Structural Analysis with Finite Elements

Friedel Hartmann Casimir Katz

Structural Analysis with Finite Elements

With 408 Figures and 26 Tables

 Springer

Friedel Hartmann
University of Kassel Structural Mechanics
Kurt-Wolters-Str. 3
34109 Kassel
Germany
friedelhartmann@uni-kassel.de

Casimir Katz
SOFiSTiK AG
Bruckmannring 38
85764 Oberschleissheim
Germany
casimir.katz@sofistik.de

Library of Congress Control Number: 2006937296

ISBN-10 3-540-49698- x Springer Berlin Heidelberg New York
ISBN-13 978-3-540-49698-4 Springer Berlin Heidelberg New York

Springer is a part of Springer Science+Business Media.

springer.com

Typesetting by authors using a Springer LaTeX macro package
Cover design: deblik, Berlin

Printed on acid-free paper   SPIN 11940241   62/3100/SPi   5 4 3 2 1 0

Preface

The finite element method has become an indispensible tool in structural analysis, and tells an unparalleled success story. With success, however, came criticism, because it was noticeable that knowledge of the method among practitioners did not keep up with success. Reviewing engineers complain that the method is increasingly applied without an understanding of structural behavior. Often a critical evaluation of computed results is missing, and frequently a basic understanding of the limitations and possibilities of the method are nonexistent.

But a working knowledge of the fundamentals of the finite element method *and* classical structural mechanics is a prerequisite for any sound finite element analysis. Only a well trained engineer will have the skills to critically examine the computed results.

Finite element modeling is more than preparing a mesh connecting the elements at the nodes and replacing the load by nodal forces. This is a popular model but this model downgrades the complex structural reality in such a way that—instead of being helpful—it misleads an engineer who is not well acquainted with finite element techniques.

The object of this book is therefore to provide a foundation for the finite element method from the standpoint of structural analysis, and to discuss questions that arise in modeling structures with finite elements.

What encouraged us in writing this book was that—thanks to the intensive research that is still going on in the finite element community—we can explain the principles of finite element methods in a new way and from a new perspective by making ample use of influence functions. This approach should appeal in particular to structural engineers, because influence functions are a genuine engineering concept and are thus deeply rooted in classical structural mechanics, so that the structural engineer can use his engineering knowledge and insight to assess the accuracy of finite element results or to discuss the modeling of structures with finite elements.

Just as a change in the elastic properties of a structure changes the Green's functions or influence functions of the structure so a finite element mesh effects a shift of the Green's functions.

We have tried to concentrate on ideas, because we considered these and not necessarily the technical details to be important. The emphasis should be on structural mechanics and not on programming the finite elements, and therefore we have also provided many illustrative examples.

Finite element technology was not developed by mathematicians, but by engineers (Argyris, Clough, Zienkiewicz). They relied on heuristics, their intuition and their engineering expertise, when in the tradition of medieval craftsmen they designed and tested elements without fully understanding the exact background. The results were empirically useful and engineers were grateful because they could suddenly tackle questions which were previously unanswerable. After these early achievements self-confidence grew, and a second epoch followed that could be called baroque: the elements became more and more complex (some finite element programs offered 50 or more elements) and enthusiasm prevailed. In the third phase, the epoch of "enlightment" mathematicians became interested in the method and tried to analyze the method with mathematical rigor. To some extent their efforts were futile or extremely difficult, because engineers employed "techniques" (reduced integration, nonconforming elements, discrete Kirchhoff elements) which had no analogy in the calculus of variations. But little by little knowledge increased, the gap closed, and mathematicians felt secure enough with the method that they could provide reliable estimates about the behavior of some elements. We thus recognize that mathematics is an essential ingredient of finite element technology.

One of the aims of this book is to teach structural engineers the theoretical foundations of the finite element method, because this knowledge is invaluable in the design of safe structures.

This book is an extended and revised version of the original German version. We have dedicated the web page `http://www.winfem.de` to the book. From this page the programs WINFEM (finite element program with focus on influence functions and adaptive techniques), BE-SLABS (boundary element analysis of slabs) and BE-PLATES (boundary element analysis of plates) can be downloaded by readers who want to experiment with the methods. Additional information can also be found on `http://www.sofistik.com`.

FriedelHartmann@uni-kassel.de   Casimir.Katz@sofistik.de

Kassel   *Friedel Hartmann*
Munich August 2003   *Casimir Katz*

Acknowledgement. We thank Thomas Graetsch, who wrote the program WINFEM and provided many illustrative examples for the approximation of influence functions with finite elements, and Marc Damashek and William J. Gordon for their help in preparing the manuscript. The permission of Oxford University Press to reprint the picture on page 145 is greatly acknowledged.

Preface to the second edition

One of the joys of writing a book is that the authors learn more about a subject. This does not stop after a book is finished. So we have added additional sections to the text

- The Dirac energy
- How to predict changes
- The influence of a single element
- Retrofitting structures
- Generalized finite element methods (X-FEM)
- Cables
- Hierarchical elements
- Sensitivity analysis
- Weak form of influence functions

in the hope that these additional topics will also attract the readers' interest.

Kassel
Munich October 2006

Friedel Hartmann
Casimir Katz

Contents

1. What are finite elements?

1.1 Introduction

In this introductory chapter various aspects of the FE method are studied, initially highlighting the key points.

1.2 Key points of the FE method

- FE method = restriction

Analyzing a structure with finite elements essentially amounts to constraining the structure (see Fig. 1.1), because the structure can only assume those shapes that can be represented by shape functions.

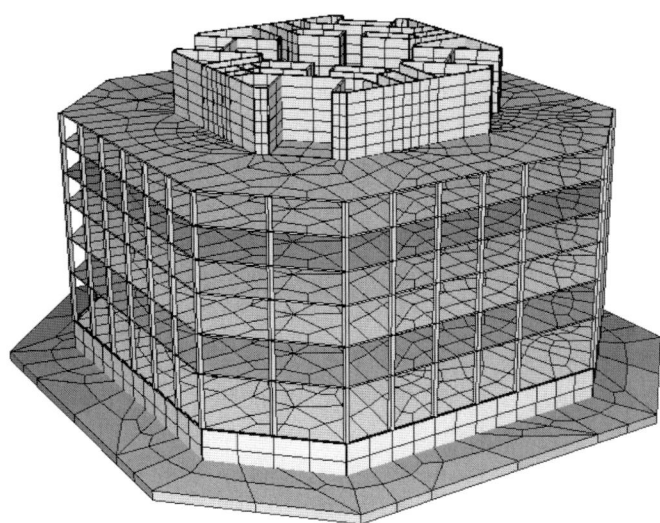

Fig. 1.1. The building can only execute movements that can be represented by shape functions

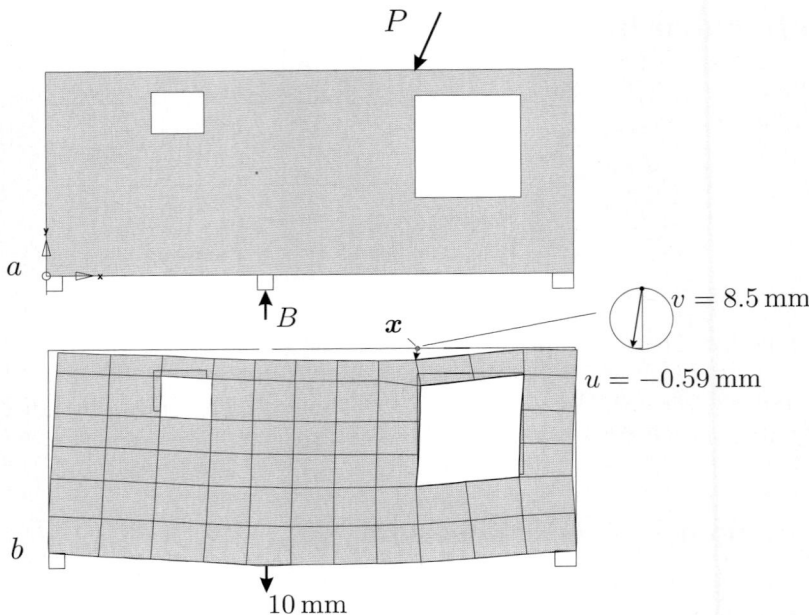

Fig. 1.2. Shear wall: **a)** support reaction B; **b)** the displacements observed at x if the support B moves in the vertical direction are a direct measure of the influence a (nearly concentrated point) load $\boldsymbol{P} = [P_x, P_y]^T$ has on the support B. About 85% of P_y and 6% of P_x will contribute to B. The better an FE program can model the movement of the support B, the better the accuracy

This is an important observation, because the accuracy of an FE solution depends fundamentally on how accurately a program can approximate the influence functions for stresses or displacements. Influence functions are displacements: they are the response of a structure to certain point loads. The more flexible an FE structure is, the better it can react to such point loads, and hence the better the accuracy of the FE solution; see Fig. 1.2.

- FE method = method of substitute load cases

It is possible to interpret the FE method as a method of substitute loadings or load cases, because in some sense all an FE program does is to replace the original load with a work-equivalent load, and solve that load case exactly. The important point is that structures are designed for these substitute loads not for the original loads.

- FE method = projection method

The shadow of a 3-D vector is that vector in the plane with the shortest distance to the tip of the vector.

The FE method is also a projection method, because the FE solution is the shadow of the exact solution when it is projected onto the trial space V_h, where

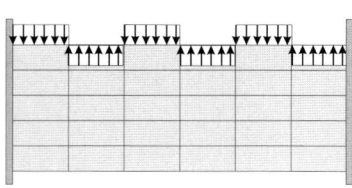
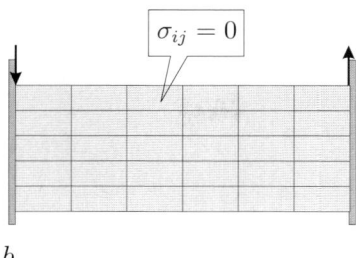

a   b

Fig. 1.3. Plate with alternating edge load: **a)** system and load; **b)** equivalent nodal forces

V_h contains all the deformations the FE structure can undergo. The metric applied in the projection is the strain energy: one chooses that deformation u_h in V_h whose distance to the exact solution u measured in units of strain energy is a minimum.

Let $\boldsymbol{u}$ denote the exact equilibrium position of a plate (subjected to some load), and let $\boldsymbol{u}_h$ be the FE approximation of this position. Now to correct the FE position, that is, to force the plate into the correct shape, a displacement field $\boldsymbol{e} = \boldsymbol{u} - \boldsymbol{u}_h$ must be added to $\boldsymbol{u}_h$.

Let σ_{ij}^e and ε_{ij}^e denote the stresses and strains caused by this displacement field $\boldsymbol{e}$. The FE solution guarantees that the energy needed to correct the FE solution is a minimum

$$a(\boldsymbol{e}, \boldsymbol{e}) = \int \left(\sigma_{xx}^e \, \varepsilon_{xx}^e + \sigma_{xy}^e \, \gamma_{xy}^e + \sigma_{yy}^e \, \varepsilon_{yy}^e \right) d\Omega \quad \rightarrow \quad \text{minimum} . \quad (1.1)$$

This is equivalent to saying[1] that the work needed to force the plate from its position $\boldsymbol{u}_h$ into the correct position $\boldsymbol{u}$ is a minimum. The effort cannot be made any smaller on the given mesh.

In a vertical projection the length of a shadow is always less than the length of the original vector (see *Bessel's inequality* [232]); this implies that the strain energy of the FE solution is always less than the strain energy of the exact solution. An engineer would say that the FE solution overestimates the stiffness of the structure.

The situation is different if a support of a structure is displaced. Then the FE projection is a skew projection (see Sect. 1.38, p. 187), that is, the shadow is longer than the original vector. This means that a greater effort is needed to displace a support of a more rigid structure than of a more flexible structure. But it will be seen later that even then a minimum principle still applies.

Because the FE solution is the shadow of the true solution, it cannot be improved on the same mesh. This is also why some load cases cannot be solved on an FE mesh. Each projection has a blind spot; see Fig. 1.3. The equivalent nodal forces at the free nodes cancel and so $\boldsymbol{K}\boldsymbol{u} = \boldsymbol{0}$.

---

[1] $a(\boldsymbol{u}, \boldsymbol{u}) = a(\boldsymbol{u}_h, \boldsymbol{u}_h) - 2\,a(\boldsymbol{u}_h, \boldsymbol{e}) + a(\boldsymbol{e}, \boldsymbol{e})$ and $a(\boldsymbol{u}_h, \boldsymbol{e}) = 0$

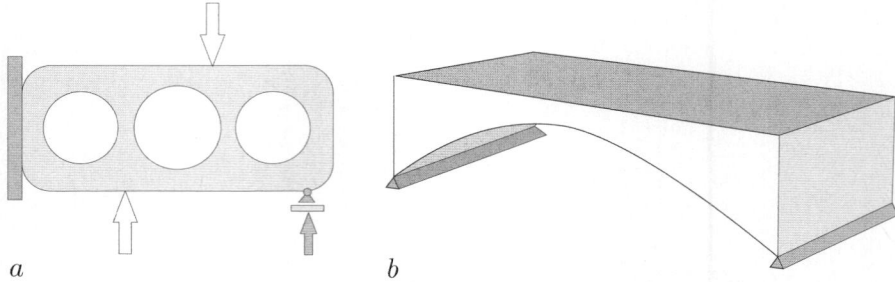

Fig. 1.4. Theoretically these load cases cannot be solved with the FE method because the strain energy is infinite: **a)** Concentrated forces acting on a plate; **b)** concrete block placed on line supports

- FE method = energy method

An FE program thinks in terms of work and energy. Loads that contribute no work do not exist for an FE program. Nodal forces represent *equivalence classes* of loads. Loads that contribute the same amount of work are identical for an FE program.

In modern structural analysis, zero is replaced by vanishing work. In classical structural analysis a distributed load $p(x)$ is identical to a second load $p_h(x)$ if at each point $0 < x < l$ of the beam the load is the same:

$$p(x) = p_h(x) \qquad 0 < x < l \qquad \text{strong equal sign}. \tag{1.2}$$

In contrast, identity is based on a weaker concept in modern structural analysis. Two loads are considered identical if the virtual work is the same for any virtual displacement $\delta w(x)$:

$$\int_0^l p(x)\, \delta w(x)\, dx = \int_0^l p_h(x)\, \delta w(x)\, dx \qquad \text{for all } \delta w(x). \tag{1.3}$$

This is the *weak equal sign*. If *all* really means *all* then of course the weak equal sign is identical to the strong equal sign. But in all other cases there remains a specific difference, in that equivalence is established only with regard to a finite set of virtual displacements δw.

Because the FE method is an energy method, problems in which the strain energy is infinite—theoretically at least—cannot be solved with this method; see Fig. 1.4.

- FE method = method of approximate influence functions

We will see that a mesh is only as good as the influence functions that can be generated on that mesh. According to Betti's theorem, the displacement $u(x)$ or the stress $\sigma_x(x)$ at a point x is the L_2-scalar product of the applied load p and the corresponding influence function (*Green's function*)

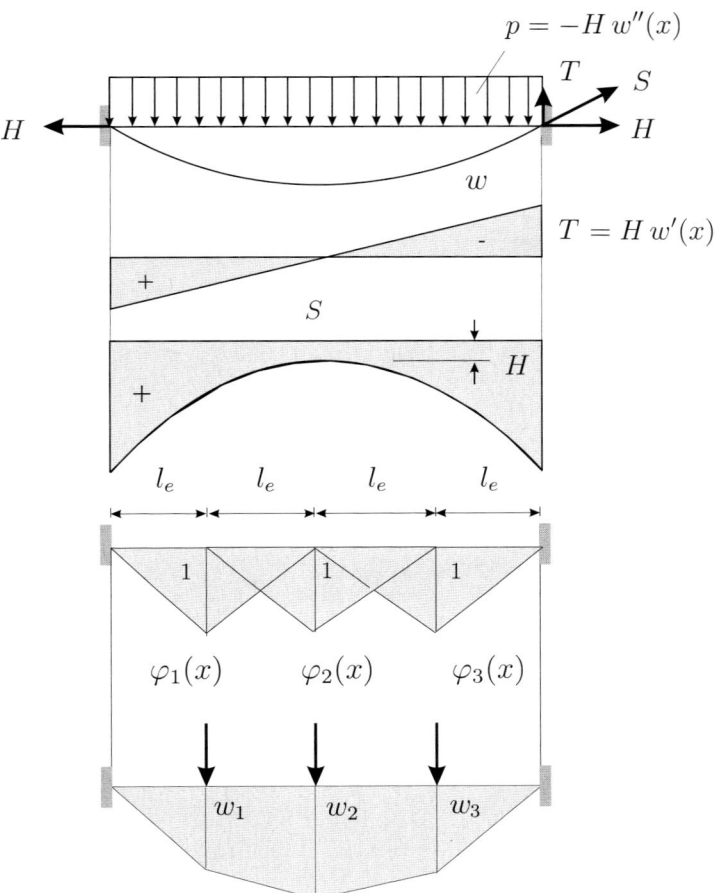

$$p = -H\, w''(x)$$

$$w$$

$$T = H\, w'(x)$$

Fig. 1.5. FE analysis of a taut rope

$$u(x) = \int_0^l G_0(y,x)\, p\,(y)\, dy\,, \qquad \sigma_x(x) = \int_0^l G_1(y,x)\, p\,(y)\, dy\,. \qquad (1.4)$$

All an FE program does is to replace the exact Green's functions with approximate Green's functions G_0^h and G_1^h, respectively. Therefore the error in an FE solution is proportional to the distance between the approximate and the exact Green's function:

$$u(x) - u_h(x) = \int_0^l \left[G_0(y,x) - G_0^h(y,x)\right]\, p\,(y)\, dy\,, \qquad (1.5)$$

$$\sigma_x(x) - \sigma_x^h(x) = \int_0^l \left[G_1(y,x) - G_1^h(y,x)\right]\, p\,(y)\, dy\,. \qquad (1.6)$$

1.3 Potential energy

To see these principles applied, we analyze a very simple structure, a taut rope (see Fig. 1.5).

Imagine that the rope is pulled taut by a horizontal force H and that it carries a distributed load p. The distribution of the vertical force V within the rope and the deflection w of the rope are to be calculated. The deflection w is the solution of the boundary value problem

$$- Hw''(x) = p(x) \qquad 0 < x < l \qquad w(0) = w(l) = 0 \,. \tag{1.7}$$

The vertical (or transverse) force T is proportional to the slope w'

$$T = Hw' \,, \tag{1.8}$$

and the vector sum of H and T is the tension S in the rope

$$S = \sqrt{H^2 + T^2} \,. \tag{1.9}$$

The potential energy of the rope is the expression

$$\varPi(w) = \frac{1}{2} \int_0^l H(w')^2 \, dx - \int_0^l p\, w \, dx = \frac{1}{2} \int_0^l \frac{T^2}{H} - \int_0^l p\, w \, dx \,. \tag{1.10}$$

For completeness we also note Green's first identity for the operator $-H\, w''$:

$$G(w, \hat{w}) = \int_0^l -H\, w'' \, \hat{w} \, dx + [T\, \hat{w}]_0^l - \int_0^l \frac{T\hat{T}}{H} \, dx = 0 \tag{1.11}$$

because it encapsulates the structural mechanics of the rope.

To approximate the deflection $w(x)$ of the rope, the rope is subdivided into four linear elements: see Fig. 1.5. The first and the last node are fixed so that only the three internal nodes can be moved. Between the nodes the deflection is linear, that is the rope is only allowed to assume shapes that can be expressed in terms of the three unit displacements $\varphi_i(x)$ of the three internal nodes (see Fig. 1.5)

$$w_h(x) = w_1 \cdot \varphi_1(x) + w_2 \cdot \varphi_2(x) + w_3 \cdot \varphi_3(x) \,. \tag{1.12}$$

The nodal deflections, w_1, w_2, w_3, play the role of weights. They signal how much of each unit deflection is contained in w_h.

All these different shapes—let the numbers w_1, w_2, w_3 vary from $-\infty$ to $+\infty$—constitute the so-called *trial space* V_h.

The space V_h itself is a subset of a greater space, the *deformation space* V of the rope. The space V contains all deflection curves $w(x)$ that the rope can possibly assume under different loadings during its lifetime. It is obvious that the piecewise linear functions w_h in the subset V_h represent only a very small fraction of V.

The next question then is: what values should be chosen for the three nodal deflections w_1, w_2, w_3 of the FE solution? What is the optimal choice?

According to the principle of minimum potential energy, the true deflection w results in the lowest potential energy on V

$$\Pi(w) = \frac{1}{2}\int_0^l H(w')^2 dx - \int_0^l p\, w\, dx\,. \tag{1.13}$$

But if the exact solution w wins the competition on the big space V, it seems a good strategy to choose the nodal deflections w_i in such a way that the FE solution

$$w_h(x) = \sum_{i=1}^3 w_i\, \varphi_i(x) \tag{1.14}$$

wins the competition on the small subset $V_h \subset V$. Then $\Pi(w_h)$ is as close as possible to $\Pi(w)$ on V_h.

Because each function w_h in V_h is uniquely determined by the nodal deflections w_i at the three interior nodes, i.e. the vector $\boldsymbol{w} = [w_1, w_2, w_3]^T$, the potential energy on V_h is a function of these three numbers only

$$\begin{aligned}
\Pi(w_h) = \Pi(\boldsymbol{w}) &= \frac{1}{2}\,\boldsymbol{w}^T \boldsymbol{K}\boldsymbol{w} - \boldsymbol{f}^T \boldsymbol{w} \\
&= \frac{1}{2}\,[w_1, w_2, w_3]\,\frac{4H}{l}\begin{bmatrix} 2 & -1 & 0 \\ -1 & 2 & -1 \\ 0 & -1 & 2 \end{bmatrix}\begin{bmatrix} w_1 \\ w_2 \\ w_3 \end{bmatrix} - [f_1, f_2, f_3]\begin{bmatrix} w_1 \\ w_2 \\ w_3 \end{bmatrix} \\
&= \frac{4H}{l}[w_1^2 - w_1 w_2 + w_2^2 - w_2 w_3 + w_3^2] - f_1 w_1 - f_2 w_2 - f_3 u_3\,,
\end{aligned} \tag{1.15}$$

where the matrix $\boldsymbol{K}$ and the vector $\boldsymbol{f}$ have the elements

$$k_{ij} = \int_0^l H\varphi_i'\,\varphi_j'\, dx \qquad f_i = \int_0^l p\,\varphi_i\, dx = p\,l_e = p\,\frac{l}{4}\,. \tag{1.16}$$

Finding the minimum value of Π on V_h is therefore equivalent to finding the vector $\boldsymbol{w}$—the "address" of $w_h \in V_h$—for which the *function* $\Pi(\boldsymbol{w})$ becomes a minimum. A necessary condition is, that the first derivatives of the function $\Pi(\boldsymbol{w})$ vanish at this point $\boldsymbol{w}$:

$$\frac{\partial \Pi}{\partial w_i} = \sum_{j=1}^3 k_{ij}\, w_j - f_i = 0\,, \qquad i = 1, 2, 3\,, \tag{1.17}$$

which leads to the system of equations

$$\boldsymbol{K}\boldsymbol{w} = \boldsymbol{f} \tag{1.18}$$

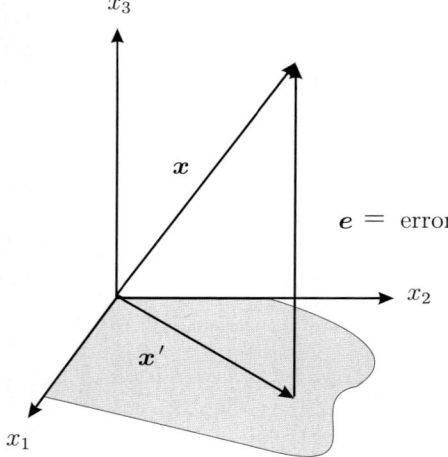

x_3

$\boldsymbol{x}$

$\boldsymbol{e} =$ error

x_2

$\boldsymbol{x}'$

x_1

Fig. 1.6. The error $\boldsymbol{e}$ is orthogonal to the plane

or

$$\frac{4\,H}{l}\begin{bmatrix} 2 & -1 & 0 \\ -1 & 2 & -1 \\ 0 & -1 & 2 \end{bmatrix}\begin{bmatrix} w_1 \\ w_2 \\ w_3 \end{bmatrix} = \frac{p\,l}{4}\begin{bmatrix} 1 \\ 1 \\ 1 \end{bmatrix}, \tag{1.19}$$

which has the solution $w_1 = w_3 = 1.5\,p\,l^2/(16\,H)$, $w_2 = 2.0\,p\,l^2/(16\,H)$. Hence the deflection

$$w_h(x) = \frac{p\,l^2}{16\,H}\,[1.5 \cdot \varphi_1(x) + 2.0 \cdot \varphi_2(x) + 1.5 \cdot \varphi_3(x)] \tag{1.20}$$

is the best approximation on V_h.

1.4 Projection

Work is a scalar quantity, as are temperature and pressure. This is nearly the most important statement that can be made about work. Work is *force ×
displacement*. Work and energy are the same. The integral

$$\frac{1}{2}\int_0^l \frac{T^2}{H}\,dx\,, \qquad T = Hw'\,, \tag{1.21}$$

is the internal energy of the rope. It measures the strain energy stored in the rope.

Energy can also serve as a scale. It is the scale FE methods work with. Having a scale means having a topology, which in turn defines "far away" and "nearby". To measure the length of a vector the *Euclidean norm* is used:

$$|\boldsymbol{x}| = \sqrt{x_1^2 + x_2^2 + x_3^2}\,. \tag{1.22}$$

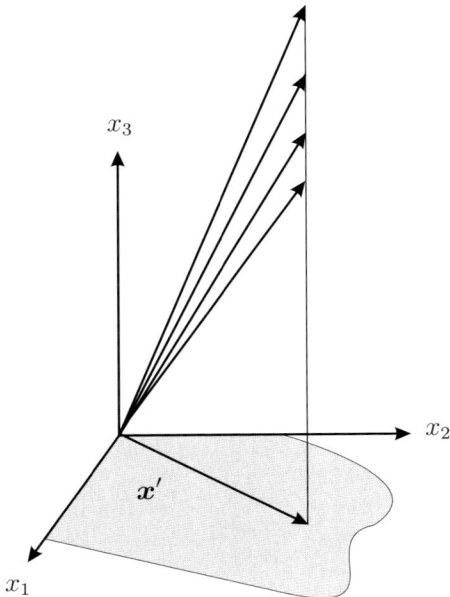

Fig. 1.7. All vectors have the same shadow $\boldsymbol{x}'$

In this *topology* two cities A and B are close neighbors if the difference between their position vectors $\boldsymbol{a}$ and $\boldsymbol{b}$ (with reference to the origin of a map) is small:

$$|\boldsymbol{a} - \boldsymbol{b}| \quad \text{``small''} \quad \Longrightarrow \quad \text{A and B are neighbors}. \quad (1.23)$$

Projections only make sense if distances can be measured. The shadow $\boldsymbol{x}'$ of a 3-D vector $\boldsymbol{x}$ is the vector in the plane which has the smallest distance to the tip of $\boldsymbol{x}$; see Fig. 1.6. The distance between the original vector and its shadow is the length of the vector

$$\boldsymbol{e} = \boldsymbol{x} - \boldsymbol{x}', \quad (1.24)$$

which points from the tip of the shadow to the tip of the vector $\boldsymbol{x}$. The shadow $\boldsymbol{x}'$ renders this distance a minimum

$$|\boldsymbol{e}| = \sqrt{(x_1 - x_1')^2 + (x_2 - x_2')^2 + (x_3 - 0)^2} = \text{minimum}. \quad (1.25)$$

Any other vector $\tilde{\boldsymbol{x}}'$ in the plane has a greater distance from the vector $\boldsymbol{x}$

$$|\tilde{\boldsymbol{e}}| = |\boldsymbol{x} - \tilde{\boldsymbol{x}}'| > |\boldsymbol{e}| = |\boldsymbol{x} - \boldsymbol{x}'|. \quad (1.26)$$

This is the *first* feature of a projection: the shadow solves a minimum problem.

The *second* feature is that the residual vector, the error $\boldsymbol{e}$, is orthogonal to the $x_1 - x_2$-plane (assuming that the sun shines from straight above), because the scalar product between the error and the shadow is zero:

$$e^T x' = 0 .$$

(1.27)

This is equivalent to saying that the shadow of the error e has no physical extent, but only if the line of sight coincides with the direction of the projection! Seen from any other direction the length of e is not zero. Hence a projection method is *blind with respect to errors which lie in the line of sight*. All vectors $\tilde{x}$ that lie "above" the vector x, which differ from x only by an additive term parallel to the line of sight (i.e., projection), have the same shadow; see Fig. 1.7.

The *third* feature is that the result of a projection cannot be improved. Repeating a projection changes nothing: the shadow of the shadow is the shadow. Which means that a projection method freezes after the first step, while other operations, such as squaring a number, can be repeated infinitely often.

The *fourth* feature of a projection is that the length of the shadow is shorter than the length of the original vector; see Fig. 1.7. This is not only true for vectors, but also for functions: the *Fourier series* $f_n(x)$ of a function $f(x)$ is the projection of $f(x)$ onto the trigonometric functions in the sense of the L_2-scalar product, and according to *Bessel's inequality* the length ($= L_2$-norm) of the Fourier series f_n is less than the L_2-norm of f:

$$||f_n||_0 = [\int_0^l f_n^2(x)\,dx]^{1/2} \le [\int_0^l f^2(x)\,dx]^{1/2} = ||f||_0 .$$

(1.28)

All this applies now to the FE method as well: the exact deflection curve $w \in V$ is projected onto a subspace V_h, and the shadow w_h is the FE solution.

In the case of the rope the space V_h contains all the deformations which are expansions in terms of the three unit displacements $\varphi_i(x)$,

$$w_h(x) = w_1 \cdot \varphi_1(x) + w_2 \cdot \varphi_2(x) + w_3 \cdot \varphi_3(x) ,$$

(1.29)

and the FE solution is the solution of the following minimum problem:

Find the deflection

$$w_h(x) = w_1 \cdot \varphi_1(x) + w_2 \cdot \varphi_2(x) + w_3 \cdot \varphi_3(x)$$

(1.30)

in V_h which has the shortest distance (= strain energy) from the exact deflection w.

In FE analysis the strain energy is usually expressed

$$a(w,w) := \int_0^l H\,(w')^2\,dx = \int_0^l \frac{T^2}{H}\,dx .$$

(1.31)

If

$$e(x) = w(x) - w_h(x)$$

(1.32)

is the error of the FE solution, then the FE solution is that function in V_h for which the strain energy of the error $e(x)$ becomes a minimum:

$$a(e, e) = \frac{1}{2} \int_0^l \frac{(T - T_h)^2}{H} \, dx = \text{minimum} \, . \tag{1.33}$$

Any other function w_h in V_h has a larger distance—in terms of energy—than the FE solution. This property of the FE solution w_h can also be expressed as follows, see (7.413) p. 572,

$$a(e, e) \leq a(w - v_h, w - v_h) \qquad \text{for all } v_h \in V_h \, . \tag{1.34}$$

We also know that the strain energy of the FE solution is always less than the strain energy of the exact solution:

$$a(w_h, w_h) = \int_0^l \frac{T_h^2}{H} \, dx < \int_0^l \frac{T^2}{H} \, dx = a(w, w) \, , \tag{1.35}$$

i.e., the shadow w_h has a shorter length (= strain energy) than w. This inequality follows directly from

$$0 < a(w, w) = a(w_h + e, w_h + e)$$
$$= a(w_h, w_h) + 2 \underbrace{a(e, w_h)}_{=0} + \underbrace{a(e, e)}_{>0} \, , \tag{1.36}$$

where

$$a(e, w_h) = \int_0^l \frac{(T - T_h) \, T_h}{H} \, dx = 0 \tag{1.37}$$

is a consequence of the *Galerkin orthogonality*

$$a(e, \varphi_i) = 0 \qquad i = 1, 2, 3 \tag{1.38}$$

i.e., the fact that the error e is orthogonal in terms of the strain energy to all unit displacements φ_i, and therefore also to $w_h = w_1 \cdot \varphi_1 + w_2 \cdot \varphi_2 + w_3 \cdot \varphi_3$.

Hence the strain energy or internal energy is the metric FE methods work with. Distance is measured in this metric and therefore also convergence.

The internal energy induces a topology on the space V which is even a *norm* on this space, because it *separates* the elements of V. Two functions w_1 and w_2 are identical if and only if their distance in terms of the strain energy is zero:

$$\frac{1}{2} \int_0^l \frac{(T_1 - T_2)^2}{H} \, dx = \frac{1}{2} \int_0^l H \, (w_1' - w_2')^2 \, dx = 0 \quad \Leftrightarrow \quad w_1 = w_2 \tag{1.39}$$

that is if $w_1 - w_2$ has zero energy.

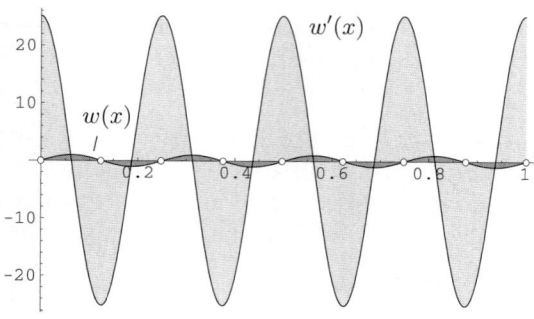

Fig. 1.8. A small deflection curve can hide a large strain energy

A function w is small in this metric if its energy (essentially the square of the first derivative) is small, and the exact deflection w and the FE solution w_h are close in this metric if the strain energy of the error

$$e(x) = w(x) - w_h(x) \qquad (e = \text{error}) \tag{1.40}$$

is small

$$\frac{1}{2} \int_0^l \frac{T_e^2}{H} \, dx = \frac{1}{2} \int_0^l H(w' - w_h')^2 \, dx = \text{small} \implies e(x) = \text{small} . \tag{1.41}$$

This energy metric makes more sense than a naive metric that considers a function such as $w(x) = \sin(8\pi x)$ a "small" function (see Fig. 1.8), while for the FE method it is a "large" function, because the strain energy due to the rapid oscillations is large

$$\int_0^1 w(x)^2 \, dx = 0.5 , \qquad \frac{1}{2} \int_0^1 Hw'(x)^2 \, dx = 316 \cdot H . \tag{1.42}$$

Hence from an engineering standpoint it makes more sense to classify functions with regard to the strain energy than their amplitude or their L_2-norm.

A better strategy would it be to base the metric on both components, the zero-order and the first-order derivative. This leads to the so-called *Sobolev norms*, which, depending on the index n, measure the derivatives up to order n

$$\|w\|_n = \left[\int_0^l \left[w(x)^2 + w'(x)^2 + \ldots + w^{(n)}(x)^2 \right] dx \right]^{1/2} \tag{1.43}$$

and classify functions according to this metric. By increasing the index n different topologies can be generated on V. In the same way the distance between two vectors does not depend on the difference of the first two components alone, $|\boldsymbol{a} - \boldsymbol{b}| = \sqrt{(a_1 - b_1)^2}$ (which would be a very crude topology) but on the difference of *all* components

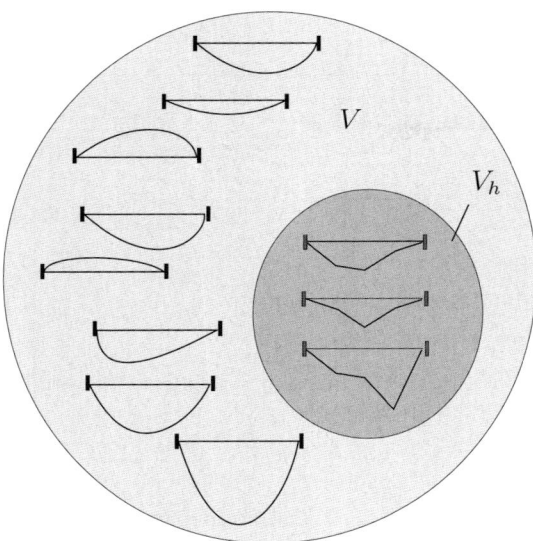

Fig. 1.9. The shapes which the FE program can represent constitute a subset V_h of V

$$|\boldsymbol{a} - \boldsymbol{b}| = \sqrt{(a_1 - b_1)^2 + (a_2 - b_2)^2 + \ldots + (a_n - b_n)^2}\,. \qquad (1.44)$$

This metric generates the finest possible topology, just as in a lottery the prize money increases, the more figures on a ticket agree with the number drawn.

Remark 1.1. Later it will be seen that in so-called load cases δ when displacements are prescribed the projection is no longer orthogonal but "skew" this implies that the length of the shadow (the strain energy) will be *greater* than the strain energy of the exact solution; see Sect. 1.38, p. 187. This is to be expected: the stiffer a structure the greater the strain energy developed by displacing a support.

1.5 The error of an FE solution

The FE method is an approximate method, see Fig. 1.9. As such it must approximate *three* functions:

- the deflection w
- the vertical force $T = Hw'$
- the load $p = -Hw''$

i.e., the zero-order, first-order, and second-order derivative of the deflection w. All three derivatives of w are relevant to the structural analysis, and hence it is legitimate to ask which of the three errors

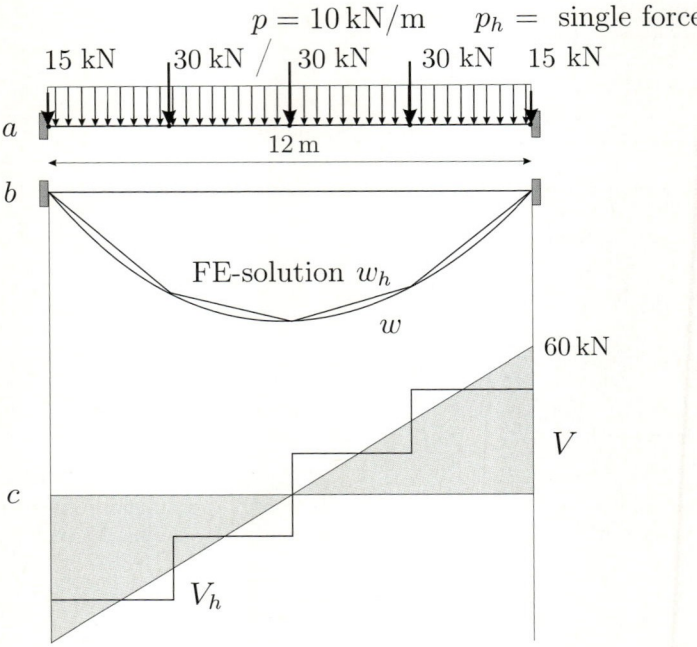

$p = 10 \, \mathrm{kN/m}$ $p_h = $ single forces

15 kN   30 kN / 30 kN   30 kN   15 kN

a

12 m

b

FE-solution w_h

w

60 kN

V

c

V_h

Fig. 1.10. The error in the displacement is zero at the nodes, while the error in the stresses is zero at the midpoints of the elements. This is a typical pattern in FE analysis

$w - w_h$	error in the deflection
$T - T_h$	error in the internal action
$p - p_h$	error in the load

is to be minimized? In principle we have already given the answer. The FE solution aims at minimizing the square of the error of the internal action $T - T_h$,

$$\int_0^l \frac{(T - T_h)^2}{H} \, dx = \int_0^l H(w' - w_h')^2 \, dx \quad \rightarrow \quad \text{minimum} . \quad (1.45)$$

Hence an FE solution does not tend to win a beauty contest by imitating the original shape w as closely as possible nor does it aim to simulate the loading; rather, the solution tends to minimize the error in the strain energy (the internal energy).

The load case p_h

A closer study of the FE solution reveals that w_h is the equilibrium position of the rope if the distributed load were concentrated at the nodes, $f_i = p \, l_e$. This load case is called the FE load case p_h, (see Fig. 1.10).

Of course we would like to know what the consequences are. How far are the results of the load case p_h (= nodal forces) from p (= distributed load)? Stated otherwise: given the error in the load

$$r := p - p_h \qquad \text{(residual forces)} \qquad (1.46)$$

how large is the error in the vertical force

$$T_e := T - T_h \qquad (1.47)$$

and the difference in the deflection

$$e := w - w_h ? \qquad (1.48)$$

In other words what can be said about the error in the first-order, $T - T_h = H(w - w'_h)$, and zero-order derivative, $w - w_h$, if the error in the second-order derivative $p - p_h$ is known?

The normal procedure is to differentiate the deflection w, yielding the vertical force T, and to differentiate T to find the load p

$$w \qquad \Rightarrow \qquad T = H w' \qquad \Rightarrow \qquad p = -H w''. \qquad (1.49)$$

In a reverse order, the functions must be integrated

$$w = \iint -\frac{p}{H} \, dx \, dx \quad \Leftarrow \quad T = \int -p \, dx \quad \Leftarrow \quad p = -H w'' \qquad (1.50)$$

and integration smoothes the wrinkles; see Fig. 1.10.

But is there a reliable method to make predictions about the distance in the first-order derivatives by looking at the distance in the second-order derivative? The answer is *no*. Otherwise it would suffice to calculate an approximate solution on a coarse mesh, and extrapolate from this solution to the exact solution. In general this seems not to be possible, certainly not in one step. There exist only different techniques which provide upper or lower bounds for the error. The development of such *error estimators* is the subject matter of adaptive methods.

1.6 A beautiful idea that does not work

- An FE solution cannot be improved on the same mesh.

Once it is understood that the error of an FE solution can be traced back to deviations in the load, could the situation not be improved by applying the residual forces $p - p_h$, solving this load case again with finite elements, and repeating this loop as long as the error is greater than a preset error margin ε?

This idea does not work, because the residual forces

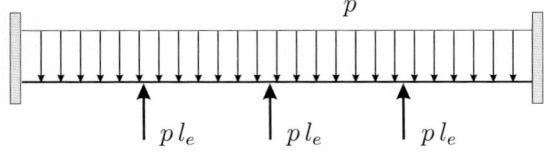

Fig. 1.11. The FE solution of this load case is zero

$$r = p - p_h \tag{1.51}$$

leave no traces on the mesh, i.e., all the equivalent nodal forces f_i^r vanish,

$$f_i^r = \int_0^l p\,\varphi_i\,dx - \sum_{j=1}^{3} f_j \cdot \varphi_i(x_j) = f_i - f_i = 0 \qquad \text{for all } \varphi_i\,, \tag{1.52}$$

so that the rope will not deflect, because zero nodal forces mean zero deflection:

$$\boldsymbol{K}\boldsymbol{u} = \boldsymbol{0} \qquad \Rightarrow \qquad \boldsymbol{u} = \boldsymbol{0}\,. \tag{1.53}$$

This riddle is easily solved by recalling that the exact curve w is projected onto the trial space V_h. But because the error $w - w_h$ is orthogonal (in the energy sense) to the space V_h,

$$\int_0^l H(w' - w_h')\,\varphi_i'\,dx = 0 \qquad \text{for all } \varphi_i\,, \tag{1.54}$$

it casts no shadow, i.e., $e = 0$.

It follows that there are load cases which cannot be solved on an FE mesh (see Fig. 1.11) namely all load cases where the load p is so arranged that it contributes no work. This is the case if all equivalent nodal forces f_i are zero:

$$f_i = \delta W_e(p, \varphi_i) = 0\,, \qquad i = 1, 2, \ldots n\,. \tag{1.55}$$

Loads that happen to be parallel to the line of sight have a "null shadow".

1.7 Set theory

In their lowest level, many systems are at their most stable position. Many processes in physics are governed by a minimum principle. The same holds in beam analysis: the deflection curve w of a continuous beam minimizes the potential energy of the beam

$$\Pi(w) = \frac{1}{2}\int_0^l \frac{M^2}{EI}\,dx - \int_0^l p\,w\,dx \qquad \rightarrow \qquad \text{minimum} \tag{1.56}$$

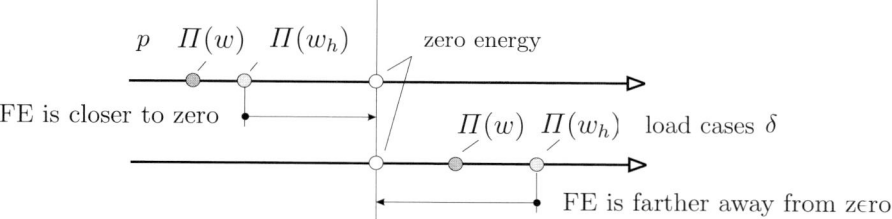

Fig. 1.12. The potential energy $\Pi(w_h)$ of the FE solution always lies to the right of the exact potential energy $\Pi(w)$

on V, which is the set of all functions w that satisfy the support conditions, i.e., that have zeros, $w = 0$, at all supports. All such functions w compete for the minimum value of $\Pi(w)$.

The winner is the deflection curve w of the continuous beam. According to Green's first identity, $G(w, w) = 0$ (see Sect. 7.2, p. 508)

$$\int_0^l \frac{M^2}{EI}\, dx = \int_0^l p\, w\, dx \qquad (2\, W_i = 2\, W_e)\,, \tag{1.57}$$

hence the minimum of the potential energy is

$$\Pi(w) = \frac{1}{2} \int_0^l \frac{M^2}{EI}\, dx - \int_0^l p\, w\, dx = -\frac{1}{2} \int_0^l p\, w\, dx\,. \tag{1.58}$$

Obviously the potential energy is *negative* in the equilibrium position, because the integral (p, w) itself is positive. It is the work done by the distributed load p inducing its own deflections, and such work (*eigenwork*) is always positive.

If no load p is applied, but instead displacements δ are prescribed at one or more supports then the potential energy is

$$\Pi(w) = \frac{1}{2} \int_0^l \frac{M^2}{EI} dx > 0\,, \tag{1.59}$$

(support displacements δ never enter into the potential energy—they only appear in the definition of the space V), i.e., the minimum value of Π must be greater than zero, because the integral of M^2 is positive. Hence the two types of load cases differ by the sign of the potential energy:

- load cases p $\Pi < 0$
- load cases δ $\Pi > 0$.

Now if a continuous beam is placed on additional supports as in Fig. 1.13, the set V "shrinks" because the candidates—the deflection curves w that compete for the minimum value of $\Pi(w)$—must have zeros, $w = 0$, at additional points. In contrast if supports are removed, then V increases, because the numbers of constraints $w = 0$ shrinks. Therefore the "size" of V is proportional to

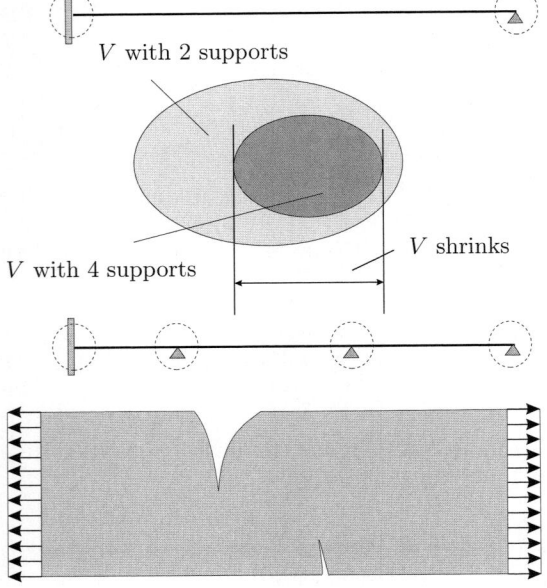

V with 2 supports

V shrinks

V with 4 supports

Fig. 1.13. The more supports, the smaller the space V

Fig. 1.14. With each crack, both the space V and $|\Pi(\boldsymbol{u})|$ increase

the number of constraints and consequently the absolute value $|\Pi(w)|$ of the potential energy must decrease (V shrinks) or increase (V grows).

Or imagine that a crack develops in a plate; see Fig. 1.14. Then the space V increases, because then also those displacement fields that are discontinuous at the faces of the crack can compete for the minimum value of $\Pi(\boldsymbol{u})$ whereupon the minimum value of $\Pi(\boldsymbol{u})$ decreases, which actually means that $|\Pi(\boldsymbol{u})|$ increases [115].

The opposite tendency is observed in FE analysis where one seeks the minimum value of $\Pi(\boldsymbol{u})$ only on a subset V_h of V. On the subset the minimum value cannot be less than the minimum on the whole space V.

A second observation can be added to this: in a load case p, the strain energy of the FE solution is always less than the strain energy of the exact solution, see (1.36),

$$\int_0^l \frac{M_h^2}{EI}\, dx \le \int_0^l \frac{M^2}{EI}\, dx \qquad \text{(load case } p), \qquad (1.60)$$

while in a load case δ the situation is just the opposite, because the strain energy of the FE solution exceeds the strain energy of the exact solution

$$\int_0^l \frac{M^2}{EI}\, dx \le \int_0^l \frac{M_h^2}{EI}\, dx \qquad \text{(load case } \delta). \qquad (1.61)$$

Both effects suggest that an FE solution tends to overestimate the stiffness of a structure.

The potential energy of the exact solution is always less than the potential energy of the FE solution:

$$\Pi(w) < \Pi(w_h) \qquad \text{because } V_h \subset V \tag{1.62}$$

or if we identify Π with numbers on the x-axis, the point $\Pi(w_h)$ will *always* lie to the right of the point $\Pi(w)$; see Fig. 1.12.

This implies that in a load case p the potential energy of the FE solution will not be as low as the potential energy of the true solution and the structure will not deflect as much—the displacements will be smaller.

The fact that $\Pi(w_h)$ lies to the right of $\Pi(w)$ means in a load case δ that more strain energy is "stored" in the FE solution than the true solution. Obviously because more energy must be supplied, to displace the support of a stiffer structure. To sum it up we have:

• in a load case p $\Pi(w_h)$ is closer to zero than $\Pi(w)$
• in a load case δ $\Pi(w_h)$ lies farther from zero than $\Pi(w)$

But these observations do not imply that FE displacements are smaller than the exact displacements! This certainly will be true for some nodes, but in general it cannot be guaranteed to be true for all nodes.

There is only one example where this conclusion—at least for one node—holds, namely if a single force P acts at a point $\boldsymbol{x}_P$ of a Kirchhoff plate. In the equilibrium position the potential energy is just the (negative) work done by the force P

$$-\frac{1}{2}P\,w(\boldsymbol{x}_P) = \Pi(w) < \Pi(w_h) = -\frac{1}{2}P\,w_h(\boldsymbol{x}_P) \tag{1.63}$$

and this inequality can only be true if the FE deflection at $\boldsymbol{x}_P$ is less than the exact value, $w_h < w$.

A similar result can be observed in a beam which is loaded at midspan, $x = l/2$, with a single force P, so that

$$\Pi(w) = -\frac{1}{2}P\,w(\frac{l}{2})\,. \tag{1.64}$$

What happens next is exactly what is predicted by set theory. The more supports that are added (see Fig. 1.15), the smaller the deflection $w(l/2)$ at the center of the beam. Then V decreases, as does the absolute value $|\Pi(w)|$ of the potential energy and thus the deflection $w(l/2)$.

The same effect can be observed if the beam is placed on one or two additional *elastic supports*. Springs are different, in that they do not change the size of V, because springs have no hard supports such as $w(0) = 0$.

Braces and diaphragms also enable the absolute value of the potential energy of a structure to decrease. The more plates, beams, columns and slabs a structure contains per cubic meter, the closer the absolute value of the potential energy of the structure in a load case p will be to zero, while in a

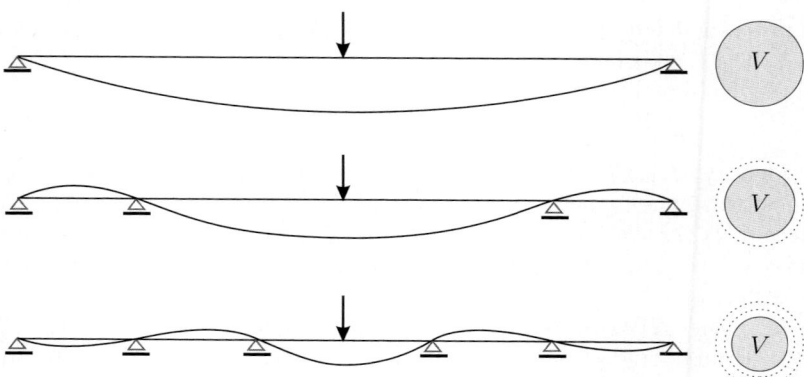

Fig. 1.15. The greater the number of supports, the smaller the value of $|\Pi|$, the smaller the deflection w, and the smaller the size of the space V

load case δ the opposite will be true. If in addition such a complex structure is modeled with just the bare minimum of elements, the structure will be very stiff.

Minimum or maximum ?

In some sense the principle of minimum potential energy could also be called a maximum principle—at least for load cases p. Calling it a minimum principle is attractive, because many processes in nature follow a principle of *least action*, but in reality the load p on a beam tends to push the beam downwards as far as possible, transforming positional energy into potential energy:

$$\Pi(w) = -\frac{1}{2} \int_0^l p\, w\, dx, \qquad \text{at } w = \text{equilibrium point} \qquad (1.65)$$

in mathematical terms, it pushes the point $|\Pi(w)|$ as far away from zero as possible.

The movement stops at the equilibrium point. This is the point at which the external work W_e equals the internal energy W_i,

$$W_e = \frac{1}{2} \int_0^l p\, w\, dx = \frac{1}{2} \int_0^l \frac{M^2}{EI}\, dx = W_i \quad \text{at the equilibrium point } w.$$

$$(1.66)$$

The more the load presses the beam down (W_e increases), the more resistance the load feels because the beam bends; the bending moments increase, thereby increasing the internal energy W_i; see Fig. 1.16. The equilibrium point is the point at which the two trends balance.

Only in load cases δ does the minimum keep its original meaning. Then the structure tries to avoid any excess strain energy, and follows with as little

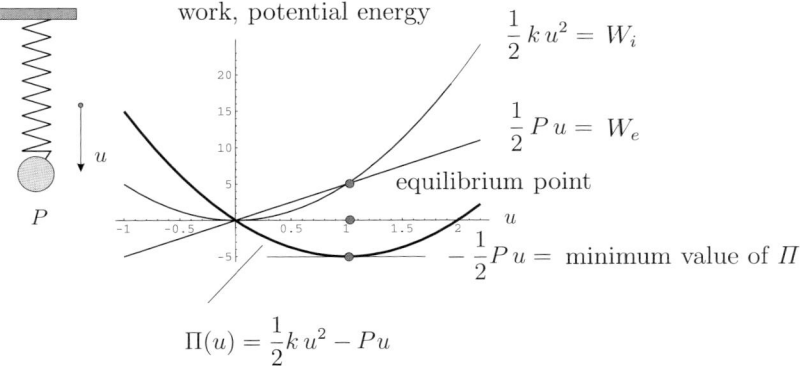

work, potential energy

$$\frac{1}{2}\,k\,u^2 = W_i$$

$$\frac{1}{2}\,P\,u = W_e$$

equilibrium point

$$-\frac{1}{2}P\,u = \text{ minimum value of } \Pi$$

$$\Pi(u) = \frac{1}{2}k\,u^2 - P\,u$$

Fig. 1.16. Because the internal energy W_i increases quadratically with u, while the external work W_e increases only linearly, W_i always catches up with W_e, and there will always be an equilibrium point where $W_i = W_e$

resistance as possible, i.e., $\Pi(w)$ is as close to zero as possible the movements imposed by the displaced supports.

But regardless of whether the minimum is positive or negative the potential energy is in any case a concave-up parabola meaning that energy must be added to the structure to move the structure out of its equilibrium position. In both types of load cases, p and δ, the equilibrium is stable.

Structural mechanics in a nutshell

Figure 1.16 is a nice illustration of the so-called *ellipticity* of a spring. Because the stiffness k of the spring is greater than zero, the strain energy is positive definite

$$a(u,u) = k\,u^2 > 0 \qquad u \neq 0, \tag{1.67}$$

and if P is bounded, the external work $P \cdot u$ is a *continuous, linear function* of u. This guarantees that there is always a solution $u = P/k$, because the parabola $1/2\,k\,u^2$ will ultimately rise faster than the straight line $1/2\,P\,u$. The parabola will catch up with the straight line; that is, there is always a balance between internal energy and external work:

$$W_i = \frac{1}{2}\,k\,u^2 = \frac{1}{2}\,P\,u = W_e \qquad \text{at } u = P/k. \tag{1.68}$$

Note that at first the external work $1/2\,P\,u$ grows faster (and *must* grow faster!) than the parabola $1/2\,k\,u^2$ of the internal energy. If that were not the case, we would see no movement! Hence in some sense—let $P = k = 1$— structural mechanics is rooted in the fact that $u > u^2$ in the interval $(0,1)$, and that beyond the end point the opposite is true. The transition point $u = 1$ is the equilibrium point.

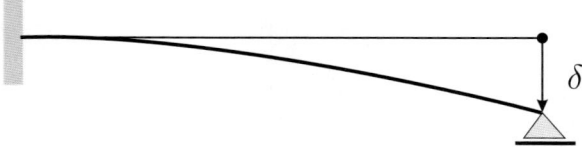

δ **Fig. 1.17.** The support settles and takes the beam with it

Skew projection

The attentive reader will recall our remark that in a load case δ the strain energy of the FE solution exceeds the energy of the exact solution. When the potential energy is minimized in a load case δ

$$\Pi(w) = \frac{1}{2} \int_0^l \frac{M^2}{EI}\,dx \qquad \rightarrow \qquad \text{minimum} \qquad (1.69)$$

the solution is sought in the solution space S. This space contains all deflections w which at the proper point, say the end of the beam, exhibit the correct deflection $w(l) = \delta$; see Fig. 1.17.

In contrast to the space S the test or trial space V consists of all those functions w which have zero displacement, $w(l) = 0$, at the end of the beam, and which of course also satisfy the other support conditions $w(0) = w'(0) = 0$ of the beam.

Normally, that is in load cases p, the two spaces coincide, $S = V$, because normally support conditions are of homogeneous type, $w(0) = 0$ or $w'(0) = 0$ etc.. Only in load cases δ the two spaces are different. Then S is simply a shift of V in a certain direction.

The setup of this space S can be illustrated as follows: *one* deflection w_δ with the property $w(l) = \delta$ is chosen and a curve from V is repeatedly added to this curve until the whole space S is generated from w_δ plus V. This may be denoted by

$$S = w_\delta \oplus V . \qquad (1.70)$$

In one regard, the space S is different from V. The sum of two curves w_1 and w_2 from S is a curve with double the deflection at the end of the beam, $2\,\delta$; that is, the sum $w_1 + w_2$ does not lie in S. Hence, if a test function $\hat{w}$ is added to w to see whether the value of $\Pi(w)$ can be reduced, $\Pi(w + \hat{w}) < \Pi(w)$, then $\hat{w}$ must be from V if the property $w(l) = \delta$ is to be retained.

Hence the minimum problem can be formulated as follows: given a fixed but otherwise arbitrary function w_δ, find the minimum of

$$\Pi(w_\delta + w) = \frac{1}{2} \int_0^l \frac{(M_\delta + M)^2}{EI} \, dx$$

$$= \frac{1}{2} \int_0^l \frac{M_\delta^2}{EI} \, dx + \int_0^l \frac{M_\delta \, M}{EI} \, dx + \frac{1}{2} \int_0^l \frac{M^2}{EI} \, dx$$

$$= \underbrace{\frac{1}{2} \int_0^l \frac{M_\delta^2}{EI} \, dx}_{I_1} + \underbrace{\int_0^l p_\delta \, w \, dx + \frac{1}{2} \int_0^l \frac{M^2}{EI} \, dx}_{\tilde{\Pi}(w)} \qquad (1.71)$$

by varying the additive term $w \in V$, where $p_\delta = EI \, w_\delta^{IV}$ is the load which be-longs to w_δ. (Depending on the character of w_δ, the work integral (p_δ, w) must eventually be supplemented with contributions of single forces or moments.)

Now the FE solution w_h of the subproblem: find the extreme point w_{ex} of $\tilde{\Pi}(w)$ on V

$$\tilde{\Pi}(w) = \frac{1}{2} \int_0^l \frac{M^2}{EI} \, dx - \int_0^l (-p_\delta) \, w \, dx \qquad \rightarrow \qquad \text{minimum} \qquad (1.72)$$

satisfies the inequality $\tilde{\Pi}(w_{ex}) < \tilde{\Pi}(w_h)$. Hence, given that

$$0 < \Pi(w_\delta + w_{ex}) = I_1 + \tilde{\Pi}(w_{ex}) < I_1 + \tilde{\Pi}(w_h) = \Pi(w_\delta + w_h) \quad (1.73)$$

the strain energy of the FE solution of a load case δ exceeds the strain energy of the exact solution, precisely because the "FE shadow" (of the homogeneous part) has a shorter length than the original function. The respective contributions at $w_{ex} \in V$ and $w_h \in V_h$ are negative:

$$\tilde{\Pi}(w_{ex}) < \tilde{\Pi}(w_h) < 0 \,, \qquad (1.74)$$

because they are solutions of a load case p.

In 2-D and 3-D problems, things are a little more complicated. To satisfy the geometric boundary condition $w = \bar{w} \neq 0$ on a part Γ_D of the boundary, a deflection surface w_δ must be constructed (mainly out of the nodal unit deflections of the nodes x_i which happen to lie on Γ_D) which interpolates $\bar{w}$ on Γ_D. Now if $\bar{w}$ is too complex, w_δ^h will only be an approximation, and the trial space will have a slightly different focus, because $S^h = w_\delta^h \oplus V$ is different from $S = w_\delta \oplus V$.

1.8 Principle of virtual displacements

The preceding text began with the principle of minimum potential energy and derived the FE method and the structural system $K \, u = f$ from this principle. But there are other variational formulations which lead to the same equations.

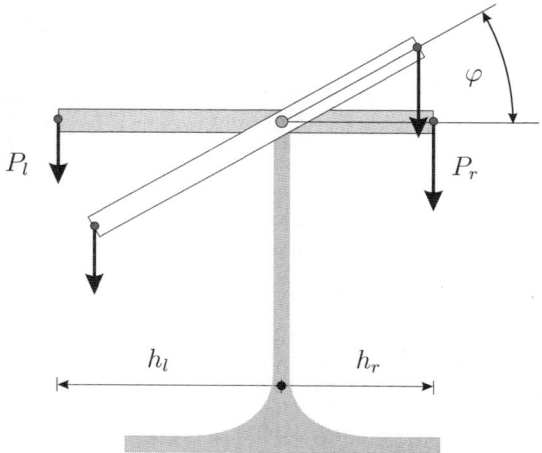

Fig. 1.18. A shopkeeper checks the equilibrium of the arm of a scale with the principle of virtual displacements

The most prominent of these is probably the principle of virtual displacements. It is more general, because there are situations where a potential energy expression does not exist, while a variational statement can still be formulated. Furthermore, the concept of an equivalent nodal force, or more generally, of the *equivalence of different load cases*, is based on the principle of virtual displacements, and therefore this principle is central to the FE method.

From a formal perspective the principle of virtual displacements is trivial. If $3 \times 4 = 12$, then $\delta u \times 3 \times 4 = 12 \times \delta u$ for arbitrary δu. This basically is the principle of virtual displacements. Things become more exciting if a) integration by parts (with functions) is performed and b) the equation $\delta u \times 3 \times u = 12 \times \delta u$ is read as a variational problem to find the "strong solution" $u = 4$.

Consider first a spring with stiffness $k = 3$ kN/m to which a force $f = 12$ kN is applied. The elongation u of the spring satisfies the equation $3\,u = 12$. But if $3\,u = 12$ then $\delta u \times 3\,u = 12 \times \delta u$ as well, whatever the value of δu.

Or if the vector $\boldsymbol{u}$ of the nodal displacements of a truss solves the system $\boldsymbol{K}\,\boldsymbol{u} = \boldsymbol{f}$, then $\boldsymbol{\delta u}^T \boldsymbol{K}\,\boldsymbol{u} = \boldsymbol{\delta u}^T \boldsymbol{f}$, for any vector $\boldsymbol{\delta u}$.

Or if the deflection curve w of a beam solves the differential equation $EI\,w^{IV} = p$, then $(\delta w, EI\,w^{IV}) = (\delta w, p)$ for arbitrary virtual deflections δw.

$$
\begin{aligned}
3\,u = 12 \quad &\Rightarrow \quad \delta u \times 3\,u = 12 \times \delta u && \text{for all } \delta u \\
\boldsymbol{K}\,\boldsymbol{u} = \boldsymbol{f} \quad &\Rightarrow \quad \boldsymbol{\delta u}^T \boldsymbol{K}\,\boldsymbol{u} = \boldsymbol{\delta u}^T \boldsymbol{f} && \text{for all } \boldsymbol{\delta u} \\
EI\,w^{IV}(x) = p(x) \quad &\Rightarrow \quad \int_0^l \delta w\, EI\,w^{IV}\, dx = \int_0^l \delta w\, p\, dx && \text{for all } \delta w\,.
\end{aligned}
$$

$$(1.75)$$

If the virtual displacements δw of the beam are *admissible* virtual displacements, these are displacements which are compatible with the support conditions, and if for simplicity all the supports are assumed to be rigid (no springs), then[2]

$$\int_0^l \delta w\, EI\, w^{IV}\, dx = \int_0^l \frac{M\,\delta M}{EI}\, dx \qquad (1.76)$$

and the third equation in (1.75) becomes

$$EI\, w^{IV}(x) = p(x) \qquad \Rightarrow \qquad \underbrace{\int_0^l \frac{M\,\delta M}{EI}\, dx}_{\delta W_i} = \underbrace{\int_0^l \delta w\, p\, dx}_{\delta W_e} . \qquad (1.77)$$

The equations on the left-hand side in (1.75) are the *Euler equations*. The equations on the right-hand side formulate the principle of virtual displacements, i.e., equilibrium in the *weak sense*:

> If a structure is in a state of equilibrium, then for any virtual displacement δu, the virtual internal work δW_i is the same as the virtual external work δW_e.

In classical structural mechanics conclusions are drawn from left to right, while in modern structural mechanics the opposite is true:

Euler equation	$\Rightarrow$	$\delta W_i = \delta W_e$	classical structural mechanics
Euler equation	$\Leftarrow$	$\delta W_i = \delta W_e$	modern structural mechanics

Today the search for the equilibrium position is cast in the form of a *variational problem*. The elongation u of the spring, the vector $\boldsymbol{u}$ of the nodal displacements of the truss, the deflection w of the beam are solutions of the following variational problems: find a number u, a vector $\boldsymbol{u}$, a function $w \in V$ such that

$$\delta u \times 3\, \underset{\uparrow}{u} = 12 \times \delta u \qquad \text{for all } \delta u \,,$$

$$\boldsymbol{\delta u}^T \boldsymbol{K} \underset{\uparrow}{\boldsymbol{u}} = \boldsymbol{\delta u}^T \boldsymbol{f} \qquad \text{for all } \boldsymbol{\delta u} \,,$$

$$\int_0^l \frac{\overset{\downarrow}{M}\,\delta M}{EI}\, dx = \int_0^l \delta w\, p\, dx \qquad \text{for all } \delta w \in V \,.$$

The next question then is: under what conditions is a variational solution also a classical solution?

$$3\, u = 12 \qquad \Leftarrow \qquad \delta u \times 3\, u = 12 \times \delta u \,, \qquad (1.78)$$

$$\boldsymbol{K}\, \boldsymbol{u} = \boldsymbol{f} \qquad \Leftarrow \qquad \boldsymbol{\delta u}^T \boldsymbol{K}\, \boldsymbol{u} = \boldsymbol{\delta u}^T \boldsymbol{f} \,, \qquad (1.79)$$

$$EI\, w^{IV}(x) = p(x) \qquad \Leftarrow \qquad \int_0^l \frac{M\,\delta M}{EI}\, dx = \int_0^l \delta w\, p\, dx \,. \qquad (1.80)$$

---

[2] This is Green's first identity $G(w, \delta w) = 0$ with $M(0) = M(l) = 0$; see Sect. 7.2, p. 508.

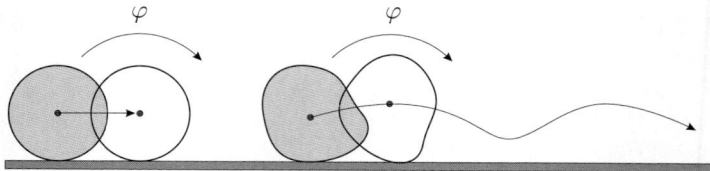

Fig. 1.19. A mechanic checks the eccentricity of a cylinder by rolling the cylinder across a flat surface

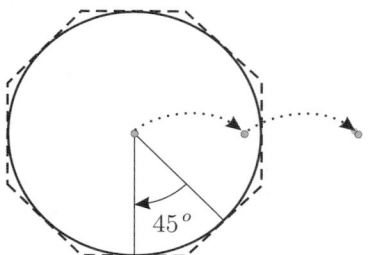

Fig. 1.20. The approximate shape is equivalent to a cylinder for all rotations which are multiples of 45^0

How many times must the spring be moved or a virtual displacement $\boldsymbol{\delta u}$ be applied to the truss or a virtual deflection δw to the beam before it is correct to say that the variational solution is also a solution in the classical sense?

The spring has one degree of freedom, so a test with one virtual displacement $\delta u \neq 0$ suffices. If the truss has n degrees of freedom, then $\delta W_i = \delta W_e$ must be verified for at least (and at most) n linear independent virtual displacements $\boldsymbol{\delta u}$ to draw the conclusion that equilibrium holds in the classical sense, $\boldsymbol{K\,u} = \boldsymbol{f}$. But a beam has infinitely many degrees of freedom, so $\delta W_i = \delta W_e$ must be verified for infinitely many virtual deflections δw before we can claim that the variational solution $w(x)$ satisfies the differential equation $EI\,w^{IV}(x) = p(x)$ at any point $0 < x < l$.

Elementary examples

When a shopkeeper checks the equilibrium of the arm of a scale

$$P_l\,h_l = P_r\,h_r\,,\tag{1.81}$$

she lightly tips the arm with her finger (Fig. 1.18), and if this slight disturbance does not start a hefty rotation of the arm then she gathers that the work done by the two loads P_l and P_r must be the same, and she concludes that (1.81) must hold:

$$P_l\,h_l = P_r\,h_r \quad \Longleftarrow \quad P_l\,h_l \tan\varphi = P_r\,h_r \tan\varphi \quad \text{for all } \varphi\,.\tag{1.82}$$

The same principle is applied by a mechanic who checks the eccentricity of a cylinder by rolling the cylinder back and forth on a flat surface. He knows that

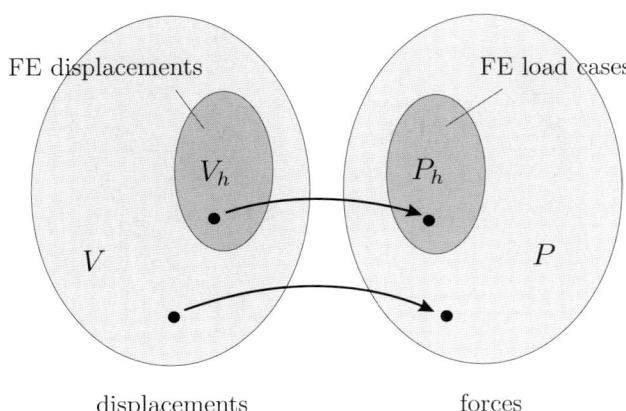

FE displacements

FE load cases

V_h

P_h

V

P

displacements

forces

Fig. 1.21. The trial space $V_h \subset V$ and its "dual" $P_h \subset P$

for any rotation angle φ of a perfect cylinder the axis maintains its (vertical) position, (see Fig. 1.19),

perfect cylinder $\quad \Rightarrow \quad$ vertical deviation vanishes for all rotations φ

and he concludes that if he senses no vertical movement with his fingers, the cylinder must be perfect:

perfect cylinder $\quad \Leftarrow \quad$ vertical deviation vanishes for all rotations φ.

Like the mechanic, the shopkeeper uses the principle of virtual displacements, which originally had the direction

$$P_l\, h_l = P_r\, h_r \qquad \Longrightarrow \qquad \delta W_e = \delta W_i \qquad (1.83)$$

in the opposite direction. If the two loads P_l and P_r satisfy the variational statement $\delta W_e = \delta W_i$ then they must also satisfy the equilibrium conditions in the classical sense

$$P_l\, h_l = P_r\, h_r \qquad \Longleftarrow \qquad \delta W_e = \delta W_i. \qquad (1.84)$$

This is also the approach of modern structural analysis. The FE method begins with the principle of virtual displacements, and it constructs an *equivalent load case* p_h which is work-equivalent to the original load case p with respect to a *finite number* of virtual displacements.

Likewise the mechanic would start with an iron bar with a quadratic cross section double the radius R of the cylinder, $(2\,R \times 2\,R)$. This shape is equivalent to the cylinder (= maintains the vertical position of its center) with regard to all rotations which are a multiple of 90^0, that is $\varphi_i = i \times (360/4)^0, i = 1, 2, \ldots$. By refining the shape, $4 \rightarrow 8 \rightarrow 16 \rightarrow \ldots$ (sides), the mechanic enlarges the "test and trial space V_h" and consequently the shape more and more begins to resemble a true cylinder; see Fig. 1.20.

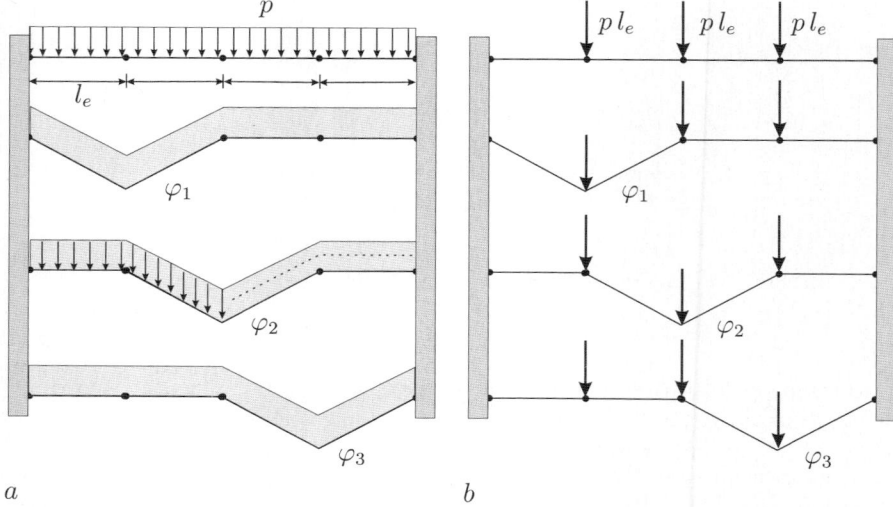

Fig. 1.22. The virtual work done by the distributed load p and the equivalent nodal forces is the same for any unit displacement φ_i: $\delta W_e(p_h, \varphi_i) = \delta W_e(p, \varphi_i)$

Remark 1.2. The virtual internal work δW_i of the arm of the balance is zero, because the arm is a rigid body, but this does not change the logic. A rigid body is in a state of equilibrium if and only if the virtual external work is zero:

$$P_l\, h_l = P_r\, h_r \qquad \Longleftrightarrow \qquad \delta W_e = \delta W_i = 0\,. \tag{1.85}$$

Equivalence

The set of shapes w that a taut rope can assume over its lifetime constitute the deformation space V. The shapes w_h that the shape functions φ_i can generate constitute a small subspace V_h within V.

Now with any shape w of the rope (Fig. 1.21), we can associate a load case p. If $w \in C^2(0, l)$, the load is simply the right-hand side corresponding to w, i.e., $p = -H\, w''$. If w is not that smooth, point forces can appear where w' is discontinuous. The set of all these load cases constitutes the "dual space" P.

By this mapping

$$\text{deflection curve} \quad w \quad \Rightarrow \quad \text{load case} \quad p \tag{1.86}$$

the subspace V_h is also mapped onto a subspace P_h of P, and true to the nature of the shape functions φ_i the load cases in the subspace $P_h \subset P$ consist of nodal forces *only*.

Because the original load case p (Fig. 1.5), is not among these load cases in P_h, the FE method chooses a substitute load case p_h in P_h, that can be solved on V_h.

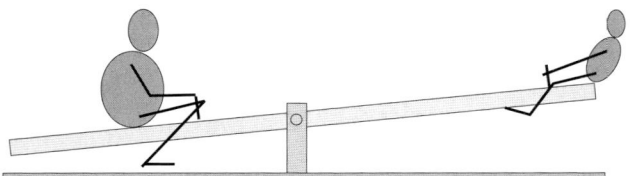

Fig. 1.23. The work done by father and son is the same over every cycle turn of the see-saw

- That load case p_h in P_h is chosen which is work-equivalent to the original load case with regard to the virtual displacements $\varphi_i \in V_h$:

$$\delta W_e(p_h, \varphi_i) = \delta W_e(p, \varphi_i) \qquad \text{for all} \qquad \varphi_i \in V_h \,. \qquad (1.87)$$

In other words, the substitute loads, the nodal forces, must—upon acting through any virtual displacement $\varphi_i \in V_h$—contribute the same virtual work as the original load p; see Fig. 1.22.

The simplest application of this idea can be seen in Fig. 1.23. Over every cycle of the see-saw, father (load case p) and son (load case p_h) (or is it vice versa?) contribute the same virtual work. With regard to all possible cycles of the see-saw, father and son represent two equivalent load cases. They are indistinguishable on V_h.

1.9 Taut rope

In the following we will show how this idea (1.87) is applied to a taut rope. The FE solution is a linear combination of the three unit deflections

$$w_h(x) = w_1 \cdot \varphi_1(x) + w_2 \cdot \varphi_2(x) + w_3 \cdot \varphi_3(x) \,. \qquad (1.88)$$

Each unit deflection φ_i represents a particular load case $p_i \in P_h$, i.e., a particular arrangement of nodal forces. Take for example the unit deflection φ_1 of the first node (Fig. 1.24). The rope assumes this shape if at the first node a force $f_1 = -P = H/l_e$ pointing upward is applied, at the next node a force double that size pointing downward $f_2 = 2P$, and at the third node a force $f_3 = -P$ again pointing upward:

$$f_1 = -\frac{H}{l_e} \ \uparrow \qquad f_2 = 2\frac{H}{l_e} \ \downarrow \qquad f_3 = -\frac{H}{l_e} \ \uparrow \qquad (1.89)$$

where H is the horizontal force that pulls the rope taut. In the same sense, two load cases p_2 and p_3 can be associated with the other two unit deflections φ_2 and φ_3. Hence, given any shape

$$w_h = w_1 \cdot \varphi_1(x) + w_2 \cdot \varphi_2(x) + w_3 \cdot \varphi_3(x) \qquad (1.90)$$

there is a load case p_h

3 unit deflections

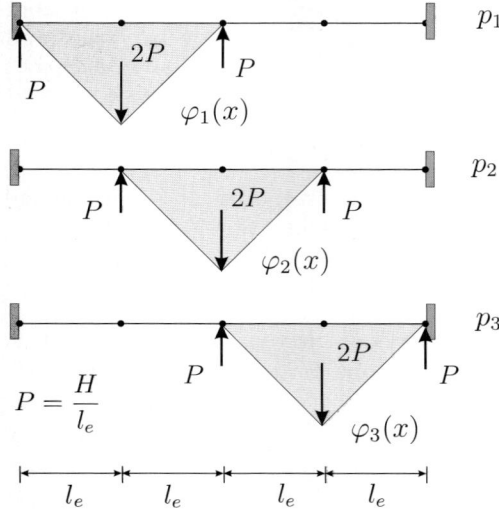

Fig. 1.24. The three unit load cases p_1, p_2, p_3 and the three unit deflections $\varphi_1, \varphi_2, \varphi_3$

$$p_h = w_1 \cdot p_1 + w_2 \cdot p_2 + w_3 \cdot p_3$$
$$= w_1 \cdot (\uparrow \quad \downarrow \quad \uparrow)_{1,2,3} + w_2 \cdot (\uparrow \quad \downarrow \quad \uparrow)_{2,3,4} + w_3 \cdot (\uparrow \quad \downarrow \quad \uparrow)_{3,4,5} \,,$$
$$(\ldots)_{1,2,3} = (\ldots) \text{ at the nodes } 1,2,3 \tag{1.91}$$

that produces the shape w_h.

By an appropriate choice of the weights w_i (the nodal deflections!), the FE load case p_h can be scaled in such a way that it is work-equivalent to the distributed load p in the sense of the principle of virtual displacements. This is the basic idea.

Because the substitute load case p_h consists of only three nodal forces, the balance between those forces and the original load p cannot be maintained with regard to *all* possible virtual deflections of the rope. Rather, the number of tests must also be restricted to three, and therefore the three unit deflections of the three nodes are chosen as test functions (virtual deflections).

The virtual work done by the distributed load p acting through unit deflection φ_i is the integral

$$\delta W_e(p, \varphi_i) = \int_0^l p \, \varphi_i \, dx \,, \tag{1.92}$$

and the virtual work of the three nodal forces f_i at the three nodes x_1, x_2, x_3 acting through the same unit deflection is the sum

$$\delta W_e(p_h, \varphi_i) = f_1 \cdot \varphi_i(x_1) + f_2 \cdot \varphi_i(x_2) + f_3 \cdot \varphi_i(x_3) \,. \tag{1.93}$$

The virtual work must be the same for any virtual deflection φ_i, hence

$$\delta W_e(p, \varphi_i) = \delta W_e(p_h, \varphi_i) \,, \quad i = 1, 2, 3 \,. \tag{1.94}$$

Next comes an important idea:

- The FE solution w_h is itself an *equilibrium solution*, and therefore it too satisfies the principle of virtual displacements.

  Hence, for any φ_i, the virtual internal work of the FE solution w_h

$$\delta W_i(w_h, \varphi_1) = \int_0^l \frac{T_h T_1}{H} \, dx \qquad T_h = H w_h', \quad T_1 = H \varphi_1', \tag{1.95}$$

is equal to the virtual external work done by the nodal forces:

$$\delta W_i(w_h, \varphi_1) = \delta W_e(p_h, \varphi_1) \,. \tag{1.96}$$

This means that the internal virtual energy of the FE solution can be added to the string of equations (1.94)

$$\delta W_e(p, \varphi_1) = \underbrace{\delta W_e(p_h, \varphi_1) = \delta W_i(w_h, \varphi_1)}_{\text{principle of virtual displacements}} \tag{1.97}$$

or if the term in the middle is dropped

$$\delta W_e(p, \varphi_1) = \ldots = \delta W_i(w_h, \varphi_1) \,, \tag{1.98}$$

and the whole equation turned around

$$\delta W_i(w_h, \varphi_1) = \delta W_e(p, \varphi_1) \tag{1.99}$$

and the original notation used, then the result is

$$\int_0^l \frac{T_h T_1}{H} \, dx = f_1 \,. \tag{1.100}$$

If these steps are repeated with φ_2 and φ_3, a system of three equations

$$\boldsymbol{K} \boldsymbol{w} = \boldsymbol{f} \,, \tag{1.101}$$

is obtained, where $\boldsymbol{K}$ is just the stiffness matrix of the rope

$$\boldsymbol{K} = \frac{4H}{l} \begin{bmatrix} 2 & -1 & 0 \\ -1 & 2 & -1 \\ 0 & -1 & 2 \end{bmatrix} \,. \tag{1.102}$$

The element k_{ij} of $\boldsymbol{K}$ is the *strain energy product* between the two unit deflections φ_i and φ_j

$$k_{ij} = \int_0^l H \varphi_i' \varphi_j' \, dx \,, \tag{1.103}$$

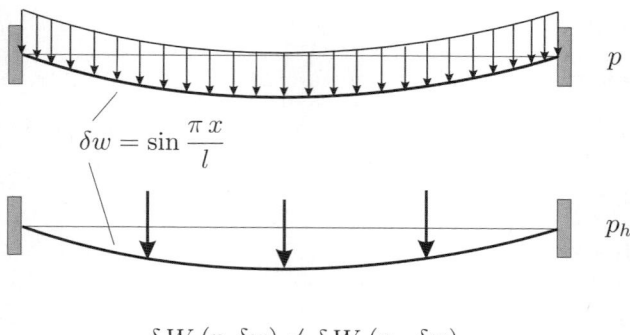

$$\delta w = \sin \frac{\pi x}{l}$$

$$\delta W_e(p, \delta w) \neq \delta W_e(p_h, \delta w)$$

Fig. 1.25. Taut rope: with regard to a sine-wave, the FE load case p_h is not work-equivalent to the load case p

while the component f_i of the vector $\boldsymbol{f}$ is the work done by the load p acting through φ_i

$$f_i = \int_0^l p\,\varphi_i\,dx\,. \tag{1.104}$$

Because the distributed load p is constant, all the equivalent nodal forces are the same

$$f_1 = f_2 = f_3 = \frac{p\,l}{4}\,. \tag{1.105}$$

Hence the nodal deflections are

$$w_1 = 1.5\,\frac{p\,l^2}{16\,H}\,, \qquad w_2 = 2.0\frac{p\,l^2}{16\,H}\,, \qquad w_3 = 1.5\frac{p\,l^2}{16\,H}\,, \tag{1.106}$$

and the solution is

$$w_h = \frac{p\,l^2}{16\,H}\left[1.5\cdot\varphi_1(x) + 2.0\cdot\varphi_2(x) + 1.5\cdot\varphi_3(x)\right], \tag{1.107}$$

which coincides with (1.20).

What we did in the end is that we replaced the original load case p with a load case p_h. A reviewing engineer who checks the FE solution with the unit deflections would not find any difference between the two load cases p and p_h. On each test with one of the unit deflections, he would recognize that the response of the rope, measured in units of virtual work, is the same.

Only by refining the tools and testing the FE solution with a sine wave

$$\delta w = \sin\frac{\pi x}{l} \tag{1.108}$$

will he realize that the two load cases cannot be the same, because the virtual work is not the same, (see Fig. 1.25),

$$\int_0^l p\sin\frac{\pi x}{l}\,dx \neq \sum_{i=1}^3 f_i\sin\frac{\pi x_i}{l} \qquad x_i = \text{location of } f_i\,. \tag{1.109}$$

Force method

To complete the picture, note that the *force method* of structural mechanics is based on the *principle of minimum complementary energy*

$$\Pi_c(w) = -\frac{1}{2} \int_0^l \frac{M^2}{EI}\, dx + V \cdot \delta \qquad \rightarrow \qquad \text{minimum}\,. \qquad (1.110)$$

(The term δ denotes a possible displacement of a support.) This principle is essentially the opposite of the principle of minimum potential energy. The minimum value of Π_c is sought among all functions $w = w_0 + X_1\, w_1 + \ldots X_n\, w_n$ that are particular solutions of the equation $EI\, w^{IV} = p$. The solution consists of a curve w_0, which is a particular solution of the equation $EI\, w_0^{IV} = p$, and the deflections w_i caused by the redundants X_i, $EI\, w_i^{IV} = 0$. Better known is the bending moment distribution M of this solution

$$M = M_0 + X_1\, M_1 + X_2\, M_2 + \ldots + X_n\, M_n\,. \qquad (1.111)$$

The condition $\partial\, \Pi_c / \partial\, X_i = 0$ leads to the system of equations

$$\boldsymbol{F}\,\boldsymbol{x} = -\,\boldsymbol{\delta}_0 \qquad (1.112)$$

where $\boldsymbol{F}$ is the *flexibility matrix* with elements $f_{ij} = (M_i, M_j)$, the vector $\boldsymbol{x} = [X_1, X_2, \ldots, X_n]^T$ contains the redundants, and the vector $\boldsymbol{\delta}_0$ embodies the interaction between M_0 and the M_i, because in the principle of minimum complementary energy the "equivalent nodal forces" are just the scalar product[3] of the bending moment M_0 (moment distribution of the statically determinate structure) with the curvature M_i/EI of the redundants

$$\delta_{i0} = \int_0^l \frac{M_0\, M_i}{EI}\, dx\,. \qquad (1.113)$$

Theoretically one could write an FE program that starts with a particular state w_0 enriched with n redundants w_i. In frame analysis, where n is equal to the degree of static indeterminacy this method would always yield the exact solution, while in plate or shell analysis infinitely many X_i would be needed.

1.10 Least squares

In FE analysis, the movements of a structure are constrained, because the structure is only allowed to assume shapes that can be expressed as piecewise linear, piecewise quadratic, or similar shape functions.

We then find that the FE solution is that deformation of the structure which among all remaining movements renders the potential energy a minimum. This is equivalent to the statement that the associated load case p_h is

---

[3] From now on we simply say scalar product instead of L_2-scalar product.

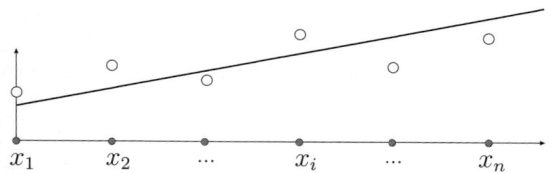

Fig. 1.26. Best fit with a straight line

work-equivalent to the original load case p. Finally the distance (in terms of energy) between the exact solution u and the FE solution u_h is a minimum:

$$a(e, e) = a(u - u_h, u - u_h) \leq a(u - v_h, u - v_h) \qquad v_h \in V_h. \quad (1.114)$$

No other function $v_h \in V_h$ comes closer to u in this sense than u_h. But a minimum in energy means (in the case of a beam for example)

$$a(e, e) = \int_0^l \frac{(M - M_h)^2}{EI}\, dx \qquad \rightarrow \qquad \text{minimum}, \quad (1.115)$$

that the least-squares error of the internal actions is a minimum.

Least squares is a concept of numerical analysis. When a straight line $M_h(x) = a\,x + b$ is drawn through a number of points (Fig. 1.26)

$$\begin{aligned} a\,x_1 + b &= M(x_1), \\ a\,x_2 + b &= M(x_2), \\ \cdots &= \cdots, \\ a\,x_n + b &= M(x_n), \end{aligned} \quad (1.116)$$

as to minimize the sum of squared errors

$$F = \sum_{i=1}^n (M_h(x_i) - M(x_i))^2 \qquad \rightarrow \qquad \text{minimum}, \quad (1.117)$$

$$\frac{\partial F}{\partial a} = 0, \qquad \frac{\partial F}{\partial b} = 0, \quad (1.118)$$

we obtain a least-squares solution. Solving (1.118) is equivalent to solving the 2×2 system of equations

$$\boldsymbol{A}^T_{(2 \times n)} \boldsymbol{A}_{(n \times 2)} \begin{bmatrix} a \\ b \end{bmatrix} = \boldsymbol{A}^T_{(2 \times n)} \boldsymbol{m}_{(n)}, \quad (1.119)$$

the *normal equations*, where $\boldsymbol{A}$ is the coefficient matrix of (1.116) and the vector $\boldsymbol{m} = [M(x_1), M(x_2)\ldots]$ is the right-hand side of (1.116).

In FE analysis, the bending moment is a function

$$M_h(x) = \sum_i w_i\, M_i(x), \qquad M_i(x) = -EI\, \varphi_i''(x), \quad (1.120)$$

and because the FE solution is minimized with respect to all points $0 < x < l$ of a beam, the sum is replaced by an integral (see (1.115)).

The equation $\boldsymbol{K}\boldsymbol{u} = \boldsymbol{f}$ is the associated normal equation [232]. This equation is obtained if the infinitely many equations $M_h(x) = M(x)$ (there are infinitely many points $x_1, x_2, \ldots$ in the interval $[0, l]$) basically a matrix $\boldsymbol{A}_{\infty \times n}$ is multiplied by the transposed matrix $\boldsymbol{A}_{n \times \infty}^T$ from the left, and the diagonal matrix $\boldsymbol{C}_{\infty \times \infty}$ with weights $C_{ii} = EI$ is placed in between:

$$\boldsymbol{A}_{(n \times \infty)}^T \, \boldsymbol{C}_{(\infty \times \infty)} \boldsymbol{A}_{(\infty \times n)} = \boldsymbol{K}_{(n \times n)} \,. \tag{1.121}$$

Weighted least squares

The term EI comes from the strain energy product

$$a(w, \hat{w}) = \int_0^l \frac{M \, \hat{M}}{EI} \, dx \,, \tag{1.122}$$

because in FE analysis the bending moments are scaled in such a way that in the *weighted* least-squares sense the discrepancies between M_h and the exact bending moment M are minimized

$$F := \int_0^l \frac{(M - M_h)^2}{EI} \, dx = \int_0^l (M - M_h) \, (\kappa - \kappa_h) \, dx \;\; \rightarrow \;\; \min. \tag{1.123}$$

which means that the error in the bending moments is multiplied by the error in the curvature, and that this product is minimized.

Of course if the bending stiffness EI is constant, then there is no difference between least squares and weighted least squares.

Global and local

In least squares the *global picture*—the error over the whole interval $[0, l]$—is studied. But the algorithm also incorporates a *local match*, as we will see in the following. To simplify the derivation it is assumed that $EI = 1$, and the short-hand notation for integrals (see Chap. 7) is used.

For the integral

$$\begin{aligned} F &= (M - M_h, M - M_h) = (M, M) - 2 \, (M, M_h) + (M_h, M_h) \\ &= (M, M) - 2 \sum_i (M, M_i) \, w_i + \sum_{i,j} (M_i, M_j) \, w_i \, w_j \end{aligned} \tag{1.124}$$

to attain a minimum at $\boldsymbol{w}$ it is necessary that the gradient of F vanishes at this point with respect to the nodal values w_i:

$$\frac{\partial F}{\partial w_i} = 2 \sum_j (M_i, M_j) \, w_j - 2 \, (M, M_i) = 2 \, (M_h, M_i) - 2(M, M_i) = 0 \,. \tag{1.125}$$

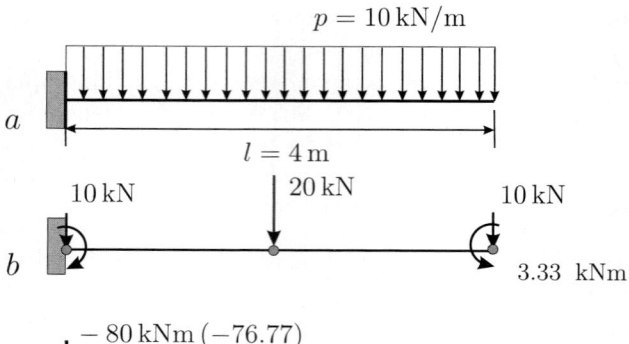

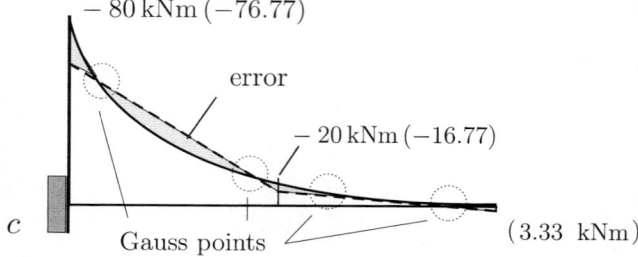

bending moments $M(x)$ and $(M_h(x))$

Fig. 1.27. Cantilever beam and FE load case. At the Gauss points the error in the bending moment is zero

Hence the solution of (1.123) is equivalent to making the error $M - M_h$ orthogonal to the functions M_i, with EI reinserted

$$\int_0^l \frac{(M - M_h)\, M_i}{EI}\, dx = 0\,, \qquad i = 1, 2 \ldots \qquad \text{(local match)}. \quad (1.126)$$

Test range

The test range for the local match between M and M_h is not a single element, but the support of the individual hat function M_i. (This is what the bending moments of the unit deflections φ_i look like.) Outside of the support Ω_i the hat function is zero, and therefore the local test is an integral over the interval $\Omega_i = [x_{i-1}, x_{i+1}]$:

$$\int_0^l \frac{(M - M_h)\, M_i}{EI}\, dx = \int_{x_{i-1}}^{x_{i+1}} \frac{(M - M_h)\, M_i}{EI}\, dx = 0\,. \quad (1.127)$$

In the case of a continuous beam the test range of an individual hat function M_i consists of two consecutive elements. And the positive and negative errors $M - M_h$ must be distributed in such a way that in the weighted average sense the error over any two neighboring elements disappears, see Fig. 1.27.

1.11 Distance inside = distance outside

The central equation of the FE method—in the case of a beam for example, is

$$a(e, e) = \int_0^l \frac{(M - M_h)^2}{EI} \, dx \qquad \rightarrow \qquad \text{minimum}. \qquad (1.128)$$

The FE solution is scaled in such a way that the mean squared error attains the smallest possible value; see Fig. 1.28. This is achieved by projecting the exact solution onto the subspace V_h in such a way that the error in the bending moments is orthogonal to all φ_i in V_h:

$$a(e, \varphi_i) = \int_0^l \frac{(M - M_h)\, M_i}{EI} \, dx = 0, \qquad i = 1, 2, \ldots n. \qquad (1.129)$$

But how can the error be controlled if the exact bending moment distribution $M(x)$ is not known? How does the program measure $M - M_h$? The answer is simple: the virtual internal energy is equal to the virtual external work. Hence if the error in the bending moments is orthogonal to the curvature produced by each φ_i

$$\delta W_i = \int_0^l \underbrace{(M - M_h) \frac{M_i}{EI}}_{unknown} dx = \int_0^l \underbrace{(p - p_h)}_{known} \varphi_i \, dx = \delta W_e = 0, \quad (1.130)$$

the residual forces $p - p_h$ must be orthogonal to each φ_i as well. The terms on the right-hand side are

$$\int_0^l (p - p_h)\, \varphi_i \, dx = \int_0^l p\, \varphi_i \, dx - \int_0^l p_h\, \varphi_i \, dx$$

$$= f_i - \sum_{j=1}^n k_{ij}\, u_j = f_i - f_i^h = 0, \qquad (1.131)$$

i.e., from the equivalent nodal force f_i of the load case p, the equivalent nodal forces f_i^h of the load case p_h are subtracted

$$\int_0^l p_h\, \varphi_i \, dx = \int_0^l \left(\sum_{j=1}^n w_j\, p_j \right) \varphi_i \, dx = \sum_{j=1}^n \int_0^l p_j\, \varphi_i \, dx\; w_j$$

$$= \sum_{j=1}^n k_{ij}\, w_j = f_i^h \qquad (1.132)$$

and because $f_i = f_i^h$ it follows that $\delta W_e(e, \varphi_i) = \delta W_i(e, \varphi_i) = 0$.

In a plate, the same equations are

distributed load $p = 10\,\mathrm{kN/m}$

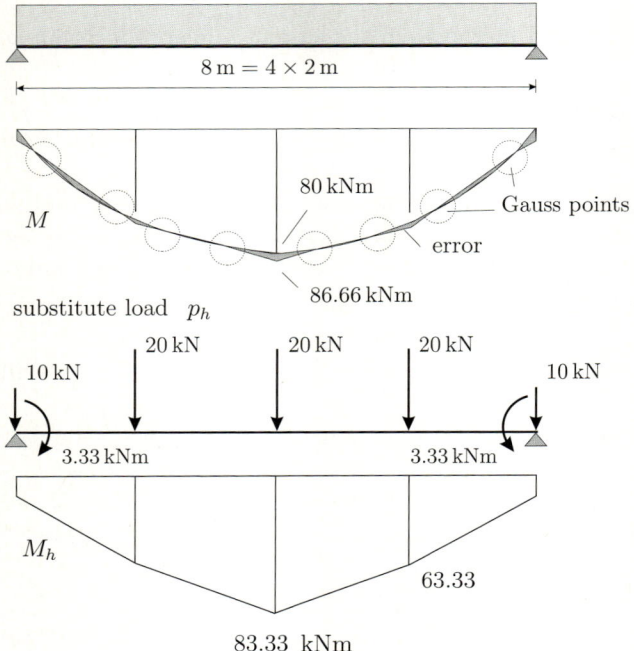

$$8\,\mathrm{m} = 4 \times 2\,\mathrm{m}$$

M

$80\,\mathrm{kNm}$

Gauss points

error

$86.66\,\mathrm{kNm}$

substitute load p_h

$20\,\mathrm{kN}$ $20\,\mathrm{kN}$ $20\,\mathrm{kN}$

$10\,\mathrm{kN}$ $10\,\mathrm{kN}$

$3.33\,\mathrm{kNm}$ $3.33\,\mathrm{kNm}$

M_h

63.33

$83.33\ \mathrm{kNm}$

Fig. 1.28. Original load and substitute load

$$\delta W_i = \int_\Omega \underbrace{(\boldsymbol{\sigma} - \boldsymbol{\sigma}_h)}_{unknown} \bullet \varepsilon_i \, d\Omega = \int_\Omega \underbrace{(\boldsymbol{p} - \boldsymbol{p}_h)}_{computable} \bullet \boldsymbol{\varphi}_i \, d\Omega$$

$$+ \int_\Gamma \underbrace{(\boldsymbol{t} - \boldsymbol{t}_h)}_{computable} \bullet \boldsymbol{\varphi}_i \, ds + \sum_k \int_{\Gamma_k} \underbrace{\boldsymbol{t}_\Delta}_{computable} \bullet \boldsymbol{\varphi}_i \, ds = \delta W_e \quad (1.133)$$

where the integrals

$$\sum_k \int_{\Gamma_k} \boldsymbol{t}_\Delta \bullet \boldsymbol{\varphi}_i \, ds \qquad\qquad (1.134)$$

are the virtual work done by line loads $\boldsymbol{t}_\Delta$, which represent the jumps in the stresses on the interelement boundaries Γ_k.

The orthogonality in the stresses is equivalent to the fact that the unit load cases $\boldsymbol{p}_i$—the *action* behind the displacement fields $\boldsymbol{\varphi}_i$—contribute no work on acting through the error $e(\boldsymbol{x}) = \boldsymbol{u}(\boldsymbol{x}) - \boldsymbol{u}_h(\boldsymbol{x})$

$$\delta W_i = \int_\Omega (\boldsymbol{\sigma} - \boldsymbol{\sigma}_h) \bullet \varepsilon_i \, d\Omega = \int_\Omega (\boldsymbol{p} - \boldsymbol{p}_h) \bullet \boldsymbol{\varphi}_i \, d\Omega + \ldots = \delta W_e = 0 \,. (1.135)$$

This is the right occasion to recall how we argue in the *force method*. The bending moment distribution $M = M_0 + X_1\,M_1 + X_2\,M_2 + \ldots$ of a continuous beam is orthogonal to all redundants X_i

$$\int_0^l \frac{M\,M_i}{EI}\,dx = 0 \quad \Rightarrow \quad w(x_i) = 0 \quad \text{or} \quad \Delta w'(x_i) = 0, \quad \text{etc.} \quad (1.136)$$

which means that the previously eliminated constraints at the supports. i.e., conditions such as $w(x_i) = 0$ or $\Delta w'(x_i) = 0$ (relative rotation), are satisfied by the exact solution.

Now it might be assumed that the orthogonality (1.133) means just this: that the error in the displacements is zero at the nodes

$$\int_\Omega (\boldsymbol{\sigma} - \boldsymbol{\sigma}_h) \bullet \varepsilon_i \, d\Omega = 0 \quad \overset{?}{\Rightarrow} \quad \boldsymbol{u}(\boldsymbol{x}_i) - \boldsymbol{u}_h(\boldsymbol{x}_i) = \boldsymbol{0}. \quad (1.137)$$

But this is not true. The FE solution does *not* interpolate the exact solution at the nodes.

Only in one-dimensional problems, as in beam problems, is the condition

$$\delta W_i = \int_0^l \frac{(M - M_h)\,M_i}{EI}\,dx = 0, \quad (1.138)$$

equivalent to the fact that the error in the deflection is zero at the nodes. Here M_i is the bending moment corresponding to the unit deflection φ_i, and the conclusion is correct, because (due to $\delta W_e = \delta W_i$) the orthogonality (1.138) is equivalent to

$$0 = \delta W_i = \delta W_e = (w(x_i) - w_h(x_i)) \cdot P = 0 \cdot P. \quad (1.139)$$

The nodal force P is the force that causes the unit nodal deflection at x_i. (If the distance from the node to the neighboring nodes is not the same, then an additional moment M can appear, but this moment does not contribute any work, because the rotation of the node is zero, $\varphi_i'(x_i) = 0$.)

In plates, slabs and shells, it is not possible to associate a single point force with a unit nodal displacement φ_i. Instead such a displacement is generated by a diffuse *cloud* of surface loads and line forces in the neighborhood of the nodes, and therefore the sharp point condition $\boldsymbol{u}(\boldsymbol{x}_i) = \boldsymbol{u}_h(\boldsymbol{x}_i)$ is not guaranteed.

It is not the intention of an FE program to interpolate the exact displacement field $\boldsymbol{u}$ at the nodes, but rather to minimize the error in the stresses. From an engineering point of view, this certainly makes more sense than to interpolate the true displacement field $\boldsymbol{u}$ at the nodes. Only in 1-D problems do we get both: *interpolation at the nodes + minimal distance in terms of energy.*

Remark 1.3. By 1-D problems are meant here and in the following the classical differential equations $-H\,w''$ (rope), $-EA\,u''$ (bar), and $EI\,w^{IV}$ (beam) of structural mechanics. With regard to extended equations such as $-EA\,u'' + c\,u$ or $EI\,w^{IV} + c\,w$, see the remark at the end of Sect. 3.1 on p. 292.

1.12 Scalar product and weak solution

In classical structural mechanics the deflection curve w of a beam is determined by solving the differential equation $EI\,w^{IV} = p$ and adjusting the solution to the boundary conditions. According to the principle of virtual displacements (Green's first identity), the classical solution is also a solution of a variational problem: find a function w such that

$$\int_0^l \frac{M\,\delta M}{EI}\,dx = \int_0^l p\,\delta w\,dx \qquad \text{for all } \delta w \in V\,. \tag{1.140}$$

The variational form and the differential equation are equivalent formulations. The differential equation $EI\,w^{IV} = p$ is the *Euler equation* of the variational principle. The variational solution is called a *weak solution*, because for the variational statement

$$\int_0^l \frac{M_h\,M_i}{EI}\,dx = \int_0^l p\,\varphi_i\,dx\,, \qquad i = 1, 2, \ldots n\,, \tag{1.141}$$

to make sense the solution must only have square-integrable second derivatives, $M_i = -EI\,\varphi_i''$, while the Euler equation requires the solution w to have fourth-order derivatives.

This is the official (?) version. But we think that the person who first spoke of a weak solution had more in mind than counting derivatives.

In mathematics there is the concept of *weak convergence*, and this concept is closely related to the scalar product (or principle of virtual displacements), and ultimately to the way the shopkeeper checks the arm of a balance and modern structural engineers argue.

To determine the mass of a brick we throw it in the air. Sensing the force f, the acceleration a and knowing that $f = m\,a$ we guess the mass m of the brick. *Basically we draw our conclusion indirectly*[4].

And this is how an FE program proceeds. To judge the load on a structure an FE program "shakes" the structure. It applies virtual displacements and it measures the virtual work done by the load. This is what the scalar product is for.

With the scalar product *duality* enters the stage, and therewith the distinction between *displacements* and *forces*. An A is tested by holding it against a B, where $A(= p)$ might be a distributed load and $B(= \delta w)$ a virtual displacement, and the work done by p acting through the displacement δw provides a measure to judge p.

If we drive a truck over a bridge and then shake the bridge by applying a series of virtual deflections δw, the truck performs virtual work. If in this scalar product

---

[4] According to a quote in [74] p. 172 Germain [93] expressed similar ideas: 'When we wish to see if a suitcase is heavy, we lift it. To estimate the tension in a (stationary) transmission belt, we try to draw it aside from its equilibrium position. The essential underlying mathematical idea is that of "duality"'.

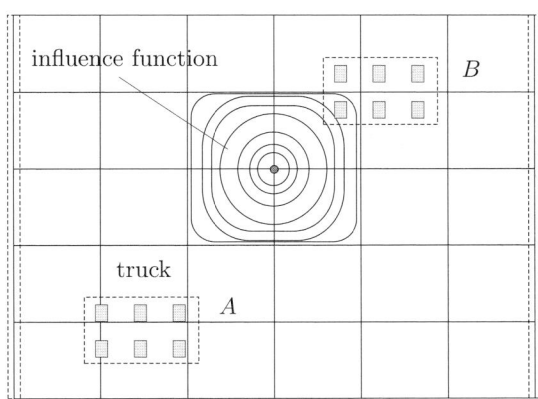

Fig. 1.29. The equivalent nodal force (= work) of truck B is the wheel load $\times$ the deflection under the wheel. The influence of truck A is zero

$$\int_{\Omega} p\,\delta w\,d\Omega =: p\,(\delta w) \qquad p = \text{truck} \qquad (1.142)$$

the load p is kept fixed and the virtual displacement δw is varied, the scalar product becomes a *functional* $p\,(\delta w)^5$. This is an expression into which a function δw is substituted and which returns a number. Any truck and any load case p constitutes a functional in this sense.

If p is the original truck and p_h the FE truck, then the FE method consists in replacing the functional $p\,()$ on V_h by a functional $p_h\,()$ in such a way that the real truck $p()$ and the pseudo-truck $p_h()$, the two functionals, are equivalent with respect to all virtual displacements $\varphi_i \in V_h$:

$$p\,(\varphi_i) = p_h\,(\varphi_i)\,, \qquad i = 1, 2, \ldots, n\,, \qquad (1.143)$$

and the FE truck p_h eventually will converge to the real truck p (if the mesh size h tends to zero) if in the limit the functional p_h agrees with the functional p with respect to *all* virtual displacements:

$$\lim_{h \to 0} p_h\,(\delta w) = p\,(\delta w) \qquad \text{for all } \delta w \text{ of the structure}\,. \qquad (1.144)$$

This is what *weak convergence* means, and in this sense the FE solution is a *weak solution*.

The distance between p and p_h, the original truck and the FE truck, is not judged directly, i.e., by comparing the pressure per square inch on the bridge $|p_h(\boldsymbol{x}) - p(\boldsymbol{x})|$, but by studying the effects which the two trucks p_h and p trigger with regard to the same virtual displacements. Our judgement is based on the belief that *if the effects are the same then the agents behind these effects must be the same.*

This conclusion is—if the reader will allow this remark—typical of our time where *substance* has been replaced by *function*. We no longer care what something is, but are only interested in how it interacts with other objects.

---

[5] Usually we denote the functional by the same letter as the load.

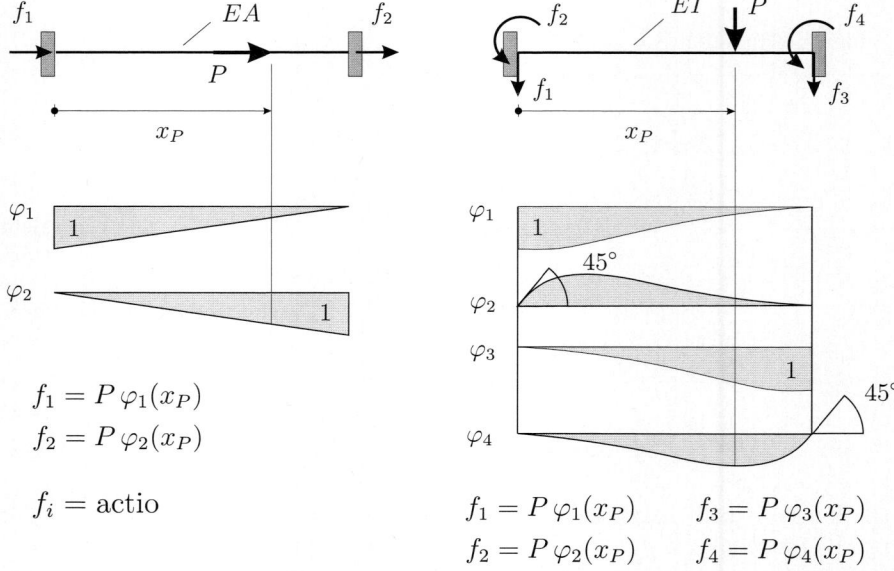

$$f_1 = P\,\varphi_1(x_P)$$
$$f_2 = P\,\varphi_2(x_P)$$

$$f_i = \text{actio}$$

$$f_1 = P\,\varphi_1(x_P) \qquad f_3 = P\,\varphi_3(x_P)$$
$$f_2 = P\,\varphi_2(x_P) \qquad f_4 = P\,\varphi_4(x_P)$$

Fig. 1.30. Reduction of the load into the nodes. The equivalent nodal forces are equal to the work which the two forces P contribute acting through the unit displacements

1.13 Equivalent nodal forces

No concept better expresses the nature of the FE method than the notion of an equivalent nodal force, because an FE program does not think in terms of forces but in terms of work. This is the medium whereby an FE program establishes contact with the outside world, and forces that contribute the same work when acting through the same displacement are the same for an FE program. They all belong to the same *equivalence class*, and the representatives of these *equivalence classes* are the equivalent nodal forces:

$$f_i = \int_{\Omega} p\,\varphi_i\,d\Omega \qquad (\text{kN m}\,) = (\text{kN/m}^2)(\text{m})(\text{m}^2)\,. \tag{1.145}$$

How much a load contributes to an equivalent nodal force depends on how much of the movement of the node is felt at the location of the load. *The influence of a node extends precisely as far as the nodal unit displacements;* see Fig. 1.29. Hence nodal unit displacements are *influence functions*. They are influence functions for equivalent nodal forces (Fig. 1.30).

Now virtual work is a fuzzy measure, because given any load p there is obviously a *second, (third, fourth, ...)*, load $\tilde{p}$ not identical to p, that contributes the same amount of work as p:

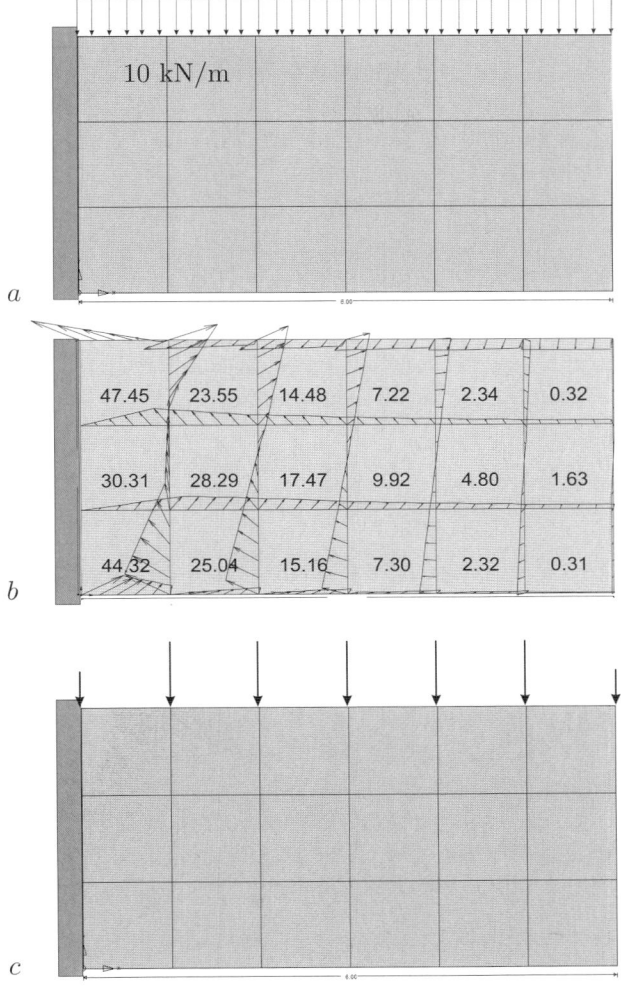

Fig. 1.31. All three load cases are equivalent: **a)** The original load case p; **b)** the FE-load case p_E; **c)** the equivalent nodal forces represent the equivalence class to which the two load cases belong

$$\int_0^l p\,\varphi_i\,dx = f_i = \int_0^l \tilde{p}\,\varphi_i\,dx\,. \qquad (1.146)$$

Hence a single nodal force f_i represents a whole class of loads, namely all the loads that contribute the same work acting through φ_i. Because they are all equivalent with respect to φ_i, we call f_i an *equivalence class* of loads (see Fig. 1.31) and we come to understand that the accuracy of the FE results cannot exceed the resolution of the FE mesh.

Reverse engineering

In soil mechanics the soil is not stress free and so we must associate with the stress state S^s (stress tensor) of the soil a set of equivalent nodal forces. But this is easy because the equivalent nodal forces are simply

$$f_i = a(u^s, \varphi_i) = \int_\Omega S^s \bullet E_i \, d\Omega \qquad (1.147)$$

where E_i is the strain tensor of the nodal unit displacement φ_i and u^s (which actually is not required) is the deformation of the soil.

This example also demonstrates that there is an "external" or an "internal" approach to calculating equivalent nodal forces. According to Green's first identity which is here formulated for an elastic solid

$$G(u, \varphi_i) = \underbrace{\int_\Omega p \bullet \varphi_i \, d\Omega + \int_\Gamma t \bullet \varphi_i \, ds}_{f_i = \delta W_e} - \underbrace{a(u, \varphi_i)}_{\delta W_i} = 0 \qquad (1.148)$$

both approaches yield the same result. If the volume forces p and surface tractions t are known then $f_i = p(\varphi_i)$ but because of (1.148) we have as well $f_i = a(u, \varphi_i)$. This is what we do in reverse engineering.

1.14 Concentrated forces

A Kirchhoff plate (which ignores transverse shear strains) sustains the attack of a concentrated force, while a Reissner–Mindlin plate does not: the force simply cuts through the plate. The same happens if a plate is put on a point support (an infinitely sharp needle). The plate simply ignores the support. Why this happens and why structures react differently to point forces will be discussed in the following.

Assume a concentrated force at the middle of a plate. If we draw a circle Γ with radius r around the force, the horizontal stresses t_x on the circle Γ must tend to infinity as $1/(2\pi r)$ (Fig. 1.32 a), because only this behavior guarantees that in the limit as r tends to zero, the stresses balance the horizontal force $P = 1$:

$$\lim_{r \to 0} \int_{\Gamma_r} t_x \, ds = \lim_{r \to 0} \int_0^{2\pi} t_x \, r \, d\varphi = \lim_{r \to 0} \int_0^{2\pi} \frac{1}{2\pi r} \overset{\downarrow}{r} \, d\varphi = 1 . \qquad (1.149)$$

The more the circles close in on the force, the tighter the lines of force are packed, the more lines pass through each square inch of the cross section of the plate, and the more singular the stresses become.

In a Kirchhoff plate (Fig. 1.32 b) the Kirchhoff shear v_n exhibits the same behavior, and for the same reason:

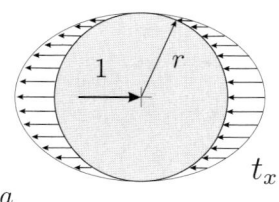
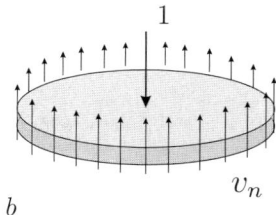

Fig. 1.32. The edge stresses must maintain the balance with the point load: **a)** plate; **b)** slab

$$\lim_{r \to 0} \int_{\Gamma_r} v_n \, ds = \lim_{r \to 0} \int_0^{2\pi} v_n \, r \, d\varphi = \lim_{r \to 0} \int_0^{2\pi} \frac{1}{2\pi r} r \, d\varphi = 1 \,. \quad (1.150)$$

If this experiment is done in a 3-D elastic solid, the stresses must tend to infinity as $1/r^2$, because the integration is carried out over a sphere and the surface S of a sphere shrinks as $S = 4\pi r^2$ as r tends to zero.

The spatial dimension

The rate at which stresses tend to infinity thus depends on the dimension of the continuum. In physics, everything that tends to a point source, the electrical forces that converge on a point charge e, the gravitational forces that converge on a point mass m, the stresses that converge on a point load P must be consistent with the dimension n of the continuum, or rather the size of the sphere that surrounds the target, see Fig. 1.33,

$$S = 2\pi r \quad (\text{circle}) \,, \qquad S = 4\pi r^2 \quad (\text{sphere}) \,, \qquad (1.151)$$

and therefore must counterbalance the rate at which the sphere shrinks, i.e., the fields must behave as $1/(2\pi r)$ or $1/(4\pi r^2)$, respectively to reach the target[6] [242]. (The factors 2π and 4π are the magnitude of the unit sphere in $\mathbb{R}_2$ and $\mathbb{R}_3$ respectively).

But if the strains $\varepsilon_{xx} = \sigma_{xx}/E$ in a plate (we take $\nu = 0$) behave as $1/r$, then the horizontal displacement u behaves as $\ln r$, the anti-derivative of $1/r$. Hence the displacement at the foot of the concentrated force also becomes infinite, the point disappears from the screen. But if the force $P = 1$ has infinite range then the work done is also infinite, $W_e = (1/2)\, P \times \infty$, and because of $W_i = W_e$ the strain energy is infinite as well. Hence in the load case $P = 1$, the stress and strain field must have infinite energy.

The reason why a Kirchhoff plate sustains the impact of a concentrated force though the Kirchhoff shear v_n (the third derivatives) also tends to infinity as $1/r$ is that the deflection w is the *triple* indefinite integral of v_n, and if $1/r$ is integrated three times then the result is the function $w = 0.5\, r^2 \ln r - 3/4\, r^2$

---

[6] Only the $1/r^2$-law of gravitation makes it possible to concentrate the mass of the Earth at its center. Given any other law say $1/r$ or $1/r^3$ the center of gravity would not lie at the center of the Earth, [232].

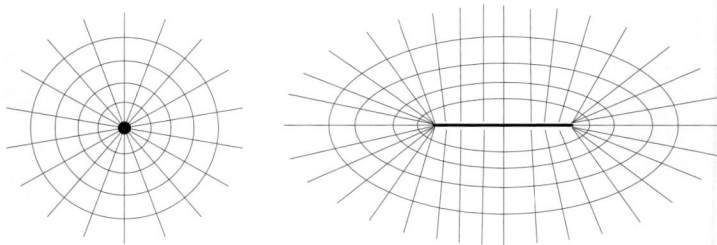

Fig. 1.33. Near a single point source the line of forces are packed so tightly that a plastic zone develops. If the load is spread over a short distance then the singularity is weaker and no plastic zone will develop

which is zero at $r = 0$ (in the limit as $r \to 0$), that is, the deflection is bounded. Note that the total deflection is not zero, because the plate deflection is the sum of this singular function and a regular function; see Equ. (2.5) in Sect. 2.1, p. 242.

Everything hinges on three numbers

$$
\begin{aligned}
i &= \text{order of the singularity, the point source} \\
n &= \text{dimension of the continuum} \\
m &= \text{order of the strain energy}
\end{aligned}
\tag{1.152}
$$

The force, or in more general terms the singularity, that deflects the plate contributes external work which is just the product of the *action* and its conjugate quantity (we may neglect the factor 1/2 typical of eigenwork, because it is irrelevant in this context)

$$
\begin{array}{ll}
W_e = \text{force} \times \text{deflection} & W_e = \text{moment} \times \text{rotation} \\
W_e = \text{rotation} \times \text{moment} & W_e = \text{dislocation} \times \text{force} .
\end{array}
\tag{1.153}
$$

Because of the principle of conservation of energy, $W_e = W_i$, the internal energy W_i is bounded if and only if the external work $W_e = action \times conjugate\ quantity$ is bounded. Because the action, the force P, the moment M, etc., is always finite—for simplicity it can be assumed that the source has magnitude 1.0—the question of whether W_e is infinite or not depends on the magnitude of response of the structure, i.e., the magnitude of the conjugate quantity. This comprises the subject of the following section.

Sobolev's Embedding Theorem

If Ω is a bounded domain in $\mathbb{R}^n$ with a smooth boundary and if $2\,m > n$, then

$$
H^{i+m}(\Omega) \subset C^i(\bar{\Omega})
\tag{1.154}
$$

and there exist constants $c_i < \infty$ such that for all $u \in H^{i+m}(\Omega)$

Table 1.1. Admissible (ok) and inadmissible (no) loads

	$n = 1$	$n = 2$	$n = 3$
singularity $m = 1$	rope, bar, Timoshenko beam	plate, Reissner–Mindlin	3-D
$i = 0:$	ok	no	no
$i = 1:$	no	no	no

singularity $m = 2$	Euler–Bernoulli beam	Kirchhoff plate
$i = 0:$	ok	ok
$i = 1:$	ok	no
$i = 2:$	no	no
$i = 3:$	no	no

$$||u||_{C^i(\bar{\Omega})} \leq c_i \, ||u||_{H^{i+m}(\Omega)} \,. \tag{1.155}$$

The norm of a function u

$$||u||_{C^i(\bar{\Omega})} := \max_{0 \leq |\boldsymbol{j}| \leq i} \left| \frac{\partial^{|j|} u(\boldsymbol{x})}{\partial x^j} \right| \tag{1.156}$$

is the maximum absolute value of u and its derivatives up to the order i on $\bar{\Omega}$.

This theorem implies that the strain energy due to a point load is bounded and the conjugate quantity is finite (and continuous) if the three numbers in (1.152) satisfy the inequality [115]

$$m - i > \frac{n}{2} \,. \tag{1.157}$$

The order of the energy is

$m = 1$ Timoshenko beams, Reissner–Mindlin plates, Elasticity theory

$m = 2$ Euler–Bernoulli beams, Kirchhoff plates

and the index of the singularity for second-order equations ($2\,m = 2$)

$i = 0$ force $i = 1$ dislocation

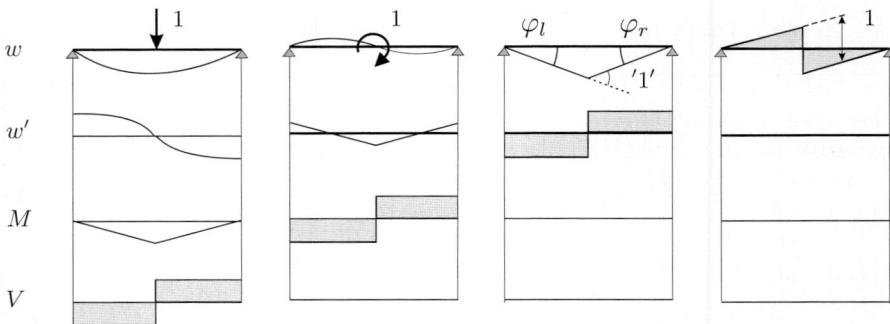

Fig. 1.34. The four singularities of a beam, $'1' \equiv \tan \varphi_l + \tan \varphi_r = 1$

and fourth-order equations $(2\,m = 4)$

$$
\begin{array}{llll}
i = 0 & \text{force} & i = 1 & \text{moment} \\
i = 2 & \text{rotation} & i = 3 & \text{dislocation}
\end{array}
$$

The spatial dimensions are $n = 1$ for ropes, bars, beams, $n = 2$ for plates, and $n = 3$ for elastic solids. Table 1.1 summarizes the inequality (1.157).

If the action is a force, then the shear forces (the third derivatives of a slab) must behave as $1/r$, the bending moments (the second derivatives) as $\ln r$, the rotations w_n (first derivative) as $r \ln r$, and the deflection w finally as $1/2\,r^2 \ln r - 3/4\,r^2$.

If this is done systematically for all possible point loads (actions), the following list with the derivatives and antiderivatives respectively of the two singular functions $1/r$ (2D) and $1/r^2$ (3D) is obtained:

$$\longrightarrow \qquad \text{differentiate}$$

$$r^2 \ln r \;\star\; r \ln r \;\star\; \ln r \;\star\; r^{-1} \;\star\; r^{-2} \;\star\; r^{-3} \;\star\; r^{-4}$$

$$\longleftarrow \qquad \text{integrate}$$

To better concentrate on the essential parts, we have dropped all constant factors and all non-essential parts.

Hence the characteristic singularities in a Kirchhoff plate are the following:

	force	moment	rotation	dislocation
w	$r^2 \ln r$	$r \ln r$	$\ln r$	$\mathbf{r^{-1}}$ $\leftarrow$
$w_{,i}$	$r \ln r$	$\ln r$	$\mathbf{r^{-1}}$ $\leftarrow$	r^{-2}
m_{ij}	$\ln r$	$\mathbf{r^{-1}}$ $\leftarrow$	r^{-2}	r^{-3}
q_i	$\mathbf{r^{-1}}$ $\leftarrow$	r^{-2}	r^{-3}	r^{-4}

and we learn that for example in the neighborhood of a single moment the shear forces q_i behave as r^{-2}, the moments m_{ij} as r^{-1}, the rotations $w_{,i}$ as ln r (the moment rotates infinitely often, a Kirchhoff plate does not sustain the attack of a moment), while the deflection, $w \simeq r \ln r$, is bounded.

Traversing the table from left to right, the characteristic 2-D singularity $1/r$ rises one level higher with each step to the right. The same tendency can be observed in a beam (Fig. 1.34). In a beam the $1/r$ discontinuity wanders from the lower left (influence function for w) to the upper right (influence function for V).

In a Reissner–Mindlin plate the deformations are w, θ_x, θ_y, and the shear forces are defined to be

$$q_x = K \frac{1-\nu}{2} \bar{\lambda}^2 (\theta_x + w_{,x}), \qquad q_y = K \frac{1-\nu}{2} \bar{\lambda}^2 (\theta_y + w_{,y}), \quad (1.158)$$

and because these shear forces must behave as $1/r$ in the neighborhood of a concentrated force, the deflection will behave as $\ln r$, that is w will be infinite at $r = 0$, and the rotations θ_x, θ_y will be infinite too.

Remark 1.4. Sobolev's Embedding Theorem deserves some remarks. Let $i = 0$, then this theorem states that

$$H^m(\Omega) \subset C^0(\bar{\Omega}) \qquad ||u||_{C^0(\bar{\Omega})} \leq c_0 ||u||_{H^m(\Omega)}, \quad (1.159)$$

which means that functions u with the property $||u||_m < \infty$ are continuous, and so is the embedding of the space $H^m(\Omega)$ into $C^0(\bar{\Omega})$—this is the meaning of the second part of (1.159).

Hence for each spatial dimension n there is a certain index m beyond which all functions in $H^m(\Omega)$ are continuous, $H^m(\Omega) \subset C^0(\bar{\Omega})$, and this index m only depends on the dimension n of the space, namely m must be greater than $n/2$.

That is if the strain energy of the structure is bounded, $||u||_m < \infty$, then the displacements are continuous—no cracks. But for this conclusion to be true it must be $m > n/2$, i.e., it is true for Kirchhoff plates, $2 > 2/2$, but not for Reissner–Mindlin plates, $1 \not> 2/2$ or elastic solids, $1 \not> 3/2$.

Let $\mathbb{R}^3$ be all vectors $\boldsymbol{x} = [x_1, x_2, x_3]^T$. The embedding of $\mathbb{R}^3$ into $\mathbb{R}^2$—simply the vertical projection of the vectors $\boldsymbol{x}$ onto the plane—is continuous because

$$||\boldsymbol{x}||_2 = \sqrt{x_1^2 + x_2^2} \leq ||\boldsymbol{x}||_3 = \sqrt{x_1^2 + x_2^2 + x_3^2}. \quad (1.160)$$

So if two vectors $\boldsymbol{x}$ and $\hat{\boldsymbol{x}}$ are close in $\mathbb{R}^3$ then they are also close in $\mathbb{R}^2$. That is a continuous embedding preserves the topological structure of the original space.

For additional remarks about Sobolev's Embedding Theorem and its consequences for structural mechanics, see Sect. 7.10, p. 552.

Working with point forces

As structural engineers we are well versed in the art of extracting information from a structure by applying point loads $P = 1$ or similar singularities. Formally this information gathering is an application of Green's first or second identity

$$G(G_0, u) = 0 \qquad B(G_0, u) = 0 \qquad \text{etc.} \qquad (1.161)$$

or stated differently, an application of the principle of virtual forces or Betti's theorem.

In light of Sobolev's Embedding Theorem, it might seem that care must be taken if these techniques are applied to 2-D and 3-D solids. But the situation is not so dramatic. One must distinguish between the formulation of Green's first identity on the diagonal, $G(u, u) = 0$, and formulations $G(u, \hat{u}) = 0$ where $u \neq \hat{u}$ that are on the secondary diagonal.

If $\boldsymbol{u} = \boldsymbol{G}_0(\boldsymbol{y}, \boldsymbol{x})$ is the displacement field due to a point load in an elastic 3-D solid the stresses behave as $1/r^2$. When $G(\boldsymbol{u}, \boldsymbol{u})$ is formulated with this field $\boldsymbol{u}$ then the strain energy density $\sigma_{ij}\,\varepsilon_{ij}d\Omega$ at $\boldsymbol{x}$ has double that singularity, and therefore the strain energy in a ball with radius $R = 1$ is infinite

$$\int_\Omega \sigma_{ij}\,\varepsilon_{ij}d\Omega \equiv \int_0^1 O(\frac{1}{r^2})\,O(\frac{1}{r^2})\,O(r^2)\,dr = \int_0^1 \frac{1}{r^2}\,dr = \infty\,. \quad (1.162)$$

But if $\hat{\boldsymbol{u}} \neq \boldsymbol{u}$ and the field $\hat{\boldsymbol{u}}$ has bounded stresses, the strain energy density at $\boldsymbol{x}$ is of the order $O(1/r^2)\,O(1)\,O(r^2)$, and therefore the strain energy product between the field $\boldsymbol{G}_0$ and the field $\hat{\boldsymbol{u}}$ is finite.

Similar considerations hold in the case of Betti's theorem. It is possible to apply a point load $P = 1$ to extract information about a regular displacement field $\hat{\boldsymbol{u}}$ as in $B(\boldsymbol{G}_0, \hat{\boldsymbol{u}}) = 0$, but this would fail if we try to formulate $B(\boldsymbol{G}_0, \boldsymbol{G}_0) = 0$, because then the singularities would cancel each other and we would be left with two meaningless boundary integrals ($\boldsymbol{\tau}_0 = $ traction vector of the field $\boldsymbol{G}_0$)

$$\lim_{\varepsilon \to 0} B(\boldsymbol{G}_0, \boldsymbol{G}_0)_{\Omega_\varepsilon} = \int_\Gamma \boldsymbol{\tau}_0 \bullet \boldsymbol{G}_0\,ds - \int_\Gamma \boldsymbol{G}_0 \bullet \boldsymbol{\tau}_0\,ds = 0\,. \qquad (1.163)$$

Hence every situation is different, and the presence of singularities requires a careful study of the limit [115]

$$\lim_{\varepsilon \to 0} G(\boldsymbol{G}_0, \hat{\boldsymbol{u}})_{\Omega_\varepsilon} = 0 \qquad \lim_{\varepsilon \to 0} B(\boldsymbol{G}_0, \hat{\boldsymbol{u}})_{\Omega_\varepsilon} = 0. \qquad (1.164)$$

Remark 1.5. Not all is well with point forces. There is one prominent victim of Sobolev's Embedding Theorem: Castigliano's Theorem, which states that the derivative of the strain energy is the displacement in the direction of the point load P_i, makes no sense in elastic solids,

$$\frac{\partial}{\partial P_i} \, a(\boldsymbol{u}, \boldsymbol{u}) \stackrel{?}{=} u_i \tag{1.165}$$

because both the strain energy and the displacement are infinite [115].

Energy estimates involving Green's functions

In this book we will operate freely with energy estimates involving Green's functions with infinite energy. One such inequality is for example

$$|u_x(\boldsymbol{x}) - u_x^h(\boldsymbol{x})| \leq ||\, \boldsymbol{G}_0[\boldsymbol{x}] - \boldsymbol{G}_0^h[\boldsymbol{x}]||_E \, ||\boldsymbol{u} - \boldsymbol{u}_h||_E \tag{1.166}$$

which is an estimate for the (horizontal) displacement error of the FE solution in a plate. (The $[\boldsymbol{x}]$ is to denote that $\boldsymbol{G}_0$ is the Green's function for $u_x(\boldsymbol{x})$ at the point $\boldsymbol{x}$). Theoretically this equation makes no sense because the strain energy of the exact Green's function $\boldsymbol{G}_0$ is infinite (we drop the $[\boldsymbol{x}]$ in the following because it is not essential here)

$$||\boldsymbol{G}_0||_E^2 := a(\boldsymbol{G}_0, \boldsymbol{G}_0) = \int_\Omega \sigma_{ij} \cdot \varepsilon_{ij} \, d\Omega \tag{1.167}$$

and so the distance of the FE Green's function $\boldsymbol{G}_0^h$ from $\boldsymbol{G}_0$ in terms of the strain energy

$$||\boldsymbol{G}_0 - \boldsymbol{G}_0^h||_E^2 := a(\boldsymbol{G}_0 - \boldsymbol{G}_0^h, \boldsymbol{G}_0 - \boldsymbol{G}_0^h) = \int_\Omega (\sigma_{ij} - \sigma_{ij}^h) \cdot (\varepsilon_{ij} - \varepsilon_{ij}^h) \, d\Omega$$
$$\tag{1.168}$$

would be infinite as well—regardless of how close $\boldsymbol{G}_0^h$ is to $\boldsymbol{G}_0$. But if we read $u_x(\boldsymbol{x})$ as the average value of the horizontal displacement over a small disk centered at $\boldsymbol{x}$ the corresponding Green's function $\boldsymbol{G}_0$ would have finite energy and then (1.166) would make sense. So in this book whenever we use an expression such as (1.166) we understand this as a statement about the average value of $u(\boldsymbol{x})$ (or other terms) over a small disk Ω_ε with a radius ε very close to zero.

We should not be deterred too much by the fact that most Green's functions have infinite energy. The FE method is surprisingly good at approximating Green's functions. Most output we see on the screen is based on the solution of ill posed problems...

1.15 Green's functions

Point solutions or *Green's functions* represent the response of a structure to point loads $P = 1$. The best known point solution is perhaps the triangular shape $G_0(y, x)$ of a guitar string (Fig. 1.35 a) when the string is plucked:

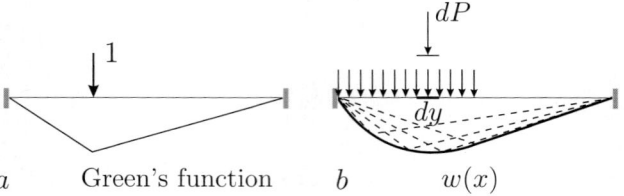

Fig. 1.35. a) Green's function; **b)** the curve $w(x)$ is the envelope of all the Green's functions of the different point loads $dP = p\,dy$

a Green's function b $w(x)$

$$G_0(y, x) = \text{deflection at } x, \text{ force } P = 1 \text{ at } y\,. \tag{1.169}$$

The importance of the point solutions is that any distributed load p can be approximated by a series of equally spaced (Δy) point loads $\Delta P_i = p(y_i)\,\Delta y$, and the envelope of all these triangles $G_0(y_i, x)\,\Delta P_i$ (as the subdivision Δy tends to zero) is the integral

$$w(x) = \int_0^l G_0(y, x)\, p(y)\, dy\,. \tag{1.170}$$

This holds for any structure, and is the reason that Green's functions play such a central albeit hidden role in structural mechanics. A Green's function is the visible embodiment of the differential equation. The structural analysis of a string or a taut rope could begin as well with the triangular shape and only in the second step would we search for the differential equation that is solved by this *unit response*.

Weak and strong influence functions

The right-hand side of (1.170) is an expression of external virtual work

$$\int_0^l G_0(y, x)\, p(y)\, dy = \delta W_e(G_0, w) \qquad w \text{ solves the load case } p \tag{1.171}$$

and because of Green's first identity

$$G(w, G_0) = \delta W_e(w, G_0) - \delta W_i(w, G_0) = \delta W_e(w, G_0) - a(w, G_0) = 0 \tag{1.172}$$

the external virtual work can be expressed as well by internal virtual work, the strain energy product between G_0 and w,

$$\delta W_e(w, G_0) = \int_0^l G_0(y, x)\, p(y)\, dy = a(G_0, w) = \delta W_i(w, G_0) \tag{1.173}$$

and so there is a "strong" and a "weak" influence function for the deflection $w(x)$ of the guitar string

$$w(x) = \underbrace{\int_0^l G_0(y,x)\, p(y)\, dy}_{strong} = \underbrace{a(w, G_0[x])}_{weak} = \int_0^l H\, w'\, G_0'(y,x)\, dy\,. \quad (1.174)$$

The proof of *Tottenham's equation* (1.210) p. 64 is based on this switch. For more on the subject see Sect. 7.7, p. 535.

Singularities of Green's functions

Green's functions being point solutions are of course subject to the conditions set forth by physics: the singularity of the stresses must be consistent with the dimension n of the continuum. To study the consequences this entails, let a force $P_j = 1$ act at a point $\boldsymbol{y} = (y_1, y_2, y_3)$ in 3-D space. A force in the y_j direction will cause the displacements

$$U_{ij}(\boldsymbol{y}, \boldsymbol{x}) = \frac{1}{8\pi G (1-\nu)\, r}\left[(3 - 4\nu)\, \delta_{ij} + r_{,i}\, r_{,j}\right] \qquad i = 1, 2, 3 \quad (1.175)$$

at a remote point $\boldsymbol{x} = (x_1, x_2, x_3)$ where the first index i indicates the direction of the displacement. The term δ_{ij} is the *Kronecker delta*, and the $r_{,i}$ are the directional derivatives of the distance $r = |\boldsymbol{y} - \boldsymbol{x}|$ with respect to the coordinates y_i of the source point

$$\delta_{ij} = \begin{cases} 1 & i = j \\ 0 & i \neq j \end{cases} \qquad r_{,i} := \frac{\partial r}{\partial y_i} = \frac{y_i - x_i}{r}\,. \quad (1.176)$$

Of course there is only one system of coordinates and therefore only one origin but to distinguish the source point $\boldsymbol{y}$ (the load) from the field point $\boldsymbol{x}$ (the observer) the coordinates are labeled differently.

As can be seen from (1.175), the displacements U_{ij} behave as $1/r$, because in 3-D the stresses and therefore the strains must tend to infinity as $1/r^2$ and the antiderivative of $1/r^2$ (the strains) is $1/r$ (the displacement).

The 3×3 functions U_{ij} form a symmetric matrix $\boldsymbol{U} = [U_{ij}]$, the columns of which are the three Green's functions (displacement fields), corresponding to the three unit forces acting at $\boldsymbol{y}$

$$\boldsymbol{U} = [\boldsymbol{G}_0^{(1)}, \boldsymbol{G}_0^{(2)}, \boldsymbol{G}_0^{(3)}] \qquad \boldsymbol{G}_0^{(i)} = \begin{bmatrix} U_{1i} \\ U_{2i} \\ U_{3i} \end{bmatrix}\,. \quad (1.177)$$

Line loads $\boldsymbol{l}(\boldsymbol{y})$ can be simulated by a succession of point forces, and volume forces $\boldsymbol{p}(\boldsymbol{y})$ can be simulated by a 3-D grid of point forces, so that the corresponding displacements can be considered the scalar product of the Green's functions and these loads:

$$u_i = \int_\Gamma \boldsymbol{G}_0^{(i)}(\boldsymbol{y}, \boldsymbol{x}) \bullet \boldsymbol{l}(\boldsymbol{y})\, ds_{\boldsymbol{y}} \equiv \int_\Gamma \frac{1}{r} \boldsymbol{l}(\boldsymbol{y})\, ds_{\boldsymbol{y}} \quad (1.178)$$

$$u_i = \int_\Omega \boldsymbol{G}_0^{(i)}(\boldsymbol{y}, \boldsymbol{x}) \bullet \boldsymbol{p}(\boldsymbol{y})\, d\Omega_{\boldsymbol{y}} \equiv \int_\Omega \frac{1}{r} \boldsymbol{p}(\boldsymbol{y})\, d\Omega_{\boldsymbol{y}} \quad (1.179)$$

Fig. 1.36. 'Dirac deltas
do exist, don't they?' (La-
Ferté pedestrian bridge in
Stuttgart, Germany)

where in the second part of each equation we have retained only the characteristic singularity $1/r$. The factor $1/r$ is the reason that a block of concrete
cannot be prestressed with a (mathematical) wire.

To explain this behavior, let us assume that on the y_1-axis (this is the
normal x-axis) and within the interval $[0, 1]$ there acts a constant line load
$l = [0, 0, l_3]^T$ pointing in the y_3 direction (the z-axis), and the observer x is
located at the origin of the coordinate system, at the front of the line load.
The observer will not be able to remain in place, because he will experience
a shift of infinite magnitude in vertical direction

$$u_3(\mathbf{0}) = \int_0^1 U_{33}(\mathbf{y}, \mathbf{0})\, l_3\, dy_1 = \frac{l_3(3 - 4\nu)}{8\,\pi\,G\,(1 - \nu)} \int_0^1 \frac{1}{y_1}\, dy_1 = \infty\,. \quad (1.180)$$

(Note that $1/r = 1/y_1$, $ds = dy_1$ and $r_{,3} = (y_3 - x_3)/r = 0$, because the load
and the observer are on the same level, $x_3 = y_3 = 0$).

This holds at any point x that happens to lie in the load path $[0, 1]$, while
outside the load path the displacements are bounded because $r > 0$.

Hence a line element $dr = O(1)$ cannot counterbalance a singularity that
goes as $1/r$. Only an area or surface element $d\Omega = r\, dr\, d\varphi = O(r)$ can cope
with such a singularity. This is why a rope that exerts only the lightest pressure
cuts through the thickest concrete, while a surface load (theoretically at least)
cannot crush a block of concrete.

In plane elasticity the situation is not so dramatic, because in 2-D the stresses are $O(1/r)$ and the displacements are $O(\ln r)$, so that a line element dr sustains the attack of the prestressing forces l within a rope

$$u_i(\boldsymbol{x}) = l \cdot \int_0^1 \ln r \, dr = O(1)\,. \qquad (1.181)$$

The more the load spreads, $point \to line \to surface$, the weaker the singularity, $r^{-1} \to \ln r \to O(1)$, as Fig. 1.35 so aptly illustrates: the peak under the point load vanishes immediately if the load is evenly spread.

The importance of Green's functions for structural mechanics probably cannot be overestimated. Engineers often claim that point loads are *imaginary quantities which do not exist in reality*, see however Fig. 1.36. This is true of a mechanical model, but it must also be recognized that line loads or surface loads are simply iterated point loads, and that therefore *any* displacement field is just a superposition of infinitely many Green's functions, each representing the influence of an infinitesimal portion dP of the total load p.

1.16 Practical consequences

Now there is no need to put up stop signs, because if a node is kept fixed, it is not a point support, and a nodal force is not a point load. This is a consequence of the inherent fuzziness of the FE method. For an FE program, a nodal force f_i is *always* an equivalent nodal force. It represents a load which upon acting through a nodal unit displacement $\boldsymbol{\varphi}_i$ contributes *work* $f_i \times 1$. The FE program neither knows nor cares whether the load is a line load, a surface load, or a point load, and therefore nodal forces lose much of their seemingly dangerous nature.

The same is true of point supports. For a node to be a point support it must not only be fixed, but the support reaction must also resemble the action of a truly concentrated force. But if the stress field near such a fixed node is studied, it soon becomes apparent that the support reaction more closely resembles a diffuse cloud of volume forces, surface loads, and line forces (jumps in the stresses at interelement boundaries) than a distinct point force f_i.

Near and far

It makes no sense to refine the mesh beyond a certain limit in the neighborhood of a point support or a point load, as this can actually force the node out of the region of interest. And it also makes no sense to design the structure for the stresses that appear in the FE output, because these numbers are "random" numbers whose magnitude indicates the presence of a hot spot in the structure but which, in and of themselves, provide no lower or upper bound for the stresses.

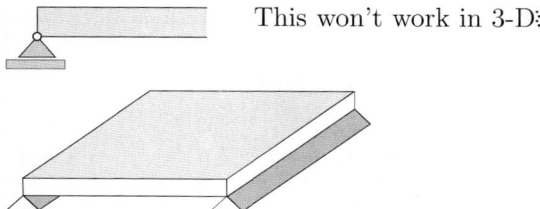

This won't work in 3-D:

Fig. 1.37. We cannot make a 3-D model of a hinged plate

But at some distance from the point load P, it no longer matters whether the applied load is a point load or a volume force p, as may be illustrated by the following (somewhat simplified) equations of 2-D elasticity:

$$u_i(\boldsymbol{x}) = \ln r_P \cdot P \qquad r_P = |\boldsymbol{y}_P - \boldsymbol{x}| \tag{1.182}$$

$$u_i(\boldsymbol{x}) = \int_\Omega \ln r \, p(\boldsymbol{y}) \, d\Omega_{\boldsymbol{y}} \simeq \ln r_P \int_\Omega p \, d\Omega = \ln r_P \cdot P. \tag{1.183}$$

In words, at some distance from the source, the effect of a point load is essentially identical to a one-point quadrature of the influence integral of the volume forces.

Capacity

The integrals

$$\int_0^\pi \sin x \, dx = 2 \qquad \int_0^\pi \int_0^\pi \sin(x\,y) \, dx \, dy = 2.90068 \tag{1.184}$$

do not change if the integrand $\sin x$ is changed at one point x_0 or if $\sin(x\,y)$ is changed along a whole curve. Similar results hold true in the theory of elasticity because

- in plates (2-D elasticity) points have zero capacity,
- in elastic solids (3-D elasticity) curves (line supports!) have zero capacity.

This means that point supports or line supports (in 3-D problems) are simply ignored by a structure. No force is necessary to displace a single point in a plate, which implies that the displacement cannot be described at a point support. And in makes no sense to specify the displacement field of an elastic solid along a curve. That is, an elastic solid ignores line supports; see Fig. 1.37.

Virtually the same can be said about Reissner–Mindlin plates. A point load effects an infinite deflection, $w = \infty$, and no force is necessary to displace a single point in vertical direction.

A Kirchhoff plate would not tolerate this. But if a single moment $M = 0$ (or almost zero) is applied at a point, the point will start to rotate infinitely rapidly. Hence if a slab is coupled with a beam via a torsional spring, the neighborhood of the spring should not be refined too much (Fig. 1.38) lest the rotational stiffness of the slab be lost.

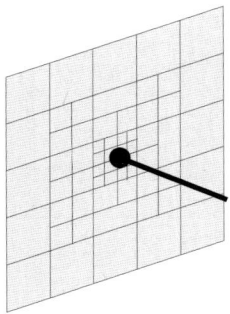 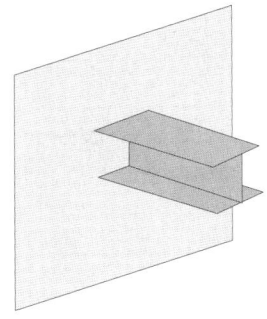

Fig. 1.38. The FE model cannot be refined too much, lest a plastic hinge develop

Summary

What is possible:

- Kirchhoff plate: point loads, line moments m kN m/m, surface loads
- Reissner–Mindlin plate: line loads, surface loads
- plates (2-D elasticity): line loads, surface/volume forces
- elastic solids (3-D elasticity): surface loads, volume forces

What is not allowed (theoretically):

- Kirchhoff plate: single moments
- Reissner–Mindlin plate: point loads, single moments
- plates (2-D elasticity): point loads
- elastic solids (3-D elasticity): point loads, line loads

Hence it is also clear which supports are admissible and which are not (theoretically).

In actual practice, an FE structure can be placed on point supports, and it is legitimate to work with point forces and single moments. Only the stresses close to such points are not reliable, and depend on the mesh size: the finer the mesh, the more singular the stresses.

1.17 Why finite element results are wrong

The reason is simply that an FE program uses the wrong influence functions.

Taut rope

Recall (Sect. 1.15) that if a force $P = 1$ is applied at a point y, the rope assumes a triangular shape denoted by the capital letter G as in *Green's function*

$$G_0(x, y), \qquad y = \text{source point of the load}, \quad x = \text{field point, variable}.$$

$$(1.185)$$

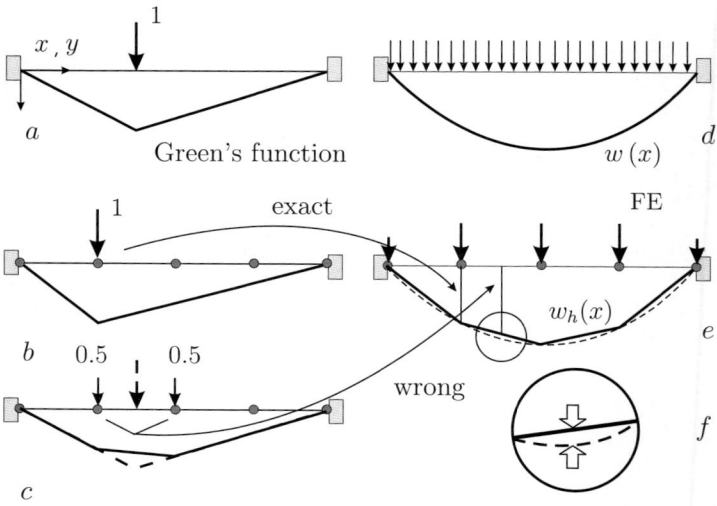

Fig. 1.39. The Green's function for the deflection w of the rope is piecewise linear. If the Green's function of a point x lies in V_h then the deflection $w_h(x)$ is exact

If a distributed load p is approximated by a sequence of small point loads $p(y_i)\,\Delta y$, the deflection is

$$w(x) \simeq \sum_i^n G_0(x, y_i)\, p(y_i)\, \Delta y \tag{1.186}$$

and in the limit the sum becomes an integral:

$$w(x) = \int_0^l G_0(x, y)\, p(y)\, dy\,. \tag{1.187}$$

Because the Green's function is symmetric, $G_0(y, x) = G_0(x, y)$, the points x and y can be exchanged, and this means that the tip of the triangle now stays fixed at x, whereas in (1.186) it moved with y_i:

$$w(x) = \int_0^l G_0(y, x)\, p(y)\, dy = \int_0^l \underline{\qquad\qquad} \boxed{}\, dy\,. \tag{1.188}$$

This influence function for the deflection $w(x)$ is the scalar product between the kernel $G_0(y, x)$ and the distributed load $p(y)$.

The finite element tries to imitate (1.188); to calculate the deflection at a point x the program (theoretically at least) proceeds as follows:

- It tries to find in V_h the Green's function for the deflection at the point x. If that is not possible, if G_0 does not lie in V_h, then it substitutes for G_0 an approximate function $G_0^h(y, x) \in V_h$.

- With this approximate kernel the program calculates the deflection

$$w_h(x) = \int_0^l G_0^h(y,x)\, p(y)\, dy\,. \qquad (1.189)$$

The result is exact if $G_0^h = G_0$, i.e., if the Green's function G_0 of the point x lies in V_h. This is true for the nodes x_k. Because if a point force $P = 1$ is applied at one of the nodes x_k the shape can be modeled with the three unit deflections φ_i exactly (see Fig. 1.39 b)

$$G_0[x_k] := G_0(y, x_k) = \sum_{i=1}^3 u_i\, \varphi_i(y) \qquad u_i = u_i(x_k) = G_0(y_i, x_k) \;(1.190)$$

and therefore (1.189) will be exact at the nodes, $w_h(x_k) = w(x_k)$.

This is a simple consequence of the *Galerkin orthogonality* of the error $e(x) = w(x) - w_h(x)$:

$$a(e, \varphi_i) = \int_0^l \frac{(T - T_h)\, T_i}{H}\, dx = 0 \qquad \text{for all } \varphi_i \in V_h \qquad (1.191)$$

where

$$T - T_h = H\, w' - H\, w_h' = H\, e' \qquad T_i = H\varphi_i'\,. \qquad (1.192)$$

Hence with $T_0(y, x_k) := H\, G_0'(y, x_k) = \sum_{i=1}^3 u_i\, T_i(y)$ and switching from the internal formulation to the external formulation, see (1.174) p. 53,

$$0 = a(e, G_0[x_k]) = \int_0^l \frac{(T(y) - T_h(y))\, T_0(y, x_k)}{H}\, dy = w(x_k) - w_h(x_k)\,. \qquad (1.193)$$

At an intermediate point x the situation is different (Fig. 1.39 c), because the Green's function of a point x that is not a node does *not* lie in V_h. The three nodal unit deflections do not allow the rope to have a peak between the nodes. Hence the FE program cannot solve the load case $P = 1$ if x is an intermediate point.

Therefore the program splits the force $P = 1$ into two equal parts and places these at the two neighboring nodes, because this is a load case it can solve. But the solution $G_0^h(y, x)$ will only be an approximation, and therefore the result of (1.189) will in general be wrong; see Fig. 1.39 e and f.

Remark 1.6. The notation $u_i = u_i(x_k)$ in (1.190) is to indicate that the coefficients u_i are different for each nodal Green's function $G_0(y, x_k)$ and so they are functions of the nodal coordinate x_k. This is the typical pattern in the FE approximation of influence functions

$$G_0^h(y, x) = \sum_i u_i(x)\, \varphi_i(y)\,. \qquad (1.194)$$

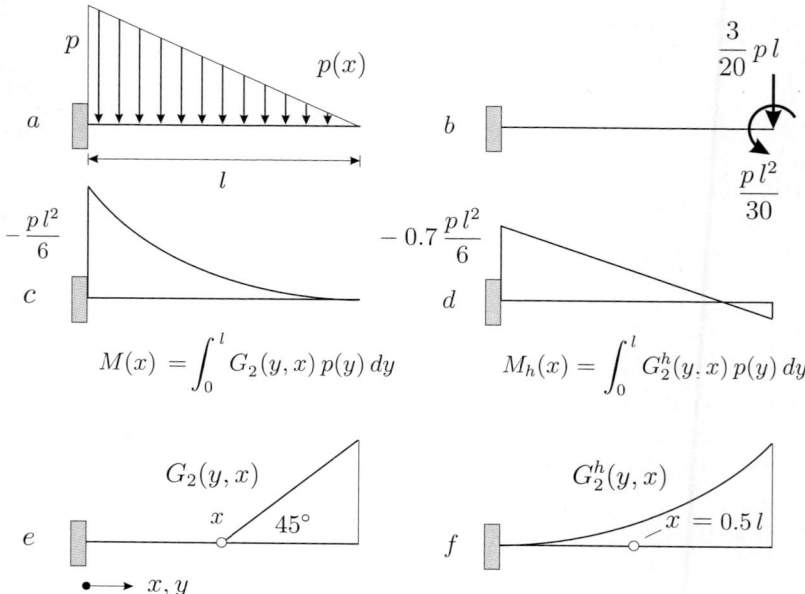

Fig. 1.40. a) Beam with triangular load; **b)** equivalent nodal forces; **c)** $M(x)$; **d)** $M_h(x)$; **e)** Green's function for $M(0.5l)$; **f)** Green's function for $M_h(0.5l)$

Beam

In the following example the logic is the same but the focus is on the error in the bending moment of a beam. The beam in Fig. 1.40 is modeled with just one element ($u_1 = u_2 = 0$ at the fixed end and u_3 and u_4 at the free end). Hence the equivalent nodal forces representing the triangular load are a vertical force $f_3 = 3p\,l/20$ and a moment $f_4 = p\,l^2/30$ at the end of the beam; see Fig. 1.40 b. The exact bending moment distribution $M(x)$ is the scalar product between the Green's function $G_2(y, x)$ and the distributed load $p(y)$:

$$M(x) = \int_0^l G_2(y, x)\, p(y)\, dy = -\frac{p\,(l - x)^3}{6\,l} \tag{1.195}$$

and the bending moment $M_h(x)$ of the FE solution is

$$M_h(x) = \int_0^l G_2^h(y, x)\, p(y)\, dy = \frac{3\,l\,p}{20}x - \frac{7\,l^2\,p}{60} \tag{1.196}$$

where the kernel

$$G_2^h(y, x) = x\left(\frac{3y^2}{l^2} - \frac{2y^3}{l^3}\right) - \frac{2y^2}{l} + \frac{y^3}{l^2} \tag{1.197}$$

is the influence function for $M_h(x)$. That is, if a point load P acts at y, the FE bending moment at x is $M_h(x) = G_2^h(y, x) \times P$.

In the next paragraph we will see that the function $G_2^h(y, x)$ is the deflection of the beam if the equivalent nodal forces $f_3 = -EI\,\varphi_3''(x)$ and $f_4 = -EI\,\varphi_4''(x)$ are applied at the end of the beam[7]. The nodal displacements $u_3 = u_3(x)$ and $u_4 = u_4(x)$ in this load case (the x is the x in $M_h(x)$) are the solutions of the system (row 3 and 4 of $\boldsymbol{K}\,\boldsymbol{u} = \boldsymbol{f}$)

$$\frac{EI}{l^3}\,(12\,u_3 + 6\,l\,u_4) = EI\,(\frac{12}{l^3}\,x - \frac{6}{l^2}) \qquad |\; \times \frac{l^3}{6\,EI} \tag{1.198}$$

$$\frac{EI}{l^3}\,(6\,l\,u_3 + 4\,l^2\,u_4) = EI\,(\frac{6}{l^2}\,x - \frac{2}{l}) \qquad |\; \times \frac{l^3}{2\,EI} \tag{1.199}$$

or if these two equations are modified as indicated

$$2\,u_3 + l\,u_4 = 2\,x - l \tag{1.200}$$

$$3\,l\,u_3 + 2\,l^2\,u_4 = 3\,l\,x - l\,. \tag{1.201}$$

The solution is $u_3(x) = x - l$ and $u_4(x) = 1$, so that $G_2^h(y, x) = u_3(x)\,\varphi_3(y) + u_4(x)\,\varphi_4(y)$; see (1.197).

The influence functions for $M(x)$ and $M_h(x)$ at $x = 0.5\,l$ are displayed in Fig. 1.40 c and d. Obviously the approximate Green's function $G_2^h(y, x)$ tries to imitate the sharp bend of $G_2(y, x)$ at $x = 0.5\,l$, but it fails, because there is no such function in V_h (= third-degree polynomials).

Summary

The preceding examples might have given the impression that the FE program employs the influence functions (1.189) and (1.196) to calculate the deflection $w_h(x)$ of the rope or the bending moment $M_h(x)$. But it is much simpler: any result at any point x has just the same magnitude it would have if it had been calculated with the approximate Green's function.

It is very important that the reader understand this concept of a Green's function. Normally we do not employ the Green's function to calculate, for example, the deflection curve of a hinged beam which carries a constant distributed load p

$$w(x) = \frac{p\,l^4}{24\,EI}\,(\frac{x}{l} - 2\,\frac{x^3}{l^3} + \frac{x^4}{l^4}) = \int_0^l G_0(y, x)\,p(y)\,dy \tag{1.202}$$

but the result is the same as if we had used the Green's function. That is, the kernel $G_0(y, x)$ stays in the background invisible for the user but it controls the exact solution, and in the same sense the approximate Green's function $G_0^h(y, x)$ controls the FE solution. Hence

- Any value in an FE solution is the scalar product of an approximate Green's function G_i^h and the applied load.

---

[7] Here the points y are the points of the beam and x acts as a parameter. The kernel $G_2^h(y, x)$ in Fig. 1.40 f was plotted by keeping $x = 0.5l$ fixed and letting y vary, $0 \le y \le l$.

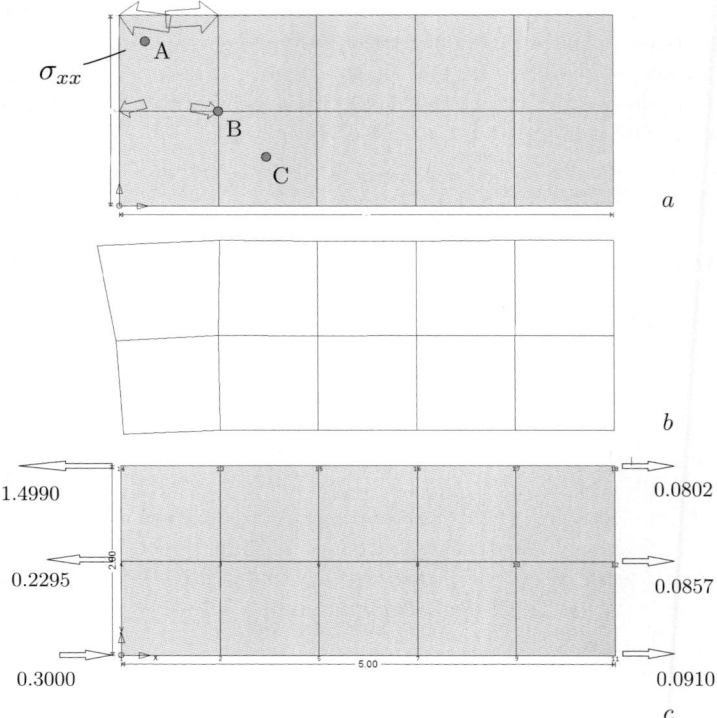

Fig. 1.41. Plate: **a)** nodal forces that generate the approximate Green's function for σ_{xx} at point A; **b)** approximate Green's function for σ_{xx} at point A; **c)** horizontal nodal displacements of the approximate Green's function for σ_{xx} at A

Values at a point

It seems then that FE stresses never can be exact,

$$\sigma(x_k) \overset{?}{=} \sigma_h(x_k)\,, \qquad (1.203)$$

because the Green's function $G_1(y, x_k)$ for the stress σ at a point x_k is a *dislocation*, and such a discontinuous displacement $u(x)$ is nonconforming and therefore it does not lie in V_h. Anything else would constitute a variational crime [230].

But it might be that the conforming approximation G_1^h in V_h accidentally fits perfectly, because the *integral* is the same, i.e., $G_1[x_k] \neq G_1^h[x_k]$, but

$$\sigma(x_k) = \int_0^l G_1(y, x_k)\, p(y)\, dy = \int_0^l G_1^h(y, x_k)\, p(y)\, dy = \sigma_h(x_k). \ (1.204)$$

This also explains why the stresses are correct if the FE solution is exact. If the plate in Fig. 1.41 a is stretched horizontally with uniform forces ± 1

kN/m^2, bilinear elements will produce the exact solution, and thus the correct horizontal stress $\sigma_{xx} = 1$ kN/m^2. But why are the stresses correct if $\boldsymbol{G}_1^{xx}$ does not lie in V_h?

To generate the shape $\boldsymbol{G}_1^{xx,h}$, i.e., the Green's function for the stress σ_{xx} at point A, for example, the equivalent nodal forces in Fig. 1.41 must be applied. Of course the shape produced by these forces (Fig. 1.41 b) is not the exact kernel $\boldsymbol{G}_1^{xx}$, but this kernel has the remarkable property that the integral of the horizontal displacements (of the kernel $\boldsymbol{G}_1^{xx,h}$) along the vertical edges on the left (Γ_L) and right (Γ_R) side yields the exact stress $\sigma_{xx} = 1.0$ (Fig. 1.41 c), because the trapezoidal rule yields

$$\sigma_{xx}^h(\boldsymbol{x}_A) = \int_{\Gamma_L} \boldsymbol{G}_1^{xx,h} \bullet \boldsymbol{t}\, ds + \int_{\Gamma_R} \boldsymbol{G}_1^{xx,h} \bullet \boldsymbol{t}\, ds \qquad (\boldsymbol{t} = \mp \boldsymbol{e}_1)$$
$$= 0.5 \cdot (-0.3000) + 0.2295 + 0.5 \cdot 1.4990$$
$$+0.5 \cdot 0.0802 + 0.0857 + 0.5 \cdot 0.0910 = 1.000\,. \qquad (1.205)$$

The same holds with regard to the influence function for σ_{xx} at the point B (of course the horizontal displacements are now different)

$$0.5 \cdot (-0.0997) + 0.5479 + 0.5 \cdot (-0.0997)$$
$$+0.5 \cdot 0.2771 + 0.2748 + 0.5 \cdot 0.2771 = 1.000 = \sigma_{xx}(\boldsymbol{x}_B) \qquad (1.206)$$

and to convince even the most skeptical reader also at point C (Fig. 1.41)

$$0.5 \cdot 0.5004 + 0.0 + 0.5 \cdot (-0.5004)$$
$$+0.5 \cdot 0.6727 + 0.5000 + 0.5 \cdot 0.3273 = 1.000 = \sigma_{xx}(\boldsymbol{x}_C)\,. \qquad (1.207)$$

This remarkable property is based on the following theorem.

- In any load case which can be solved exactly on V_h, the error in the Green's functions is orthogonal to the applied load, i.e.,

$$\int_0^l [G_1(y, x_k) - G_1^h(y, x_k)]\, p(y)\, dy = \sigma(x_k) - \sigma_h(x_k) = 0\,. \qquad (1.208)$$

Basically this theorem states that regardless of whether or not the Green's function lies in V_h, any point value calculated with an approximate Green's function is exact if the exact solution lies in V_h. This is not so trivial as it sounds. Rather it is an interesting statement about approximate Green's functions and the FE method.

In the notation of Chap. 7, the proof of (1.208) is simple:

$$(G_1 - G_1^h, p) = a(G_1 - G_1^h, u) = a(G_1 - G_1^h, u_h) = 0\,. \qquad (1.209)$$

First the principle of virtual displacements is invoked, then the fact that $u = u_h$ and finally the Galerkin orthogonality of the error in the Green's function. Of course (1.209) holds for any Green's function, not just for G_1.

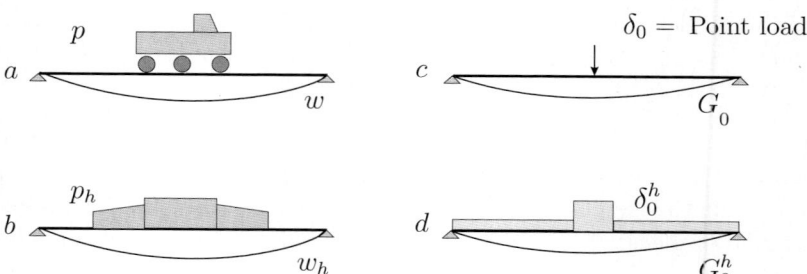

Fig. 1.42. A truck on a bridge: **a)** The truck is the load case p; **b)** the load case p_h, which is equivalent to the truck; **c)** the load case δ_0 (= single force); **d)** the load case δ_0^h, the work-equivalent substitute for the single force

It would seem that if the exact solution u does not lie in V_h, there is little chance for $\sigma_{xx}^h = \sigma_{xx}$, because (i) $\boldsymbol{G}_1^{xx,h} \neq \boldsymbol{G}_1^{xx}$ and (ii) the orthogonality (1.209) does not hold. But if the stress field is simple, then it might happen that the value of σ_{xx} at the centroid of the element coincides with the average value of σ_{xx} over the element, so that there might be a good chance to catch this stress, because the Green's functions for average values of stresses are much easier to approximate; see Sect. 1.21, p. 86.

Remark 1.7. We should also not be too critical of the FE method on account of a certain mismatch between values at a point and the strain energy (= *integral*). In the end we want to have results *at points*, but can we expect such sharp results from an energy method? Can we expect that unbounded point functionals yield accurate results if the primary interest of an FE program is to minimize the *error in the energy* and not to achieve high accuracy at a single point? Though one could object that nature too only minimizes the strain energy but nevertheless each single value is exact ...

1.18 Proof

It is now time to prove that the FE solution is indeed the scalar product of the approximate Green's function G_0^h (which is never calculated explicitly) and the applied load p. The proof fits on one line[8].

Tottenham's equation

$$w_h(\boldsymbol{x}) = \underbrace{(\delta_0, w_h)}_{\delta W_e(\delta_0, w_h)} = \underbrace{a(G_0^h, w_h)}_{\delta W_i(G_0^h, w_h)} = \underbrace{(p, G_0^h)}_{\delta W_e(G_0^h, p)} , \qquad (1.210)$$

---

[8] A paper published by Tottenham in 1970 [245] is the earliest reference to this equation known to the authors.

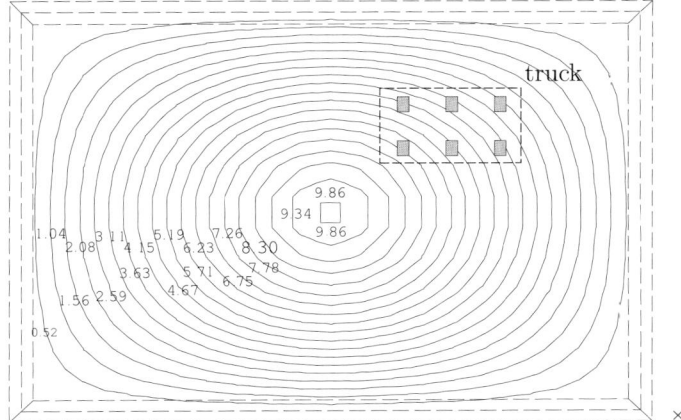

$\times 10^{-1}$

Fig. 1.43. Influence function for the force in the column in the middle of the plate, and the truck

or in a more conservative notation

$$w_h(\boldsymbol{x}) = \int_\Omega \delta_0(\boldsymbol{y} - \boldsymbol{x})\, w_h(\boldsymbol{y})\, d\Omega_{\boldsymbol{y}} = a(G_0^h, w_h) = \int_\Omega G_0^h(\boldsymbol{y}, \boldsymbol{x})\, p(\boldsymbol{y})\, d\Omega_{\boldsymbol{y}}\,.$$
$$(1.211)$$

First w_h is considered to be the FE solution of the load case p and $G_0^h \in V_h$ assumes the role of a virtual displacement:

$$a(G_0^h, w_h) = \int_\Omega G_0^h(\boldsymbol{y}, \boldsymbol{x})\, p(\boldsymbol{y})\, d\Omega_{\boldsymbol{y}} \qquad (1.212)$$

then G_0^h is considered to be the FE solution of the load case δ_0, and w_h assumes the role of a virtual displacement:

$$w_h(\boldsymbol{x}) = \int_\Omega \delta_0(\boldsymbol{y} - \boldsymbol{x})\, w_h(\boldsymbol{y})\, d\Omega_{\boldsymbol{y}} = a(G_0^h, w_h) \qquad (1.213)$$

which explains the left-hand side. The symmetric strain energy $a(G_0^h, w_h)$ plays the role of a turnstile.

Because (1.210) contains so much structural analysis, it is perhaps best to repeat the proof in single steps.

Assume a truck is parked in the middle of a bridge; see Fig. 1.42. The truck constitutes the load case p. Next two FE solutions are calculated: a) the FE solution w_h of the load case p; b) the FE solution G_0^h, which simulates a single force δ_0 in the middle of the bridge. Both solutions, w_h and G_0^h, lie in V_h. Because G_0^h lies in V_h, it follows that

$$a(G_0^h, w_h) = p(G_0^h) \qquad G_0^h \text{ is a virtual displacement in the load case } p$$
$$(1.214)$$

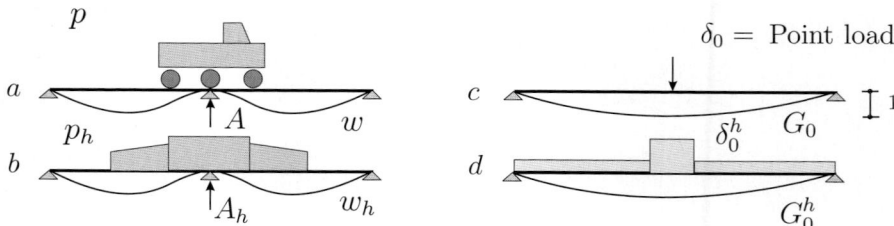

Fig. 1.44. The truck causes the force A in the column while the substitute load p_h causes the force A_h. The force A_h is obtained if the truck is placed on the influence function G_0^h

but it must also be true that

$$a(G_0^h, w_h) = (\delta_0, w_h) \qquad w_h \text{ is a virtual displacement in the load case } \delta_0 \tag{1.215}$$

and because $(\delta_0, w_h) = w_h(\boldsymbol{x})$, the proof is complete.

Next assume the truck traverses a plate supported in the middle by a column; see Fig. 1.43. The influence function for the force A in the column is the deflection $P \cdot G_0$ of the plate if the column is removed and instead a concentrated force P is applied (for simplicity in the following it is assumed that $P = 1$) which pushes the plate down by one unit of deflection. If p denotes the truck, the force in the column is

$$A = \int_\Omega G_0(\boldsymbol{y}, \boldsymbol{x}_A)\, p(\boldsymbol{y})\, d\Omega_{\boldsymbol{y}} \,. \tag{1.216}$$

Concentrated forces are out of reach for an FE program, so the program replaces the concentrated force with an equivalent load δ_0^h that is an aggregate of surface and line loads, which in the sense of the principle of virtual work is equivalent to the concentrated force P acting at $\boldsymbol{x}_A$ (see Fig. 1.44):

$$\underbrace{\delta_0^h(\varphi_i)}_{\delta W_e(p_h,\varphi_i)} = f_i = \underbrace{(\delta_0, \varphi_i)}_{\delta W_e(p,\varphi_i)} = P \cdot \varphi_i(\boldsymbol{x}_A) \quad i = 1, 2, \ldots n\,. \tag{1.217}$$

The deflection surface G_0^h of this equivalent load case δ_0^h is an approximate influence function, and if now the truck is placed on a contour plot of this deflection surface and the contour lines covered by the wheels of the truck are traced with a planimeter, the result is exactly the support reaction A_h of the FE program:

$$A_h = \int_\Omega G_0^h(\boldsymbol{y}, \boldsymbol{x}_A)\, p(\boldsymbol{y})\, d\Omega_{\boldsymbol{y}} \,. \tag{1.218}$$

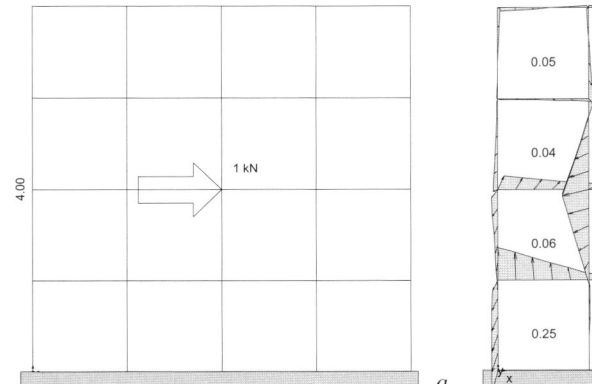
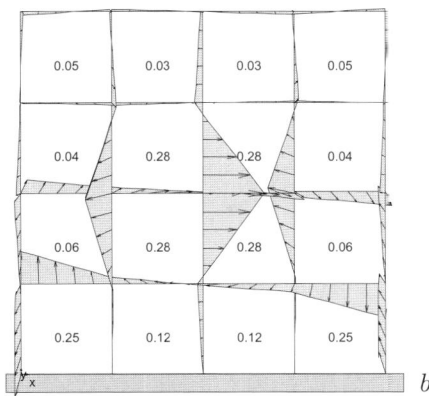

Fig. 1.45. Plate: **a)** point load $\boldsymbol{\delta}_0$; **b)** FE approximation $\boldsymbol{\delta}_0^h$ with bilinear elements. In V_h the two deltas yield the same result $u_h(\boldsymbol{x}) = (\boldsymbol{\delta}_0, \boldsymbol{u}_h) = (\boldsymbol{\delta}_0^h, \boldsymbol{u}_h)$

A converse statement

Consider a plate as in Fig. 1.45 which is subjected to a volume force $\boldsymbol{p}$ (not shown). In a second load case a horizontal force $P = 1$ is applied at a point $\boldsymbol{x}$; see Fig. 1.45 a. Let $\boldsymbol{p}_0^h$ and $\boldsymbol{j}_0^h$ be the element residual forces and jump terms, respectively, of the FE solution $\boldsymbol{G}_0^h$ of this second load case (Fig. 1.45 b). According to Betti's theorem

$$W_{1,2} = \int_{\Omega} \boldsymbol{G}_0^h \bullet \boldsymbol{p} \, d\Omega_{\boldsymbol{y}} = \int_{\Omega} \boldsymbol{p}_0^h \bullet \boldsymbol{u} \, d\Omega_{\boldsymbol{y}} + \sum_k \int_{\Gamma_k} \boldsymbol{j}_0^h \bullet \boldsymbol{u} \, ds_{\boldsymbol{y}} = W_{2,1}$$

$$(1.219)$$

and because of

$$\int_{\Omega} \boldsymbol{G}_0^h \bullet \boldsymbol{p} \, d\Omega_{\boldsymbol{y}} = u_x^h(\boldsymbol{x}) \tag{1.220}$$

we have as well

$$u_x^h(\boldsymbol{x}) = \int_{\Omega} \boldsymbol{p}_0^h \bullet \boldsymbol{u} \, d\Omega_{\boldsymbol{y}} + \sum_k \int_{\Gamma_k} \boldsymbol{j}_0^h \bullet \boldsymbol{u} \, ds_{\boldsymbol{y}} =: (\boldsymbol{\delta}_0^h, \boldsymbol{u}) \tag{1.221}$$

which means that the horizontal displacement $u_x^h(\boldsymbol{x})$ is equal to the work done by the "approximate Dirac delta" $\boldsymbol{\delta}_0^h = \{\boldsymbol{p}_0^h, \boldsymbol{j}_0^h\}$ acting through the true displacement field $\boldsymbol{u}$. In the limit the volume forces more and more resemble a true Dirac delta, $\boldsymbol{p}_0^h \to \boldsymbol{\delta}_0$, and the stress jumps vanish, $\boldsymbol{j}_0^h \to \boldsymbol{0}$, so that

$$u_x(\boldsymbol{x}) = \int_{\Omega} \boldsymbol{\delta}_0(\boldsymbol{y} - \boldsymbol{x}) \bullet \boldsymbol{u}(\boldsymbol{y}) \, d\Omega_{\boldsymbol{y}} . \tag{1.222}$$

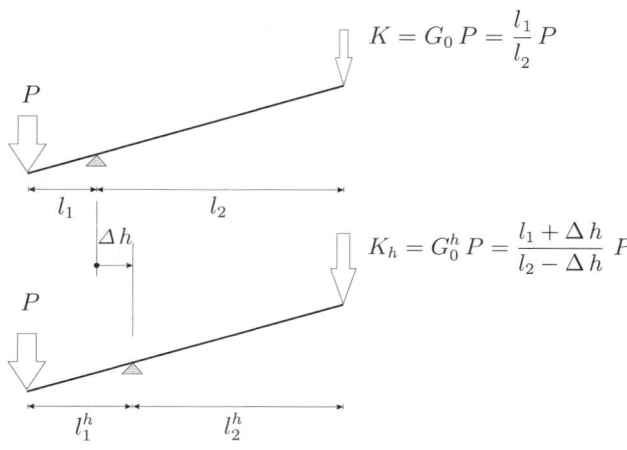

$$K = G_0 P = \frac{l_1}{l_2} P$$

$$K_h = G_0^h P = \frac{l_1 + \Delta h}{l_2 - \Delta h} P$$

Fig. 1.46. Effect of a small perturbation Δh of the fulcrum on the influence coefficient $G_0 = l_1/l_2$ for the leverage K

Hence the residual forces $\{\boldsymbol{\delta}_0 - \boldsymbol{p}_0^h, \boldsymbol{j}_0^h\}$ are to be minimized. Note that (1.221) holds for any FE solution. The statement

$$u_h(\boldsymbol{x}) = (\boldsymbol{\delta}_0^h, \boldsymbol{u}) \tag{1.223}$$

where $\boldsymbol{\delta}_0^h$ stands for $\{\boldsymbol{p}_0^h, \boldsymbol{j}_0^h\}$ is just the converse of

$$u_h(\boldsymbol{x}) = \boldsymbol{p}(\boldsymbol{G}_0^h). \tag{1.224}$$

Proxies

More will be said about this subject in Chap. 7. For now it suffices to state that

- the approximate Green's function G_0^h is a proxy for the exact function G_0 on P_h = the set of all load cases based on the unit load cases p_i;
- the approximate Dirac delta δ_0^h is a proxy for δ_0 on V_h.

In other words the Galerkin orthogonality $a(u - u_h, \varphi_i) = 0$ also holds for the Green's functions $a(G_0 - G_0^h, \varphi_i) = 0$, or after integration by parts (Green's first identity)

$$a(G_0 - G_0^h, \varphi_i) = p_i(G_0) - p_i(G_0^h) = \varphi_i(\boldsymbol{x}) - p_i(G_0^h) = 0 \tag{1.225}$$

which implies that G_0^h on P_h is a perfect replacement for the kernel G_0, and because

$$(\delta_0, \varphi_i) = (\delta_0^h, \varphi_i) = \varphi_i(x) \tag{1.226}$$

the same holds with regard to δ_0^h and δ_0 on V_h. This means that to calculate the horizontal displacement $u_x^h(\boldsymbol{x})$ of *any* displacement field $\boldsymbol{u}_h \in V_h$, instead of the exact Dirac delta in Fig. 1.45 a, the substitute Dirac delta in Fig. 1.45 b may be invoked. This is a remarkable—but obvious—result. It is obvious because in the FE method the original Dirac delta δ_0 (the load case p) is replaced by a substitute Dirac delta δ_0^h (a load case p_h) that is equivalent with respect to all $\varphi_i \in V_h$ and in a load case such as $p = \delta_0$ work equivalence just means that

$$\varphi_i(x) = (\delta_0, \varphi_i) = (\delta_0^h, \varphi_i) = \varphi_i(x). \tag{1.227}$$

Summarizing all these results, we can state that the FE solution $u_h(x) \in V_h$ can be written in six different ways

$$u_h(x) = p_h(G_0) = p_h(G_0^h) = p(G_0^h) = (\delta_0, u_h) = (\delta_0^h, u_h) = (\delta_0^h, u). \tag{1.228}$$

To better pick up the pattern in these equations we use the short hand notation (p_h, G_0) instead of $p_h(G_0)$ for the virtual external work so that

$$u_h(x) = (p_h, G_0) = (L\,u_h, G_0) = (u_h, L^*G_0) = (u_h, \delta_0) \tag{1.229}$$
$$u_h(x) = (p_h, G_0^h) = (L\,u_h, G_0^h) = (u_h, L^*G_0^h) = (u_h, \delta_0^h) \tag{1.230}$$

where L is the differential operator and $L^* = L$ is the adjoint operator which is the same operator as L because in structural analysis L is (most often) self-adjoint.

1.19 Influence functions

As in many engineering problems (Fig. 1.46) the accuracy of an FE solution depends on how well the influence functions for the displacements, stresses, or support reactions can be approximated (Fig. 1.47). The nature of these Green's functions is therefore to be discussed next.

All influence functions are displacements or deflections. In a beam the influence functions for $w(x), w'(x), M(x)$, or $V(x)$ at a point x are the deflection curves of the beam if a *dual load*, i.e., a force $P = 1$, a moment $M = 1$, a sharp bend $w'(x_+) - w'(x_-) = 1$ or a dislocation $w(x_+) - w(x_-) = 1$, is applied at x (see Fig. 1.34, p. 48).

In an energy method the focus is not on the loads themselves, but rather on the work done by the loads acting through the virtual displacements. The characteristic property of a single force $P = 1$ at x is that it contributes work $\delta w(x)$ upon acting through a virtual displacement δw.

A convenient symbol to describe such an action (point load) is the *Dirac delta*

$$\delta_0(y - x) = 0 \quad \text{for all } y \neq x, \qquad \int_0^l \delta_0(y - x)\varphi_i(y)\,dy = \varphi_i(x). \tag{1.231}$$

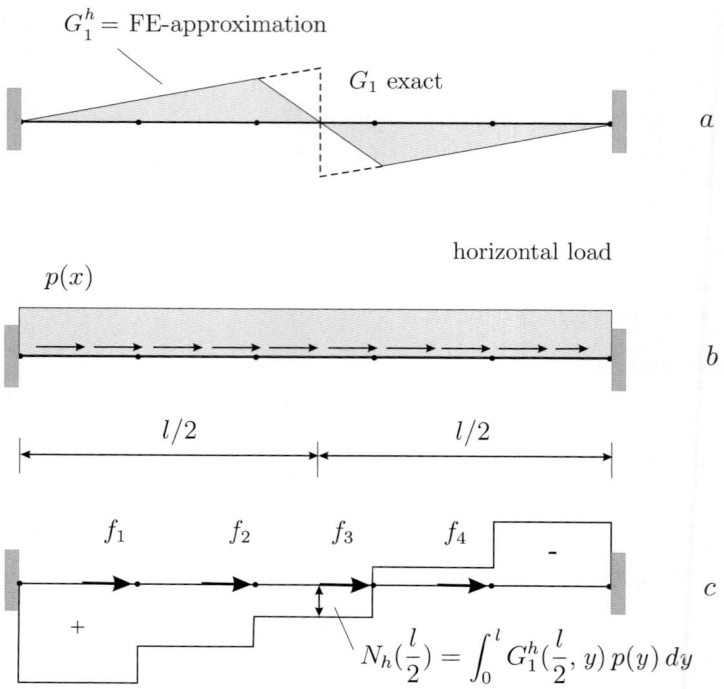

$N_h(x) = $ normal force of FE-solution

Fig. 1.47. The scalar product of the load p and the approximate Green's function G_1^h is the normal force N_h of the FE solution

This is a function that is zero almost everywhere—up to the point x—and that acting through a nodal unit displacement $\varphi_i(y)$ or any other virtual displacement contributes work $\varphi_i(x)$ (= the value of $\varphi_i(y)$ at x). This is what a single force looks like in energy methods.

To have symbols for the other dual quantities, higher Dirac deltas are introduced:

$$\delta_1(y-x) \qquad \text{moment} \qquad \int_0^l \delta_1(y-x)\varphi_i(y)\,dy = \varphi_i'(x)$$

$$\delta_2(y-x) \qquad \text{sharp bend} \qquad \int_0^l \delta_2(y-x)\varphi_i(y)\,dy = M_i(x)$$

$$\delta_3(y-x) \qquad \text{dislocation} \qquad \int_0^l \delta_3(y-x)\varphi_i(y)\,dy = V_i(x)\,,$$

where $M_i(x)$ and $V_i(x)$ are the moment and shear force, respectively, of φ_i at the point x.

The equivalent nodal forces f_i that belong to a Dirac delta δ_0 are

$$f_i = \int_0^l \delta_0(y - x)\, \varphi_i(y)\, dy = \varphi_i(x) \qquad (1.232)$$

and the same holds for the higher Dirac deltas. Because each Dirac delta extracts from the virtual displacement φ_i just the term conjugate to δ_i, the f_i are just the values of φ_i at x conjugate to δ_i. Hence if the influence function for a quantity $Q(x)$ is to be calculated, the equivalent nodal forces f_i are just $f_i = Q(\varphi_i)(x)$.

Table 1.2. Equivalent nodal forces for influence functions in beams and slabs

Beam	dual	quantity	f_i	Kirchhoff plate	f_i
w	δ_0	force	$\varphi_i(x)$	w	$\varphi_i(\boldsymbol{x})$
w'	δ_1	moment	$\varphi_i'(x)$	$w_{,x}$, $w_{,y}$	$\varphi_{i,x}(\boldsymbol{x})$, $\varphi_{i,y}(\boldsymbol{x})$
M	δ_2	sharp bend	$M_i(x)$	m_{xx}, m_{xy}, m_{yy}	$m_{xx}^{(i)}(\boldsymbol{x}), m_{xy}^{(i)}(\boldsymbol{x}), m_{yy}^{(i)}(\boldsymbol{x})$
V	δ_3	dislocation	$V_i(x)$	q_x, q_y	$q_x^{(i)}(\boldsymbol{x}), q_y^{(i)}(\boldsymbol{x})$

Table 1.2 lists the equivalent nodal forces f_i for Euler–Bernoulli beams and Kirchhoff plates; quantities for second-order equations are listed in Table 1.3.

Table 1.3. Equivalent nodal forces for influence functions of bars and plates

Bar	dual	quantity	f_i	plate	f_i
u	δ_0	force	$\varphi_i(x)$	u_x, u_y	$\varphi_i(\boldsymbol{x})$
N	δ_1	dislocation	$N_i(x)$	$\sigma_{xx}, \sigma_{xy}, \sigma_{yy}$	$\sigma_{xx}^{(i)}(\boldsymbol{x}), \sigma_{xy}^{(i)}(\boldsymbol{x}), \sigma_{yy}^{(i)}(\boldsymbol{x})$

Deflection of a taut rope

To calculate the influence function for the deflection w of the rope in Fig. 1.39 p. 58 at a node x_k, a single force $P = 1$ is applied at x_k. The equivalent nodal forces f_i are

$$f_i = \int_0^l \delta_0(x_k - y)\, \varphi_i(y)\, dy = \varphi_i(x_k) = \begin{cases} 1 & i = k \\ 0 & i \neq k \end{cases} \qquad (1.233)$$

so that $\boldsymbol{f}$ is identical to the unit vector $\boldsymbol{e}_k$, and with $\boldsymbol{u} = \boldsymbol{K}^{(-1)}\boldsymbol{e}_k$ the Green's function becomes

$$G_0^h(x_k, y) = \sum_i u_i\, \varphi_i(y) = \sum_i (\boldsymbol{K}^{(-1)}\boldsymbol{e}_k)_i\, \varphi_i = \sum_i k_{ik}^{(-1)}\, \varphi_i(y)\,. \quad (1.234)$$

In a load case p the deflection $w_h(x_k)$ at the node x_k is therefore

influence function for $N(l/2)$

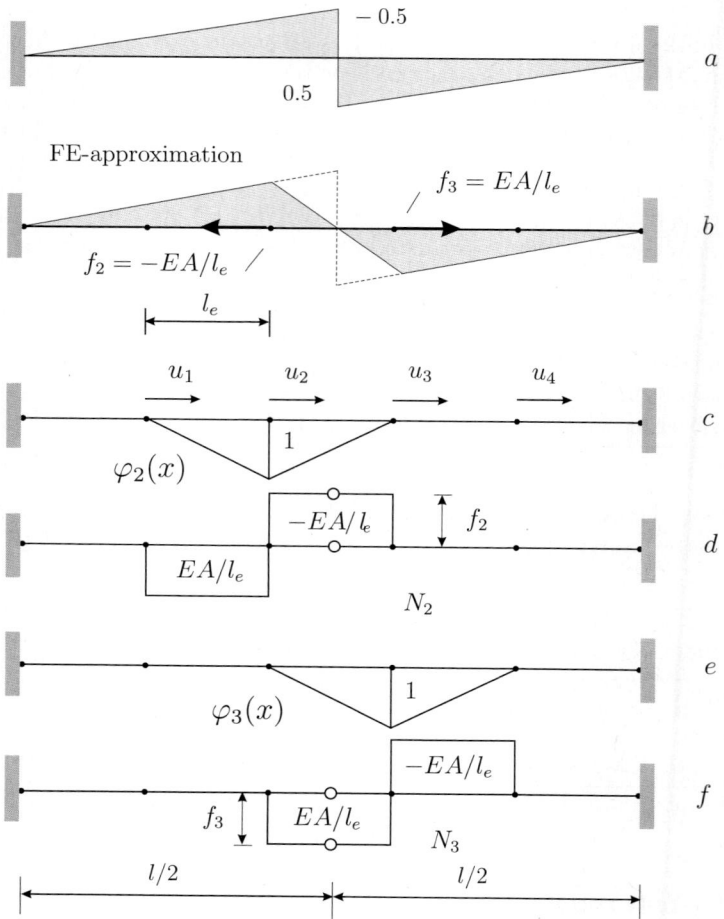

Fig. 1.48. a) Influence function for the normal force at the center of the bar; **b)** FE approximation. The equivalent nodal forces f_i are the normal forces of the nodal unit displacements at the point $x = l/2$

$$w_h(x_k) = \int_0^l G_0^h(x_k, y)\, p(y)\, dy = \sum_i k_{ik}^{(-1)} \int_0^l \varphi_i(y)\, p(y)\, dy$$

$$= \sum_i k_{ik}^{(-1)} f_i \tag{1.235}$$

where the f_i are now the equivalent nodal forces belonging to the distributed load p.

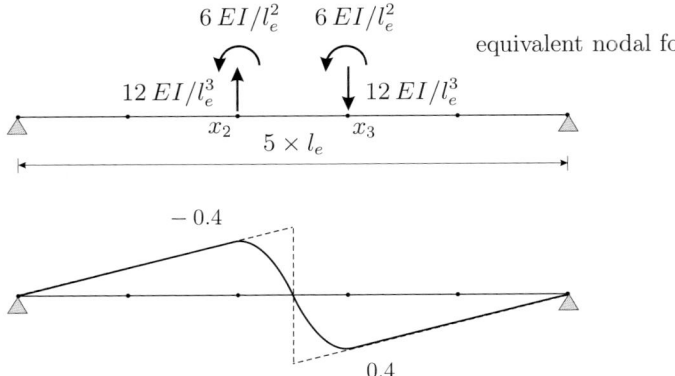

Fig. 1.49. Influence function for the FE shear force V_h at the center of the beam

Normal force in a bar

The influence function for the normal force N at the center of the bar in Fig. 1.48 is the longitudinal displacement of the bar if the bar is split at the center, $u(x_+) - u(x_-) = 1$; see Fig. 1.48. The equivalent nodal forces f_i that belong to this load case are

$$f_i = \int_0^l \delta_1(\frac{l}{2} - y)\, \varphi_i(y)\, dy = EA\, \varphi_i'(\frac{l}{2}) = N_i(\frac{l}{2}). \tag{1.236}$$

Evidently the FE solution of this load case, the shape in Fig. 1.48 b, is not the exact influence function for $N(l/2)$.

But it is the exact influence function for the *average* value N_a of the normal force $N(x)$ in the center element having end points x_2 and x_3 and length $l_e = x_3 - x_2$. To see this, note that

$$\begin{aligned}
N_a &= \frac{1}{l_e}\int_{x_2}^{x_3} N(x)\, dx = \frac{1}{l_e}\int_{x_2}^{x_3}\int_0^l G_1(y,x)\, p(y)\, dy\, dx \\
&= \frac{1}{l_e}\int_{x_2}^{x_3}\int_0^l EA\frac{d}{dx} G_0(y,x)\, p(y)\, dy\, dx \\
&= \frac{EA}{l_e}\int_0^l [G_0(y,x_3) - G_0(y,x_2)]\, p(y)\, dy
\end{aligned} \tag{1.237}$$

and this kernel $EA/l_e[G_0(y,x_3) - G_0(y,x_2)]$ is just the shape in Fig. 1.48 b. Namely to reproduce this kernel in V_h the following equivalent nodal forces must be applied

$$f_i = \frac{EA}{l_e}\int_0^l [\delta_0(y - x_3) - \delta_0(y - x_2)]\, \varphi_i(y)\, dy = \frac{EA}{l_e}[\varphi_i(x_3) - \varphi_i(x_2)] \tag{1.238}$$

which are just the forces, $f_2 = -EA/l_e$, $f_3 = EA/l_e$, that were applied previously to approximate G_1.

Remark 1.8. Because the normal force is defined as $N(x) = EA\,u'(x)$, the influence function $G_1(y, x)$ for $N(x)$ is

$$G_1(y, x) = EA\,\frac{d}{dx}\,G_0(y, x)\,. \tag{1.239}$$

So if $\sigma_{ij} = \mathrm{Op}(u)$, where $\mathrm{Op}()$ is some differential operator and G_0 is the Green's function for u, then the Green's function for σ_{ij} is $\mathrm{Op}(G_0)$ and differentiation is carried out with respect to x.

Shear force in a beam

To obtain the influence function for the shear force $V(l/2)$ in the middle of the beam (see Fig. 1.49) a dislocation $w_+ - w_- = 1$ must be applied so that the equivalent nodal forces

$$f_i = \int_0^l \delta_3(\frac{l}{2} - y)\varphi_i(y)\,dy = V_i(\frac{l}{2}) \tag{1.240}$$

are the shear forces of the nodal unit displacement φ_i at the point $x = l/2$.

Obviously the FE influence function is not correct. Could the situation be saved by claiming that the FE approximation is the exact influence function for the average value of $V(x)$? No, this time the previous logic fails by a narrow margin. The influence function for the average value V_a of $V(x)$ in the center element is

$$
\begin{aligned}
V_a &= \frac{1}{l_e} \int_{x_2}^{x_3} V(x)\,dx = \frac{1}{l_e} \int_{x_2}^{x_3} \int_0^l G_3(y, x)\,p(y)\,dy\,dx \\
&= \frac{1}{l_e} \int_{x_2}^{x_3} \int_0^l \frac{d}{dx}\,G_2(y, x)\,p(y)\,dy\,dx = \frac{1}{l_e} \int_0^l [G_2(y, x_3) - G_2(y, x_2)]\,p(y)\,dy
\end{aligned}
\tag{1.241}
$$

where $1/l_e[G_2(y, x_3) - G_2(y, x_2)]$ is almost the figure in Fig. 1.49. "Almost" because the influence functions for bending moments have a sharp bend at x that is not to be seen at the end points x_3 and x_2. But it is obviously possible to come very close to this shape in V_h. In other words, FE solutions gain in accuracy if averages are studied, rather than point values.

To approximate the shape of the kernel $1/l_e[G_2(y, x_3) - G_2(y, x_2)]$ in V_h, the same equivalent nodal forces must be applied as in Fig. 1.49, because the third-order difference quotient and the third-order derivative of a third-degree polynomial φ_i at any point $\bar{x} \in (x_2, x_3)$ are the same:

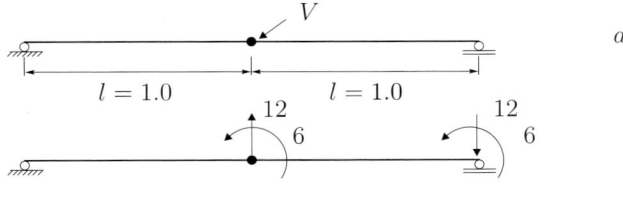

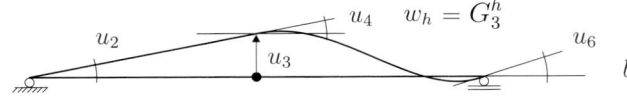

a

b

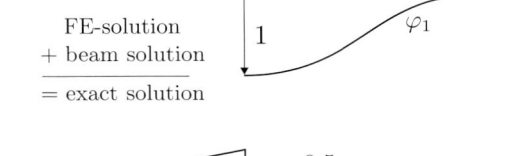

FE-solution
+ beam solution
———————
= exact solution

c

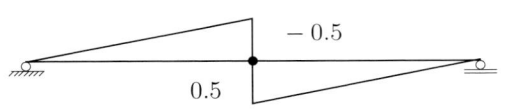

d

Fig. 1.50. How a commercial FE program calculates the influence function for the shear force V at the center of a beam ($EI = 1$), and how it obtains the correct result

$$f_i = \frac{1}{l_e} \int_0^l [\delta_2(y - x_3) - \delta_2(y - x_2)]\, \varphi_i(y)\, dy = \int_0^l \delta_3(y - \bar{x})\, \varphi_i(y)\, dy$$

$$= \frac{M_i(x_3) - M_i(x_2)}{l_e} = V_i(\bar{x})\,. \tag{1.242}$$

Remark 1.9. A commercial FE program calculates the influence function for $V(l/2)$ as follows. First it keeps all nodes fixed and it applies the dislocation ($= \varphi_1$) (see Fig. 1.50 c) to the second element. Next it applies the fixed end forces ($\times(-1)$) of this load case to the structure and adds to the resulting deflection curve which is just G_3^h—the deflection caused by the dislocation in element 2. The result is the exact solution (Fig. 1.50).

The deflection in the first element agrees with the exact solution

$$w_h^{(1)} = -0.5\, x = w_{exact} = G_3 \tag{1.243}$$

but to the deflection in the second element the program adds the beam solution φ_1 (= dislocation at the left end of the fixed beam)

$$w_h^{(2)} = -0.5 - 0.5\, x + 3\, x^2 - 2\, x^3$$
$$+ 1.0 \qquad\quad - 3\, x^2 + 2\, x^3 \qquad (= \varphi_1)$$
$$w_{exact} = +0.5 - 0.5\, x \tag{1.244}$$

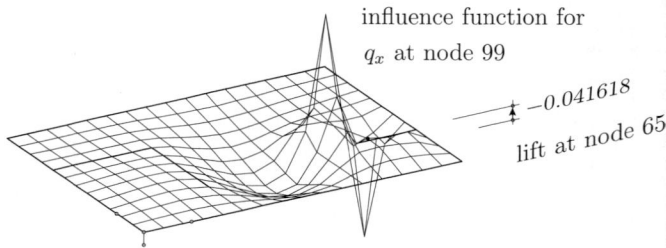

influence function for
q_x at node 99

-0.041618

lift at node 65

a

$$q_x = -0.041618 = w \times P = -0.041618 \times 1$$

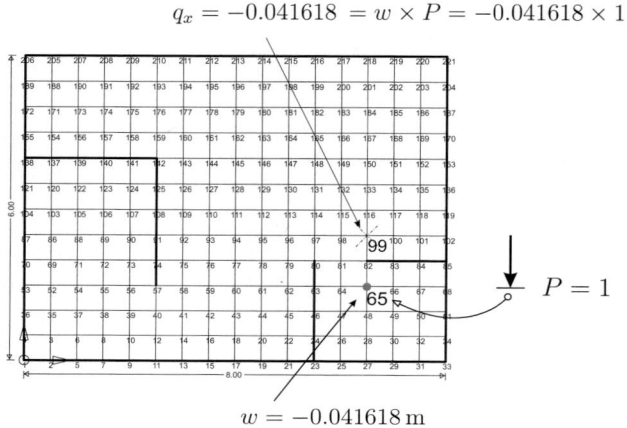

$P = 1$

b $w = -0.041618\,\mathrm{m}$

Fig. 1.51. a) FE influence function for the shear force q_x^h at node 99; **b)** FE mesh and deflection at node 65, $E = 3.0\ \mathrm{E}^7\ \mathrm{kN/m^2}$, $\nu = 0.16$, thickness $d = 0.20$ m, element size $= 0.5$ m $\times$ 0.5 m

In 2-D and 3-D problems the displacement field cannot be split into a homogeneous and a particular displacement field, so this technique is not applicable to such problems.

Note that in the FE model of the beam we must distinguish between the shear force V^h on the left-hand side (V_l^h) and on the right-hand side (V_r^h) of the center node. G_3^h is the influence function for V_r^h. This explains the asymmetry of G_3^h (see Fig. 1.50 b).

Shear force in a slab

The influence function $G_{3,x}^h$ for the shear force q_x at node 99 of the slab in Fig. 1.51 was calculated with conforming Kirchhoff elements by applying a unit dislocation at that node. The deflection at node 65 was -0.041618 m, which is exactly the value of q_x^h at node 99 if a unit force $P = 1$ is placed at node 65.

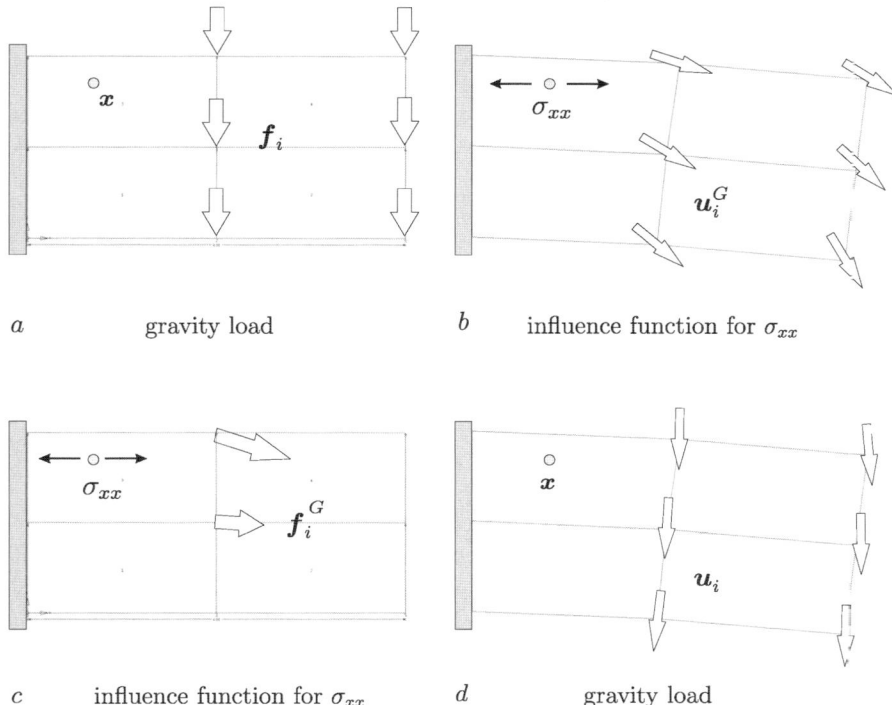

a   gravity load   b   influence function for σ_{xx}

c   influence function for σ_{xx} d   gravity load

Fig. 1.52. Nodal forces and nodal displacements of two load cases: gravity load, **a)** and **d)**, and Green's function for σ_{xx}, **c)** and **b)**. The scalar product of the nodal vectors **a)** $\boldsymbol{f}_i$ (gravity load) and **b)** $\boldsymbol{u}_i^G$ (Green's function) gives the stress σ_{xx}^h of the FE solution or—alternatively—of the nodal vectors **c)** $\boldsymbol{f}_i^G$ (Green's function) and **d)** $\boldsymbol{u}_i$ (gravity load)

Nodal influence functions

Nodal influence functions is short for *nodal form of influence functions* and by this we mean that in FE analysis the evaluation of the two equivalent influence functions

$$u_h(x) = \int_0^l G_0^h(y, x)\, p(y)\, dy = \int_0^l u_h(y)\, \delta_0(y - x)\, dy \qquad (1.245)$$

can be done by summing over the nodes. This is a key point of the FE method. We have

$$u_h(x) = \int_0^l G_0^h(y,x)\, p(y)\, dy = \int_0^l \sum_i \varphi_i(y)\, u_i^G(x)\, p(y)\, dy = \sum_i u_i^G(x)\, f_i$$

$$= \boldsymbol{u}_G^T\, \boldsymbol{f} = \boldsymbol{u}_G^T\, \boldsymbol{K}\, \boldsymbol{u} = \boldsymbol{u}_G^T\, \boldsymbol{K}^T\, \boldsymbol{u} = \boldsymbol{f}_G^T\, \boldsymbol{u} = \sum_i f_i^G(x)\, u_i$$

$$= \sum_i \varphi_i(x)\, u_i = \int_0^l \sum_i u_i\, \varphi_i(y)\, \delta(y-x)\, dy = \int_0^l u_h(y)\, \delta_0(y-x)\, dy\,.$$

$$(1.246)$$

So the displacement $u_h(x)$ is the scalar product between the nodal displacements of the Green's function and the equivalent nodal forces of the load case p or—vice versa—between the nodal displacements of the FE solution u_h and the nodal forces f_i^G of the Green's function

$$u_h(x) = \begin{cases} \displaystyle\sum_i \varphi_i(x)\, u_i = \int_0^l u_h(y)\, \delta_0(y-x)\, dy = \boldsymbol{f}_G^T\, \boldsymbol{u} \\[2mm] \displaystyle\int_0^l G_0^h(y,x)\, p(y)\, dy = \boldsymbol{u}_G^T\, \boldsymbol{f}\,. \end{cases} \qquad (1.247)$$

So when we evaluate $u_h(x)$ or $\sigma_h(x)$ then we apply the corresponding Dirac delta to each shape function φ_i and multiply the result $(= f_i^G)$ with the weight u_i of the shape function. That is the equivalent nodal forces f_i^G of the Green's functions are simply the displacements or stresses, etc., of the shape functions at the point x. This is clearly seen in the middle of the long chain of equations (1.246) because the identity

$$\sum_i f_i^G(x)\, u_i = \sum_i \varphi_i(x)\, u_i \qquad \Rightarrow \qquad f_i^G(x) = \varphi_i(x) \qquad (1.248)$$

just means that.

For example the stress $\sigma_{xx}^h(\boldsymbol{x})$ of the plate in Fig. 1.52 is equal to the work done by the equivalent nodal forces $\boldsymbol{f}_i$ on acting through the nodal displacements $\boldsymbol{u}_i^G$ of the Green's function (a and b in Fig. 1.52)

$$\sigma_{xx}^h(\boldsymbol{x}) = \int_\Omega \boldsymbol{G}_1^h(\boldsymbol{y},\boldsymbol{x}) \bullet \boldsymbol{p}(\boldsymbol{y})\, d\Omega_{\boldsymbol{y}} = \sum_i \boldsymbol{u}_i^G \bullet \boldsymbol{f}_i \equiv b \times a\,. \qquad (1.249)$$

Because of Betti's theorem ($\boldsymbol{K} = \boldsymbol{K}^T$) this result is equivalent to

$$\sigma_{xx}^h(\boldsymbol{x}) = \sum_i \boldsymbol{f}_i^G \bullet \boldsymbol{u}_i = \sum_j \sigma_{xx}(\boldsymbol{\varphi}_j)(\boldsymbol{x})\, u_j \equiv c \times d\,. \qquad (1.250)$$

The first sum extends over all nodes $i = 1, 2, \dots N$ of the structure while the second sum extends over all degrees of freedom $j = 1, 2, \dots 2 \times N$ of the nodes.

This result means that in FE methods we calculate stresses as in finite difference methods, see Sect. 7.6 p. 533.

The inverse of K

The entries in the inverse K^{-1} of the stiffness matrix of a structure are the coefficients g_{ij} of the projections (= FE approximations) of the n nodal Green's functions $G_0[x_i], i = 1, 2, \ldots n$ onto V_h:

$$G_0^h(x_i, y) = \sum_{j=1}^{n} g_{ij} \, \varphi_j(y) \qquad K^{(-1)} = [g_{ij}]. \qquad (1.251)$$

That is to each node i belongs a Green's function $G_0(x_i, y)$ (= influence function for the displacement $u_i = u(x_i)$ at x_i) and the entries g_{ij} in row i of K^{-1} describe the expansion of the FE Green's function in terms of the φ_j. In 1-D problems the g_{ij} would be just the nodal values of the Green's function $G_0(x_i, y)$ that is $g_{ij} = G_0(x_i, y_j)$, $j = 1, 2, \ldots n$, where $y_1, y_2, \ldots$ are the coordinates of the nodes.

Equation (1.251) is easily verified if the analytic result

$$u_i = u_h(x_i) = \int_\Omega G_0^h(x_i, y) \, p(y) \, d\Omega_y = \sum_j \int_\Omega g_{ij} \, \varphi_j(y) \, p(y) \, d\Omega_y$$

$$= \sum_j g_{ij} \, f_j \qquad (1.252)$$

is compared with the computer output

$$u_i = \sum_j k_{ij}^{(-1)} f_j \qquad (1.253)$$

and if the n unit vectors, $f = e_j$ are substituted consecutively for f. This establishes $k_{ij}^{(-1)} = g_{ij}$.

Linear algebra provides the same result: let u and $\hat{u}$ any two vectors; then $\hat{u}^T K u = u K \hat{u}$ which implies that if $K u = f$, $\hat{u}^T f = u K \hat{u}$. Next let g_i be the solution of $K g_i = e_i$; then it follows that $u_i = g_i^T f$. If this is compared with (1.252), it follows that the coefficients g_{ij} are the solutions of $K g_i = e_i$, which is equivalent to saying that $K^{(-1)} = [g_{ij}]$.

Commercial codes

Commercial codes normally provide no routines for to calculate influence functions but a diligent user can circumvent this problem. The issue is to determine the equivalent nodal forces f_i which will produce the influence function.

Let us assume that the influence function for the shear stress σ_{xy} at the center of a bilinear element which is part of a larger structure is to be calculated.

The analysis is done on a single bilinear element which has the same shape and size as the original element. In this case the element has eight degrees of

point force   dislocation

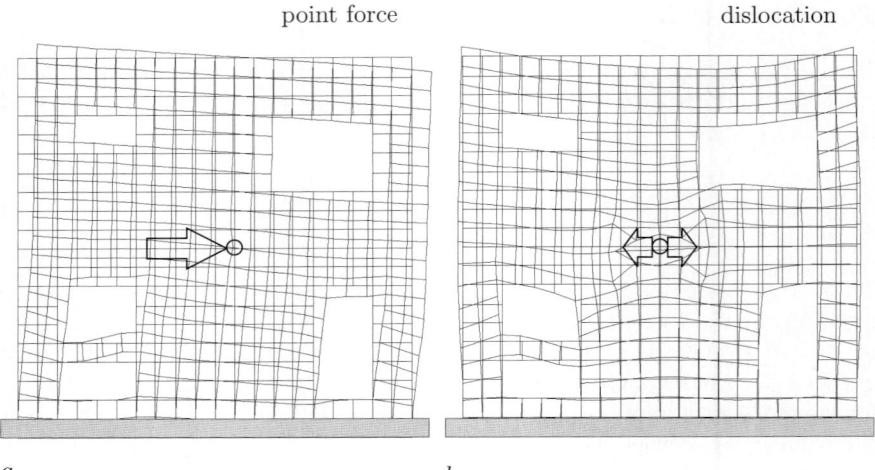

a b

Fig. 1.53. Influence functions in a plate for **a)** the horizontal displacement u_x, **b)** the stress σ_{xx}

freedom. In the first load case we let $u_1 = 1$ and all other $u_i = 0$. The shear stress σ_{xy} at the center of the element in this load case is the equivalent nodal force f_1, etc. So by solving eight different load cases $\boldsymbol{u} = \boldsymbol{e}_i, i = 1, 2, \ldots 8$, we can calculate the eight stresses $\sigma_{xy}(\boldsymbol{\varphi}_i)(\boldsymbol{x}) = f_i$ which—as equivalent nodal forces f_i—produce the FE influence function for $\sigma_{yx}^h(\boldsymbol{x})$.

Because the equivalent nodal forces depend only on the element which contains the point $\boldsymbol{x}$ the formulas in Sect. 4.8 p. 357 would have provided the stresses more easily but if the elements which are implemented are of an unknown type or if the elements are curved or not affine to the master element then this technique can help.

The influence functions for nodal stresses, which are usually average values

$$\sigma_{xy}^h = \frac{\sigma_{xy}^{(1)} + \sigma_{xy}^{(2)} + \sigma_{xy}^{(3)} + \sigma_{xy}^{(4)}}{4} \qquad (1.254)$$

are obtained in the same way: it is only that the 8×4 nodal forces—for each of the four element stresses $\sigma_{xy}^{(i)}$ at the node—must be applied simultaneously and must be weighted with $1/4$.

1.20 Accuracy

Each displacement u, v, w, and each stress or stress resultant

$$\sigma_{xx}, \sigma_{xy}, \sigma_{yy} \qquad m_{xx}, m_{xy}, m_{yy}, \qquad q_x, q_y \qquad (1.255)$$

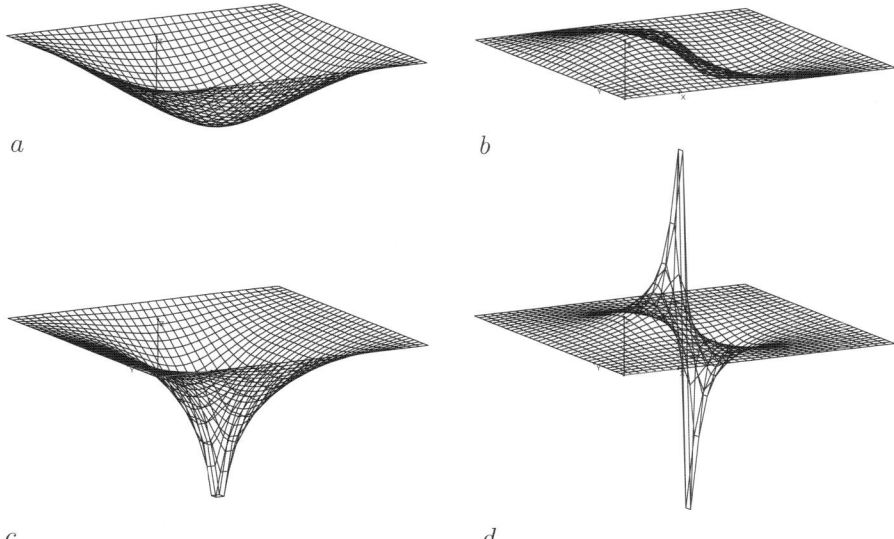

a b

c d

Fig. 1.54. Influence functions for **a)** the deflection w, **b)** the slope $w_{,x}$, **c)** the bending moment m_{xx}, and **d)** the shear force q_x at the center of the slab

at a node or a Gauss point (see Fig. 1.53) is the scalar product of the associated Green's function G_j (the index j corresponds to the index j of the Dirac delta δ_j) and the applied load p

$$m_{xx}(\boldsymbol{x}) = \int_\Omega G_2(\boldsymbol{y}, \boldsymbol{x}) \, p(\boldsymbol{y}) \, d\Omega_{\boldsymbol{y}} \,. \qquad (1.256)$$

The FE program replaces—as shown previously—the exact Green's function with an approximation G_2^h, with what it "considers" to be the exact Green's function, and therefore the error in the bending moments is proportional to the distance between G_2 and G_2^h:

$$m_{xx}(\boldsymbol{x}) - m_{xx}^h(\boldsymbol{x}) = \int_\Omega [G_2(\boldsymbol{y}, \boldsymbol{x}) - G_2^h(\boldsymbol{y}, \boldsymbol{x})] \, p(\boldsymbol{y}) \, d\Omega_{\boldsymbol{y}} \,. \qquad (1.257)$$

This holds true for any other value as well.

Hence the *real task* of an FE program is not the solution of a single load case, but an optimal approximation of the *Green's functions*, because normally more than one load case is solved on the same mesh. The truck drives on, but the Green's functions, being *mesh-dependent*, are invariant with respect to the single load cases that are solved on the mesh, although the accuracy may depend on where the truck is parked because the error in the Green's function varies locally.

The shift

In plate bending (Kirchhoff) problems the Green's functions for

$$w, w_{,x}, w_{,y}, m_{xx}, m_{xy}, m_{yy}, q_x, q_y \tag{1.258}$$

at a specific point $\boldsymbol{x}$ are generated if a point source conjugate to $w, w_{,x}, w_{,y}$ etc. is applied at $\boldsymbol{x}$. A source conjugate to w is a single force, a source conjugate to $w_{,x}$ is a single moment, etc.

As can be seen in Fig. 1.54, the complexity of the deflection surfaces generated by these sources increases with the order of the derivative. The higher the derivative the more narrow and focused the influence function and therefore the more difficult it is to approximate these surfaces with simple shape functions.

This tendency of the influence functions has to do with the *shift* of the kernel functions.

The surface load p that acts on a slab is (in somewhat simplified terms) the fourth-order derivative of the deflection surface w. The influence function for the deflection

$$w(\boldsymbol{x}) = \int_{\Omega} \underbrace{G_0(\boldsymbol{y}, \boldsymbol{x})}_{kernel} p(\boldsymbol{y}) \, d\Omega_{\boldsymbol{y}} \tag{1.259}$$

extracts from p—which is the "fourth-order derivative"—the zeroth order derivative. Thus the kernel G_0 *integrates* four times. Its shift is of order -4.

Table 1.4. The kernels G_i and their shifts

Magnitude	Derivative	Kernel	Action	Shift
$w(\boldsymbol{x})$	0	G_0	force	-4
$w_{,i}(\boldsymbol{x})$	1	G_1	moment	-3
$m_{ij}(\boldsymbol{x})$	2	G_2	kink	-2
$q_i(\boldsymbol{x})$	3	G_3	dislocation	-1

Hence, influence functions are *integral operators*. They transform the load p into the deflection w, the normal (or more general directional) derivative $w_{,n}$ etc. The more they achieve, the more they integrate p, the more negative the shift, and the more these functions spread in all directions; see Fig. 1.54.

Therefore if the deflection at a point $\boldsymbol{x}$ is to be calculated very precisely the mesh must have the same quality everywhere, because the integral operator integrates four times, although because the operator is a very smooth function a coarse mesh probably is sufficient. But if the focus is on the shear force q_x, the mesh in the neighborhood of the point is critical, because the integral

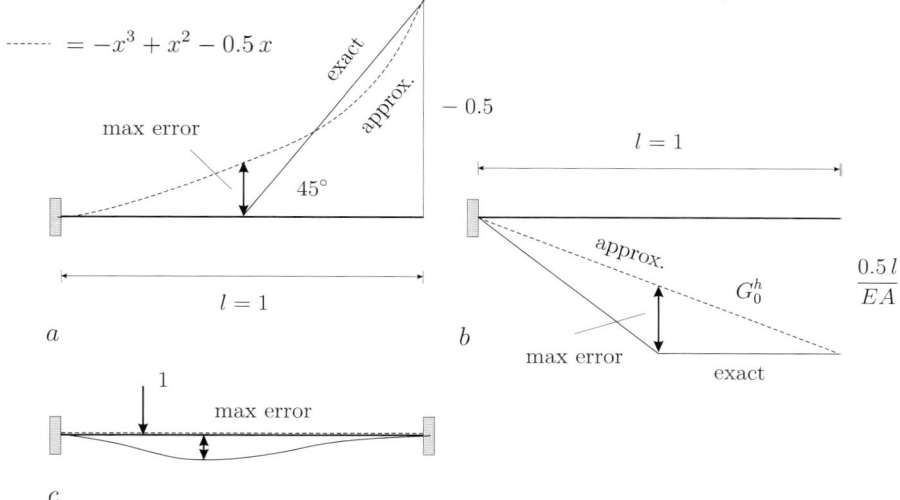

Fig. 1.55. a) Influence function for the bending moment at $x = l/2$, and **b)** the horizontal displacement at the same point. **c)** Influence function for the deflection at the quarter point $x = l/4$. The dashed curves are the FE approximations

operator in the influence functions for the shear force has the low shift -1. It integrates only once. This makes it a nearly local operator.

An operator which does nothing (it has shift zero) is the Dirac delta, which is usually identified with a point force. Now it must be considered a displacement, because all influence functions are displacements. In the case of a Kirchhoff plate it would be the deflection surface $w(\boldsymbol{x}) = K \, \Delta\Delta \, g_0(\boldsymbol{y}, \boldsymbol{x})$ with $g_0 = 1/(8\pi K)r^2 \ln r$, which in a 3-D picture would be a lone *peak* hovering over the mesh. This peak neither integrates nor differentiates what is placed under the integral sign:

$$p(\boldsymbol{x}) = \int_{\Omega} \delta_0(\boldsymbol{y} - \boldsymbol{x}) \, p(\boldsymbol{y}) \, d\Omega_{\boldsymbol{y}} \, . \tag{1.260}$$

It just reproduces the function. Its shift is of order zero. It is truly a local operator.

There are also integral operators that differentiate, which have a positive shift. If a bar is stretched by u units of displacement, then the normal force $N(0)$ is

$$N(0) = EA \frac{u}{l} \, . \tag{1.261}$$

The operator $u \to N = EA \, u'$ is an integral operator which differentiates. Actually, to see an integral sign would require a 2-D model of the bar, but evidently u/l is a difference quotient.

The bending moment in a beam fixed on the left-hand side and simply supported on the right-hand side is, if the simple support moves downward by δ units,

$$M(0) = -\frac{3\,EI}{l^2}\,\delta\,. \tag{1.262}$$

This integral operator differentiates twice as can be seen from the l^2 in the denominator.

The maximum error in the Green's functions

It is not easy to predict where the maximum error in the displacements or stresses will occur, because the accuracy depends on the accuracy of the approximate Green's function *and* the nature of the load which is applied. It is only guaranteed that if the exact Green's function lies in V_h, then the value will be exact. But the opposite need not be true: recall that if the exact solution lies in V_h, then the stresses are exact even though the Green's function does not lie in V_h.

We thus concentrate on the error in the Green's function alone. In Figure 1.55 two influence functions are plotted, one for the bending moment of a beam at $x = l/2$ and the other for the longitudinal displacement at the same point if the bar is stretched or compressed. The dashed curves are the approximate Green's functions if just one element is used. Obviously the maximum error occurs at the source point $x = l/2$ itself. It is easy to see why this happens: outside of the element $G_i = G_i^h$, and within the element G_i^h is essentially the curve obtained if G_i is interpolated at the nodes with a third-degree polynomial (a homogeneous solution), while the essential part w_p, which contains the peak, $EI\,w_p^{IV} = \delta_i(y - x)$, is neglected. Obviously the distance between the smooth interpolating function G_i^h and the peak is at its maximum at the source point x. This is even more evident in plate bending problems. Recall the infinite peaks in the influence functions for the bending moments m_{ij} in a slab! Clearly the maximum error will occur at the source point.

Hence the stresses at the foot of a single force are the least reliable, independent of the problematic nature of single forces. Can we say then that the more a load is spread, the better the accuracy of the FE results? Do gravity loads therefore have an advantage over traffic loads?

It is not guaranteed that the maximum error *always* occurs at the source point. One counterexample is the influence function for the deflection at the first quarter point of a beam with fixed ends; see Fig. 1.55 c. The FE approximation G_0^h is zero, so that the maximum error is identical to the maximum deflection of G_0, which occurs at some distance from the quarter point. But it can be assumed that this "mismatch" occurs only in influence functions for displacements, and not resultant stresses, because the peaks in the latter functions are more pronounced. But note that the influence functions for support reactions are also of displacement type.

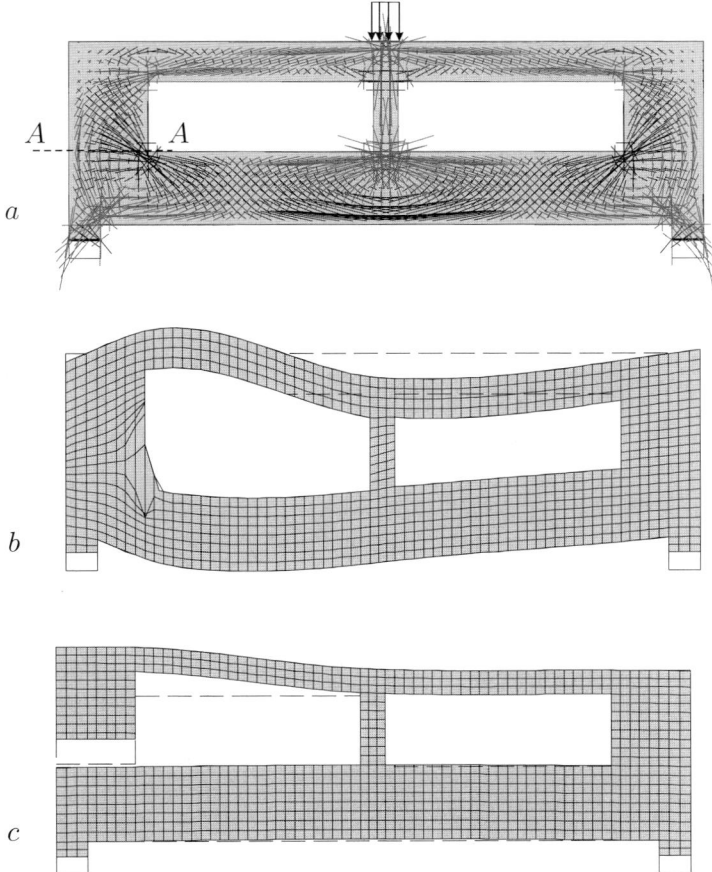

Fig. 1.56. Plate: **a)** system and loading; **b)** influence function for the stress σ_{yy} near the corner point; **c)** influence functions for the internal action N_y in cross-section A–A

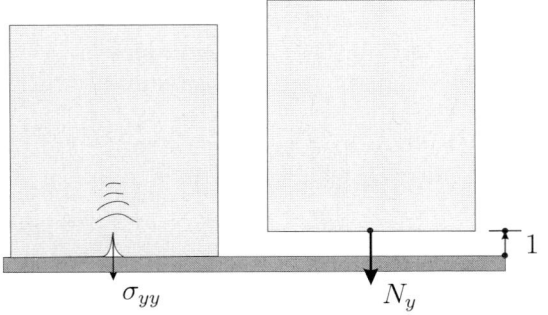

Fig. 1.57. Influence function for σ_{yy} and N_y

1.21 Why resultant stresses are more accurate

Near points where stresses become singular it is better to concentrate on resultant stresses than single values:

$$N_y = \int_0^l n_y \, dx = \int_0^l \sigma_{yy} \, d \, dx \qquad M_y = \int_0^l m_{yy} \, dy \,. \qquad (1.263)$$

The lower left corner point of the left opening in the plate in Fig. 1.56 is just such a singular point. An influence function for the singular stress σ_{yy} at this point does not exist, see Sect. 7.6, p. 532.

The stress σ_{yy} at the corner point increases steadily when the mesh is refined while the resultant stress N_y in the cross section $A - A$ is much more stable [146]. The reason is that the influence function for N_y has a simple shape. It is the displacement field of the plate if all the points in the cross section are spread simultaneously; see Fig. 1.56 c. Even a coarse mesh suffices to approximate this shape, and this is why N_y changes little when the mesh is refined.

Next consider the plate in Fig. 1.57. The influence function for the stress σ_{yy} at a single point, for example at the lower edge of the plate, certainly does not lie in V_h, but the influence function for the normal force N_y must lie in V_h^+ ($= V_h$ plus the rigid-body motions of the plate), because it represents a rigid-body motion of the plate, and therefore the FE program finds (as it must!) the correct result for the stress resultant N_y.

In plate-bending problems the situation is the same, as can be seen in Fig. 1.58. The influence function for a resultant bending moment is much easier to approximate than for a single value.

Equilibrium

The resultant force R_h of the FE solution in a cross section will balance the external load if the Green's function for R lies in V_h, see Sect. 1.37, p. 184. The Green's function for the sum of the horizontal forces and the vertical forces are simple movements, $u_x = 1$ and $u_y = 1$ respectively. In the case of the plate in Fig. 1.59 the influence function for N_x in the cross section $A - A$ is a rigid-body motion of the part to the right of $A - A$, i.e., a unit dislocation of *all points* on the line $A - A$.

The equivalent nodal forces f_i that try to generate this shape are the integrals of the stresses of the nodal unit displacements fields φ_i along the line $A - A$:

$$f_i = \int_{A-A} \int_\Omega \delta_1^{xx}(\boldsymbol{y} - \boldsymbol{x}) \cdot \boldsymbol{\varphi}_i(\boldsymbol{y}) \, d\Omega_{\boldsymbol{y}} \, ds = \int_{A-A} \sigma_{xx}^{(i)}(\boldsymbol{x}) \, ds \,. \qquad (1.264)$$

The FE solution (Fig. 1.59 c) is not exact, because displacement fields φ_i in V_h cannot model step functions.

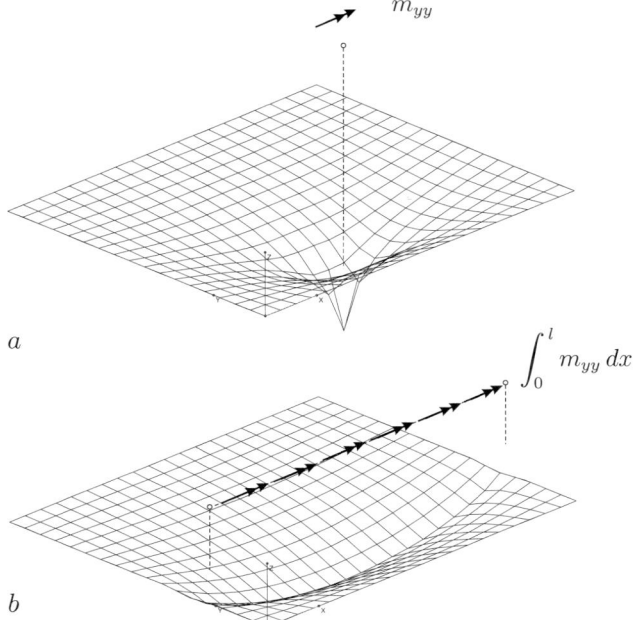

$$m_{yy}$$

$$\int_0^l m_{yy}\, dx$$

a

b

Fig. 1.58. Slab: **a)** influence function for m_{yy}; **b)** influence function for the integral of m_{yy}

The equilibrium condition $\sum H = 0$ is a simple condition because the influence function is simple, $u_x = 1$. The condition that the sum of the moments is zero, $\sum M = 0$, is more difficult because it involves rotations. If a vertical force $P = 1$ kN is applied at the end of the plate in Fig. 1.60, the maximum bending stresses of the FE solution in the cross section $A - A$ are ± 10 kN/m^2, so the bending moment is

$$M^h = \int_{-0.5}^{+0.5} z \cdot \underbrace{(-20\,z)}_{\sigma_{xx}}\, dz = -1.\bar{6}\,\text{kN m} \qquad (1.265)$$

which is less than the exact value of -2.5 kN m. The reason is that the FE program cannot model the exact influence function (see Fig. 1.60 b) so it operates with the shape in Fig. 1.60 c instead, which it obtains when it applies the moments of the nodal unit displacement φ_i in cross section $A - A$ as equivalent nodal forces:

$$f_i = \int_{-h/2}^{+h/2} \int_\Omega \delta_1^{xx}(\boldsymbol{y} - \boldsymbol{x}) \bullet \varphi_i(\boldsymbol{y})\, d\Omega_{\boldsymbol{y}} \cdot z\, dz$$

$$= \int_{-h/2}^{+h/2} \sigma_{xx}^{(i)}(z) \cdot z\, dz\,. \qquad (1.266)$$

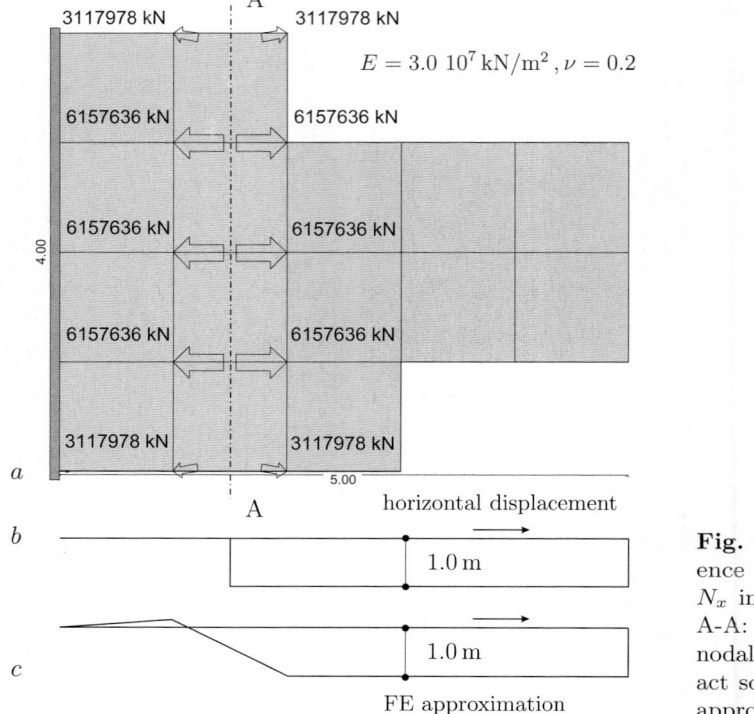

Fig. 1.59. Influence function for N_x in cross section A-A: **a)** equivalent nodal forces, **b)** exact solution, **c)** FE approximation

This pseudorotation lifts node 10 vertically by $u_y = -1.\bar{6}$ m, which is exactly the bending moment $M_h = 1 \cdot (-1.\bar{6})$ kNm of the FE solution in section $A - A$ (see Equ. (1.265)).

1.22 Why stresses at midpoints are more accurate

The simpler a Green's function the better the chance that the Green's function lies either in V_h or at least not too far away from it so that the error in the FE results will either be zero or small. This holds in particular for the Green's function of average values.

To calculate the average stress σ_{xx} in a region Ω_e of a plate, the stress is integrated and divided by the size Ω_e of the region[9]:

$$\sigma_{xx}^a = \frac{1}{\Omega_e} \int_{\Omega_e} \sigma_{xx} \, d\Omega = \frac{E}{\Omega_e} \int_{\Omega_e} (\varepsilon_{xx} + \nu \, \varepsilon_{yy}) \, d\Omega \,. \tag{1.267}$$

Because the strains are derivatives, $\varepsilon_{xx} = u_{x,x}$ and $\varepsilon_{yy} = u_{y,y}$, the domain integral can be transformed into a contour integral

---

[9] We simply write Ω_e instead of $|\Omega_e|$ or similar expressions.

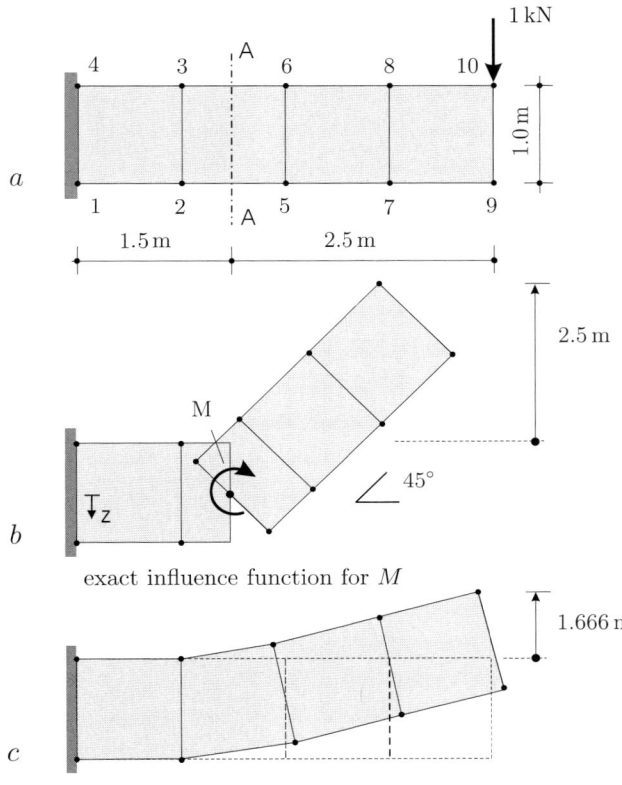

exact influence function for M

FE-approximation on V_h ($=$ bilinear elements)

Fig. 1.60. a) Cantilever plate; **b)** exact influence function for the moment M; **c)** FE approximation

$$\sigma_{xx}^a = \frac{E}{\Omega_e} \int_{\Omega_e} (\varepsilon_{xx} + \nu\,\varepsilon_{yy})\,d\Omega = \frac{E}{\Omega_e} \int_{\Gamma_e} (u_x\,n_x + \nu\,u_y\,n_y)\,ds \quad (1.268)$$

over the boundary Γ_e of Ω_e. To keep the formulas short let us assume in the following that $\nu = 0$, so that

$$\sigma_{xx}^a = \frac{E}{\Omega_e} \int_{\Omega_e} \varepsilon_{xx}\,d\Omega = \frac{E}{\Omega_e} \int_{\Gamma_e} u_x\,n_x\,ds\,. \quad (1.269)$$

The influence function for u_x at a point $\boldsymbol{x} \in \Gamma_e$ is the displacement field due to a single force $P_x = 1$ acting at $\boldsymbol{x}$. Hence the influence function for the integral

$$\frac{E}{\Omega_e} \int_{\Gamma_e} u_x\,n_x\,ds \quad (1.270)$$

is the displacement field of the plate if distributed horizontal forces $E/\Omega_e \times n_x$ are acting along the edge Γ_e; see Fig. 1.61.

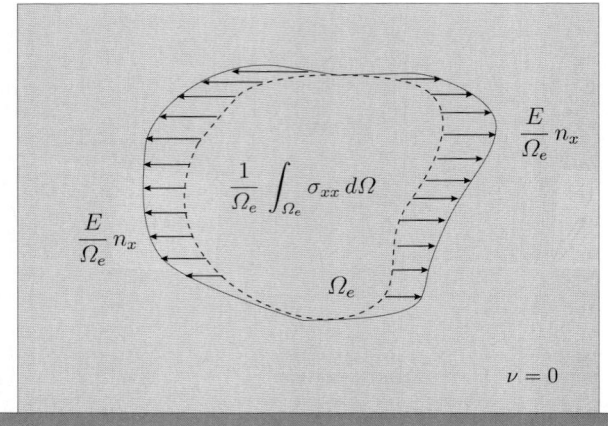

Fig. 1.61. The "Dirac delta" for the average value of σ_{xx} in a region Ω_e is a boundary layer of horizontal forces on the edge of the region

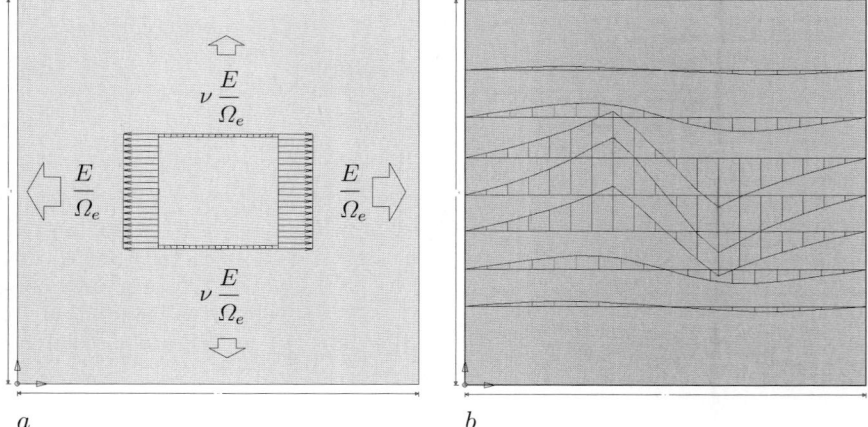

a b

Fig. 1.62. Influence function for the average value of σ_{xx} in the boxed region. The plate is fixed on all sides, **a)** the "Dirac delta", **b)** horizontal displacements

In Sect. 1.19, p. 69, we saw that the FE influence function for the normal force $N(x)$ at the mid-point of an element (Fig. 1.47) is the exact influence function for the average value N_a of the normal force $N(x)$ in that element.

A similar result holds, as we will see, for rectangular bilinear plane elements; there is no difference between the FE influence function for the stress σ_{xx} at the centroid $\boldsymbol{x}_c$ of an element and the FE influence function for the average value of σ_{xx} over the element.

To obtain the influence function for $\sigma_{xx}(\boldsymbol{x}_c)$, the stresses $\sigma_{xx}(\boldsymbol{\varphi}_i)(\boldsymbol{x}_c)$ of the nodal unit displacements at the center $\boldsymbol{x}_c$ of the element must be applied as equivalent nodal forces. The stresses σ_{xx} in a bilinear element are (Sect. 4.8, p. 357),

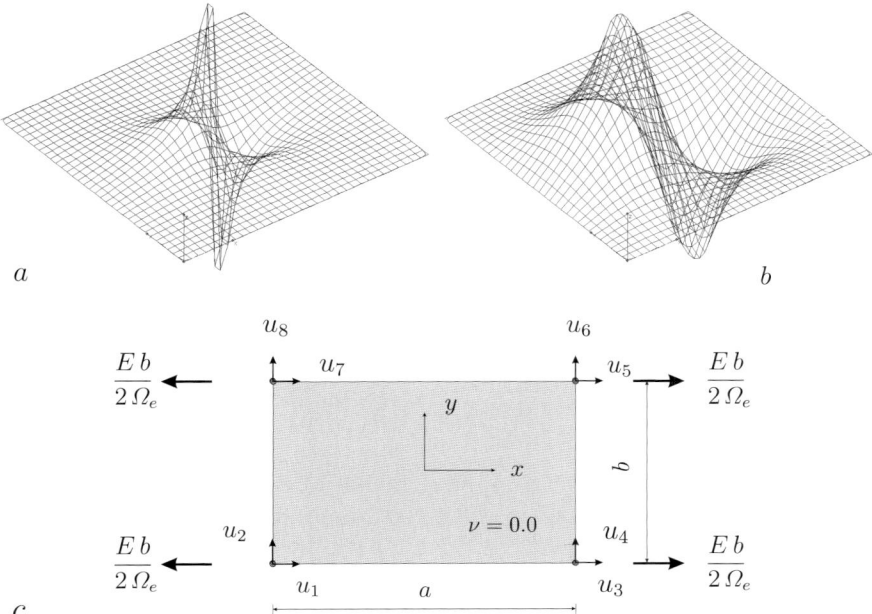

Fig. 1.63. A bilinear element does not distinguish between **a)** the influence function for σ_{xx} at the center and **b)** the influence function for the average value σ_{xx}^a in the element (3-D visualization of the horizontal displacements). The equivalent nodal forces and therefore the approximate influence functions are the same

$$\sigma_{xx}(x,y) = \frac{E}{a\,b\,(-1+\nu^2)} \times \left[b\,(u_1 - u_3) + a\,\nu\,(u_2 - u_8) + \right.$$

$$\left. + x\,\nu\,(-u_2 + u_4 - u_6 + u_8) + y\,(-u_1 + u_3 - u_5 + u_7) \right] \quad (1.271)$$

so that (Fig. 1.63)

$$f_3 = \sigma_{xx}(\boldsymbol{\varphi}_3)(\boldsymbol{x}_c) = f_5 = \sigma_{xx}(\boldsymbol{\varphi}_5)(\boldsymbol{x}_c) = +\frac{E\,b}{2\,\Omega_e} \quad (1.272)$$

$$f_1 = \sigma_{xx}(\boldsymbol{\varphi}_1)(\boldsymbol{x}_c) = f_7 = \sigma_{xx}(\boldsymbol{\varphi}_7)(\boldsymbol{x}_c) = -\frac{E\,b}{2\,\Omega_e} \quad (1.273)$$

where $\Omega_e = a\,b$ is the area of the element. Note that $f_2 = f_4 = f_6 = f_8 = 0$, because $\nu = 0$.

The average value of σ_{xx} is (see (1.269))

$$\sigma_{xx}^a = \frac{1}{\Omega_e} \int_{\Omega_e} \sigma_{xx}\,d\Omega = \frac{E}{\Omega_e} \int_{\Gamma_e} u_x\,n_x\,ds = \frac{E}{\Omega_e} \left[\int_{\Gamma_R} u_x\,ds - \int_{\Gamma_L} u_x\,ds \right]$$

$$(1.274)$$

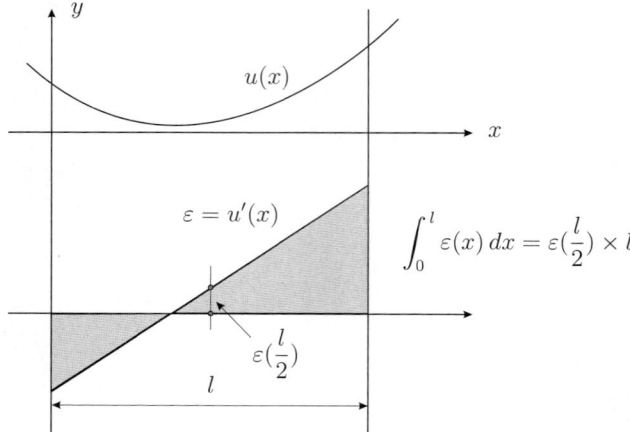

Fig. 1.64. The integral of a linear function is l times the midpoint value

where Γ_L and Γ_R are the left and the right side of the element—at the upper and lower edge $n_x = 0$—and u_x is the horizontal displacement of the edge. Hence to generate the influence function for σ_{xx}^a we must apply as equivalent nodal forces $f_i = \sigma_{xx}^a(\varphi_i)$ the average stresses of the unit displacement fields φ_i.

Let us do this for example with the unit displacement field $\varphi_3(x, y)$ that describes the horizontal displacement of the lower right corner, $u_3 = 1$ (see Fig. 1.63 b). The shape function of this corner point (see Eq. (4.30) p. 338)

$$\psi_2^e(x, y) = \frac{1}{4\,\Omega_e}(a + 2\,x)\,(b - 2\,y)\,, \tag{1.275}$$

is the u_x in the nodal unit displacement field $\varphi_3(x, y) = [u_x, 0]^T$. The shape function is zero on Γ_L, so that the integral (1.274) is

$$\frac{E}{\Omega_e}\int_{\Gamma_R} u_x\,ds = \frac{E}{\Omega_e}\int_{-b/2}^{b/2}\psi_2(\frac{a}{2}, y)\,dy = \frac{E\,b}{2\,\Omega_e} = f_3\,,$$

$$\tag{1.276}$$

which is the same f_3 as in (1.272).

From these formulas, it follows that an element has the property

FE influence function for $\sigma_{xx}(\boldsymbol{x}_c)$ = FE influence function for σ_{xx}^a

if the average value of the strains is equal to the value at the centroid $\boldsymbol{x}_c$ of the element (see Fig. 1.64):

$$\frac{1}{\Omega_e}\int_{\Omega_e}\varepsilon_{xx}(\boldsymbol{\varphi}_i)\,d\Omega = \varepsilon_{xx}(\boldsymbol{\varphi}_i)(\boldsymbol{x}_c)\qquad i = 1, 2, \ldots n \tag{1.277}$$

and likewise for ε_{yy} and ε_{xy}.

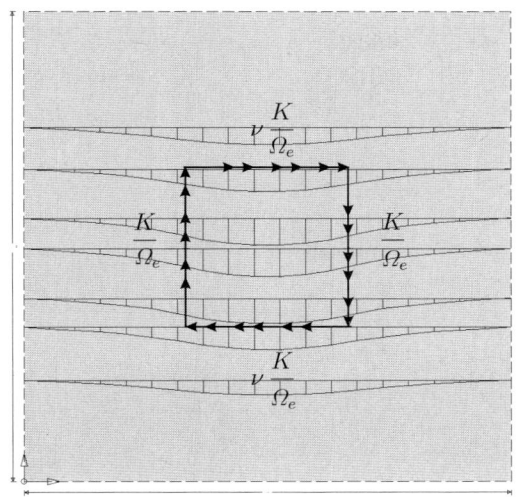

Fig. 1.65. Influence function for the average value of m_{xx} in the boxed region

Hence a CST element has this property, as does a quadratic triangle (straight sides, midside nodes, and six quadratic shape functions), because quadratic displacements imply linear strains, and the integral of a linear function ε_{xx} is $\varepsilon_{xx}(\boldsymbol{x}_c) \cdot \Omega_e$ where $\boldsymbol{x}_c$ is the centroid of the element.

The strains of the six shape functions corresponding to $u_1, u_2, \ldots, u_6$ are[10]

$$\frac{1}{2\Omega_e}\,(4\xi_1 - 1)\,y_{23} \qquad \frac{1}{2\Omega_e}\,(4\xi_2 - 1)\,y_{31} \qquad \frac{1}{2\Omega_e}\,(4\xi_3 - 1)\,y_{12} \quad (1.278)$$

$$\frac{2}{\Omega_e}\,(\xi_2\,y_{23} + \xi_1\,y_{31}) \qquad \frac{2}{\Omega_e}\,(\xi_3\,y_{31} + \xi_2\,y_{12}) \qquad \frac{2}{\Omega_e}\,(\xi_1\,y_{12} + \xi_3\,y_{23}) \quad (1.279)$$

The centroid has the natural coordinates $\xi_1 = \xi_2 = \xi_3 = 1/3$, and integration is done with the formula

$$\int_{\Omega_e} \xi_1^k\,\xi_2^l\,\xi_3^m\,d\Omega = 2\,\Omega_e\,\frac{k!\,l!\,m!}{(2 + k + l + m)!} \qquad k, l, m \geq 0. \quad (1.280)$$

Hence

$$\varepsilon_{xx}(\boldsymbol{\varphi}_1)(\boldsymbol{x}_c) = \frac{1}{2\Omega_e}\Big(\frac{4}{3} - 1\Big)\,y_{23} \quad (1.281)$$

is the same as

$$\frac{1}{\Omega_e}\int_{\Omega_e} \varepsilon_{xx}(\boldsymbol{\varphi}_1)\,d\Omega = \frac{1}{2\Omega_e^2}\int_{\Omega_e} (4\,\xi_1 - 1)\,y_{23}\,d\Omega = \frac{1}{2\Omega_e}\Big(\frac{4}{3} - 1\Big)\,y_{23}$$

$$(1.282)$$

---

[10] [70] p. 158

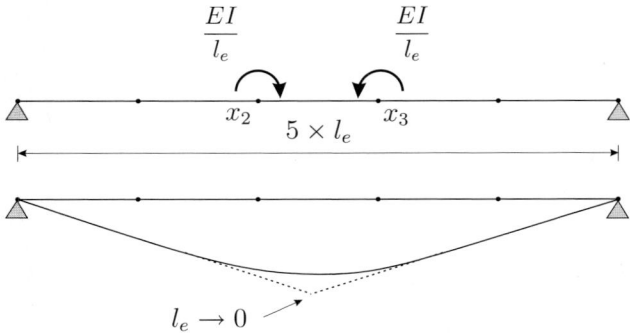

Fig. 1.66. Influence function for the average value of the bending moment $M(x)$ in the center element

or

$$\varepsilon_{xx}(\boldsymbol{\varphi}_4)(\boldsymbol{x}_c) = \frac{2}{\Omega_e}(\frac{1}{3}y_{23} + \frac{1}{3}y_{31}) \tag{1.283}$$

is the same as

$$\frac{1}{\Omega_e}\int_{\Omega_e}\varepsilon_{xx}(\boldsymbol{\varphi}_4)\,d\Omega$$
$$= \frac{2}{\Omega_e^2}\int_{\Omega_e}(\xi_2 y_{23} + \xi_1\,y_{31})\,d\Omega = \frac{2}{\Omega_e}(\frac{1}{3}\,y_{23} + \frac{1}{3}\,y_{31})\,. \tag{1.284}$$

Of course the same can be said about the strains ε_{yy} and ε_{xy}, so the generalization to the case $\nu \neq 0$ is obvious.

Kirchhoff plates

The average value of the bending moment m_{xx} can be expressed by a boundary integral,

$$\frac{1}{\Omega_e}\int_{\Omega_e}m_{xx}\,d\Omega = \frac{K}{\Omega_e}\int_{\Omega_e}(w_{,xx} + \nu\,w_{,yy})\,d\Omega = \frac{K}{\Omega_e}\int_{\Gamma_e}(w_{,x}\,n_x + \nu\,w_{,y}\,n_y)\,ds \tag{1.285}$$

and if we let $\nu = 0$,

$$m_{xx}^a = \frac{1}{\Omega_e}\int_{\Omega_e}m_{xx}\,d\Omega = \frac{K}{\Omega_e}\int_{\Gamma_e}w_{,x}\,n_x\,ds \tag{1.286}$$

which means that the influence function for the average value of m_{xx} inside the boxed region (Fig. 1.65) is the deflection surface of the slab under the action of pairs of opposing line moments (kN m/m), which rotate the plate inwards.

In a beam, two single moments would even yield the exact influence function for the average value M_a because (see Fig. 1.66)

$$M_a = \frac{1}{l_e} \int_{x_2}^{x_3} M(x)\, dx = \frac{1}{l_e} \int_{x_2}^{x_3} \int_0^l G_2(y,x)\, p(y)\, dy$$

$$= \frac{1}{l_e} \int_{x_2}^{x_3} \int_0^l -EI \frac{d}{dx} G_1(y,x)\, p(y)\, dy$$

$$= \frac{-EI}{l_e} \int_0^l [G_1(y,x_3) - G_1(y,x_2)]\, p(y)\, dy \qquad (1.287)$$

where G_1 is the Green's function for the rotation.

Note that as the length of the element tends to zero, $l_e \to 0$, the moments $M = EI/l_e$ tend to ∞, while at the same time their distance shrinks to zero. Hence, a sharp bend develops at the center point x_c, i.e., the influence function for the average bending moment becomes the influence function for $M(x)$ at the midpoint of the element.

These two moments are also the equivalent nodal forces for the influence function of the FE bending moment $M_h(0.5\, l_e)$ at the center of the element, as follows from

$$M(\varphi_1^e)(0.5\, l_e) = 0 \qquad M(\varphi_2^e)(0.5\, l_e) = \frac{-EI}{l_e} \qquad (1.288)$$

$$M(\varphi_3^e)(0.5\, l_e) = 0 \qquad M(\varphi_4^e)(0.5\, l_e) = \frac{+EI}{l_e}. \qquad (1.289)$$

Hence, in beam analysis as well the two influence functions for $M_h(0.5\, l_e)$ and the average value M_h^a coincide. This is simply a consequence of the fact that the nodal unit displacements are cubic polynomials, and therefore the bending moments are piecewise linear.

More generally the condition

$$\frac{1}{\Omega_e} \int_{\Omega_e} m_{xx}(\varphi_i)\, d\Omega = m_{xx}(\varphi_i)(x_c) \qquad i = 1, 2, \ldots n \qquad (1.290)$$

(and for m_{yy} likewise) is sufficient for the FE influence functions of $m_{xx}(x_c)$ and the average value m_{xx}^a to be the same. In conforming elements where the bending moments m_{ij} are linear functions, this condition is satisfied.

In a Reissner–Mindlin plate the bending moments are defined as

$$m_{xx} = K\, (\theta_{x,x} + \nu\, \theta_{y,y}) \qquad (1.291)$$

and therefore the rotations θ_x and θ_y must be linear polynomials.

Summary

To give credit to our claim that the stresses at midpoints are more accurate, we argue as follows:

in any load case $\displaystyle\int_0^l N(x)\,dx = 0$ in any load case $\displaystyle\int_0^l M(x)\,dx = 0$

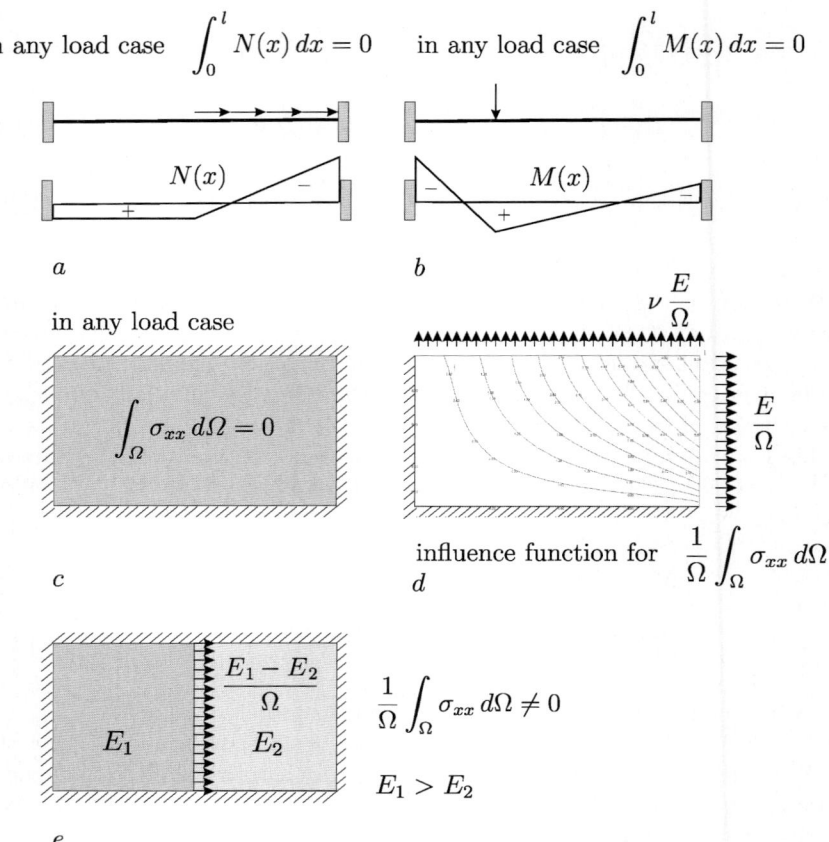

a   b

in any load case

c   d   influence function for $\displaystyle\frac{1}{\Omega}\int_\Omega \sigma_{xx}\,d\Omega$

e

Fig. 1.67. The average values of the stresses or moments are zero in structures with fixed or clamped edges: **a)** bar, **b)** beam, **c)** plate, **d)** but not in a plate with free edges; **e)** in plates with different elastic properties the influence functions also are not zero

- The stresses at the centroid of an element are approximately identical to the average stresses in the element. This follows from a simple Taylor expansion

$$\sigma_{xx}(\boldsymbol{x}) = \sigma_{xx}(\boldsymbol{x}_c) + \nabla\sigma_{xx}(\boldsymbol{x}_c)\,(\boldsymbol{x} - \boldsymbol{x}_c) + O(h^2) \qquad (1.292)$$

because

$$\frac{1}{\Omega_e}\int_{\Omega_e}\sigma_{xx}(\boldsymbol{x})\,d\Omega = \sigma_{xx}(\boldsymbol{x}_c) + \frac{1}{\Omega_e}\int_{\Omega_e}O(h^2)\,d\Omega\,. \qquad (1.293)$$

Hence if the element gets small enough the stresses at the centroid are nearly identical to the average value, $\sigma_{xx}(\boldsymbol{x}_c) \simeq \sigma_{xx}^a$.

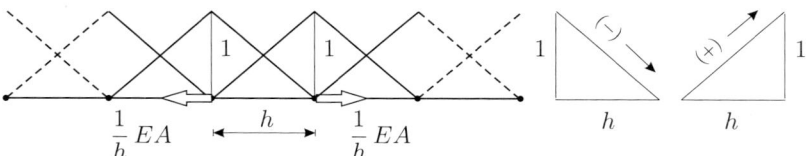

Fig. 1.68. Equivalent nodal forces for the influence function of $N = \sigma A$ at the center of the element

- If the element stresses are linear, the FE influence function for the stress at the centroid and for the average stress are the same so that $\sigma_{xx}^h(\boldsymbol{x}_c) = \sigma_{xx}^{h,a}$.
- Because the influence function for the average stress σ_{xx}^a is relatively simple (see Fig. 1.62 p. 90) the error $\sigma_{xx}^a - \sigma_{xx}^{h,a}$ will be small, and because $\sigma_{xx}(\boldsymbol{x}_c) \simeq \sigma_{xx}^a$ also the error $\sigma_{xx}(\boldsymbol{x}_c) - \sigma_{xx}^h(\boldsymbol{x}_c)$.

Remark 1.10. It follows that the average stresses are zero over any structural element with fixed or clamped edges, because the edge forces or edge moments which generate the influence functions will effect nothing when they are applied to fixed edges; see Fig. 1.67. An elementary analysis shows that this is no longer true if the material properties change; see Fig. 1.67 e. In that case the influence function is generated by placing tractions E_i/Ω separately around the edges of the two subdomains and the resulting action $(E_1 - E_2)/\Omega$ at the interface makes that the plate deforms and so the influence function is not zero.

Remark 1.11. Many interesting things can be learned about the nature of influence functions if the transition from average values to point values is studied. Basically the action behind the influence function for σ_{xx} involves two opposing forces $\pm 1/\Delta x$ which point in opposite directions which become infinite if the distance Δx between the two forces becomes zero. In physics this is called a *dipole*, [221]. At distances far from the source point, the effects cancel, but near the source point the material is stretched in two opposing directions, so that two opposing peaks in the horizontal displacement u as in the influence function for q_x in a slab develop. This dipole nature of the kernel is the reason why the influence function for the stress has this local character.

We can actually see how the tendency $1/\Delta x \to \infty$ develops. To calculate the influence function for σ_{xx} at the center of an element, the stresses $\sigma_{xx}(\boldsymbol{\varphi}_i)$ are applied as equivalent nodal forces. Now the more the element size $h \, (= \Delta x)$ shrinks, the more the derivative of the nodal unit displacement

$$\frac{d}{dx}\, \varphi_i(x) = \frac{1}{h} \tag{1.294}$$

tends to infinity. This is best seen in the 1-D problem of a bar: the smaller the elements, the better the FE program can simulate the action of a dipole at the center of an element with two opposing forces at the neighboring nodes

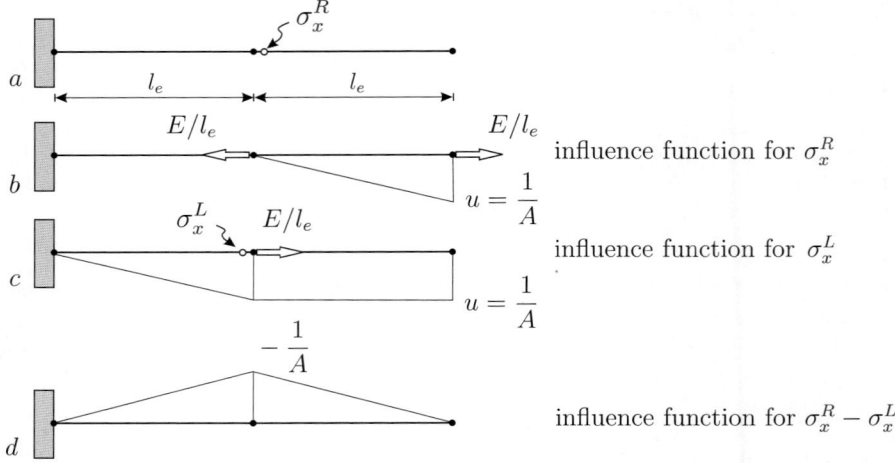

Fig. 1.69. Construction of the influence function for the stress discontinuity at the center node

$$f = \pm \frac{1}{h} \times EA \qquad (1.295)$$

where the change in sign is due to the fact that the unit displacement of the node on the left-hand side has a negative slope at the center of the element while for the opposing node the opposite is true; see Fig. 1.68.

Influence functions for average values of displacements

These influence functions follow the same logic. If a single force $P = 1$ is placed at a point $\boldsymbol{x}$ of a prestressed membrane Ω (Poisson equation $-\Delta u = p$), the average deflection u in a region Ω_e is

$$u^a = \frac{1}{\Omega_e} \int_{\Omega_e} u \, d\Omega = \frac{1}{\Omega_e} \int_{\Omega_e} G_0(\boldsymbol{y}, \boldsymbol{x}) \, d\Omega_{\boldsymbol{y}} . \qquad (1.296)$$

Hence the Green's function for u^a is simply the shape of the membrane if a constant pressure $p = 1$ is applied to the region Ω_e, so that the equivalent nodal forces for the Green's function are

$$f_i = \frac{1}{\Omega_e} \int_{\Omega_e} \varphi_i \times 1 \, d\Omega = \frac{1}{3} \qquad (1.297)$$

where $1/3$ is the result for the node of a CST element Ω_e. So that if the size Ω_e shrinks to zero, the action of the three nodal forces increasingly resembles the action of a true point load $P = 1/3 + 1/3 + 1/3 = 1$.

1.23 Why stresses jump

This seems a trivial question. Stresses jump, because the displacements are only C^0, so the derivatives are discontinuous at interelement boundaries. But perhaps it is worthwhile to study the phenomenon from the perspective of influence functions. To keep things simple we consider a bar (Fig. 1.69). The influence function for the jump

$$\sigma_x^R - \sigma_x^L \tag{1.298}$$

in the stress at the center node is the influence function for σ_x^R minus the influence function for σ_x^L, and because this compound influence function is not zero, the stress jumps.

At an interior point of an element the two influence functions are identical, and therefore they cancel. In other words jumps in the stresses do not occur at any interior point.

But this is no surprise; rather, it smacks of a circular argument. Because the equivalent nodal forces for the influence functions of the stress σ_x are, up to the factor E, the first derivative of the shape functions, the jump in the stresses will always be zero if the first derivative is the same on both sides of the point, i.e., if the stress is continuous ...

But there are two interesting points to make looking at Fig. 1.69. Obviously the maximum jump occurs if the load is applied directly at either side of the node and the jump will be zero if the load alternates, $+p$ in the first element and $-p$ in the second element. This is probably also true in 2-D and 3-D problems. Checkerboard loads leave few traces in V_h, i.e., the equivalent nodal forces f_i are relatively small.

1.24 Why finite element support reactions are relatively accurate

When the same problem is solved with various FE programs, it is often found that agreement in the support reactions is quite good, and it is soon recognized that they change little when the mesh is refined. The reason is that influence functions for support reactions have a particularly simple shape. They are the deflection curves or surfaces if the support settles by one unit of deflection; see Fig. 1.70. These simple shapes are easy to approximate, even on a coarse mesh.

Things are different if the wall is not isolated, but is instead in contact with other walls. Now if the wall settles, the slab cracks in the transition zone between the rigid supports, $w = 0$, and the sagging wall, $w = 1$; see Fig. 1.71. In practice the neighboring walls will not be perfectly rigid, but will move too, so the discontinuity in the transition zone will not be so sharp. But accuracy will certainly suffer. Short intermediate supports will be affected more than longer walls.

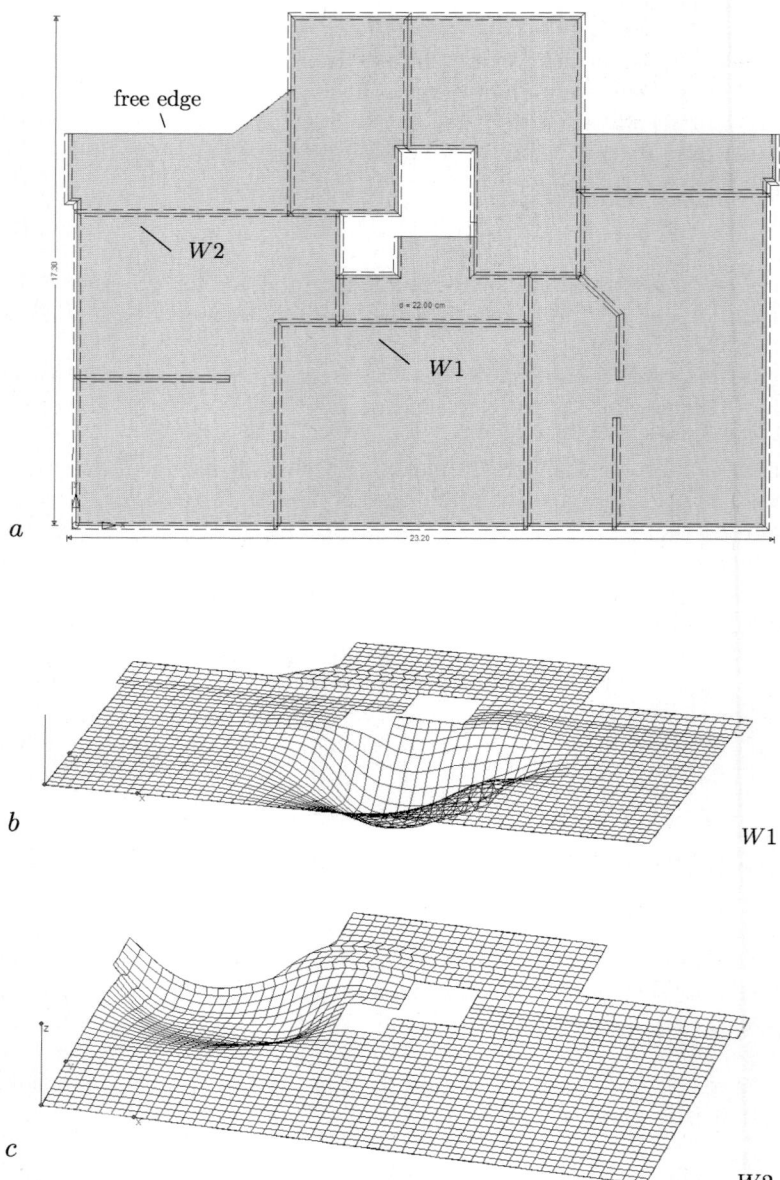

free edge

$W2$

$W1$

b

$W1$

c

$W2$

Fig. 1.70. Slab: **a)** plan view; **b)** influence function for the sum of the support reactions in wall $W1$; **c)** in wall $W2$

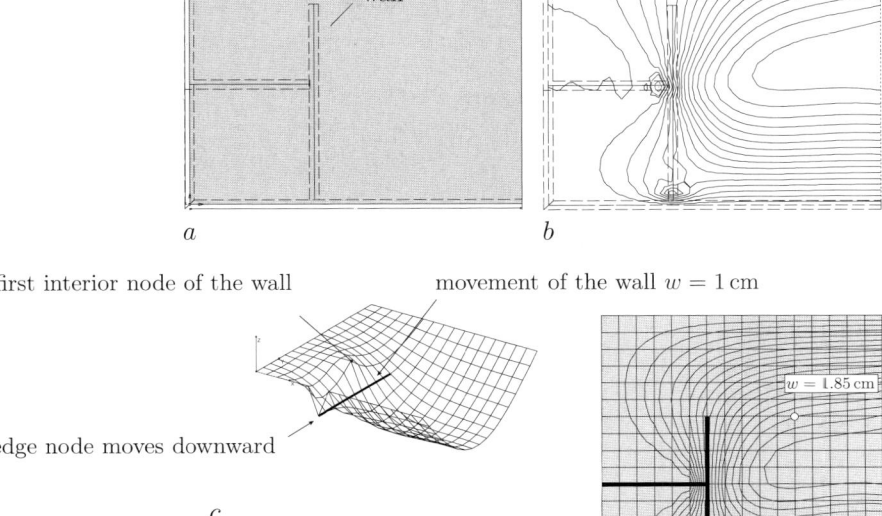

first interior node of the wall   movement of the wall $w = 1$ cm

edge node moves downward

Fig. 1.71. Floor plate: **a)** plan view; **b)** influence function for the support reaction in the wall extending vertically; **c)** 3-D view of the FE approximation; **d)** contour lines of the influence function

Example

Consider the slab in Fig. 1.71. The exact influence function G_0 for the support reaction of the wall (the one extending vertically) is displayed in Fig. 1.71 b, and the FE approximation is shown in Fig. 1.71 d. The latter is the shape of the slab if the nodes of the FE mesh that lie on the wall are pushed down by $w = 1$ cm. This produces the deflection $w = 1.85$ cm at the distant node $\boldsymbol{y}_k$; see Fig. 1.71 d. This number is exactly the sum of the equivalent nodal forces of the wall if a force $P = 1$ kN is placed at this node

$$\sum_i f_i = \int_\Omega G_0^h(\boldsymbol{y})\, p(\boldsymbol{y})\, d\Omega = G_0^h(\boldsymbol{y}_k) \cdot 1\,\text{kN} = 1.85 \ \text{kN} . \qquad (1.299)$$

The exact value (on a very fine mesh) for the support reaction is 2.2 kN, so the FE solution underestimates the support reaction on this mesh by about 16%. When a uniform load is applied, the error in the support reaction is about 5%.

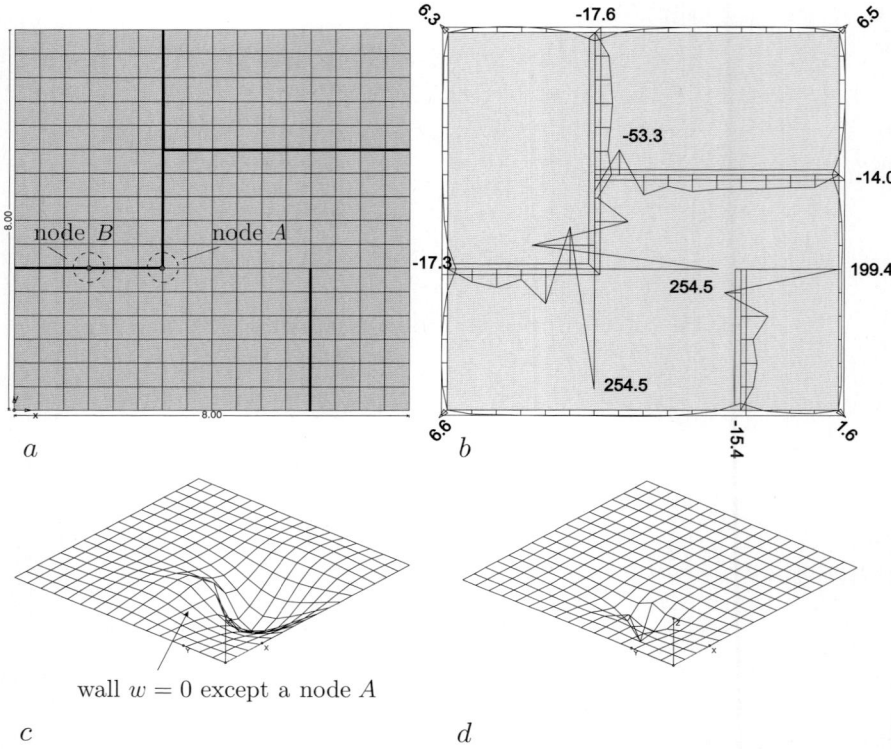

Fig. 1.72. Floor plate on rigid supports: **a)** plan view and FE mesh; **b)** support reactions under gravity load; **c)** influence function for the nodal force A, and **d)** for the nodal force B

Peaks

Support reactions tend to oscillate and end in high *peaks*—in particular near the ends of free-standing walls and at corner points; see Fig. 1.72. This is easily understood by looking at the influence functions for the nodal force directly at the end of the wall, node A, and at a node further back, node B. The node up front has a much larger influence area than the node behind it, and the force at node B is most often negative, simply because a movement $w = 1$ of this node lifts the part in front upwards.

Point supports

A Kirchhoff plate is about the only 2-D structure that can safely be placed on a point support. A Reissner–Mindlin plate or a plate (shear wall) simply ignores point supports, because a single point can travel freely in any direction; see Fig. 1.73. A Kirchhoff plate will finally succumb to even the smallest moment

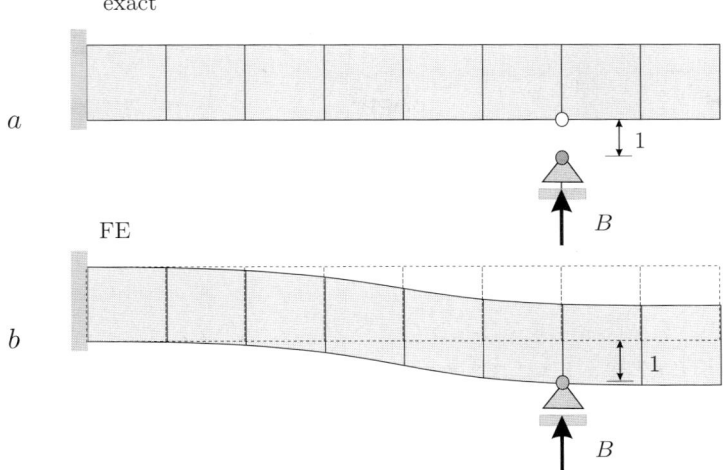

Fig. 1.73. a) The exact influence function for the support reaction B is zero. One single point of the plate (= the support) can move downward by one unit length without disturbing the plate **b)** The FE approximation to the influence function closely follows the beam solution

M, and will not try to prevent the point of attack from rotating. Rather it will let it loose so that it can rotate freely. Polynomial shape functions cannot accomplish such remarkable feats. If one point moves, the whole neighborhood follows suit, and therefore such influence functions always turn out wrong.

But the situation can be saved if the mathematical idea of a point support is abandoned, and instead the supports are allowed to have finite extent. Then the support reaction acts over a small surface area, and it may be assumed that the influence function for such a patch of forces comes close to the shape if the node is moved by one unit of displacement.

Cantilever plate

According to the theory of elasticity, the support reaction of the cantilever plate in Fig. 1.74 should be zero. The reason that it is not zero is that an FE program cannot generate the exact influence function; see Fig. 1.73 a. Instead it produces the shape in Fig. 1.73 b which closely follows the deflection curve of a beam. This is also why the FE support reaction B is identical to the beam solution. At least for bilinear elements (Q4), this tendency prevailed even when the mesh was refined ($8 \rightarrow 32 \rightarrow 128 \rightarrow 512$ elements) as illustrated by the results in Table 1.5. While the support reaction hardly changed, the stresses near the point support increased with each refinement step.

Hence, point supports can be freely used, and the results are also reasonable, in the sense of beam analysis, only stresses in the neighborhood of such supports are meaningless.

gravity load

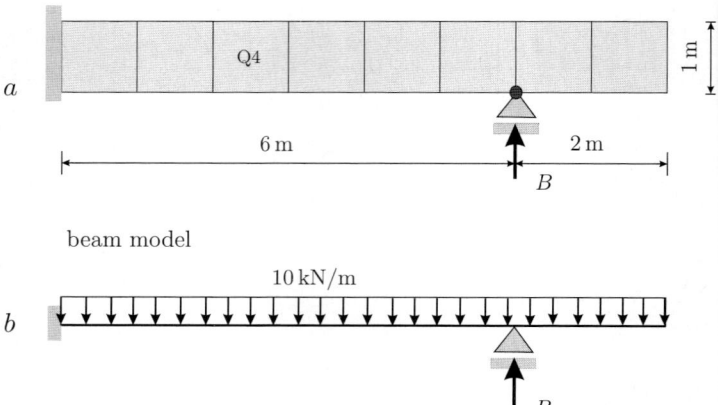

a

6 m   2 m

B

beam model

10 kN/m

b

B

Fig. 1.74. Cantilever plate with a point support and equivalent beam model

Table 1.5. Results for the plate (beam) in Fig. 1.74

Elements	Support reaction B (kN)	Min σ (kN/m^2) near the support
8	47.72	-480
32	47.54	-963
128	47.49	-1960
512	47.46	-3854

1.25 Gauss points

It is often found that the accuracy of the FE solution is superior at the Gauss points. To understand this phenomenon, it is best to begin with 1-D problems. In 1-D problems the FE solution agrees with the exact solution at the nodes. Hence also the approximate Green's function G_0^h coincides with G_0 at the nodes. We then have: (i) the unit nodal displacements φ_i are piecewise homogeneous solutions; (ii) the Green's functions are homogeneous solutions (except at x); and (iii) homogeneous solutions are determined by their nodal values. Hence it follows that the error in the Green's function G_0^h is zero outside the element that contains the source point x (see Fig. 1.75) so a 1-D FE solution is exact at all points x that happen to lie on a load-free element.

Now what happens if x lies on an element that carries, say, a uniform load p? The exact deflection curve w in each element can be split into a homogeneous solution w_0 and a particular solution w_p (corresponding to fixed ends):

$$w(x) = w_0(x) + w_p(x) \qquad EI\, w_0^{IV}(x) = 0 \qquad EI\, w_p^{IV} = p\,. \qquad (1.300)$$

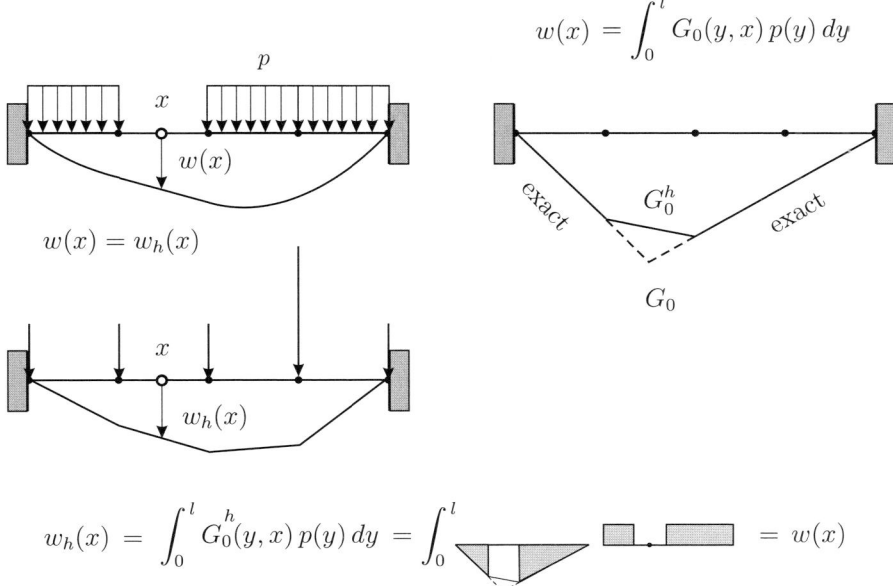

$$w(x) = \int_0^l G_0(y, x)\, p(y)\, dy$$

$$w_h(x) = \int_0^l \overset{h}{G_0}(y, x)\, p(y)\, dy = \int_0^l \;\; = w(x)$$

Fig. 1.75. In 1-D problems the error is zero in any element that carries no load

Because of the special nature of the trial space V_h and the FE method, this string of homogeneous solutions (element per element) is identical to the FE solution: $w_h = w_0$ on every element. The error $e(x)$ in the FE solution is therefore $e(x) = w(x) - w_h(x) = w_p(x)$, so the error in the bending moment within an element is just the bending moment distribution in an element with fixed ends:

$$M_p(x) = -EI\, w_p''(x) = \frac{p\, l_e^2}{2} \left(\frac{1}{6} - \frac{x}{l_e} + \frac{x^2}{l_e^2} \right) \qquad l_e = \text{length}. \quad (1.301)$$

Now we are in for a surprise! Evidently the error is zero where $M_p(x) = 0$, and (we let $l_e = 1$) these two points $x_1 = 0.21132$ and $x_2 = 0.78868$ are just the Gauss points! How does this happen?

(i) The integral of M_p is zero because the ends are fixed, i.e., because $w_p'(0) = w_p'(l_e) = 0$:

$$\int_0^{l_e} M_p(x)\, dx = \int_0^{l_e} -EI\, w_p''(x)\, dx = -EI\, [w_p'(l_e) - w_p'(0)] = 0. \quad (1.302)$$

This is the key to the problem.

(ii) $M_p(x)$ is a symmetric second-degree polynomial. Therefore the function must vanish at the $n = 2$ Gauss points of a $2n - 1 = 3$ formula,

$$\int_0^{l_e} M_p(x)\, dx = w_1\, M_p(x_1) + w_2\, M_p(x_2) = 2\, w_1\, M_p(x_1) = 0 \quad (1.303)$$

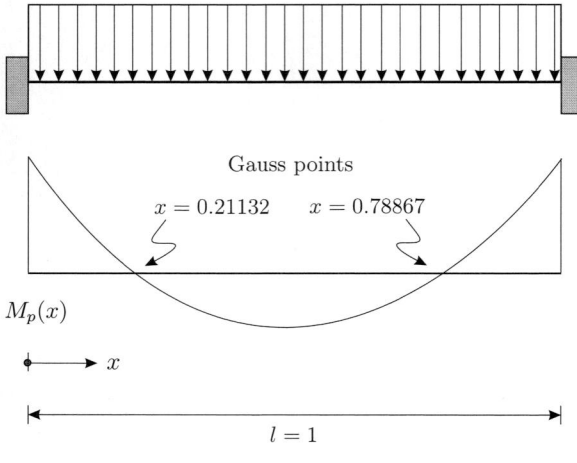

Fig. 1.76. The Gauss points coincide with the zeros of $M_p(x)$

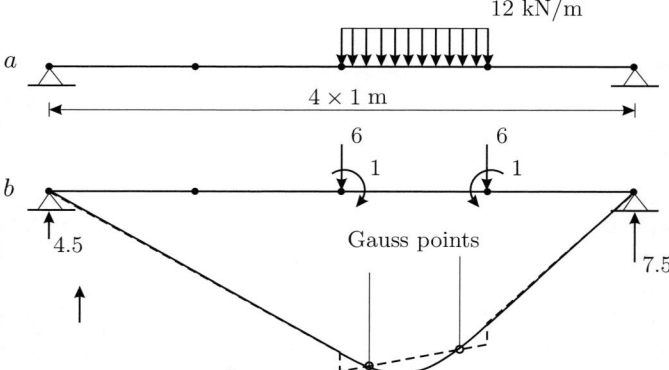

Fig. 1.77. The error in the bending moments is zero at the Gauss points

where the w_i are the weights at the Gauss points x_i.

When other types of loads are applied, it is not guaranteed that M_p is zero at the Gauss points. Then the super-convergent points (the zeros of M_p) must be found by studying the graph of M_p in engineering handbooks. Some hidden symmetries still apply, however. If for example the distributed load in the interval $[-1, 1]$ were a triangular load, $p(x) = (1 + x)/2$, then because of

$$\int_{-1}^{+1} M_p(x)\,dx = \int_{-1}^{+1} [-\frac{1}{12} - \frac{x}{20} + \frac{x^2}{4} + \frac{x^3}{12}]\,dx$$
$$= 1.0 \cdot M_p(-0.5775) + 1.0 \cdot M_p(0.5775) = 0 \quad (1.304)$$

M_p must have opposite values at the two Gauss points.

Note also that the error in the shear force $V(x) = -EI\,w'''(x)$ is zero at the center of the element if the distributed load is constant, $p = c$. The reason is basically the same as before: the integral of $V(x) = M'(x)$ is zero because the bending moments at the fixed ends are the same, so that $M(l_e) - M(0) = 0$, and because $V(x)$ is a linear function, which can be integrated with one Gauss point exactly.

Next let us study a rectangular slab clamped along its edge. Because the slope $w_n = \nabla w \cdot \boldsymbol{n} = 0$ and the tangential derivative $w_t = \nabla w \cdot \boldsymbol{t} = 0$ are zero on the boundary, the gradient $\nabla w = [w,_x , w,_y]^T$ must be zero too so that the integral of m_{xx} vanishes:

$$\int_{\Omega_e} m_{xx} \, d\Omega = \int_{\Omega_e} (w,_{xx} + \nu \, w,_{yy}) \, d\Omega = \int_{\Gamma_e} (w,_x \, n_x + \nu \, w,_y \, n_y) \, ds = 0$$

$$(1.305)$$

as do the integrals of m_{yy} and m_{xy}. Note that

$$\int_{\Omega_e} w,_i \, d\Omega = \int_{\Gamma_e} w \, n_i \, ds \qquad \text{(integration by parts)}. \qquad (1.306)$$

In a beam the bending moment $M(x)$ would be a quadratic polynomial if a constant load $p = 1$ were applied. If the same were true of a slab, the bending moments m_{ij} assuming that they were perfect symmetric second-order polynomials would vanish at the four Gauss points. But this is only approximately true; see Fig. 1.78.

In a plane rectangular element with fixed edges, $\boldsymbol{u} = \boldsymbol{0}$, the integral of the stress $\sigma_{xx} = \varepsilon_{xx} + \nu \, \varepsilon_{yy} = u_{x,x} + \nu \, u_{y,y}$ must vanish,

$$\int_{\Omega_e} \sigma_{xx} \, d\Omega = E \int_{\Omega_e} (u_{x,x} + \nu \, u_{y,y}) \, d\Omega = E \int_{\Gamma_e} (u_x \, n_x + \nu \, u_y \, n_y) \, ds = 0$$

$$(1.307)$$

as must the integral of σ_{yy} and σ_{xy} as well. When a horizontal volume force $\boldsymbol{p} = [1,0]^T$ is applied the horizontal stresses σ_{xx} are approximately linear functions and the vertical stresses σ_{yy} quadratic functions; see Fig. 1.79. Perfectly linear stresses σ_{xx} would have a zero at the center of the element, and perfectly quadratic stresses σ_{yy} would have zeros at the four Gauss points.

To summarize, if the edges of an element are kept fixed, the integrals of the stresses (and bending moments) must be zero. If the stresses σ_{ij} are linear, symmetry conditions imply that they vanish at the centroid of the element, and if the stresses are quadratic, they vanish at the four Gauss points of a rectangular element.

The relevance of this insight is best illustrated by studying a triangular plane element (Fig. 1.80) with fixed edges subjected to a horizontal volume force $\boldsymbol{p} = [1,0]^T$. Notice that the principal stresses vanish near the Gauss point. This means that if the edges are first kept fixed and then released—so that the effects of the load can spill over into the neighboring elements—and if it is assumed that the exact displacement field within the element is the sum

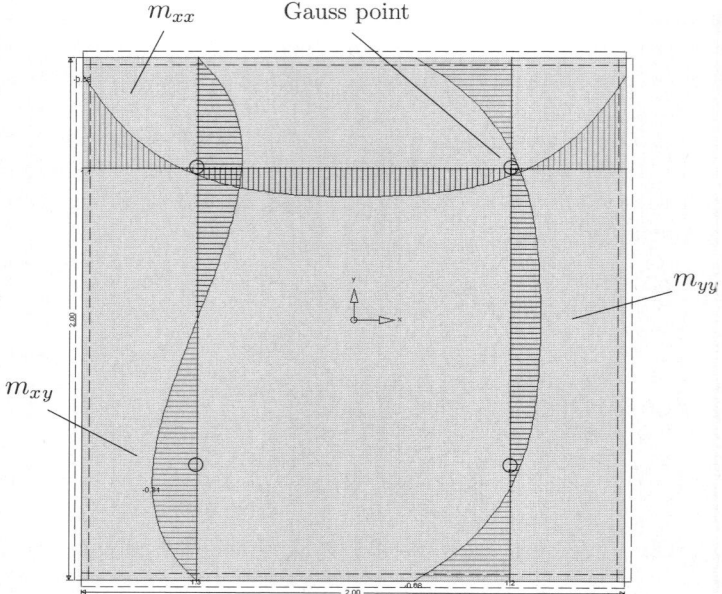

Fig. 1.78. Clamped plate, the position of the four Gauss points, and plots of m_{xx}, m_{xy}, and m_{yy}

of a linear displacement field (CST element) and a particular solution $\boldsymbol{u}_p$ of the load case $\boldsymbol{p} = [1, 0]^T$ then the error in the principal stresses will be zero at the Gauss point. These are very many if's, but obviously these assumptions are more often true than not.

If we let in (1.307) for simplicity $\nu = 0$

$$\int_{\Omega_e} \sigma_{xx}\, d\Omega = E \int_{\Omega_e} \varepsilon_{xx}\, d\Omega = E \int_{\Gamma_e} u_x\, n_x\, ds\,, \qquad (1.308)$$

and if we apply it to an unconstrained element Ω_e (not fixed at its edges), then the equation states that the integral of σ_{xx} equals the integral of the edge displacement of the element in the horizontal direction. Because the average bending stress σ_{xx} in the plane element in Fig. 1.81 a is zero, the overall extension of the element—the amount it stretches to the right and to the left—must be zero as well. In its simplest form this equation states that the integral of the normal force N in a bar equals the relative displacement $u(l) - u(0)$ of the end points.

One is tempted to employ these equations as well to study the error of an FE solution. Let $\sigma_{xx}^e = \sigma_{xx} - \sigma_{xx}^h$ denote the error in the stresses, then a non-vanishing error

$$\int_{\Omega_e} \sigma_{xx}^e\, d\Omega = \int_{\Gamma_e} (u_x - u_x^h)\, n_x\, ds \neq 0 \qquad (\nu = 0) \qquad (1.309)$$

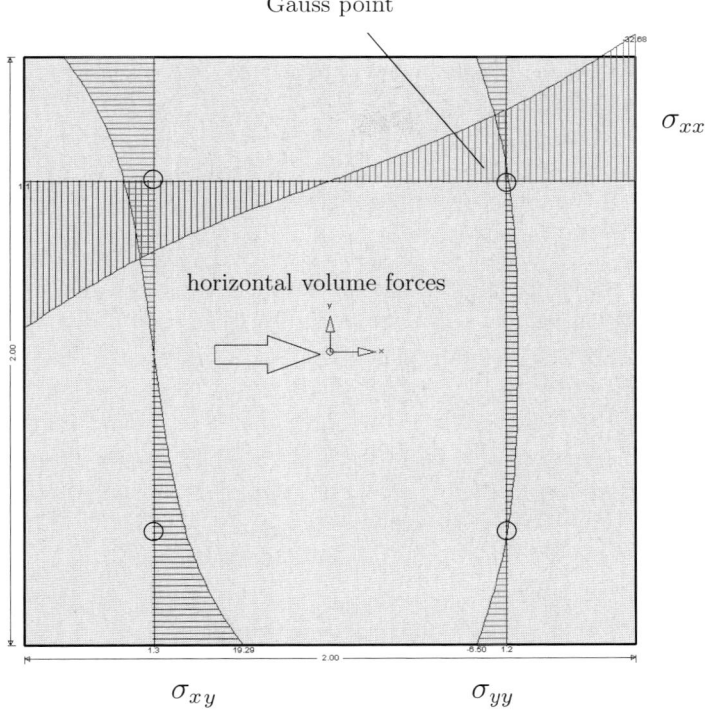

Fig. 1.79. Plane element with fixed edges subjected to a horizontal volume force $\boldsymbol{p} = [1,0]^T$; σ_{xx} and σ_{xy} are approximately linear functions, and σ_{yy} a quadratic function

implies that the *shape* of the deformed finite element differs from the true shape of the deformed region Ω_e of the structure (*shape* discounts rigid-body motions). The non-averaged stresses $\sigma_{xx} - \sigma_{xx}^h$ are responsible for the (horizontal) offset of an element. But care must be taken: (i) it cannot be said that the element as a whole is displaced, only that there is an excess or lack of horizontal movement; (ii) two elements with the same average stresses σ_{xx}^a must not have the same shape. When the bending stresses in the element in Fig. 1.81 are doubled, the average stress remains zero, but the shape certainly changes. Only the converse is true: if two elements differ in σ_{xx}^a, then they must also differ in shape.

The same holds for σ_{yy}^a and σ_{xy}^a, and the extension to plate bending problems is also straightforward. In plate bending problems, the integral of the bending moment m_{xx} equals the boundary integral of the slope on Γ in the x-direction $w,_x n_x$

$$\int_\Omega m_{xx} \, d\Omega = \int_\Gamma w,_x \, n_x \, ds \qquad (\nu = 0) . \tag{1.310}$$

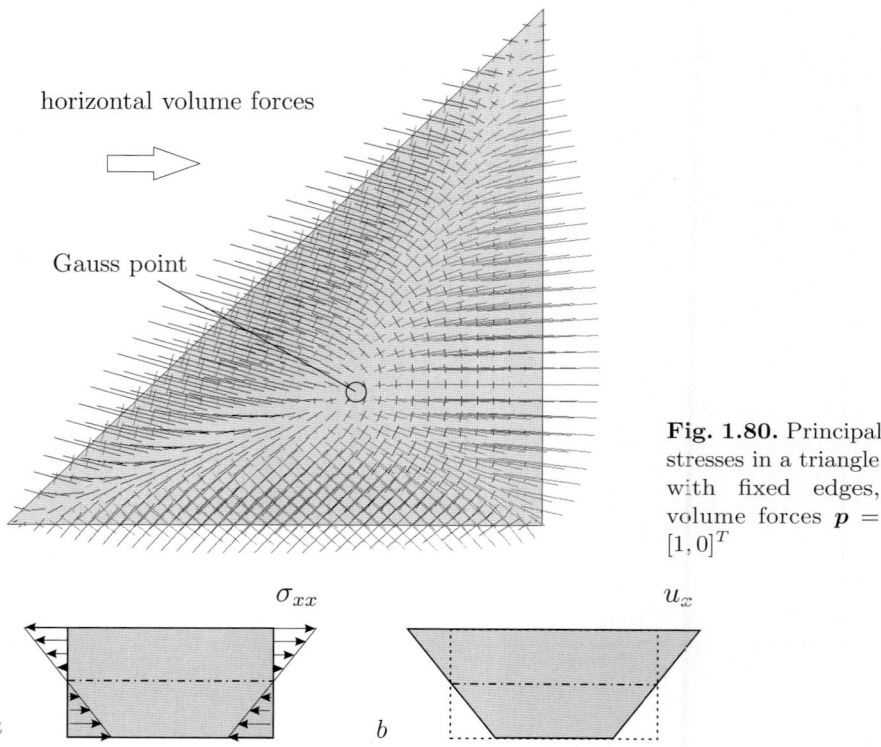

horizontal volume forces

Gauss point

Fig. 1.80. Principal stresses in a triangle with fixed edges, volume forces $\boldsymbol{p} = [1, 0]^T$

σ_{xx}

u_x

a b

Fig. 1.81. The average value of σ_{xx} is zero, and therefore the average horizontal edge displacement u_x is zero as well ($\nu = 0$)

Imagine a unit vector fixed to the edge of the slab and pointing outward. Under load, the slab curves inward or outward, so that the vector rotates about an angle $\psi = \arctan(w_{,x}\, n_x + w_{,y}\, n_y)$. If upon circling the slab, the excursions of this indicator $\times$ the arc length ds are counted, and if the count is zero, then the integral of $m_{xx} + m_{yy}$, or more appropriately of the curvature terms $\kappa_{xx} + \kappa_{yy}$, will be zero as well.

1.26 The Dirac energy

Energy plays a fundamental role in mechanics. We speak of the strain energy, $U = \boldsymbol{u}^T \boldsymbol{K} \boldsymbol{u}$, that is stored in a truss and we know that it must be equal to the work done by the exterior load, $\boldsymbol{u}^T \boldsymbol{f}$,

$$U = \boldsymbol{u}^T \boldsymbol{K} \boldsymbol{u} = \boldsymbol{u}^T \boldsymbol{f}. \tag{1.311}$$

Here we want to show that also each single displacement $u(x)$, each single stress $\sigma(x)$ is an energy quantum which represents a specific amount of energy—the Dirac energy[11]

The pulley in Fig. 1.82 is momentarily at rest and also the scale of the market woman in Fig. 1.18, p. 24, has come to a stop. But how is the balance maintained when the forces are not the same, when $G \neq H$ or $P_l \neq P_r$? The answer is: because each side knows that it cannot win. If the weight P_l on the left side of the scale moves down by δu units P_r moves up by $h_r/h_l \, \delta u$ units and so the total effort $P_l \cdot \delta u - P_r \cdot (h_r/h_l \, \delta u) = 0$ is zero. In the classical sense equilibrium means $actio = reactio$ or $ut\ tenso\ sic\ vis$ (Hooke). The force that pulls at a rod or a spring is equal to the force that holds the rod or proportional to the elongation of the spring (Hooke). But in a more precise sense equilibrium is defined by zero virtual work, which means that the forces must be orthogonal to the rigid-body motions r, these are the functions r such that $a(r, u) = 0$ for all u,

$$G(u, r) = \int_0^l -EA\, u''\, r\, dx + [N\, r]_0^l - a(u, r) = \int_0^l -EA\, u''\, r\, dx + [N\, r]_0^l = 0\,.$$

(1.312)

So we may conclude that $work = force \times displacement$ is the common denominator in mechanics.

Sure, we say we calculate displacements or stresses, but what we actually calculate is work

$$u(x) \cdot 1 = \ldots \qquad \sigma_{xx} \cdot 1 = \ldots \qquad (1.313)$$

because behind each quantity stands an influence function which is based on energy principles and so the result is of the dimension $work = force \cdot displacement$.

To repeat: for to calculate the shear force $V(x)$ in a beam we apply a dislocation $\delta_3 = 1$ at x. The beam tries its best to lessen the strain by assuming the shape $G_3(y, x)$. According to Betti's theorem the work done by the two shear forces $V_l(x_-)$ and $V_r(x_+)$ on both sides of x, which equals $V(x) \cdot 1$, plus the work done by the applied load p on acting through G_3 is zero

$$\delta W_e = -V(x) \cdot 1 + \int_0^l G_3(y, x)\, p(y)\, dy = 0 \qquad (1.314)$$

and so $V(x)$ must be equal to the work done by the load on acting through the Green's function G_3. We call this work or energy the *Dirac energy*

$$V(x) \cdot 1 = Dirac\ energy = \int_0^l G_3(y, x)\, p(y)\, dy\,. \qquad (1.315)$$

---

[11] We understand that there is also a Dirac energy in quantum mechanics but because structural mechanics operates on a very different length scale we think there is very little danger of getting things mixed up.

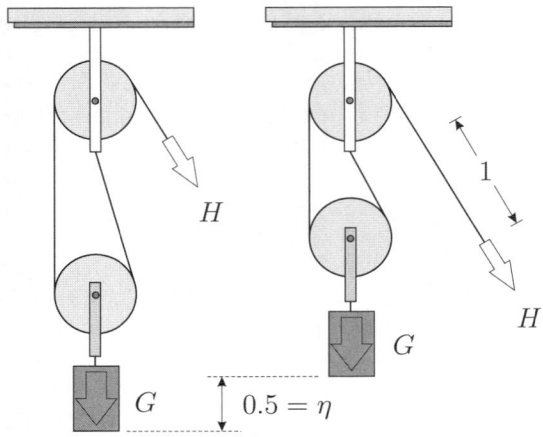

Fig. 1.82. Pulley

So to each displacement $u(x)$, each stress $\sigma(x)$, belongs a certain mechanism which, if released, induces a certain displacement in the structure and the work done by the load on acting through this displacement is $u(x)$, is $\sigma(x)$, is the Dirac energy. The Dirac energy is specific for each point x and each value $u(x), \sigma(x)$.

This energy balance $\delta W_e = 0$—which is a natural extension of Newton's law

$$\text{actio} = \text{reactio} \quad\Rightarrow\quad \delta\,u \cdot \text{actio} = \text{reactio} \cdot \delta\,u \qquad (1.316)$$

is a basic law of structural mechanics and a simple application of the idea behind a pulley. A pulley is characterized by its transmission ratio. The hand H that pulls the rope downward by one unit length moves the weight G upward by η units

$$\delta W_e = H \cdot 1 - G \cdot \eta = 0\,. \qquad (1.317)$$

It is only that in structures the ratio, $\eta = G_3(y, x)$, is not constant but depends on y, that is where the load is applied.

In a well designed structure the maximum values of the influence functions for the support reactions are less than one, $|G_R(y, x)| \leq 1$, because otherwise the structure *amplifies* the load. Archimedes knew this: *'Give me a place to stand and I will move the Earth'*.

Statics is not static

'Statics is static, isn't it? Nothing is supposed to move. Otherwise it would be called dynamics'. 'No—statics is not static, it is kinematics'.

The tourist gazing in wonder at the Eiffel tower does not realize this. The mighty tower does not move. How then should the tourist understand that the forces in each member are well tuned, that they reflect the kinematics of the tower from the foundation up to the very top. If the tourist drives up to the uppermost platform then each frame element of the tower will support her or his weight G with a fraction f which is equal to the movement of the platform in vertical direction if a corresponding hinge is introduced in the frame element and spread by one unit in vertical direction and obviously all members have decided to bear the load jointly and in fair shares because in each cross section the sum of all factors f_i equals 1.0

$$G = (f_1 + f_2 + \ldots f_n) \cdot G. \tag{1.318}$$

That is a structure consists of infinitely many mechanisms—all bolted and fixed so that the structure can carry the load. But if we release one mechanism, apply a unit rotation or dislocation, then this will induce a movement in the structure and the work done by the load on acting through this movement is equal to $M(x) \cdot 1$ or $V(x) \cdot 1$, is equal to the Dirac energy. In FE analysis we hinder the movements of the structure and so the mechanism gets the wrong signal of how large the Dirac energy really is, and consequently $M_h(x) \neq M(x)$ and $V_h(x) \neq V(x)$.

So the kinematics of a mesh determines the accuracy of an FE solution

- mesh = kinematics = accuracy of influence functions = quality of results.

Remark 1.12. Actually, once a structure has found the equilibrium position we could remove the bolts and nuts from all the internal mechanisms without having to fear that the structure would collapse because for any movement that is compatible with the kinematics of the structure δW_e would be zero, so the structure would be "safe".

1.27 How to predict changes

How will cracks in a beam, that is changes in the stiffness of single elements, affect the stress distribution in a structure? How will the Dirac energy change? This is the topic of this section.

Betti would say that the effects of the cracks can only be predicted by meticulously evaluating at each integration point y (that is effectively the whole structure if $p =$ gravity load) the changes in the influence function for, say, the bending moment $M(x)$

$$M_c(x) - M(x) = \int_0^l [G_2^c(y, x) - G_2(y, x)] \, p(y) \, dy. \tag{1.319}$$

Here G_2 and G_2^c are the influence functions for $M(x)$ in the uncracked and $M_c(x)$ in the cracked structure respectively.

But we want to show that there is a better approach which effectively restricts the analysis to those members which change. This approach may be

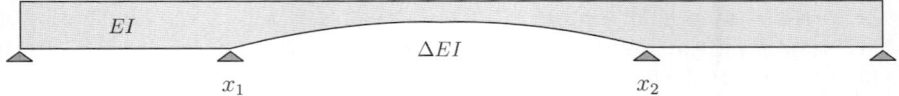

Fig. 1.83. Bridge structure

summarized as follows: To determine the change in a quantity $u(x)$, $\sigma(x)$, ... due to a change $EI \rightarrow EI + \varDelta EI$ in a certain member it suffices to evaluate the increase/decrease in the strain energy product in that member only

$$
M_c(x) - M(x) = -\int_{x_1}^{x_2} \varDelta EI \, w'' \, (G_2^c)''(y, x) \, dy = -\int_{x_1}^{x_2} \frac{\varDelta EI}{EI} \frac{M \, M_2^c}{EI} \, dy \, .
$$
$$(1.320)$$

Here G_2^c is the Green's function for $M_c(x)$ in the structure *after* the beam has cracked and w is the deflection of the element *before* the beam cracked. Note that x can be any point of the structure, that is we can effectively predict the changes in any member by integrating over the cracked element *only*. Of course the trick is that the distant member is present under the integral sign via the Green's function.

Remark 1.13. The bending moments of a Green's function are very large— even when they are "small"—but because we divide by the stiffness EI, see (1.320), the effects cancel.

Localization

In the following we want to do the localization more systematically.

Assume the stiffness in the center span of a bridge deviates from the stiffness EI in the outer spans by a term $\varDelta EI$, see Fig. 1.83.

The weak formulation of the original problem (uniform EI)

$$
\underbrace{\int_0^l EI \, w'' \, v'' \, dx}_{a(w,v)} = \underbrace{\int_0^l p \, v \, dx}_{(p,v)} \qquad v \in V \tag{1.321}
$$

and the weak formulation of the changed problem

$$
\underbrace{\int_0^l EI \, w_c'' \, v'' \, dx}_{a(w_c,v)} + \underbrace{\int_{x_1}^{x_2} \varDelta EI \, w_c'' \, v'' \, dx}_{d(w_c,v)} = \underbrace{\int_0^l p \, v \, dx}_{(p,v)} \qquad v \in V \tag{1.322}
$$

differ only by one additional term, the integral from x_1 to x_2.

Hence in an abstract setting modifications of the stiffness of a structure lead to weak formulations with an additional symmetric term $d(u, v)$

$$u_c \in V : \qquad a(u_c, v) + d(u_c, v) = (p, v) \qquad v \in V. \qquad (1.323)$$

The notation we adopted here is short for: *find $u_c \in V$ such that ... for all $v \in V$*. In FE methods we restrict the search to the FE functions in $V_h \subset V$.

How then does the solution u of the original problem

$$u \in V : \qquad a(u, v) = (p, v) \qquad v \in V \qquad (1.324)$$

differ from the solution u_c of the changed/cracked model (1.323)? Or more to the point: how does $J(u_c)$ differ from $J(u)$ where $J(.)$ is any output functional, that is the displacement or the stress or ...

$$J(u) = u(x) \qquad J(u) = \sigma(x) \qquad J(w) = M(x) \qquad \text{etc.} \qquad (1.325)$$

at a specific point?

Let $e_u = u_c - u$ the difference between the two solutions. Recall that— where applicable—

$$J(u) = a(G, u) \quad \equiv \quad w(x) = \int_0^l EI\, G''(y, x)\, w''(y)\, dy \qquad (1.326)$$

and hence

$$J(e_u) = a(G, e_u) \quad \equiv \quad w_c(x) - w(x) = \int_0^l EI\, G''(y, x)\, (w_c''(y) - w''(y))\, dy \qquad (1.327)$$

as well.

Obviously we have, subtract (1.324) from (1.323),

$$a(e_u, v) = -d(u_c, v) \qquad v \in V \qquad (1.328)$$

and therefore also, choose $v = G$,

$$J(e_u) = a(e_u, G) = -d(u_c, G) \qquad (1.329)$$

or in terms of the beam

$$w_c(x) - w(x) = -\int_{x_1}^{x_2} \Delta\, EI\, G''(y, x)\, w_c''(y)\, dy. \qquad (1.330)$$

$$\underbrace{}_{J(e_w)} \qquad \underbrace{\phantom{\int_{x_1}^{x_2} \Delta EI G''(y,x) w_c''(y) dy}}_{d(w_c, G)}$$

Eq. (1.329) is the central equation.

To express the same result with u and G_c, the Green's function of the linear functional $J(u)$ in the cracked model, we note that

$$G_c \in V : \qquad a(G_c, v) + d(G_c, v) = J(v) \qquad v \in V. \qquad (1.331)$$

Gerber beam #1

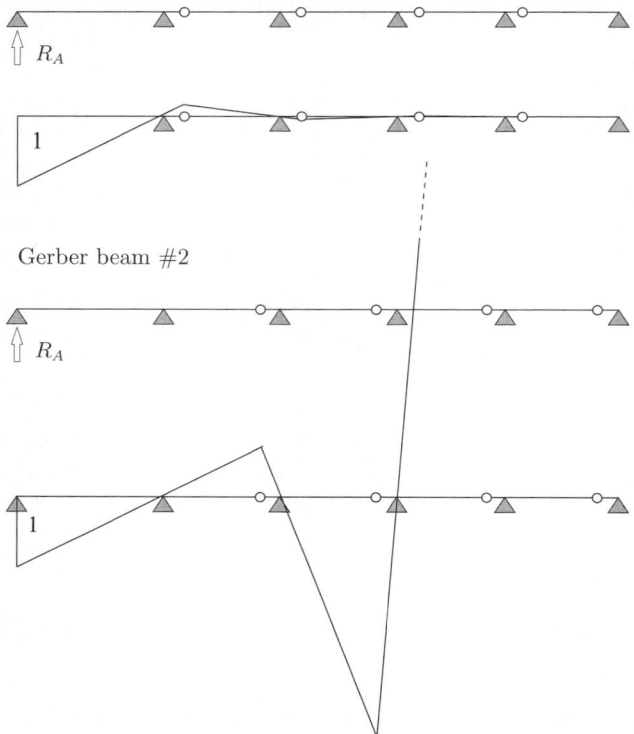

Gerber beam #2

Fig. 1.84. Influence
function for support
reaction R_A. Not
all Green's functions
decay!

Hence the error in the output functional is as well

$$J(e_u) = a(e_u, G_c) + d(e_u, G_c) = -d(u_c, G_c) + d(e_u, G_c) = -d(u, G_c) \tag{1.332}$$

so we can express $J(e_u) = -d(u_c, G) = -d(u, G_c)$ both ways. Either combination $u_c \times G$ or $u \times G_c$ will do.

Remark 1.14. By applying the same arguments to the FE equations the results can be extended to $e_u^h = u_c^h - u_h$

$$J(e_u^h) = -d(u_c^h, G_h) = -d(u_h, G_c^h) \tag{1.333}$$

Summary

The change in any output value

$$J(e_u) := J(u_c) - J(u) \tag{1.334}$$

can be expressed by an energy integral, the d-scalar product between u and G_c or—vice versa—between u_c and G. Because $d(G_c, u)$ only extends over the elements affected by the change, see (1.320), this formula is superior to Betti's formula

$$J(e_u) = \underbrace{-d(G_c, u)}_{local\ analysis} = \underbrace{\int_0^l [G_c - G]\, p(y)\, dy}_{global\ analysis\ =\ Betti} \tag{1.335}$$

or

$$w_c(x) - w(x) = -\int_{x_1}^{x_2} \Delta\, EI\, G''_c\, w''\, dy = \int_0^l [G_c - G]\, p(y)\, dy \tag{1.336}$$

where the second formulation requires that we trace the deviations between G_c and G over the whole structure—at least in the load case $p =$ dead weight.

Recall that all quantities in linear mechanics are energies, $u(\boldsymbol{x}) \times 1, \sigma(\boldsymbol{x}) \times 1$, see Sect. 1.26 p. 110, and so when the stresses and displacements change then the *Dirac energy*, $(u(\boldsymbol{x}) + \Delta u) \times 1$—that is the work done by the exterior load on acting through the Green's function—changes and this change in energy, $\Delta u \times 1$, is just

$$-d(u, G_c) = \text{ante} \times \text{post} = -d(u_c, G) = \text{post} \times \text{ante} \tag{1.337}$$

so we only need to look at the single spring, the single member, the single plate or slab element where the change occurs

$$-J(e_u) = d(u_c, G) = \Delta k\, G(l, x)\, w_c(l) \qquad\qquad \text{spring } k \tag{1.338}$$

$$= \Delta EI \int_{x_1}^{x_2} G''\, w''_c\, dy \qquad\qquad \text{beam} \tag{1.339}$$

$$= \Delta t \int_{\Omega_e} \sigma^G_{ij}\, \varepsilon^c_{ij}\, d\Omega_{\boldsymbol{y}} \qquad\qquad \text{plate} \tag{1.340}$$

$$= t \int_{\Omega_e} \Delta C^{ijkl} \varepsilon^G_{kl}\, \varepsilon^c_{ij}\, d\Omega_{\boldsymbol{y}} \qquad\qquad \text{plate} \tag{1.341}$$

$$= \Delta K \int_{\Omega_e} m^G_{ij}\, \kappa^c_{ij}\, d\Omega_{\boldsymbol{y}} \qquad\qquad \text{slab} \tag{1.342}$$

to assess the change in $u(\boldsymbol{x})$ or $\sigma(\boldsymbol{x})$ at an arbitrary point $\boldsymbol{x}$ of the structure.

The stresses in the element induced by the Dirac delta—which may be located at some very distant point $\boldsymbol{x}$—act like weights. That is if the concrete cracks under tension in a slab element, $K \to K + \Delta K$, then typically the effects will be scaled by some negative power $r^{-1}, r^{-2}, \ldots$ of the distance r between the element and the point $\boldsymbol{x}$.

But please note that not all Green's functions decay. If rigid-body motions are involved then the opposite may be true; see Fig. 1.84 and also Fig. 1.60 p. 89.

Simplification

To apply the formula (1.332) for $J(e_u)$ we must know the solution u of the original problem and the Green's function G_c of the changed problem (or vice versa). This is not very practical because once we have set up the equations for both systems we could compare the two solutions u and u_c directly.

So let us try a different approach. With the Green's function G of the original problem

$$G \in V : \qquad a(G,v) = J(v) \qquad v \in V \tag{1.343}$$

we obtain the formula

$$J(e_u) = -d(u,G_c) = -d(u,G) - d(u,G_c - G) \tag{1.344}$$

or if we drop the second term

$$J(e_u) \simeq -d(u,G) . \tag{1.345}$$

This approximate formula has the advantage that all terms come from the original model.

Force terms

In Sect. 7.7 p. 535 we will see that if the Green's function G is the influence function for a force term (a stress or any other internal action) at a point x then the formula

$$J(w) = a(G,w) \qquad \equiv \qquad M(x) \stackrel{?}{=} \int_0^l \frac{M_2\,M}{EI}\,dy = 0 \tag{1.346}$$

makes no sense—at least not in the naive sense. We cannot calculate $M(x)$ or $V(x)$ at a point x with Mohr's integral. Rather what (1.346) means is this: if there is a sequence of Green's functions $\{G_h\}$ which converges to G then we have

$$\lim_{h \to 0} a(G_h, w) = a(G, w) + J(w) = 0 + J(w) \tag{1.347}$$

that is in the limit out of $a(G_h, w)$ pops $J(w)$ but $a(G, w)$ itself is zero, that is a computer cannot calculate $J(w)$ by evaluating $a(G, w)$ *a posteriori* that is when all is done. Rather the computer must follow the action from the start only then will it have a chance to catch $J(w)$—as the limit of all expressions $a(G_h, w)$.

But luckily in FE analysis we have Tottenham's equation which guarantees that the weak formulation and Betti are identical

$$J(w_h) = \int_0^l G_h(y, x)\,p(y)\,dy = a(G_h, w_h) \tag{1.348}$$

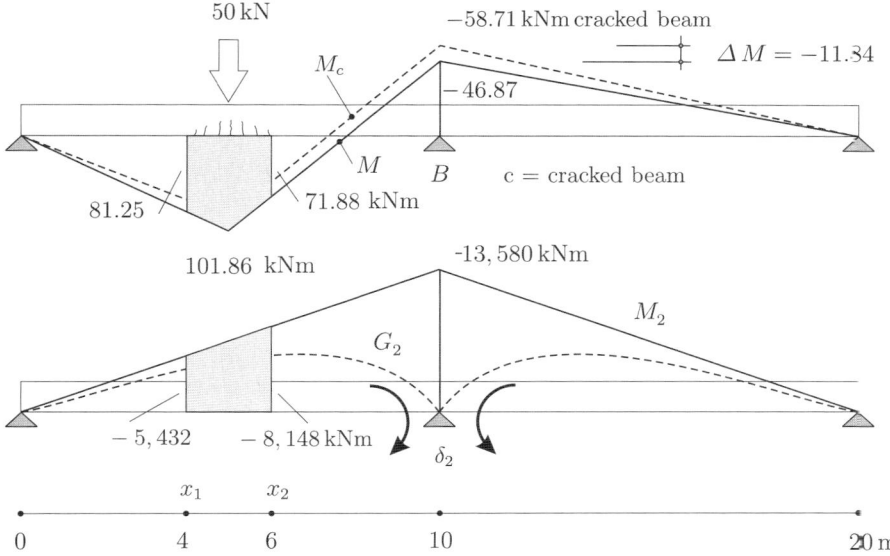

Fig. 1.85. Cracks in a continuous beam. The integral over the shaded areas, $(M, M_2) \times (EI)^{-1} = -13.3$ kNm, is approximately $\Delta M = -11.84$ kNm

whatever the type of Green's function G—even if $J(w_h)$ is a force term. So in FE analysis we must not be concerned with such "subtile" distinctions. The equation

$$M_h(x) = \int_0^l \frac{M_2^h \, M_h}{EI} \, dy \tag{1.349}$$

simply works—because we make it work, see (7.221) p. 535. This guarantees also that

$$J(e_u^h) = -d(u_c^h, G_h) = -d(u_h, G_c^h) \tag{1.350}$$

for any functional $J(.)$.

Example—cracks in a beam

Under the action of the point load the concrete cracks (see Fig. 1.85) so that the bending stiffness drops from $EI = 90,625$ kNm^2 to $EI + \Delta EI = 46,400$ kNm^2 which is a drop of nearly 50 % ($\Delta EI = -44,225$).

To predict the change in the hogging moment at support B we use the approximation

$$J(e_w) = -d(w, G_2^c) \simeq -d(w, G_2) \tag{1.351}$$

that is we substitute for the exact Green's function G_2^c (cracked beam) the original Green's function G_2 (sans cracks) so that the change in M is approximately the d-scalar product between the two deflections, w and G_2, of the uncracked beam

$$M_c - M = -\int_{x_1}^{x_2} \frac{\Delta\,EI}{EI} \frac{M\,M_2}{EI}\,dy = -13.3\,\text{kNm} \qquad (1.352)$$

while the true value is -11.84 kNm.

Remark 1.15. The study of the equation

$$J(e_w) = M_c(x) - M(x) = -\int_{x_1}^{x_2} \frac{\Delta\,EI}{EI} \frac{M\,M_2^c}{EI}\,dy = -d(w, G_2^c)$$

$$(1.353)$$

makes for an interesting subject.

If in a continuous beam the change, $EI \to EI + \Delta EI$, is the same in all cross sections then $[x_1, x_2]$ is the whole beam $[0, l]$ and so $d(w, G_2^c)$ essentially coincides with $a(w, G_2^c)$

$$J(e_w) = \frac{\Delta\,EI}{EI} \int_0^l \frac{M\,M_2^c}{EI}\,dy = \frac{\Delta\,EI}{EI}\,a(w, G_2^c) = 0 \qquad (1.354)$$

but because $a(w, G_2^c)$ is zero, see Sect. 7.7 p. 535, the change in the bending moment must be zero too. Hence the bending moment distribution in a continuous beam—with a uniform EI—is independent of the magnitude of EI.

Vice versa, if the bending stiffness EI in a continuous beam varies locally then the bending moment distribution is sensitive to such variations in EI.

In a statically determinate beam the Green's function for $M(x)$ is piecewise linear so that $M_2 = 0$ and so $J(e_w) = 0$ for any ΔEI, that is in a statically determinate beam $M(x)$ does not depend on EI.

Nodal form of $J(e_u)$

The nodal form of a functional $J(.)$ is

$$J(u_h) = u_G^T\,K\,u \qquad (1.355)$$

where u_G is the nodal vector of the Green's function and the nodal form of the change $J(e_u) = J(u_c^h - u_h)$ in a functional is—as we will show

$$J(e_u) \simeq -u_G^T\,K_\Delta\,u\,. \qquad (1.356)$$

To start we observe that the vector-and-matrix form of the two equations (1.323) and (1.324) is

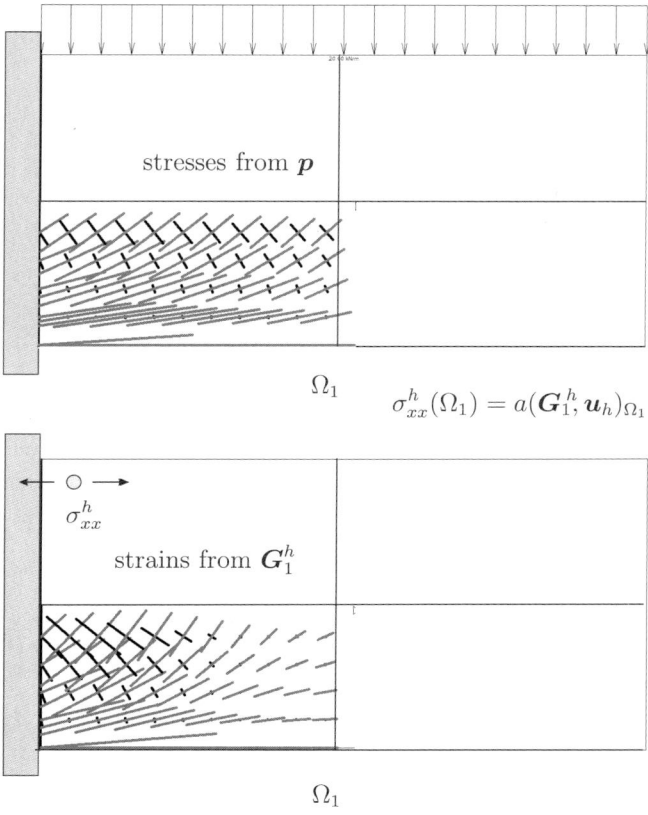

$$\Omega_1 \qquad \sigma^h_{xx}(\Omega_1) = a(\boldsymbol{G}^h_1, \boldsymbol{u}_h)_{\Omega_1}$$

$$\Omega_1$$

Fig. 1.86. Contributions of element Ω_1 to the stress σ_{xx} **a)** stresses from the edge load **b)** strains from $\boldsymbol{G}_1$

.

$$\boldsymbol{u}_c \in \mathbb{R}^n \qquad \boldsymbol{v}^T \boldsymbol{K} \boldsymbol{u}_c + \boldsymbol{v}^T \boldsymbol{K}_\Delta \boldsymbol{u}_c = \boldsymbol{v}^T \boldsymbol{f} \qquad \text{for all } \boldsymbol{v} \in \mathbb{R}^n \quad (1.357)$$

and

$$\boldsymbol{u} \in \mathbb{R}^n \qquad \boldsymbol{v}^T \boldsymbol{K} \boldsymbol{u} = \boldsymbol{v}^T \boldsymbol{f} \qquad \text{for all } \boldsymbol{v} \in \mathbb{R}^n \qquad (1.358)$$

where the matrix $\boldsymbol{K}_\Delta$ encapsulates the change in the stiffness matrix. It corresponds to the d-scalar product.

By subtracting these two equations we obtain

$$\boldsymbol{v}^T \boldsymbol{K}(\boldsymbol{u}_c - \boldsymbol{u}) = -\boldsymbol{v}^T \boldsymbol{K}_\Delta \boldsymbol{u}_c \qquad \boldsymbol{v} \in \mathbb{R}^n \qquad (1.359)$$

or if we let $\boldsymbol{v} = \boldsymbol{u}_G$, the nodal vector of a Green's function which belongs to a functional $J(.)$

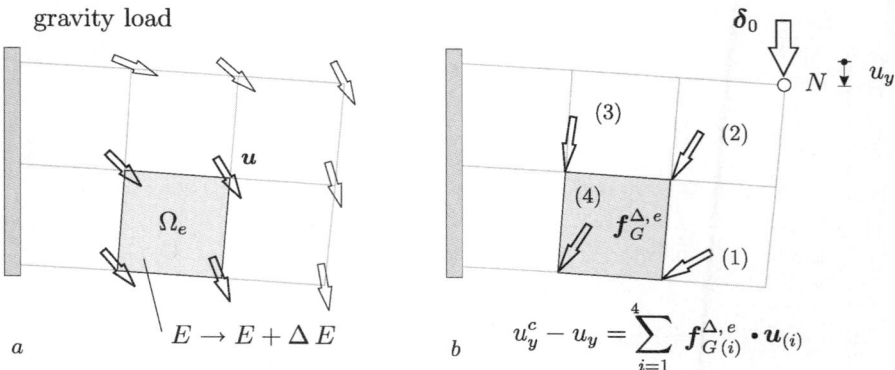

Fig. 1.87. Change of Young's modulus in a single element and influence on the vertical displacement of node N: **a)** nodal displacements under gravity load; **b)** incremental equivalent nodal element forces $f_{G(i)}^{\Delta,\,e}$ in element Ω_e due to the Dirac delta at node N

$$J(u_c^h - u_h) = u_G^T \, K(u_c - u) = -u_G^T \, K_\Delta \, u_c \simeq -u_G^T \, K_\Delta \, u. \quad (1.360)$$

In most cases the stiffness of an element Ω_e changes, $E \to E + \Delta E$, and then the d-scalar product is

$$d(u_G^h, u_h) = \frac{\Delta E}{E} \, a(u_G^h, u_h)_{\Omega_e} \quad (1.361)$$

so that the stiffness matrix of the modified structure is (E is a common factor of all entries k_{ij}^e)

$$K_{mod} = K + K_\Delta \qquad (K_\Delta)_{ij} = \frac{\Delta E}{E} \, k_{ij}^e \quad (1.362)$$

where the additional matrix K_Δ contains only contributions (k_{ij}^e) from the element Ω_e weighted with $\Delta E / E$ —in all other elements $\Delta E = 0$—and so

$$J(u_c^h - u_h) \simeq -d(G_h, u_h) = -u_G^T K_\Delta \, u = -u^T K_\Delta \, u_G. \quad (1.363)$$

Because of the "local" character of K_Δ this triple product can be identified with the work done by the equivalent element nodal forces $f_G^{\Delta,\,e} = K_\Delta^e \, u_G$ of the change of the Green's function on Ω_e on acting through the nodal displacements u or vice versa, that is the d-scalar product reduces to

$$- d(G_h, u_h) = -u_{(8)}^T (K_\Delta^e)_{(8\times 8)} \, (u_G)_{(8)} \quad (1.364)$$

(e.g. bilinear element with eight degrees of freedom) or (see Fig. 1.87)

$$- d(G_h, u_h) = -u_{(8)}^T (f_G^{\Delta,\,e})_{(8)}. \quad (1.365)$$

If only one entry in the stiffness matrix changes, say the entry $k_{11\,11} \rightarrow k_{11,11} + \Delta k$ of an elastic support, then the change in the output functional is simply

$$J(\boldsymbol{u}_c^h - \boldsymbol{u}^h) = f_{11}^G \cdot \Delta k \cdot u_{11} \qquad (1.366)$$

where f_{11}^G is the nodal force (= support reaction) in the spring due to the action of the Dirac delta (located somewhere else) and u_{11} is the nodal displacement under load.

Nonlinear problems

In nonlinear problems there are no Green's functions and so the equation

$$J(\boldsymbol{u}_c) - J(\boldsymbol{u}) = -d(\boldsymbol{z}, \boldsymbol{u}_c) \qquad (1.367)$$

is not applicable. But if we substitute for the missing $\boldsymbol{z}$ the Green's function $\boldsymbol{z}$ at the linearization point,

$$a_T(\boldsymbol{u}; \boldsymbol{z}, \boldsymbol{v}) = J'(\boldsymbol{u}; \boldsymbol{v}) \qquad \boldsymbol{v} \in V \qquad (1.368)$$

—in an FE setting this would be

$$\boldsymbol{K}_T(\boldsymbol{u})\,\boldsymbol{z} = \boldsymbol{j} \qquad (1.369)$$

where $\boldsymbol{K}_T$ is the tangential stiffness matrix—it follows, [50], that

$$J(\boldsymbol{u}_c) - J(\boldsymbol{u}) = -d(\boldsymbol{u}, \boldsymbol{z}) - \frac{1}{2}\{d(\boldsymbol{u}, \boldsymbol{e}_z) + d'(\boldsymbol{u})(\boldsymbol{e}_u, \boldsymbol{z}) - R\} \quad (1.370)$$

where d' is the Gateaux derivative of d and R is a remainder that is cubic in $\boldsymbol{e}_u = \boldsymbol{u}_c - \boldsymbol{u}$ and $\boldsymbol{e}_z = \boldsymbol{z}_c - \boldsymbol{z}$.

The error in the output functional of a nonlinear problem can be expressed as, see Sect. 7.5, p. 526,

$$J(\boldsymbol{u}) - J(\boldsymbol{u}_h) = \frac{1}{2}\rho(\boldsymbol{u}_h)(\boldsymbol{z} - \boldsymbol{z}_h) + \frac{1}{2}\rho^*(\boldsymbol{u}_h, \boldsymbol{z}_h)(\boldsymbol{u} - \boldsymbol{u}_h) + R_h^{(3)} \quad (1.371)$$

where the first term

$$\rho(\boldsymbol{u}_h)(\boldsymbol{z} - \boldsymbol{z}_h) = (\boldsymbol{p} - \boldsymbol{p}_h, \boldsymbol{z} - \boldsymbol{z}_h) \qquad (1.372)$$

is the work done by the residual forces on acting through the error, $\boldsymbol{z} - \boldsymbol{z}_h$, in the Green's function and the second term is—in a somewhat symbolic notation

$$\rho^*(\boldsymbol{u}_h, \boldsymbol{z}_h)(\boldsymbol{u} - \boldsymbol{u}_h) = (\boldsymbol{\delta}_0 - \boldsymbol{\delta}_0^h, \boldsymbol{u} - \boldsymbol{u}_h) \qquad (1.373)$$

the error in the functional $J(.)$ evaluated at $\boldsymbol{u} - \boldsymbol{u}_h$ and $R_h^{(3)}$ is a cubic remainder.

By combining the discretization error (1.371) with the model error (1.370) it follows that

$$J(\boldsymbol{u}_c) - J(\boldsymbol{u}_h) = -d(\boldsymbol{u}_h, \boldsymbol{z}_h)$$
$$+ \frac{1}{2}\{(\boldsymbol{p} - \boldsymbol{p}_h, \boldsymbol{z} - \boldsymbol{z}_h) + (\boldsymbol{\delta}_0 - \boldsymbol{\delta}_0^h, \boldsymbol{u} - \boldsymbol{u}_h)\}$$
$$- \frac{1}{2}\{d(\boldsymbol{u}_h, \boldsymbol{e}_z) + d'(\boldsymbol{u}_h)(\boldsymbol{e}_u, \boldsymbol{z}_h)\} + \frac{1}{2}R \qquad (1.374)$$

or if we neglect higher order terms

$$J(\boldsymbol{u}_c) - J(\boldsymbol{u}_h) = -d(\boldsymbol{u}_h, \boldsymbol{z}_h)$$
$$+ \frac{1}{2}\{(\boldsymbol{p} - \boldsymbol{p}_h, \boldsymbol{z} - i_h\,\boldsymbol{z}) + (\boldsymbol{\delta}_0 - \boldsymbol{\delta}_0^h, \boldsymbol{u} - i_h\,\boldsymbol{u})\}.$$
$$(1.375)$$

Here $\boldsymbol{u}_c$ is the exact solution of the modified problem, $E \to E + \Delta E$, and $\boldsymbol{u}_h$ is the FE solution of the original (or simplified) problem. The terms $i_h\boldsymbol{u}$ and $i_h\,\boldsymbol{z}$ signal that $\boldsymbol{u}_h$ and $\boldsymbol{z}_h$ respectively can be replaced by *any* two functions that interpolate $\boldsymbol{u}$ and $\boldsymbol{z}$ on V_h—or come close to $\boldsymbol{u}$ and $\boldsymbol{z}$ in what sense ever—so that they provide a tight upper bound on the error.

To make this formula applicable the unknown functions $\boldsymbol{z}$ and $\boldsymbol{u}$ are approximated by higher-order interpolations of the FE solutions $\boldsymbol{z}_h$ and $\boldsymbol{u}_h$

$$\boldsymbol{z} - i_h\,\boldsymbol{z} \simeq i_{2h}^{(2)}\boldsymbol{z}_h - \boldsymbol{z}_h \qquad (1.376)$$
$$\boldsymbol{u} - i_h\,\boldsymbol{u} \simeq i_{2h}^{(2)}\boldsymbol{u}_h - \boldsymbol{u}_h \qquad (1.377)$$

that is if V_h exists for example of piecewise linear functions then a quadratic interpolation may be used. For additional details see [50].

Remark 1.16. To put these results in a better perspective we add some remarks: In some sense analysis is all about inequalities, about bounds, about estimates or—as we might say as well—about errors. The error in a linear interpolation u_I of a regular function u is bounded by the maximum value of the second derivative $u''(x)$ and the element length h

$$|u(x) - u_I(x)| \le h^2 \cdot \max|u''|. \qquad (1.378)$$

That is these two parameters *control* the error. The aim of any numerical analysis must be the search for such estimates.

With the influence function for $u(x)$

$$u(x) - u_I(x) = -\int_0^h G_0(y, x)\,(u'' - u_I'')\,dy = -\int_0^h G_0(y, x)\,(u'' - 0)\,dy$$
$$= -\int_0^h \overline{}\,u''\,dy. \qquad (1.379)$$

it should not be too difficult to prove that $h^2 \cdot \max |u''|$ is a bound for the interpolation error. The proof never mentions the influence function explicitly but it relies of course on the properties of the triangle $G_0(y, x)$, see Sect. 7.11, p. 560.

Now in a nonlinear problem there is no Green's function $\boldsymbol{z}$ such that

$$J(u) = \int_\Omega \boldsymbol{z}(\boldsymbol{x}, \boldsymbol{y}) \cdot p(\boldsymbol{y}) \, d\Omega_{\boldsymbol{y}} . \tag{1.380}$$

But we know that the error

$$J(u) - J(u_h) \tag{1.381}$$

of the FE solution is small if the error $\boldsymbol{z} - \boldsymbol{z}_h$ in the Green's function at the linearization point can be made small. This is the important observation. The Green's function $\boldsymbol{z}$ at the linearization point is of no direct use—in the sense of (1.380)—but it seems reasonable to assume that if the error $\boldsymbol{z} - \boldsymbol{z}_h$ is small (and the error $\boldsymbol{u} - \boldsymbol{u}_h$) then also the error $J(u) - J(u_h)$ will be small. Obviously the error in the Green's function is—strangely enough—correlated with the discretization error $J(u) - J(u_h)$ and also, as shown above, with the modeling error.

For more on the discretization error $J(u) - J(u_h)$ in nonlinear problems, see Sect. 7.5 p. 526.

Linear versus nonlinear

The variational formulation of a nonlinear problem, see Sect. 4.21 p. 401,

$$a(\boldsymbol{u}, \boldsymbol{v}) := \int_\Omega \boldsymbol{E}_{\boldsymbol{u}}(\boldsymbol{v}) \bullet \boldsymbol{S}(\boldsymbol{u}) \, d\Omega = (\boldsymbol{p}, \boldsymbol{v}) \tag{1.382}$$

and a linear problem

$$a(\boldsymbol{u}, \boldsymbol{v}) := \int_\Omega \boldsymbol{E}(\boldsymbol{v}) \bullet \boldsymbol{S}(\boldsymbol{u}) \, d\Omega = (\boldsymbol{p}, \boldsymbol{v}) \tag{1.383}$$

differ only in the term

$$\begin{aligned}
d(\boldsymbol{u}, \boldsymbol{v}) &= \int_\Omega \left(\boldsymbol{E}_{\boldsymbol{u}}(\boldsymbol{v}) - \boldsymbol{E}(\boldsymbol{v}) \right) \bullet \boldsymbol{S}(\boldsymbol{u}) \, d\Omega \\
&= \int_\Omega \left(\nabla \boldsymbol{u}^T \nabla \boldsymbol{v} + \nabla \boldsymbol{v}^T \nabla \boldsymbol{u} \right) \bullet \boldsymbol{S}(\boldsymbol{u}) \, d\Omega
\end{aligned} \tag{1.384}$$

and so it should not be too far-fetched to assume that nonlinear effects can be predicted by the formula

$$J(\boldsymbol{u}_c) - J(\boldsymbol{u}) \simeq -d(\boldsymbol{z}, \boldsymbol{u}_c) = -\int_\Omega \left(\nabla \boldsymbol{u}_c^T \nabla \boldsymbol{z} + \nabla \boldsymbol{z}^T \nabla \boldsymbol{u}_c \right) \bullet \boldsymbol{S}(\boldsymbol{u}_c) \, d\Omega \tag{1.385}$$

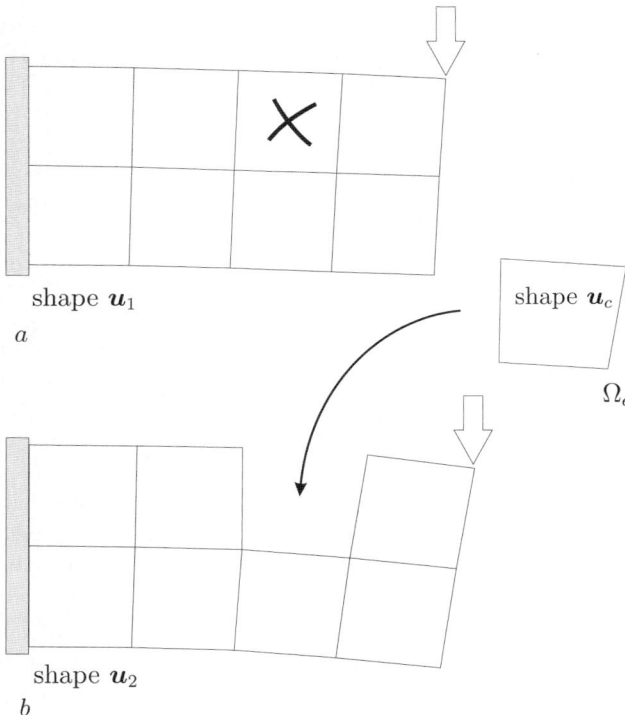

shape $\boldsymbol{u}_1$

a

shape $\boldsymbol{u}_c$

Ω_e

shape $\boldsymbol{u}_2$

b

Fig. 1.88. Influence of an element: The shape $\boldsymbol{u}_c$ of the single element Ω_e is different from the shape it originally had in Figure **a)**. The new shape $\boldsymbol{u}_c$ would fill the void in Figure **b)**

where the Green's function $\boldsymbol{z}$ is taken from the linear model and $\boldsymbol{u}_c$ is the solution of the nonlinear problem.

In FE analysis we would substitute also for $\boldsymbol{u}_c$ the FE solution of the linear problem and so

$$J(\boldsymbol{u}_c) - J(\boldsymbol{u}_h) \approx -d(\boldsymbol{z}_h, \boldsymbol{u}_h) = -\int_\Omega (\nabla \boldsymbol{u}_h^T \nabla \boldsymbol{z}_h + \nabla \boldsymbol{z}_h^T \nabla \boldsymbol{u}_h) \bullet \boldsymbol{S}(\boldsymbol{u}_h) \, d\Omega \,.$$

$$(1.386)$$

1.28 The influence of a single element

The influence of a single element Ω_e on, say, the displacement $u_x^h(\boldsymbol{x})$ at a point $\boldsymbol{x}$ in a plate is the contribution

$$\int_{\Omega_e} \boldsymbol{G}_0^h(\boldsymbol{y}, \boldsymbol{x}) \bullet \boldsymbol{p}(\boldsymbol{y}) \, d\Omega_{\boldsymbol{y}} \qquad (1.387)$$

of the element to the sum total

$$u_x^h(\boldsymbol{x}) = \sum_e \int_{\Omega_e} \boldsymbol{G}_0^h(\boldsymbol{y}, \boldsymbol{x}) \bullet \boldsymbol{p}(\boldsymbol{y}) \, d\Omega_{\boldsymbol{y}} \,. \tag{1.388}$$

If we employ a weak form of the influence function ("Mohr's integral"), see Sect. 7.7 p. 535,

$$u_x^h = a(\boldsymbol{G}_0^h, \boldsymbol{u}_h) = \int_\Omega [\sigma_{xx} \cdot \varepsilon_{xx} + 2\,\sigma_{xy} \cdot \varepsilon_{xy} + \sigma_{yy} \cdot \varepsilon_{yy}] \, d\Omega_{\boldsymbol{y}} \tag{1.389}$$

then the contribution of a single element (see Fig. 1.86)

$$\int_{\Omega_e} [\sigma_{xx} \cdot \varepsilon_{xx} + 2\,\sigma_{xy} \cdot \varepsilon_{xy} + \sigma_{yy} \cdot \varepsilon_{yy}] \, d\Omega_{\boldsymbol{y}} \tag{1.390}$$

is the strain energy product between the stress field of the Green's function $\boldsymbol{G}_0^h$ and the strains of the FE solution $\boldsymbol{u}_h$ in this element or vice versa, because the strain energy product is symmetric, $a(\boldsymbol{G}_0^h, \boldsymbol{u}_h) = a(\boldsymbol{u}_h, \boldsymbol{G}_0^h)$.

The important point to note is that influence depends on *two* quantities. The strains (or stresses) from the load case $\boldsymbol{p}^h$ are weighted with the stresses (or strains) from $\boldsymbol{G}_i^h$. Only if both quantities are large will the contribution be significant. And typically influence depends on the distance $r = |\boldsymbol{y} - \boldsymbol{x}|$ between the two points, $G_0(\boldsymbol{y}, \boldsymbol{x}) = G_0(\boldsymbol{y} - \boldsymbol{x})$, (and the angular orientation between the two points $\boldsymbol{x}/|\boldsymbol{x}|$ and $\boldsymbol{y}/|\boldsymbol{y}|$ on the unit sphere) so that influence functions act like *convolutions*.

But the influence of a single element could also be understood in the following sense: how would the results change if the element were removed from the structure? Or stated otherwise: how important is a certain element for a structure?

This question can be answered with the same formulas as before, it is only that the displacement field $\boldsymbol{u}_h$ of the element must be replaced by the field $\boldsymbol{u}_h^c$ which is the shape of the element if it were drained of all its stiffness.

If a frame element $[x_1, x_2]$ is removed from a structure then the change in any output functional $J(w)$ at any point x is, see Sect. 3.8,

$$J(e_w) := J(w_c) - J(w) = \int_{x_1}^{x_2} \frac{M_c \, M_i}{EI} \, dx \tag{1.391}$$

where

- M_i is the bending moment of the influence function G_i for $J(w)$.
- M_c is the bending moment in the spline w_c that reconnects the released nodes.

The spline w_c is that curve that bridges the gap after the frame has found its new equilibrium position. It attaches *seamlessly* to the two released nodes.

Equation (1.391) holds true for other structures as well; see Fig. 1.88. The change in any output functional $J(\boldsymbol{u})$ due to the loss of an element Ω_e is

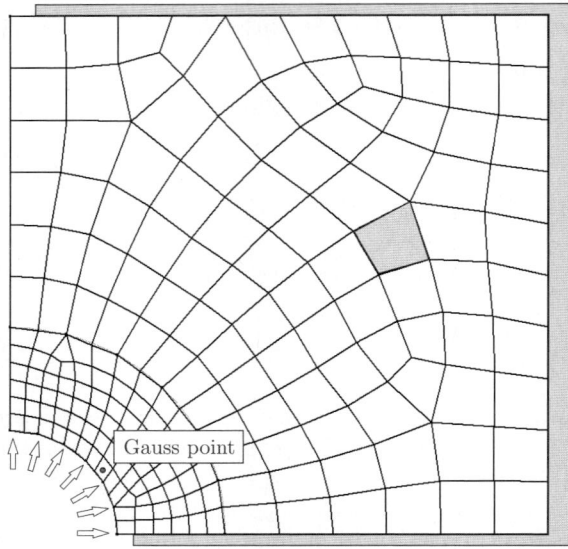

Fig. 1.89. The influence of the element on the stresses at the Gauss point depends (1) on the shape $\boldsymbol{u}_c$ of the void the element leaves if it is removed and (2) the magnitude of the stresses in the element caused by the dislocations at the Gauss point

$$J(\boldsymbol{e}_u) := J(\boldsymbol{u}_c) - J(\boldsymbol{u}) = a(\boldsymbol{u}_c, \boldsymbol{G}_i)_{\Omega_e} \tag{1.392}$$

where $\boldsymbol{u}_c$ is the shape of the "phantom element" that bridges seamlessly the void left by the original element Ω_e.

Because of

$$|J(\boldsymbol{e}_u)| = |a(\boldsymbol{u}_c, \boldsymbol{G}_i)_{\Omega_e}| \leq |a(\boldsymbol{u}_c, \boldsymbol{u}_c)_{\Omega_e}| \cdot |a(\boldsymbol{G}_i, \boldsymbol{G}_i)_{\Omega_e}| \tag{1.393}$$

we conclude that the strain energy of the displacement field $\boldsymbol{u}_c$ and the strain energy of the influence function $\boldsymbol{G}_i$ (which also is a displacement field) provide an upper bound for the change.

To repeat: the displacement field $\boldsymbol{u}_c$ is that displacement field which reconnects the edges of the void *after* the structure has found its new equilibrium position. It is the shape the element would assume if it were slowly drained of all its stiffness but would cling to the structure. Alternatively one could imagine a single element with the original stiffness which by prestressing forces is bent into the shape $\boldsymbol{u}_c$ so that it gives the impression as if it would bridge the gap.

The energy $a(\boldsymbol{u}_c, \boldsymbol{u}_c)$ is just the strain energy in this element. If the prestressing forces on the edge of the element would be applied in opposite direction on the edge of the void the structure would assume the shape it had before the element was removed.

Hence the importance of a single element Ω_e for a single value $J(\boldsymbol{u})$ depends on the strain energy of the fields $\boldsymbol{u}_c$ and $\boldsymbol{G}_i$ inside that element. The more the phantom element must stretch to fill the void and the more intense the strain energy of the Green's function G_i is in the element the more important is Ω_e for $J(\boldsymbol{u})$, see Fig. 1.89.

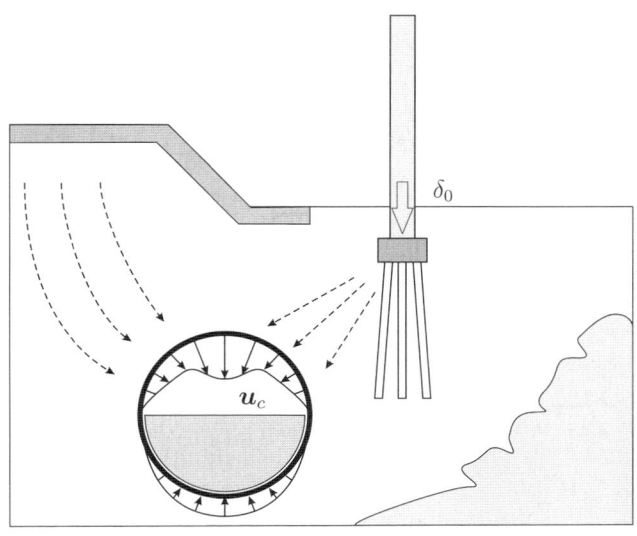

Fig. 1.90. Planned excavation of a tunnel and influence on the settlement of a pier

Every structural engineer knows that the elements that get stretched or bent the most are the most important for a structure. But with (1.392) we can quantify this feeling and give it a precise mathematical expression. And as it turns out it is not exactly the shape $\boldsymbol{u}$ of the element we see on the screen but the shape $\boldsymbol{u}_c$ of the element when it is drained of all its stiffness which is decisive. The larger the gap the drained element must bridge the more important the element is.

The logic can also be applied to a planned excavation if we want to know how much the cavity will affect the foundation of a nearby pier; see Fig. 1.90. Do the following:

1. Apply a vertical point load $P = 1$ at the foot of the foundation and calculate the strain energy $a(\boldsymbol{G}_0, \boldsymbol{G}_0)$ of the region Ω_X which is to be excavated with a one-point quadrature that is

$$a(\boldsymbol{G}_0, \boldsymbol{G}_0) \simeq [\sigma_{xx}(\boldsymbol{x}_c)\,\varepsilon_{xx}(\boldsymbol{x}_c) + \ldots + \sigma_{yy}(\boldsymbol{x}_c)\,\varepsilon_{yy}(\boldsymbol{x}_c)] \times \Omega_X \tag{1.394}$$

   where $\boldsymbol{x}_c$ is the center of the cavity and Ω_X is the area of the cross section.
2. Excavate Ω_X, determine the edge displacement ($= \boldsymbol{u}_c$) of the cavity under load and apply these displacements to a plate which has the same extension as Ω_X.
3. Calculate as before the strain energy in this plate. The product of these two energies provides a rough upper bound for the additional settlement of the pier due to the excavation.

Of course this is purely theoretical and impractical because in step 2 we very nearly have the answer. But these steps may provide a clue as to how we

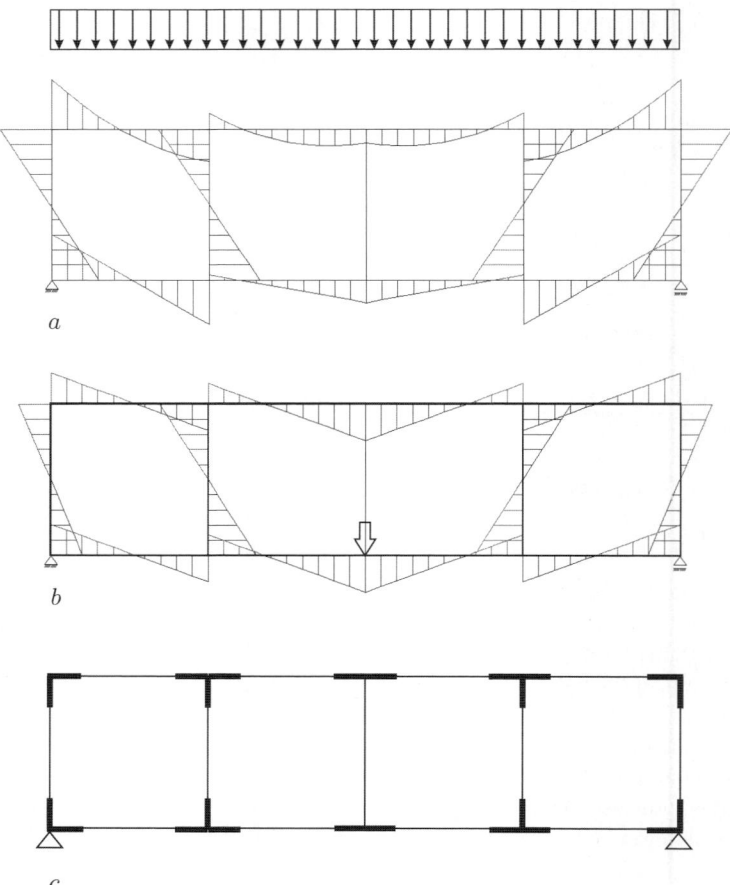

Fig. 1.91. Retrofitting a Vierendeel girder **a)** bending moments from the traffic load **b)** bending moments from the point load $P = 1$ (Dirac delta δ_0) **c)** proposed modifications

argue when we try to make predictions. Foremost it is the distance between the foundation and the cavity which interests us—the energy $a(\boldsymbol{G}_0, \boldsymbol{G}_0)$ depends on this distance—and the shape $\boldsymbol{u}_c$ of the cavity when the load is applied to the surface of the halfspace.

1.29 Retrofitting structures

Green's functions are also an ideal tool to find the zones in a structure where retrofit measures will be the most effective. The Vierendeel girder in Fig. 1.91 may serve as an introductory example. Which parts of the girder should be

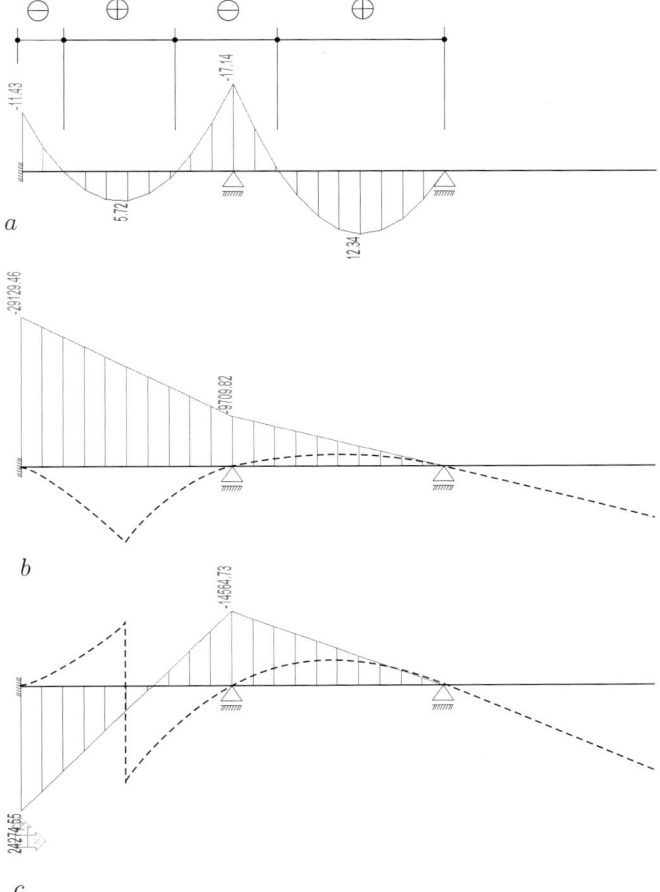

Fig. 1.92. Continuous beam, fixed on the left side **a)** moment $M(x)$ from distributed load **b)** moment $M_2(x)$ of the influence function for $M(0.5\,l)$ at the center of the first span **c)** moment $M_3(x)$ of influence function G_3 for the shear force $V(0.5\,l)$ at the same point; $\oplus$ and $\ominus$ indicate where ΔEI must be positive or negative if the bending moment $M(0.5\,l)$ is to be increased by a change $EI + \Delta EI$ in the stiffness

retrofitted, $EI \rightarrow EI + \Delta EI$, to reduce the deflection (at the bottom of the girder) due to a uniform load on the upper part of the girder?

The bending moment distribution caused by the load is plotted in Fig. 1.91 a and the bending moment of the Dirac delta is plotted in Fig. 1.91 b. According to the equation

$$J(e_w) = w_c(x) - w(x) = -\frac{\Delta EI}{EI} \int_{x_1}^{x_2} \frac{M_c\, M_0}{EI}\, dy \qquad M_0 = -EI\, \frac{d^2}{dy^2}\, G_0''$$

$$(1.395)$$

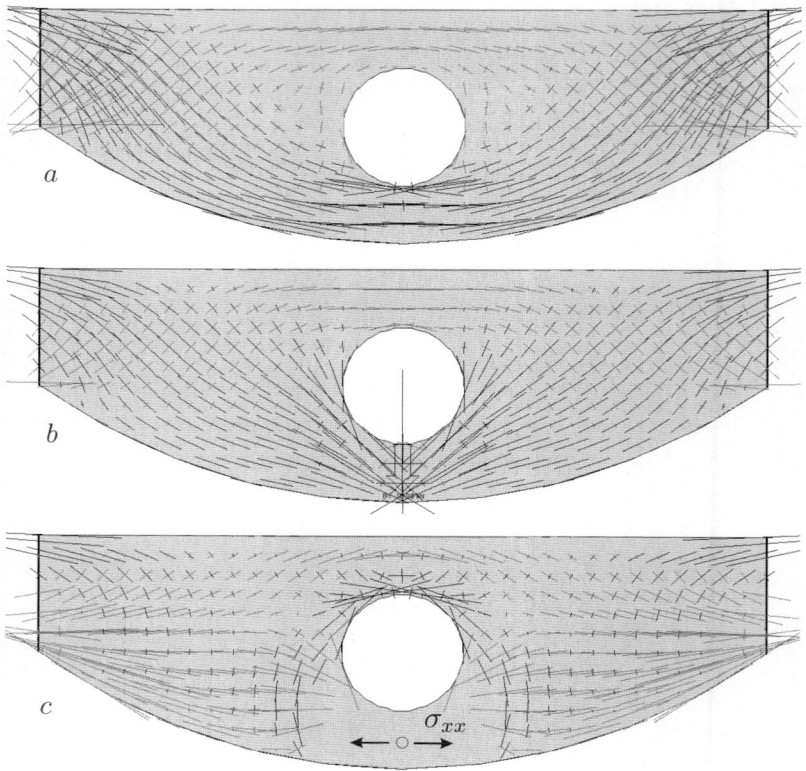

Fig. 1.93. Retrofitting a plate to reduce the vertical displacement and the horizontal stress σ_{xx} at the bottom of the plate. Depicted are the principal stresses **a)** of the original load case (gravity load), **b)** of the Dirac delta δ_0 and **c)** of the Dirac delta δ_1. The plate is fixed on both sides. In the last figure the stresses near the source point have been masked out because they would outshine all other stresses

the parts where both contributions are large should be retrofitted and these are obviously the joints of the frame; see Fig. 1.91 c. (As is customary we approximate M_c by the bending moment distribution M of the unmodified frame).

This result is typical for frame analysis. In a clamped one-span beam which carries a uniform load p the bending moment at mid-span is $p\,l^2/24$ while the bending moment at the ends is double that value, $p\,l^2/12$. So that retrofitting measures at the ends of frame elements will be more effective in general.

But note that the change has a also direction which depends on the sign of $M_c \times M_i$

$$J(w_c) - J(w) = -\frac{\Delta EI}{EI} \int_{x_1}^{x_2} \frac{M_c(y)\, M_i(y,x)}{EI}\, dy \qquad (1.396)$$

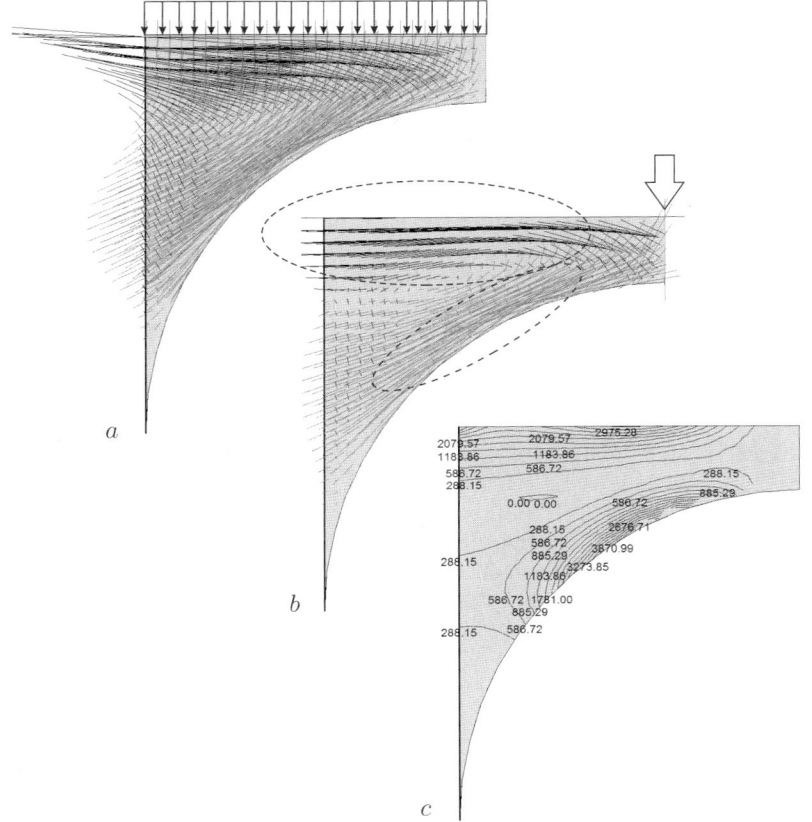

Fig. 1.94. Cantilever plate, fixed on the left side. To reduce the deflection the thickness t should be increased where indicated **a)** main stresses from edge load **b)** main stresses from point load (δ_0) **c)** contour lines of element strain energy product $a(\boldsymbol{G}_0, \boldsymbol{u})_{\Omega_e}$

Where the sign is negative an increase ΔEI will effect a positive change, $J(w_c) - J(w) > 0$ and vice versa.

In the case of the Vierendeel girder things are simple because the correlation between M and M_0 is very strong. In the case of the continuous beam in Fig. 1.92, which carries a uniform load in the first two spans, things are a bit more complicated. Plotted are the bending moment distribution M from the traffic load, Fig. 1.92 a, the moment M_2 of the influence function G_2 for the bending moment at the center of the first span, Fig. 1.92 b, and the moment M_3 of the influence function G_3 for the shear force $V(0.5\,l)$ at the same point, Fig. 1.92 c. Assume the goal is to increase the bending moment $M(0.5\,l)$ at the center of the first span

$$M_c(0.5\,l) - M(0.5\,l) = -\frac{\Delta EI}{EI} \int_{x_1}^{x_2} \frac{M_c(y)\,M_2(y, 0.5\,l)}{EI}\,dy \qquad (1.397)$$

then EI must be increased, $\Delta EI > 0$, where the product of $M_c \times M_2$ is negative and EI must be decreased, $\Delta EI < 0$, where $M_c \times M_2$ is positive. The symbols $\oplus$ and $\ominus$ respectively indicate these regions.

Though it is almost too trivial to mention: a change ΔEI in the last span, the cantilever beam, would effect nothing in the first span because the indicator function $M_2 = 0$ is zero in that part of the beam.

Note also that these predictions are based on the simplifying assumptions $G_i^c \sim G_i$ and therewith $M_i^c \sim M_i$. So if the change ΔEI becomes too large the indicator functions M_i may be too far off from the true M_i^c.

As a third example we consider the plate in Fig. 1.93. A local (Ω_e) change in the thickness $t \to t + \Delta t$ effects a change

$$J(e_u) = -\Delta t \int_{\Omega_e} \sigma_{ij}^G \cdot \varepsilon_{ij}^c \, d\Omega_{\boldsymbol{y}} \qquad (1.398)$$

in any quantity $J(\boldsymbol{u})$ of the plate in Fig. 1.93. Hence the regions where the strain energy product of the field $\boldsymbol{u} \sim \boldsymbol{u}^c$ and the Green's function is the largest are the most important. According to Fig. 1.93 these are the regions near the fixed edges and the bottom of the plate. Similar considerations hold true for the cantilever plate in Fig. 1.94 where the aim is the reduction of the deflection at the end of the plate.

To evaluate the strain energy product, (we let $\boldsymbol{u}^c = \boldsymbol{u}$),

$$\Delta t \int_{\Omega_e} \sigma_{ij}^G \cdot \varepsilon_{ij} \, d\Omega_{\boldsymbol{y}} = \Delta t \cdot \boldsymbol{f}^e \bullet \boldsymbol{u}_G \qquad (1.399)$$

in a single element Ω_e we could either use Gaussian quadrature or we could calculate the vector of element nodal forces $\boldsymbol{f}^e = \boldsymbol{K}^e \boldsymbol{u}^e$ and multiply this vector with the vector $\boldsymbol{u}_G$ of the nodal displacements of the Green's function.

In some cases the solution $\boldsymbol{u}$ itself is the Green's function for the output functional. Consider for example a cantilever plate to which an edge load $\boldsymbol{p} = \{0, 1\}^T$ is applied at the upper edge Γ_u so that

$$\boldsymbol{u} \in V : \; a(\boldsymbol{u}, \boldsymbol{v}) = \int_{\Gamma_u} \boldsymbol{p} \bullet \boldsymbol{u}\,ds = \int_{\Gamma_u} u_y\,ds \qquad \boldsymbol{v} \in V. \qquad (1.400)$$

If the output functional is the average value of the edge displacement

$$J(\boldsymbol{u}) = \int_{\Gamma_u} u_y\,ds \qquad (1.401)$$

then $J(\boldsymbol{u}) = a(\boldsymbol{u}, \boldsymbol{u})$ and so to reduce the deflection

$$J(e_u) = \int_{\Gamma_u} u_y^c\,ds - \int_{\Gamma_u} u_y\,ds = -\sum_e \Delta t \cdot a(\boldsymbol{u}, \boldsymbol{u})_{\Omega_e} \qquad (1.402)$$

$$A(x) = A_0 + A_1 x$$

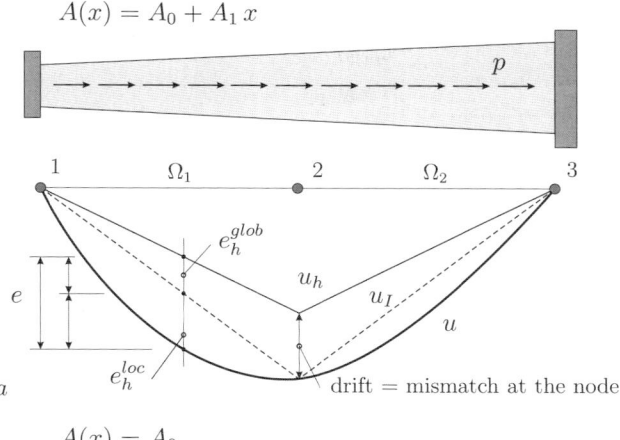

$$A(x) = A_0$$

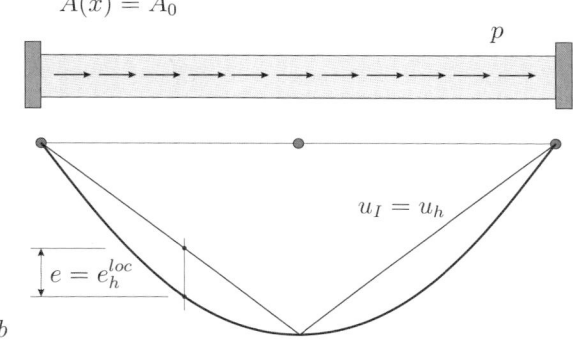

Fig. 1.95. Local error and global error in two shafts, u is the axial displacement **a)** the FE solution does not interpolate the exact solution at the nodes while in **b)** the drift is zero and so $e_h^{glob} = u_I - u_h = 0$

the thickness, $t \to t + \Delta t$, should be increased in all those elements Ω_ϵ where $a(\boldsymbol{u}, \boldsymbol{u})$ is large.

We can generalize this: the displacement field $\boldsymbol{u}$ of a structure is the Green's function for the *weighted* average—the load $\boldsymbol{p}$ is the weight—of the displacement field taken over the region Ω_p where the load $\boldsymbol{p}$ is applied

$$J(\boldsymbol{u}) = \int_{\Omega_p} \boldsymbol{u} \bullet \boldsymbol{p} \, d\Omega = \int_{\Omega_p} (u_x \cdot p_x + u_y \cdot p_y) \, d\Omega = a(\boldsymbol{u}, \boldsymbol{u}). \quad (1.403)$$

Because in this case the indicator function $\boldsymbol{G}$ is identical with $\boldsymbol{u}$, the correlation between $\boldsymbol{G} = \boldsymbol{u}$ and $\boldsymbol{u}$ is of course optimal. But also in the first two examples—the Vierendeel girder and the cantilever plate—the Green's function and the original displacement field are of similar type and so in these two cases $a(\boldsymbol{u}, \boldsymbol{u})$ would be nearly as good an indicator as $a(\boldsymbol{G}, \boldsymbol{u})$: simply where the stresses are large the thickness t should be increased.

1.30 Local errors and pollution

It seems intuitively clear that the error $e_h(x) = u(x) - u_h(x)$ of an FE solution can be split into a *local* and a *global* error:

$$e_h(x) = u(x) - u_I(x) + u_I(x) - u_h(x) = e_h^{loc}(x) + e_h^{glob}(x). \quad (1.404)$$

The local error is that part of the solution remaining after, say, a linear interpolation $u_I \in V_h$ and the *drift* of the element—the mismatch between u and u_h at the nodes—is the global error, s. Fig. 1.95,

$$e_h^{loc}(x) = u(x) - u_I(x) \qquad e_h^{glob}(x) = u_I(x) - u_h(x). \quad (1.405)$$

Closely related to this splitting of the error in two parts is the concept of *the local solution*. In 1-D problems the local solution u_h^{loc} is the function on any element Ω_e that minimizes the error in the strain energy of the element under the side condition that it agrees with the exact solution at the nodes of the element.

If the FE solution interpolates the exact solution at the nodes, then the FE solution u_h is also the local solution, $u_h = u_h^{loc}$, and the local error

$$e_h^{loc} = u - u_h^{loc} \quad (1.406)$$

within an element Ω_e (a bar element) simply the particular solution $-EA\,u_p'' = p$ if both sides of the element are fixed.

Because in standard 1-D problems the interpolating function is identical with the FE solution, $u_I(x) = u_h(x)$, the global error is zero while in 2-D and 3-D problems we observe a drift at the nodes. The drift in Fig. 1.95 a is due to the fact that the cross section $A = A_0 + A_1 \cdot x$ of the shaft changes, (see Chap. 3 p. 292).

In some 2-D and 3-D problems the exact nodal displacement is infinite, for example $u = \ln(\ln 1/r)$ at $r = 0$, (double the logarithm for u to have bounded strain energy $a(u, u) = (\nabla u, \nabla u)$) and so the solution cannot be interpolated at such a node that is u_I in (1.404) must be replaced by a slightly different function, [19], but for our purposes we may neglect these special cases.

Interest often focuses on the error of the solution in a certain patch Ω_p of the mesh, and then local and global may refer to contributions to the error from sources inside or outside the patch, respectively. And in this context the local error is also termed the *near-field error* and the global error is referred to as the *far-field error*.

If for example linear triangles are used in the FE analysis of a plate that carries an edge load only, then given any patch Ω_P we may consider the displacement field due to the line forces j_h within the patch (the jump in the traction vectors at interelement boundaries) the local error and the displacement field due to the line forces outside the patch the global error.

Or imagine a beam element with a local solution w^{loc} which produces the exact curvature in the beam, but which is saddled onto an FE solution with

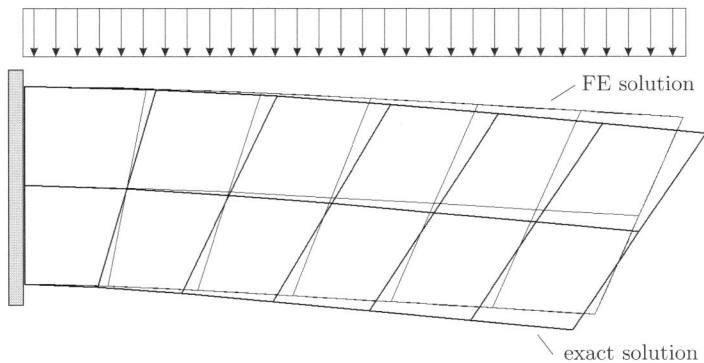

Fig. 1.96. Drift of an FE solution = mismatch at the nodes. Because the FE structure is too stiff the Dirac deltas δ_0 at the nodes (= influence functions for the nodal displacements) effect too little—that is the edge deflects too little—so that the nodes get the wrong message from the edge of how large the load really is and consequently also the nodes deflect less than necessary

a large drift at the nodes. This is the situation most often found in 2-D and 3-D FE analysis. Locally an FE solution fits relatively well because often the load is either only applied at the edge or in a small region of the problem domain Ω so that the FE solution—in most parts—must "only" approximate a homogeneous displacement field but the stress discontinuities between the elements produce a drift which spoils the picture; see Fig. 1.96.

Pollution

Hence we can give pollution a name, it is the drift at the nodes caused by the element residuals and jump terms on the element edges. Because we know that the dip (displacement) caused by a single force ebbs away as

$\ln r$	2-D elasticity	$\dfrac{1}{r}$	3-D elasticity
$r^2 \ln r$	point load, slab	$r \ln r$	moment, slab (1.407)

and because a one-point quadrature rule, $\boldsymbol{x}_p$ = center of Ω_p, $r = |\boldsymbol{x} - \boldsymbol{x}_p|$,

$$w(\boldsymbol{x}) = \int_{\Omega_P} \frac{1}{8\,\pi\,K}\,r^2\,\ln r\,p\,d\Omega\boldsymbol{y} \simeq \frac{1}{8\,\pi\,K}\,r^2\,\ln r \cdot p(\boldsymbol{x}_p) \cdot \Omega_p$$

(1.408)

will not change the picture much the disturbances introduced by *one* non-matching edge load alone and similar errors would ebb away rather quickly but the sheer multitude of these edge loads causes a noticeable drift. (For a more detailed picture see Sect. 1.32 p. 172).

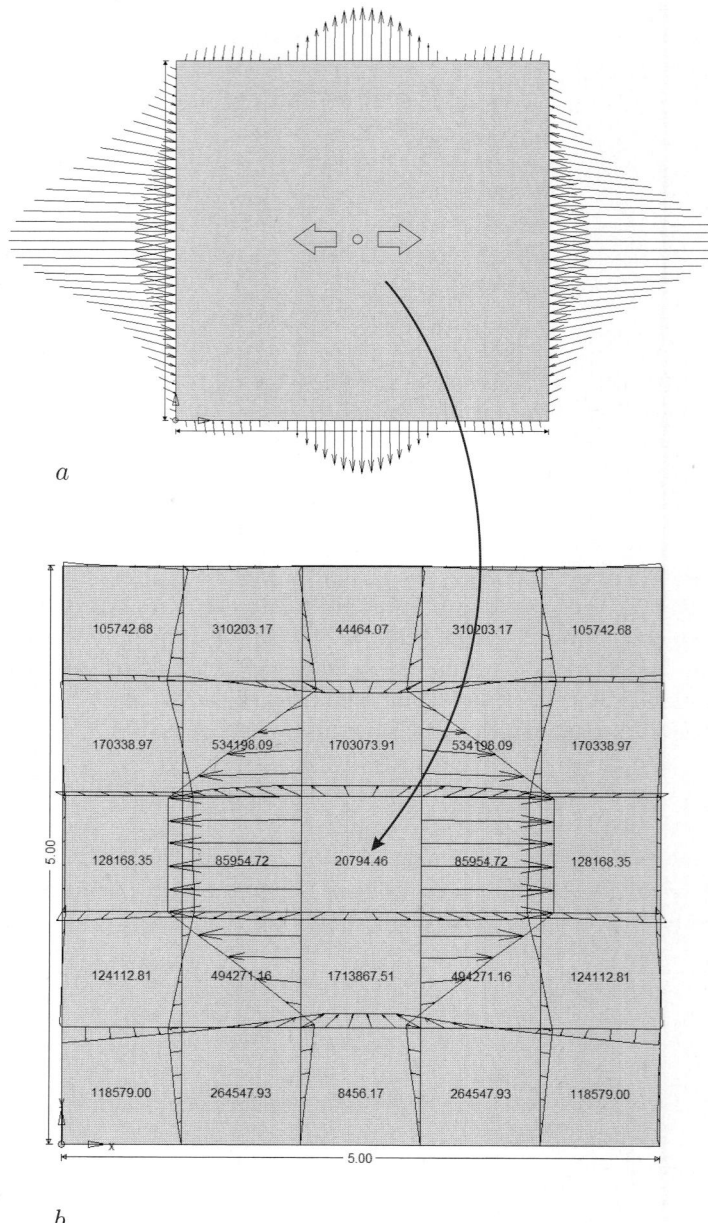

Fig. 1.97. FE influence function for $\sigma_x(\boldsymbol{x}_c)$ **a)** fixed end forces in a single element (BE solution) **b)** FE loads $\boldsymbol{p}_h$; these loads are equivalent to the Dirac delta $\boldsymbol{\delta}_1$ on V_h that is $(\boldsymbol{p}_h, \boldsymbol{u}_h) = (\boldsymbol{\delta}_1, \boldsymbol{u}_h) = \sigma_{xx}^h(\boldsymbol{x}_c)$; the single values are the resultants of the volume forces $((p_x, p_x) + (p_y, p_y))^{1/2}$ in each element

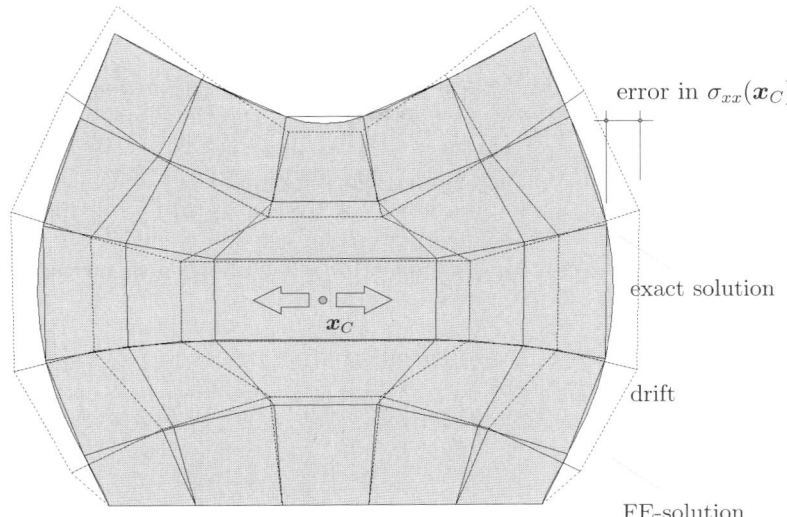

Fig. 1.98. The drift of the FE influence function for $\sigma_{xx}(\boldsymbol{x}_c)$ on the rectangular mesh in the previous Fig. 1.97. The "exact solution" is a BE solution. The displacements are greatly exaggerated

The drift of influence functions

Influence is measured in units of *work = force × displacement*. Where the displacement is the displacement in the structure due to the action of the Dirac delta δ_i, say δ_1 at the Gauss point $\boldsymbol{x}$ (= influence function for $\sigma_{xx}(\boldsymbol{x})$). If we get the displacement caused by the Dirac delta at the foot $\boldsymbol{y}$ of a point load wrong then the influence of the point load on the stress $\sigma_{xx}(\boldsymbol{x})$ at the Gauss point $\boldsymbol{x}$ will be wrong, that is $\sigma_{xx}^h \neq \sigma_{xx}$. It is as simple as that.

Imagine that we apply a dislocation $\delta_1 = 1$ in horizontal direction at the center $\boldsymbol{x}_C$ of a bilinear element (which is part of a larger structure) to calculate the influence function for the stress σ_{xx} at the center of the element; see Fig. 1.97. To be as accurate as possible we remove the element from the structure, generate a very fine mesh on this isolated element (we keep the edges fixed) and solve this load case "exactly"; see Fig. 1.97 a. Then we apply the fixed end actions to the edge of the cut-out in the opposite direction.

Because we cannot reproduce all the fine details of the fixed end actions on the edge of the cut-out there will be a mismatch between the original edge loads $\boldsymbol{t}$ and the FE edge loads $\boldsymbol{t}_h$. If we neglect for a moment the element residuals and jumps in the stresses of the FE solution at elements farther away then these "parasitic" edge loads $\boldsymbol{r}_h = \boldsymbol{t} - \boldsymbol{t}_h$ are (mainly) responsible for the drift at the nodes, s. Fig. 1.98.

The magnitude of the drift is exactly equal to the magnitude of the error in the stress σ_{xx} at the center of the element when we apply a point load at any of the nodes. And in this example the drift is negative at all nodes so that in the FE model the stress σ_{xx}^h at the center is too large.

Things become really bad if parts of a structure can perform nearly rigid body motions because then even a small erroneous rotation can cause large drifts.

Recall the problem of the cantilever plate in Fig. 1.60, p. 89, where the drift of the FE solution, -2.5 m - (-1.66 m) = -0.84 m, is nearly one meter or three yards! A purely local error analysis in, say, the last element Ω_e would probably yield only a small L_2 residual (the integral of the square of the errors)

$$||\boldsymbol{p} - \boldsymbol{p}_h||_{\Omega_e}^2 + ||\boldsymbol{t} - \boldsymbol{t}_h||_{\Gamma_e}^2 \tag{1.409}$$

because at the far end the real plate and the FE plate essentially only perform rotations, $\boldsymbol{u} = \boldsymbol{a} + \boldsymbol{x} \times \boldsymbol{\omega}$, which are stress free so that $\boldsymbol{p} \simeq \boldsymbol{p}_h \simeq \boldsymbol{0}$ and also $\boldsymbol{t} \simeq \boldsymbol{t}_h \simeq \boldsymbol{0}$ and so we would be led to believe that the error is small—while the opposite is true.

By adding elements to the plate that is by extending the plate in horizontal direction we could even make the error (= the vertical displacements at the end nodes) in the FE solution arbitrarily large and at the same time the residual (1.409) in the last element arbitrarily small. A truly paradoxical situation—we would easily win any contest "for the worst of all FE solutions", [18].

Okay, measuring only the residual in the last element is not fair. The real estimate for the displacement error at the end nodes is the inequality

$$|u_y(\boldsymbol{x}) - u_y^h(\boldsymbol{x})| \le ||\boldsymbol{G}_M - \boldsymbol{G}_M^h||_E \cdot ||\boldsymbol{G}_0||_E \tag{1.410}$$

—in a beam problem this equation would read

$$|w(x) - w_h(x)| \le \left[\int_0^l EI\,(G_M'' - (G_M^h)'')^2\,dx\right]^{1/2} \cdot \left[\int_0^l EI\,(G_0'')^2\,dx\right]^{1/2} \tag{1.411}$$

—where the field $\boldsymbol{G}_M$ is the influence function for the bending moment—it effects a rotation of the cross section by $\tan \varphi = 1$—while $\boldsymbol{G}_0$ is the influence function for the vertical displacement at the end nodes, that is if a point load $P = 1$ is applied at one of the end nodes[12].

The inequality (1.410) is based on

$$u_y(\boldsymbol{x}) - u_y^h(\boldsymbol{x}) = \int_\Omega [\boldsymbol{G}_0(\boldsymbol{y}, \boldsymbol{x}) \bullet (\boldsymbol{\delta}_M - \boldsymbol{\delta}_M^h)]\,d\Omega_{\boldsymbol{y}} = a(\boldsymbol{G}_0, \boldsymbol{G}_M - \boldsymbol{G}_M^h) \tag{1.412}$$

---

[12] We tacitly assume that the fields $\boldsymbol{G}_M$ and $\boldsymbol{G}_0$ have finite strain energies, see the remark on p. 51

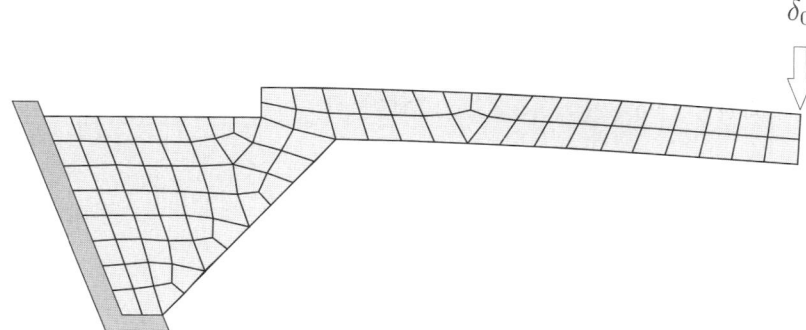

Fig. 1.99. The more the structure extends outward the larger the displacement error gets at the far end—in practically all load cases

where the "Dirac delta" $\boldsymbol{\delta}_M$ is the action that effects the rotation $\tan \varphi = 1$ and $\boldsymbol{\delta}_M^h$ is its FE approximation. So the Dirac deltas are loads, $(\boldsymbol{\delta}_M - \boldsymbol{\delta}_M^h) \equiv (\boldsymbol{p} - \boldsymbol{p}_h)$. The fields $\boldsymbol{G}_M$ and $\boldsymbol{G}_M^h$ respectively are the response of the structure to these actions.

The integral

$$\int_\Omega \boldsymbol{G}_0(\boldsymbol{y}, \boldsymbol{x}) \bullet \boldsymbol{\delta}_M \, d\Omega_{\boldsymbol{y}} \qquad (= -2.5\,\text{kNm}) \qquad (1.413)$$

which is the lift $(u_y(\boldsymbol{x}) \cdot 1\,\text{kN} = \text{work})$ produced at the far end by the Dirac delta $\boldsymbol{\delta}_M$ is according to Betti's theorem equal to the bending moment M in cross section $A - A$ (see Fig. 1.60) produced by the point load $\boldsymbol{\delta}_0$ $(P = 1)$ sitting at the far end of the plate and

$$\int_\Omega \boldsymbol{G}_0(\boldsymbol{y}, \boldsymbol{x}) \bullet \boldsymbol{\delta}_M^h \, d\Omega_{\boldsymbol{y}} \qquad (= -1.66\,\text{kNm}) \qquad (1.414)$$

is an approximation of this integral—with a relative large error.

After applying the Cauchy-Schwarz inequality to (1.412) Equ. (1.410) follows directly

$$|a(\boldsymbol{G}_0, \boldsymbol{G}_M - \boldsymbol{G}_M^h)| \le a(\boldsymbol{G}_0, \boldsymbol{G}_0)^{1/2} \cdot a(\boldsymbol{G}_M - \boldsymbol{G}_M^h, \boldsymbol{G}_M - \boldsymbol{G}_M^h)^{1/2}$$
$$= ||\boldsymbol{G}_0||_E \cdot ||\boldsymbol{G}_M - \boldsymbol{G}_M^h||_E . \qquad (1.415)$$

Now it is evident what happens: with each element that we add to the plate the lever arm of the point load $P = 1$ increases that is with each element the maximum stress of the field $\boldsymbol{G}_0$ will increase and therewith the strain energy and the energy norm

$$||\boldsymbol{G}_0||_E = a(\boldsymbol{G}_0, \boldsymbol{G}_0)^{1/2} = \left\{ \int_\Omega [\sigma_{xx} \cdot \varepsilon_{xx} + 2\,\sigma_{xy} \cdot \varepsilon_{xy} + \sigma_{yy} \cdot \varepsilon_{yy}] \, d\Omega \right\}^{1/2} .$$
$$(1.416)$$

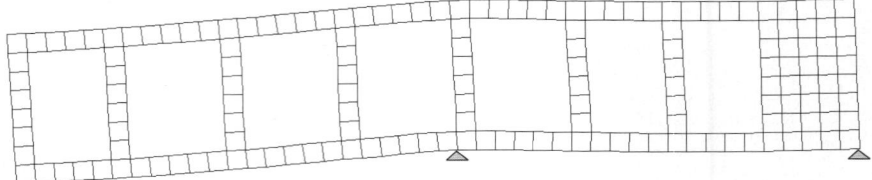

Fig. 1.100. Gravity load in a Vierendeel girder. In structures with large overhanging parts the results must be checked very carefully

At the same time the energy of the error, that is $||\boldsymbol{G}_M - \boldsymbol{G}_M^h||_E$, remains the same because the elements added are stress free in the load case $\tan \varphi = 1$—they will only perform rigid body rotations. Hence the estimate (1.410) will deteriorate with each element that we add to the plate.

So if a part of a structure can perform large rigid body motions (or very nearly such motions) then it pays for this "liberty" with large lever arms for the Dirac delta δ_0 which automatically deteriorate the estimates for the nodal displacements (see Fig. 1.99) which means that also the accuracy of the numerical influence functions suffers.

But we must also keep an eye on the equilibrium conditions in such structures. The left part of the Vierendeel girder in Fig. 1.100 behaves like a cantilever beam so that the displacements in that part will be relatively large. Hence the FE results should be checked very carefully. Because the equilibrium conditions, which play an important role here, are not guaranteed—rather the error probably will be quite pronounced—a frame analysis with beam elements would be a much better choice.

In 1-D problems only loads which act on elements through which the cut passes contribute wrongly to the equilibrium conditions (see Fig. 1.75 p. 105) so the possible error is much smaller than in 2-D analysis. Commercial FE codes eliminate also this small error by adding the fixed end actions to the FE results and so the equilibrium conditions are satisfied exactly.

Pollution due to singularities on the boundary

The drift, so far, stems from the mismatch on the right-hand side, $\boldsymbol{p} - \boldsymbol{p}_h$. An additional source are singular points on the boundary. At such points the exact solution, for example,

$$u(r, \varphi) = k\, r^{0.5} f(\varphi) + \text{smooth terms} \qquad k = \text{stress intensity factor}$$

$$(1.417)$$

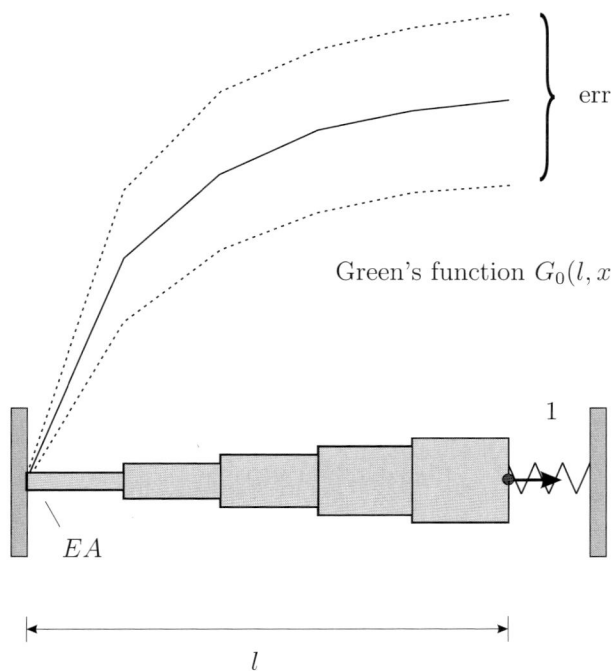

error

Green's function $G_0(l, x)$

EA

1

l

Fig. 1.101. The slope of the Green's function for the end displacement $u(l)$ is proportional to $1/EA$

typically sharply drops to zero as in Fig. 1.122 a, p. 175 (or any other value, if the edge with the corner point itself is displaced), so that the steep slope, $\varepsilon = r^{-0.5}$, leads to infinite stresses at the corner point.

Why such a corner singularity has a negative influence on the accuracy of the solution in *other regions* of the mesh is best illustrated by the one-dimensional example of a stepped bar; see Fig. 1.101. The left end of the bar is fixed, and the other end abuts a spring with a certain stiffness α_2. The Green's function for the end displacement $u(l)$ is displayed in Fig. 1.101. Note that the slope in the Green's function is proportional to $1/EA$. Because of the steep slope at the beginning, a slight change or error in the stiffness EA_1 of the first bar element Ω_1 will lead to a rather large error at the other end of the bar.

This problem was studied by Babuška and Strouboulis [19] where it was assumed that the stiffness of the bar varies according to the rule

$$E\,A(x) = E\,x^{\vartheta} \quad 0 < \vartheta = 0.75 < 1 \quad E = 1, \quad \alpha_2 = \frac{5}{2}, \quad l = 1 \quad (1.418)$$

which gives the bar the shape in Fig. 1.102. The first bar element is now infinitely thin and infinitely short, so to speak, because the stiffness has a zero at $x = 0$. Note that the normal stress $\sigma = P/A$ becomes infinite at $x = 0$, but that the stress resultant $N = \sigma A = P = 1$ remains bounded.

$EA = x^{0.75}$

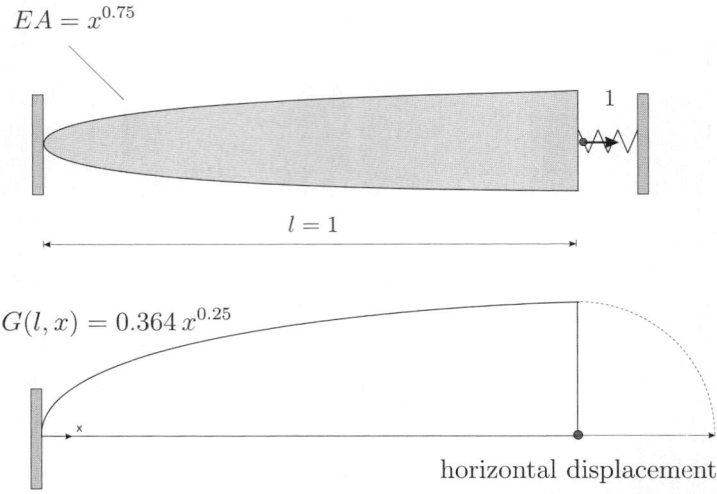

$G(l, x) = 0.364\, x^{0.25}$

horizontal displacement

Fig. 1.102. Model problem with $EA = x^{0.75}$. The Green's function for $u(l)$ is proportional to $x^{0.25}$, and therefore the gradient has a singularity at $x = 0$. The drawings are not to scale, [19] p. 8

The Green's function for the end displacement $u(l) = u(1)$ is $G(l, x) = 0.364\, x^{0.25}$ (see Fig. 1.102 b). Its slope is infinite at $x = 0$.

The local and the global error of a piecewise linear FE approximation of the Green's function on a uniform mesh is displayed in Fig. 1.103 (The annotations in the figure are due to the present authors). Only in the first element is the local error larger than the pollution error. This is the important observation and it directly contradicts St. Venant's principle, which seemingly guarantees that given a large enough distance from the disturbance all negative effects vanish. *This is not true if pollution is a problem.* St. Venant's principle applies so to speak only if the FE solution is close to the exact solution. But what is displayed here is the numerical solution of a singular problem.

Note that averaging techniques would scarcely address this problem, because the global error that dominates the problem is relatively smooth. The L_2-projections would only have an effect on the small local error; see Fig. 1.103 b. But if the "ground wave" is large, it will not help much to smooth out the wrinkles.

Or to cite Babuška and Strouboulis, [19] p. 278,

- *'Nevertheless any local refinements of the mesh in an area of interest reduce only the local error, and the magnitude of the pollution error is determined by the density of the mesh in the area of interest compared with the density of the mesh in the rest of the domain. Thus, unless special care is taken to design the global mesh and the local meshes to balance the pollution and local error in the area of interest, the pollution error may be the dominant component of the error and, if this is the case, only minimal gains*

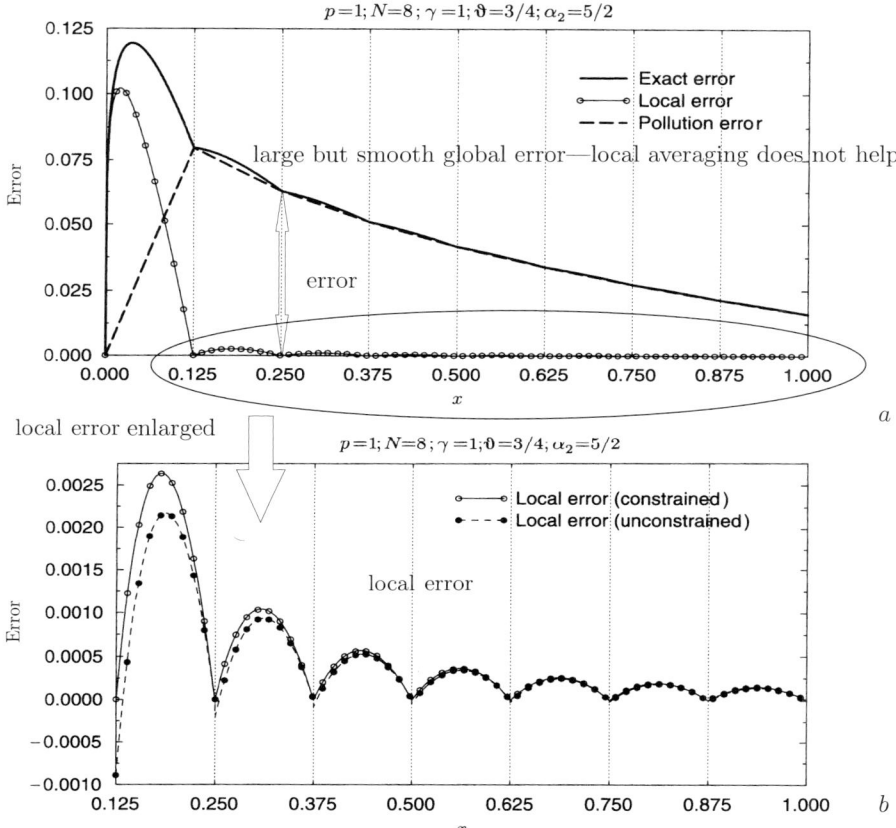

Fig. 1.103. FE results with linear elements ([19], p. 275)

in accuracy may be achieved by local refinements. The justification of the global/local or zooming approaches used in engineering is based on an intuitive understanding of Saint Venant's principle for the error. The error is the exact solution of the model problem loaded by the residuals in the finite element solution ... and since the residuals are oscillatory, it is argued that they influence the error only locally and hence the pollution error cannot [be] significant. This intuitive understanding of Saint Venant's principle could be misleading, and a quantitative analysis is needed.' [end of quote]

Other possible sources of pollution are

- sudden changes in the right-hand side p, the applied load (a mild problem in statics)

- discontinuities in the coefficients of the differential equation when, for example, the stiffness changes—this produces a kink in the Green's function; see Fig. 1.101.
- a non-uniform mesh, i.e., a very large element followed by a series of very small elements, may also cause pollution in the numerical solution. The small elements never recover from the gross error—the drift—they inherit from the large element.

Because our problems are usually set in domains with corners and changing boundary conditions, one simply must expect the solution to have singularities, and therefore most often these disturbances aggravate the pollution additionally. The pollution error can be successfully controlled by refining the mesh in the neighborhood of the singularities. But because *one cannot detect pollution by any local analysis* adaptive refinement must employ energy error measures for the *whole* mesh.

Each node and each Gauss point is—implicitly—a source of pollution

On a given mesh we are not just finding the equilibrium position of the structure but implicitly we solve n additional load cases δ_i for the n Green's functions of the n values which the program outputs at the nodes and Gauss points. Most often the singularities of these Green's functions are *much stronger* than the singularities at the corner points. That is pollution is a major problem for the numerical Green's functions and the source of these adverse effects are not primarily the corner points but the innocent looking nodes and Gauss points of the mesh.

Verification and validation

- Verification asks whether the equations were solved correctly; validation asks whether the right equations were solved.

As these examples demonstrate, any error in the coefficients of the governing equations (the elastic constants) will cause a drift, a global error, in the FE solution. Simple examples of this phenomenon are displayed in Fig. 1.104. A change in the elastic constants EA, EI or the spring stiffness c_φ will directly affect the solutions:

$$u(l) = \frac{P\,l}{EA} \qquad w(l) = \frac{P\,l^2}{c_\varphi} + \frac{P\,l^3}{3\,EI}\,. \tag{1.419}$$

Note that the error in the internal actions, $N = \sigma\,A$, V, and M is zero because the structures are statically determinate. If it were otherwise changes in the elastic constants would most often also lead to changes in the internal actions.

From an engineering standpoint, the choice of a correct model is at least as important as an asymptotic error analysis. How sensitive is a solution to

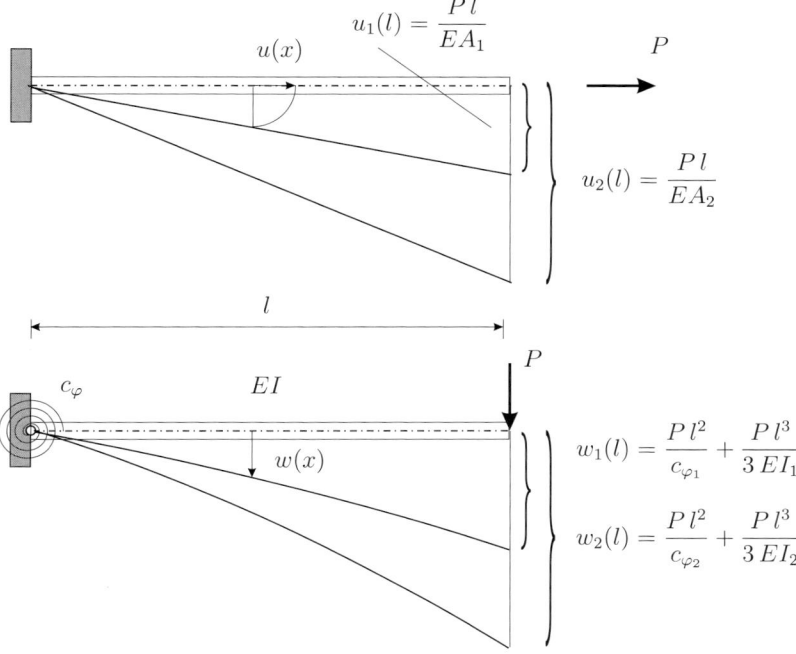

Fig. 1.104. The solutions depend on the parameters of the model

the modeling assumptions? How do the Green's functions change if the elastic constants are altered?

Model deficiencies can only be detected by measuring the response of the actual structure and comparing it to the computed results. This process is called *parameter identification*. Such a calibration can be done, for example, by studying the eigenfrequencies of a slab. Any deviation between the first three or four eigenfrequencies of the FE model and the slab is an indication of a modeling error. The problem the analyst faces is that FE models can be very complex, and the sheer number of parameters that might possibly affect the solution make it difficult to identify those parameters to which the eigenfrequencies are most sensitive.

1.31 Adaptive methods

In a plate the residual forces $\boldsymbol{r} = \boldsymbol{p} - \boldsymbol{p}_h$ and the jumps $\boldsymbol{t}_h^{\Delta}$ of the traction vector at interelement boundaries are assumed to be an appropriate measure of the quality of an FE solution; see Fig. 1.105. Where these *a posteriori error indicators* are large, the size of the elements is diminished or the order of the polynomial shape functions is increased. Repeating this loop for many cycles,

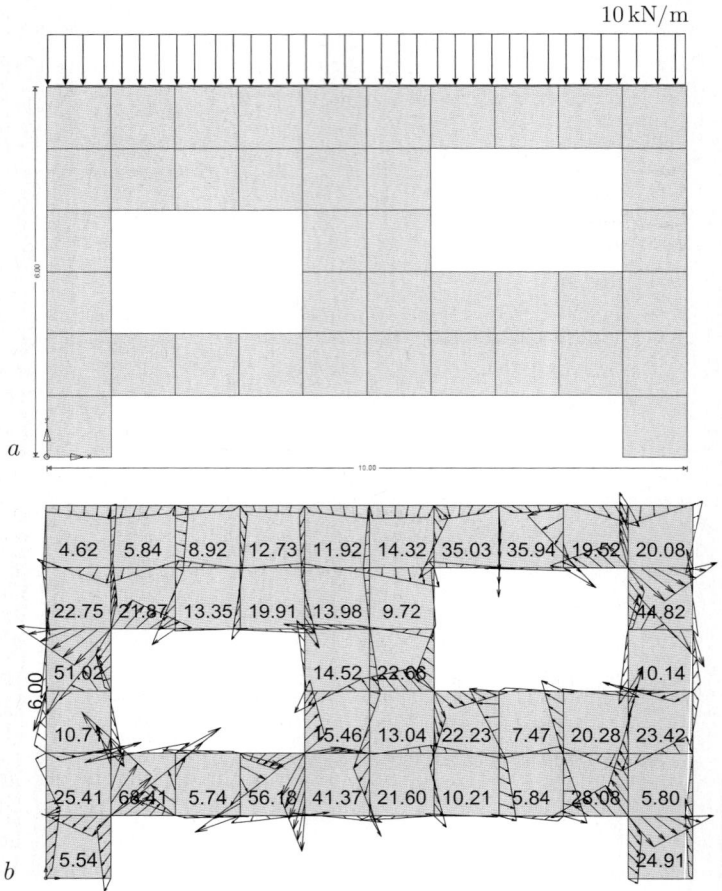

Fig. 1.105. Plate analysis with bilinear elements. **a)** System and load case p (edge load); **b)** load case p_h; the numbers are the resulting volume forces (kN)

one hopes the solution improves; see Fig. 1.106. This is the idea of adaptive methods.

Although this whole process seems rather straightforward, it is not quite clear how to weight the various errors. Would it suffice to look only at the residual forces r_i and neglect the jumps t_h^Δ, or do the jumps carry the message? In a beam which is subjected to a uniform load p, the nodal forces $f_k = p\,l/(n+1)$ (shear force discontinuities) get smaller and smaller the more elements n are used, but because $p_h = 0$ the element residuals remain the same throughout the refinement, see Fig. 1.107,

$$\int_0^l (p-0)^2 \, dx = ||p||_0^2, \tag{1.420}$$

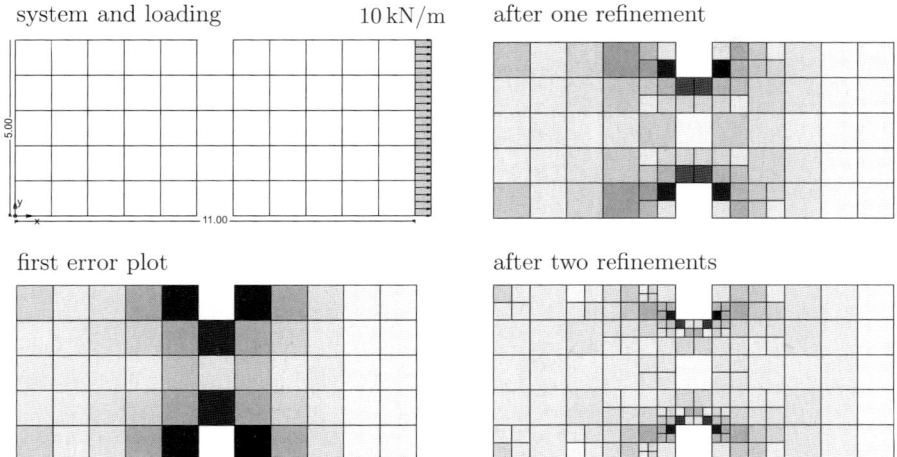

system and loading   10 kN/m   after one refinement

first error plot   after two refinements

Fig. 1.106. Adaptive refinement

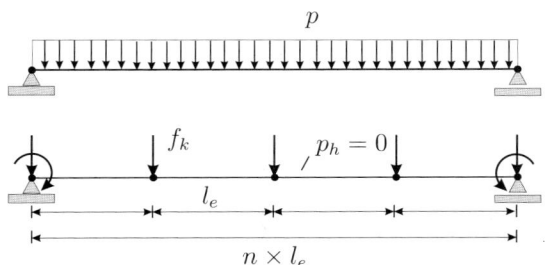

Fig. 1.107. The residual forces in each element remain the same, although the FE solution converges to the exact solution for $n \to \infty$

although we are convinced that the FE solution w_h tends to the exact solution w. This is an indication that in this example the jump terms—which tend to zero as the mesh is refined—are a more sensitive error indicator than the element residuals. In a beam, the jump terms are the discontinuities in the second- and third-order derivatives, $f_i = M(x_i^+) - M(x_i^-)$ and $f_j = V(x_j^+) - V(x_j^-)$, while the element residuals measure the discrepancies in the fourth-order derivative $r = EI\, w^{IV} - EI\, w_h^{IV}$.

The relevance of the two errors depends on the degree p of the shape functions. For even order, $p = 2, 4, \ldots$, the residual r dominates while for odd order, $p = 1, 3, \ldots$, the jump terms dominate (*"dichotomy"*), see [19] p. 424.

Plate analysis

Following these heuristic remarks, we now concentrate on the adaptive refinement of plates for the details. The model problem is a plate which is subjected to volume forces $\boldsymbol{p}$ and to traction boundary conditions $\boldsymbol{t}$ on the part $\varGamma_N$ of the boundary $\varGamma = \varGamma_D + \varGamma_N$. The boundary value problem consists of finding the solution $\boldsymbol{u} = [u_1, u_2]^T$ of the system

$$- \boldsymbol{L}\boldsymbol{u} := -\mu\Delta\boldsymbol{u} - \frac{\mu}{1 - 2\nu}\nabla\operatorname{div}\boldsymbol{u} = \boldsymbol{p} \tag{1.421}$$

where $\boldsymbol{u}$ is subject to the following boundary conditions

$$\boldsymbol{u} = \boldsymbol{0} \quad \text{on } \Gamma_D, \qquad \boldsymbol{Sn} = \bar{\boldsymbol{t}} \quad \text{on } \Gamma_N. \tag{1.422}$$

Let $\Omega = \sum_i \Omega_i$ be a suitable partition of Ω into finite elements Ω_i with diameters h_i and with edges Γ_i. The FE solution $\boldsymbol{u}_h \in V_h$ is the solution of the variational problem

$$a(\boldsymbol{u}_h, \boldsymbol{v}_h) = \boldsymbol{p}(\boldsymbol{v}_h) := \int_\Omega \boldsymbol{p} \bullet \boldsymbol{v}_h \, d\Omega + \int_{\Gamma_N} \bar{\boldsymbol{t}} \bullet \boldsymbol{v}_h \, ds \quad \text{for all } \boldsymbol{v}_h \in V_h. \tag{1.423}$$

Evidently the error $\boldsymbol{e} = \boldsymbol{u} - \boldsymbol{u}_h$ satisfies on V the equation

$$a(\boldsymbol{e}, \boldsymbol{v}) = \boldsymbol{p}(\boldsymbol{v}) - a(\boldsymbol{u}_h, \boldsymbol{v}) = \boldsymbol{p}(\boldsymbol{v}) - \boldsymbol{p}_h(\boldsymbol{v}) =: \boldsymbol{r}(\boldsymbol{v}) \quad \boldsymbol{v} \in V \tag{1.424}$$

with

$$\boldsymbol{r}(\boldsymbol{v}) = \sum_{\Omega_i} \left\{ \int_{\Omega_i} \boldsymbol{r} \cdot \boldsymbol{v} \, d\Omega + \int_{\Gamma_i} \boldsymbol{j} \cdot \boldsymbol{v} \, ds \right\} \tag{1.425}$$

where $\boldsymbol{r}$ are the element residual forces,

$$\boldsymbol{r} := \boldsymbol{p} - \boldsymbol{p}_h = \boldsymbol{p} + \boldsymbol{L}\boldsymbol{u}_h \quad \text{on } \Omega_i \tag{1.426}$$

and $\boldsymbol{j}$ are the (evenly distributed) jumps of the tractions at the element boundaries

$$\boldsymbol{j} = \begin{cases} \frac{1}{2}(\boldsymbol{S}_h^+ \boldsymbol{n}^+ + \boldsymbol{S}_h^- \boldsymbol{n}^-) & \text{on } \Gamma_i \not\subseteq \Gamma \\ \bar{\boldsymbol{t}} - \boldsymbol{S}_h \boldsymbol{n} & \text{on } \Gamma_N \\ \boldsymbol{0} & \text{on } \Gamma_D. \end{cases} \tag{1.427}$$

Recall that for test functions $\boldsymbol{v}_h \in V_h$ the fundamental Galerkin orthogonality

$$a(\boldsymbol{e}, \boldsymbol{v}_h) = \boldsymbol{r}(\boldsymbol{v}_h) = 0 \quad \boldsymbol{v}_h \in V_h \tag{1.428}$$

holds. The single error terms in an element

$$||\boldsymbol{r}_i||_0^2 = \int_{\Omega_i} \boldsymbol{r} \bullet \boldsymbol{r} \, d\Omega \qquad ||\boldsymbol{j}_i||_0^2 = \int_{\Gamma_i} \boldsymbol{j} \bullet \boldsymbol{j} \, ds \tag{1.429}$$

are weighted with h_i^2 and h_i respectively to produce a resulting error term

$$\eta_i^2 := h_i^2 \, ||\boldsymbol{r}_i||_0^2 + h_i \, ||\boldsymbol{j}_i||_0^2. \tag{1.430}$$

Under suitable assumptions this measure η_i^2 is an upper bound for the strain energy of the field $\boldsymbol{e}$ in a single element Ω_i (see Sect. 7.11, p. 565)

$$||\boldsymbol{e}_i||_E^2 := a(\boldsymbol{e}, \boldsymbol{e})_{\Omega_i} \le c\,\eta_i^2 \tag{1.431}$$

where c is an unknown constant. Evidently

$$||\boldsymbol{e}||_E^2 = a(\boldsymbol{e}, \boldsymbol{e}) \le c\,\eta^2 := c\sum_i \eta_i^2 \tag{1.432}$$

and so the ratio between η and the energy norm of the exact solution $\boldsymbol{u}$ defines a global error indicator

$$\eta^{rel} := \frac{\eta}{||\boldsymbol{u}||_E}\,. \tag{1.433}$$

Because of $\boldsymbol{u} = \boldsymbol{u}_h + \boldsymbol{e}$ and the Galerkin orthogonality, $a(\boldsymbol{u}_h, \boldsymbol{e}) = 0$, we have, (see Sect. 7.12, p. 568, and the "Pythagorean theorem")

$$\begin{aligned} ||\boldsymbol{u}||_E^2 &= a(\boldsymbol{u}_h + \boldsymbol{e}, \boldsymbol{u}_h + \boldsymbol{e}) = a(\boldsymbol{u}_h, \boldsymbol{u}_h) + 2\,a(\boldsymbol{u}_h, \boldsymbol{e}) + a(\boldsymbol{e}, \boldsymbol{e}) \\ &= a(\boldsymbol{u}_h, \boldsymbol{u}_h) + a(\boldsymbol{e}, \boldsymbol{e}) = ||\boldsymbol{u}_h||_E^2 + ||\boldsymbol{e}||_E^2 \end{aligned} \tag{1.434}$$

so that

$$\eta^{rel} = \frac{\eta}{\sqrt{||\boldsymbol{u}_h||_E^2 + ||\boldsymbol{e}||_E^2}}\,. \tag{1.435}$$

The relative error in a single element is defined to be the ratio between η_i and the energy norm $||\boldsymbol{u}_i||_E := a(\boldsymbol{u}, \boldsymbol{u})_{\Omega_i}^{1/2}$ of the exact solution in that element:

$$\eta_i^{rel} := \frac{\eta_i}{||\boldsymbol{u}_i||_E}\,. \tag{1.436}$$

Locally the Galerkin orthogonality must not be true, $a(\boldsymbol{e}, \boldsymbol{u}_h)_{\Omega_i} \ne 0$, so that

$$\bar{\eta}_i^{rel} = \frac{\eta_i}{\sqrt{||\boldsymbol{u}_{hi}||_E^2 + ||\boldsymbol{e}_i||_E^2}} \tag{1.437}$$

is not the same as (1.436), but it is an obvious approximation and to make it computable $||\boldsymbol{e}_i||_E^2$ is replaced by η_i^2. Hence the idea is to refine all those elements where the (so-modified) relative error

$$\tilde{\eta}_i^{rel} = \frac{\eta_i}{\sqrt{||\boldsymbol{u}_{h_i}||_E^2 + \eta_i^2}} \tag{1.438}$$

exceeds a certain threshold value η_0 [145]. Normally only the first 30% of all elements on the list are refined; otherwise the problematic zones will not crystallize so well. Often the element residual forces $\boldsymbol{r}$ are also neglected because the contributions of the jump terms are more significant.

How the energy error is typically distributed can be seen in Fig. 1.108, where dark regions symbolize high errors and light regions small errors in the strain energy. The critical regions are simply those with stress concentrations, so that eventually an engineer can easily predict where the program will refine the mesh.

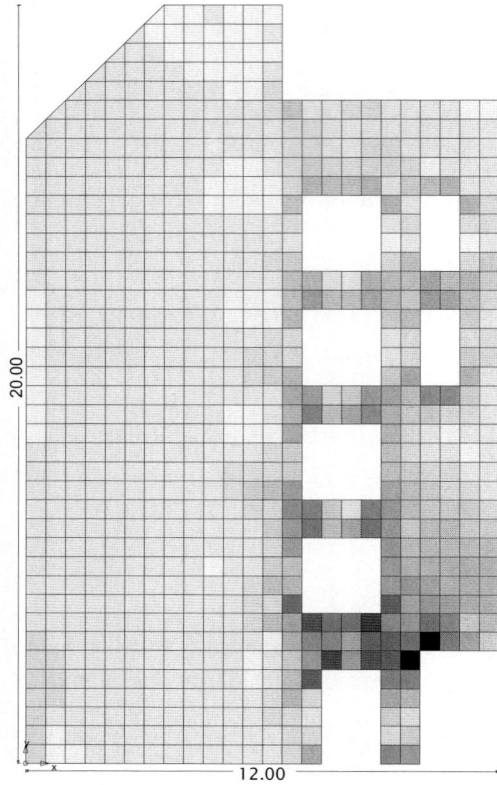

Fig. 1.108. Distribution of error indicators in a plate (gravity load)

L_2-projections

Another popular way to quantify the error is to compare the raw FE stresses with averaged stresses. This is known as the method of Zienkiewicz and Zhu or simply as Z^2 [261].

The averaged stresses are written in terms of the shape functions ψ_i,

$$\bar{\boldsymbol{\sigma}}_h(\boldsymbol{x}) = \sum_i^n \bar{\boldsymbol{\sigma}}_i\,\psi_i(\boldsymbol{x}) \quad \text{or} \quad \begin{bmatrix} \bar{\sigma}_{xx}^h(\boldsymbol{x}) \\ \bar{\sigma}_{yy}^h(\boldsymbol{x}) \\ \bar{\sigma}_{xy}^h(\boldsymbol{x}) \end{bmatrix} = \sum_i^n \begin{bmatrix} \bar{\sigma}_{xx_i} \\ \bar{\sigma}_{yy_i} \\ \bar{\sigma}_{xy_i} \end{bmatrix} \psi_i(\boldsymbol{x}) \quad (1.439)$$

and the nodal values $\bar{\sigma}_{xx_i}$ (analogously for $\bar{\sigma}_{yy_i}, \bar{\sigma}_{xy_i}$) are determined in such a way that the square of the error $||\sigma_{xx}^h - \bar{\sigma}_{xx}||_0^2$ is minimized (L_2-projection), which leads to the system of equations

$$\sum_j \int_\Omega \psi_i\,\psi_j\,d\Omega\,\bar{\sigma}_{xx_j} = \int_\Omega \psi_i\,\sigma_{xx}^h\,d\Omega, \qquad i = 1, 2, \ldots n. \quad (1.440)$$

Often this system is only solved approximately, by diagonalizing the mass matrix. The energy error in each element is then

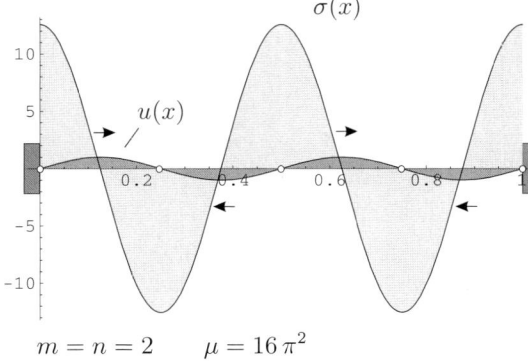

$m = n = 2$ $\mu = 16\,\pi^2$

Fig. 1.109. Oscillating longitudinal displacements produce oscillating stresses

$$\eta_i^2 = \int_{\Omega_i} (\bar{\boldsymbol{\sigma}}_{h_i} - \boldsymbol{\sigma}_{h_i}) \bullet (\bar{\boldsymbol{\varepsilon}}_{h_i} - \boldsymbol{\varepsilon}_{h_i})\, d\Omega\,. \tag{1.441}$$

The method can be improved by interpolating not at the nodes, but at the superconvergent points [262], [263]. In many cases this method is very successful.

The weak point of the Z^2-method though is the implicit assumption that

- oscillations indicate errors
- smooth stresses mean accurate stresses

and so the method breaks down if the false stresses are smooth as in Fig. 1.103, p. 136, where pollution produced a large but smooth error in the stresses.

In the following problem [2]

$$- EA\,u''(x) = \mu\,\sin(2^m\,\pi\,x) \qquad u(0) = u(1) = 0\,, \quad m > 0\,, \tag{1.442}$$

a bar is stretched and compressed by rapidly oscillating longitudinal forces having an amplitude μ. If this problem is solved on a uniform mesh of $2^n, n \leq m$, linear elements so that the nodes are located at the points

$$x_k = \frac{k}{2^n} \qquad k = 0, 1, \ldots, 2^n\,, \tag{1.443}$$

the piecewise linear FE solution is $u_h = 0$, because we know that it interpolates the exact solution (we let $EA = 1$)

$$u(x) = \frac{\mu}{4^m\,\pi^2}\,\sin(2^m\,\pi\,x)\,, \quad \sin(2^m\,\pi\,x_k) = \sin(2^{(m-n)}\,\pi\,k) = 0 \tag{1.444}$$

at the nodes x_k, where the exact solution is zero; see Fig. 1.109. Hence the FE stress $\sigma_h = 0$ is infinitely smooth—so all seems well—while in truth the exact stress oscillates rapidly

$$\sigma(x) = E\,\varepsilon(x) = \frac{\mu}{2^m\,\pi}\,\cos(2^m\,\pi\,x)\,. \tag{1.445}$$

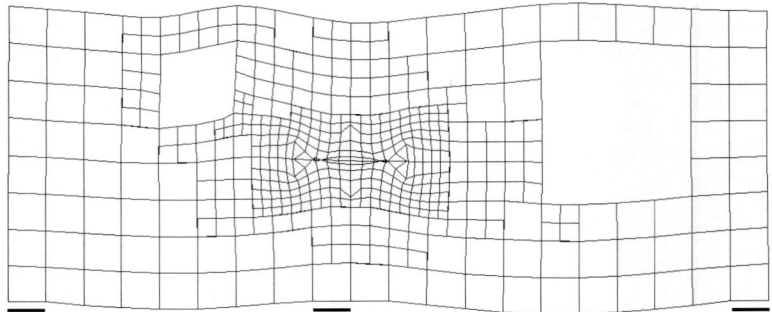

Fig. 1.110. Green's function for σ_{xx} at a node after an adaptive refinement

Of course, the fact that all $f_i = 0$ are zero should have warned us, but this problem is apt to illustrate, that deformations caused by oscillating loads are hard to catch with finite elements; see Fig. 1.3, p. 3. The contributions to the f_i tend to cancel, and also the influence functions for the jumps in the stresses at the nodes are symmetric (see Fig. 1.69, p. 98) so that oscillating loads—paradoxically enough—produce smooth FE stress fields.

Additional methods

In FE adaptivity we essentially distinguish between

- *h-methods*, which reduce the element size
- *p-methods*, which increase the degree p of the polynomial shape functions .

A combination of the two methods is the *hp-method*, where an increase in the order p is combined with a refinement of the mesh.

There are other methods, such as *r-adaptivity* or *remeshing*, where the layout of a mesh is changed without changing the number of elements, *d-adaptivity* (dimension), and *m-adaptivity* (model), [227].

The latter two methods acknowledge that it makes no sense to make the solution error smaller than the modeling error. In transition zones, there might occur effects which cannot be handled with a reduced model (3D → 2D), but where a switch back to the original 3-D model is necessary. The term *m*-adaptivity means that the model is changed and the constitutive equations are amended, while *d*-adaptivity means that the dimension of the problem is changed.

In particular, an increase in the order of the polynomials should probably be reserved for regions where the solution is smooth. In regions where the model does not fit well, it often makes more sense to shrink the elements and surmount thus the hurdles. In zones were the thickness changes, it is often much more effective to increase the number of elements than to increase the degree of the polynomials.

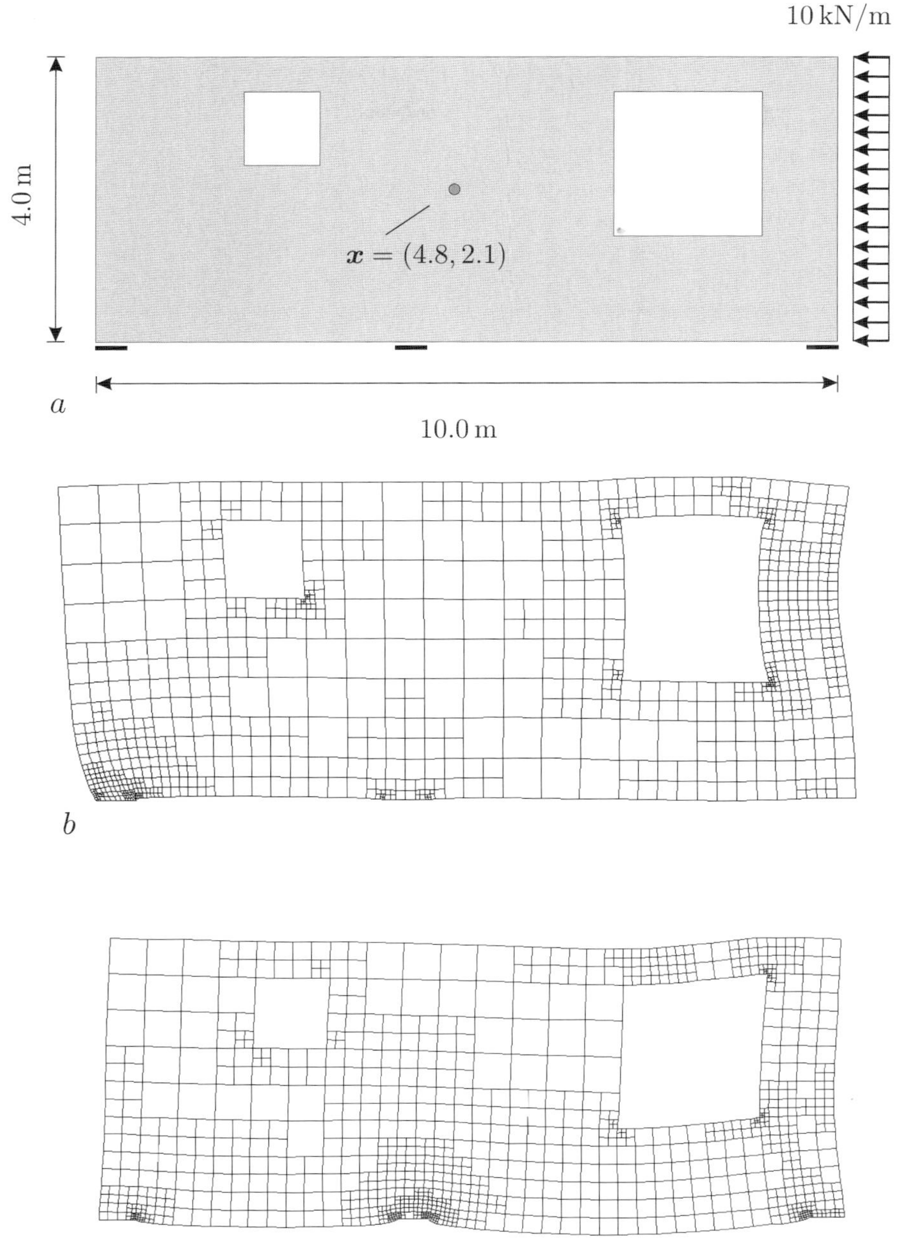

10 kN/m

4.0 m

$x = (4.8, 2.1)$

a

10.0 m

b

c

Fig. 1.111. Conventional adaptive refinement of a plate (bilinear elements), two load cases wind load and gravity load **a)** system and wind load; **b)** optimal mesh for $\sigma_{xx}(x)$ (wind load); **c)** for $\sigma_{yy}(x)$ (gravity load)

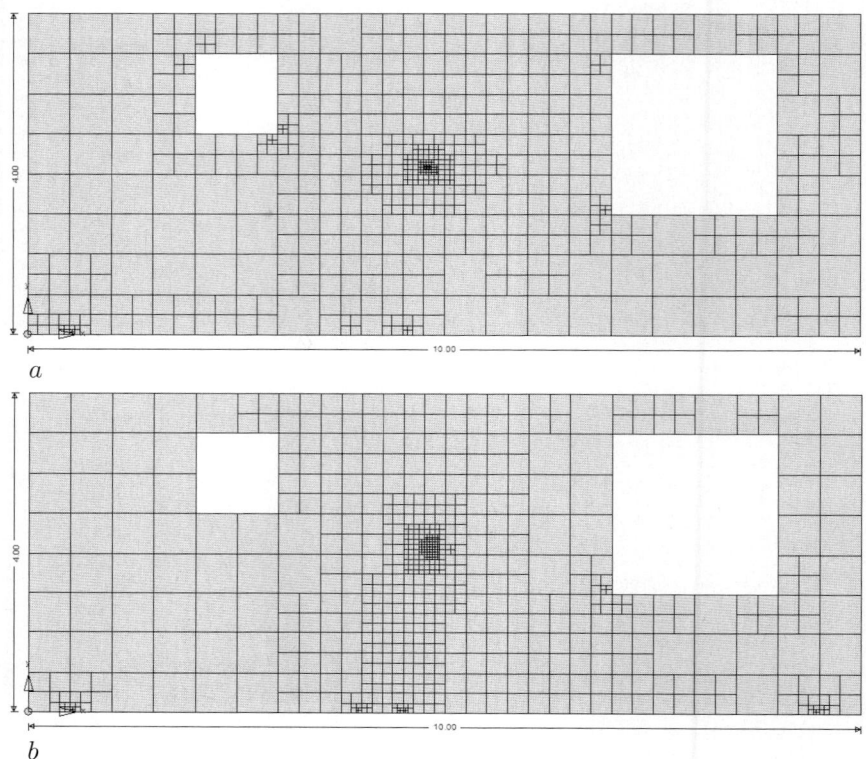

Fig. 1.112. Goal-oriented refinement (duality techniques), optimal meshes for
a) $\sigma_{xx}(\boldsymbol{x})$ under wind load; **b)** $\sigma_{yy}(\boldsymbol{x})$ under gravity load

Duality techniques

Let us assume that the displacement at a particular point $\boldsymbol{x}$ is to be calculated to high accuracy. We know that the distance between the exact and
approximate Green's function is responsible for the error of the FE solution:

$$e(\boldsymbol{x}) = u(\boldsymbol{x}) - u_h(\boldsymbol{x}) = \int_\Omega [\, G_0(\boldsymbol{y}, \boldsymbol{x}) - G_0^h(\boldsymbol{y}, \boldsymbol{x})\,]\, p(\boldsymbol{y})\, d\Omega_{\boldsymbol{y}}$$

$$= a(G_0 - G_0^h, u)\,. \tag{1.446}$$

So what is needed is a good approximation of the Green's function. This is
the strategy of duality techniques: We optimize the mesh in such a way that
the error in the Green's function G_0 becomes small. To achieve this goal first
a point load $P = 1$ is placed at the point $\boldsymbol{x}$ where $u(\boldsymbol{x})$ is to be calculated, and
the mesh is refined for this load case by a conventional adaptive procedure.
The solution of this load case is the approximate Green's function G_0^h for

displacement caused by $P_1 = 1$

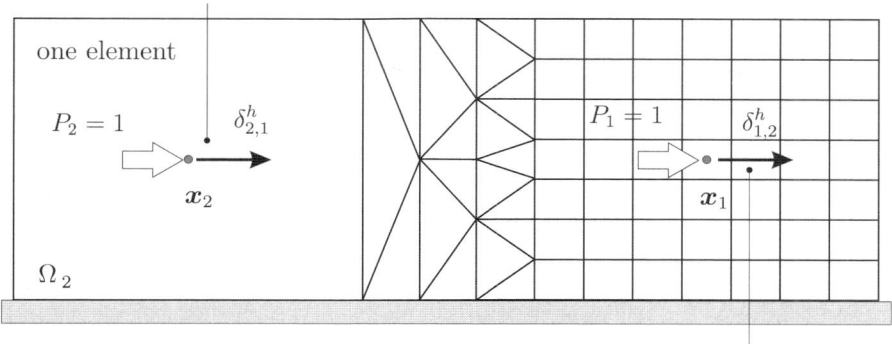

displacement caused by $P_2 = 1$

Fig. 1.113. A very uneven mesh but Maxwell's theorem implies that the displacements are the same, $\delta^h_{1,2} = \delta^h_{2,1}$

.

$u(\boldsymbol{x})$. Then in the second step the original load case is solved on this optimized mesh. And because the mesh refinement in the first step has reduced the error $G_0 - G_0^h$ (which is an intrinsic property of the mesh) the displacement error $e(\boldsymbol{x})$ in the original load case at the point $\boldsymbol{x}$ should be markedly reduced. This is the basic idea, see Fig. 1.110, Fig. 1.111 and Fig. 1.112.

But in doing so we also incorporate information about the error in the original problem, $u - u_h$, or—as it is called—the error in the *primal problem*. That is the mesh is refined where the error $u - u_h$ is large and where the error $G_0 - G_0^h$ is large. Both errors steer the refinement process.

To understand this strategy recall that influence functions are based on Betti's principle, $W_{1,2} = W_{2,1}$. In Fig. 1.113 a single force $P_1 = 1$ is applied at the point $\boldsymbol{x}_1$. We want to calculate the horizontal displacement at the point $\boldsymbol{x}_2$ caused by this force

$$u_x(\boldsymbol{x}_2) = \int_\Omega \boldsymbol{G}_0(\boldsymbol{y}, \boldsymbol{x}_2) \bullet \boldsymbol{p}(\boldsymbol{y}) \, d\Omega_{\boldsymbol{y}} = P_1 \cdot G_0^x(\boldsymbol{x}_1, \boldsymbol{x}_2) \,. \qquad (1.447)$$

G_0^x is the horizontal component of the displacement field $\boldsymbol{G}_0 = [G_0^x, G_0^y]^T$; in the following we write simply G_0 and also u instead of u_x.

The Green's function takes only samples where the load $\boldsymbol{p}$ is applied and if the load happens to be a point load then it samples at only *one* point! Hence to minimize the error in the displacement at the point $\boldsymbol{x}_2$

$$e(\boldsymbol{x}_2) = P_1 \cdot [G_0(\boldsymbol{x}_1, \boldsymbol{x}_2) - G_0^h(\boldsymbol{x}_1, \boldsymbol{x}_2)] = P_1 \cdot [\delta_{1,2} - \delta^h_{1,2}] \qquad (1.448)$$

we have to minimize—so it seems—only the error [...] in the Green's function at the point $\boldsymbol{x}_1$ where the load is applied.

Now the Green's function $G_0(\boldsymbol{y}, \boldsymbol{x}_2)$ takes off from the point $\boldsymbol{x}_2$ where the Dirac delta, the single force $P_2 = 1$ in Fig. 1.113, presses against the plate to generate the influence function for $u(\boldsymbol{x}_2)$. So would it not make sense to refine the mesh near this point too so that the effect felt at the distant point $\boldsymbol{x}_1$ becomes more accurate? The answer is yes! The two points (or rather the displacements at the two points) are like twins, are adjoint, because according to Maxwell's theorem (see p. 550) we have

$$\delta^h_{1,2} = \delta^h_{2,1} \, . \tag{1.449}$$

Assume we turn the whole situation around: the point load acts at the center of the "Gargantuan" element $\varOmega_2$ and we inquire about the displacement $u(\boldsymbol{x}_1)$ at the point $\boldsymbol{x}_1$. According to the previous logic the result should not be too good because we failed to refine the neighborhood of the point load P_2. But then $\delta^h_{1,2} = \delta^h_{2,1}$ and so if the second problem is poorly solved then also the first!

Hence to carry the information from a point A (where the action is) to a point B where the observer is standing the neighborhood of *both* points must be refined. This is an inherent requirement in self-adjoint problems.

But note that the refinement at the two points must not be of the same order, see Fig. 1.114. It depends on the strength of the singularity at A and at B, that is what the action is at A (a point force, a single moment, a bend, a dislocation, a twist) and what we measure at B (the deflection, the slope, the bending moment, etc.).

If the load is not a point load but spread over a region A then the integration

$$\partial^i u(\boldsymbol{x}_B) = \int_A \boldsymbol{G}_i(\boldsymbol{y}, \boldsymbol{x}_B) \bullet \boldsymbol{p}(\boldsymbol{y}) \, d\varOmega_{\boldsymbol{y}} \tag{1.450}$$

will lower the order of the singularity and then only the refinement at the point B will be noticeable. ($\partial^i u$ is short for $u, \sigma, \dots$).

The incorporation of the defect $\boldsymbol{p} - \boldsymbol{p}_h$, the error in the primal problem, is easy, because due to the Galerkin orthogonality (1.446) is equivalent to[13]

$$
\begin{aligned}
e(\boldsymbol{x}) = u(\boldsymbol{x}) - u_h(\boldsymbol{x}) &= \int_\varOmega [\, \boldsymbol{G}_0(\boldsymbol{y}, \boldsymbol{x}) - \boldsymbol{G}^h_0(\boldsymbol{y}, \boldsymbol{x}) \,] \bullet (\boldsymbol{p}(\boldsymbol{y}) - \boldsymbol{p}_h(\boldsymbol{y})) \, d\varOmega_{\boldsymbol{y}} \\
&= a(\boldsymbol{G}_0 - \boldsymbol{G}^h_0, \boldsymbol{u} - \boldsymbol{u}_h) \, . \tag{1.451}
\end{aligned}
$$

We essentially have added a zero to (1.446), because the error in the Green's function is orthogonal to each $\boldsymbol{v}_h \in V_h$ and therefore also to $\boldsymbol{u}_h$

$$0 = \boldsymbol{p}_h(\boldsymbol{G}_0[\boldsymbol{x}]) - \boldsymbol{p}_h(\boldsymbol{G}^h_0[\boldsymbol{x}]) = a(\boldsymbol{G}_0[\boldsymbol{x}] - \boldsymbol{G}^h_0[\boldsymbol{x}], \boldsymbol{u}_h) \, . \tag{1.452}$$

---

[13] Our tacit assumption is that the load case $\boldsymbol{p}_h$ consists of volume forces $\boldsymbol{p}_h$ only. We shall make this assumption whenever appropriate to simplify the notation.

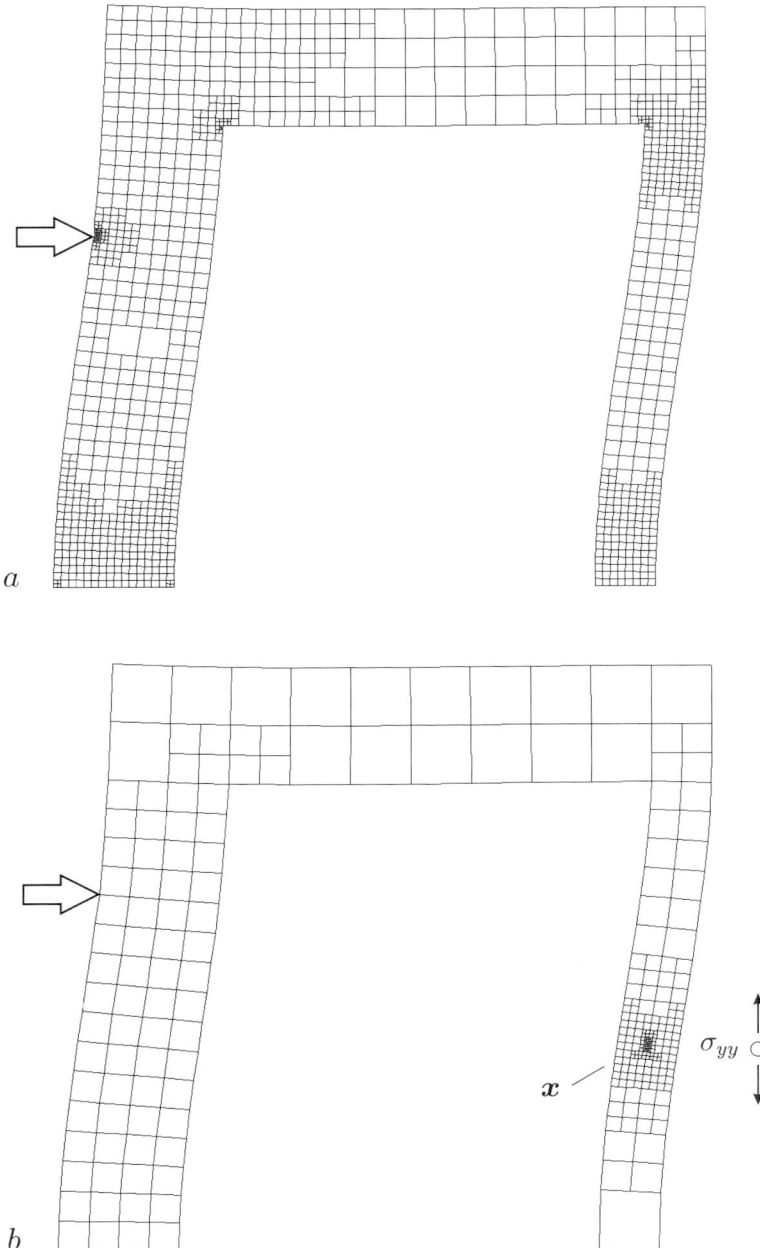

Fig. 1.114. Adaptive refinement **a)** standard refinement $\eta_p \leq \varepsilon_{TOL}$ **b)** goal oriented refinement $\eta_p \times \eta_G \leq \varepsilon_{TOL}$

After applying the Cauchy-Schwarz inequality to (1.451) we obtain the fundamental estimate

$$|u(\boldsymbol{x}) - u_h(\boldsymbol{x})| \leq || \, \boldsymbol{G}_0[\boldsymbol{x}] - \boldsymbol{G}_0^h[\boldsymbol{x}]||_E \, ||\boldsymbol{u} - \boldsymbol{u}_h||_E \qquad (1.453)$$

where $||.||_E$ is the standard energy norm, $||\boldsymbol{u}||_E^2 := a(\boldsymbol{u}, \boldsymbol{u})$.

This result is motivation to minimize the error in the Green's function *and* the error in the solution u_h simultaneously. That is, at each refinement step an error indicator η_G^e for the Green's function and an error indicator η_p^e for the original problem is calculated on each element and the combined error indicator on each element is then $\eta^e = \eta_G^e \cdot \eta_p^e$.

This last step is based on

$$|u(\boldsymbol{x}) - u_h(\boldsymbol{x})| = |\sum_e a(\boldsymbol{G}_0 - \boldsymbol{G}_0^h, \boldsymbol{u} - \boldsymbol{u}_h)_{\Omega_e}|$$

$$\leq \sum_e |a(\boldsymbol{G}_0 - \boldsymbol{G}_0^h, \boldsymbol{u} - \boldsymbol{u}_h)_{\Omega_e}|$$

$$\leq \sum_e \underbrace{a(\boldsymbol{G}_0 - \boldsymbol{G}_0^h, \boldsymbol{G}_0 - \boldsymbol{G}_0^h)_{\Omega_e}^{1/2}}_{\eta_G^e} \cdot \underbrace{a(\boldsymbol{u} - \boldsymbol{u}_h, \boldsymbol{u} - \boldsymbol{u}_h)_{\Omega_e}^{1/2}}_{\eta_p^e}$$

$$(1.454)$$

so that the sum of the local errors $\eta^e = \eta_G^e \cdot \eta_p^e$ provides an upper bound for the error

$$|u(\boldsymbol{x}) - u_h(\boldsymbol{x})| \leq \sum_e \eta_G^e \cdot \eta_p^e \, . \qquad (1.455)$$

The energy norms of the fields $\boldsymbol{e}_G = \boldsymbol{G}_0 - \boldsymbol{G}_0^h$ and $\boldsymbol{e}_u = \boldsymbol{u} - \boldsymbol{u}_h$ are calculated by measuring the eigenwork done by the element residual forces and jump terms on the element edges as on page 149, so that for example

$$\eta_p = a(\boldsymbol{e}_u, \boldsymbol{e}_u)_{\Omega_e}^{1/2} = r(\boldsymbol{e}_u)^{1/2} \, . \qquad (1.456)$$

Because this technique consists in applying a quantity (a Dirac delta) conjugate to the original value, we speak of duality techniques. In particular the method is well adapted to problems where the focus is on one or two values that are approximated to a high degree of accuracy [22], [67], [132]. Which is why the technique is also called *goal-oriented recovery*.

The interesting point about the duality technique is that the error in the Green's function serves as a weight to the error in the primal problem. The consequences of this can be seen in Fig. 1.114. The first mesh, the mesh in Fig. 1.114 a, is the result of a standard adaptive refinement

$$\eta_p^e \leq \varepsilon_{TOL} \, . \qquad (1.457)$$

To push the error below the preset error margin the program has to refine all those elements—in practice only the first, say, 30%—where the error η_p^e of the primal problem exceeds this margin. It has no other choice.

The mesh in Fig. 1.114 b is based on weighting the primal error η_p with the error η_G of the Green's function elementwise

$$\eta_G^e \cdot \eta_p^e \leq \varepsilon_{TOL} \, . \tag{1.458}$$

In most parts of the mesh the error η_G^e of the numerical Green's function is so low that this inequality is *automatically* satisfied. That is many of the refinements in Fig. 1.114 a are not necessary if we are only interested in the stress σ_{yy} at $\boldsymbol{x}$.

For a more detailed analysis of duality techniques and the concept of generalized Green's functions see Sect. 7.4, p. 519.

Remark 1.17. Clearly the duality technique is based on Tottenham's equation. But historically the duality technique was developed independently, see [19] and [22]. Only latter it was discovered that the fundamental equation had already been published by Tottenham, [67]. Probably there are also other precursors as, for example, the L^*L-method of Kato-Fujita where to obtain bounds for a solution at a point a similar technique is applied [174].

Nonlinear problems

Although the duality technique is motivated by Betti's theorem, it can be applied successfully to nonlinear problems as well, [22], [67], [200]. Near an equilibrium point a nonlinear structure essentially exhibits a linear behavior with regard to load increments and this suffices to establish a connection between the error in the functional, $J(u) - J(u_h)$, and the error in the Green's function, $\boldsymbol{z} - \boldsymbol{z}_h$, at the linearization point.

By doing a "Taylor expansion" of Green's first identity it follows, see Sect. 4.21 p. 403, that the displacement increment $\boldsymbol{u}_\Delta$ and the load increment $\boldsymbol{p}_\Delta$ satisfy—to first order—the variational statement

$$a_T(\boldsymbol{u}; \boldsymbol{u}_\Delta, \boldsymbol{v}) = (\boldsymbol{p}_\Delta, \boldsymbol{v}) \qquad \boldsymbol{v} \in V \tag{1.459}$$

where a_T is the Gateaux derivative of the strain energy product, that is the gradient of $a(\boldsymbol{u}, \boldsymbol{v})$ at $\boldsymbol{u}$ in the direction of $\boldsymbol{u}_\Delta$.

The Newton-Raphson algorithm for the solution of the nonlinear system of equations $\boldsymbol{k}(\boldsymbol{u}) = \boldsymbol{f}$ is based on this expansion

$$\boldsymbol{K}_T(\boldsymbol{u}_i)\,\boldsymbol{u}_{i+1} = \boldsymbol{f} - \boldsymbol{k}(\boldsymbol{u}_i) \tag{1.460}$$

where the tangential stiffness matrix $\boldsymbol{K}_T$ corresponds to a_T.

If the Gateaux derivative a_T is linear in the second and third argument the associated Euler equation

$$\boldsymbol{L}_\Delta \boldsymbol{u}_\Delta = \boldsymbol{p}_\Delta \tag{1.461}$$

is linear which means that near an equilibrium point the response of the structure to load increments is linear and so for each functional $J(\boldsymbol{u}_\Delta)$ there is a (generalized) Green's function z such that the effect of a load increment on any functional value of the displacement increment can be predicted as in the linear theory, $J(u_\Delta) = (p_\Delta, z)$.

Now the key point is—we have discussed this before in Sect. 1.27—that the error in the output functional of a nonlinear problem can be expressed as

$$J(u) - J(u_h) = \frac{1}{2}\rho(u_h)(z - z_h) + \frac{1}{2}\rho^*(u_h, z_h)(u - u_h) + R_h^{(3)} \quad (1.462)$$

where the first term

$$\rho(u_h)(z - z_h) = (p - p_h, z - z_h) \quad (1.463)$$

is the work done by the residual forces on acting through the error, $z - z_h$, in the Green's function and the second term is

$$\rho^*(u_h, z_h)(u - u_h) = (\delta_0 - \delta_0^h, u - u_h) \quad (1.464)$$

the error in the functional $J(.)$ evaluated at $u - u_h$ and $R_h^{(3)}$ is a cubic remainder.

So the error $z - z_h$ in the Green's function at the linearization point is one part of the error $J(u) - J(u_h)$ or else: if the error $z - z_h$ can be made small and the error $u - u_h$ then *automatically* also the error $J(u) - J(u_h)$ will be small—at least that is what we hope. This is the idea; see also the Remark on p. 124.

To repeat: the Green's function z at the linearization point can not be used to calculate

$$J(u) = \int_\Omega z(y, x) \cdot p(y) \, d\Omega y \,. \quad (1.465)$$

This does not work. But the study of the error $z - z_h$ allows to adapt the mesh so that eventually the error $J(u) - J(u_h)$ will become small.

Hence basically what is done is, see Box 1.1, that at the actual equilibrium point the tangential stiffness matrix is solved for the generalized Green's function,

$$K_T(u) \, z = j \quad (1.466)$$

corresponding to

$$a_T(u_h; z_h, \varphi_i) = J'(u_h; \varphi_i) \qquad \varphi_i \in V_h \quad (1.467)$$

and the error in the primal problem is weighted with the error in this dual problem. For an application see the example in Sect. 4.21, p. 412 and for a more detailed analysis of goal-oriented refinement applied to nonlinear problems see Sect. 7.5 p. 526.

Start with an initial mesh $\mathcal{T}_k$ and let $k = 0$

(A) For $t = 1$ to M (loop over all load increments M)

$$\lambda = t \cdot \frac{1}{M} \quad \text{actual load parameter}$$

1. Newton-Raphson: Let $i = 0$, $\boldsymbol{u}_{\Delta_k} = 0$, $\boldsymbol{u}_k^{(0)} = \boldsymbol{u}_{k-1}$ and solve the equations

$$\boldsymbol{K}_T(\boldsymbol{u}_k^{(i)}) \, \boldsymbol{u}_{\Delta_k}^{(i+1)} = \lambda \cdot \boldsymbol{f} - \boldsymbol{k}(\boldsymbol{u}_k^{(i)})$$

$$\boldsymbol{u}_k^{(i+1)} = \boldsymbol{u}_{\Delta_k}^{(i+1)} + \boldsymbol{u}_k^{(i)}$$

   Let $i = i + 1$ and repeat iteration till convergence is achieved.
2. Calculate the error indicators of this primal problem $\eta_e^{(p)}$.
3. Formulate and solve the dual problem at the actual equilibrium point $\boldsymbol{u}_k$:

$$\boldsymbol{K}_T(\boldsymbol{u}_k) \, \boldsymbol{z}_k = \boldsymbol{j}_k$$

4. Calculate the error indicators of the dual problem $\eta_e^{(z)}$.
5. Mesh refinement:
   - Determine the *weighted* error indicator

$$\eta_e = \eta_e^{(p)} \cdot \eta_e^{(z)} \,.$$

   - Calculate the error estimator

$$J(\boldsymbol{e}) \approx \eta = \sum_{\Omega_e} \eta_e \,.$$

   - IF $|\eta| \leq TOL$ (global error margin) $\rightarrow$ $t = t + 1$, GOTO (A).
   - IF $\eta_e > TOL_e$ (local error margin) $\rightarrow$ refine element Ω_e.
   - Generate a new mesh $\mathcal{T}_{k+1}$, transfer data, let $k = k + 1$.
   - GOTO 1.

Box 1.1: Goal oriented refinement of nonlinear problems

Model adaptivity

Often the FE model of a structure is based on simplifying assumptions and so we would like to have estimates for the modeling error and also indicators which could steer an adaptive process by which the model can be updated if necessary[14]. The equation

$$J(u_c) - J(u) = -d(u_c, G) \simeq -d(u, G) \,, \tag{1.468}$$

_____

[14] 'One should make a model as simple as possible but not simpler' (A. Einstein).

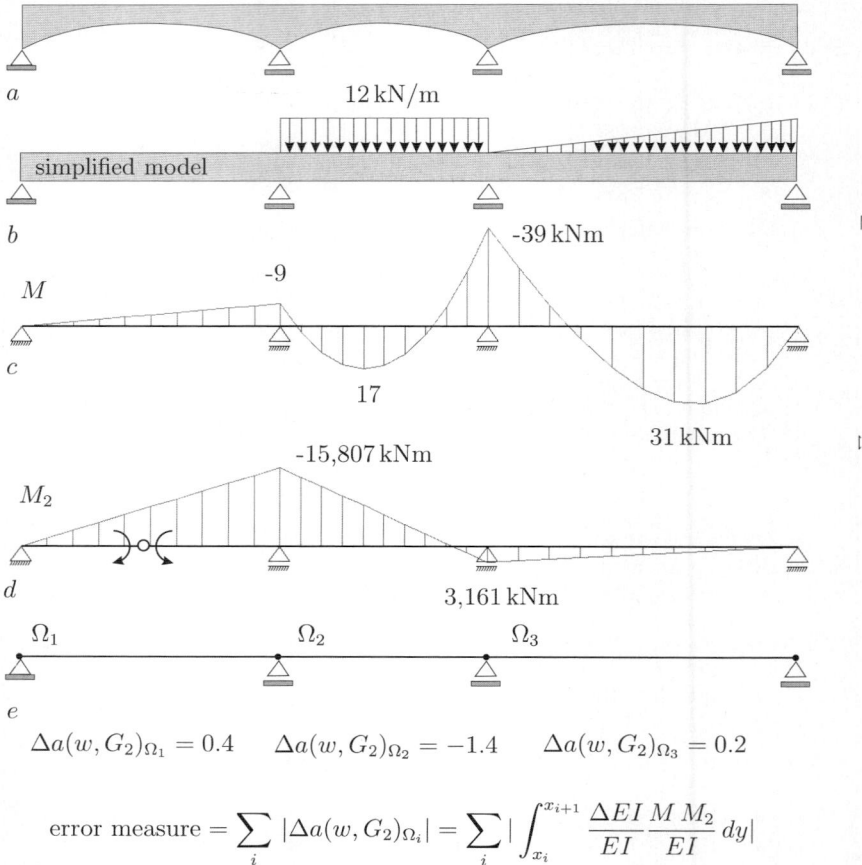

$$\Delta a(w, G_2)_{\Omega_1} = 0.4 \qquad \Delta a(w, G_2)_{\Omega_2} = -1.4 \qquad \Delta a(w, G_2)_{\Omega_3} = 0.2$$

$$\text{error measure} = \sum_i |\Delta a(w, G_2)_{\Omega_i}| = \sum_i |\int_{x_i}^{x_{i+1}} \frac{\Delta EI}{EI} \frac{M\, M_2}{EI}\, dy|$$

Fig. 1.115. Effect of the modeling error on the bending moment $M(x)$ at the center of the first span **a)** original design **b)** simplified model **c)** $M(x)$ **d)** bending moment of the influence function G_2 for $M(x)$ **e)** elements

which allows to asses changes in any output functional $J(u)$ due to changes in the stiffness of a structure serves just this purpose. That is Green's functions and the d-scalar product allow to estimate how sensitive the results in a part Ω_A of a structure are to the choice of the elastic parameters in a part Ω_B of the structure.

An introductory problem, the bridge in Fig. 1.115 a, may illustrate the basic idea. The simplified FE model consists of three beam elements with a uniform stiffness EI. The goal is the evaluation of the bending moment $M(x)$ at the center of the first span. The error in $M(x)$ is approximately

$$M_{ex}(x) - M(x) \simeq -\int_0^l \frac{\Delta EI}{EI} \frac{M\, M_2}{EI}\, dy = -(0.4 - 1.4 + 0.2) \quad (1.469)$$

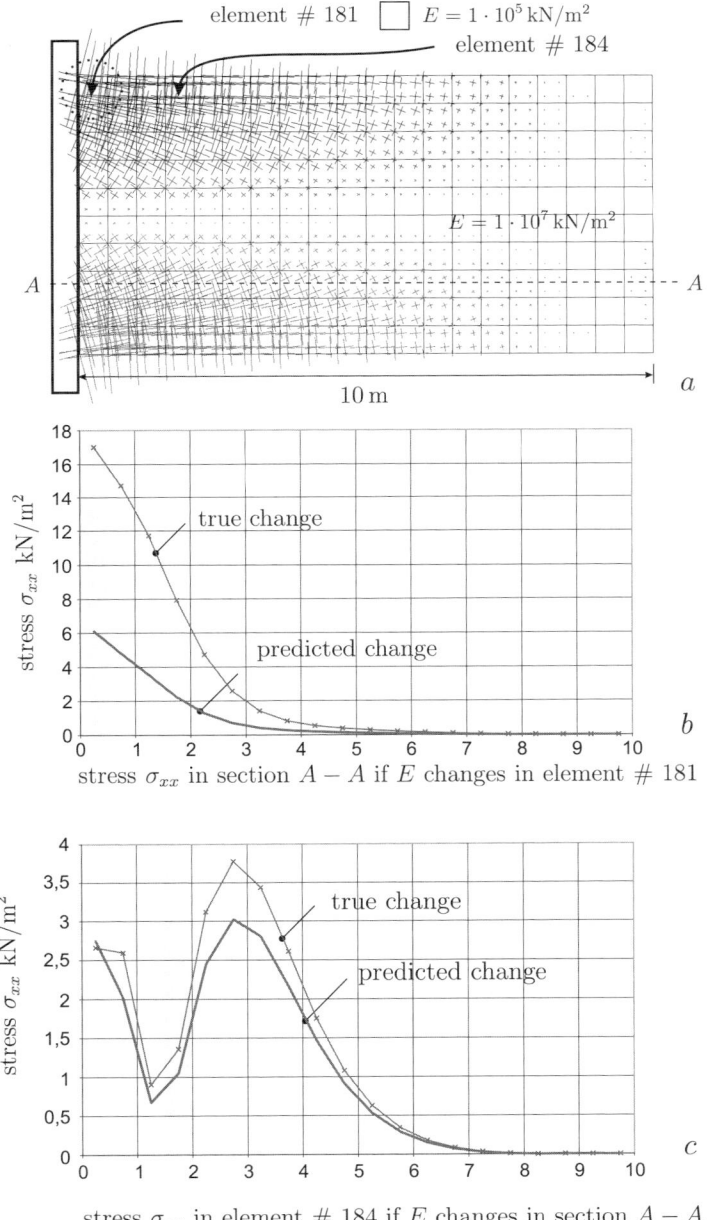

Fig. 1.116. Change of Young's modulus, $E = 1 \cdot 10^7 \rightarrow 1 \cdot 10^5$ kN/m^2 **a)** stress distribution under gravity load, uniform E, **b)** change in σ_{xx} along section $A - A$ if E changes in element # 181 **c)** change of σ_{xx} in element # 184 if E changes—one element at a time—in section $A - A$

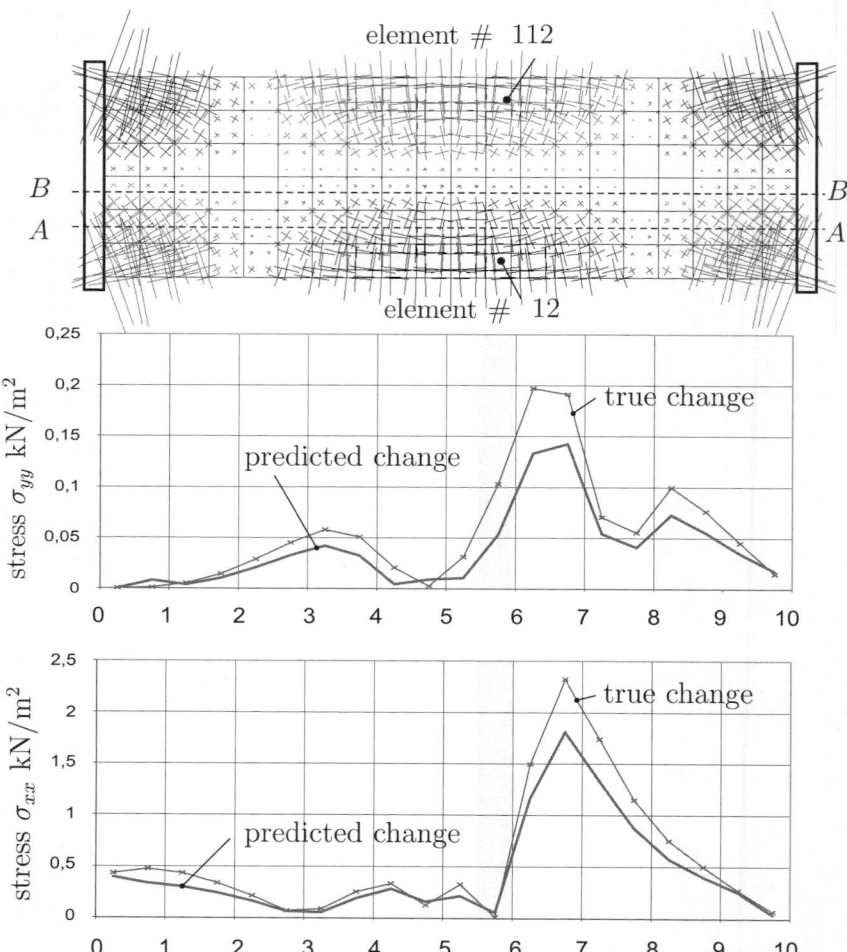

Fig. 1.117. Shear wall, 10 m × 3 m. Change of Young's modulus, $E = 1 \cdot 10^7 \rightarrow 1 \cdot 10^5$ kN/m^2 **a)** stress distribution under gravity load, uniform E, **b)** change in σ_{yy} in element # 112 if E changes elementwise in section $A - A$ **c)** change of σ_{xx} in element # 12 if E changes elementwise in section $B - B$

where the three numbers are the contributions of the three spans to the error. Because the error is largest in the second span in the next step this span should be subdivided into, say, three separate elements with more realistic values of EI and then the analysis should be repeated.

Of course the modeling error

$$\eta_m = \sum_e |d(u_c, G)_{\Omega_e}| \simeq \sum_e |d(u, G)_{\Omega_e}| \qquad (1.470)$$

should be combined with the discretization errors η_p and η_G of the primal and dual problem so that—eventually after some proper scaling—the expression

$$\eta = \eta_p \cdot \eta_G + \eta_m \tag{1.471}$$

may serve as an engineering error indicator.

In Figure 1.116 this technique was applied to a cantilever plate. In the element # 181, which gets stressed the most, the modulus of elasticity was reduced from $E = 1 \cdot 10^7$ to $E = 1 \cdot 10^5$ kN/m^2. Plotted in Fig. 1.116 b are the true change of the horizontal stress σ_{xx} in cross section $A - A$ and the change as predicted by the d-scalar product

$$\sigma_{xx}^c - \sigma_{xx} \simeq -d(\boldsymbol{G}_1^h, \boldsymbol{u}_h) \tag{1.472}$$

where both fields, $\boldsymbol{G}_1^h$ and $\boldsymbol{u}_h$, were taken from the original unmodified model.

Next the modulus of elasticity was changed in cross-section $A - A$—one element at a time—and the change in the stress σ_{xx} at the center of element # 184 was studied. The effects on σ_{xx} are plotted in Fig. 1.116 c.

Similar modifications were done to the plate in Fig. 1.117 and also the results are similar.

It is clearly visible (1) that the influence of local changes, $E \rightarrow E + \Delta E$, on the stress field is limited and that (2) in most of the cases the d-scalar product $d(\boldsymbol{G}_1^h, \boldsymbol{u}_h)$ with both fields taken from the unmodified structure is capable of predicting the effects of the change.

Nonlinear problems

Model adaptivity can be applied to nonlinear problems as well because the change in any output functional can be approximately predicted by applying Equ. 1.375 in Sect. 1.27, p. 123.

Fundamental solutions

The formula

$$u_h(\boldsymbol{x}) = \int_\Omega G_0^h(\boldsymbol{y}, \boldsymbol{x}) \, p(\boldsymbol{y}) \, d\Omega_{\boldsymbol{y}} \tag{1.473}$$

suggests, that all that is needed is a good approximation to $G_0(\boldsymbol{y}, \boldsymbol{x})$. To ease the burden for the FE program, to approximate the function G_0 it can be split into a *fundamental solution* g_0 and a regular part u_R,

$$G_0(\boldsymbol{y}, \boldsymbol{x}) = g_0(\boldsymbol{y}, \boldsymbol{x}) + u_R(\boldsymbol{y}, \boldsymbol{x}) \tag{1.474}$$

and then only the regular part u_R needs to be approximated with finite elements. The regular part u_R is a homogeneous solution of the differential

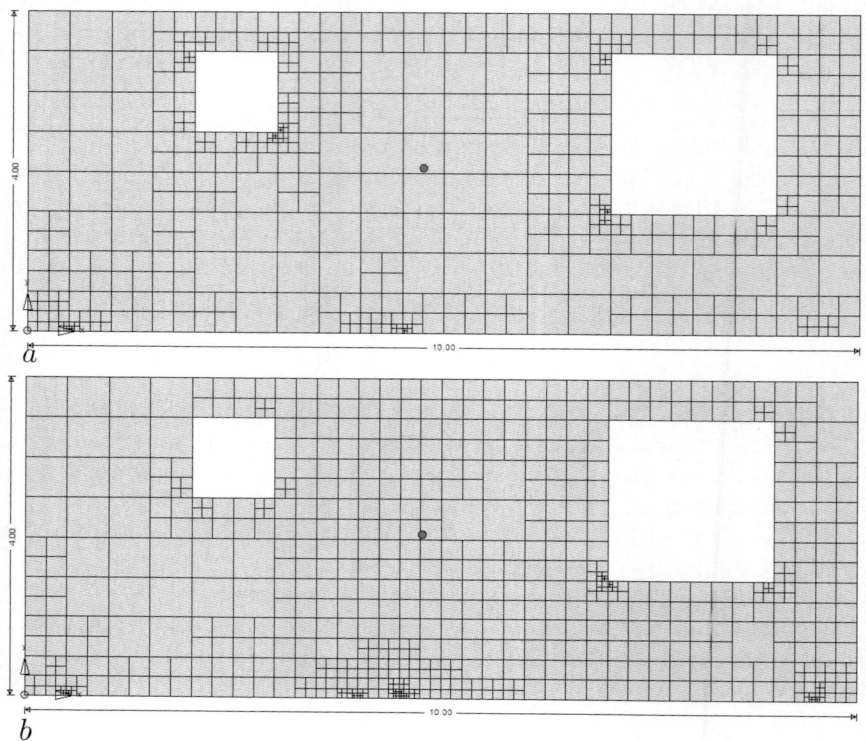

Fig. 1.118. Improved goal-oriented refinement with fundamental solutions, optimal meshes for **a)** $\sigma_{xx}(\boldsymbol{x})$ under wind load; **b)** $\sigma_{yy}(\boldsymbol{x})$ under gravity load

equation, and its boundary values are such that $g_0 + u_R$ satisfies the same boundary conditions as G_0.

In the following, this split will be exemplified with the kernels for the stresses σ_{ij} in a plate:

$$\boldsymbol{G}_1^{ij}(\boldsymbol{y}, \boldsymbol{x}) = \boldsymbol{u}_R^{ij}(\boldsymbol{y}, \boldsymbol{x}) + \boldsymbol{g}_1^{ij}(\boldsymbol{y}, \boldsymbol{x}) \,. \tag{1.475}$$

The three fundamental solutions form a matrix

$$\boldsymbol{D}(\boldsymbol{y}, \boldsymbol{x}) = [\boldsymbol{g}_1^{11}(\boldsymbol{y}, \boldsymbol{x}), \boldsymbol{g}_1^{12}(\boldsymbol{y}, \boldsymbol{x}), \boldsymbol{g}_1^{22}(\boldsymbol{y}, \boldsymbol{x})] \tag{1.476}$$

with the elements [116]

$$D_{kij}(\boldsymbol{y}, \boldsymbol{x}) = \frac{1}{r}\{(1 - 2\,\nu)[\delta_{ki}r_{,j} + \delta_{kj}r_{,i} - \delta_{ij}r_{,k}] + 2r_{,i}\,r_{,j}\,r_{,k}\}\frac{1}{4\pi(1 - \nu)}\,. \tag{1.477}$$

The index k is the vector component, and the indices $ij \equiv \{11, 12, 22\}$ denote the stresses. The traction vectors of these fields have the components (k)

$$S_{kij}(\boldsymbol{y},\boldsymbol{x}) = \frac{2\mu}{r^2}\{2\frac{\partial r}{\partial n}[(1-2\,\nu)\delta_{ij}r,_k + \nu(\delta_{ik}r,_j + \delta_{jk}r,_i) - 4r,_i\,r,_j\,r,_k]$$

$$+ 2\,\nu\,(n_i r,_j\,r,_k + n_j r,_i\,r,_k) + (1-2\,\nu)(2n_k r,_i\,r,_j + n_j\delta_{ik}$$

$$+ n_i\delta_{jk}) - (1-4\,\nu)n_k\delta_{ij}\}\frac{1}{4\pi(1-\nu)}\,. \tag{1.478}$$

If the boundary conditions are assumed to be homogeneous on Dirichlet (Γ_D) and Neumann (Γ_N) boundaries, the regular solution must satisfy the equations

$$\boldsymbol{G}_1^{ij} = \boldsymbol{u}_R^{ij} + \boldsymbol{g}_1^{ij} = \boldsymbol{0} \quad \text{on } \Gamma_D, \tag{1.479}$$

$$\boldsymbol{t}(\boldsymbol{G}_1^{ij}) = \boldsymbol{t}_R^{ij} + \boldsymbol{t}_1^{ij} = \boldsymbol{0} \quad \text{on } \Gamma_N\,. \tag{1.480}$$

The approximate regular part is the solution of the variational problem

$$a(\boldsymbol{u}_{R,h}^{ij}, \boldsymbol{\varphi}_i) = -\int_{\Gamma_N} \boldsymbol{t}_1^{ij} \cdot \boldsymbol{\varphi}_i\, ds \quad \boldsymbol{\varphi}_i \in V_h\,. \tag{1.481}$$

The improved Green's function is

$$\boldsymbol{G}_{1,h}^{ij+}(\boldsymbol{y},\boldsymbol{x}) = \boldsymbol{u}_{R,h}^{ij}(\boldsymbol{y},\boldsymbol{x}) + \boldsymbol{g}_1^{ij}(\boldsymbol{y},\boldsymbol{x})\,, \tag{1.482}$$

and the stresses are

$$\sigma_{ij,h}^+(\boldsymbol{x}) = \int_\Omega \boldsymbol{G}_{1,h}^{ij+}(\boldsymbol{y},\boldsymbol{x}) \cdot \boldsymbol{p}(\boldsymbol{y})\, d\Omega_{\boldsymbol{y}}\,. \tag{1.483}$$

Note that the error in the Green's function is replaced in (1.453) by the error in the regular solution u_R

$$|u(\boldsymbol{x}) - u_h(\boldsymbol{x})| \le ||\,u_R[\boldsymbol{x}] - u_R^h[\boldsymbol{x}]||_E\,||u - u_h||_E\,, \tag{1.484}$$

which will speed up the convergence, because at points not too close to the boundary the error in the regular part will be much smaller than the error in the Green's function. As before two error indicators η_R (indicating the error in the regular part) and η_p are calculated and the combination of both indicators, $\eta = \eta_R \cdot \eta_p$, steers the adaptive refinement.

Although there is one subtle difference between the new approach and the previous technique. In the new approach we actually *construct* for, say, the stress σ_{xx} at a point $\boldsymbol{x}$ an approximate Green's functions $\boldsymbol{G}_{1,h}^+$ and we calculate the stress with this function, see (1.483). While in the previous method we simply calculate the FE stress $\sigma_{xx}^h(\boldsymbol{x}) = \sum_i \sigma_{xx}(\boldsymbol{\varphi}_i)(\boldsymbol{x})\,u_i$ (after the mesh has been improved) knowing that in so doing we inherently form the scalar product between the improved kernel $\boldsymbol{G}_1^h$ and the applied load.

Comparative study

The plate in Fig. 1.111 on p. 155 was analyzed with all three methods

- conventional energy-norm refinement
- goal oriented refinement
- goal oriented refinement based on fundamental solutions

The plate was subject to a wind load and gravity load. The aim of the refinement was to predict the horizontal stress σ_{xx} at the point $\boldsymbol{x} = (4.8, 2.1)$ under wind load and the vertical stress σ_{yy} at the same point under gravity load to high accuracy. Hence each load case required a different refined mesh. The "exact" values in Tables 1.6 and 1.7 are the results of a BE analysis.

As is typical for goal-oriented methods the adaptive refinement concentrates on the neighborhood of the point $\boldsymbol{x}$; see Fig. 1.112. While if the Green's function is split into a fundamental solution and a regular part then the refinement spares the neighborhood of $\boldsymbol{x}$ because the fundamental solution is already the quasi-optimal solution near this point (see Fig. 1.118), and all effort goes into minimizing the energy error of the regular part, see (1.484).

Table 1.6. Gravity load, σ_{yy}, exact $= -87.126$ kN/m^2

d.o.f.	Energy norm	d.o.f.	Goal-or.	d.o.f.	Fund. sol.
350	-73.9557	350	-73.9557	350	-85.1865
558	-74.2296	528	-84.0145	576	-85.8080
900	-74.9279	823	-87.7160	870	-86.0676
1462	-74.6239	1206	-85.5208	1235	-86.2783
2325	-85.0523	1619	-85.0156	1694	-86.4994

Table 1.7. Wind load, σ_{xx}, exact $= -43.634$ kN/m^2

d.o.f.	Energy norm	d.o.f.	Goal-or.	d.o.f.	Fund. sol.
350	-45.2505	350	-45.2505	350	-43.7053
538	-45.2789	518	-42.5768	537	-43.7753
858	-45.0542	716	-43.4873	766	-43.6761
1370	-45.0911	982	-44.0888	1065	-43.6685
2079	-45.0155	1254	-43.8960	1378	-43.6596
$-$	$-$	1565	-43.6094	$-$	$-$

Changes in the elastic parameters

Even if the material is not homogeneous, the Green's function can still be split into a fundamental solution and a regular part. It is merely necessary that additional forces must be applied at the interface of the different zones.

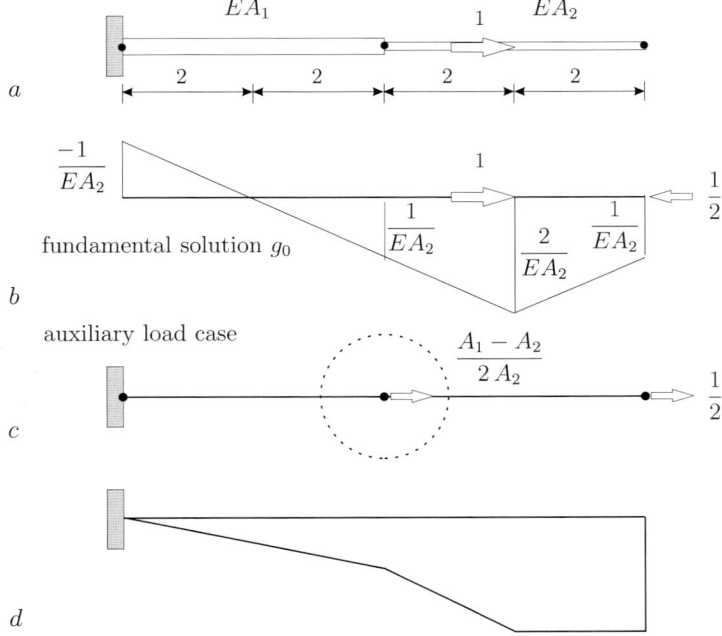

Fig. 1.119. The Green's function is split into a fundamental solution and a regular solution, which consists of two parts

These forces account for the discontinuities in the material parameters. The 1-D problem of a simple bar will suffice to illustrate the technique.

The bar in Fig. 1.119 consists of two elements with different cross sections A_1 and A_2. The Green's function for the longitudinal displacement at the center of the second element is to be calculated. The displacement g_0 in Fig. 1.119 a is an appropriate fundamental solution because it corresponds to the application of a single force $P = 1$ at the center of the second element. Now the Green's function has to have (i) zero displacement at $x = 0$, (ii) a zero normal force $N(l) = 0$ at the free end of the bar and (iii) the jump in the normal force must be zero at the transition point x_1 between the two zones. Note that the fundamental solution does not and cannot satisfy this last condition because the slope g_0' of the fundamental solution is continuous at x_1. Its value is $1/(2\,EA_2)$, so

$$N^+(x_1) - N^-(x_1) = EA_2\, g_0'(x_1) - EA_1\, g_0'(x_1)$$
$$= \frac{1}{2}(1 - \frac{A_1}{A_2}) = -\frac{A_1 - A_2}{2}. \qquad (1.485)$$

Because $A_1 > A_2$, the net effect is that a force $(A_1 - A_2)/(2A_2)$ pulls the node x_1 to the left:

$$\frac{1}{2} \frac{A_1}{A_2} \quad \leftarrow \bullet \rightarrow \quad \frac{1}{2}. \tag{1.486}$$

(Multiply the right-hand side by A_2/A_2.) Now to correct this defect and enforce $N(l) = 0$ at the end of the bar (currently $N(l) = -1/2$) the regular solution must solve the load case in Fig. 1.119 c, and in addition a constant term $1/EA_2$ must be added to correct the nonzero displacement, $g_0(0, x_c) = -1/EA_2$, of the fundamental solution at $x = 0$:

$$G_0 = g_0 + u(\text{Fig 1.119 c}) + 1/EA_2. \tag{1.487}$$

1.32 St. Venant's principle

According to St. Venant's principle, the difference between the stresses due to statically equivalent loads becomes insignificant at distances greater than the largest dimension of the area over which the loads are spread.

St. Venant's principle is valid for *elliptic differential equations*, i.e., for most of the equations of structural mechanics. Typically, for static or harmonic loads, the solutions decay very rapidly outside of the loaded region, as can be seen for example, from the influence function for the bending moment m_{xx} of a hinged slab:

$$m_{xx}(\boldsymbol{x}) = \int_\Gamma \left[g_2 \cdot v_\nu + m_\nu \cdot (g_2) \frac{\partial w}{\partial \nu} \right] ds\boldsymbol{y} + \int_\Omega g_2 \cdot p \, d\Omega\boldsymbol{y}$$
$$+ \sum_c g_2(\boldsymbol{y}^c) \cdot F_c. \tag{1.488}$$

Note that the subscript ν indicates that the functions depend on the normal vector $\boldsymbol{\nu} = [\nu_1, \nu_2]^T$ at the integration point $\boldsymbol{y}$. This vector must be distinguished from the normal vector $\boldsymbol{n}$ at the observation point $\boldsymbol{x}$.

The contributions to this influence function come from the support reaction v_ν, the slope $\partial w / \partial \nu$, the surface load p, and the corner forces F_c; the influence decays as $\ln r$ or r^{-2}:

$$g_2(\boldsymbol{y}, \boldsymbol{x}) = O(\ln r) \qquad m_\nu(g_2(\boldsymbol{y}, \boldsymbol{x})) = O(r^{-2}). \tag{1.489}$$

In a typical FE solution, many more sources contribute to the bending moment:

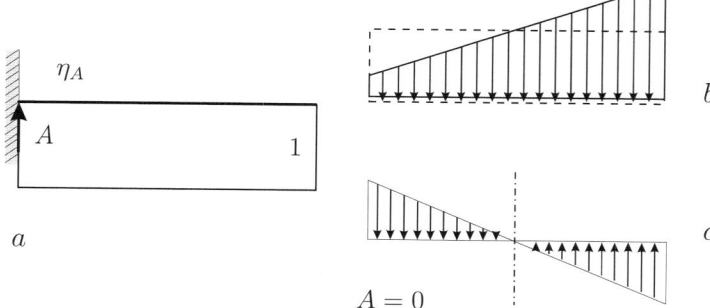

Fig. 1.120. a) Influence functions for the support reaction A; **b)** only the average value matters; **c)** antisymmetric loads are orthogonal to the kernel of the influence functions

$$m_{xx}^h(\boldsymbol{x}) = \int_\Gamma \underbrace{[\, g_2 \cdot v_\nu^h + m_\nu(g_2) \cdot \frac{\partial w^h}{\partial \nu}\,]\, ds\boldsymbol{y}}_{influence\ of\ boundary\ values}$$

$$+ \underbrace{\sum_e \int_{\Omega_e} g_2 \cdot p_h\, d\Omega\boldsymbol{y} + \sum_i \int_{\Gamma_i} [\, g_2 \cdot v_h^\Delta - \frac{\partial g_2}{\partial \nu} \cdot m_h^\Delta\,]\, ds\boldsymbol{y}}_{influence\ of\ sources\ in\ the\ domain}$$

$$+ \underbrace{\sum_{k=1}^n g_2(\boldsymbol{y}_k) \cdot F_k^h}_{influence\ of\ nodal\ forces} + \underbrace{\sum_c g_2(\boldsymbol{y}^c) \cdot F_c^h}_{influence\ of\ corner\ forces} \qquad (1.490)$$

namely the element load p_h, the jumps in the Kirchhoff shear, v_h^Δ, the discontinuities m_h^Δ in the bending moments, the nodal forces F_k^h (due to the corner discontinuities of the twisting moment m_{xy}^h) and the corner forces F_c^h. All these forces together constitute the load case p_h.

This influence function looks very complicated, but in the end it is the same polynomial that is obtained when the shape functions are differentiated directly:

$$m_{xx}^h = 3.14 + 2.72\, x + 9.81\, y = \int_\Gamma [\, g_2\, v_\nu^h - \frac{\partial}{\partial \nu} g_2\, m_\nu^h - \dots \qquad (1.491)$$

The strange thing is that n data cells (n = number of degrees of freedom of the plate element) obviously suffice to store all the influence that the distant sources have on a single element.

If the two expressions (1.488) and (1.490) are subtracted, a representation of the FE error is obtained:

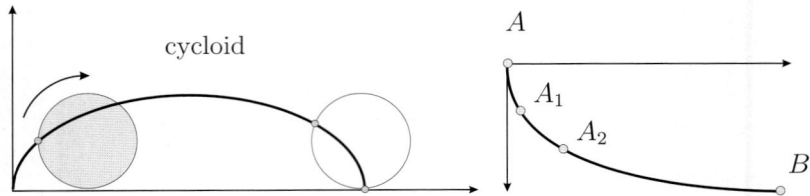

Fig. 1.121. The fastest connection between two points A and B is a cycloid. Because the initial acceleration at the points A_1 or A_2 is less than at A, it takes the same time to travel to B from A_1 or A_2 as from A [231]

$$m_{xx}(\boldsymbol{x}) - m_{xx}^h(\boldsymbol{x}) = \int_{\Gamma} [\, g_2(v_\nu - v_\nu^h) + m_\nu(g_2)\frac{\partial}{\partial\nu}(w - w_h)\,]\, ds_{\boldsymbol{y}} - \dots$$

$$(1.492)$$

The error cannot be calculated because the support reactions v_ν and the slope on the edge, $\partial w/\partial \nu$, are not known, but the formula provides a glimpse into how the error propagates, which depends on the nature of the kernel functions:

$$g_2 = O(\ln r) = \text{deflection surface due to } M_x = 1 \text{ at } \boldsymbol{x}$$

$$\frac{\partial}{\partial\nu}\, g_0 = O(r^{-1}) = \text{slope at the edge}$$

$$m_\nu(g_2) = O(r^{-2}) = \text{bending moment at the edge}$$

$$v_\nu(g_0) = O(r^{-3}) = \text{Kirchhoff shear}$$

These kernel functions decay very rapidly. The later increase of the logarithm comes too late to be of any significance ($\ln 100 = 4.6$).

St. Venant's principle depends on this rapid decay of the kernels and the averaging effect of integration, as can be illustrated by a simple example.

The support reaction A of the cantilever beam in Fig. 1.120 is the scalar product of the influence function $\eta_A = 1$ and the distributed load p:

$$A = \int_0^l \eta_A(x)\, p(x)\, dx\,.$$

$$(1.493)$$

Because η_A is constant the support reaction A is simply the *average value* p_a of the distributed load p times the length l:

$$A = \int_0^l p(x)\, dx = p_a \times l\,.$$

$$(1.494)$$

That is, the kernel $\eta_A = 1$ eliminates all "harmonics" of p which are antisymmetric with respect to the center $x = l/2$ of the beam (see Fig. 1.120 c) because they are *orthogonal* to the kernel η_A.

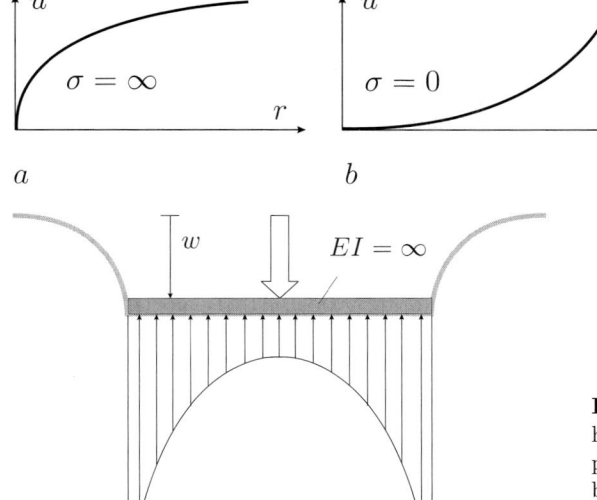

Fig. 1.122. Depending on how the displacements tend to zero, the stresses are either infinite or bounded

Fig. 1.123. Rigid punch on a half-space. At the edge of the punch the stresses in the soil become infinite because the strains are infinite

1.33 Singularities

Stresses are infinite where the strains are infinite, $\varepsilon = du/dx = \infty$, i.e., where the displacements change "on the spot". Why this happens is best explained by the problem of the *brachistochrone*, the problem to find the fastest route between two points A and B. The solution of this famous problem is a *cycloid*; see Fig. 1.121.

"It is better to start out vertically and pick up speed early, even if the path is longer" [231]. This is also the tendency we observe in structures. The material tries to escape as fast as possible from the dangerous zones by starting with an infinite slope $u'(0) = \infty$. Such an abrupt growth where the displacements change *stante pede*, on the spot, (see Fig. 1.122 a) is described by a function as

$$u = r^\alpha \qquad \alpha < 1 \qquad \Rightarrow \qquad \sigma = \frac{E}{r^{1-\alpha}}, \qquad (1.495)$$

whose derivative du/dr for values of $\alpha < 1$ is infinite at the start. If the displacement decays in a soft slope as in Fig. 1.122 b, then α is greater than one and the stresses remain bounded.

The best known example for abruptly changing deformations of type b is the rigid punch (Fig. 1.123). Outside the compression zone the displacement of the soil shoots straight up to taper off very rapidly. This abrupt decrease in the settlement is the reason for the infinite stresses at the edge of the punch.

In traffic accident research it is said *if the braking distance is zero then the force is infinite*. The same holds in structural mechanics. What for a speeding

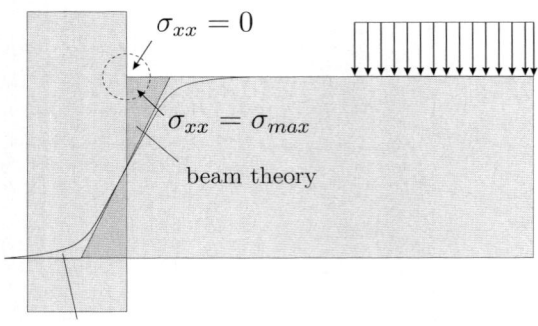

$$\sigma_{xx} = 0$$

$$\sigma_{xx} = \sigma_{max}$$

beam theory

theory of elasticity

Fig. 1.124. At the same point the stresses must be zero and they must assume the maximum value

car is the acceleration $a = dv/dt$[15] is the strain $\varepsilon = du/dx$ or the curvature $\kappa = d^2w/dx^2$ for a structure. If a plate cracks, then the strain is infinite, because in the uncracked concrete the two faces of the crack had the distance $dx = 0$ and even an infinitely small crack opening du will result in infinite strains, $du/dx = du/0 = \infty$. The same holds for a slab. At a sharp bend the radius R is zero, and therefore the curvature $\kappa = 1/R$ is infinite.

Stress singularities occur primarily at the edge, at reentrant corners, or at points where the boundary conditions change. Some singularities are simply the result of contradictory boundary conditions. Above the point where the cantilever beam intersects the wall (Fig. 1.124), the horizontal stress σ_{xx} must be zero, while directly below that point the bending stress $\sigma = M/W$ attains its maximum value.

This conflict is not the result of a "discretization error", which could be circumvented with a simple trick, but the treatment of the problem is not adequate. Each abrupt change in the boundary conditions is not in agreement with the fact that *partial differential equations* are to be solved.

All abrupt changes in the boundary conditions should theoretically be replaced by more "blurred" formulations, were it not that an FE program has its own interpretation of boundary conditions: geometric boundary conditions are satisfied exactly, but static boundary conditions only in the L_2-sense.

In the vicinity of a singularity, the displacement field of a plate consists of a "non-polynomial" singular part $\boldsymbol{u}_S$ and a regular "polynomial" part $\boldsymbol{u}_R$,

$$\boldsymbol{u}(x, y) = k\, r^\alpha \begin{bmatrix} u(\varphi) \\ v(\varphi) \end{bmatrix} + \boldsymbol{u}_R(x, y) = \boldsymbol{u}_S(x, y) + \boldsymbol{u}_R(x, y)\,. \qquad (1.496)$$

The factor k is the so-called *stress intensity factor*, and the exponent $\alpha < 1$ depends on the angle of the corner point and the boundary conditions. Because $\alpha < 1$ the stresses become singular:

$$\sigma_{ij} = k\, \frac{1}{\sqrt{r}} \ \ldots \qquad (\text{for } \alpha = 0.5)\,. \qquad (1.497)$$

---

[15] If a car hits the wall with a speed $v = 100$ km/h and is brought to a halt in 0 seconds, the negative acceleration is $a = \Delta v/\Delta t = -100/0 = -\infty$.

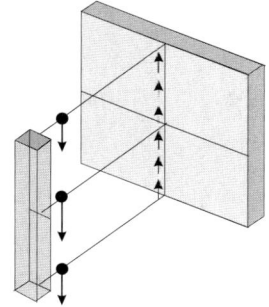

Fig. 1.125. Column and shear wall

The idea to handle only the regular part $\boldsymbol{u}_R$ with finite elements implies that the exact shape of the singular part is known, because when, say, instead of the exact function $r^{0.5}$ the function $r^{0.4}$ is subtracted, not much is gained, as the FE program still must approximate the missing singular part $r^{0.1}$ (actually things are a bit more complicated we cannot simply add and subtract exponents).

If the solution cannot be split into these two parts, the FE program must also approximate the singular part, and then one must be careful. One can then make *snapshots* of the stress state, which are "correct" for one mesh but which—in the neighborhood of the singularity—bear no resemblance to the subsequent stress states as soon as the mesh is refined adaptively.

1.34 Actio = reactio?

We expect that the stresses on the two faces of a cut are the same. But this must not be true in FE analysis. This does not contradict Newton's principle, because equilibrium in the FE sense means only that the virtual work done by the stresses on the two faces *and* the surface loads or volume forces on the left- and right-hand side of the cut is the same: $\delta W_e^+ + \delta W_e^- = 0$.

Any load in the neighborhood which senses the movement contributes to the virtual work and thereby blurs the picture, so that $\delta W_e^+ + \delta W_e^- = 0$ in general does not imply that the resultant stresses on the two faces are the same, $\boldsymbol{R}_+ = \boldsymbol{R}_-$.

Consider for example the masonry wall and the column in Fig. 1.125. If the column is modeled with linear elements, concentrated forces will act at the nodes and line loads at the opposing face if the wall is modeled with CST elements; see Fig. 1.125. What these different forces though have in common is that they are work-equivalent with respect to the nodal unit displacements of the interface nodes (volume forces are absent from this model).

Or imagine that a slab is modeled with Kirchhoff plate elements, and a T beam (Fig. 1.126) with beam elements. Evidently it is not possible, to simply

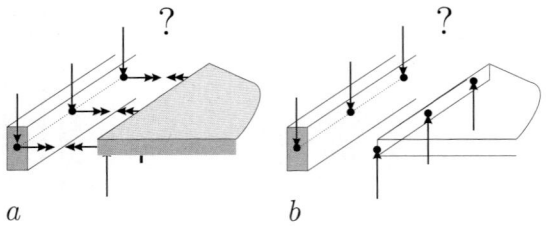

Fig. 1.126. The coupling between a beam and a slab is a work-equivalent coupling but not a mechanical coupling

a   *b*

transfer the nodal moments from the beam to the slab. Reissner–Mindlin elements would not even tolerate a transfer of the nodal forces.

Add to this that usually the displacements on the two faces are not the same because the shape functions on the two sides are different. Such inconsistencies are much more common than users are aware of. But they should only occur at the interfaces of different structural elements, because a structure can hardly be modeled with a series of gaps.

Equivalent nodal forces

What is the same though are the equivalent nodal forces at the interface between two structural components because at any such node N

$$\boldsymbol{f}_L + \boldsymbol{f}_R = \boldsymbol{f}_N \qquad (1.498)$$

where $\boldsymbol{f}_N$ is the equivalent nodal force applied at the node. The components f_i of the two nodal forces (L) and (R) are the strain energy products between the stress field $\boldsymbol{S}_h$ of the FE solution $\boldsymbol{u}_h$ and the nodal unit displacements $\boldsymbol{\varphi}_i$ of the node on the left and on the right

$$f_i = a(\boldsymbol{u}_h, \boldsymbol{\varphi}_i) = \int_\Omega \boldsymbol{S}_h \bullet \boldsymbol{E}(\boldsymbol{\varphi}_i)\, d\Omega = \int_\Omega [\sigma_{11} \cdot \varepsilon_{11} + 2\,\sigma_{12} \cdot \varepsilon_{12} + \sigma_{22} \cdot \varepsilon_{22}]\, d\Omega.$$

$$(1.499)$$

This is also the technique how equivalent nodal forces can be assigned to the nodes of interelement boundaries if a structure is split into different parts; see Fig. 1.127.

If no force is applied at the node, $\boldsymbol{f}_N = \boldsymbol{0}$, then the sum of the two forces, the two "energy quanta", is zero. Note that this result is independent of the shape of the elements on both sides of the interface. Large elements bordering on small elements possess the same nodal forces as the small elements.

What the small elements miss in size they make good in strains (not the strains from the FE solution $\boldsymbol{u}_h$ but from the fields $\boldsymbol{\varphi}_i$) because the smaller an element gets the larger the strains from the unit displacements of the nodes will be, this is the $1/h$ effect (see Fig. 1.68 p. 97) so that

$$f_i^L = \int_{small} \boldsymbol{S}_h \bullet \underbrace{\boldsymbol{E}(\boldsymbol{\varphi}_i^L)}_{large}\, d\Omega = \int_{large} \boldsymbol{S}_h \bullet \underbrace{\boldsymbol{E}(\boldsymbol{\varphi}_i^R)}_{small}\, d\Omega = f_i^R. \quad (1.500)$$

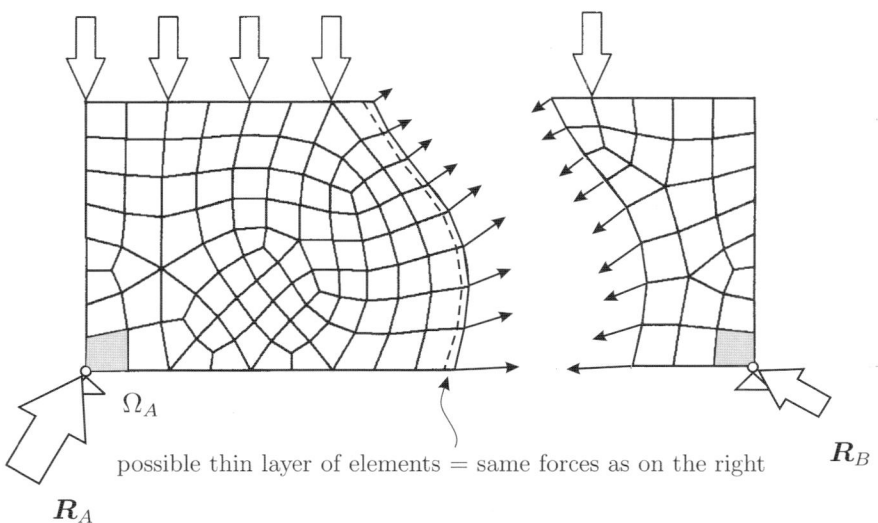

possible thin layer of elements = same forces as on the right

Fig. 1.127. The equivalent nodal forces on both sides are the same. This is independent of the shape of the elements on both sides. Even a thin layer of elements on the left would produce the same forces as the large elements on the right

Note also that the support reactions $\boldsymbol{R}_A$ and $\boldsymbol{R}_B$ of the plate are uniquely determined by the stress state in the two shaded elements bordering on the supports. For example the horizontal component $(R_A)_x$ of $\boldsymbol{R}_A$ is the scalar product between the stress state $\boldsymbol{S}_h$ in the element Ω_A and the strains of the unit displacement in horizontal direction of the node that attaches to the support

$$(R_A)_x = \int_{\Omega_A} \boldsymbol{S}_h \bullet \boldsymbol{E}(\boldsymbol{\varphi}_1^A)\, d\Omega\,. \tag{1.501}$$

The more the mesh is refined the smaller Ω_A gets but the loss in size is easily balanced by the increase in the stresses $\boldsymbol{S}_h$—the elements gets closer to the hot spot.

Example

The plate in Fig. 1.128 a carries a horizontal edge load of magnitude $2P$ and is modeled with two bilinear elements. The equivalent nodal forces are $f_1 = f_2 = P$ while all other f_i (also the vertical components) are zero. The FE solution is the solution of the load case in Fig. 1.128 g. The stress state $\boldsymbol{S}^h$ of the FE solution on acting through the strains $\boldsymbol{E}(\boldsymbol{\varphi}_i)$ of the unit displacement fields $\boldsymbol{\varphi}_i$ yields the same equivalent nodal forces f_i as the original load

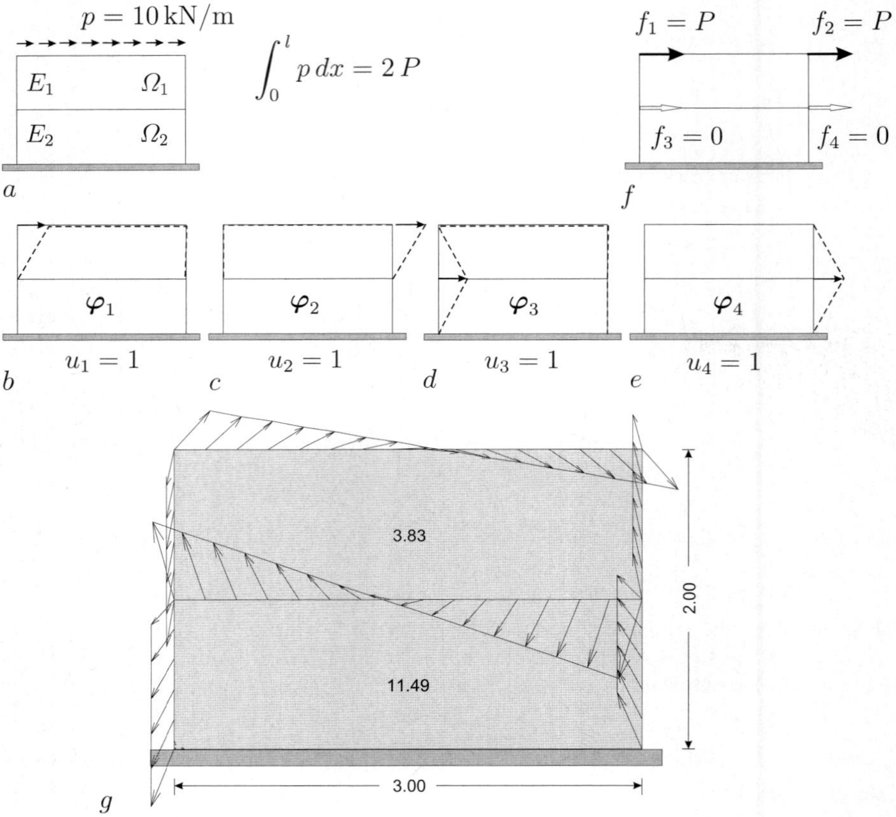

$p = 10 \, \text{kN/m}$

$$\int_0^l p \, dx = 2P$$

| E_1 | Ω_1 |
| E_2 | Ω_2 |

a

$f_1 = P$ $f_2 = P$

$f_3 = 0$ $f_4 = 0$

f

φ_1 φ_2 φ_3 φ_4

$u_1 = 1$ $u_2 = 1$ $u_3 = 1$ $u_4 = 1$

b   c   d   e

3.83

11.49

2.00

3.00

g

Fig. 1.128. Plate and shear forces: **a)** system and loading, **b)** - **e)** horizontal nodal unit displacements, **f)** equivalent horizontal nodal forces, **g)** FE solution

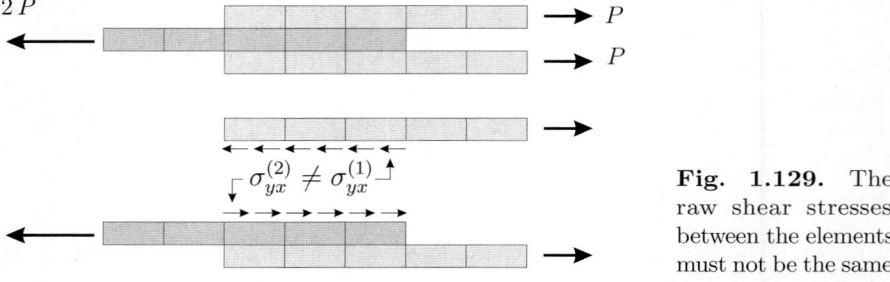

$2P$ P

P

$\ulcorner \, \sigma_{yx}^{(2)} \neq \sigma_{yx}^{(1)} \, \urcorner$

Fig. 1.129. The raw shear stresses between the elements must not be the same

$$f_i = a(\boldsymbol{u}_h, \boldsymbol{\varphi}_i) = \int_{\Omega} \boldsymbol{S}_h \bullet \boldsymbol{E}(\boldsymbol{\varphi}_i) \, d\Omega = \begin{cases} P & i = 1, 2 \\ 0 & i = 3, 4 \end{cases} . \tag{1.502}$$

(Of course the plate has more than four degrees of freedom). At the interface between the two elements the sum of the shear stresses is not zero ($d =$ thickness)

$$N_{yx}^{(1)} + N_{yx}^{(2)} = d \int_0^l \sigma_{yx}^{(1)} \, dx + d \int_0^l \sigma_{yx}^{(2)} \, dx \neq 0 , \tag{1.503}$$

because the shear forces must balance the horizontal component of the line force that acts at the interface. The obvious remedy (see also Fig. 1.129) is to work with averaged stresses. Let

$$\boldsymbol{j} = \boldsymbol{t}_h^{(1)} + \boldsymbol{t}_h^{(2)} \quad (= \downarrow + \uparrow) \qquad \text{on } \partial\Omega_1 \cap \partial\Omega_2 \tag{1.504}$$

the jump in the tractions, the improved ($\hat{}$) averaged tractions are

$$\hat{\boldsymbol{t}}^{(1)} = \boldsymbol{t}_h^{(1)} - \frac{1}{2}\boldsymbol{j} \qquad \hat{\boldsymbol{t}}^{(2)} = \boldsymbol{t}_h^{(2)} - \frac{1}{2}\boldsymbol{j} . \tag{1.505}$$

Unlike the resultant stresses $N_{yx}^{(i)}$ the equivalent nodal forces $f_3^{(i)}$ on the two sides of the interface balance

$$f_3 = f_3^{(1)} + f_3^{(2)} = \int_{\Omega_1} \boldsymbol{S}_h \bullet \boldsymbol{E}(\boldsymbol{\varphi}_3) \, d\Omega + \int_{\Omega_2} \boldsymbol{S}_h \bullet \boldsymbol{E}(\boldsymbol{\varphi}_3) \, d\Omega = 0 . \tag{1.506}$$

1.35 The output

To assess the accuracy of FE results correctly, it must be understood how an FE program processes the raw output and how it displays it on the screen.

The load case p_h

In general the equivalent load case p_h is not displayed on the screen, because a user not well-acquainted with FE techniques would be irritated.

Support reactions

One would assume that an FE program outputs the support reactions of the FE load case p_h. These forces plus the forces that have been reduced to the supports at the start would be the true support reactions. But instead what is displayed on the screen are the *equivalent support reactions*, the equivalent nodal forces spread along the supports to simulate a continuous support.

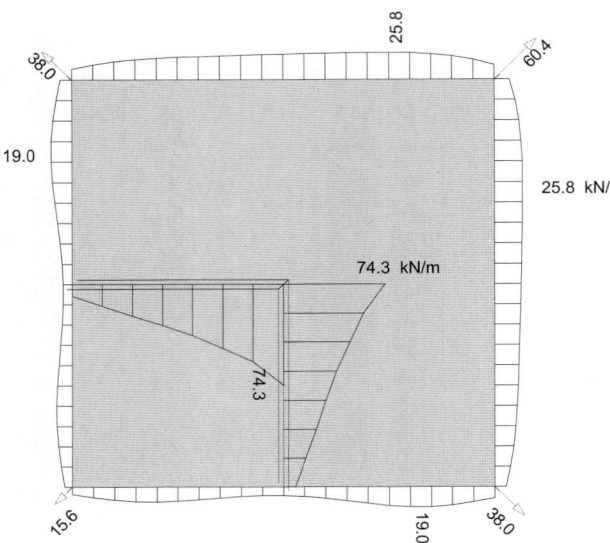

Fig. 1.130. On-screen appearance. The support reactions are the evenly spread equivalent nodal forces f_i

Formally, what happens is that the program converts the element volume loads p_h and interelement line loads l_h into equivalent nodal support reactions by letting these loads act through the nodal unit displacements of the supports:

$$f_i = \int_\Omega p_h \, \varphi_i \, d\Omega + \int_\Gamma l_h \, \varphi_i \, ds$$

φ_i = unit displacement of a support node.

Because in the neighborhood of supports there are probably more loads p_h and l_h pointing upward (having a negative sign) the net result will be a series of equivalent nodal forces that point upward, i.e., which support the slab.

Basically all this was already done when the global stiffness matrix was assembled. Hence the stiffness matrix $\boldsymbol{K}$ must only be multiplied by the nodal unit displacements:

$$\boldsymbol{K}\,\boldsymbol{u} = \boldsymbol{f} \qquad \leftarrow \qquad \text{list of equivalent nodal forces}. \qquad (1.507)$$

These equivalent nodal forces f_i (kN m) are then transformed into equivalent line forces (kN/m). Assuming a linear distribution between two nodes, this would result in a distribution such as

$$l(x) = \frac{1}{2} \times \left[\frac{f_i}{l_e}(1 - \frac{x}{l_e}) + \frac{f_{i+1}}{l_e}\frac{x}{l_e} \right] \qquad 0 < x < l_e. \qquad (1.508)$$

In Fig. 1.131 and Fig. 1.130 the two versions can be seen side by side. The first figure shows the distribution of the support reactions as they appear

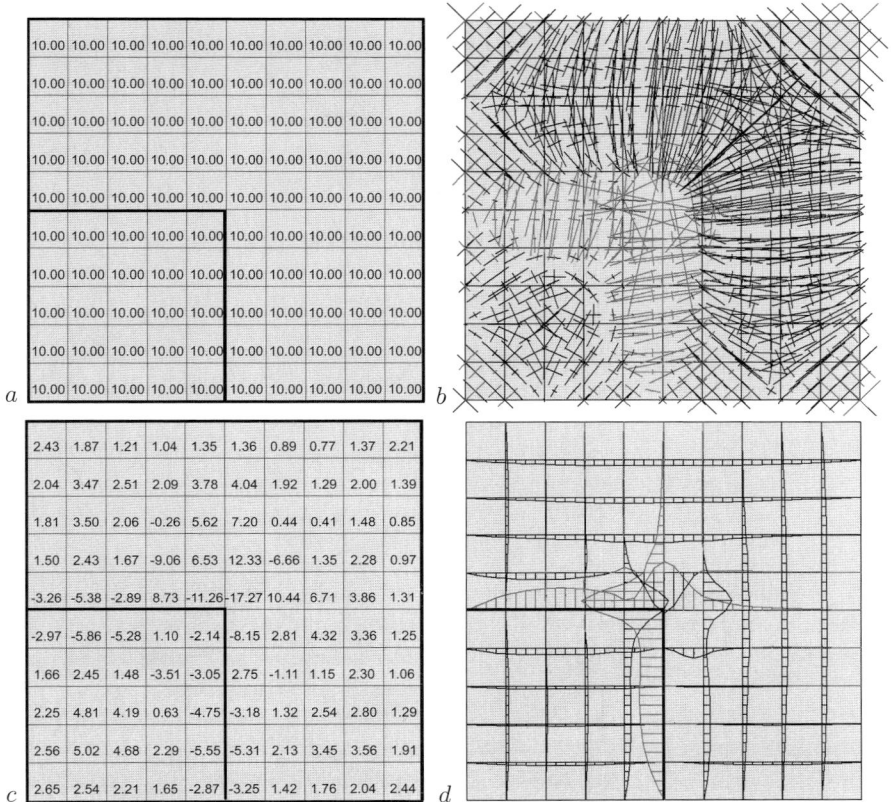

Fig. 1.131. Slab **a)** system and loading, **b)** principal moments, **c)** element surface loads, **d)** vertical forces along the interelement boundaries

on-screen—these are the transformed equivalent nodal forces f_i—while the second figure displays the "true" support reactions, where it is seen that the slab is not only supported by the walls but by negative element surface loads as well. Note also that the support reactions do not end abruptly at the ends of the walls, but continue beyond these points.

1.36 Support conditions

An FE solution satisfies geometric boundary conditions such as

$$\text{plate:} \quad u = 0, \quad v = 0 \qquad \text{slab:} \quad w = 0 \qquad w_{,n} = 0 \qquad (1.509)$$

pointwise, while static boundary conditions

$$\text{plate:} \quad \boldsymbol{S}\,\boldsymbol{n} = \bar{\boldsymbol{t}} \qquad \text{slab:} \quad m_n = \bar{m} \qquad v_n = \bar{v} \qquad (1.510)$$

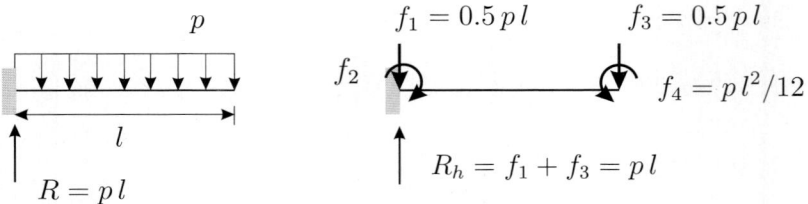

Fig. 1.132. The resultant forces are the same, $R = R_h$

are only satisfied in the weak sense, i.e., along a free edge of a slab it is only guaranteed that the support reaction v_n together with the surface load p_h and interelement jump terms v_n^Δ in the neighborhood of the edge (see Sect. 1.34, p. 177) contribute no work through any of the nodal unit displacements of the edge nodes:

$$\int_\Omega p_h\,\varphi_i\,d\Omega + \sum_k \int_{\Gamma_k} v_n^\Delta\,\varphi_i\,ds + \int_\Gamma v_n\,\varphi_i\,ds = 0\,. \qquad (1.511)$$

The same holds for the bending moment m_n^h, which is nonzero along a free or hinged edge. The distribution of m_n^h is skillfully balanced in such a way that m_n^h annihilates the work done by the other terms:

$$f_i = \int_\Omega p_h\,\varphi_i\,d\Omega + \sum_k \int_{\Gamma_k} m_n^\Delta\,\frac{\partial\varphi_i}{\partial n}\,ds + \int_\Gamma m_n^h\,\frac{\partial\varphi_i}{\partial n}\,ds = 0\,. \quad (1.512)$$

The same logic applies of course to edge loads. The substitute FE edge loads are only weakly equivalent to the true load.

1.37 Equilibrium

Statements such as

- *Global equilibrium is satisfied*
- *Equilibrium is usually not satisfied within an element*
- *Equilibrium is usually not satisfied across interelement boundaries*
- *Equilibrium of nodal forces and moments is satisfied*   (?)

do not boost our confidence in the FE method, but they lose much of their alarmism if we recall that in the FE method the original load case p is replaced by a work-equivalent load case p_h and it is therefore quite natural for the stress resultants of the load case p_h not to maintain equilibrium with the forces of the load case p.

An FE program commits only *one* error, and this at the very start: it replaces the original load case by a work-equivalent load case. Everything else

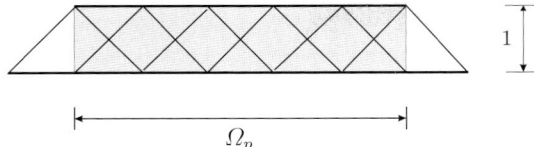

Ω_p

Fig. 1.133. Because the footprint of the nodal unit displacements extends beyond the patch, equilibrium is not maintained

that follows is classical structural analysis. An FE program solves the work-equivalent load case *exactly*. Therefore both the whole structure and every individual part maintains equilibrium—with the forces of the work-equivalent load case.

Global equilibrium

This term has a very special meaning in the FE method. After the global stiffness matrix $\boldsymbol{K}$ is assembled it contains the contributions of the nodal unit displacement of all the nodes of the structure. Next, those rows and columns are deleted that belong to fixed degrees of freedom ($u_i = 0$), and the so-called *reduced stiffness matrix* is obtained. The entries $k_{ij} = a(\varphi_i, \varphi_j)$ in the reduced stiffness matrix are the strain energy products of the nodal unit displacements in V_h, while the entries in the full matrix $\boldsymbol{K}$ are the strain energy products of the φ_i in the space V_h^+, which is the space V_h plus all nodal unit displacements of the fixed nodes, the support nodes.

Because all rigid-body motions $u_0 = a \times x + b$ of the structure do lie in V_h^+, the resultant force $\boldsymbol{R}_h$ of the substitute load p_h coincides in size, direction, and position with the resultant force $\boldsymbol{R}$ of the original load distribution p. This follows simply from the fact that $p(\varphi_i) = p_h(\varphi_i)$, $i = 1, 2, \ldots n$, and that any rigid-body motion u_0 can be written in terms of the φ_i. Hence *global equilibrium* in FE terminology means $\boldsymbol{R}_h = \boldsymbol{R}$.

If $\boldsymbol{A}$ is the resultant support reaction in the load case p, then $\boldsymbol{R} + \boldsymbol{A} = \boldsymbol{0}$, and the same holds for the FE solution $\boldsymbol{R}_h + \boldsymbol{A}_h = \boldsymbol{0}$. Because $\boldsymbol{R}_h = \boldsymbol{R}$, it follows that $\boldsymbol{A}_h + \boldsymbol{R} = \boldsymbol{0}$ must be true as well. In this last equation we are comparing apples ($\boldsymbol{A}_h$ is from the load case p_h) with oranges ($\boldsymbol{R}$ is from the load case p), but because of global equilibrium, $\boldsymbol{R}_h = \boldsymbol{R}$, it makes no difference.

Local equilibrium

Why is it that the Kirchhoff shear v_n^h of an FE solution integrated along the edge of an arbitrary patch Ω_p of elements does not balance the original load acting on that patch? The reason is that the rigid-body motions of the patch extend beyond the patch, see Fig. 1.133.

Let us consider a patch Ω_p of a slab which is subjected to a uniform surface load p that vanishes outside the patch. Let $\boldsymbol{R}^p$ the resultant force and let $\boldsymbol{R}_h^p$ the resultant force of the FE load p_h acting on this patch. If the resultant forces were the same, $\boldsymbol{R}^p = \boldsymbol{R}_h^p$, the integral of the Kirchhoff shear along the perimeter of the patch would also be the same, $(v_n - v_n^h, 1)_\Gamma = 0$ (we neglect the corner forces). The equation $\boldsymbol{R}^p = \boldsymbol{R}_h^p$ is valid if and only if

$$\int_{\Omega_p} p\, u_0\, d\Omega = \int_{\Omega_p} p_h\, u_0\, d\Omega \tag{1.513}$$

for all rigid-body motions u_0 of the patch. The problem is that the rigid-body motions of the patch do not lie in V_h. For to lift the patch by one unit of displacement the movement

$$u_0(\boldsymbol{x}) = \begin{cases} 1 & \boldsymbol{x} \in \Omega_p \\ 0 & \text{else} \end{cases} \tag{1.514}$$

would have to lie in V_h. But such a discontinuous function is nonconforming.

Imagine that a tablecloth is spread over the slab and that the patch is lifted while we try to hold down the rest of the tablecloth. This shape is as close as we can get on V_h to the lift of the patch.

Because both load cases p and p_h are equivalent with respect to all φ_i they are also equivalent with respect to the shape of the "tablecloth". In a somewhat symbolic notation this means

$$\int_{\Omega_{p+1}} p\, (\triangle\!\Box\!\triangle\,)\, d\Omega = \int_{\Omega_{p+1}} p_h\, (\triangle\!\Box\!\triangle\,)\, d\Omega, \tag{1.515}$$

where Ω_{p+1} denotes that part of the slab where the height of the tablecloth is not zero. This is Ω_p plus one row of elements (probably). According to our assumptions p is zero outside of Ω_p—this simplifies the derivation—and therefore equilibrium is "almost" established

$$\int_{\Omega_p} p\, \Box\, d\Omega = \int_{\Omega_{p+1}} p_h\, (\triangle\!\Box\!\triangle\,)\, d\Omega. \tag{1.516}$$

Specifically the weight of the load p on the patch Ω_p is the same as the weight of the load "$1/2 p_h + p_h + 1/2 p_h$" on the patch Ω_{p+1}, and therefore R_h^p (the integral of v_n^h along the edge of the patch Ω_p) cannot be the same as the weight of p on Ω_p, that is

$$R^p = \int_{\Omega_p} p\, d\Omega \neq \int_{\Gamma_p} v_n^h\, ds = R_h^p. \tag{1.517}$$

The reviewing engineer would like to have

$$\int_{\Omega_p} p\, \Box\, d\Omega = \int_{\Omega_p} p_h\, \Box\, d\Omega, \tag{1.518}$$

which would only be true if p_h were zero in the neighboring elements, which is very improbable. We can only hope that the smaller the elements become, the smaller the ramp becomes, and the closer we come to a true local equilibrium.

Nodal forces

The structural equation $\boldsymbol{Ku} = \boldsymbol{f}$ is not an equilibrium condition. To speak of nodal equilibrium is misleading because the layperson takes this statement literally. Rather it expresses the equivalence of the FE load case p_h with the original load case p,

$$p_h(\varphi_i) = f_i^h = \sum_j k_{ij}\, u_j = f_i = p(\varphi_i) \qquad i = 1, 2, \ldots n, \qquad (1.519)$$

i.e., virtual work (kN m) is equated, not forces (kN); see Eq. (7.92), p. 515.

1.38 Temperature changes and displacement of supports

To get things straight let us repeat the constitutive equations for a bar $[0, l]$ with mechanical and thermal loading, [115],

$$\begin{aligned} u' - \varepsilon &= \alpha_T\, \Delta T \\ EA\,\varepsilon - N &= 0 \\ -N' &= p\,. \end{aligned} \qquad (1.520)$$

The longitudinal displacement of the bar is, see Fig. 1.134,

$$u = u_{el} + u_T \qquad (1.521)$$

where $u_{el} = u - u_T$ is the part that corresponds to the mechanical loading. We assume that the bar is fixed on the left side, $u(0) = 0$, and that it carries additionally a single force P at the other end at $x = l$. Green's first identity $G(u_{el}, v) = (p, v) + P \cdot v(l) - a(u_{el}, v) = 0$ then implies

$$a(u_{el}, v) = \int_0^l p\, v\, dx + P \cdot v(l) \qquad v \in V \qquad (1.522)$$

so that with $u_{el}^h = u_h - u_T$

$$a(u_{el}^h, \varphi_i) = a(u_h - u_T, \varphi_i) = \int_0^l p\, \varphi_i\, dx + P \cdot \varphi_i(l) \qquad \varphi_i \in V_h \quad (1.523)$$

or

$$a(u_h, \varphi_i) = a(u_T, \varphi_i) + f_i \qquad \varphi_i \in V_h \qquad (1.524)$$

it follows

$$\boldsymbol{Ku} = \boldsymbol{f}_T + \boldsymbol{f} \qquad (1.525)$$

where

$$u = \alpha_t \, \Delta T \, x$$

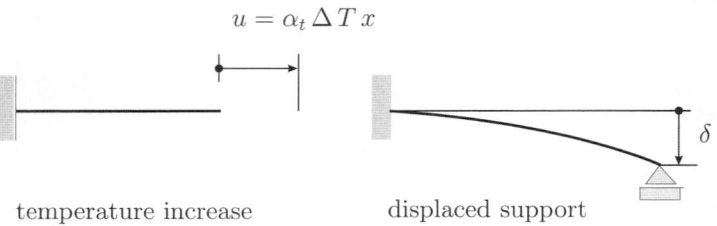

temperature increase   displaced support

Fig. 1.134. Change of temperature and a displaced support

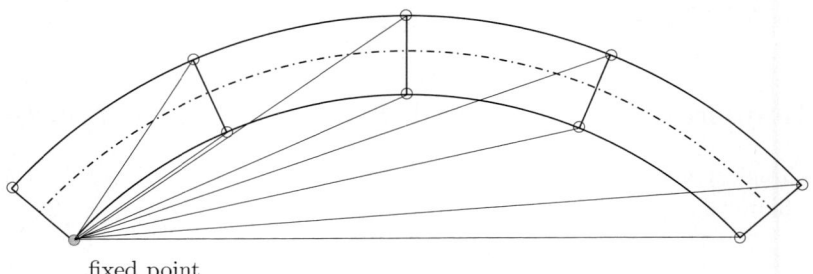

fixed point

Fig. 1.135. Temperature changes should leave the bridge stress-free

$$k_{ij} = a(\varphi_i, \varphi_j) \qquad f_{T_i} = a(u_T, \varphi_i) \qquad f_i = \int_0^l p\,\varphi_i\,dx + P \cdot \varphi_i(l)\,.$$

$$(1.526)$$

In the following we employ a two-node bar element and we assume for simplicity that the mechanical load is zero, $p = 0, P = 0$.

When the thermal loading is constant as in (1.520) then $u_T = \alpha \, \Delta T \, x$ and consequently the equivalent nodal forces are

$$f_{T1} = a(u_T, \varphi_1) = EA \int_0^l u_T' \, \varphi_1' \, dx = EA \int_0^l \alpha_T \, \Delta T \cdot \frac{-1}{l} \, dx$$
$$= -EA \, \alpha_T \, \Delta T = -f_{T2} = -a(u_T, \varphi_2)\,. \qquad (1.527)$$

Hence the system

$$\frac{EA}{l} \begin{bmatrix} 1 & -1 \\ -1 & 1 \end{bmatrix} \begin{bmatrix} u_1 \\ u_2 \end{bmatrix} = \begin{bmatrix} f_{T1} \\ f_{T2} \end{bmatrix} \qquad (1.528)$$

has the solution $u_1 = 0$, $u_2 = \alpha_T \, \Delta T \, l$ and so the elastic part u_{el} is—as we expect—zero

$$u_{el}^h = u_h - u_T = \alpha_T \, \Delta T \, l \cdot \varphi_2(x) - \alpha_T \, \Delta T \, x = \alpha_T \, \Delta T (\frac{x}{l} \cdot l - x) = 0$$

$$(1.529)$$

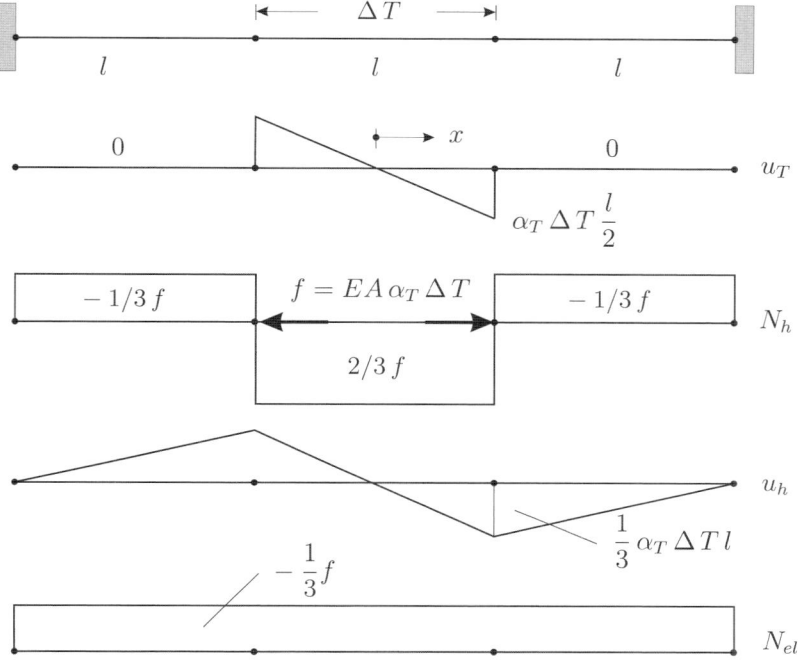

Fig. 1.136. Change of temperature in one element only

which means that the bar is stress free $\sigma = N/A = E\,\varepsilon = E\,(u^h_{el})' = 0$.

Next let the thermal loading increase with the distance x from the fixed support, $\alpha_T\,\Delta T\,x$, then

$$u_T = \frac{1}{2}\,\alpha_T\,\Delta T\,x^2 \tag{1.530}$$

and so

$$f_{T1} = -\frac{1}{2}EA\,\alpha_T\,\Delta T\,l = -f_{T2}. \tag{1.531}$$

Now the system (1.528) has the solution $u_1 = 0$, $u_2 = 1/2\,\alpha_T\,\Delta T\,l^2$. But because u_h is linear and u_T is quadratic the elastic solution is *not* zero

$$u^h_{el} = u_h - u_T = \frac{1}{2}\,\alpha_T\,\Delta T(x\,l - x^2) \tag{1.532}$$

and so spurious stresses $\sigma = E\,(u^h_{el})'$ appear in the bar which should be stress-free.

This means that temperature fields must have the same polynomial order as the strains of the shape functions. If necessary higher order fields must be interpolated by lower order functions: if the shape functions are quadratic the temperature fields must be linear, if they are linear the temperature fields must be constant, etc., see Fig. 1.135.

Example

In Fig. 1.136 the temperature increases in the second element only, so that $u_T = \alpha_T \, \Delta T \, x$ in the second element and $u_T = 0$ in the first and the third element. In these elements the elastic solution $u_{el} = u_h - u_T$ coincides with u_h and in the second element

$$u_{el} = u_h - u_T = \alpha_T \, \Delta T \left(\frac{2}{3} x - x\right) = -\alpha_T \, \Delta T \, \frac{x}{3} \qquad (1.533)$$

so that $N = -\alpha_T \, \Delta T \, (EA/3)$ in that element.

Supports

If the support of a beam settles, $w(l) = \delta$, the procedure is virtually the same. The solution is

$$w(x) = w_\delta(x) + w_{el}(x), \qquad (1.534)$$

where w_δ is a deflection curve with the property $w_\delta(l) = \delta$ and $w_{el}(x)$ corresponds to the mechanical load p.

As before we have

$$a(w_{el}^h, \varphi_i) = a(w_h - w_\delta, \varphi_i) = \int_0^l p \, \varphi_i \, dx \qquad (1.535)$$

and so

$$\boldsymbol{Ku} = \boldsymbol{f}_\delta + \boldsymbol{f} \qquad (1.536)$$

where

$$k_{ij} = a(\varphi_i, \varphi_j) \qquad f_{\delta_i} = a(w_\delta, \varphi_i) \qquad f_i = \int_0^l p \, \varphi_i \, dx \,. \qquad (1.537)$$

In both cases the equivalent nodal forces

$$f_{T_i} = a(u_T, \varphi_i) = \delta W_e(p_T, \varphi_i) \qquad (1.538)$$
$$f_{\delta_i} = a(w_\delta, \varphi_i) = \delta W_e(p_\delta, \varphi_i), \qquad (1.539)$$

can be identified, via Green's first identity

$$G(u_T, \varphi_i) = \underbrace{\int_0^l -EA \, u_T'' \, \varphi_i \, dx + [N_T \, \varphi_i]_0^l}_{\delta W_e(p_T, \varphi_i)} - a(u_T, \varphi_i) = 0, \qquad (1.540)$$

with the virtual work done by external forces p_T and p_δ, respectively. Note that $N_T(x) = EA \, u_T'(x)$. In 1-D problems the loads p_T are just the fixed end forces $\times(-1)$ due to the change in temperature $\alpha \, \Delta T$,

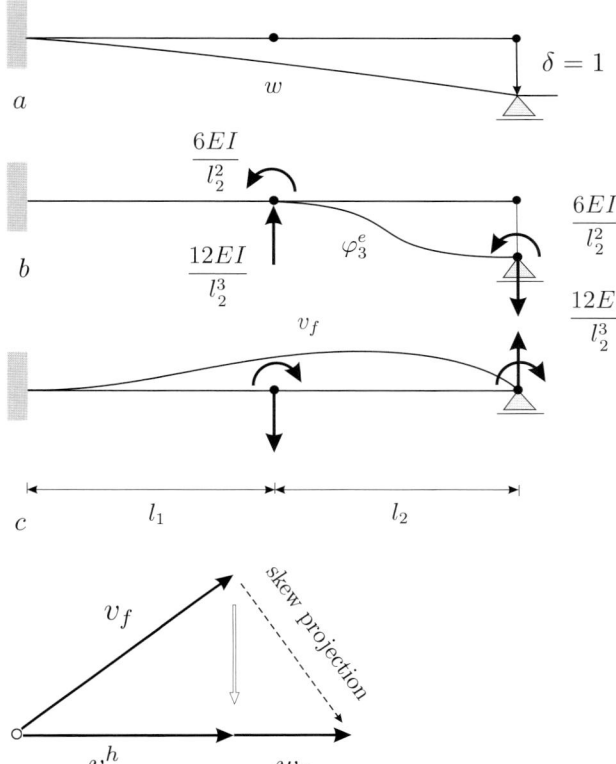

Fig. 1.137. Even in load cases δ where $\delta W_i(w_{el}, \varphi_i) = 0$, projections are applied

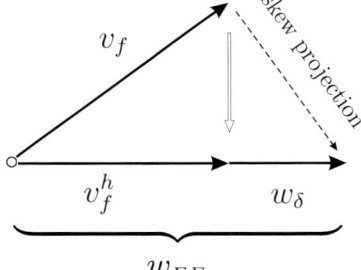

W_{FE}

Fig. 1.138. The FE solution $w_{FE} = v_f^h + w_\delta$

$$a(u_T, \varphi_i) = \int_0^l -EA \left(\alpha \, \Delta T \, x \right)'' \varphi_i \, dx + [N_T \, \varphi_i]_0^l = N_T(l) \, \varphi_i(l) - N_T(0) \, \varphi_i(0)$$

$$(1.541)$$

that is the forces ($\leftarrow \rightarrow$) the bar would exert on the confining walls if it were fixed on both sides

$$\leftarrow \quad -f_1 = N_T(0) = EA \, \alpha \, \Delta T \quad \rightarrow \quad f_2 = N_T(l) = EA \, \alpha \, \Delta T \quad (1.542)$$

and the forces p_δ result from the movement of the displaced node $w_i = \delta$, because w_δ is constructed by picking an appropriate nodal unit displacement φ_i of the structure.

The solution technique can be summarized as follows:

- First all nodes are kept fixed, and the fixed end forces f_{Ti} and $f_{\delta i}$ due to the temperature change or the movement of a node are calculated.

- The system $\boldsymbol{Ku} = \boldsymbol{f} + \boldsymbol{f}_\delta$ is solved, and with the nodal displacements u_i the elastic displacements and the elastic stresses are calculated.
- The stresses caused by the temperature change and the displaced support respectively are added to the elastic stresses.

Projection

Also load cases δ are solved by a projection method, even when (seemingly) $p = 0$. To see this, note that the strain energy product of the part w_{el}^h of the FE solution (1.534) with any $\varphi_i \in V_h$ must be zero:

$$a(w_{el}^h, \varphi_i) = a(w_h, \varphi_i) - a(w_\delta, \varphi_i) = 0 \,. \tag{1.543}$$

To achieve this we proceed as follows: for the beam to assume the shape $w_\delta = \delta \cdot \varphi_3^e$ (we let $\delta = 1$), the nodal forces in Fig. 1.137 b must be applied. These forces, so to speak, are the fixed end actions of the load case δ. Next the opposite of these forces are applied to the *original* beam (see Fig. 1.137 c) i.e., when $w(l) = 0$. Let this load case be called $-p_\delta^0$ and the associated deflection curve v_f. When this load case $-p_\delta^0$ is solved with finite elements, the deflection curve v_f is projected onto the set V_h (see Fig. 1.138), and to this function v_f^h the deflection curve w_δ is added. The result is the FE solution $w_{FE} = w_\delta + v_f^h$.

Note that here—as in all standard 1-D problems—$v_f^h = v_f$ because v_f is a piecewise third-order polynomial which lies in V_h.

Influence functions

If a support of a beam settles by 5 mm the Green's function for the deflection $w(x)$ at any point x is

$$w(x) \cdot 1 = R_0(x) \cdot 5 \, \text{mm} \tag{1.544}$$

where $R_0(x)$ is the support reaction due to the point load $P = 1$, the Dirac delta δ_0, acting at x, s. Sect. 7.3 p. 516. For any other quantity, $w'(x), M(x), V(x), R_0$ must be replaced by the appropriate support reaction R_i corresponding to δ_i.

If the temperature in a frame element changes ($\alpha\, T$ = temperature strain) or if an element is prestressed (N^+) the influence function for the axial displacement is

$$u(x) = \int_0^l [N_0\, \alpha\, T + \varepsilon_0\, N^+]\, dx \,, \tag{1.545}$$

where N_0 and ε_0 are the normal force and the strain respectively due to the Dirac delta δ_0. This result is based on a mixed formulation, see Sect. 4.19 p. 399, which provides the theoretical background for such problems. Though in practice it is much simpler to think in terms of equivalent nodal forces and to apply the negative end fixing forces to the structure and to follow their effects with the standard influence functions for load cases p.

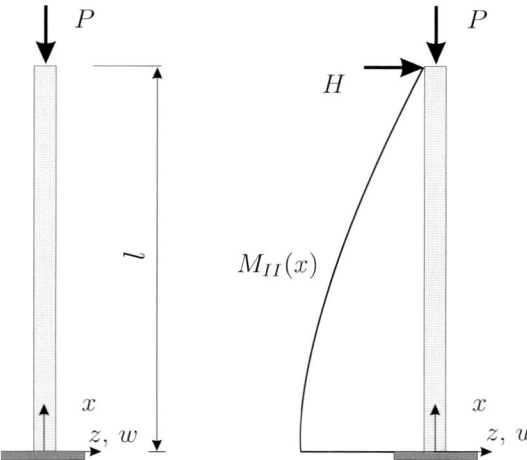

Fig. 1.139. Stability problem and stress problem

1.39 Stability problems

In practice there are no stability problems, because even in "perfect" structures we find eccentricities. But then also in stress problems, failure occurs if the critical load level is reached, as in the example of the Euler beam I in Fig. 1.139. If the horizontal force H is absent, it is a stability problem, and with the force H it becomes a stress problem, but the critical load

$$P_{crit} = \frac{\pi^2}{4} \frac{EI}{l^2} \qquad (1.546)$$

also dominates the stress problem, because when the load reaches P_{crit}, which corresponds to $\varepsilon = \pi/2$, the bending moment at the base of the column becomes infinite

$$M = -\frac{H\,l}{\varepsilon} \tan \varepsilon\,, \qquad \varepsilon^2 = l^2 \frac{|P|}{EI}\,, \qquad (1.547)$$

because $\tan \varepsilon = \infty$ for $\varepsilon = \pi/2$.

   In a true stability problem there are no lateral loads p. The only external load, the compressive force P, enters the problem via the differential equation. Formally it does not count as an external load.

   In stability problems the potential energy Π consists only of the internal energy $\Pi(w) = 1/2\,a(w,w)$, and Π is zero when the structure buckles (!)

$$\Pi(w_{crit}) = \frac{1}{2}\,a(w_{crit}, w_{crit}) = \frac{1}{2} \int_0^l [\frac{M_{crit}^2}{EI} - P\,(w'_{crit})^2]\,dx = 0 \ (1.548)$$

so that w_{crit} cannot be found by minimizing the potential energy. It also makes no sense to search for a work-equivalent load case p_h, because in stability problems $p = 0$. Instead *Galerkin's method (weighted residual method)* is applied. The buckled shape w_{crit} of the beam must satisfy the differential equation

$$EIw^{IV}(x) + P\,w''(x) = 0 \qquad (1.549)$$

and homogeneous boundary conditions such as $w(0) = 0$ and/or $w'(0) = 0$, etc. All elastic curves w which satisfy the geometric boundary conditions form the space V.

The beam is subdivided into m finite elements and allowed to assume under compression only those shapes that can be expressed by the n nodal unit displacements of the free nodes, $\sum_i u_i \varphi_i$, where the φ_i are usually the nodal unit displacements of the first-order beam theory (!). These shape functions form the basis of the subspace $V_h \subset V$.

Because of (1.549), the right-hand side of the exact deflection curve $w = w_{crit}$ is orthogonal to all shape functions $\varphi_i \in V_h$:

$$\int_0^l [EIw^{IV}(x) + P\,w''(x)] \cdot \varphi_i\, dx = 0\,. \qquad (1.550)$$

After integration by parts, it follows—because the shape functions $\varphi_i \in V_h$ satisfy the boundary conditions—that the strain energy product between w and the shape functions must also be zero:

$$a(w, \varphi_i) = \int_0^l [EIw''\varphi_i'' - P\,w'\varphi_i']\, dx = 0 \qquad i = 1, 2, \ldots, n\,. \qquad (1.551)$$

The FE solution w_h tries to imitate this property of the true solution. That is, the nodal displacements u_i must satisfy the system

$$(\boldsymbol{K} - P \times \boldsymbol{K}_G)\,\boldsymbol{u} = \boldsymbol{0} \qquad (1.552)$$

where

$$k_{ij} = \int_0^l EI\,\varphi_i''\,\varphi_j''\, dx \qquad k_{ij}^G = \int_0^l \varphi_i'\,\varphi_j'\, dx\,. \qquad (1.553)$$

The trivial solution would be $\boldsymbol{u} = \boldsymbol{0}$, which is the neutral position of the beam. Because the right-hand side is zero, a solution $\boldsymbol{u} \neq \boldsymbol{0}$ can only exist if the determinant of the matrix is zero:

$$\det (\boldsymbol{K} - P \times \boldsymbol{K}_G) = 0\,. \qquad (1.554)$$

The smallest positive number $P > 0$, for which this holds is the approximate *buckling load* P_{crit}^h.

We know that the pitch of a guitar string will increase with the tension in the string. The opposite tendency we observe in a column. The frequency will decrease if the compression increases and if the column finally buckles

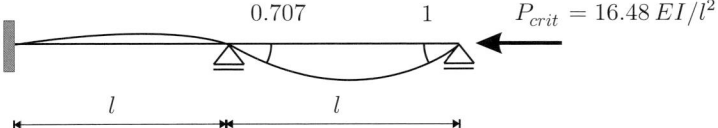

Fig. 1.140. The buckling load and the first eigenmode

the return frequency has reached its lowest possible value, namely zero, which means that it takes the column infinitely long to perform one full swing.

Not all stability problems possess a distinctive lowest eigenvalue. In some cases a geometric nonlinear analysis with proper imperfections is not only more concise but sometimes also the only possible way to predict the limit state of a structure.

Rayleigh quotient

In FE analysis the buckling load P^h_{crit} is an overestimate. This follows from the fact that the buckled shape w_{crit} minimizes the Rayleigh quotient on V, and that the minimum value is just P_{crit}:

$$P_{crit} = \frac{\int_0^l EI(w''_{crit})^2 \, dx}{\int_0^l (w'_{crit})^2 \, dx} . \tag{1.555}$$

But because the minimum on a subspace V_h is always greater than the minimum on the whole space V, it follows that $P^h_{crit} \geq P_{crit}$.

Usually the eigenvector $\boldsymbol{u}$ that belongs to the eigenvalue P^h_{crit} is normalized in the sense that $|u_i| \leq 1$. If the associated shape

$$w_h = \sum_i u_i \, \varphi_i \tag{1.556}$$

is substituted element-wise into the differential equation $EIw^{IV}(x) + Pw''(x) = 0$ and the associated nodal forces and moments are studied, the FE load case p_h is recovered. The latter is an expansion in terms of the unit load cases p_i

$$p_h = \sum_i u_i \, p_i . \tag{1.557}$$

Because the nodal unit displacements φ_i of first-order beam theory as for example

$$\varphi_1(x) = 1 - \frac{3x^2}{l^2} + \frac{2x^3}{l^3} , \tag{1.558}$$

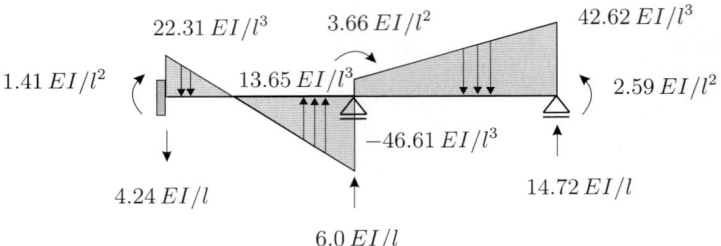

Fig. 1.141. The load case p_h solved by the first eigenmode

are not homogeneous solutions of the differential equation of second-order beam theory

$$(EI\frac{d^4}{dx^4} + P\frac{d^2}{dx^2})\,\varphi_1(x) = P\,(\frac{12\,x}{l^3} - \frac{6}{l^2}) \qquad (1.559)$$

lateral loads hold the "buckled" beam in place. That is, the FE solution is the solution of a stress problem, even though it shares with the exact curve the property that it is orthogonal to all $\varphi_i \in V_h$. In normal FE analysis such functions would be called *spurious modes*, because they do not interact with the other shape functions φ_i.

In FE analysis to the "buckled" shape of a beam or a plate belongs a load case $p_h \neq 0$ which is orthogonal to all nodal unit displacements.

If the homogeneous solution of second-order beam theory were used as shape functions, the FE program would be an implementation of the second-order slope-deflection method, and the buckled shape would be exact because then w_{crit} would lie in V_h.

Example. An FE analysis of the continuous beam in Fig. 1.140 with two elements yielded for the buckling load the value

$$P^h_{crit} = \frac{16.48\,EI}{l^2} > \frac{12.7\,EI}{l^2} = P_{crit} \qquad (1.560)$$

and the buckled shape

$$\begin{bmatrix} u_4 \\ u_6 \end{bmatrix} = \begin{bmatrix} -0.707 \\ 1 \end{bmatrix}. \qquad (1.561)$$

If the FE solution w_h is substituted into the differential equation and the jumps in the shear force V and the bending moment M are measured at the nodes, then this gives an impression of the load case p_h (see Fig. 1.141). But note that this arrangement is only a snapshot because the load case p_h can be scaled in an arbitrary way, since any multiple of the "buckling mode" w_h is also a possible solution.

As can be seen in Fig. 1.141 forces are necessary to hold the buckled beam in place. This is equivalent to saying that the opposing forces prevent

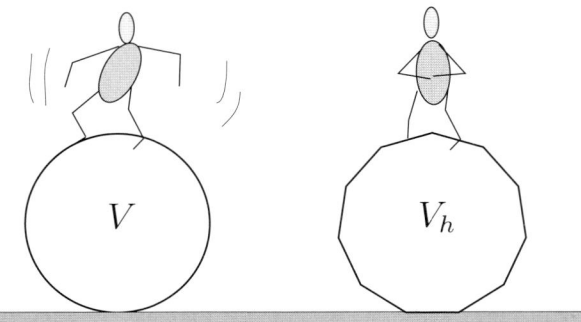

Fig. 1.142. A polygon is a more stable object than a perfect circle

the beam from buckling. This explains why the approximate buckling load is greater than the exact load. The same phenomenon is experienced by two acrobats in Fig. 1.142. The acrobat on the perfect circle finds himself in an unstable position, while his colleague profits from the fact that the vertices of the polygon hamper rotation, and are marginally stabilizing his position.

1.40 Interpolation

In some sense the FE method is structural analysis with polynomials, i.e., functions such as

$$u(x) = x + 3\,x^2 \qquad u(x, y) = 1 + x\,y + x^5\,y^7 \,. \tag{1.562}$$

Polynomials are very versatile functions, and easy to handle, but if the displacement u is assumed to be zero in the first span and to increase linearly in the second span, two distinct polynomials are needed. Interpolation with *piecewise polynomials*, as in Fig. 1.143, is therefore the basic procedure of FE analysis.

For a mathematician these hat functions, or more generally these nodal unit displacements φ_i, are the real finite elements.[16] The structural elements are only considered a convenient tool to generate the nodal unit displacements, the "real" finite elements.

Indeed the term *finite element* is not unique. When we speak of linear elements we mean the shape functions. But when we speak of plate or shell elements we mean the structural element.

The characteristic feature of the FE method is that the shape functions have a *finite support*, because they are nonzero only over a small region of the structure while the basis functions of a Fourier series such as

---

[16] "The use of the concept finite *element* may seem deceptive. In principle we subdivide the domain into *elements*, that is geometric objects, while by *finite elements* we mean functions." [51]

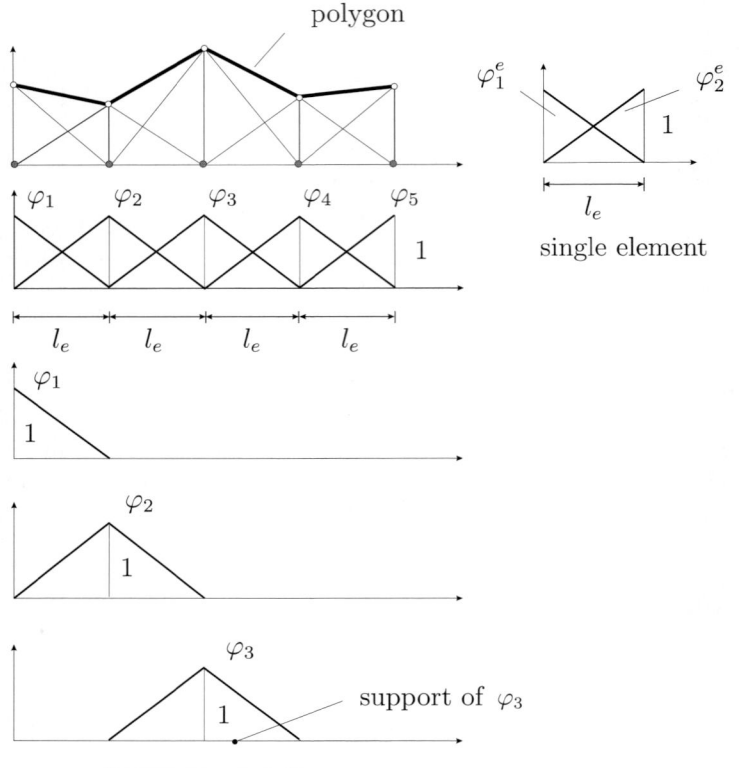

Fig. 1.143. Construction of a polygon from hat functions φ_i

$$w_n(x) = \sum_{k=1}^{n} \left(a_k \cdot \sin \frac{k\pi x}{l} + b_k \cdot \cos \frac{k\pi x}{l} \right), \tag{1.563}$$

are everywhere oscillatory.

Hence, in this sense the FE method is actually a method of finite functions. This is similar to the *three-moment equation*, where by a smart choice of redundants the bandwidth of the flexibility matrix $\boldsymbol{F} = [\delta_{ij}]$ can be kept small (Fig. 1.144 a). If instead all interior supports were removed (Fig. 1.144 b), the structure would be statically determinate as well but the flexibility matrix

$$\delta_{ij} = \int_0^l \frac{M_i \, M_j}{EI} \, dx \tag{1.564}$$

would be fully populated and certainly ill-conditioned, and therefore susceptible to rounding errors, because—given that the number of spans is large—the moments M_i and M_j, and thus the numbers δ_{ij} of adjacent nodes, would be nearly identical.

a

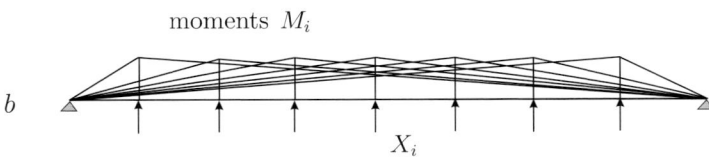

b

moments M_i

X_i

c

X_i

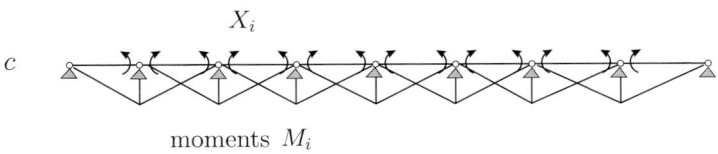

moments M_i

Fig. 1.144. Continuous beam: **a)** system, **b)** unfavorable choice of redundants, **c)** optimal choice

Shape functions φ_i that extend over the whole structure—like the deflection curves of the redundants $X_i = 1$ in Fig. 1.144 b—are not a good choice for the FE method. The overlap between the shape functions must be kept small. In this sense the nodal unit displacements are a better choice, because they are *almost orthogonal*, and hence lead to much better-conditioned systems of equations.

The FE method could rightly be called an interpolation method if there were not the problem that the function to be interpolated is not known. It would not be claiming too much to say that

The whole theory of finite elements is only concerned with the question of what the best choice for the unknown nodal deflections w_i might be?

Here *best* does not necessarily mean that the interpolating function passes through the nodes of the original curve, just that the difference between the FE stresses and the true stresses is minimized. This is the difference between a "normal" interpolation and an FE interpolation.

1.41 Polynomials

Each function can be expanded in a *Taylor series*

$$u(x) = u(0) + u'(0)\,x + u''(0)\,\frac{x^2}{2} + u'''(0)\,\frac{x^3}{3!} + \ldots \qquad (1.565)$$

and in the same fashion the displacements within an element can be approximated by constant, linear, or quadratic functions. The shape functions of the

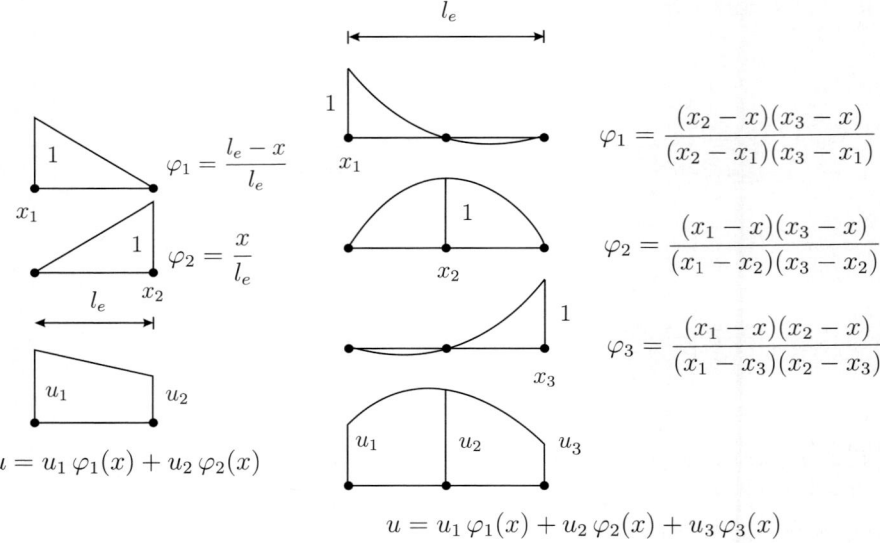

$$\varphi_1 = \frac{(x_2 - x)(x_3 - x)}{(x_2 - x_1)(x_3 - x_1)}$$

$$\varphi_2 = \frac{(x_1 - x)(x_3 - x)}{(x_1 - x_2)(x_3 - x_2)}$$

$$\varphi_3 = \frac{(x_1 - x)(x_2 - x)}{(x_1 - x_3)(x_2 - x_3)}$$

$$u = u_1 \varphi_1(x) + u_2 \varphi_2(x)$$

$$u = u_1 \varphi_1(x) + u_2 \varphi_2(x) + u_3 \varphi_3(x)$$

Fig. 1.145. Linear and quadratic shape functions

single nodes x_j

$$\varphi_i(x_j) = \delta_{ij} = \begin{cases} 1 & i = j \\ 0 & i \neq j \end{cases} \qquad (\delta_{ij} = \text{Kronecker delta}) \qquad (1.566)$$

are polynomials of degree $\leq n$. A *Lagrange element* has internal nodes and edge nodes, while *serendipity elements* only have edge nodes. Lagrange elements are based on *Lagrange polynomials*; see Fig. 1.145.

It is not guaranteed that the shape functions form a complete set, i.e., that they can represent all possible polynomials of degree n on the element

$$f(x) = a_0 + a_1 x + a_2 x^2 + \ldots + a_n x^n \overset{?}{=} \sum_{i=1}^{n+1} u_i \varphi_i(x). \qquad (1.567)$$

The number of terms needed for a complete polynomial of degree n in the $x{-}y$-plane increases rapidly, as can be seen from *Pascal's triangle*:

$$1$$
$$x \quad y$$
$$x^2 \quad xy \quad y^2$$
$$x^3 \quad x^2y \quad xy^2 \quad y^3$$
$$x^4 \quad x^3y \quad x^2y^2 \quad xy^3 \quad y^4$$
$$x^5 \quad x^4y \quad x^3y^2 \quad x^2y^3 \quad xy^4 \quad y^5$$

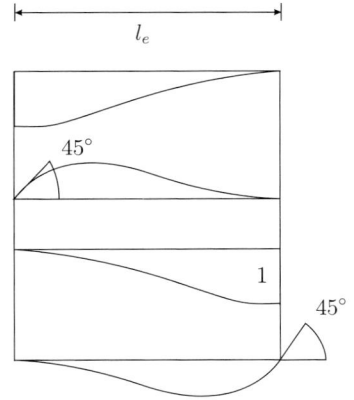

$$\varphi_1 = 1 - \frac{3x^2}{l_e^2} + \frac{2x^3}{l_e^3}$$

$$\varphi_2 = -x + \frac{2x^2}{l_e} - \frac{x^3}{l_e^2}$$

$$\varphi_3 = \frac{3x^2}{l_e^2} - \frac{2x^3}{l_e^3}$$

$$\varphi_4 = \frac{x^2}{l_e} - \frac{x^3}{l_e^2}$$

Fig. 1.146. The nodal unit displacements of a beam element are third-degree polynomials

A complete polynomial of order zero, one, two, three, four, or five (last row in the triangle) must have 1, 3, 6, 10, 15, or 21 terms which means that only elements with 1, 3, 6, 10, 15, or 21 nodes are complete.

In Euler–Bernoulli beams, *Hermite polynomials* are used, which enable one to interpolate the deflection and the first derivative at the nodes; see Fig. 1.146.

In the sense of *backward error analysis*, the shape an element assumes tells us which load the element carries, as in Fig. 1.147. If a rope is slung around a wheel, then the pressure p is inversely proportional to the radius R of the wheel

$$p = -Hw'' = -H\frac{1}{R} \, . \tag{1.568}$$

And if $w_h(x) = (1 + 0.2\,x + 3\,x^2 - 5\,x^3 + 3\,x^5 - x^6)/EI$ is the deflection of an element, the element obviously carries the distributed load

$$p_h(x) = EI\,w_h^{IV}(x) = 360\,(x - x^2)\ \text{kN/m}\,, \tag{1.569}$$

which is balanced by the shear forces V and moments M at the ends of the beam element (see Fig. 1.148, p. 203) because

- Each polynomial satisfies the equilibrium conditions.

This is true for *all* elements. The resultant stresses at the edge of an element *always* balance the distributed load to which the element is subjected. The proof is based on Green's first identity: for any smooth function u—not just polynomials (!)—$G(u, r) = \delta W_e - \delta W_i = \delta W_e = 0$, where $r = a + x\,b$ is a rigid-body motion.

Mapped polynomials

In FE analysis mostly *isoparametric elements* are used, i.e., each element Ω_e is generated by mapping a master element onto the region Ω_e of the structure

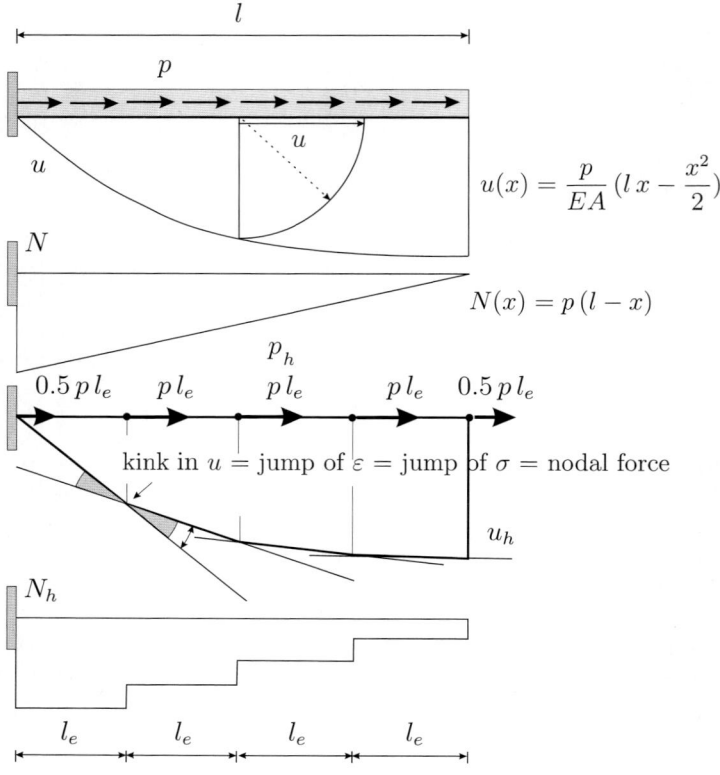

$$u(x) = \frac{p}{EA}\left(l\,x - \frac{x^2}{2}\right)$$

$$N(x) = p\,(l - x)$$

Fig. 1.147. Piecewise linear shape functions in a bar

where the element is located and the polynomials that define the mapping are the same polynomials that define the nodal unit displacements.

Let us assume that on the master element $\Omega_M = [-1, 1]$ two nodal unit displacements are defined,

$$\hat{\varphi}_1(\xi) = \frac{1 - \xi}{2} \qquad \hat{\varphi}_2(\xi) = \frac{1 + \xi}{2}\,, \tag{1.570}$$

and that this master element is mapped onto the interval $\Omega_e = [3,7]$ of the x-axis:

$$x(\xi) = 3 \cdot \varphi_1(\xi) + 7 \cdot \varphi_2(\xi) = 5 + 2\,\xi\,. \tag{1.571}$$

Now to map the nodal unit displacements onto the element Ω_e, the inverse $\xi(x) = 0.5\,x - 2.5$ of this mapping function

$$\varphi_1(x) = \frac{1 - \xi(x)}{2} = \frac{3.5 - 0.5\,x}{2} \qquad \varphi_2(x) = \frac{1 + \xi(x)}{2} = \frac{0.5\,x - 1.5}{2}$$

$$\tag{1.572}$$

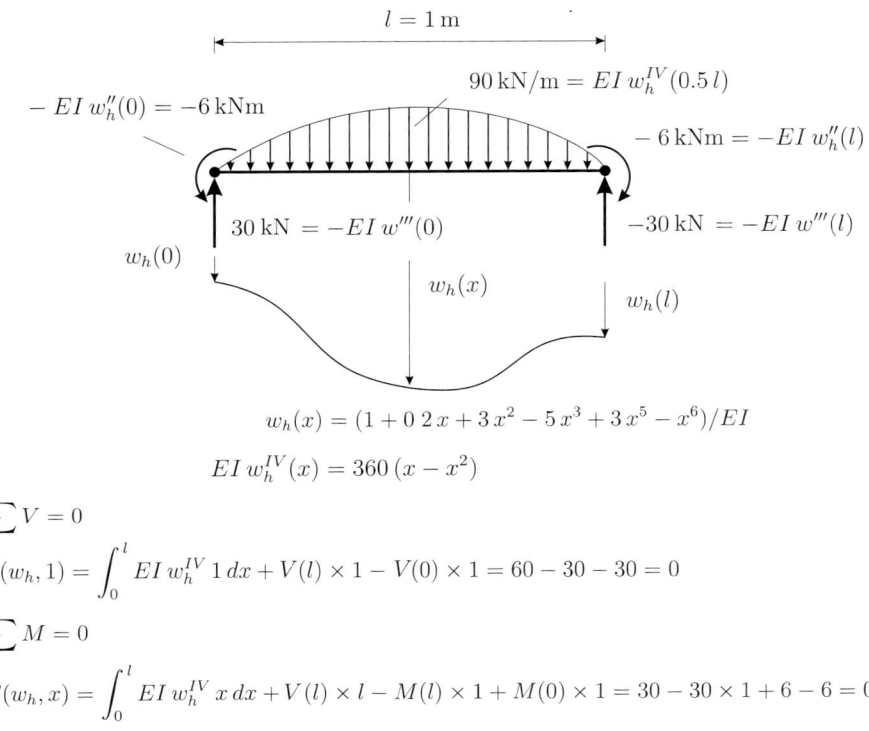

$$w_h(x) = (1 + 0\,2\,x + 3\,x^2 - 5\,x^3 + 3\,x^5 - x^6)/EI$$

$$EI\,w_h^{IV}(x) = 360\,(x - x^2)$$

$$\sum V = 0$$

$$G(w_h, 1) = \int_0^l EI\,w_h^{IV}\,1\,dx + V(l) \times 1 - V(0) \times 1 = 60 - 30 - 30 = 0$$

$$\sum M = 0$$

$$G(w_h, x) = \int_0^l EI\,w_h^{IV}\,x\,dx + V(l) \times l - M(l) \times 1 + M(0) \times 1 = 30 - 30 \times 1 + 6 - 6 = 0$$

Fig. 1.148. Each polynomial satisfies the equilibrium conditions

must be applied. These functions are called *mapped polynomials*. Formally the mapped polynomials are compositions of the "pullback" function $\xi(x)$ and the original shape functions $\hat{\varphi}_i(\xi)$:

$$\varphi_i = \hat{\varphi}_i \circ \xi\,. \tag{1.573}$$

In isoparametric elements all nodal unit displacements are such mapped polynomials. The interesting question then is: When are the mapped polynomials actually polynomials? When does the transformation $\xi \to x$ leave the nature of the shape functions invariant? This is true if the master element Ω_M and the actual finite element Ω_e are affine, that is, if the finite element Ω_e is simply a blow-up of the master element. To stretch an element, linear mapping functions

$$x(\xi, \eta) = a_0 + a_1\,\xi + a_2\,\eta \qquad y(\xi, \eta) = b_0 + b_1\,\xi + b_2\,\eta \tag{1.574}$$

suffice. Therefore in such elements the determinant of the Jacobi matrix is constant, that is the ratio $d\Omega/d\Omega_M$ is at all points the same and it is simply a scaling factor. In a mesh consisting of simple triangular or rectangular elements with linear or bilinear shape functions, this is guaranteed. But if

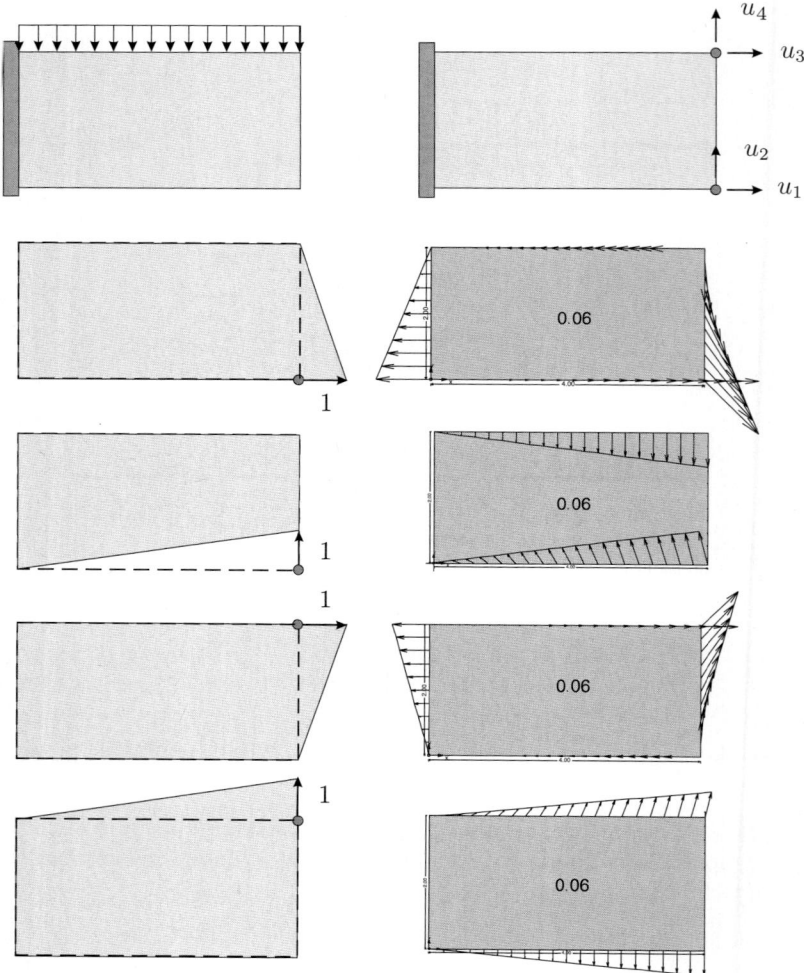

Fig. 1.149. FE analysis of a plate with bilinear elements. The four nodal unit displacements of the two nodes and the associated unit load cases. The element loads are not displayed

a rectangular bilinear element is mapped onto a skew-edged element, or if a single node is intentionally displaced, then it is not.

What is more important, though is that the mapping between the master element Ω_M and the element Ω_e is one-to-one and onto so that every point in Ω_e can be uniquely identified with a point in Ω_M and vice versa. That being the case, the mapped polynomials, the composition of the pullback functions $\xi(x,y), \eta(x,y)$, and the master-element shape functions $\varphi_i(\xi,\eta)$, are all smooth functions, even though they might not be polynomials [121].

Unit load cases

Typically the unit load cases p_i which can be associated with the nodal unit displacements have the same polynomial character as the displacement fields—only some orders lower; see Fig. 1.149.

Interpolation

At first glance the FE method can be considered interpolation of an unknown function with *piecewise polynomials* (or mapped polynomials). But caution is in order here, because the right strategy is not to interpolate but to minimize! Suppose the deflection surface w of a slab were interpolated at the nodes:

$$w_I(\boldsymbol{x}) = w(\boldsymbol{x}_1)\,\varphi_1(\boldsymbol{x}) + w_{,x}\,(\boldsymbol{x}_1)\,\varphi_2(\boldsymbol{x}) + \ldots + w_{,y}\,(\boldsymbol{x}_{3n})\,\varphi_{3n}(\boldsymbol{x})\,.(1.575)$$

This interpolating function w_I would then be an inferior solution, as its distance from the exact solution in terms of potential energy

$$\Pi(w) < \Pi(w_h) < \Pi(w_I) \quad \leftarrow \quad w_I \text{ is not as close to } w \text{ as } w_h \quad (1.576)$$

and also in terms of strain energy

$$a(e,e) = a(w - w_h, w - w_h) < a(w - w_I, w - w_I) = a(e_I, e_I) \quad (1.577)$$

would exceed the distance of the FE solution (see Eq. (7.412), p. 572).

Many asymptotic error estimates are based on this property and on *Céa's lemma* which states that

$$||w - w_h||_m \le c \inf_{v_h \in V_h} ||w - v_h|| \qquad (\text{inf} = \text{infimum})\,. \qquad (1.578)$$

This lemma essentially means that the error in the FE solution is proportional to the minimum distance of w from V_h and so the problem of estimating the error $||w - w_h||_m$ is reduced to a problem in approximation theory. Because the strain energy product $a(w,w)$ and $||w||_m^2$ are equivalent norms we may write as well

$$a(e,e) = a(w - w_h, w - w_h) \le \tilde{c} \inf_{v_h \in V_h} ||w - v_h||\,. \qquad (1.579)$$

Hence if the interpolation error on the space V_h is of order

$$||w - w_I||_m \le h^{t-m}\,||w||_t \qquad (1.580)$$

then this automatically provides an upper bound of the error in the FE solution because the FE solution is closer *in the sense of the strain energy* to the exact solution than the interpolating function w_I

$$a(e,e) = a(w - w_h, w - w_h) \le \hat{c}\,h^{t-m}\,||w||_t\,. \qquad (1.581)$$

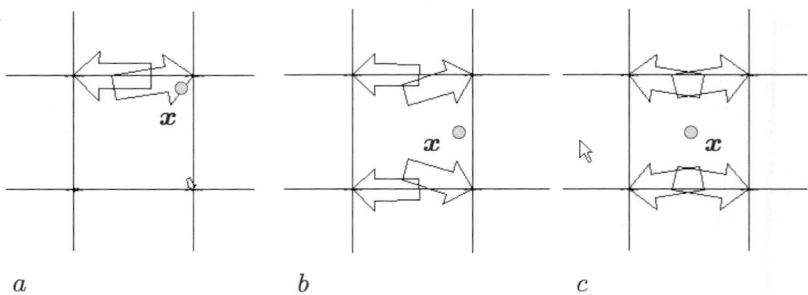

a   b   c

Fig. 1.150. Weights $\boldsymbol{f}_i^G$ for σ_{xx} in a bilinear element, $\sigma_{xx} = \sum_{i=1}^4 \boldsymbol{f}_i^G \bullet \boldsymbol{u}_i$

So if the mesh is well qualified to interpolate the deflection surface then we might have a good mesh.

The difference between the interpolating function w_I and the FE solution w_h lies only in the coefficients w_i, because the shape functions φ_i are the same. The coefficients w_i^I of the interpolating function w_I are the nodal values of $w(\boldsymbol{x})$, while the FE coefficients w_i^h are the solution of the system $\boldsymbol{K}\boldsymbol{w} = \boldsymbol{f}$, which guarantees that the FE solution minimizes the distance in the strain energy. If the interpolating function w_I were a better solution the nodal values of the exact solution would solve the system $\boldsymbol{K}\boldsymbol{w} = \boldsymbol{f}$. Because this is not true in 2-D and 3-D problems, the interpolating function must be an inferior solution.

Superconvergence

The nodes (with regard to displacements) and the Gauss points or the mid-points (with regard to stresses) of the elements are called superconvergent points because the accuracy of an FE solution is often superior at these points.

From an engineering standpoint it seems clear why the displacements are superior at the nodes. Simply because the dip caused by a point load δ_0 (Green's function) can best be represented at a node, while a dislocation can best be modeled, so it seems, if the source point lies halfway between the nodes.

The latter point is best understood by looking at the nodal influence function for the stresses, say,

$$\sigma_{xx}(\boldsymbol{x}) = \sum_i \boldsymbol{f}_i^G \bullet \boldsymbol{u}_i = \int_\Omega \boldsymbol{G}_1(\boldsymbol{y}, \boldsymbol{x}) \bullet \boldsymbol{p}(\boldsymbol{y}) \, d\Omega_{\boldsymbol{y}} \,. \qquad (1.582)$$

Recall that the equivalent nodal forces $\boldsymbol{f}_i^G$ are the stresses $\sigma_{xx}(\varphi_j)(\boldsymbol{x})$ of the shape functions at $\boldsymbol{x}$. If the mesh consists of a regular array of bilinear elements of size $h \times h$ then at the node $\boldsymbol{x}$ itself the vector $\boldsymbol{f}_i^G$ is zero, if we let $\sigma_{xx}(\boldsymbol{x})$ the average stress at the node, because the stresses of the four neighboring element shape functions on the four sides of the node cancel, similar to

$$\frac{1}{h} - \frac{1}{h} + \frac{1}{h} - \frac{1}{h} = 0 \tag{1.583}$$

and so only the $\boldsymbol{f}_i^G$ at the edge of the larger patch $[2\,h \times 2\,h]$ are non-zero, but this means that by averaging the stresses at the nodes we artificially double the mesh width, $h \to 2 \cdot h$, we loose accuracy.

If the point $\boldsymbol{x}$ lies inside an element then only the four nodes of the element $[h \times h]$ contribute to the formula (1.582); see Fig. 1.150. If $\boldsymbol{x}$ is the center of the element then the weights in the finite difference scheme for σ_{xx} are all the same (see Fig. 1.150 c) but if $\boldsymbol{x}$ wanders away from the center the node with the shortest distance to the source point $\boldsymbol{x}$ gains the upper hand, it contributes the most to $\sigma_{xx}(\boldsymbol{x})$, and if the point $\boldsymbol{x}$ crosses the line that separates two elements then the weights change abruptly, which explains the typical jumps in FE stresses.

Ideally the weights in the finite difference scheme for $\sigma_{xx}(\boldsymbol{x})$ should be the same at each point in Ω. That they are not the same is a simple consequence of the fact that the FE solution is an expansion in terms of nodal basis functions and the derivative of u_h is simply the sum of the derivatives of the shape functions

$$u_h'(\boldsymbol{x}) = u_1\,\varphi_1'(\boldsymbol{x}) + u_2\,\varphi_2'(\boldsymbol{x}) + \ldots = u_1 \cdot \text{weight}_1 + u_2 \cdot \text{weight}_2 + \ldots \tag{1.584}$$

that is the weights are the slopes of the shape functions at the point $\boldsymbol{x}$.

Note that the weights for a displacement, say $u_x(\boldsymbol{x})$, are not that sensitive to the question of which element contains the point. When the point $\boldsymbol{x}$ is close to a node then 90% of the weight is concentrated in that node—regardless of on which side of the node the point $\boldsymbol{x}$ lies.

In narrower terms, superconvergence means that in some cases the FE solution w_h approximates the interpolating function $w_I \in V_h$ of the exact solution (w_I = the function which agrees with w at the nodes) with a higher rate of convergence than the solution w itself. This is no surprise given that both approximate solutions are based on the same functions φ_i and so the error

$$e_{I-h}(\boldsymbol{x}) = w_I(\boldsymbol{x}) - w_h(\boldsymbol{x}) = \sum_i e_i\,\varphi_i(\boldsymbol{x}) \tag{1.585}$$

can be traced back to the error in the output of the approximate nodal Green's functions

$$e_i = w_i^I - w_i^h = \int_\Omega (G_0(\boldsymbol{y}, \boldsymbol{x}_i) - G_0^h(\boldsymbol{y}, \boldsymbol{x}_i))\,p(\boldsymbol{y})\,d\Omega_{\boldsymbol{y}} \tag{1.586}$$

(for rotational degrees of freedom G_0 would have to be replaced by G_1) so that if the error at the nodes is small the two solutions will also be close between the nodes.

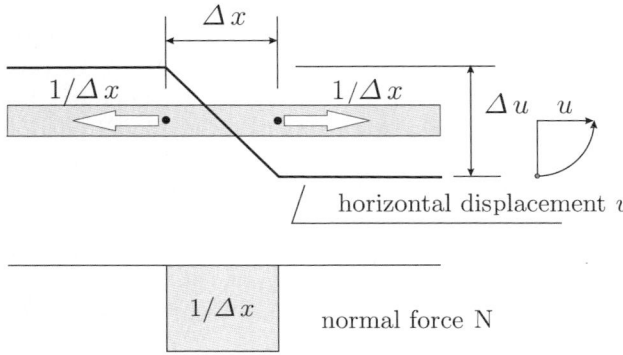

Fig. 1.151. Effect of two opposing forces $1/\Delta x$. The horizontal displacements are plotted vertically downwards (+) and upwards (-) respectively

1.42 Infinite energy

If a bar is stretched by two opposing forces which increase in magnitude with decreasing distance Δx between the two forces (Fig. 1.151)

$$P = \frac{1}{\Delta x} \qquad P = -\frac{1}{\Delta x} \qquad (1.587)$$

then as the distance Δx tends to zero the horizontal displacement u of the bar becomes discontinuous; see Fig. 1.151. At the collision point of the two forces a gap of size $u(x_+) - u(x_-) = 1/EA$ opens up. It is no surprise that in the final stage of this experiment the strain energy becomes infinite ($EA = 1$):

$$\frac{1}{2}\int_0^l \frac{N^2}{EA}dx = \frac{1}{2}\int_0^l \frac{1}{\Delta x^2}dx = \frac{1}{2}\frac{1}{\Delta x^2}\Delta x = \frac{1}{2}\frac{1}{\Delta x} = \infty \qquad \Delta x \mapsto 0.$$

$$(1.588)$$

Similar things happen in a beam, see Fig. 1.152. If two opposing moments are applied to a beam, and if these moments increase in magnitude as the distance Δx between the two moments shrinks,

$$M = \frac{1}{\Delta x} \qquad M = -\frac{1}{\Delta x} \qquad (1.589)$$

then in the limit $\Delta x \mapsto 0$ a plastic hinge will form at the collision point and the strain energy will become infinite:

$$\frac{1}{2}\int_0^l \frac{M^2}{EI}dx = \frac{1}{2}\int_0^l \frac{1}{\Delta x^2}dx = \frac{1}{2}\frac{1}{\Delta x} = \infty \qquad \Delta x \mapsto 0. \qquad (1.590)$$

What these examples are saying is that infinite forces are necessary to tear a bar apart or to form a plastic hinge in a beam, and by virtue of the energy balance, $W_i = W_e = \infty \times gap$, the strain energy W_i must also be infinite.

In mathematical terms, a fracture or a plastic hinge is a discontinuity in a displacement, and the message is that discontinuous displacements mean infinite energy, see Fig. 1.153.

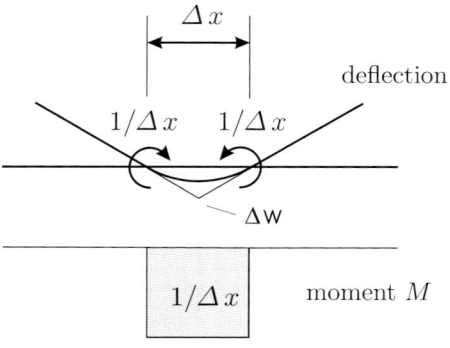

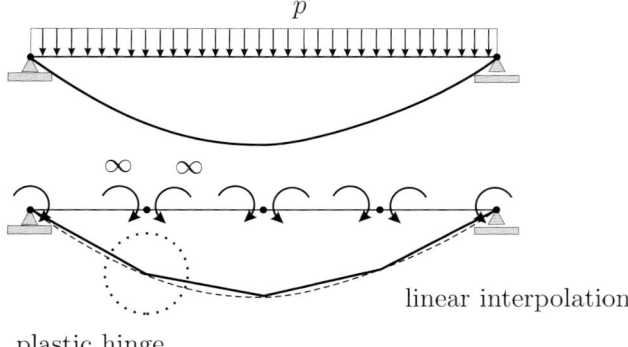

Fig. 1.152. The closer the moments, the larger they become

linear interpolation

plastic hinge

Fig. 1.153. A linear interpolation requires plastic hinges

Remark 1.18. Note that when the work done by these forces or moments $1/\Delta x$ via functions u or w is calculated and we take the limit, we are actually differentiating u or w', because

$$\lim_{\Delta x \to 0} \frac{u(x + 0.5\,\Delta x) - u(x - 0.5\,\Delta x)}{\Delta x} = u'(x) \qquad (1.591)$$

$$\lim_{\Delta x \to 0} \frac{w'(x + 0.5\,\Delta x) - w'(x - 0.5\,\Delta x)}{\Delta x} = w''(x)\,. \qquad (1.592)$$

1.43 Conforming and nonconforming shape functions

Elements are called *conforming* if the functions φ_i—more accurately the displacement terms of the φ_i—are continuous across interelement boundaries.

What counts as a displacement term depends on the order of the differential equation. In a second-order equation such as $-EA\,u''$, the zero-th order derivative u is a displacement and the first-order derivative $N = EA\,u'$ is a force. In a fourth-order equation like $EI\,w^{IV}$, the deflection w and the slope w' are displacement terms and the second- and third-order derivatives

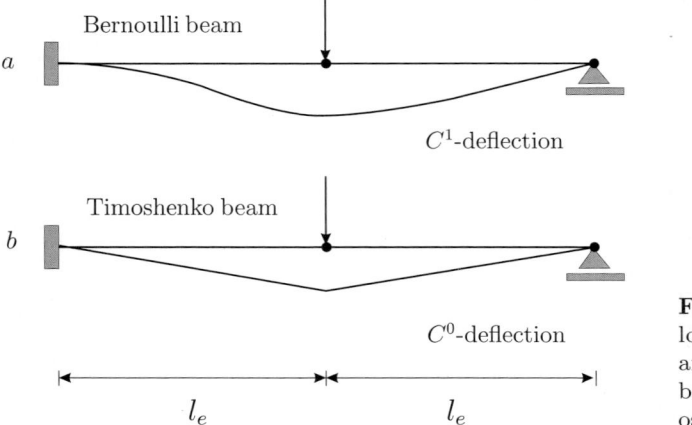

Fig. 1.154. Point load applied to **a)** an Euler–Bernoulli beam and **b)** a Timoshenko beam

$M = -EI\,w''$ and $V = -EI w'''$ are force terms. Accordingly, conforming elements are classified as C^0 or C^1 elements, (see Fig. 1.154),

$$C^0 - \text{elements}: \text{ for second-order equations}$$
$$C^1 - \text{elements}: \text{ for fourth-order equations}$$

Because discontinuous displacements imply infinite energy, it might seem that nonconforming shape functions could not be a proper choice for an energy method, because only the finite part of the energy can be considered in the analysis. But a whole range of nonconforming elements are successfully employed in FE analysis. There are different reasons for this:

- The elements are basically conforming elements, and only by enriching the shape functions with additional terms do they become nonconforming (Wilson's element).
- Hybrid variational principles are used or other modifications are applied.

One member of the first class is Wilson's plane element, which is based on a conforming bilinear element to which two internal modes are added; see Sect. 4.4, p. 338. Because these two internal modes ensure that the deformed elements overlap, the element is nonconforming. But the element is superior to a standard bilinear element, and if the element size tends to zero $h \to 0$, the element becomes conforming.

The second class of nonconforming elements is based on hybrid or modified variational principles, where the "defect" of the element, i.e., the jump in the displacements at the interelement boundaries, is built into the functional with the help of Lagrange multipliers. Instead of the principle of minimum potential energy

$$\Pi(\boldsymbol{u}) = \frac{1}{2} \int_{\Omega} \boldsymbol{E} \cdot \boldsymbol{S}\, d\Omega - \int_{\Omega} \boldsymbol{p} \cdot \boldsymbol{u}\, d\Omega, \qquad (1.593)$$

a hybrid form of this principle is employed:

$$\Pi(\boldsymbol{u}, \boldsymbol{t}_\sigma) = \frac{1}{2} \sum_e \int_{\Omega_e} \boldsymbol{E} \cdot \boldsymbol{S} \, d\Omega - \int_\Omega \boldsymbol{p} \cdot \boldsymbol{u} \, d\Omega + \sum_i \int_{\Gamma_i} \boldsymbol{t}_\sigma \cdot (\boldsymbol{u}^+ - \boldsymbol{u}^-) \, ds$$

$$(1.594)$$

where $\boldsymbol{u}^+ - \boldsymbol{u}^-$ are the jump terms of the displacement field at the interelement boundaries Γ_i, and the traction vector $\boldsymbol{t}_\sigma$ plays the role of a Lagrange multiplier.

Now it is of no concern that the displacement field is discontinuous. Hence, what is a conforming element and what is not depends on the variational principle employed. The error committed if nonconforming shape functions are used in the standard functional (1.593) is that the penalty terms at the interelement boundaries Γ_i are neglected.

The message is that the "measuring device" that is the strain energy in the functional $\Pi(\boldsymbol{u})$, must be compatible with the peculiarities of the shape functions; see Eq. (1.594).

The so-called *spurious modes* also belong in this context. These are shape functions $\varphi_i(x) \neq 0$ with zero strain energy:

$$\delta W_i(\varphi_i, \varphi_i) = \int_0^l EI \, (\varphi_i'')^2 \, dx = 0 \qquad \text{but } \varphi_i \neq 0. \qquad (1.595)$$

The entries on the main diagonal of the stiffness matrix vanish for such shape functions:

$$k_{ii} = \delta W_i(\varphi_i, \varphi_i) = 0. \qquad (1.596)$$

Spurious modes normally only occur if a program author tinkers with the basic algorithm, if the author reduces the sensitivity of an FE program by using, for example, reduced integration.

But in mixed formulations or multi-physics problems spurious modes are not that seldom observed. They are an indication that either the implementation is not adequate or that the mathematical model is very sensitive to the physical parameters as for example in the analysis of a nearly incompressible fluid.

1.44 Partition of unity

There is a logic built into influence functions: the influence functions for the support reaction A and B of the beam in Fig. 1.155 add to 1 at every point x:

$$\eta_A(x) + \eta_B(x) = 1.0 \qquad \text{at all points } x. \qquad (1.597)$$

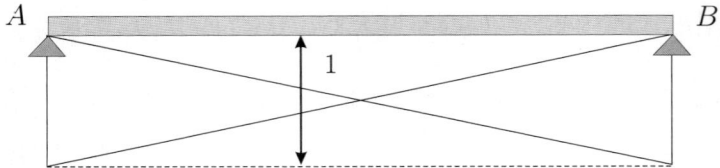

Fig. 1.155. The sum of the two influence functions for the support reactions R_A and R_B is 1.0 at *every* point

It cannot be 1.10 or 0.90; it must be exactly 1 otherwise the beam would be a wonderful machine that could increase or decrease the load at will.

The same logic applies to the nodal unit displacements, because they are influence functions for the equivalent nodal forces. To perform a unit transla-tion $w = 1$ of a slab, the value $w = 1$ is assigned to the translational degrees of freedom. The virtual work done by the surface load representing the weight g is the integral

$$\sum_i \int_\Omega g \, \varphi_i \, d\Omega \,. \tag{1.598}$$

This integral is only equal to the work done by the total weight G

$$G \cdot 1 = \sum_i \int_\Omega g \, \varphi_i \, d\Omega \,, \tag{1.599}$$

if the nodal unit displacements add to 1 at every point $\boldsymbol{x}$:

$$\sum_i \varphi_i(\boldsymbol{x}) = 1 \qquad \text{for all } \boldsymbol{x} \,. \tag{1.600}$$

In other words if the value $u_i = 1$ is assigned to all translational degrees of freedom of a structure, the structure must undergo a true rigid-body motion and no point may lag or may rush ahead, i.e., the shape functions must form a *partition of unity* in the domain Ω. This is a very important property, because without it there would be no *global equilibrium*.

In Euler–Bernoulli beams and Kirchhoff plates it is also required that the nodal unit displacements represent (pseudo) rotations such as

$$w(x) = a \, x \quad \text{(beam)} \quad w(x, y) = a \, x + b \, y \quad \text{(slab)} \tag{1.601}$$

exactly, because without that property the resulting moments would not be the same, $\sum M_h = \sum M$.

In plates the symmetry of the stress tensor $\sigma_{xy} = \sigma_{yx}$ implies $M = 0$. This is why in 2-D elasticity only translations such as $\boldsymbol{r} = [1, 0]^T$ or $\boldsymbol{r} = [0, 1]^T$ count as rigid-body motions.

A mesh inherits the ability to represent rigid-body motions from the indi-vidual elements. This means that if an individual element has this property, then so does the whole mesh if the nodal unit displacements are C^0 or C^1 respectively (beams and slabs), that is, if the elements are conforming.

The logic behind all this is the following:

1. If rigid-body motions can be represented exactly on the mesh, they also lie in V_h^+, because V_h^+ contains *all* deformations that can be "reached" by the nodal unit displacements—with a proper choice of the coefficients u_i this is possible ($V_h^+ = V_h +$ rigid-body motions).
2. Because the two load cases $\boldsymbol{p}$ and $\boldsymbol{p}_h$ are equivalent with respect to the $\varphi_i \in V_h^+$ they are also equivalent with respect to all rigid-body motions

$$\delta W_e(\boldsymbol{p}, \boldsymbol{r}) = \delta W_e(\boldsymbol{p}_h, \boldsymbol{r}) \qquad \boldsymbol{r} = \text{rigid-body motion}. \qquad (1.602)$$

3. We have

$$\delta W_e(\boldsymbol{p}, \boldsymbol{r}) = \sum H \quad \ldots = \sum V \quad \ldots = \sum M \qquad (1.603)$$

depending on what kind of rigid-body motion $\boldsymbol{r} = \boldsymbol{a} + \boldsymbol{\omega} \times \boldsymbol{x}$ is.
4. But (1.602) and (1.603) imply that

$$\rightarrow \sum H = \sum H_h \quad \downarrow \sum V = \sum V_h \quad \curvearrowright \sum M = \sum M_h$$
$$(1.604)$$

which is just the statement that $\boldsymbol{R} = \boldsymbol{R}_h$.

Next to rigid-body motions, constant stress states are the most important stress fields which an FE program must be able to represent exactly. Only then will an FE program have a chance to come close to the exact solution if the element size h shrinks to zero.

1.45 Generalized finite element methods

In the past decade the FE method has been extended in various directions. Most of these extensions, as the Element Free Galerkin (EFG) method or the X-FEM method, can be characterized in the framework of the Generalized Finite Element Method (GFEM) or as it was called previously the Partition of Unity Method (PUM).

According to the GFEM common to all these techniques is that the domain Ω is divided into different regions ω_j and that to each region belongs a local space V_j of functions, not necessarily polynomials, that match the (assumed) character of the solution and thus ensure good local approximation. Then a partition of unity is used to "paste" these spaces together to form the trial space V_h. The partition of unity of the domain Ω may be based on a simple triangulation and so it offers more freedom when compared to standard FE methods, [20], where the shape functions and the mesh are closely linked.

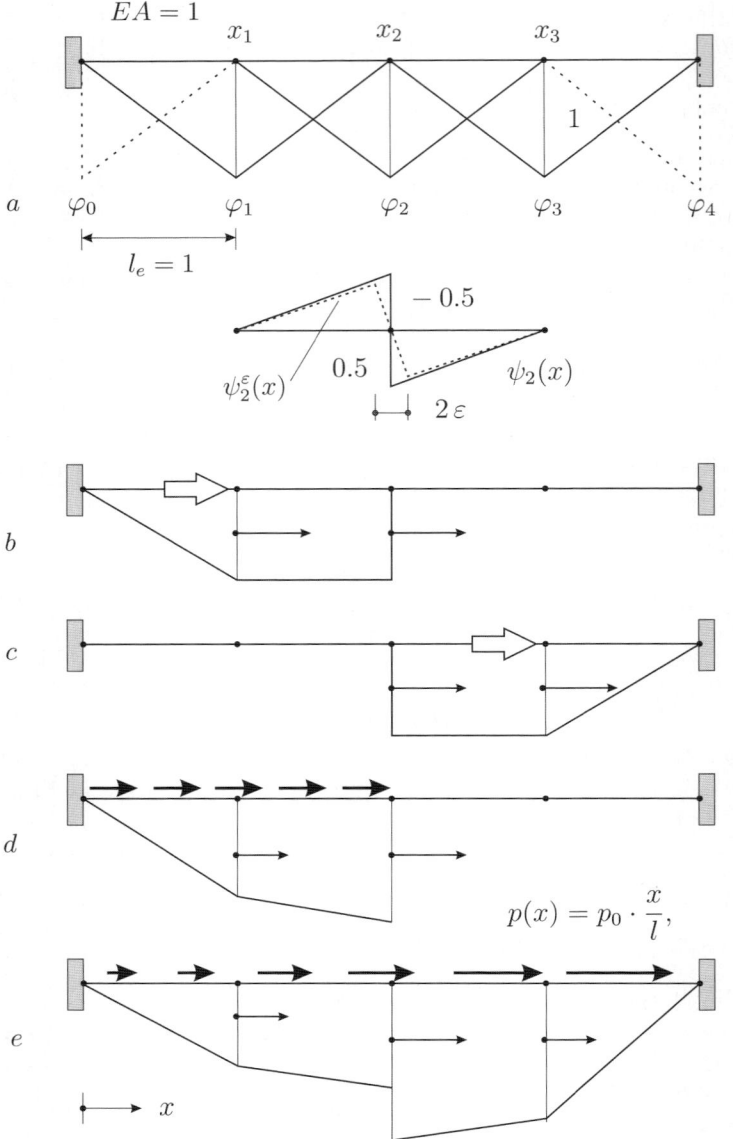

Fig. 1.156. X-FEM and horizontal displacement in a cracked bar: **a)** regular mesh and discontinuous shape function; **b)** point load on the left side and **c)** on the right side of the crack; **d)** half-sided load p **e)** linear load $p(x) = p_0 \cdot x/l$

Meshless methods

A significant stimulus in the development of meshless methods comes from the problem of moving boundaries as in crack growth. The Element Free Galerkin (EFG) method and the Particle or the Finite Point method attempt to overcome this problem.

The EFG is based on a moving least squares scheme that is a method of reconstructing continuous functions from a set of "arbitrarily" distributed point samples via the calculation of a weighted least squares measure which is biased towards the region around the point at which the reconstructed value is requested.

The advantages of meshfree methods compared to finite elements are the higher order of the shape functions, simpler incorporation of h- and p-adaptivity and certain advantages in crack problems. The most important drawback of meshfree methods though is their higher computational cost. The calculation of the strain energy products k_{ij} of "free floating" meshless shape functions is anything but trivial.

X-FEM

The idea of the so-called extended FE method (X-FEM) is to model cracks and other discontinuities by locally enriching the trial space V_h with discontinuous shape functions through a partition of unity method.

To fix ideas we consider the bar in Fig. 1.156 a which is subdivided into four linear elements. We want to model a crack—or rather a fracture—of the bar at node 2. To this end we introduce at this node an additional shape function

$$\psi_2(x) = J(x_2) \cdot \varphi_2(x) \qquad J(x_2) := \begin{cases} 0.5 & x > x_2 \\ -0.5 & x < x_2 \end{cases} \qquad (1.605)$$

which is the product of a step function and the unit displacement $\varphi_2(x)$ of the node x_2.

The five standard linear shape functions $\varphi_i(x), i = 0, 1, 2, 3, 4$, form a partition of unity

$$\sum_i \varphi_i(x) = 1 \qquad 0 \leq x \leq l, \qquad (1.606)$$

that is an assemblage of *nodal influence functions*, and by multiplying the step function $J(x_2)$ with $\varphi_2(x)$ the step function is restricted to the immediate neighborhood of the node x_2. That is any additional function which is introduced to enrich the trial space V_h lives under the umbrella of one of the nodal shape functions. In the X-FEM terminology this is written as

$$u_h(x) = \sum_i \varphi_i(x) \cdot (u_i + \sum_j \psi_j(x) u_j^i)$$

$$= \sum_i \varphi_i(x) \cdot u_i + \sum_i \varphi_i(x) \cdot (\psi_1(x) u_1^i + \psi_2(x) u_2^i)). \qquad (1.607)$$

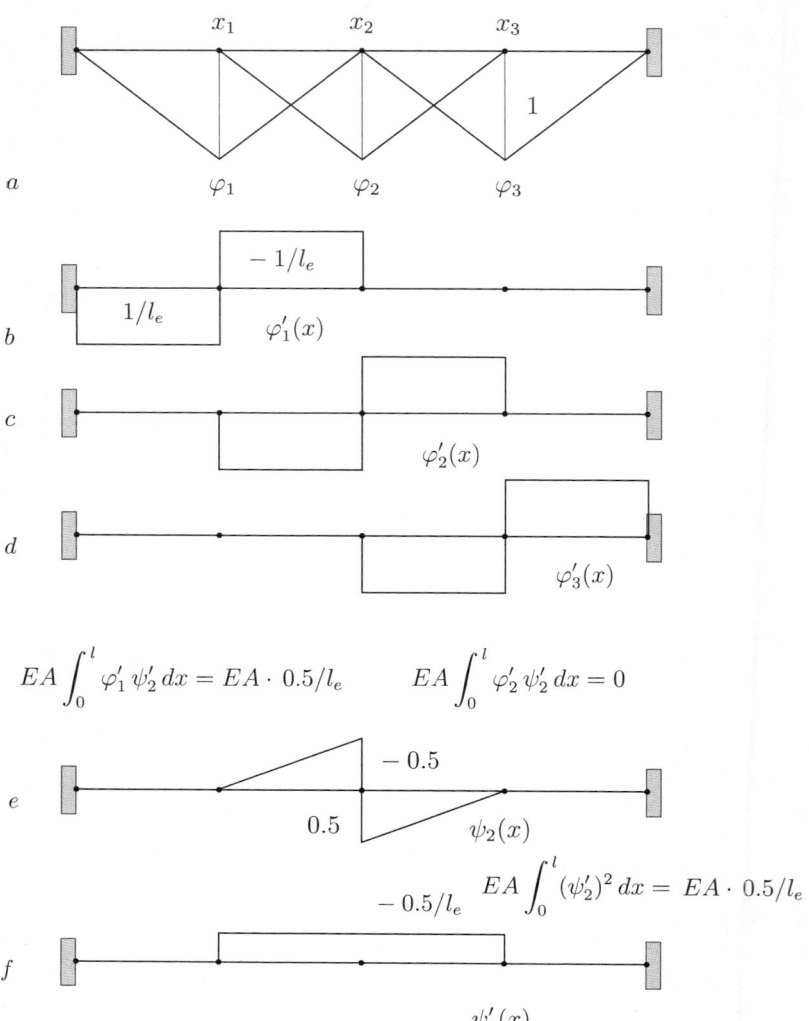

Fig. 1.157. Shape functions and derivatives of the shape functions

The extra functions are the functions $\psi_j(x)$ and the coefficients u_j^i are the additional degrees of freedom. At nodes i where no functions are added the u_j^i are zero. In (1.607) we assumed that two extra functions are added at each node. Of course the $\psi_j(x)$ can vary from node to node and some of the u_j^i can be zero.

Our numbering scheme is simply the following:

$$u_h(x) = u_1 \cdot \varphi_1(x) + u_2 \cdot \varphi_2(x) + u_3 \cdot \psi_2(x) + u_4 \cdot \varphi_3(x). \qquad (1.608)$$

The problem we face is that the new function $\psi_2(x)$ is discontinuous at x_2 which means that the strain energy product

$$k_{33} = \int_0^l EA\,(\psi_2'(x))^2\,dx = \infty \tag{1.609}$$

is infinite. Now we could simply ignore the infinite slope and split the integral into two parts

$$k_{33} = \int_0^{x_2-\varepsilon} \ldots\, dx + \int_{x_2+\varepsilon}^l \ldots\, dx = \left(\frac{0.5}{l_e}\right)^2 \cdot l_e + \left(\frac{0.5}{l_e}\right)^2 \cdot l_e = EA \cdot 0.5\, l_e \tag{1.610}$$

or we could argue as follows: the discontinuous shape function $\psi_2(x)$ can be considered the limit of the zig-zag function $\psi_2^\varepsilon(x)$ in Fig. 1.156 if ε tends to zero. And if we take the strain energy of the zig-zag function the integral in the middle

$$k_{33} = \int_0^{x_2-\varepsilon} \ldots\, dx + \int_{x_2-\varepsilon}^{x_2+\varepsilon} EA\,[(\psi_2^\varepsilon(x))']^2\,dx + \int_{x_2+\varepsilon}^l \ldots\, dx \tag{1.611}$$

explodes if $\varepsilon \to 0$

$$\int_{x_2-\varepsilon}^{x_2+\varepsilon} EA\,[(\psi_2^\varepsilon(x))']^2\,dx \simeq EA \cdot \frac{0.5^2}{\varepsilon^2} \cdot 2\,\varepsilon\,. \tag{1.612}$$

To prevent this from happening we let $EA = 0$ in that part of the bar. The result is of course the same as before but our treatment of the singularity seems better justified because that is what we want to model: a bar where $EA = 0$ at one point. We will see in the following that the structure follows *exactly* this argument.

The stiffness matrix of the bar is calculated as usually, (see Fig. 1.157), by forming the strain energy products of the shape functions—the fact that $EA = 0$ at x_2 has no influence on the other values k_{ij}

$$\boldsymbol{K} = \frac{EA}{l_e} \begin{bmatrix} 2 & -1 & 0.5 & 0 \\ -1 & 2 & 0 & -1 \\ 0.5 & 0 & 0.5 & -0.5 \\ 0 & -1 & -0.5 & 2 \end{bmatrix}\,. \tag{1.613}$$

For a first try a point load $P = 1$ is applied at the node x_1 and the FE program promptly produces the shape in Fig. 1.156 b. Obviously does the program understand what we want. Simply by adding a discontinuous shape function the structure develops a crack! And if the point load acts on the other side of the "crack" the situation is simply reversed, see Fig. 1.156 c. Also the response to a half-sided constant load p, Fig. 1.156 d, and a load $p(x) = p_0 \cdot x/l$, Fig. 1.156 e, is correct.

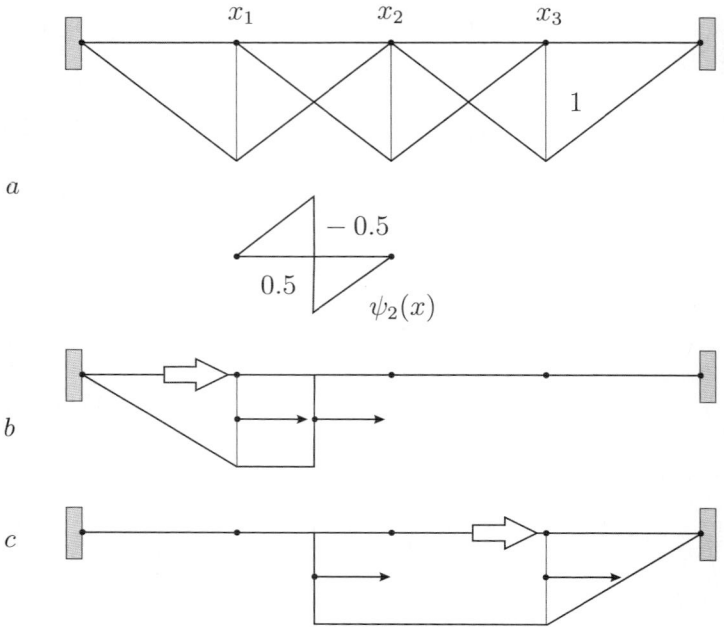

Fig. 1.158. X-FEM and crack between two nodes: **a)** regular mesh and discontinuous shape function; **b)** horizontal displacement due to the point load on the left side and **c)** on the right side of the crack

If the crack lies between two nodes (see Fig. 1.158) the discontinuous shape function extends only from node x_1 to x_2, ($H(\xi)$ = Heaviside function, which is zero for $\xi < 0$ and $+1$ for $\xi > 0$),

$$\psi_2(x) = \varphi_1(x) \cdot H(x - 1.5) + \varphi_2(x) \cdot (H(x - 1.5) - 1) \qquad (1.614)$$

$(H(\xi) - 1)$ = Heaviside on the negative axis) so that

$$u_h(x) = u_1 \cdot \varphi_1(x) + u_2 \cdot \psi_2(x) + u_3 \cdot \varphi_2(x) + u_4 \cdot \varphi_3(x). \qquad (1.615)$$

Now the stiffness matrix is

$$\boldsymbol{K} = \frac{EA}{l_e} \begin{bmatrix} 2 & 1 & -1 & 0 \\ 1 & 1 & -1 & 0 \\ -1 & -1 & 2 & -1 \\ 0 & 0 & -1 & 2 \end{bmatrix}. \qquad (1.616)$$

And as before do the actions of the point loads stop at the crack, see Fig. 1.158 b and c.

The extension of this technique to 2-D and 3-D problems is straightforward, see Fig. 1.159,

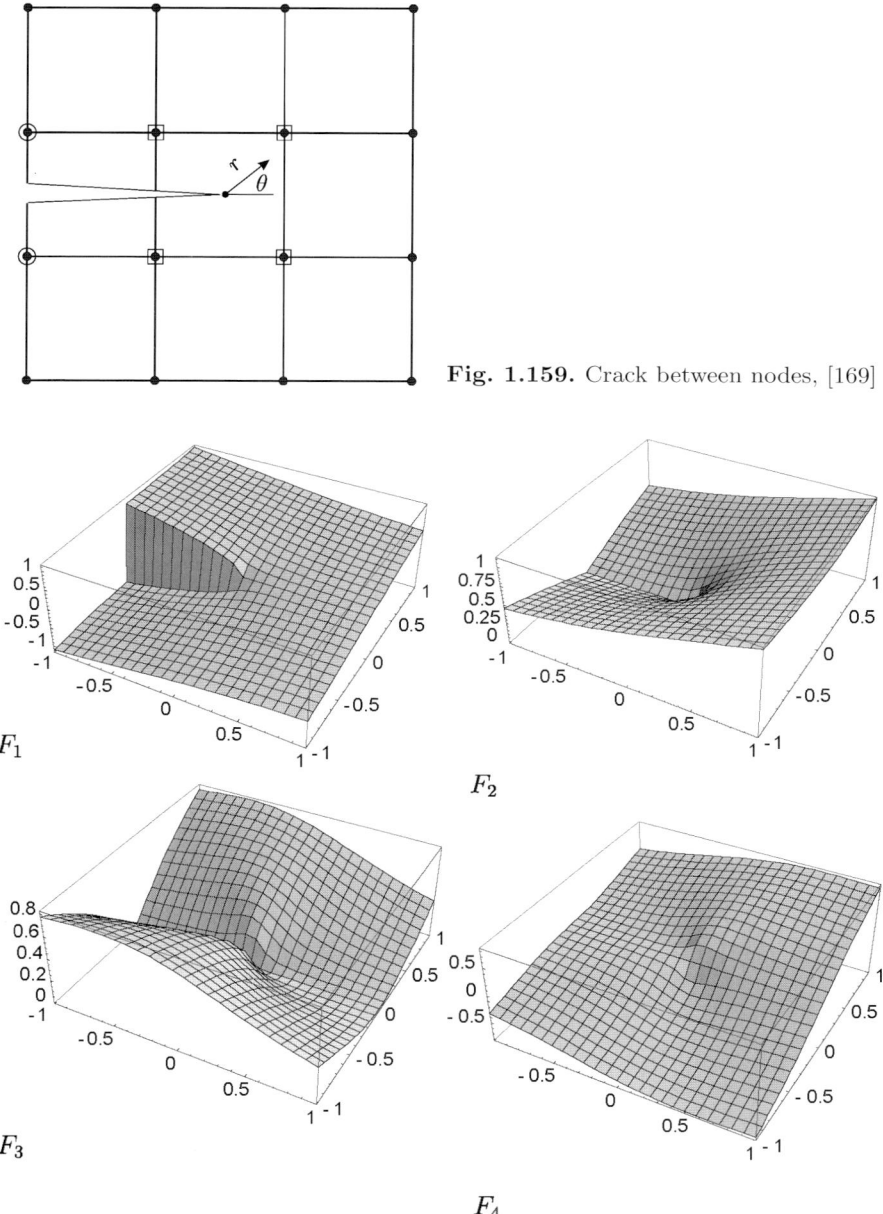

Fig. 1.159. Crack between nodes, [169]

F_1

F_2

F_3

F_4

Fig. 1.160. Crack-tip functions

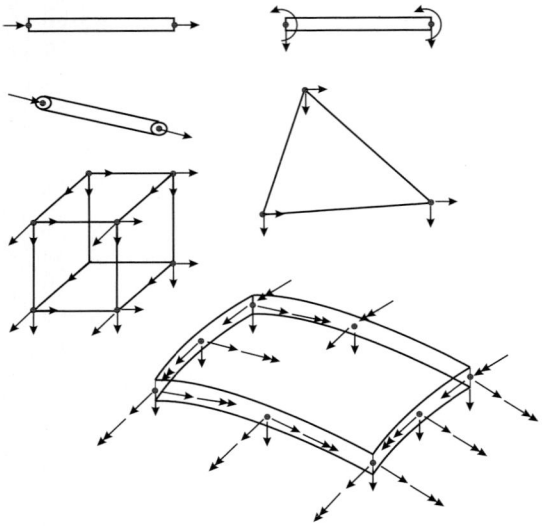

Fig. 1.161. Various elements

$$\boldsymbol{u}_h(\boldsymbol{x}) = \sum_i \boldsymbol{u}_i \, \varphi_i(\boldsymbol{x}) + \sum_{j \in J} \boldsymbol{u}_j^H \, \varphi_j(\boldsymbol{x}) \, H(\boldsymbol{x}) + \sum_{k \in K} \varphi_k(\boldsymbol{x}) \, (\sum_{l=1}^{4} \boldsymbol{c}_k^l \, F_l(\boldsymbol{x}))$$

(1.617)

where the Heaviside function $H(\boldsymbol{x})$ which models the discontinuity at the crack is multiplied with the shape functions of the circled nodes, set J, and the asymptotic crack-tip displacement field which consists of four modes (see Fig. 1.160)

$$F_1(\boldsymbol{x}) = \sqrt{r} \sin \frac{\theta}{2} \qquad F_2(\boldsymbol{x}) = \sqrt{r} \cos \frac{\theta}{2} \qquad (1.618)$$

$$F_3(\boldsymbol{x}) = \sqrt{r} \sin \frac{\theta}{2} \cdot \sin \theta \qquad F_4(\boldsymbol{x}) = \sqrt{r} \cos \frac{\theta}{2} \cdot \sin \theta \qquad (1.619)$$

$$r = \sqrt{x^2 + y^2} \qquad \theta = \arctan \frac{y}{x} \qquad (1.620)$$

is multiplied with the shape functions of the set K of squared nodes, [77], [169].

1.46 Elements

The type of an element, (see Fig. 1.161), is determined by the strains and stresses which result if the element is deformed; that is, the type depends foremost on the definition of the strain energy of the element.

Element displacements are represented in general by polynomials. The higher the degree of the polynomials, the more flexible an element is, the more stress and strain states it can represent, and flexibility—think of the Green's functions—is very important in the FE method. There are three requirements which any element must meet:

- *Rigid-body motions:* the element must be able to represent rigid-body motions and constant strain states exactly, that is, it must be capable of following the element through the first two terms in the *Taylor series* of the displacement field, $\boldsymbol{u}(x) = \boldsymbol{u}(0) + \nabla \boldsymbol{u}(0)\, \boldsymbol{x}$.
- *Isotropy and rotational invariance:* Theoretically a solution should not depend on the orientation of the element, that is the elements should not prefer particular directions. This is guaranteed if the polynomials are *complete*.
- *Continuity:* At the interelement boundaries the displacements must be continuous. Such elements are called C^0-elements. If two neighboring elements have the same displacements at the common nodes then in general the displacements along the interelement boundary are the same. In plate theory (Kirchhoff plates, $K\,\Delta\Delta\,w$) and beam theory (Euler–Bernoulli beam, $EI\,w^{IV}$) the first-order derivatives must be C^1 across interelement boundaries.

The requirement that the polynomial shape functions be complete can be relaxed: to have isotropy and rotational invariance it suffices that all terms which are symmetric to the diagonal of Pascal's triangle be included.

1.47 Stiffness matrices

A stiffness matrix $\boldsymbol{K}$ is quadratic and symmetric. The number of rows and columns is equal to the number of degrees of freedom of the element. The elements of a stiffness matrix, as for example of a bar element or a beam element, see Fig. 1.162,

$$\boldsymbol{K} = \frac{EA}{l}\begin{bmatrix} 1 & -1 \\ -1 & 1 \end{bmatrix}, \qquad \boldsymbol{K} = \frac{EI}{l^3}\begin{bmatrix} 12 & -6l & -12 & -6l \\ -6l & 4l^2 & 6l & 2l^2 \\ -12 & 6l & 12 & 6l \\ -6l & 2l^2 & 6l & 4l^2 \end{bmatrix},$$

$$(1.621)$$

are the strain energy products between the nodal unit displacements:

$$k_{ij} = a(\varphi_i, \varphi_j) = \int_0^l EA\,\varphi_i'\,\varphi_j'\,dx = \int_0^l \sigma_i\,\varepsilon_j\,A\,dx \qquad \text{bar} \quad (1.622)$$

$$k_{ij} = a(\varphi_i, \varphi_j) = \int_0^l EI\,\varphi_i''\,\varphi_j''\,dx = \int_0^l m_i\,\kappa_j\,dx \qquad \text{beam}. \quad (1.623)$$

The same holds for plates and slabs (Kirchhoff):

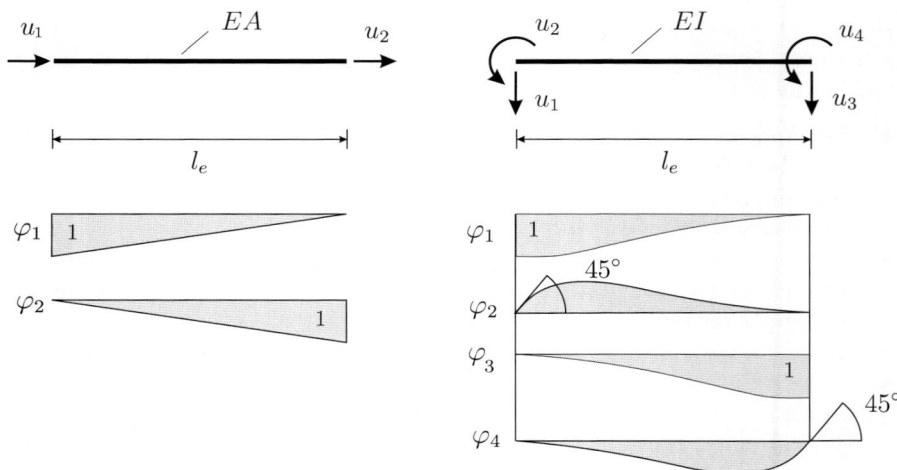

Fig. 1.162. A bar and a beam element, and the nodal unit displacements

$$k_{ij} = a(\boldsymbol{\varphi}_i, \boldsymbol{\varphi}_j) = \int_\Omega \boldsymbol{S}_i \cdot \boldsymbol{E}_j \, d\Omega = \int_\Omega \boldsymbol{\sigma}_i \cdot \boldsymbol{\varepsilon}_j \, d\Omega \qquad (1.624)$$

$$k_{ij} = a(\varphi_i, \varphi_j) = \int_\Omega \boldsymbol{M}_i \cdot \boldsymbol{K}_j \, d\Omega = \int_\Omega \boldsymbol{m}_i \cdot \boldsymbol{\kappa}_j \, d\Omega \,, \qquad (1.625)$$

where the corresponding terms are the scalar product between the stress tensor of the field $\boldsymbol{\varphi}_i$ and the strain tensor of the field $\boldsymbol{\varphi}_j$, or the moment tensor $\boldsymbol{M}_i$ of φ_i and the curvature tensor $\boldsymbol{K}_j$ of φ_j.

If the stress vectors $\boldsymbol{\sigma}_i$ and strain vectors $\boldsymbol{\varepsilon}_i$ of the element nodal unit displacement fields $\boldsymbol{\varphi}_i$ are written as row vectors, then

$$\boldsymbol{B}_{(n\times3)} = [\boldsymbol{\varepsilon}_1^T, \boldsymbol{\varepsilon}_2^T, \boldsymbol{\varepsilon}_3^T, \ldots, \boldsymbol{\varepsilon}_n^T]^T \qquad \boldsymbol{S}_{(n\times3)} = [\boldsymbol{\sigma}_1^T, \boldsymbol{\sigma}_2^T, \boldsymbol{\sigma}_3^T, \ldots, \boldsymbol{\sigma}_n^T]^T \,,$$
$$(1.626)$$

and the stiffness matrix becomes

$$\boldsymbol{K}_{(n\times n)} = \int_\Omega \boldsymbol{B}_{(n\times3)} \, \boldsymbol{S}_{(3\times n)}^T \, d\Omega = \int_\Omega \boldsymbol{B}_{(n\times3)} \, \boldsymbol{D}_{(3\times3)} \, \boldsymbol{B}_{(3\times n)}^T \, d\Omega \,,$$
$$(1.627)$$

where $\boldsymbol{D}$ is a 3×3-matrix which transforms the strains into stresses, $\boldsymbol{\sigma}_i = \boldsymbol{D}\,\boldsymbol{\varepsilon}_i$:

$$\begin{bmatrix} \sigma_{xx} \\ \sigma_{yy} \\ \sigma_{xy} \end{bmatrix} = \frac{E}{1-\nu^2} \begin{bmatrix} 1 & \nu & 0 \\ \nu & 1 & 0 \\ 0 & 0 & (1-\nu)/2 \end{bmatrix} \begin{bmatrix} \varepsilon_{xx} \\ \varepsilon_{yy} \\ 2\,\varepsilon_{xy} \end{bmatrix} \,. \qquad (1.628)$$

This matrix $\boldsymbol{D}$ is for plane stress problems. For plane strain problems it has the form

$$\boldsymbol{D} = \frac{E}{(1+\nu)(1-2\nu)} \begin{bmatrix} (1-\nu) & \nu & 0 \\ \nu & (1-\nu) & 0 \\ 0 & 0 & (1-2\nu)/2 \end{bmatrix}. \qquad (1.629)$$

In plate bending the procedure is analogous.

The symmetric triple product in (1.627) has its origin in the structure of the differential equation[17]

$$G(u, \hat{u}) = \int_0^l -\frac{d}{dx} EA \frac{d}{dx} u \cdot \hat{u} \, dx + [EA \frac{d}{dx} u \cdot \hat{u}]_0^l$$
$$- \int_0^l \frac{d}{dx} u \cdot EA \cdot \frac{d}{dx} \hat{u} \, dx = 0. \qquad (1.630)$$

Virtual work

The setup of a stiffness matrix can also be explained as follows. First all nodes are kept fixed and the first degree of freedom, $u_1 = 1$, (only) is activated. The forces necessary to force the structure into this particular shape, $\boldsymbol{u} = \boldsymbol{e}_1$, constitute the unit load case $\boldsymbol{p}_1$. Next we let these forces consecutively act through the nodal unit displacements $\boldsymbol{\varphi}_1, \boldsymbol{\varphi}_2, \ldots, \boldsymbol{\varphi}_n$, and each time the virtual work is calculated. These n numbers

$$\delta W_e(\boldsymbol{p}_1, \boldsymbol{\varphi}_i) = \text{work done by } \boldsymbol{p}_1 \text{ on acting through } \boldsymbol{\varphi}_i \qquad i = 1, 2, \ldots n \qquad (1.631)$$

form the first column of the stiffness matrix. Then the next degree of freedom, u_2, is activated (all other u_i are zero) and the procedure is repeated. The result is the stiffness matrix $\boldsymbol{K}$.

Because the product $\boldsymbol{K} \boldsymbol{u} = u_1 \boldsymbol{c}_1 + u_2 \boldsymbol{c}_2 + \ldots + u_n \boldsymbol{c}_n$ is the weighted (u_i) sum of the columns $\boldsymbol{c}_i$ of $\boldsymbol{K}$ and because

$$\boldsymbol{K} \boldsymbol{e}_i = \boldsymbol{c}_i \qquad \boldsymbol{e}_i = \text{i-th unit vector}, \qquad (1.632)$$

the columns $\boldsymbol{c}_i$ of $\boldsymbol{K}$ are just the equivalent nodal forces that belong to the single nodal unit displacements.

Three properties

Each stiffness matrix has the following three properties:

$$\hat{\boldsymbol{u}}^T \boldsymbol{K} \boldsymbol{u} = \boldsymbol{u}^T \boldsymbol{K} \hat{\boldsymbol{u}} \qquad \text{symmetry}$$
$$\boldsymbol{u}_0^T \boldsymbol{K} \boldsymbol{u} = 0 \qquad \text{equilibrium}$$
$$\boldsymbol{K} \boldsymbol{u}_0 = \boldsymbol{0} \qquad \boldsymbol{u}_0 = \text{rigid-body motion}.$$

---

[17] Gilbert Strang has written nearly a whole book—and very readable book—about this subject, [232].

Because of the latter property the stiffness matrix of an unconstrained structure or element is singular. If the rows and columns in $\boldsymbol{K}$ which correspond to fixed degrees of freedom are deleted, then a regular matrix, the so-called *reduced stiffness matrix* $\boldsymbol{K}$, is obtained.

Energy

The elastic energy stored in the FE structure is

$$W_i = \frac{1}{2}\,\boldsymbol{u}^T \boldsymbol{K} \boldsymbol{u} \tag{1.633}$$

and if a virtual displacement φ_i is applied then the virtual work is the scalar product of row $\boldsymbol{r}_i$ of the stiffness matrix and the vector $\boldsymbol{u}$

$$\delta W_e(p_h, \varphi_i) = \boldsymbol{r}_i\, \boldsymbol{u} = \sum_{j=1}^{n} k_{ij}\, u_j \, . \tag{1.634}$$

1.48 Coupling degrees of freedom

If two elements are joined at a node, the displacements must be the same at the node. Conversely this implies that a force that acts at the node must work against the stiffness of both elements. Hence if two degrees of freedom are coupled, their stiffness adds as in springs working in parallel.

- This simple coupling of even the most diverse elements is the real advantage of the FE method with regard to other numerical methods.

To understand why the stiffness adds let us first recall two rules of matrix algebra.

a) If the columns of a unit matrix $\boldsymbol{I}$ are permuted in an arbitrary fashion $\boldsymbol{I} \to \boldsymbol{I}_P$, and if a matrix $\boldsymbol{K}$ (having the same size) is multiplied from the right by this matrix $\boldsymbol{I}_P$, the *columns* of $\boldsymbol{K}$ are permuted in the same way. If the matrix $\boldsymbol{K}$ is multiplied from the left by the transposed matrix $\boldsymbol{I}_P^T$, the *rows* of $\boldsymbol{K}$ are interchanged in the same way.

b) If a 2×2 matrix is multiplied from the right by a vector $[1, 1]^T$ the columns of the matrix are added. If the same is done from the left, the rows are added:

$$\begin{bmatrix} a & b \\ c & d \end{bmatrix} \begin{bmatrix} 1 \\ 1 \end{bmatrix} = \begin{bmatrix} a+b \\ c+d \end{bmatrix} \qquad [1, 1] \begin{bmatrix} a & b \\ c & d \end{bmatrix} = [\, a+c,\, b+d\,] \, . \tag{1.635}$$

Next consider the bar in Fig. 1.163, which consists of two elements, so that

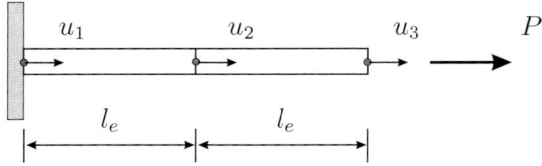

Fig. 1.163. Bar consisting of two elements

$$
\begin{bmatrix}
k_1 & -k_1 & 0 & 0 \\
-k_1 & k_1 & 0 & 0 \\
0 & 0 & k_2 & -k_2 \\
0 & 0 & -k_2 & k_2
\end{bmatrix}
\begin{bmatrix}
u_1^{(1)} \\
u_2^{(1)} \\
u_1^{(2)} \\
u_2^{(2)}
\end{bmatrix}
=
\begin{bmatrix}
f_1^{(1)} \\
f_2^{(1)} \\
f_1^{(2)} \\
f_2^{(2)}
\end{bmatrix}
\quad \text{or} \quad \boldsymbol{K}_l \, \boldsymbol{u}_l = \boldsymbol{f}_l \,. \quad (1.636)
$$

In matrix algebra the coupling between the global, u_i, and local degrees of freedom, $u_i^{(j)}$, can be written as

$$
\begin{bmatrix}
u_1^{(1)} \\
u_2^{(1)} \\
u_1^{(2)} \\
u_2^{(2)}
\end{bmatrix}
=
\begin{bmatrix}
1 & 0 & 0 \\
0 & 1 & 0 \\
0 & 1 & 0 \\
0 & 0 & 1
\end{bmatrix}
\begin{bmatrix}
u_1 \\
u_2 \\
u_3
\end{bmatrix}
\quad \text{or} \quad \boldsymbol{u}_l = \boldsymbol{A}\,\boldsymbol{u}\,, \quad (1.637)
$$

and because

$$
\boldsymbol{u}_l^T \, \boldsymbol{K}_l \, \boldsymbol{u}_l = \boldsymbol{u}^T \, \boldsymbol{A}^T \, \boldsymbol{K}_l \, \boldsymbol{A}\,\boldsymbol{u} = \boldsymbol{u}^T \, \boldsymbol{K}\,\boldsymbol{u} \quad (1.638)
$$

we have

$$
\begin{bmatrix}
1 & 0 & 0 & 0 \\
0 & 1 & 1 & 0 \\
0 & 0 & 0 & 1
\end{bmatrix}
\begin{bmatrix}
k_1 & -k_1 & 0 & 0 \\
-k_1 & k_1 & 0 & 0 \\
0 & 0 & k_2 & -k_2 \\
0 & 0 & -k_2 & k_2
\end{bmatrix}
\begin{bmatrix}
1 & 0 & 0 \\
0 & 1 & 0 \\
0 & 1 & 0 \\
0 & 0 & 1
\end{bmatrix}
=
\begin{bmatrix}
k_1 & -k_1 & 0 \\
-k_1 & k_1 + k_2 & -k_2 \\
0 & -k_2 & k_2
\end{bmatrix}
= \boldsymbol{K}\,.
$$

$$(1.639)$$

Due to the multiplication from the right, columns 2 and 3 are added; due to the multiplication from the left, rows 2 and 3 are added. This is the algebra which governs the assemblage of the global stiffness matrix.

Rigid elements should be modeled by formulating coupling conditions, and not by raising the stiffness. The displacements u_x, u_y, u_z of a node $\boldsymbol{x} = (x, y, z)$ in a rigid element can easily be expressed in terms of displacements $u_{x,ref}, u_{y,ref}, u_{z,ref}$ and rotations $\varphi_x, \varphi_y, \varphi_z$ of a reference node

$$
u_z = u_{z,ref} - (x - x_{ref}) \cdot \varphi_{y,ref} + (y - y_{ref}) \cdot \varphi_{x,ref} \,. \quad (1.640)
$$

Implicit formulations as in the case of a skew roller support

$$
\boldsymbol{u} \bullet \boldsymbol{n} = u_x\,n_x + u_y\,n_y + u_z\,n_z = 0 \quad (1.641)
$$

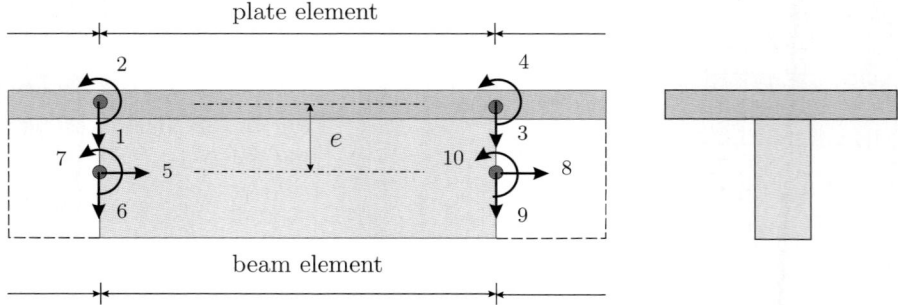

Fig. 1.164. Coupling of a beam and a slab

must be transformed into an explicit form to be able to distinguish between master and slave degrees of freedom. A master degree of freedom is a genuine degree of freedom which has its place in the structural system of equations, while a slave is eliminated either at the element level or later, when the global system of equations is assembled. Elimination at the element level only makes sense if the coupling condition is a genuine property of the element, and if the number of equations in the global stiffness matrix does not increase. Repeated application of the explicit form can raise the rank of the matrix.

T beams

If a girder is modeled by a beam as in Fig. 1.164, the movements of the beam must follow the movements of the slab:

$$
\begin{bmatrix} u_5 \\ u_6 \\ u_7 \\ u_8 \\ u_9 \\ u_{10} \end{bmatrix} = \begin{bmatrix} 0 & -e & 0 & 0 \\ 1 & 0 & 0 & 0 \\ 0 & 1 & 0 & 0 \\ 0 & 0 & 0 & e \\ 0 & 0 & 1 & 0 \\ 0 & 0 & 0 & 1 \end{bmatrix} \begin{bmatrix} u_1 \\ u_2 \\ u_3 \\ u_4 \end{bmatrix} \qquad \text{or} \qquad \boldsymbol{u}^B_{(6)} = \boldsymbol{A}_{(6\times4)} \boldsymbol{u}^S_{(4)} . \quad (1.642)
$$

Correspondingly a modified beam element matrix is obtained

$$
\boldsymbol{A}^T_{(4\times6)} \, \boldsymbol{K}_{(6\times6)} \, \boldsymbol{A}_{(6\times4)} = \boldsymbol{K}_{(4\times4)} , \qquad (1.643)
$$

which can be incorporated directly into the global stiffness matrix of the slab.

A different approach to formulating coupling conditions between degrees of freedom is provided by *Lagrange multipliers*. Their application is simple, but it is often difficult to obtain stable solutions with this technique.

1.49 Numerical details

When it comes to the solution of the structural equation $\boldsymbol{K} \boldsymbol{u} = \boldsymbol{f}$, two things can happen: the system of equations cannot be solved because the matrix $\boldsymbol{K}$

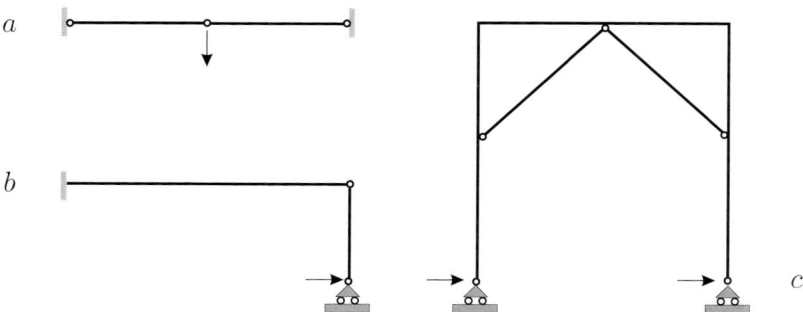

Fig. 1.165. All of these structures are statically underdetermined, and therefore the reduced stiffness matrix K is singular

is either *singular* or *ill-conditioned*. In the first case no solution exists, and in the second the solution may be sensitive to rounding errors.

Singular matrices

If the whole structure or parts of it can perform rigid-body motions, (see Fig. 1.165), then no equilibrium position u can be found, because the stiffness matrix K is singular. The computer stops the triangular decomposition of $K\,u = f$ when it is instructed to divide by zero.

If a matrix is singular, exist vectors $u \neq 0$, which are mapped by K onto the zero-vector. In general these are the rigid-body motions u_0 which do not cause any strains in the structure:

$$a(u_0, u_0) = u_0^T K\, u_0 = u_0^T\, 0 = 0 \quad u_0 = \text{rigid-body motion}\,. \quad (1.644)$$

The opposite of singular matrices are regular matrices. For any right-hand side f there is a unique solution u. In particular, zero strain energy implies zero displacement:

$$a(u, u) = u^T K\, u = 0 \quad \Rightarrow \quad u = 0\,. \quad (1.645)$$

Because singular matrices have at least one eigenvalue $\lambda = 0$, an inspection of the distribution of eigenvalues can provide clues to the "stability" of a structure. Nevertheless the calculation of the first three, four eigenvalues is rather expensive, and in addition it is not at all evident whether an eigenvalue is "nearly zero" or definitely greater than zero. Solvers are very sensitive to even hidden or infinitesimal movements.

Element matrices are singular because they let rigid-body motions pass through. The stiffness matrix of a bar ignores simple translations such as $u_0 = [1, 1]^T$, i.e., the equivalent nodal forces f are zero:

$$\frac{EA}{l}\begin{bmatrix} 1 & -1 \\ -1 & 1 \end{bmatrix}\begin{bmatrix} 1 \\ 1 \end{bmatrix} = \begin{bmatrix} 0 \\ 0 \end{bmatrix}\,. \quad (1.646)$$

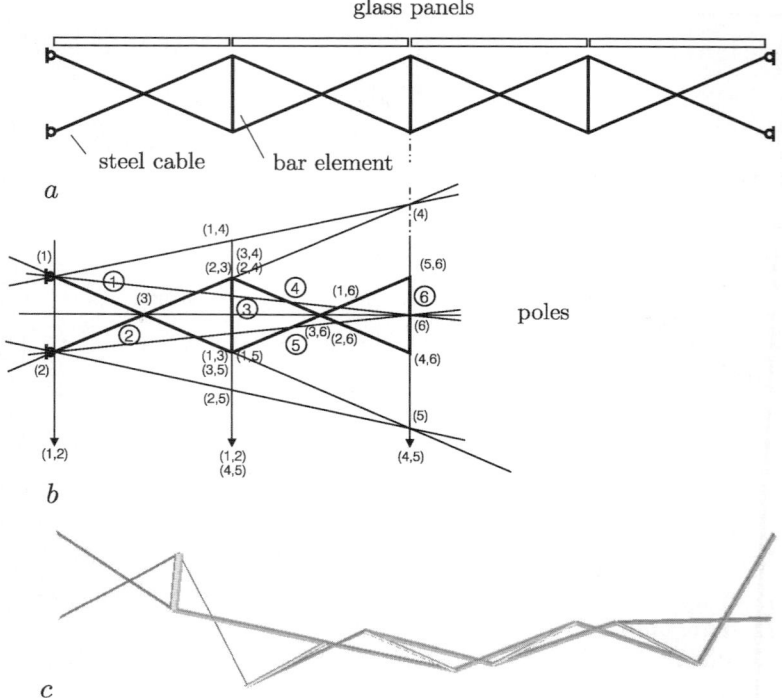

Fig. 1.166. Steel cables carry the weight of the glass panels and bear the force of the wind: **a)** system, **b)** position of the poles, **c)** deformations under dynamical load

The global assembled stiffness matrix of a structure is singular. Only if the structure is constrained, and rows and columns deleted which belong to fixed degrees of freedom—the support nodes—is the so-called *reduced global stiffness matrix* obtained, which normally is a regular matrix.

The structure in Fig. 1.166 consists of two ropes held apart by rigid bar elements. This structure was to bear the weight of glass panels and a wind load. The whole structure is kinematically unstable (see Fig. 1.166 b), even though the results of a first-order analysis seemed plausible. Obviously, the FE program assumed that the ropes were (mildly) prestressed. This seems to be a common approach, because other FE programs rendered similar results. Only a second-order analysis was sensitive enough to raise concerns. A dynamic analysis (Fig. 1.166 c) finally revealed the instability of the whole structure.

Reduced integration

A stiffness matrix is regular if and only if zero strain energy ($a(\boldsymbol{u}, \boldsymbol{u}) = \boldsymbol{u}^T \boldsymbol{K} \boldsymbol{u} = 0$) implies $\boldsymbol{u} = \boldsymbol{u}_0$. Energy is an integral, but in FE programs this integral is calculated by evaluating the strain energy density $\sigma_{ij}\, \varepsilon_{ij}\, d\Omega$

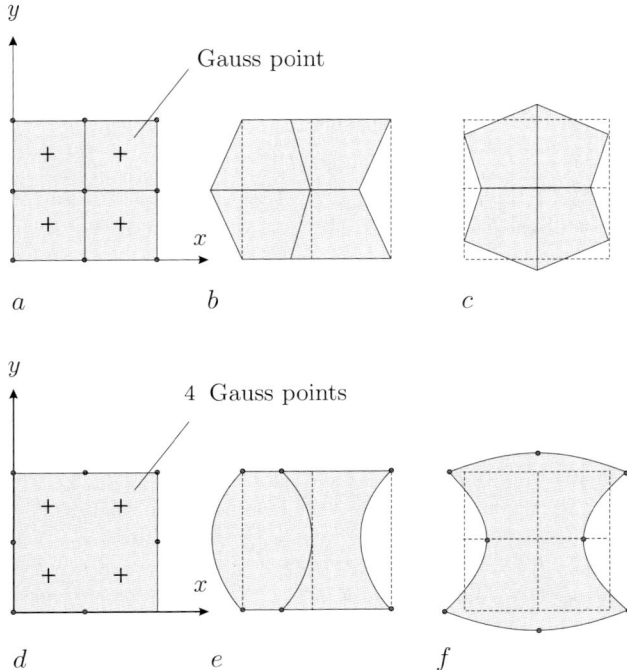

Fig. 1.167. Hourglass modes caused by reduced integration: **a)** cluster of four bilinear elements with one Gauss point each; **b)** and **c)** shapes with zero energy; **d)** quadratic 8-node-element with four Gauss point (2× 2); **e)** and **f)** shapes with zero energy

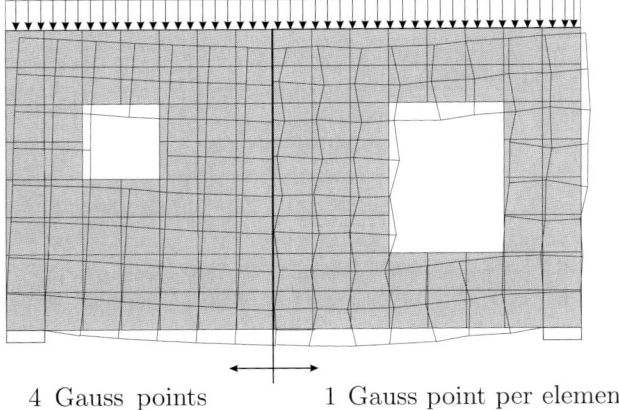

4 Gauss points   1 Gauss point per element

Fig. 1.168. Analysis of a plate with bilinear elements, regular integration, and reduced integration

at n Gauss points $\boldsymbol{x}_k$ and multiplying the point values by the integration weights w_k:

$$a(\boldsymbol{u}, \boldsymbol{u}) = \int_\Omega \sigma_{ij} \, \varepsilon_{ij} \, d\Omega = \sum_k \sigma_{ij}(\boldsymbol{x}_k) \, \varepsilon_{ij}(\boldsymbol{x}_k) \, d\Omega(\boldsymbol{x}_k) \, w_k \,. \quad (1.647)$$

If reduced integration is used it can happen that there are certain modes $\boldsymbol{u} \neq \boldsymbol{0}$ whose strain energy density happens to be zero at the reduced set of Gauss points. These modes $\boldsymbol{u} \neq \boldsymbol{0}$ (but $a(\boldsymbol{u}, \boldsymbol{u}) = 0$) are called *zero-energy hourglass modes* because of their shape, see Fig. 1.167 and Fig. 1.168.

Rounding errors

A vector $\boldsymbol{u}$ is an eigenvector of the matrix $\boldsymbol{K}$ if $\boldsymbol{K}\boldsymbol{u}$ is a multiple of $\boldsymbol{u}$, $\boldsymbol{K}\boldsymbol{u} = \lambda\boldsymbol{u}$. The scalar λ is called the eigenvalue. An $n \times n$ matrix has exactly n eigenvalues, which can all be different. Likewise some may coincide, and sometimes they can even all be the same, as in the case of a unit matrix $\boldsymbol{I}$, where all $\lambda_i = 1$. If a matrix is symmetric and positive definite, then all eigenvalues are positive, $\lambda_i > 0$.

   The condition number c of the reduced stiffness matrix $\boldsymbol{K}$ is the ratio of the largest eigenvalue to the smallest eigenvalue:

$$c = \frac{\lambda_{max}}{\lambda_{min}} \,. \quad (1.648)$$

The greater this ratio, the worse the condition of the matrix. The condition number of the unit matrix $\boldsymbol{I}$ is $c = 1/1 = 1$. If the matrix is singular, as for example the matrix

$$\boldsymbol{K} = \frac{EA}{l} \begin{bmatrix} 1 & -1 \\ -1 & 1 \end{bmatrix} \quad \text{eigenvalues} \quad \lambda_1 = 0 \,, \quad \lambda_2 = 2\frac{EA}{l} \,, \quad (1.649)$$

then the condition number is $c = \infty$, because $\lambda_{min} = 0$. This is the worst case.

   The worse the condition number of the matrix $\boldsymbol{K}$ in the system $\boldsymbol{K}\boldsymbol{u} = \boldsymbol{f}$, the more susceptible the solution $\boldsymbol{u}$ is to rounding errors. The condition number of a stiffness matrix is always large if the structure or parts of it can perform movements which come close to rigid-body motions, i.e., if the structure sits on very soft supports, or if some parts of the structure are very stiff compared to others. The reason for this behavior is the principle of conservation of energy or equivalently Green's first identity

$$\frac{1}{2} \, G(w, w) = W_e - W_i = 0 \,. \quad (1.650)$$

   Let us study the nearly rigid beam in Fig. 1.169. The internal energy of the beam is

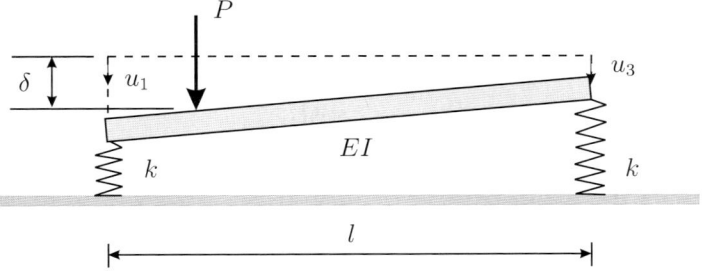

Fig. 1.169. The greater the stiffness EI, the more the deflection curve assumes the shape of a rigid-body motion. This is a consequence of the principle of conservation of energy

$$W_i = \frac{1}{2}\,\boldsymbol{u}^T\{\underbrace{\frac{EI}{l^3}\begin{bmatrix} 12 & -6l & 6l & -6l \\ -6l & 4\,l^2 & 6l & 2\,l^2 \\ -12 & 6l & 12 & 6\,l \\ -6\,l & 2\,l^2 & 6\,l & 4\,l^2 \end{bmatrix}}_{\boldsymbol{K}_B} + \underbrace{\begin{bmatrix} k & 0 & 0 & 0 \\ 0 & 0 & 0 & 0 \\ 0 & 0 & k & 0 \\ 0 & 0 & 0 & 0 \end{bmatrix}}_{\boldsymbol{S}}\}\boldsymbol{u}$$

$$= \frac{1}{2}\,\{\boldsymbol{u}^T\,\boldsymbol{K}_B\,\boldsymbol{u} + \boldsymbol{u}^T\,\boldsymbol{S}\,\boldsymbol{u}\} = \frac{1}{2}\,\boldsymbol{u}^T\,\boldsymbol{K}\,\boldsymbol{u} = \frac{1}{2}\,a(w,w)\,, \qquad (1.651)$$

because the stiffness matrix $\boldsymbol{K}$ is the sum of the beam matrix $\boldsymbol{K}_B$ plus the "spring matrix" $\boldsymbol{S}$. If the beam stiffness becomes very large, $EI \to \infty$, W_i too will become very large. On the other hand, the internal energy W_i can never, because of (1.650), exceed the external eigenwork $W_e = 1/2 \cdot P \cdot \delta$. The beam avoids this dilemma by performing mostly a rotation, i.e., the vector $\boldsymbol{u}$ more and more resembles a rigid-body motion, $\boldsymbol{u} \to \boldsymbol{u}_0$, and this preserves the energy balance, because rotations lie in the *kernel*[18] of the beam matrix $\boldsymbol{K}_B$, and the beam can thus (even in the case $EI = \infty$) preserve the energy balance:

$$\frac{1}{2}\,\boldsymbol{u}_0^T\,\boldsymbol{K}\,\boldsymbol{u}_0 = \underbrace{\frac{1}{2}\,[u_1^2\,k + u_3^2\,k]}_{W_i} = \underbrace{\frac{1}{2}\,P\,\delta}_{W_e}\,. \qquad (1.652)$$

Clearly the transition must be smooth. Obviously, the more a vector $\boldsymbol{u}$ resembles a vector $\boldsymbol{u}_0$, the smaller the strain energy $1/2\,\boldsymbol{u}^T\,\boldsymbol{K}_B\,\boldsymbol{u}$.

It is evident that for large values of EI the *difference* between the end rotations of the beam, $w'(0) - w'(l)$, is small but on this difference depends the energy of the nearly rigid beam.

An elementary example of the difficulties that can result from large differences in element stiffness is that of two bar elements that differ in longitudinal

---

[18] The kernel of a matrix $\boldsymbol{K}$ contains all vectors that are mapped onto the null vector by $\boldsymbol{K}$.

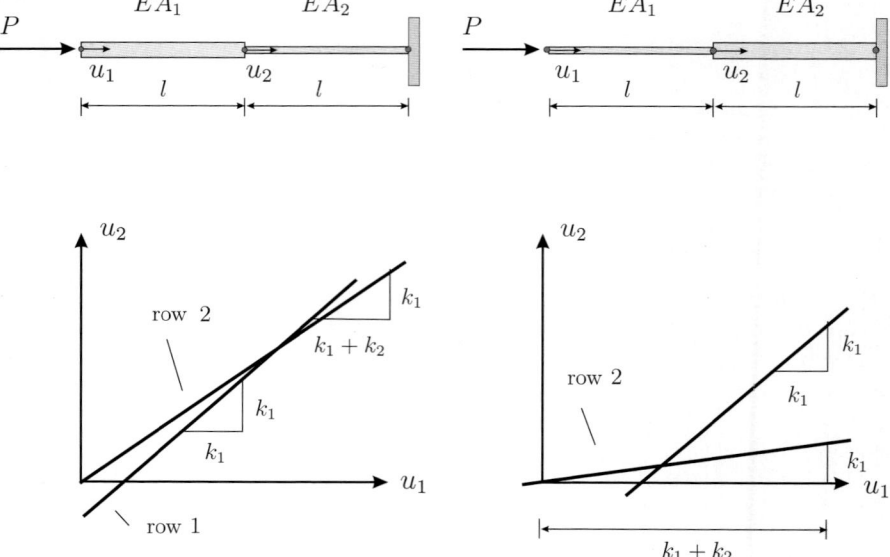

Fig. 1.170. Favorable and unfavorable arrangement

stiffness, $k_i = EA_i/l_i$, $k_1 = 100$ and $k_2 = 1$, as in Fig. 1.170. If the stiffer element is supported by the weaker element, the system of equations for the u_i is

$$\begin{bmatrix} k_1 & -k_1 \\ -k_1 & k_1 + k_2 \end{bmatrix} \begin{bmatrix} u_1 \\ u_2 \end{bmatrix} = \begin{bmatrix} P \\ 0 \end{bmatrix} \qquad \begin{bmatrix} 100 & -100 \\ -100 & 100 + 1 \end{bmatrix} \begin{bmatrix} u_1 \\ u_2 \end{bmatrix} = \begin{bmatrix} P \\ 0 \end{bmatrix}. \tag{1.653}$$

The solution of this system of equations is the point in the u_1–u_2-plane where the two straight lines $k_1 u_1 - k_1 u_2 = P$ and $-k_1 u_1 + (k_1 + k_2) u_2 = 0$ intersect; see Fig. 1.170. Because of the nearly identical slopes, the intersection is hard to localize.

A computer adds these two equations and thus eliminates the unknown u_1

$$[(k_1 + k_2) - k_1]\, u_2 = P\,. \tag{1.654}$$

If the computer uses only three decimal places, the result is $(k_1 + k_2) - k_1 = 100.01 - 100 = 1.001 \times 10^2 - 1 \times 10^2 = 0.000$ or $u_2 = 0$.

If the two bar elements are rearranged so that the weaker element is supported by the more rigid element

$$\begin{bmatrix} 1 & -1 \\ -1 & 1 + 100 \end{bmatrix} \begin{bmatrix} u_1 \\ u_2 \end{bmatrix} = \begin{bmatrix} P \\ 0 \end{bmatrix}, \tag{1.655}$$

i.e., if the structure is better "grounded" then the problem is well conditioned, and the intersection of the two lines is easy to spot.

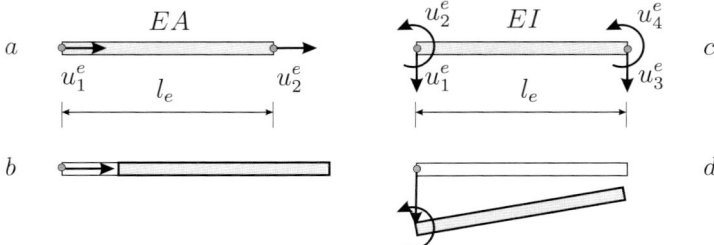

Fig. 1.171. Rigid-body motions

A sounder approach is to replace nearly rigid zones of a structure by formulating coupling conditions. For example, if the coupling condition of the first bar element were $u_1^e = u_2^e$, the stiffness of this bar element would vanish

$$\begin{bmatrix} 1, 1 \end{bmatrix} \begin{bmatrix} k_1 & -k_1 \\ -k_1 & k_1 \end{bmatrix} \begin{bmatrix} 1 \\ 1 \end{bmatrix} = 0 \,, \tag{1.656}$$

and the structural system would reduce to the equation

$$k_2 \, u_2 = P \,, \tag{1.657}$$

which would yield the exact solution for u_2.

In the case of a beam, such a rigid-body constraint (see Fig. 1.171 c) would result in

$$\begin{bmatrix} u_1^e \\ u_2^e \\ u_3^e \\ u_4^e \end{bmatrix} = \begin{bmatrix} 1 & 0 \\ 0 & 1 \\ 1 & -l \\ 0 & 1 \end{bmatrix} \begin{bmatrix} u_1^e \\ u_2^e \end{bmatrix} \qquad \text{or} \qquad \boldsymbol{u} = \boldsymbol{T} \, \boldsymbol{u}_M \,. \tag{1.658}$$

Because the columns of the matrix $\boldsymbol{T}$ represent rigid-body motions, the modified stiffness matrix

$$\boldsymbol{T}^T \, \boldsymbol{K} \, \boldsymbol{T} = \begin{bmatrix} 0 & 0 \\ 0 & 0 \end{bmatrix} \tag{1.659}$$

is zero.

In Fig. 1.172 the coupling conditions are

$$\begin{bmatrix} u_1 \\ u_2 \\ u_3 \\ u_4 \\ u_5 \\ u_6 \end{bmatrix} = \begin{bmatrix} 1 & 0 & 0 & 0 \\ 0 & 1 & 0 & 0 \\ 0 & 0 & 1 & 0 \\ 0 & 1 & -l_2 & 0 \\ 0 & 0 & 1 & 0 \\ 0 & 0 & 0 & 1 \end{bmatrix} \begin{bmatrix} u_1 \\ u_2 \\ u_3 \\ u_6 \end{bmatrix} \qquad \text{or} \qquad \boldsymbol{u} = \boldsymbol{T} \, \boldsymbol{u}_M \,. \tag{1.660}$$

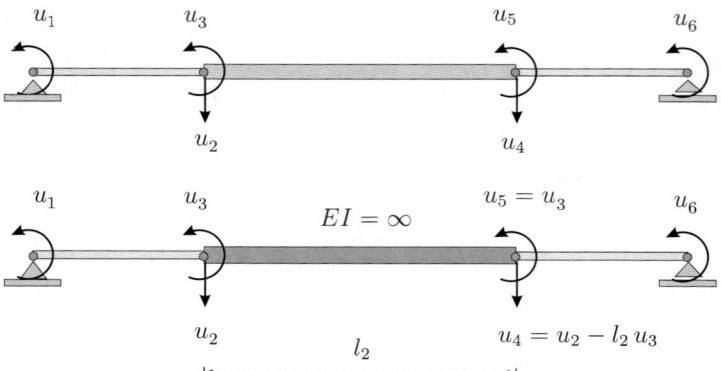

Fig. 1.172. The beam in the middle is assumed to be inflexible

If the stiffness matrix

$$
\begin{bmatrix}
k_{22}^1 & k_{23}^1 & k_{24}^1 & 0 & 0 & 0 \\
k_{32}^1 & k_{33}^1 + k_{11}^2 & k_{34}^1 + k_{12}^2 & k_{13}^2 & k_{14}^2 & 0 \\
k_{42}^1 & k_{43}^1 + k_{21}^2 & k_{44}^1 + k_{22}^2 & k_{23}^2 & k_{24}^2 & 0 \\
0 & k_{31}^2 & k_{32}^2 & k_{33}^2 + k_{11}^3 & k_{34}^2 + k_{12}^3 & k_{13}^3 \\
0 & k_{41}^2 & k_{42}^2 & k_{43}^2 + k_{21}^3 & k_{44}^2 + k_{22}^3 & k_{23}^3 \\
0 & 0 & 0 & k_{31}^3 & k_{32}^3 & k_{33}^3
\end{bmatrix}
\tag{1.661}
$$

is multiplied from the right by the matrix $\boldsymbol{T}$, then

- column 4 is added to column 2
- column 4 $\times(-l_2)$ is added to column 3
- column 5 is added to column 3
- column 6 is added to column 4,

Multiplication from the left by $\boldsymbol{T}^T$ effects the same operations with the rows. The result is a 4×4 matrix $\boldsymbol{T}^T \boldsymbol{K} \boldsymbol{T}$ which only contains contributions from the first and last beam elements:

$$
\begin{bmatrix}
k_{22}^1 & k_{23}^1 & k_{24}^1 & 0 \\
\ldots & k_{33}^1 + k_{11}^3 & k_{34}^1 - l_2\,k_{11}^3 + l_2\,k_{12}^3 & k_{13}^3 \\
\ldots & \ldots & k_{44}^1 + l_2^2\,k_{11}^3 - 2\,l_2^2 k_{12}^3 + k_{22}^3 & -l_2\,k_{13}^3 + k_{23}^3 \\
\text{sym.} & \ldots & \ldots & k_{33}^3
\end{bmatrix}. \tag{1.662}
$$

Hence if rigid-body motions are enforced, the element matrix shrinks. No other solution is possible because, even infinitesimal displacements would produce infinite strain energy in the constrained beam, and the total strain energy of the beam in Fig. 1.172

$$
a(w, w) = a(w_1, w_1) + a(w_2, w_2) + a(w_3, w_3) < \infty \tag{1.663}
$$

remains bounded for $EI_2 \to \infty$ only if the center element performs a rigid-body motion $w_2(x) = a\,x + b$.

Note that the reduction of the equivalent nodal forces obeys the rule $T^T_{(4\times6)}\, f_{(6)} = f_{(4)}$, as follows from

$$K_{(6\times6)}u_{(6)} = f_{(6)} \quad \rightarrow \quad T^T_{(4\times6)}\, K_{(6\times6)}T_{(6\times4)}u_{(4)} = T^T_{(4\times6)}f_{(6)} = f_{(4)}\,. \tag{1.664}$$

1.50 Warning

We want to close this introductory chapter with a warning note. A reviewing engineer who has followed our analysis up to this point is perhaps now tempted to request that in the future the design engineer document the FE load case p_h so that he, the reviewing engineer, can quickly check how close the FE solution is to the true solution. Theoretically such a proposition makes sense, but on the other hand one must be careful not to overinterpret the FE load case p_h.

In real structures the load case p_h is seemingly miles away from the original load case p. If we really take the trouble to plot the load case p_h, we are surprised how little the FE load case and the original load case have in common. This is why commercial FE programs do not show the load case p_h. Any user not well acquainted with FE theory would doubt the FE results.

The real secret of the FE method is that nonetheless the results are accurate, and if we as structural engineers want to have more faith in FE methods, we must deal with this question more intensively. Foremost this is a problem of structural mechanics and not of mathematics.

What do we mean by *near* and *far* in structural mechanics? What level of uncertainty can we allow, and when would we lose focus?[19]

What we see, if we concentrate on plate bending problems, and what we can compare are the loads, the load case p, and the load case p_h, i.e., (we simplify somewhat) the fourth-order derivatives of the two deflection surfaces w_h and w. But the bending moments are the second-order derivatives

$$m = \int\int p\,d\Omega\,d\Omega\,, \qquad m_h = \int\int p_h\,d\Omega\,d\Omega\,, \tag{1.665}$$

and because integration smoothes out the wrinkles, the bending moments of the FE solution are in relatively good agreement with the exact bending moments. No reviewing engineer has the tools to make guesses about the deviations in the bending moments by looking at the discrepancies in the load, the fourth-order derivatives.

If on the uppermost floor, the fourth floor, the walls deviate by 20 cm from their position, how large then is the deviation on the second floor? The

---

[19] Read the paper by Bürg and Schneider [55] on the design of a simple flat garage roof by 32 different professional engineers!

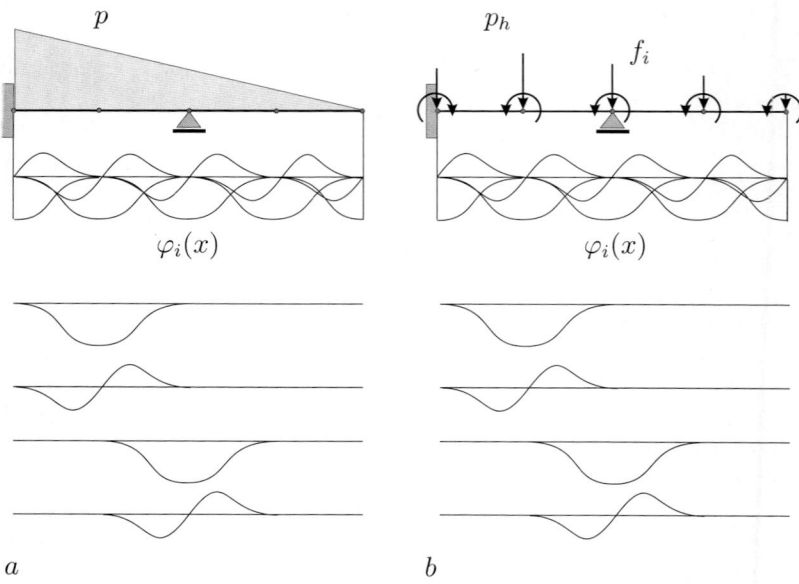

Fig. 1.173. Beam: **a)** load case p, **b)** load case p_h and nodal unit displacements

deviation on the fourth floor we can measure. The deviation on the second floor we cannot. About the latter we can only speculate. This is the problem.

There are examples in mathematics which should warn us. A linear system of equations $K\,u = f$ can only be solved approximately on a computer. It seems then that the *residual* $r = K\,u_c - f$ is an appropriate measure for the error of the computed solution u_c. But there are examples where a solution u_c with a larger residual is closer to the exact solution than a computer solution with a smaller residual.

Fortunately such ill-conditioned problems are rare in structural mechanics. Most of our problems are well behaved. We solve elliptic boundary value problems and we may assume that a small residual (probably) indicates a small error. Instinctively we also rely on St. Venant's principle.

In boundary element analysis the residual forces $p - p_h$ are zero because the BE solution satisfies the differential equation. Hence it seems that BE solutions should be more accurate than FE solutions. They are in general but only by a rather small margin. If we compare the stress resultants between a BE solution and an FE solution, then we are surprised to see how good for example the agreement between the bending moments of an FE solution and a BE solution are. This is the point where one begins to speculate: how significant are the residual forces $p - p_h$ really? Is the whole interpretation not too naive?

Yes and no. The FE solution is the solution of a load case p_h. This interpretation is correct, and as demonstrated by adaptive methods it makes sense to refine the mesh in regions where the residual forces $p - p_h$ are large. Could it then be that we are looking at the wrong data because an FE program does not focus on the distance $p - p_h$ but on the distance in the strain energy? No, the data we inspect are correct, only an FE program looks at the data from a slightly different perspective.

As engineers, we tend to say the distance is large if the distributed loads p and p_h deviate by a large margin. But not so the FE program. It does not compare the shape of p and of p_h, it does not measure the gap $|p - p_h|$, but it measures *the action*.

The FE program lets the residual forces $p - p_h$ act through the nodal unit displacements φ_i, and only if the residual forces are orthogonal

$$\int_0^l (p - p_h)\, \varphi_i \, dx = 0, \qquad i = 1, 2, \ldots, n \qquad (1.666)$$

to all functions φ_i is the distance $p - p_h$ well tuned.

Imagine that a triangular area (p) is to be covered with a stack of needles (point forces f_i). Calculus provides no tool to measure the distance between the triangular area and the area covered by the needles, because (mathematical) needles are infinitely thin. But FE analysis offers an ingenious way to give a meaning to it by applying the principle of virtual displacements.

The beam problem in Fig. 1.173 may exemplify this. In the naive sense, p_h will never converge to the triangular load p, because $p_h = 0$. But the nodal forces f_i and moments which are the real p_h do converge to p in the finite element sense, i.e., in the weak sense

$$\int_0^l p\, \varphi_i \, dx = \sum_{j=1}^{\infty} f_j \, \varphi_i(x_j) \qquad i = 1, 2, \ldots, \infty. \qquad (1.667)$$

Hence we may not take the residual forces prima facie to insist on a good match between p and p_h. In FE analysis, action is what counts—we approximate actions not functions (!)—and the action we do not see on the screen.

2. What are boundary elements?

The boundary element method (BE method) is an integral equation method, or as we could say as well an influence function method. It is based on the fact that in linear problems the boundary values uniquely determine the displacements and stresses inside a structure such as the frame in Fig. 2.1, so that it suffices to discretize the edge with boundary elements only.

In one form or the other the idea is applied each day. If for example two moments, $M(0)$ and $M(l)$, act at the ends of a beam the bending moment $M(x)$ can be generated from these boundary values with the help of a simple ruler; see Figure 2.2. If the beam carries also a distributed load p, then a curved ruler is used instead, see Fig. 2.2.

The key to this technique is that the linear bending moment $M(x) = a\,x + b$ is a homogeneous solution of the equation $-M'' = 0$, and the quadratic bending moment is a particular solution of the differential equation $-M'' = p$.

Technically the BE method applies influence functions (= integrals) to relate the displacements and tractions on the boundary with the displacements and stresses at internal points.

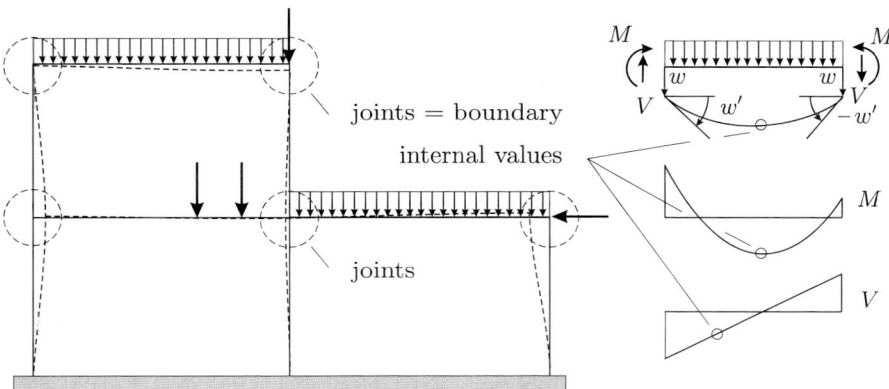

Fig. 2.1. The boundary values w, w', M, V and the distributed load p suffice to calculate any value in the interior

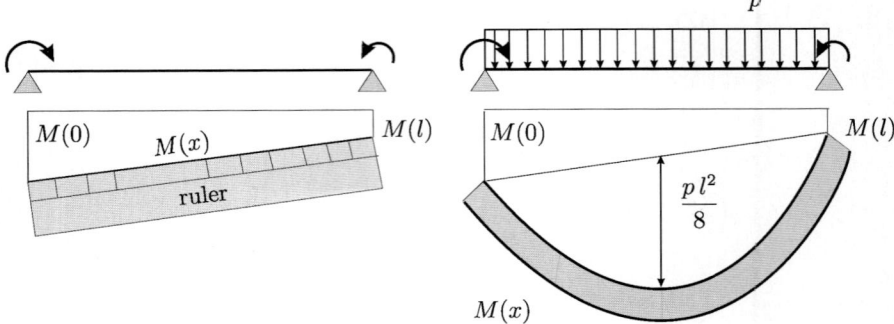

Fig. 2.2. Influence functions for bending moments

2.1 Influence functions or Betti's theorem

Beams

To calculate the deflection $w(x)$ at the center of a beam with the *principle of virtual forces* the scalar product of the two bending moments $M(x)$ and $\bar{M}(x)$ is formed (see Fig. 2.3):

$$1 \cdot w(x) = \int_0^l \frac{\bar{M}\, M}{EI}\, dx \qquad \text{Mohr's integral}. \tag{2.1}$$

But according to Betti's theorem the scalar product of the distributed load p and the deflection curve $G_0(y, x)$ (= Green's function) which belongs to the load case $\bar{P} = 1$ would provide the same result:[1]

$$W_{1,2} = 1 \cdot w(x) = \int_0^l G_0(y, x)\, p(y)\, dy = W_{2,1}. \tag{2.2}$$

Betti's theorem remains valid if the span of the second beam is larger than the span of the first beam. It is only necessary to cut off the protruding parts before Betti's theorem is formulated, and to compensate for this loss the previous internal actions must be applied as external forces.

Those external forces then contribute external work too, so that Betti's theorem becomes rather lengthy, as is seen if the reciprocal statement $W_{1,2} = W_{2,1}$ is solved for $w(x)$:

---

[1] One could also simply integrate (2.1) by parts: $(\bar{M}, M/EI) = (G_0, p)$.

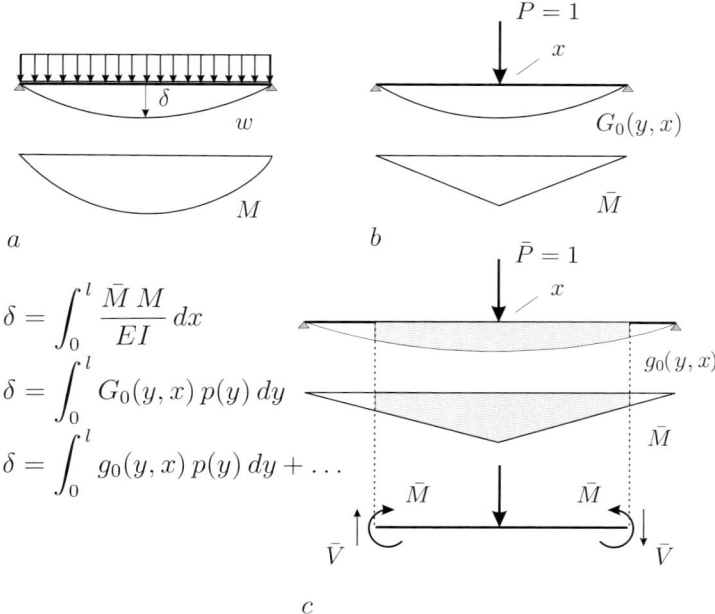

$$\delta = \int_0^l \frac{\bar{M} M}{EI} dx$$

$$\delta = \int_0^l G_0(y, x) \, p(y) \, dy$$

$$\delta = \int_0^l g_0(y, x) \, p(y) \, dy + \dots$$

Fig. 2.3. Calculation of the deflection δ **a)** with the principle of virtual forces, **b)** with Betti's theorem (the same beam), **c)** with Betti's theorem (a different beam)

$$
\begin{aligned}
w(x) = \int_0^l & g_0(y, x) \, p(y) \, dy \\
& + \underbrace{\bar{M}(0) \, w'(0) - \bar{M}(l) \, w'(l) + \bar{V}(l) \, w(l) - \bar{V}(0) \, w(0)}_{-work\ on\ the\ boundary\ W_{1,2}} \\
& + \underbrace{M(0) \, g_0'(x, 0) - M(l) \, g_0'(x, l) + V(0) \, g_0(x, 0) - V(l) \, g_0(x, l)}_{work\ on\ the\ boundary\ W_{2,1}} \, .
\end{aligned}
\tag{2.3}
$$

The deflection curve $w = g_0(y, x)$ of the second beam (see Fig. 2.3 c) is called a *fundamental solution*, because unlike the Green's function G_0 it does not satisfy the support conditions of the original beam. But both curves g_0 and G_0 share the property that the shear force jumps at the source point, $V(x_+) - V(x_-) = 1$. To construct influence functions—to have $w(x)$ emerge—only the jump in the shear force V is needed.

Kirchhoff plates

The procedure is virtually the same as before: to calculate the deflection at a point $\boldsymbol{x}$ a single force $\bar{P} = 1$ is applied at this point and the scalar product of the deflection surface $G_0(\boldsymbol{y}, \boldsymbol{x})$ and the load p is formed:

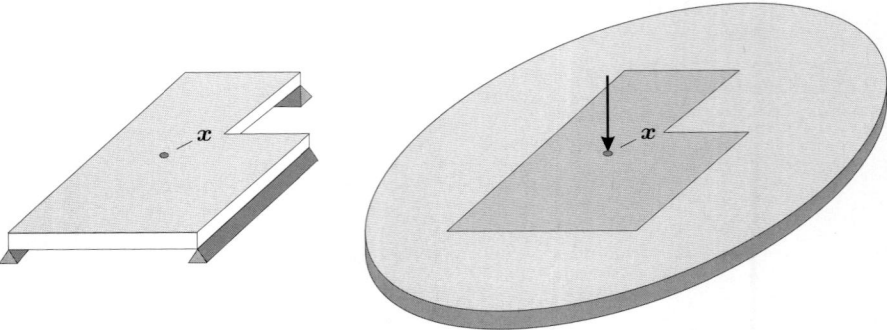

Fig. 2.4. Circular plate serving as auxiliary problem

$$W_{1,2} = 1 \cdot w(\boldsymbol{x}) = \int_{\Omega} G_0(\boldsymbol{y}, \boldsymbol{x}) \, p(\boldsymbol{y}) \, d\Omega_{\boldsymbol{y}} = W_{2,1} \,. \tag{2.4}$$

The problem with this approach is that in general the Green's function G_0 is unknown. Only if a single force $\bar{P}$ is applied directly at the center of a circular plate of radius R (see Fig. 2.4) is the deflection surface known:

$$w(r) = \frac{\bar{P}}{16 \, K \, \pi} \left[\frac{3+\nu}{1+\nu} \, R^2 \left(1 - \frac{r^2}{R^2}\right) - 2 \, r^2 \ln \frac{R}{r} \right] \,. \tag{2.5}$$

But this suffices. First the radius R of the plate is extended to infinity, $R = \infty$, so that all plates that will eventually be considered fit into the interior of the circular plate. That is, (2.5) is simply stripped of all unnecessary parts and only the fundamental solution is retained:

$$g_0(\boldsymbol{y}, \boldsymbol{x}) = \frac{1}{8 \, \pi \, K} \ln r \,. \tag{2.6}$$

Next the source point $\boldsymbol{x}$—the center of the infinite circle—is moved to the point at which the deflection $w(\boldsymbol{x})$ is to be calculated. Then that part of the circular plate which coincides in shape and position with the original plate is cut out of the circular plate, the previous internal actions are applied at the new edge as external forces, and finally the reciprocal work of the pair $\{w, g_0\}$ is calculated. The result is an expression of the form

$$w(\boldsymbol{x}) = \int_{\Omega} g_0(\boldsymbol{y}, \boldsymbol{x}) \, p(\boldsymbol{y}) \, d\Omega_{\boldsymbol{y}} - \text{work on the boundary } W_{1,2}$$
$$+ \text{work on the boundary } W_{2,1} \tag{2.7}$$

or

$$c(\boldsymbol{x})\, w(\boldsymbol{x}) = \int_{\Gamma} [\, g_0(\boldsymbol{y},\boldsymbol{x})\, \underline{v_\nu(\boldsymbol{y})} - g_{0\nu}(\boldsymbol{y},\boldsymbol{x})\, \underline{m_\nu(\boldsymbol{y})}$$

$$-v_\nu(g_0)(\boldsymbol{y},\boldsymbol{x})\, \underline{w(\boldsymbol{y})} + m_\nu(g_0)(\boldsymbol{y},\boldsymbol{x})\, \underline{w_\nu(\boldsymbol{y})}\,]\, ds_{\boldsymbol{y}}$$

$$+ \int_{\Omega} g_0(\boldsymbol{y},\boldsymbol{x})\, p(\boldsymbol{y})\, d\Omega_{\boldsymbol{y}}$$

$$+ \sum_{c} [\, g_0(\boldsymbol{y}^c,\boldsymbol{x})\, \underline{F(w)(\boldsymbol{y}^c)} - \underline{w(\boldsymbol{y}^c)}\, F(g_0)(\boldsymbol{y}^c,\boldsymbol{x})\,]\,, \qquad (2.8)$$

where the underlined terms are the boundary values of the plate:

$$v_\nu = \text{support reaction} \quad m_\nu = \text{bending moment}$$

$$w_\nu = \text{rotation} \qquad\qquad w = \text{deflection}$$

and the F are the corner forces. The influence coefficients or kernel functions

$$\downarrow\ g_0(\boldsymbol{y},\boldsymbol{x}) \qquad \curvearrowright\ g_{0\nu}(\boldsymbol{y},\boldsymbol{x}) \qquad \curvearrowright\ m_\nu(g_0)(\boldsymbol{y},\boldsymbol{x}) \qquad \downarrow\ v_\nu(g_0)(\boldsymbol{y},\boldsymbol{x})$$

$$(2.9)$$

are, in this sequence, the deflection, slope, bending moment and Kirchhoff shear of the fundamental solution g_0 on the boundary. The so-called characteristic function $c(\boldsymbol{x})$ on the left-hand side has the value 1 at all interior points, and the value $c(\boldsymbol{x}) = \Delta\varphi/2\pi$ at boundary points where $\Delta\varphi$ is the angle of the boundary point.

By replacing the kernel g_0 (single force) with the kernel $g_{0n} = \partial g_0/\partial n$ (moment), an influence function for the slope $\partial w/\partial n$ can be derived in the same way.

If the observation point $\boldsymbol{x}$ is located on the boundary, these two influence functions become—in a somewhat simplified notation—a system of two integral equations:

$$\int_{\Gamma} \boldsymbol{H}(\boldsymbol{y},\boldsymbol{x})\, \boldsymbol{u}(\boldsymbol{y})\, ds_{\boldsymbol{y}} = \int_{\Gamma} \boldsymbol{G}(\boldsymbol{y},\boldsymbol{x})\, \boldsymbol{t}(\boldsymbol{y})\, ds_{\boldsymbol{y}}$$

$$+ \int_{\Omega} \boldsymbol{U}(\boldsymbol{y},\boldsymbol{x})\, \boldsymbol{p}(\boldsymbol{y})\, d\Omega_{\boldsymbol{y}} + \boldsymbol{R}(\boldsymbol{y}_c,\boldsymbol{x})\, \boldsymbol{r}_c \qquad \boldsymbol{x} \in \Gamma, \qquad (2.10)$$

for the boundary values of a slab

$$\boldsymbol{u} = \{w(\boldsymbol{x}), w_n(\boldsymbol{x})\} \qquad \boldsymbol{t} = \{m_n(\boldsymbol{x}), v_n(\boldsymbol{x})\} \qquad \boldsymbol{r}_c = \{w_c, F_c\}. \quad (2.11)$$

The subscript c in the vector $\boldsymbol{r}_c$ indicates that these terms are the corner deflections w_c and the corner forces F_c.

To solve this system approximately, the boundary functions are interpolated with polynomial shape functions, and the unknown nodal values are determined by a collocation procedure, so that the two coupled integral equations are equivalent to a linear system of equations

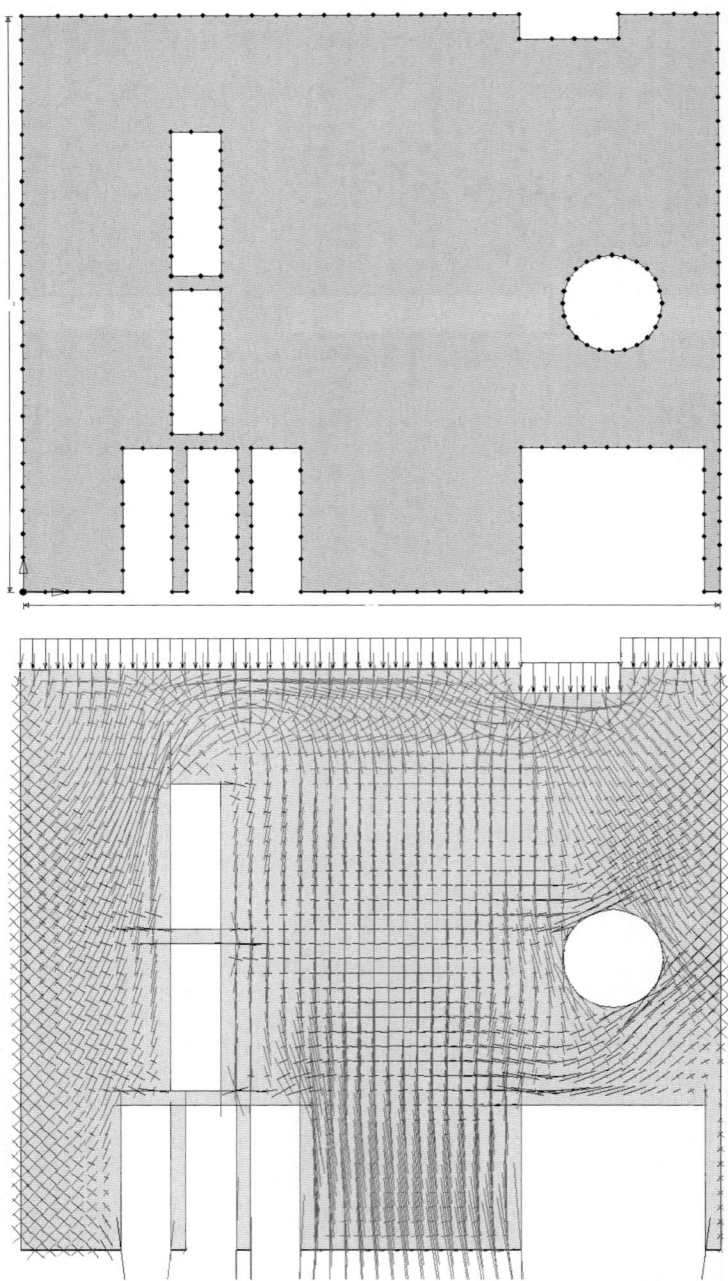

Fig. 2.5. BE analysis of a shear wall

$$\boldsymbol{H}\boldsymbol{u} = \boldsymbol{G}\boldsymbol{t} + \boldsymbol{d} \tag{2.12}$$

for the nodal values $\boldsymbol{u}$ and $\boldsymbol{t}$.

If the equation is multiplied from the left first by $\boldsymbol{G}^{-1}$ and then by the mass matrix $\boldsymbol{F}$,

$$\boldsymbol{F}\boldsymbol{G}^{-1}\boldsymbol{H}\boldsymbol{u} = \boldsymbol{F}\boldsymbol{G}^{-1}\boldsymbol{t} + \boldsymbol{F}\boldsymbol{G}^{-1}\boldsymbol{d} \tag{2.13}$$

and terms are collected, we obtain the system

$$\boldsymbol{K}\boldsymbol{u} = \boldsymbol{f} + \boldsymbol{p} \tag{2.14}$$

which is formally identical to the structural equation in FE analysis, even though the matrix $\boldsymbol{K}$ is a slightly unsymmetric stiffness matrix.

Linear elasticity

The influence functions for the displacement field $\boldsymbol{u}(\boldsymbol{x})$ of a plate as in Fig. 2.5 is

$$\boldsymbol{C}(\boldsymbol{x})\,\boldsymbol{u}(\boldsymbol{x}) = \int_{\Gamma} [\boldsymbol{U}(\boldsymbol{y},\boldsymbol{x})\,\boldsymbol{t}(\boldsymbol{y}) - \boldsymbol{T}(\boldsymbol{y},\boldsymbol{x})\,\boldsymbol{u}(\boldsymbol{y})]\,ds_{\boldsymbol{y}}$$
$$+ \int_{\Omega} \boldsymbol{U}(\boldsymbol{y},\boldsymbol{x})\,\boldsymbol{p}(\boldsymbol{y})\,d\Omega_{\boldsymbol{y}} \tag{2.15}$$

where $\boldsymbol{t} = \boldsymbol{t}(\boldsymbol{y})$ is the traction vector $\boldsymbol{S}\boldsymbol{\nu} = \boldsymbol{t}$ at the integration point $\boldsymbol{y}$ on the boundary with the normal vector $\boldsymbol{\nu} = [\nu_1, \nu_2]^T$. The fundamental solutions are

$$U_{ij}(\boldsymbol{y},\boldsymbol{x}) = \frac{1}{8\pi\mu(1-\nu)}[(3-4\nu)\ln\frac{1}{r}\,\delta_{ij} + r_{,i}\,r_{,j}] \qquad \text{(2-D)}, \tag{2.16}$$

$$U_{ij}(\boldsymbol{y},\boldsymbol{x}) = \frac{1}{8\pi\mu(1-\nu)r}[(3-4\nu)\,\delta_{ij} + r_{,i}\,r_{,j}] \qquad \text{(3-D)}, \tag{2.17}$$

and the tractions of these solutions on the boundary are

$$T_{ij}(\boldsymbol{y},\boldsymbol{x}) = -\frac{1}{4\alpha\pi(1\nu)r^{\alpha}}[\frac{\partial r}{\partial\nu}((1-2\nu)\,\delta_{ij} + \beta\,r_{,i}\,r_{,j})$$
$$-(1-2\nu)\{r_{,i}\,\nu_j(\boldsymbol{y}) - r_{,j}\,\nu_i(\boldsymbol{y})\}], \tag{2.18}$$

where $\alpha = 1$, $\beta = 2$ in 2-D, $\alpha = 2$, $\beta = 3$ in 3-D, and

$$r_{,i} := \frac{\partial r}{\partial y_i} = -\frac{\partial r}{\partial x_i} = \frac{y_i - x_i}{r}. \tag{2.19}$$

The columns i (or rows) of the symmetric matrix $\boldsymbol{U}(\boldsymbol{y},\boldsymbol{x})$ are the displacement fields of the elastic continuum if a concentrated force $\boldsymbol{P} = \boldsymbol{e}_i$ acts at a point $\boldsymbol{x}$, and the columns of the matrix $\boldsymbol{T}(\boldsymbol{y},\boldsymbol{x})$ are the associated tractions at a point $\boldsymbol{y}$ with the normal vector $\boldsymbol{\nu}$.

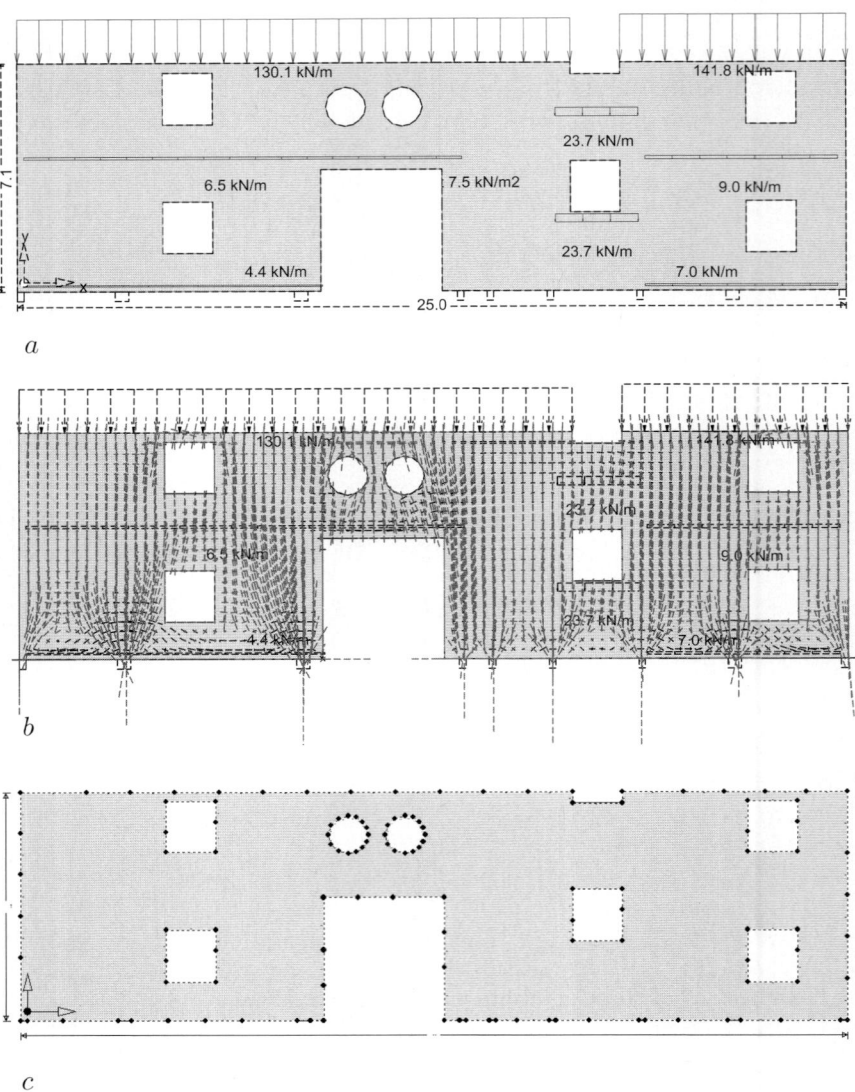

Fig. 2.6. Shear wall: **a)** system and loads, **b)** principal stresses, **c)** boundary elements

Technical details

The boundary discretization poses no difficulties. The elements should form a piecewise smooth approximation and the boundary functions should be C^0 or where necessary C^1 functions (the deflection w in plate bending). Essentially the whole technology can be carried over from the FE literature.

More important is an efficient integration of the influence integrals. If the elements are straight, then in general the integrals can be evaluated analytically. Analytic integration will not eliminate all problems because in a very small region near the boundary nodes the solution will still deteriorate, but these effects can be attributed to the kinks (higher-oder discontinuities) in the boundary displacements or tractions.

Another problem is that not all the stresses on the boundary can be determined directly from the integral equations. The traction vector t has two components, while the stress tensor S has three. In plate bending problems the moment m_t (which requires reinforcement in the tangential direction) and in some sense also the twisting moment m_{nt} must be recovered artificially from the boundary values w and $w,_n$ by finite differences.

2.2 Structural analysis with boundary elements

Some examples may serve to highlight the application of BE methods in structural analysis, and the potentials of the method.

Shear wall

The shear wall in Fig. 2.6 carries its own weight plus line loads coming from different structural elements that are connected to the wall. The wall rests on a series of narrow supports, which in FE analysis would be modeled as point supports. Here they are modeled as short boundary elements.

The outer boundary plus the edges of the openings were subdivided into a total of 123 quadratic boundary elements. Some minor inconsistencies arise at the supports, because at the transition point between the free edge and a support, the physical tractions are discontinuous, while C^0 shape functions cannot model this behavior—but these are very minor details.

The unknowns are the boundary displacements u_x, u_y along the free edges, while at the supports, some (at a roller support) or all of the tractions t_x, t_y are unknown. After the support reactions and the shape of the deformed structure are determined by solving the system $H\, u = G\, t + d$, the influence function (2.15) makes it possible to calculate the displacements and stresses in the interior.

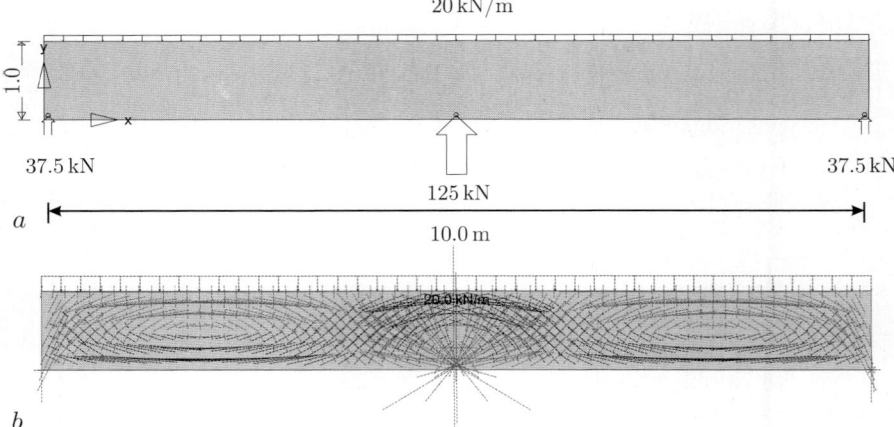

Fig. 2.7. Wall on point supports: **a)** system and loading, **b)** principal stresses

Two-span wall

The two-span beam in Fig. 2.7 was modeled as a plate. To compare the results with a beam solution, the plate was placed on point supports. Theoretically the support reactions should be zero, because the influence of a support reaction P on the displacement of the support itself is $-P \ln r = -P \ln 0 = \infty$, since the distance is zero ($r = 0$). This implies that the support reaction must be zero, $P = 0$, since in a set of equations such as

$$\begin{bmatrix} \infty & b \\ c & d \end{bmatrix} \begin{bmatrix} P \\ x \end{bmatrix} = \begin{bmatrix} e \\ f \end{bmatrix} \tag{2.20}$$

P must be zero.

But in reality the BE program replaces the function call $\ln r$ for values of $r < 10^{-3}$, (= one millimeter) with $\ln 10^{-3} = -6.9077$ to avoid overflow. This is equivalent to replacing the point support with a very short line support, which suffices to make P equal to the beam solution.

Slabs

Engineering plate-bending problems are much more complex than the biharmonic problems usually discussed in the mathematical literature. Add to this that the correct modeling of a complex floor plate depends on so many parameters, and so many assumptions must be made by the analyst, that these assumptions tend to have much more influence on the results than, say, the order of the boundary elements or other mathematical "subtleties". Hence programming the biharmonic equation $K\Delta\Delta w = p$ for the analysis of slabs is a real challenge, but the results are rewarding; see Fig. 2.8.

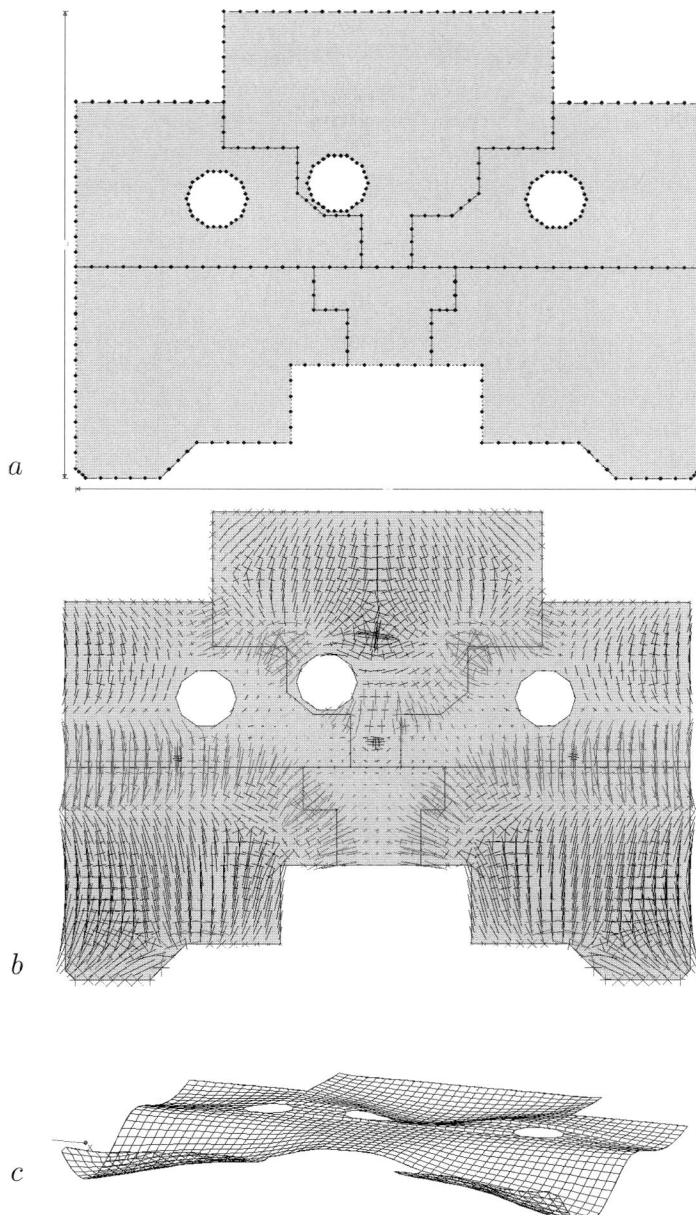

Fig. 2.8. Slab: **a)** system and 274 boundary elements, **b)** principal moments under gravity load, **c)** 3-D plot of the deflection surface

The choice of boundary elements for interpolating the boundary values w, w_n, m_n, v_n is dictated by the regularity assumptions made in the derivation of the integral equations [116]. The deflection w must be C^1 on any smooth part of the boundary, while for the other functions C^0-continuity suffices.

Support conditions are usually of mixed type, not jut $w = m_n = 0$ (hinged) or $w = w_n = 0$ (clamped). Most supports are modeled as elastic supports, so that a hinged support becomes $c\,w + v_n = m_n = 0$ and a clamped support becomes $c\,w + v_n = c_\varphi w_n + m_n = 0$, where c and c_φ are elastic constants. Add to this that at corner points, because of the continuity of the gradients ∇w, $\nabla\nabla w$ etc., boundary conditions on both sides of a corner point cannot be formulated independently of each other. If in addition the thickness of a plate changes at the corner point, then things can become really complicated. In a BE program, all the modeling is done on the boundary, and the boundary of a Kirchhoff plate is a very thin layer. But luckily the biharmonic equation—being of elliptic type—is a very patient equation, so that for engineering purposes the accuracy attainable with BE methods is more than sufficient, and at least on the same level as FE results.

Walls and T beams

Internal supports are subdivided into boundary elements (or rather line elements) to model the distribution of the support reaction s with piecewise linear functions; see Fig. 2.9. The nodal values are so determined that the deflection w of the plate is zero at the nodes of the wall (or that $c\,w + s = 0$ if the walls are elastic). As in FE analysis, it can be assumed that the calculated support reactions are relatively accurate.

The support reactions s of T beams are determined such that the deflection of the plate and the T beams are the same at the nodes of the T beams. The T beams are modeled with finite elements, and the reduced stiffness matrices $\boldsymbol{K}$ are inverted to provide flexibility matrices $\boldsymbol{F} = \boldsymbol{K}^{(-1)}$, which better fit a boundary element scheme, because $\boldsymbol{F}\,\boldsymbol{s} = \boldsymbol{w}$.

Columns

The BE method can assess the magnitude of the bending moments near columns very accurately, because the correct singularity is built into a BE program; see Fig. 2.10. If $p = P/\Omega_c$ is the pressure in the contact zone Ω_c, where P is the support reaction, the influence of the support reaction P on the bending moment $m_{xx}(\boldsymbol{x})$ is the integral

$$\int_{\Omega_c} [-K(g_{0,x_1 x_1} + \nu\, g_{0,x_2 x_2})]\, p\, d\Omega_{\boldsymbol{y}}\,, \qquad g_{0,x_1 x_1} = \frac{\partial^2 g_0}{\partial^2 x_1}(\boldsymbol{y}, \boldsymbol{x})\,, \quad (2.21)$$

and these kernel functions $(g_0(\boldsymbol{y}, \boldsymbol{x}))_{,x_i x_j}$ are part of the BE code, and must not be approximated with piecewise polynomials as in the FE method. The

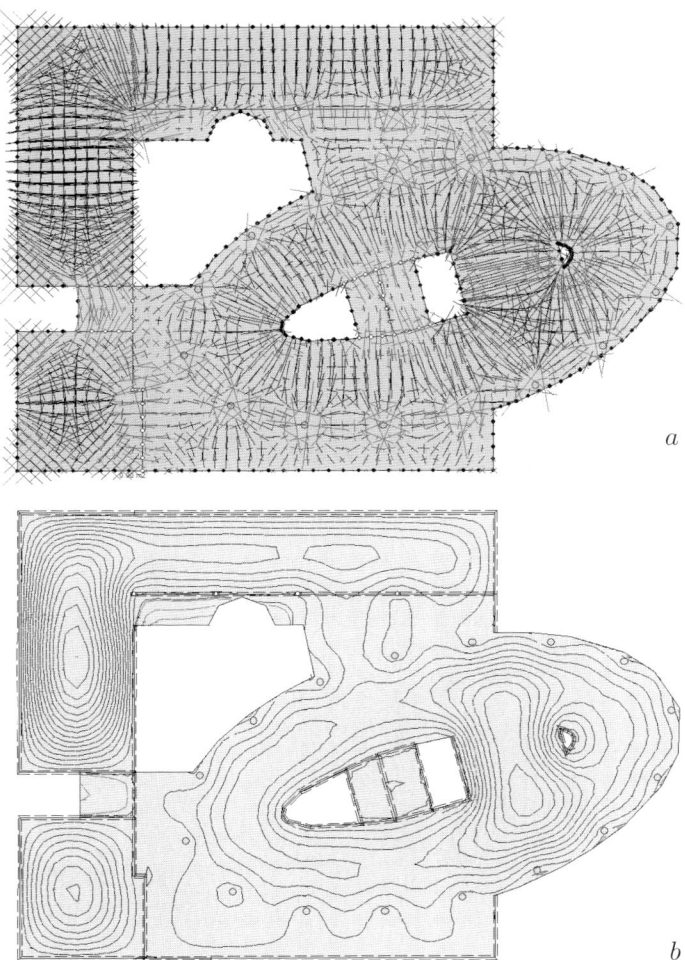

Fig. 2.9. Slab: **a)** principal moments, and **b)** contour plot of the deflection

surface integral in (2.21) also ensures that the bending moments are automatically rounded out; see Fig. 2.11.

The same can be said about the shear forces. The contribution of the support reaction in the column is the integral

$$\int_{\Omega_c} [-K(g_{0,x_1x_1x_1} + g_{0,x_2x_2x_1})] \, p \, d\Omega_{\boldsymbol{y}} , \qquad (2.22)$$

where

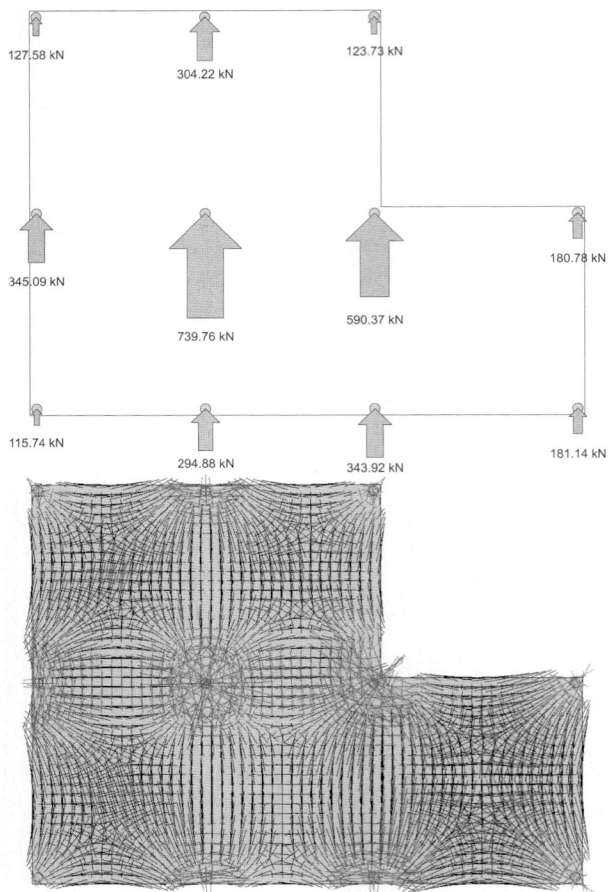

Fig. 2.10. Slab on point supports

$$(g_0(\boldsymbol{y}, \boldsymbol{x}))_{,x_1 x_1 x_1} = \frac{2}{8\pi K\, r} \left[-2\, r_{,2}^2\; r_{,1} - r_{,1}^3 - r_{,1}\; r_{,2}^2 \right]\,, \qquad (2.23)$$

$$(g_0(\boldsymbol{y}, \boldsymbol{x}))_{,x_2 x_2 x_1} = \frac{1}{8\pi K\, r} \left[-r_{,2}^3 - 2\, r_{,2}^2\; r_{,1} - r_{,1}^3 \right]\,. \qquad (2.24)$$

The resolution attainable with these influence functions depends solely on how tightly the observation points $\boldsymbol{x}$ are packed; see Fig. 2.12.

The internal actions are smooth continuous functions, because each individual value is calculated by evaluating the proper influence function. Theoretically there is a loss of accuracy close to the boundary, but for engineering purposes this loss is negligible; see Fig. 2.13.

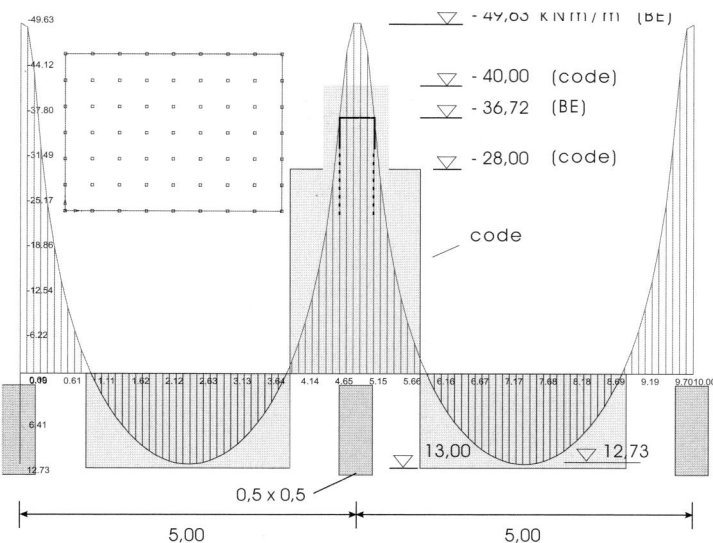

Fig. 2.11. Slab 40 m × 30 m on a grid of columns 5 m × 5 m, $p = 10$ kN/m^2; comparison of bending moments with building code

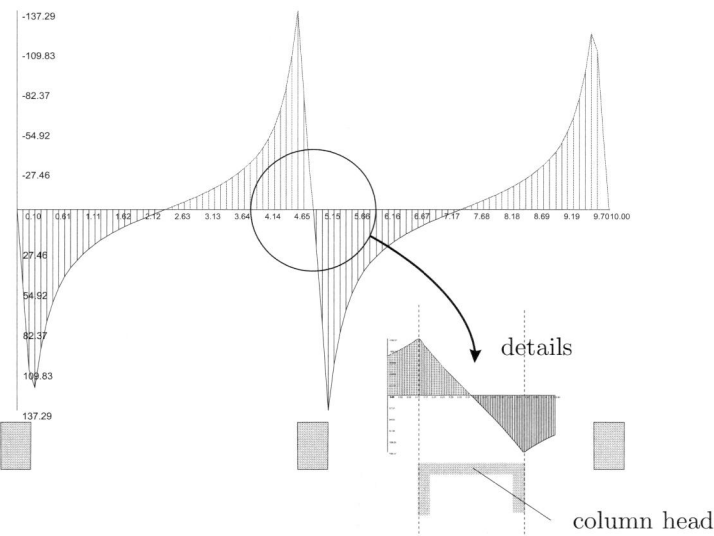

Fig. 2.12. Distribution of the shear forces q_x between the columns. The resolution can be made so fine that the variation of q_x across the column head is detailed out

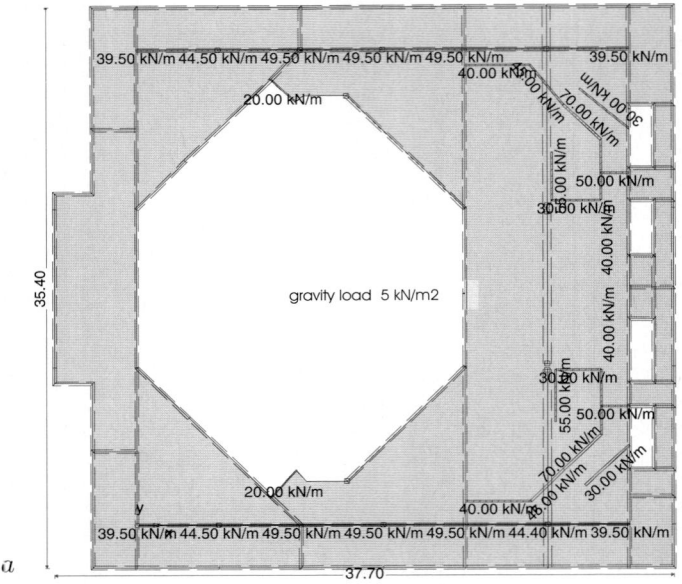

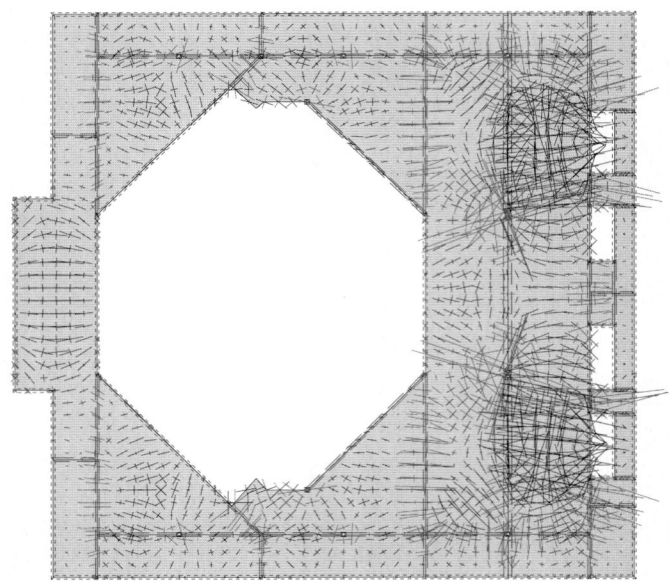

Fig. 2.13. Slab with a large opening **a)** system and loads, **b)** principal moments; the loss of accuracy close to the boundary is negligible

Loads

Formally it is of no concern which load is applied and where, because every load makes its own contribution to the influence function; see Fig. 2.13. The influence of a force P is the term $g_0(\boldsymbol{y}_P, \boldsymbol{x}) \cdot P$ + regular parts. If it is a line load, the influence is a line integral, and if it is a surface load on a patch Ω_p, the influence function is a surface integral:

$$\underbrace{g_0(\boldsymbol{y}_P, \boldsymbol{x}) \cdot P}_{single\ force} \qquad \underbrace{\int_{\Gamma_L} g_0(\boldsymbol{y}, \boldsymbol{x})\, l(\boldsymbol{y})\, ds_{\boldsymbol{y}}}_{line\ load} \qquad \underbrace{\int_{\Omega_p} g_0(\boldsymbol{y}, \boldsymbol{x})\, p(\boldsymbol{y})\, d\Omega_{\boldsymbol{y}}}_{surface\ load} \quad .(2.25)$$

In particular, a BE program is well-qualified to calculate influence functions, because it must only approximate the regular part of the solution.

Foundation slabs

In the *Winkler-model* the soil is treated as a grid of springs that can move independently, and that exert a force $c\,w$, where c (kN/m^3) is the modulus of subgrade reaction, so that

$$K\Delta\Delta w + c\,w = p\,. \tag{2.26}$$

The simplest approach to solving this equation is it to actually place the slab on a grid of springs and determine the unknown spring forces by extending the collocation equations to the springs. In many cases the accuracy attainable with this technique is sufficient. Its advantage is that it can be directly incorporated into an existing BE code with very slight modifications.

A more scientific approach is it to employ the two Fourier–Bessel integrals

$$g_0(\boldsymbol{y}, \boldsymbol{x}) = \frac{a^2}{2\pi K} \int_0^\infty \frac{t}{t^4 + \frac{k}{K}\,a^4}\, J_0(t\,\frac{r}{a})\, dt \tag{2.27}$$

and

$$g_1(\boldsymbol{y}, \boldsymbol{x}) = \frac{a\,r_n}{2\pi K} \int_0^\infty \frac{t}{t^4 + \frac{k}{K}\,a^4}\, J_1(t\,\frac{r}{a})\, dt \qquad r_n = \nabla_{\boldsymbol{x}} r \bullet \boldsymbol{n} \tag{2.28}$$

which are fundamental solutions of (2.26) [126]. Formally the influence function for the deflection $w(\boldsymbol{x})$ is nearly identical to the influence function (2.8),

$$c(\boldsymbol{x})\,w(\boldsymbol{x}) = \int_\Gamma [\,g_0(\boldsymbol{y}, \boldsymbol{x})\,\underline{v_\nu(\boldsymbol{y})} - g_{0\nu}(\boldsymbol{y}, \boldsymbol{x})\,\underline{m_\nu(\boldsymbol{y})}$$

$$-v_\nu(g_0)(\boldsymbol{y}, \boldsymbol{x})\,\underline{w(\boldsymbol{y})} + m_\nu(g_0)(\boldsymbol{y}, \boldsymbol{x})\,\underline{w_\nu(\boldsymbol{y})}\,]\, ds_{\boldsymbol{y}}$$

$$+ \int_\Omega g_0(\boldsymbol{y}, \boldsymbol{x})\,p(\boldsymbol{y})\, d\Omega_{\boldsymbol{y}} + \frac{1}{8\sqrt{c\,\hat{K}}}\, F(w)(\boldsymbol{x})$$

$$+ \sum_c [\,g_0(\boldsymbol{y}^c, \boldsymbol{x})\,\underline{F(w)(\boldsymbol{y}^c)} - \underline{w(\boldsymbol{y}^c)}\, F(g_0)(\boldsymbol{y}^c, \boldsymbol{x})\,]\,, \tag{2.29}$$

except for an extra free term $(F/8\sqrt{c\,\hat{K}})$ where $\hat{K} = a^4 c/K$.

The Fourier–Bessel functions

$$T_{i,j} := \frac{1}{2\,\pi} \int_0^\infty \frac{t^i}{t^4 + \hat{K}}\, J_j\!\left(t\,\frac{r}{a}\right) dt \tag{2.30}$$

can be represented in the form of zero-order Kelvin or Thomson functions $\mathrm{kei}(x)$ and $\mathrm{ker}(x)$ and their respective derivatives

$$T_{1,0} = -\frac{1}{2\,\pi}\,\mathrm{kei}(\frac{r}{a}) \qquad T_{3,0} = \frac{1}{2\,\pi}\,\mathrm{ker}(\frac{r}{a}), \tag{2.31}$$

which can be calculated using rapidly convergent expansions [127].

Half-space model

In the year 1885 Boussinesq [49] found the solution for an elastic half-space loaded at is surface with a vertical force P:

$$u_B := u_3(\boldsymbol{x}, \boldsymbol{y}) = \frac{1+\nu}{2\pi E}\left[\frac{(x_3 - y_3)^2}{r^3} + 2\,\frac{1-\nu}{r}\right] P. \tag{2.32}$$

If the interface between the foundation slab and the soil is subdivided into rectangular elements, the soil pressure can be expanded with regard to the n nodal shape functions giving the expression

$$w_S(\boldsymbol{x}) = \sum_i \int_\Omega u_B(\boldsymbol{y}, \boldsymbol{x})\,\varphi_i(\boldsymbol{y})\,d\Omega_{\boldsymbol{y}} \cdot p_i \tag{2.33}$$

for the deflection of the soil. The nodal values p_i are found by requiring that the deflection of the slab and the soil be the same at each node:

$$w(\boldsymbol{x}_i) = w_S(\boldsymbol{x}_i) \qquad i = 1, 2, \ldots n. \tag{2.34}$$

By this very simple technique, a coupled analysis of a foundation slab on top of an elastic half-space can be modeled; see Fig. 2.14. What is striking is how different the deformation patterns of a foundation slab are, depending on which model—the Winkler model or the half-space model—is used; see also Sect. 5.17, p. 469.

If the soil consists of different layers with moduli E_i and Poisson ratios ν_i, the total settlement beneath a point force P is the sum of all the individual contributions s_i within the single layers:

$$u_B^\Sigma(\boldsymbol{x}) = \sum_i s_i(\boldsymbol{x})\,P. \tag{2.35}$$

The contribution $s_i = s_i(E_i, \nu_i, h_i)$ of one layer can be calculated approximately as follows:

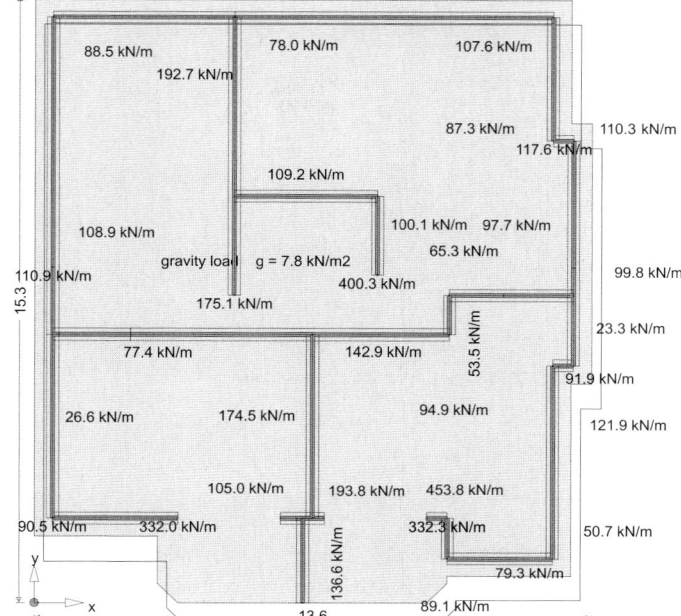

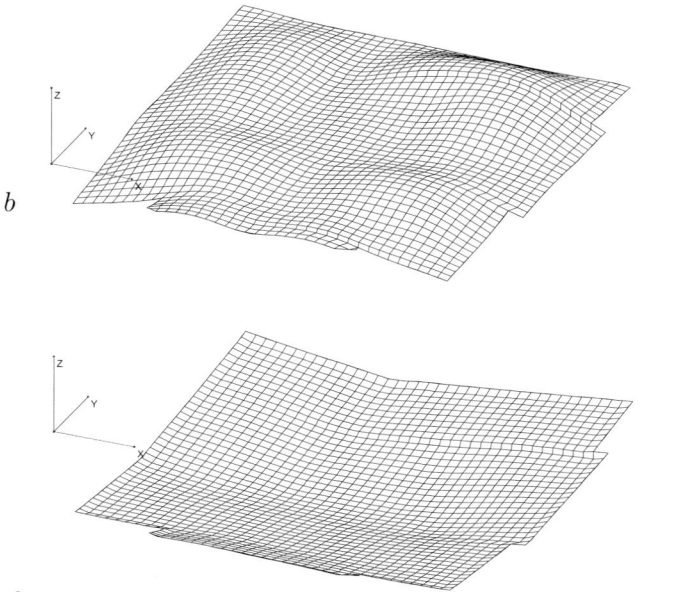

Fig. 2.14. Foundation plate: **a)** system and loads, **b)** deformation according to the Winkler model, **c)** deformation according to the half-space model

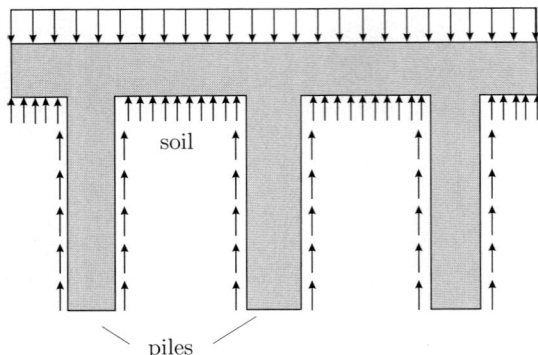

Fig. 2.15. Piled raft foundation

$$s_i(\boldsymbol{x}) = [u_B(\boldsymbol{x}, \boldsymbol{y}_u^i) - u_B(\boldsymbol{x}, \boldsymbol{y}_l^i)]|_{E_i, \nu_i, h_i} \tag{2.36}$$

$$|\boldsymbol{y}_u^i - \boldsymbol{y}_l^i| = h_i \quad \text{thickness of the layer}, \tag{2.37}$$

where the points $\boldsymbol{y}_u^i$ and $\boldsymbol{y}_l^i$ are the projections of the surface point $\boldsymbol{x}$ onto the upper and lower cover of the layer i. That is, the contribution of an individual layer is calculated as if the total half-space had the elastic properties E_i, ν_i.

Piled raft foundations

In a piled raft foundation the contribution of the piles as well as the raft is taken into account to transfer the building load. The piles transfer a part of the loads into deeper and stiffer layers of soil and thereby allow the reduction settlement and differential settlement in a very economic way; see Fig. 2.15.

To model such foundations it must be possible to predict the soil deformation if a vertical point force acts *in the interior* of the elastic half-space. Mindlin found the following solution for this problem [166]:

$$
u_M := u_3(\boldsymbol{x}, \boldsymbol{y}) = \frac{P}{16\,\pi\,E\,(1-\nu)} \left[\frac{3-4\nu}{R_1^3} + \frac{8\,(1-\nu)^2 - (3-4\nu)}{R_2} \right.
$$
$$
+ \frac{(x_3 - y_3)^2}{R_1^3} + \frac{(3-4\nu)(x_3 + y_3)^2 - 2\,y_3\,x_3}{R_2^3}
$$
$$
\left. + \frac{6\,y_3\,x_3\,(x_3 + y_3)^2}{R_2^5} \right], \tag{2.38}
$$

where

$$R_1 = \sqrt{x_1^2 + x_2^2 + (x_3 - y_3)^2} \qquad R_2 = \sqrt{x_1^2 + x_2^2 + (x_3 + y_3)^2}. \tag{2.39}$$

The coupled problem of the foundation plate and the elastic half-space can be stated as follows: on the free surface the tractions must vanish, $\boldsymbol{t} = \boldsymbol{0}$, while at the interface between the slab and the soil surface the deflection must be the same, and the vertical stresses must have opposite signs:

$$u_3 = w \qquad t_3 - p_S = 0\,.$$ (2.40)

Here p_S is the soil pressure, which also appears on the right-hand side of the plate equation

$$K\,\Delta\,\Delta\,w = p - p_S\,.$$ (2.41)

Modeling the slab

The unknowns are the boundary values w and w_n of the slab, the soil pressure distribution p_S under the slab and the shear forces s along the piles so that for example the first integral equation becomes

$$c(\boldsymbol{x})\,w(\boldsymbol{x}) = \int_\Gamma [-V_\nu(g_0)w + M_\nu(g_0)\frac{\partial w}{\partial \nu}]\,ds\boldsymbol{y} + \int_\Omega g_0(p - p_S)\,d\Omega\boldsymbol{y}$$

$$- \sum_i Q_i\,g_0[\boldsymbol{y}_i] + \sum_c [-w(\boldsymbol{y}^c)F(g_0)(\boldsymbol{y}^c, \boldsymbol{x})]$$ (2.42)

where the Q_i are the pile forces.

Modeling the soil

The soil pressure distribution p_S is interpolated with piecewise linear shape functions

$$p_S(\boldsymbol{x}) = \sum_i p_i\,\varphi_i(\boldsymbol{x})\,,$$ (2.43)

and to simplify the notation it is assumed that the same is done with the surface load p coming from the building.

When the edge of the slab is divided into boundary elements and the first and second integral equation (2.42) formulated at the collocation points, the pertinent system of equations is

$$\boldsymbol{C}\,\boldsymbol{u} = \boldsymbol{H}\,\boldsymbol{u} + \boldsymbol{D}\,(\boldsymbol{p} - \boldsymbol{p}_S) + \boldsymbol{L}\,\boldsymbol{q}$$ (2.44)

or if the unknowns are placed on the left-hand side and an index B is attached to denote either that the terms are boundary terms, $\boldsymbol{u} = \boldsymbol{u}_B$, or to indicate that the collocation points lie on the boundary

$$\boldsymbol{C}_B\,\boldsymbol{u}_B - \boldsymbol{H}_B\,\boldsymbol{u}_B + \boldsymbol{D}_B\,\boldsymbol{p}_S - \boldsymbol{L}_B\,\boldsymbol{q} = \boldsymbol{D}_B\boldsymbol{p}\,.$$ (2.45)

The vector $\boldsymbol{q} = [Q_1, Q_2, \ldots]^T$ contains the pile forces. The matrices $\boldsymbol{H}, \boldsymbol{D}, \boldsymbol{L}$ are influence matrices that describe the influence of the nodal values u_i, p_i, p_{S_i}

and the pile forces Q_i at the collocation points $\boldsymbol{x}_i$ on the boundary. The matrix $\boldsymbol{C}_B$ is nearly identical to $1/2 \times \boldsymbol{I}$. Only at corner points are the entries different.

The soil pressure p_S and frictional stress $f_{s_i} =: f_i$ along the individual piles are responsible for the deflection of the surface of the elastic half-space so that the deformation is a superposition of Boussinesq's solution u_B and Mindlin's solution u_M:

$$u_3(\boldsymbol{x}) = u(\boldsymbol{x}) = \int_\Omega u_B(\boldsymbol{y}, \boldsymbol{x})\, p_S(\boldsymbol{y})\, d\Omega_{\boldsymbol{y}} + \sum_i \int_{\Gamma_i} u_M(\boldsymbol{y}, \boldsymbol{x})\, f_i(\boldsymbol{y})\, ds_{\boldsymbol{y}} \quad (2.46)$$

where Γ_i is the surface of the individual piles $i = 1, 2, \ldots$ and f_i is the frictional stress acting along pile i.

Let $\boldsymbol{u}_I$ denote the deflection of the soil surface at the nodes of the grid, that is the points at which the soil pressure p_S is interpolated. The influence of the soil pressure p_S and the frictional stress f_i as expressed by (2.46) can then be written as

$$\boldsymbol{u}_I^{soil} = \boldsymbol{B}_I\, \boldsymbol{p}_S + \boldsymbol{M}_I\, \boldsymbol{f}\,. \qquad (2.47)$$

At the grid points the deflection of the soil (2.47) and the deflection of the slab

$$\boldsymbol{u}_I^{slab} = \boldsymbol{H}_I\, \boldsymbol{u}_B + \boldsymbol{D}_I\, (\boldsymbol{p} - \boldsymbol{p}_S) + \boldsymbol{L}_I\, \boldsymbol{q} \qquad (2.48)$$

must be the same, $\boldsymbol{u}_I^{slab} - \boldsymbol{u}_I^{soil} = \boldsymbol{0}$, which gives

$$\boldsymbol{H}_I\, \boldsymbol{u}_B - \boldsymbol{D}_I\, \boldsymbol{p}_S + \boldsymbol{L}_I\, \boldsymbol{q} - \boldsymbol{B}_I\, \boldsymbol{p}_S - \boldsymbol{M}_I \boldsymbol{f} = -\boldsymbol{D}_I \boldsymbol{p}\,. \qquad (2.49)$$

Modeling the piles

At the interface between the piles and the soil, the pile displacement u and the vertical soil displacement u_3 must be the same (slipping is neglected), which means that the deformations must be the same,

$$u_3(\boldsymbol{x}_i) = u_B(\boldsymbol{x}_i) + u_M(\boldsymbol{x}_i) = u(x_i)\,, \qquad (2.50)$$

at the k collocation points along the axis of each of the n piles. At one individual pile i therefore,

$$\boldsymbol{B}_{P_i}\, \boldsymbol{p}_S + \boldsymbol{M}_{P_i}\, \boldsymbol{f} = \boldsymbol{u}_i = \boldsymbol{K}_i^{-1}\, \boldsymbol{N}_i \boldsymbol{f}_i = \boldsymbol{F}_i\, \boldsymbol{f}_i \qquad (2.51)$$

where $\boldsymbol{F}_i$ is the flexibility matrix of the pile and the matrix $\boldsymbol{N}_i$ has the elements

$$N_{kj} = \pi d \int_0^{l_i} \psi_k\, \psi_j\, dx \qquad d = \text{diameter of the pile}\,. \qquad (2.52)$$

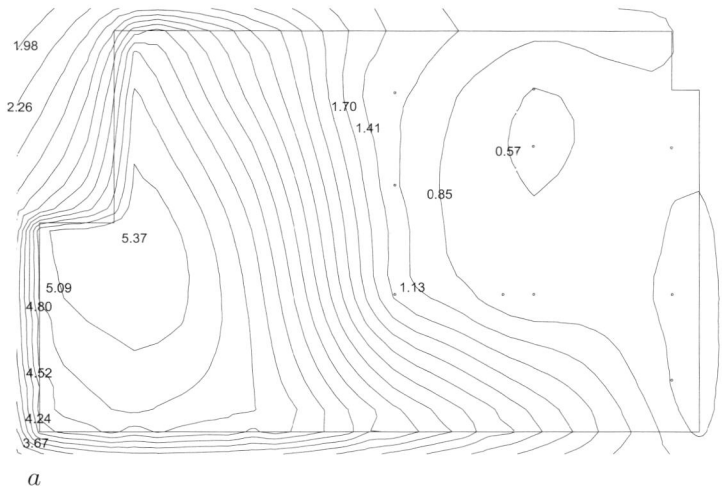

a

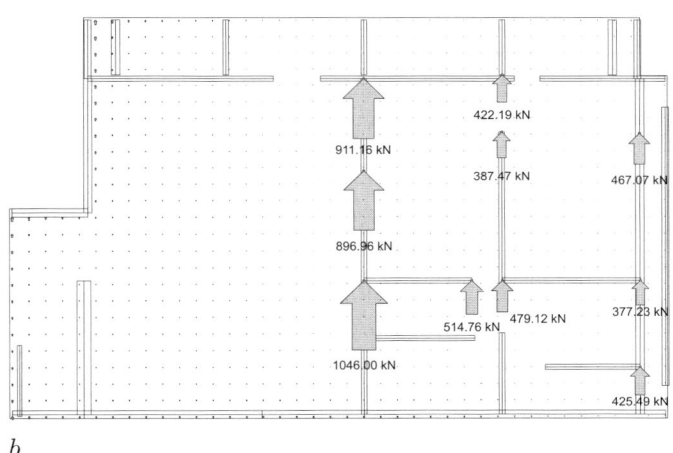

b

Fig. 2.16. Study of a piled raft foundation **a)** settlement, **b)** pile forces

The ψ_k are the shape functions of the pile. The vector $\boldsymbol{N}_i\,\boldsymbol{f}_i = \boldsymbol{s}_i$ represents the equivalent nodal forces s_i, due to the frictional stress f. Because each pile has an influence on all the others it follows that

$$\boldsymbol{B}_P\,\boldsymbol{p}_S + (\boldsymbol{M}_P - \boldsymbol{F}\,\boldsymbol{N})\,\boldsymbol{f} = \boldsymbol{0} \qquad (2.53)$$

where the diagonal entries in the matrix $\boldsymbol{F}$ are the flexibility matrices $\boldsymbol{F}_i$ of the individual piles and the subscript P on the Boussinesq and Mindlin influence matrices indicates that the control points, the collocation points, lie on the piles.

Finally at each pile the sum (the integral) of the shear forces f_i must be equal to the resultant pile force Q_i which, with an abstract matrix Σ whose somewhat simplified entries are the lengths l_i of the single bar elements, can be stated as

$$\Sigma N f = q. \tag{2.54}$$

If the results are collected, the following system of equations is obtained:

$$\begin{bmatrix} C_B - H_B & D_B & -L_B & 0 \\ H_I & -D_I - B_I & L_I & -M_I \\ 0 & B_P & 0 & M_P - F N \\ 0 & 0 & I & -\Sigma N \end{bmatrix} \begin{bmatrix} u_B \\ p_S \\ q \\ f \end{bmatrix} = \begin{bmatrix} D_B p \\ -D_I p \\ 0 \\ 0 \end{bmatrix} \begin{array}{l} \text{(slab)} \\ \text{(soil)} \\ \text{(piles)} \\ \text{(sum)} \end{array}$$

The slab in Fig. 2.16 was analyzed with this technique [117].

2.3 Comparison finite elements—boundary elements

In the BE analysis of a plate at each point the influence that all the different sources have on that particular point must be calculated separately. Typically the influence a source exerts depends on the distance between the source point y and the observation point x. This means that it typically fades away like $\ln r$ (in the near term) or like r^{-1} or r^{-2}, and the influence depends on the angle φ between the two points. A change in the position, $(r, \varphi) \to (\hat{r}, \hat{\varphi})$, makes that the influence coefficients change, and thereby also the stresses.

In the FE method the situation is basically the same. It is only that the FE method plugs all the information on how the influence changes when a point (r, φ) moves to a new location $(\hat{r}, \hat{\varphi})$ into the nodal displacements u_i of an element, and interpolates the element displacement field using linear or quadratic polynomials. Because of this technique the element is autonomous. The price the FE method pays for this technique is that it is not as sensitive as the BE method. If for example a cantilever plate is modeled with bilinear elements and Poisson's ratio is zero, $\nu = 0$, then the FE bending moment is the same in any vertical plane of an element, which is impossible. Obviously the 2×4 data cells u_i of a bilinear element are not sufficient to register the change in the influence coefficients if the observer moves from x to $\hat{x} = x + \Delta x$.

But one can also view things differently: an FE program replaces the original load case p only with a load case p_h that is compatible with its limited means.

Green's function

The FE solution, $u_h(x) = (G_0[x], p)$, is based on a mesh-dependent approximation of the Green's function. Does this also hold true for BE solutions? To answer this question let us study the Poisson equation $-\Delta u = p$ in a domain

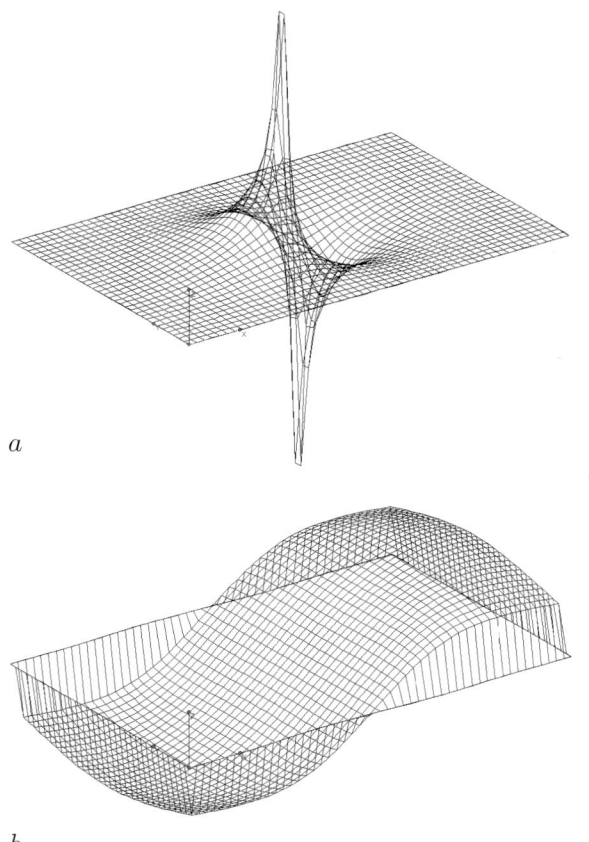

a

b

Fig. 2.17. Influence function for q_x **a)** BE solution $G_3^{xx} = g_3^{h,xx} + u_R^h$, **b)** regular part $u_R^h = G_3^{xx} - g_3^{h,xx}$

Ω with boundary conditions $u = 0$ on Γ_D (fixed edge) and $\partial u/\partial n = 0$ on Γ_N (free edge) ($\Gamma = \Gamma_D \cup \Gamma_N$). It is assumed that the solution can be represented by the formula

$$u(\boldsymbol{x}) = \int_\Omega G_0(\boldsymbol{y}, \boldsymbol{x})\, p(\boldsymbol{y})\, d\Omega_{\boldsymbol{y}} = \int_{\Gamma_D} g_0(\boldsymbol{y}, \boldsymbol{x})\, t(\boldsymbol{y})\, ds_{\boldsymbol{y}}$$

$$- \int_{\Gamma_N} \frac{\partial g_0}{\partial \nu}(\boldsymbol{y}, \boldsymbol{x})\, u(\boldsymbol{y})\, ds_{\boldsymbol{y}} + \int_\Omega g_0(\boldsymbol{y}, \boldsymbol{x})\, p(\boldsymbol{y})\, d\Omega_{\boldsymbol{y}} \qquad (2.55)$$

where $t = \partial u/\partial \nu$ is the slope (traction) on the boundary at the integration point $\boldsymbol{y}$ with normal vector $\boldsymbol{\nu}$ and Green's function G_0. The function g_0 in the second integral representation is a suitable fundamental solution, for example $g_0 = -1/(2\pi) \ln r$. If the BE solution

$$u_h(\boldsymbol{x}) = \int_{\Gamma_D} g_0(\boldsymbol{y},\boldsymbol{x})\, t_h(\boldsymbol{y})\, ds_{\boldsymbol{y}} - \int_{\Gamma_N} \frac{\partial g_0}{\partial \nu}(\boldsymbol{y},\boldsymbol{x})\, u_h(\boldsymbol{y})\, ds_{\boldsymbol{y}}$$

$$+ \int_{\Omega} g_0(\boldsymbol{y},\boldsymbol{x})\, p(\boldsymbol{y})\, d\Omega_{\boldsymbol{y}} \tag{2.56}$$

is compared with the FE solution

$$u_h(\boldsymbol{x}) = \sum_i u_i\, \varphi_i(\boldsymbol{x}) = \int_{\Omega} G_0^h(\boldsymbol{y},\boldsymbol{x})\, p(\boldsymbol{y})\, d\Omega_{\boldsymbol{y}} , \tag{2.57}$$

there seems not to be much agreement. But the BE solution (2.56) is just the formula

$$u_h(\boldsymbol{x}) = \int_{\Omega} G_0^h(\boldsymbol{y},\boldsymbol{x})\, p(\boldsymbol{y})\, d\Omega_{\boldsymbol{y}} \tag{2.58}$$

in disguise, of course with a different G_0^h. To see this note that the Green's function G_0 can be split (see Fig. 2.17) into a fundamental solution g_0 and a regular part u_R,

$$u(\boldsymbol{x}) = \int_{\Omega} G_0(\boldsymbol{y},\boldsymbol{x})\, p(\boldsymbol{y})\, d\Omega_{\boldsymbol{y}}$$

$$= \int_{\Omega} g_0(\boldsymbol{y},\boldsymbol{x})\, p(\boldsymbol{y})\, d\Omega_{\boldsymbol{y}} + \int_{\Omega} u_R(\boldsymbol{y},\boldsymbol{x})\, p(\boldsymbol{y})\, d\Omega_{\boldsymbol{y}} \tag{2.59}$$

from which it follows that the boundary integrals in the influence function (2.55) are just an equivalent expression for the work done by the distributed load p acting through the regular part u_R:

$$\int_{\Omega} u_R(\boldsymbol{y},\boldsymbol{x})\, p(\boldsymbol{y})\, d\Omega_{\boldsymbol{y}} = \int_{\Gamma_D} g_0(\boldsymbol{y},\boldsymbol{x})\, t(\boldsymbol{y})\, ds_{\boldsymbol{y}} - \int_{\Gamma_N} \frac{\partial g_0}{\partial \nu}(\boldsymbol{y},\boldsymbol{x})\, u(\boldsymbol{y})\, ds_{\boldsymbol{y}} . \tag{2.60}$$

Hence it can be assumed that the boundary integrals in the BE solution (2.56) play the same role,

$$\int_{\Omega} u_R^h(\boldsymbol{y},\boldsymbol{x})\, p(\boldsymbol{y})\, d\Omega_{\boldsymbol{y}} := \int_{\Gamma_D} g_0(\boldsymbol{y},\boldsymbol{x})\, t_h(\boldsymbol{y})\, ds_{\boldsymbol{y}} - \int_{\Gamma_N} \frac{\partial g_0}{\partial \nu}(\boldsymbol{y},\boldsymbol{x})\, u_h(\boldsymbol{y})\, ds_{\boldsymbol{y}} , \tag{2.61}$$

and so we arrive at (2.58), which makes the two methods look alike, even though the FE Green's function is only mesh dependent, while the BE Green's function is also load-case dependent.

The difference between the two methods is that the FE method must approximate both parts of the Green's function $G_0 = g_0 + u_R$ while the BE method must only approximate the regular part u_R because the fundamental solution g_0 is part of the code.

Note that the BE method uses an ingenious approach: it does not approximate $u_R(\boldsymbol{y}, \boldsymbol{x})$—this would be simply too laborious, because at every point $\boldsymbol{x}$ it would have to approximate a new function $u_R(\boldsymbol{y}, \boldsymbol{x})$—but instead substitutes for the work integral $(u_R[\boldsymbol{x}], p)$ the work done by the Cauchy data u on Γ_N and t on Γ_D via the conjugate terms of the fundamental solution g_0. That is, the program knows that it suffices to approximate the "static" data u and t leaving the effects caused by a change in the observation point $\boldsymbol{x}$ to the fundamental solution $g_0(\boldsymbol{y}, \boldsymbol{x})$. This is the essence of (2.61).

Note also that the boundary values

$$\tilde{u}_h(\boldsymbol{x}) = u_h(\boldsymbol{x}) + \varepsilon(\boldsymbol{x}) \qquad \tilde{t}_h(\boldsymbol{x}) = t_h(\boldsymbol{x}) + \eta(\boldsymbol{x}) \tag{2.62}$$

of $u_h(\boldsymbol{x})$ in (2.56) differ (slightly) from the values under the integral signs, because the solution (2.56) satisfies the integral equation

$$c(\boldsymbol{x})\, u(\boldsymbol{x}) = \int_{\Gamma_D} g_0(\boldsymbol{y}, \boldsymbol{x})\, t_h(\boldsymbol{y})\, ds_{\boldsymbol{y}} - \int_{\Gamma_N} \frac{\partial g_0}{\partial \nu}(\boldsymbol{y}, \boldsymbol{x})\, u_h(\boldsymbol{y})\, ds_{\boldsymbol{y}}$$

$$+ \int_{\Omega} g_0(\boldsymbol{y}, \boldsymbol{x})\, p(\boldsymbol{y})\, d\Omega_{\boldsymbol{y}} \qquad \boldsymbol{x} \in \Gamma \tag{2.63}$$

only at the collocation points $\boldsymbol{x}_i$. On Γ_D the left-hand side is zero, $u(\boldsymbol{x}) = 0$, while on Γ_N the left-hand side is $c(\boldsymbol{x})\, u_h(\boldsymbol{x})$ where $u_h(\boldsymbol{x})$ is the same function as under the second boundary integral.

Accuracy

Can we expect the same tendency as in the FE method, namely that the accuracy reflects the nature of the Green's function? To check this let us apply Betti's theorem to the pair $\{u_h, G_0\}$, where u_h is the BE solution (2.56) of the Poisson equation $-\Delta u = p$ with boundary conditions $u = 0$ on Γ_D and $t = 0$ on Γ_N and where G_0 is the Green's function for $u(\boldsymbol{x})$ at a point $\boldsymbol{x}$. Then

$$\int_{\Omega} G_0\, p\, d\Omega_{\boldsymbol{y}} + \int_{\Gamma_N} \tilde{t}_h\, G_0\, ds_{\boldsymbol{y}} = u_h(\boldsymbol{x}) \cdot 1 + \int_{\Gamma_D} t_0\, \tilde{u}_h\, ds_{\boldsymbol{y}} \tag{2.64}$$

and

$$u(\boldsymbol{x}) = \int_{\Omega} G_0\, p\, d\Omega_{\boldsymbol{y}} + \int_{\Gamma_N} G_0\, t\, ds_{\boldsymbol{y}} - \int_{\Gamma_D} t_0\, u\, ds_{\boldsymbol{y}}\,, \tag{2.65}$$

or

$$u(\boldsymbol{x}) - u_h(\boldsymbol{x}) = \int_{\Gamma_N} G_0\, (t - \tilde{t}_h)\, ds_{\boldsymbol{y}} - \int_{\Gamma_D} t_0\, (u - \tilde{u}_h)\, ds_{\boldsymbol{y}} \tag{2.66}$$

where t_0 are the tractions of the Green's function on Γ. Obviously the BE solution will be exact if $\tilde{u}_h = u$ on Γ_D (which is only true at the collocation points $\boldsymbol{x}_i$), and if $\tilde{t}_h = t$ on Γ_N.

In view of our previous interpretation it can be stated as well that the error is attributable to the approximation error in u_R^h:

$$u(\boldsymbol{x}) - u_h(\boldsymbol{x}) = \int_\Omega (u_R - u_R^h)\, p\, d\Omega_{\boldsymbol{y}}\,. \tag{2.67}$$

Compare this with the FE error

$$u(\boldsymbol{x}) - u_h(\boldsymbol{x}) = \int_\Omega (G_0 - G_0^h)\, p\, d\Omega_{\boldsymbol{y}}\,, \tag{2.68}$$

and the difference between the two methods becomes apparent. In the FE method the accuracy is linked to the nature of the Green's function, and in the BE method the accuracy is linked to the character of the regular solution u_R, which clearly deteriorates the closer the point comes to the boundary. (Hence "regular solution" may be an euphemism.)

This is confirmed by (2.66), because the propagation of the error is governed by the Green's function $G_0 = O(\ln r)$ and the traction $t_0 = O(1/r)$ of the Green's function. Both functions "live" on the boundary. Seemingly, the closer the point $\boldsymbol{x}$ comes to the boundary Γ, the more negative the influence of the boundary error on the results. This agrees with our observations.

The load case p_h

In FE methods p is replaced by a work-equivalent load case p_h. In BE analysis the choice of p_h is not so simple to explain. But the BE method bears some resemblance to the force method. The volume potential (G_0, p) of a BE formulation corresponds to the deflection curve w_0 of the statically determinate structure and the boundary potentials are the deflection curves w_i of the redundant forces X_i.

Because the BE solution satisfies the differential equation in any patch Ω_P—"a truck remains a truck"—local equilibrium is satisfied in any patch Ω_P. But this is not true if $\Omega_P = \Omega$ because on the boundary there are "parasitic" stresses (the η in (2.62)) that ensure that $\boldsymbol{R} \neq \boldsymbol{R}_h$, that is, that the resultant $\boldsymbol{R}_h$ of the BE load (p + parasitic stresses) deviates from the true resultant $\boldsymbol{R}$, and therefore the sum of the BE support reactions deviates from $-\boldsymbol{R}$; see Tab. 2.1.

These parasitic stresses on Γ are a result of the deviations between the tractions of the BE solution and the tractions of the exact solution. Usually the imbalance $\boldsymbol{R} - \boldsymbol{R}_h \neq \boldsymbol{0}$ is very small (less than 3%).

Theoretically the BE method is restricted to linear problems, because the scalar product is *distributive*

$$\int_\Omega G_0(p_1 + p_2)\, d\Omega = \int_\Omega G_0\, p_1\, d\Omega + \int_\Omega G_0\, p_2\, d\Omega \tag{2.69}$$

Table 2.1. FE versus BE

	FE	BE
differential equation satisfied	no	yes
geometric boundary conditions	yes	(no)
static boundary conditions	no	no
"global equilibrium"	yes	no
local equilibrium	no	yes

but in practice the BE method can also be successfully applied to many such problems by placing the nonlinear terms on the right-hand side and solving the equations iteratively [257].

The ideal application for boundary elements are dynamic problems in unbounded domains. An example of such a problem is the analysis of displacements of the surface of an elastic half-space if a (very) fast train is speeding across the surface; see Fig. 2.18. The velocity of propagation for the p- and s-waves respectively is, $c_p = 995.1$ m/s, $c_s = 300.0$ m/s, and the density of the half-space is $\rho = 2000$ kg/m^3, from which follows that the speed of propagation for *Rayleigh waves* is $c_R = 284.7$ m/s. The train is a point load simulated by a triangular stress distribution. The train moved at different speeds v across the surface:

$$0 < v < c_R \qquad \text{Figure a} \qquad c_R < v < c_s \qquad \text{Figure b}$$
$$c_s < v < c_p \qquad \text{Figure c} \qquad c_p < v < +\infty \qquad \text{Figure d}.$$

At $v < c_R$ (see Fig. 2.18 a) a symmetric depression appeared (as in the static case on both sides of the point load the deformation was the same) which moved at speed v across the surface. At speeds $c_R < v < c_s$ (see Fig. 2.18 b), a wave front builds up ahead of the moving point load. At speeds $c_s < v < c_p$ the wave front becomes wedge-like (see Fig. 2.18 c) while at the same time the impression the triangular "point load" makes become smaller. At speeds $c_p < v = 1350$ m/s $< \infty$, which exceed the p-wave velocity, deformations only occur *after* the point load has passed, i.e., they trail the point load (see Fig. 2.18 d).

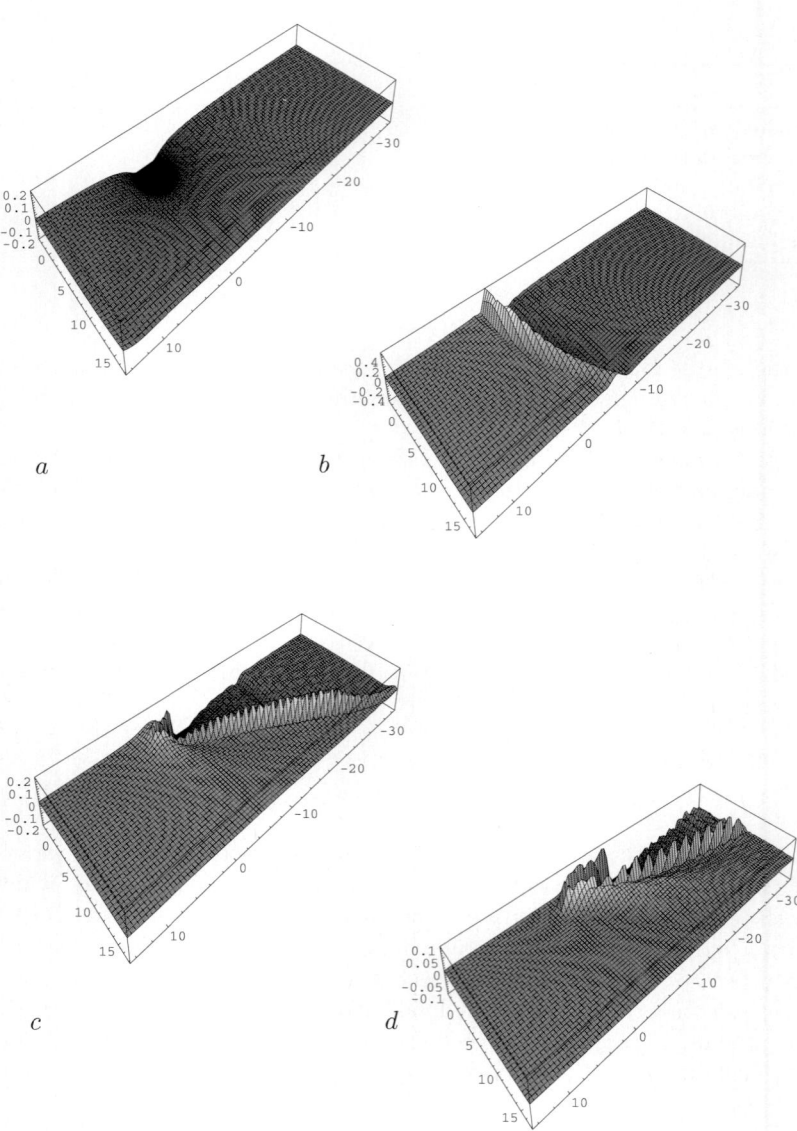

Fig. 2.18. Moving point load and surface waves: **a)** $v < c_R$, **b)** $c_R < v < c_s$, **c)** $c_s < v < c_p$, **d)** $c_p < v < \infty$

3. Frames

3.1 Introduction

In frame analysis the FE method is basically identical to the slope-deflection method. But the FE method extends beyond this method, insofar it can solve problems approximately which cannot be solved by the slope-deflection method or other classical methods, as the force method.

Frame analysis itself is an approximation as certain effects like axial ($EA = \infty$) or shear deformations ($GA = \infty$) are often neglected. An interesting example of how we mask certain effects is the example of an eccentric moment applied to the middle of a beam with fixed ends; see Fig. 3.1.

According to beam theory, the moment vector can be moved in arbitrary fashion along its axis. The bending moment in the beam will always be the same. But when we calculate influence functions, such an eccentric moment is also the derivative of the influence function of an eccentric force that generates non-constant torsional moments. This implies that torsional moments of magnitude $M \cdot a/L$ will be observed within the beam.

If the same load is applied to an FE structure (Fig. 3.2) the deflection of the bridge deck will cause a twist of the longitudinal axis and therewith a torsional moment. And indeed if the resultant stresses are summed a torsional moment of this magnitude is obtained.

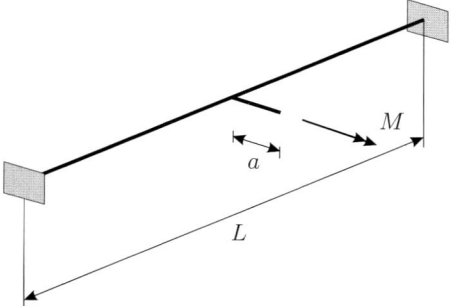

Fig. 3.1. Torsional moment

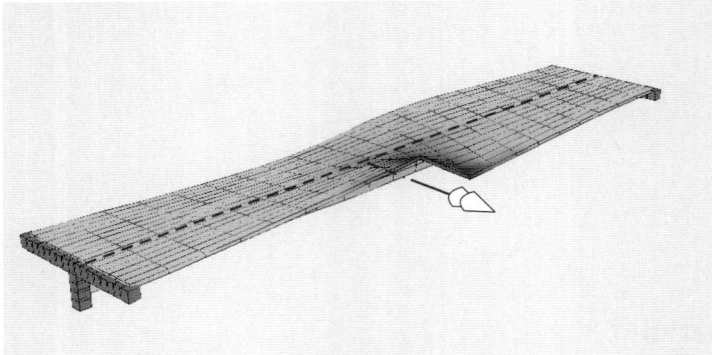

Fig. 3.2. Eccentric moment creates torsion in the bridge deck

3.2 The FE approach

Many engineers consider frame analysis settled for good because all problems of frame analysis are solved. No one expects new results, more complex structures are analyzed with finite elements, and the benefit of 1-D models seems questionable. But the main advantage of 1-D analysis lies in the clear and descriptive representation of structures because results are immediately presented in integral form. This gives many engineers the false impression that

- 1-D elements are exact
- 1-D elements are simple

But the more effects that must be considered in the analysis, the more baroque 1-D analysis becomes. It is far easier to program a 2-D finite element than to consider all the intricacies when a 2-D structure is reduced to a series of 1-D elements.

Degrees of freedom

In a typical analysis, the x-axis follows the long dimension of the element and the coordinates y and z point in orthogonal directions. Displacements, and rotations, forces and moments respectively are defined analogously; see Fig. 3.3. Simple formulations for 1-D elements are only obtained if the description concentrates on the axis. Preferably resultant internal actions refer to the centroids of the cross sections. But different construction stages possibly mean changes in the location of the centroids, and often the inclination of the neutral axis changes as well. At every stage the axes can differ in length, and the normal force and shear forces can point in different directions. Then the quantities must either be transformed, or one works with average values. Occasionally when transferring data from the CAD-model to the structural

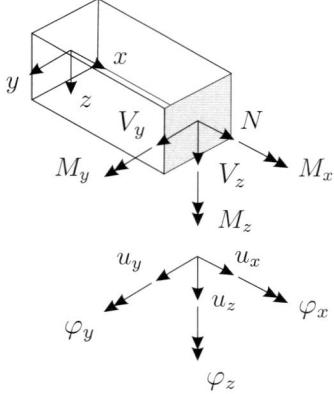

Fig. 3.3. Internal actions in a beam

model, new axes or surfaces must be generated that will not coincide with the CAD-model.

To avoid these difficulties it is a good strategy to allow from the start for eccentricities in the position of the centroid and the shear center with respect to the axes. And it is also a good approach to refer the resultant internal actions to the user-provided y and z axes, and not to the centroidal axes of the cross section, because the position of the principal axes within a tapered beam may rotate.

In such a beam (see Fig. 3.4), the internal actions can be referred to at least three different coordinate systems. The simplest choice are the main axes shown in Fig. 3.4 a. This is the classical treatment, where the normal force and shear forces follow the dashed line, indicating the longitudinal axis. These components N and V are also the components referred to when the stresses are calculated and the reinforcement is designed.

But it would be a more elegant and also more efficient approach if the longitudinal force and the transverse forces of second-order beam theory which refer to the axes of the undeformed system, were used (see Fig. 3.4 b and c).

Superposition of the stresses when the centroid changes position is only easy in case (c) because in the other two cases the position of the centroid must be known, and in case (a) the inclination of the longitudinal axis must be known as well. But the superposition of internal actions to calculate resulting stresses only makes sense if the cross section does not change. Therefore the longitudinal and transverse forces are better referred to the centroid of the cross section, as in Fig. 3.4 b, because if an arbitrary point of reference is chosen as in Fig. 3.4 c, it is much harder to interpret the results. Add to this that if the internal actions are to be plotted, then besides the bending moment the transverse forces, the longitudinal force, and the rotation of the beam axis according to second-order theory must also be known.

At each end of the element lies one node whose displacements are given in terms of the global coordinate system. It is helpful to transform these

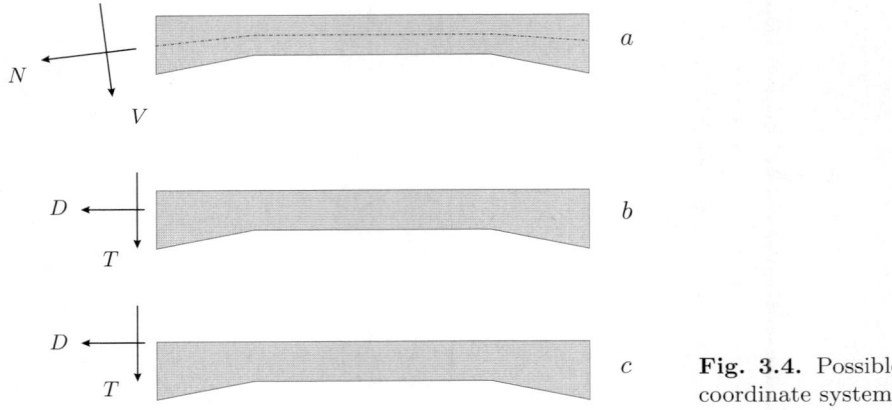

a

b

c **Fig. 3.4.** Possible coordinate systems

deformations into the local element coordinate system. In accordance with Fig. 3.4 b, we choose as internal reference nodes the centroids of the cross sections at both ends of the element. Each of these nodes can have up to seven degrees of freedom:

1. the longitudinal displacement u_x at the centroid of the cross section
2. local transverse deformations u_y and u_z as well as rotations φ_y and φ_z
3. the twist φ_x and the angle of twist per unit length φ'_x

The displacements and rotations of the two nodes are determined by the global displacements (u_X, u_Y, u_Z) and rotations $(\varphi_X, \varphi_Y, \varphi_Z)$ if the position of the local nodes with respect to the global nodes are known (see Fig. 3.5). While the rotations are identical for the local and the associated global node, the displacements must be transformed by taking the eccentricities into account. The longitudinal displacements at the centroid $\boldsymbol{x}_c$, for example, are

$$u_{x0}(\boldsymbol{x}_c) = u_x(\boldsymbol{x}_i) + \varphi_y \, \Delta z - \varphi_z \, \Delta y \qquad \boldsymbol{x}_i = \text{node}. \qquad (3.1)$$

To consider effects due to warping, matching transition conditions must be formulated at corner points and at rigid joints. The choice of components is arbitrary, insofar as also higher derivatives of the displacements could be used as nodal values, although these would be difficult to control.

Along the beam axis these displacements are interpolated as follows:

1. Linear interpolation of the axial displacement u_x
2. Coupled interpolation of the displacement u_y and the rotation φ_z
3. Coupled interpolation of the displacement u_z and the rotation φ_y
4. Coupled interpolation of the twist φ_x and the angular twist per unit length φ' (linear interpolation of φ_x if warping torsion is neglected)

The coupled interpolation can be done with cubic splines, in which case the rotations and the warping effects respectively are simply the derivatives of the

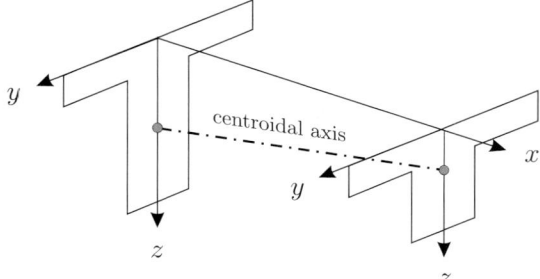

Fig. 3.5. Position of the centroidal axis

displacements and the rotations. Then it is convenient to choose the shape functions of a beam

$$H_1 = (1 - \xi^2)(1 + 2\,\xi) \qquad H_2 = l\,(1 - \xi^2)\,\xi \tag{3.2}$$

$$H_3 = \xi^2(3 - 2\,\xi) \qquad H_4 = -l\,(1 - \xi)\,\xi^2 \qquad \xi = x/l \tag{3.3}$$

so that

$$u_z(x) = \sum_i H_i(x) \cdot u_{zi} \qquad \varphi_y = \sum_i H_i'(x) \cdot u_{zi} \tag{3.4}$$

$$u_y(x) = \sum_i H_i(x) \cdot u_{yi} \qquad \varphi_z = \sum_i H_i'(x) \cdot u_{yi}\,. \tag{3.5}$$

Here u_{yi} and u_{zi} are nodal displacements in the local coordinate system at the two ends of the element.

But the coupling can also be done by considering the shearing strains according to Timoshenko:

$$\theta_y = \frac{V_z}{G\,A_z}\,, \qquad \varphi_y = u_{z,x} + \theta_y\,, \tag{3.6}$$

$$\theta_z = \frac{V_y}{G\,A_y}\,, \qquad \varphi_z = u_{y,x} + \theta_z\,. \tag{3.7}$$

The displacements within a cross section are then obtained by a product approach,

$$u_j = \sum_{i=1}^{7} N_{ij}(y,z) \cdot u_i(x) \tag{3.8}$$

or if we substitute for $u_i(x)$ the interpolating functions

$$u_j = \sum_{k=1}^{2} \sum_{i=1}^{7} N_{ij}(y,z) \cdot H_{ik}(x) \cdot u_{ik}\,. \tag{3.9}$$

In general the following functions are used to model the displacements and higher-order terms within the cross section:

$$u_x(y,z) = u_{x0} + \varphi_y \cdot (z - z_s) - \varphi_z \cdot (y - y_s)$$
$$+ U_w \cdot (\theta_x - \varphi_y' \varphi_z + \varphi_y \varphi_z') + U_y \cdot \theta_y + U_z \cdot \theta_z + U_{w2} \cdot \theta_{t2},$$
$$(3.10)$$

$$u_y(y,z) = u_{y0} - \varphi_x \cdot (z - z_m) - \frac{1}{2}(\varphi_x^2 + \varphi_z^2) \cdot (y - y_m), \tag{3.11}$$

$$u_z(y,z) = u_{z0} - \varphi_x \cdot (y - y_m) - \frac{1}{2}(\varphi_x^2 + \varphi_y^2) \cdot (z - z_m). \tag{3.12}$$

Here u_{x0}, u_{y0} and u_{z0} are the displacements of the centroid, y_s, z_s are the cross-sectional coordinates of the centroid, and y_m, z_m are the coordinates of the shear center.

The first three terms of the longitudinal displacement are determined by the requirement that the cross section remain flat, in agreement with Bernoulli's hypothesis. Next the unit warping functions U_w, U_y, U_z are added, and finally terms (U_{w2}, θ_{t2}), which take into account deformations caused by shear forces and secondary torsional moments. The latter add complex patterns of warping, which in general are not easy to investigate.

Displacements orthogonal to the axis are simple translations and rotations (rigid-body motions), that is, it is assumed that the cross section keeps its shape. To allow for changes in the cross section would require either a sophisticated 3-D model of flat shell elements—think of a double T beam—or the single terms must be decoupled.

The strains are the derivatives of the displacements. If the higher-order terms coming from second-order theory are neglected, the stresses become

$$\sigma_x = E\,\varepsilon_x = E\,u_{,x} = E\,[u_{,x} + \varphi_{y,x}(z - z_s)$$
$$-\varphi_{z,x}(y - y_s) - z_{s,x}\,\varphi_y + y_{s,x}\,\varphi_z + \sum(U_{i,x} \cdot \theta_i + U_i \cdot \theta_{i,x})\,x],$$
$$(3.13)$$

$$\tau_{xy} = G\,\gamma_{xy} = G[u_{x,y} + u_{y,x}] = G\,[(u_{y0,x} - \varphi_z)$$
$$+\sum(U_{i,y} \cdot \theta_i) - (z - z_m)\,\vartheta_{,x} + z_{m,x}\,\vartheta], \tag{3.14}$$

$$\tau_{xz} = G\,\gamma_{xz} = G[u_{x,z} + u_{z,x}] = G\,[(u_{z0,x} - \varphi_y)$$
$$+\sum(U_{i,z} \cdot \theta_i) - (y - y_m)\,\vartheta_{,x} - y_{m,x}\,\vartheta], \tag{3.15}$$

$$\sigma_y = \sigma_z = \tau_{yz} = 0. \tag{3.16}$$

Note that the last three stress components vanish, because the movements orthogonal to the axis are assumed to be only rigid-body motions.

When the derivatives are calculated, the product rule ensures that terms appear which are absent in a prismatic member, because in such members the position of the centroidal axis and the axis of the shear center do not change. In a tapered beam these terms should not be neglected, because the element will benefit from their inclusion.

With regard to the shear stresses two approaches are possible. If Timoshenko's and Mindlin's approach is applied, the terms in the first bracket effect a constant shear distribution, and the cross section remains flat.

A shear distribution which instead satisfies the equilibrium conditions locally as well (differential equation) is determined by the warping of the cross section. But this approach can also be used for the classical beam (if the first bracket becomes zero). The shear stresses then simply depend on the unit warping deformations, which must be scaled in such a way that equilibrium is satisfied with regard to the total shear force in the cross section.

In frame analysis the focus is on the resulting stresses, that is, the so-called internal actions

$$N = \int_A \sigma_x \, dA = EA \left(u_{,x} - z_{s,x} \, \varphi_y + y_{s,x} \, \varphi_z \right) + EA_z \, \varphi_{y,x} - EA_y \, \varphi_{y,x}$$

$$\tag{3.17}$$

$$M_y = \int_A z \, \sigma_x \, dA = EA_z \left(u_{,x} - z_{s,y} \, \varphi_y + y_{s,x} \, \varphi_z \right) + EA_{zz} \, \varphi_{y,x} - EA_{yz} \varphi_{y,x}$$

$$\tag{3.18}$$

$$M_z = \int_A y \, \sigma_x \, dA = EA_y \left(u_{,x} - z_{s,x} \, \varphi_y + y_{s,x} \, \varphi_z \right) + EA_{yz} \, \varphi_{y,x} - EA_{yy} \varphi_{y,x}$$

$$\tag{3.19}$$

$$V_y = \int_A \tau_{xy} \, dA \,, \qquad V_z = \int_A \tau_{xz} \, dA \tag{3.20}$$

$$M_t = \int_A \left[(y - y_m) \, \tau_{xz} - (z - z_m) \, \tau_{xy} \right] dA \tag{3.21}$$

where

$$EA = \int E \, dA \qquad EA_{zz} = \int E \, z^2 \, dA \tag{3.22}$$

$$EA_y = \int E \, y \, dA \qquad EA_{yy} = \int E \, y^2 \, dA \tag{3.23}$$

$$EA_z = \int E \, z \, dA \qquad EA_{yz} = \int E \, y \, z \, dA \,. \tag{3.24}$$

The centroid of the cross section is the point at which the area integrals EA_y and EA_z vanish. In addition, the unit warping functions U_i are scaled in such a way that their contributions to the first three integrals vanish. The cross-sectional centrifugal moment $E \, A_{yz}$ should be retained in any case, because a transformation of the cross section to the principal axes is neither appropriate nor always possible in the presence of shear deformations.

Cross sectional values

The cross-sectional values can be easily determined if the cross section is polygonal (see Fig. 3.6)

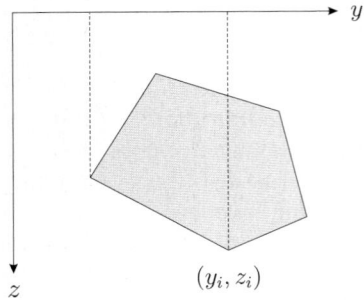

(y_i, z_i)

Fig. 3.6. Cross section

$$A = \sum_{i=0}^{n} \frac{1}{2} \left(z_{i+1} + z_i\right) \left(y_{i+1} - y_i\right) \tag{3.25}$$

$$I_y = \sum_{i=0}^{n} \frac{1}{12} \left(z_{i+1} + z_i\right) \left(y_{i+1} - y_i\right) \left(z_{i+1}^2 + z_i^2\right). \tag{3.26}$$

Sometimes a correction term in the form of an effective width is added if Bernoulli's hypothesis seems too crude an approximation. But these correction terms depend on what is to be corrected, that is, they are different if cross-sectional values and thereby the stiffnesses are to be modified, or if they modify terms which are relevant for the design. In such situations one must be careful to group the correct position of the centroidal axes with the correct cross-sectional values. For example it is a conventional approach in the design of prestressed concrete to derive the internal actions for a cross section, which is different from the approach that is later used for design purposes.

Because shear stresses do not appear in the constitutive equations of an Euler–Bernoulli beam, they are usually introduced by formulating an equilibrium condition:

$$\tau_{,s} = \sigma_{,x} , \qquad \tau = \frac{V\,Z}{I\,b} . \tag{3.27}$$

But unfortunately this formula has many drawbacks:

- The shear force V is only correct if the normal force is constant and the beam has a constant cross section (no tapered beam).
- The section modulus Z is only correct if the cross section is simply connected.
- Shear stresses must especially not be constant across the width of the beam.
- Eventually I must be replaced by Swain's formula for skew bending.

Mainly the second remark indicates the basic problem namely that the force method is not very appropriate for computer programs.

The natural choice would be a displacement method in which the primary variable is the warping of the cross section. The constitutive equation for this model is either the principle of minimum potential energy, or equivalently the set of equations that formulates the equilibrium conditions:

$$\tau_{xy} = G\left(w_{,y} - z\,\theta_{x,x}\right) \tag{3.28}$$

$$\tau_{xz} = G\left(w_{,z} + y\,\theta_{x,x}\right) \tag{3.29}$$

$$G\,\Delta w = G\left(w_{,xx} + w_{,yy}\right) = -\sigma_{x,x} \tag{3.30}$$

$$\tau_{xy}\,n_y + \tau_{xz}\,n_z = 0 \qquad \text{on the edge}. \tag{3.31}$$

This problem can either be solved in one step or divided into four sub-problems:

- The primary torsion problem: $d\theta/dx = $ warping ; $\sigma_x = 0$
- The two shear problems: $d\theta/dx = 0$; $\sigma_x = $ from the shear force V
- The secondary torsion problem: $d\theta/dx = 0$; $\sigma_x = $ from the warping moment

In the case of thin-walled cross sections where the stresses in thickness direction are nearly constant, the problem is easily solvable. Otherwise the Poisson equation must be solved numerically.

The result of such an advanced analysis is a precise shear stress distribution, which facilitates the design. By employing an equivalence principle, the internal strain energy can also be used to calculate the shear deformations, and the shear stress distribution also defines the unit warping functions U_i.

Stiffness matrix

If the interaction between the mixed terms of the stresses and the unit warping functions are neglected, the strain energy of the element can be written

$$\Pi_i = \frac{1}{2}\int E\,\varepsilon^2\,dV = \frac{1}{2}\int \big(EA\left[(u_0')^2 - 2\,u_0'\left[\varphi_y\,z_s' - \varphi_z\,y_s'\right]\right]$$
$$+ EA\left[\varphi_y^2\,(z_s')^2 + \varphi_z^2\,(y_s')^2 - 2\,\varphi_y\,z_s'\,\varphi_z\,y_s'\right]$$
$$+ EI_y\,(\varphi_y')^2 + EI_z\,(\varphi_z')^2 - 2\,EI_{yz}\,\varphi_y'\,\varphi_z'\big)\,dx\,. \tag{3.32}$$

This integral yields a stiffness matrix that not only takes into account the normal displacements and the bending stiffness, but also the stiffening effect of a tapered beam.

If the shape functions satisfy the differential equations, the stiffness matrix is exact. If this is not the case, as for example in the presence of an elastic foundation or if the beam is tapered or second-order theory effects must be considered, the beam must be subdivided into small elements to improve the approximation.

In the case of a tapered beam (see Fig. 3.7) the results obtained with different elements are displayed in the following table.

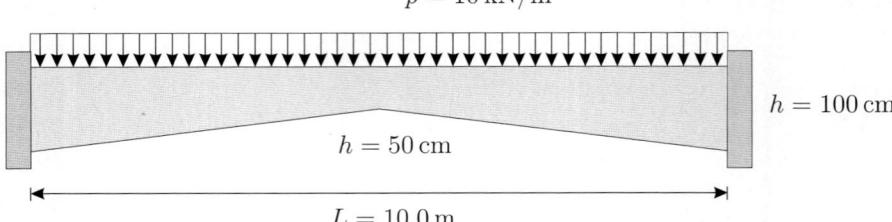

$p = 10\,\mathrm{kN/m}$

$h = 100\,\mathrm{cm}$

$h = 50\,\mathrm{cm}$

$L = 10.0\,\mathrm{m}$

Fig. 3.7. Tapered beam

	w(mm)	N_e (kN)	N_m (kN)	M_{ye} (kN m)	M_{ym} (kN m)
Inclined axis					
* 1 element	0.397	−80.5	−78	−73.58	31.91
* 8 elements	0.208	−46.3	−43.8	−94.87	19.17
** 1 element	0.172	−39.8	−37.3	−93.65	22
** 8 elements	0.206	−45.8	−43.3	−95.02	19.14
Horizontal axis					
1 element	0.168	−37.9	−37.9	−93.01	22.52
8 elements	0.204	−44.2	−44.2	−94.85	19.1

Here * indicates that the shape of $1/EI$ was interpolated, while ** indicates that the shape of EI was interpolated and $e = $ end and $m = $ middle refer to the position of the cross section.

Shear stresses and shear deformations

Shear deformations are neglected in classical beam theory. It would not be correct to simply add strain energy terms that account for the equivalent shear cross sections, because these add stiffness to the system but not flexibility. This can only be done in a formulation based on the complementary energy approach.

The simplest approach is it to reduce the bending stiffness in such a way that the deformations become the same. But this technique is limited to prismatic members, and would only achieve the intended effect for one particular load case. Also an extension to tapered beams is not so easy.

Timoshenko proposed to do the coupling of the rotations and the derivative of the deflection with a Lagrange multiplier. Hence the equations

$$M = -EI\,w'', \qquad V = -EI\,w''', \qquad (3.33)$$

are replaced by the two equations

$$M = -EI\,\varphi_{,x}\,, \qquad V = -GA\,\theta = GA\,(\varphi - w_{,x})\,. \qquad (3.34)$$

If the rotations and displacements are interpolated with linear polynomials, the bending moment is constant and the shear force is linear. This leads to two new problems that need attention:

- For large values of GA *locking* becomes a problem. A possible remedy is the introduction of a so-called Kirchhoff mode, which guarantees that at one point on the axis the derivative M' equals the shear force V. This condition relates u to φ, which enables one to modify the shape functions accordingly. If Hughes' approach is adopted [121] then

$$M = -EI\,\varphi_{,x} \qquad (3.35)$$

$$V = -GA\,\theta = GA\,[\frac{\varphi_i + \varphi_j}{2} - \frac{w_j - w_i}{L}]\,. \qquad (3.36)$$

- In addition, the element can no longer model simple problems correctly, as for example a cantilever beam that carries a single force at the free end. The rotations and the internal actions are correct, but the deflection is not. To overcome this problem the beam must be subdivided into many elements, which is hardly a sound approach from the standpoint of the user. In this case it helps to supplement the rotations with a nonconforming quadratic function φ_m so that—considering the Kirchhoff-condition—it follows

$$\varphi = \varphi_i\,(1 - \xi) + \varphi_j\,\xi + \varphi_m\,(4\,\xi\,(1 - \xi))\,, \qquad (3.37)$$

$$M = -\frac{EI}{L}\,[(\varphi_j - \varphi_i) + 1.5\,\varphi_m\,(8\,\xi - 4)]\,, \qquad (3.38)$$

$$V = -GA\,\theta = -GA\,[\frac{\varphi_i + \varphi_j}{2} + \varphi_m - \frac{w_j - w_i}{L}]\,. \qquad (3.39)$$

The function produces exactly the missing linear variation of the bending moment. The shear force stays at its maximum value, which requires a correction factor 1.5 for the bending moment.

But even such an advanced model as a Timoshenko beam cannot accommodate all effects. In numerical tests the results depended for some cross-sectional shapes on the orientation of the coordinate system, because if the shear force vector as well as the vector of the shearing strains is transformed, the tensor of the inverse shear areas is obtained

$$\begin{bmatrix} \theta_y \\ \theta_x \end{bmatrix} = \begin{bmatrix} 1/GA_y & 1/GA_{yz} \\ 1/GA_{yz} & 1/GA_z \end{bmatrix} \begin{bmatrix} V_y \\ V_z \end{bmatrix}\,. \qquad (3.40)$$

On the one hand the introduction of a mixed shear area $1/A_{yz}$ removes the inconsistency in the results; on the other hand, this additional area cannot simply be added to the Timoshenko beam as an additional stiffness, because normally it is infinite, and it would lead to a coupling of the bending moments about the two axes.

There is no easy way to account for this matrix in the classical beam element. There are only two possibilities:

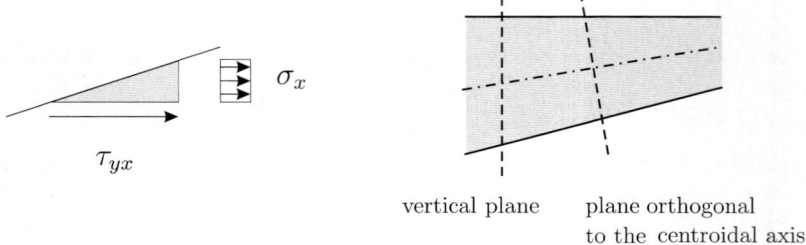

vertical plane   plane orthogonal
to the centroidal axis

Fig. 3.8. Shear stresses

- The first is to use a flexibility matrix. This may be established via numerical integration of the differential equations. Extra shear deformations are simply added and eventually included when this matrix is inverted to derive the stiffness matrix.
- The other solution based on the Timoshenko beam operates with the inverse of the above matrix, which yields the following term for the potential:

$$[V_y, V_z] \frac{1}{1-a} \begin{bmatrix} A_y & -b \\ b & A_y \end{bmatrix} \begin{bmatrix} V_y \\ V_z \end{bmatrix} \qquad a = \frac{A_y A_z}{A_{yz}} \qquad b = \frac{A_y A_y}{A_{yz}} . \quad (3.41)$$

In the Timoshenko model the shear stresses are constant within the cross section, while they have a parabolic shape in rectangular cross sections. The shear areas for the deformations and for the maximum stresses are therefore in general not identical:

$$\theta = \frac{V}{GA} \qquad (A = 0.833 \cdot b \cdot h \text{ for a rectangle}),$$

$$\tau = \frac{V}{A} \qquad (A = 0.666 \cdot b \cdot h \text{ for a rectangle}).$$

One aspect deserves our particular attention: normally the prismatic bar has a constant cross section and in calculating the shear stresses it is assumed that they vanish at the edge of the cross section. But in the case of a tapered beam this is not correct; see Fig. 3.8.

Shear stresses appear at the edge of a tapered beam (see Fig. 3.9). In a plane orthogonal to the reference axis the distribution of the shear stresses is asymmetric, (Fig. 3.10 a), while in a plane orthogonal to the centroidal axis the shear stress distribution is more harmonic (Fig. 3.10 b).

In the cross section $x = 1.0$ the transverse force is 40 kN, hence the shear force is 39.95 kN. If the cross section were rectangular with a height of 93.75 cm, the shear stress would be 63.92 kN/m². Because the bending moment in this cross section is 52.6 kN m, the shear force $M/d \cdot \tan \alpha$ is reduced by 5.6 kN, so that the real value is 54.9 kN/m². The FE shear stress is 53 kN/m² at the center and 30 kN/m² at the edge. Hence this reduction of the shear force comes close to the maximum value, but the distribution over the cross section

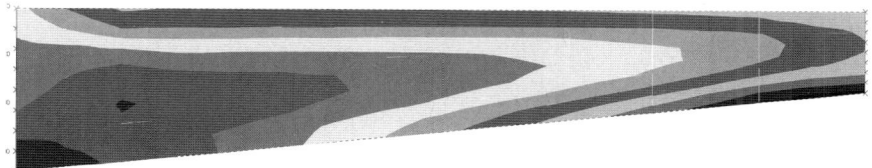

Fig. 3.9. Shear stresses τ_{xy}

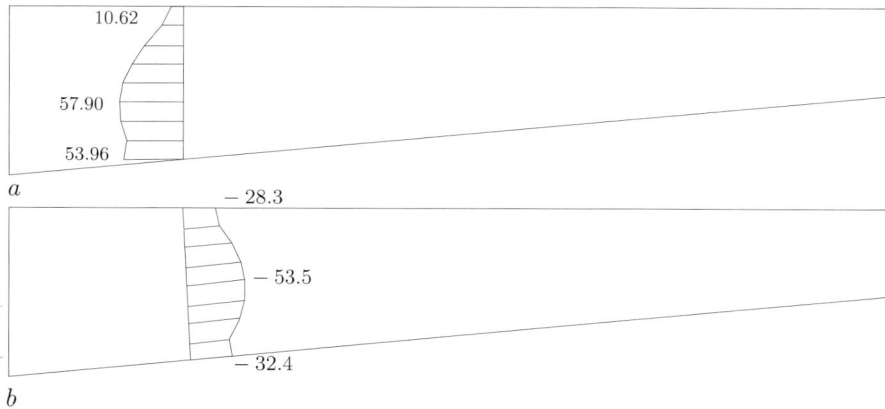

Fig. 3.10. Shear stresses: **a)** τ_{xy} in a vertical plane, **b)** τ_{nq} in a plane orthogonal to the centroidal axis

is only an approximation, so it is questionable whether the calculated principal tensile stresses or the maximum von Mises stresses are always correct.

Influence of the shear center, warping torsion

If the influence of warping torsion is to be considered as well (and assuming that the shear center axis changes its location) so many terms contribute to the strain energy integral that no complete analysis seems to have been done yet. In the simpler case that the reference axis coincides with the rotational axis, use could be made of the following formula for the strain energy:

$$
\begin{aligned}
\Pi_{i2} = \frac{1}{2} \int_0^l \big(& E\, C_M\, (\vartheta'')^2 + G\,I\,(\vartheta')^2 \\
& + N\,[2\,\vartheta'\,z_m\,v_m' + 2\,\vartheta'\,y_m\,w_m' + (v_m')^2 + (w_m')^2 + i_m\,(\vartheta')^2] \\
& + M_y\,[-2\,\vartheta\,w_m'' + r_{M_y}\,(\vartheta')^2] + M_t\,[2\,\vartheta\,v_m'' + r_{M_z}\,(\vartheta')^2] \\
& + M_b\,[r_{M_w}\,(\vartheta')^2] + M_t\,[v_m'\,w_m'' - v_m''\,w_m'] \big)\, dx\,.
\end{aligned}
\tag{3.42}
$$

Of course these terms are to be supplemented with additional terms that depend on the load.

With regard to the torsional moment, two components are now to be considered. The total moment can be split into a Saint-Venant part and a part that stems from secondary torsion:

$$M_t = M_{tv} + M_{t2} = G\,I_T\,\vartheta' - E\,C_M\,\vartheta'''. \tag{3.43}$$

Evidently a cubic polynomial is now needed to model the twist of the longitudinal axis. This is best achieved by introducing two additional degrees of freedom for the warping of the cross section. Then the same standard third-degree Hermite polynomials are also used to model the twist. As in the case of the shear force, one obtains in each element a constant and a stepped distribution of the secondary torsional moment. These formulas are also valid if no warping occurs ($C_M = 0$).

With regard to the shearing deformations due to warping, the same observations are true as for the shear deformations. They can either be incorporated directly, or one can use an approach similar to the approach used in a Timoshenko beam.

The general beam element

In the most general case—if arbitrary loads are to be applied, if the cross section can have any shape and support conditions can be rather complicated—use can be made of transfer matrices. The transfer matrix for a beam with constant values of EI, GA_s and a constant load p is

$$
\begin{bmatrix} w \\ w' \\ M \\ V \end{bmatrix}
=
\begin{bmatrix}
1 & -x & \dfrac{x^2}{2\,EI} & \dfrac{x^3}{6\,EI} - \dfrac{x}{G\,A_s} \\
0 & 1 & -\dfrac{x}{EI} & -\dfrac{x^2}{2\,EI} \\
0 & 0 & -1 & -x \\
0 & 0 & 0 & -1
\end{bmatrix}
\begin{bmatrix} w_0 \\ w'_0 \\ M_0 \\ V_0 \end{bmatrix}
+
\begin{bmatrix}
\dfrac{x^4}{24\,EI} + \dfrac{x^2}{2\,GA_s} \\
-\dfrac{x^3}{6\,EI} \\
-\dfrac{x^2}{2} \\
-x
\end{bmatrix}
p\,.
\tag{3.44}
$$

From an abstract point of view these equations can be interpreted as the result of a direct or numerical integration of the differential equations. Numerical integration is best done with Runge–Kutta methods. In simple cases the exact solution is obtained after just one step and the additional effort is minimal. In the worst case a stiff system of differential equations is obtained, and then the beam must be subdivided into many elements to control the numerical solution. The advantage of transfer matrices over variational methods is that by using transfer matrices a flexibility matrix can be derived, which corresponds to the principle of minimum complementary energy. This is an important property when it comes to a correct assessment of the shear deformations. An additional advantage is that the differential equation is easier to program.

Resistance factor design

There are other areas where a generalization of the current models opens up new possibilities. Some building codes allow to work in cross sections with plastic limit loads. Here too we have in principle the possibility of using finite elements and to apply elasto-plastic laws, or the whole analysis can be done in the cross section. In the case of thin-walled cross sections a complete description of the interaction of all resultant stresses is feasible [143]. If a general nonlinear procedure is employed, then the analysis must be done iteratively due to the changing stiffnesses, and the need to handle the interaction in the partially plasticized zones.

Frame analysis is usually based on the following assumptions:

- linear elastic distribution of the normal stresses and the stresses caused by torsional warping
- linear elastic distribution of the shear stresses due to shear forces and torsional moments
- a plane strain state plus warping
- a yield condition

To calculate the linear distribution of shear stresses in the cross section advanced methods—based either on a displacement or a force method—are necessary to determine the shear flow.

In the first step, given a certain combination of loads, a linear equivalent stress distribution can be calculated, and these stresses can then be related to the yield stresses.

In the second step one might "somehow" reduce these stresses. Besides the special case where the shear force is taken care of separately, other techniques are available.

In the third step, the stresses can be transformed by numerical integration into resultant internal actions, and these can then be used to calculate effective nonlinear stiffnesses. By an iterative procedure based on the stiffnesses, the equilibrium of the internal actions ultimately is maintained [141].

To control the yield condition it suffices to check just the normal stress σ_x and the shear stresses τ_{xy} and τ_{xz}, because the shear stresses τ_{yz} and the stresses σ_y and σ_z will only be noticeable near the loaded regions, and they are probably not influenced by changes in the stiffness of the cross section. For an investigation into the local limit load of such regions, 1-D analysis is not appropriate. Rather this would require experimental tests or an elaborate FE analysis with flat shell elements.

With regard to the intended reduction of the stresses, in principle three methods are conceivable. Common to all three methods is that each stress point is independent of the neighboring points. Plastic strains that might lead to additional stresses because the movement in y or z direction is impeded are therefore ignored, and perhaps this provides an additional safety factor with respect to experimental results. Formally this is in agreement with other engineering approaches, as for example the Winkler model. We can choose between three methods:

- **Prandtl solution** In adopting Prandtl's yield conditions, we calculate the plastic strains, which are orthogonal to the yield surface. To this end [72], [258], first the onset of the development of a plastic zone is calculated by a uniform reduction according to the first method, and then for the remaining plastic strain increments an elasto-plastic elasticity matrix is calculated by considering the elasticity matrix C:

$$\sigma = \left[C - \frac{q \cdot C \cdot q'}{q' \cdot C \cdot q} \right] \cdot \varepsilon, \qquad q = \frac{\partial F}{\partial \sigma}. \qquad (3.45)$$

- **Isotropic reduction** Shear and normal stresses are relaxed in the same ratio, so that the equivalent stress just reaches the yield stress:

$$\sigma = \left[\frac{f_y}{\sigma_{v,\,\text{elastic}}} \right] \cdot \sigma_{v,\,\text{elastic}}, \qquad \tau = \left[\frac{f_y}{\sigma_{v,\,\text{elastic}}} \right] \cdot \tau_{v,\,\text{elastic}}. \qquad (3.46)$$

- **Shear stresses** Shear stresses are absorbed fully and therefore normal stresses are reduced. This is the usual strategy in manual calculations. It remains unsatisfactory in the presence of strong shear stresses, because it can lead to situations where an increase in the curvature has no effect on the system

$$\tau = \min \left\{ \frac{f_y}{\sqrt{3}}, \tau_{\text{elastic}} \right\}, \qquad \sigma = \min \left\{ \sqrt{f_y^2 - 3\,\tau^2}, \sigma_{\text{elastic}} \right\}. \qquad (3.47)$$

With one of these methods, the resulting internal actions can be calculated by numerical integration.

If these internal actions exceed the actual internal actions, the structure is safe. If we do an elastic–plastic analysis, we have only to check the slenderness ratio (or a similar criterion) to pass the design check. Otherwise an iterative analysis must follow to allow for a redistribution of the forces.

If plastic zones develop in a cross section, the stiffnesses change. To assess the rearrangement of internal forces, these changes must be considered in the analysis. In bending-dominated problems, the equations can easily be modified if the following equation is either solved for the plastic curvature κ_0 or used to calculate a secant stiffness:

$$\begin{bmatrix} M_y \\ M_z \end{bmatrix} = \begin{bmatrix} E\,I_y & E\,I_{yz} \\ E\,I_{yz} & E\,I_z \end{bmatrix} \begin{bmatrix} \kappa_y \\ \kappa_z \end{bmatrix} + \begin{bmatrix} \kappa_{y0} \\ \kappa_{z0} \end{bmatrix}. \qquad (3.48)$$

With regard to shear strains, it is important to decide whether a rearrangement of the shear strains is possible or desirable. In some cases a reduction in the bending stiffness alone will lead to a reduction in the shear strains. But in general a static analysis must incorporate the shear deformations, because

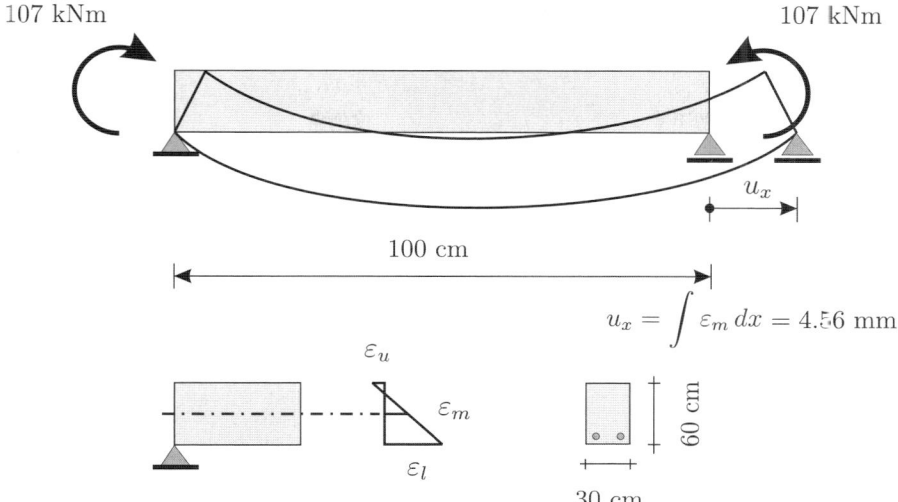

Fig. 3.11. Nonlinear analysis of a beam

only then can the shear stiffnesses be reduced. But unlike the situation in bending problems, there is no simple rule to determine nonlinear shear deformations. Nevertheless one can simply reduce the shear stiffness according to the ratio between internal and external shear force, but note that during the iteration the stiffness can increase again if the shear forces decrease.

Nonlinear analysis

In nonlinear analysis the stiffness of a structure depends on the actual strains and stresses within the structure. By introducing an equivalent *secant modulus* the nonlinear problem can be reduced to an elastic problem but then an iterative analysis must be performed to find the strains which are compatible with the internal actions. In the final step of such an iterative analysis it must be checked whether the internal actions are compatible with the iteratively determined stiffnesses. We perform this check for the beam in Fig. 3.11, grade C 20 concrete and grade S 500 reinforcement steel (Eurocode), [180].

If the origin of the system of coordinates does not coincide with the elastic centroid the matrix which describes the relation between the strains and the internal actions is fully populated

$$\begin{bmatrix} N_x \\ M_y \\ M_z \end{bmatrix} = \begin{bmatrix} EA & EA_z & -EA_y \\ EA_z & EA_{zz} & -EA_{yz} \\ -EA_y & -EA_{yz} & EA_{yy} \end{bmatrix} \begin{bmatrix} u' \\ -w'' \\ v'' \end{bmatrix} . \tag{3.49}$$

For a definition of the terms EA, EA_z, etc. see (3.22). A nonlinear analysis with an FE program rendered a value of 4.70 cm^2 for the reinforcement and

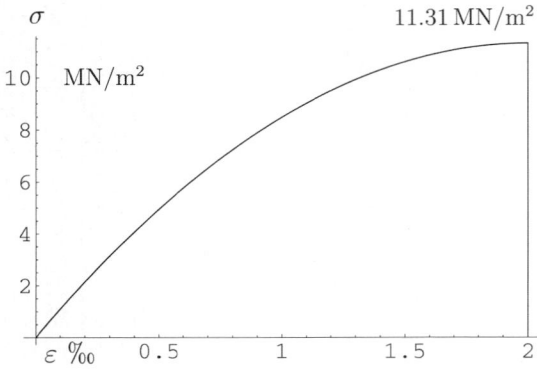

Fig. 3.12. $\sigma - \varepsilon$ diagram

for the strains the values

$$\varepsilon_u = -2.0\,\text{‰} \qquad \varepsilon_l = 11.12\,\text{‰} \qquad \varepsilon_m = \frac{11.12 - 2.0}{2} = 4.56\,\text{‰} \quad (3.50)$$

so that the curvature of the cross sections is

$$- w'' = \frac{2.0\text{‰} + 11.12\text{‰}}{0.6\,\text{m}} = 21.87 \cdot 10^{-3}\,\text{m}^{-1}. \qquad (3.51)$$

The nonlinear analysis was based on the following stress-strain law for the concrete (see Fig. 3.12)

$$\sigma_c = \varepsilon \cdot \left(1 - \frac{\varepsilon}{4}\right) \cdot \alpha \cdot f_{cd} \qquad \alpha = 0.85\,, \;\; f_{cd} = \frac{20}{1.5} = 13.3\,\text{MN/m}^2 \quad (3.52)$$

so that the secant stiffness becomes

$$E = \frac{\sigma_c}{\varepsilon} = \left(1 - \frac{\varepsilon}{4}\right) \cdot \alpha \cdot f_{cd}\,. \qquad (3.53)$$

Both the concrete $(b = 30 \text{ cm})$

$$EA_c := \int_{z=0}^{z_u} \left(1 - \frac{\varepsilon}{4}\right) \cdot \alpha \cdot f_{cd} \cdot b\,dz = 234.5\,\text{MN} \qquad (3.54)$$

and the steel

$$EA_s := \frac{\sigma_s}{\varepsilon_s} \cdot A_s = 20.65\,\text{MN} \qquad (3.55)$$

contribute to the longitudinal stiffness

$$EA := \int E\,dA = EA_c + EA_s = 234.5 + 20.65 = 255.15\,\text{MN}\,. \qquad (3.56)$$

In the same sense the statical moment is the sum of two parts

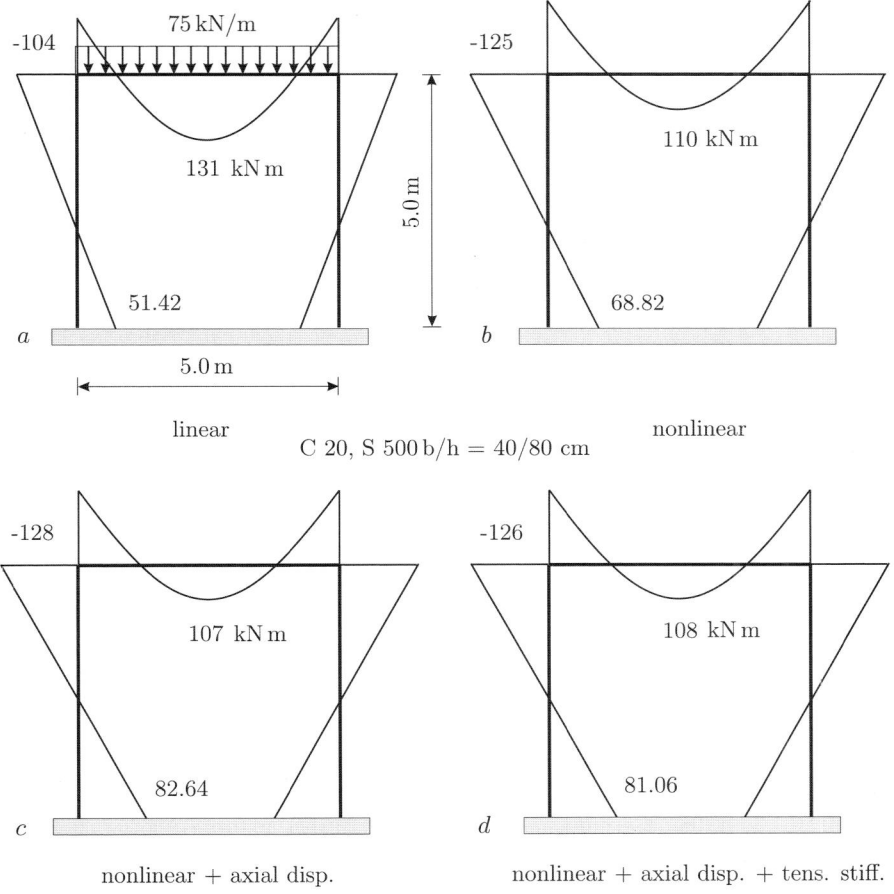

Fig. 3.13. An FE analysis of a standard frame: **a)** linear analysis; **b)** nonlinear analysis; **c)** nonlinear analysis + axial displacements; **d)** plus tension stiffening

$$EA_z := \int E\, z\, dA = EA_{z,c} + EA_{z,s} = -58.37 + 5.16 = -53.21\,\text{MN} \tag{3.57}$$

and the moment of inertia as well

$$EA_{zz} := \int E\, z^2\, dA = EA_{zz,c} + EA_{zz,s} = 14.69 + 1.29 = 15.98\,\text{MNm} \tag{3.58}$$

so that

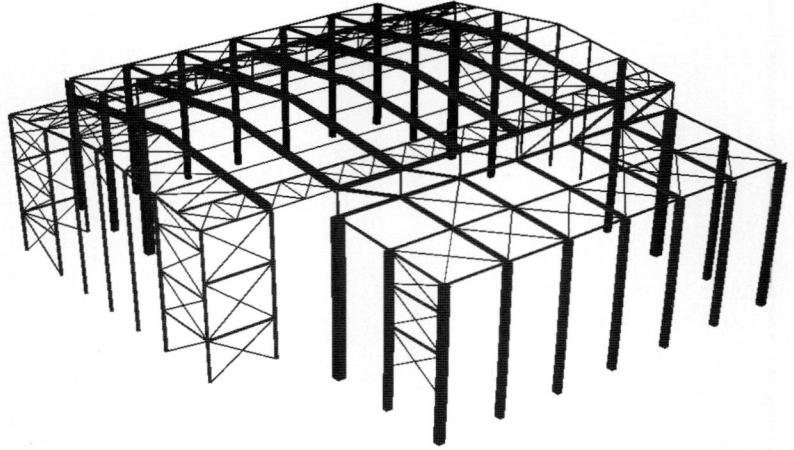

Fig. 3.14. An FE analysis of standard frames yields the exact values

$$
\begin{bmatrix} N_x \\ M_y \\ M_z \end{bmatrix} = \begin{bmatrix} 255.15 & -53.21 & -EA_y \\ -53.21 & 15.98 & 0 \\ -EA_y & 0 & EA_{yy} \end{bmatrix} \begin{bmatrix} 4.56 \cdot 10^{-3} \\ 21.87 \cdot 10^{-3} \\ 0 \end{bmatrix} = \begin{bmatrix} 0 \\ 107 \\ 0 \end{bmatrix} . \quad (3.59)
$$

The result agrees with the internal actions, $N = 0, M_y = 107\,\mathrm{kNm}, M_z = 0$.
The constant strain ε_m of the x-axis causes a horizontal displacement

$$
u_x = \int_0^l \varepsilon_m \, dx = 4.56\%_0 \cdot 1.0\,\mathrm{m} = 4.56\,\mathrm{mm} . \quad (3.60)
$$

Often the horizontal displacements are neglected but their influence on the structural reaction can be considerable and sometimes they produce quite surprising results.

Today it is standard that FE programs allow to consider shear deformations in frame analysis. It is hoped that in the future it will be possible to include the effects of nonlinear axial strains and *tension stiffening* as well in the analysis. The influence of these different effects

- linear analysis
- nonlinear analysis
- nonlinear analysis + axial displacements
- nonlinear analysis + axial displacements + tension stiffening

on the bending moment distribution of a concrete frame is depicted in Fig. 3.13.

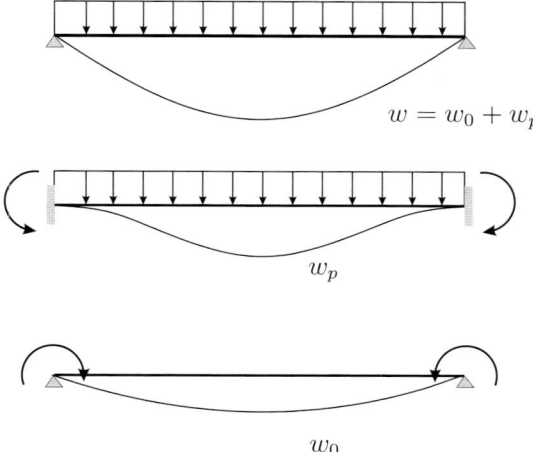

Fig. 3.15. Any deflection curve can be split into two parts

$$w = w_0 + w_p$$

$$w_p$$

$$w_0$$

3.3 Finite elements and the slope deflection method

If a frame is analyzed with an FE program, is the solution an approximation or is it exact?

The solution is exact if the frame could also have been analyzed with the slope deflection method, because in standard frames—no tapered beams, constant stiffness—the FE result is identical to the results of the slope deflection-method; see Fig. 3.14. The reason is that the reduction of the load into the nodes does not change the nodal displacements, as is explained below.

The deflection curve $w = w_0 + w_p$ of a beam can be split into two parts, a *homogeneous deflection curve* w_0 and a *particular deflection curve* w_p (see Fig. 3.15), where

$$w_0 = u_1 \cdot \varphi_1(x) + u_2 \cdot \varphi_2(x) + u_3 \cdot \varphi_3(x) + u_4 \cdot \varphi_4(x) \qquad (3.61)$$

and where w_p is a solution corresponding to fixed ends.

The homogeneous solution solves $EIw_0^{IV} = 0$, while the particular deflection curve w_p solves the equation $EIw_p^{IV} = p$. The deflection w_0 "carries" the end displacements u_i while the deflection w_p "carries" the load p. The second function is "mute" at the ends of the beam.

Correspondingly the deformation of a frame (see Fig. 3.16) can be split into two parts: deformations resulting from the movement of joints (nodal displacements), and local deformations, i.e., displacements between the nodes caused by the distributed load p. The local deformations would also occur if all nodes were fixed. The nodal displacements are the decisive terms, because they establish the *interaction* between the individual beams.

Hence, if in the FE method the movements of a frame are expanded with regard to the *unit displacements* φ_i of the nodes,

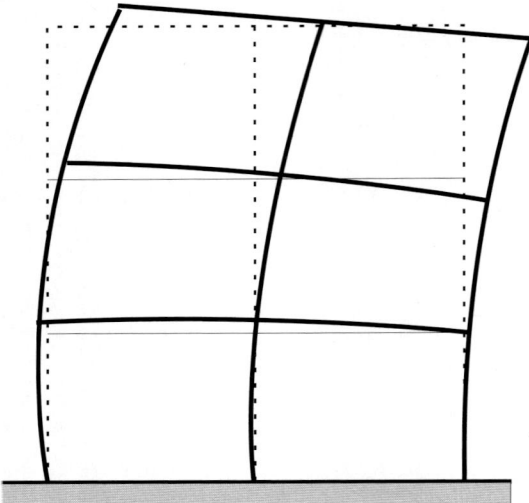

Fig. 3.16. The FE deformation is an expansion in terms of the nodal unit displacements of the joints. The expansion is exact if only nodal forces act on the frame

$$u_h(x) = \sum_i u_i \cdot \varphi_i(x) \, , \tag{3.62}$$

the local contributions w_p are neglected, i.e., it is assumed that $p = 0$.

Because such unit displacements can only yield the exact shape if the load is concentrated at the nodes, the FE method reduces all distributed loads to the nodes. To this end, it lets the distributed load p act through the unit displacements $\varphi_i(x)$, and it places nodal forces $\bar{f}_i$ at the nodes that contribute the same amount of work,

$$\delta W_e(p, \varphi_i) = \bar{f}_i \cdot 1 \, , \tag{3.63}$$

which means that the nodal displacements $\boldsymbol{u}$ satisfy

$$\boldsymbol{K}\boldsymbol{u} = \bar{\boldsymbol{f}} \, . \tag{3.64}$$

But the vector $\bar{\boldsymbol{f}}$ on the right-hand side

$$\bar{\boldsymbol{f}} = \boldsymbol{f} + \boldsymbol{p} \tag{3.65}$$

is identical to the right-hand side in the slope deflection method, because when the nodes are released and the forces (*reactio*) which previously prevented any movement of the nodes are applied in the opposite direction (*actio*), and when the equilibrium position $\boldsymbol{u}$ of the frame is determined, then this system (3.64) is solved.

The first vector $\boldsymbol{f}$ in (3.65) is the vector of the true nodal forces, i.e., the concentrated loads applied directly at the nodes, while the second vector $\boldsymbol{p}$ contains the equivalent nodal forces resulting from the distributed load. But

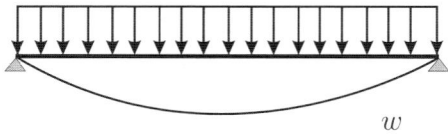

w **Fig. 3.17.** Beam and constant load

these equivalent forces are just the fixed end forces $\times(-1)$ of the distributed load, because

$$(-1) \times \text{fixed end forces} = p_i = \int_0^l p\,\varphi_i\,dx$$

$$= \int_0^l p\,\varphi_i\,dx = p_i = \text{equivalent nodal force}\,.$$

$$(3.66)$$

For this to be true, it is of course necessary that the shape functions be the exact unit displacements φ_i of the beam, because these functions are also the influence functions for the fixed end forces. This duality is the reason why the two sides in (3.66) are the same, why the equivalent nodal forces are also the fixed end forces.

Hence, an FE program which employs exact unit displacements is an implementation of the slope deflection method (in one step). Whether the distributed load is left out on the beam or reduced to the nodes makes no difference—the nodal displacements are the same. This remarkable result only holds for 1-D problems (ordinary differential equations—but see the remark on the next page).

Of course, between the joints the exact solution (distributed load) and the FE solution (equivalent nodal forces) differ, but this is of no consequence, because (3.64) is only used to calculate the nodal displacements. To calculate the stress resultants and the displacements in the individual beam elements, for each element (e) we invoke the relation

$$\boldsymbol{K}^e \boldsymbol{u}^e = \boldsymbol{f}^e + \boldsymbol{p}^e \quad \Rightarrow \quad \boldsymbol{f}^e = \boldsymbol{K}^e \boldsymbol{u}^e - \boldsymbol{p}^e\,. \quad (3.67)$$

This provides the beam end forces $\boldsymbol{f}^e$ that belong to the end displacements $\boldsymbol{u}^e$ and once the f_i^e are calculated, the internal actions between the nodes can be calculated with influence functions or transfer matrices.

A study of the displacements of a simple one span beam with a constant load will explain this (see Fig. 3.17). The beam has a length of 15 m, the bending stiffness is $EI = 34,167$ kNm^2, and the applied load is $p = 10$ kN/m. The following table shows results for an analysis with one or two elements.

	Exact	1 element	2 elements	
Max. moment	281.25	281.25	281.25	kNm
End rotations	41.147	41.147	41.147	
Center deflection	19.2876	15.4301	19.2876	mm

The maximum bending moments and the rotations are identical, but the center deflections are not. This is no surprise, given that the shape functions are cubic polynomials. The deflection curve must be a symmetric function with respect to the center of the beam, it must be a second-order polynomial! The end rotations are correct, because a reduction of the distributed load into the nodes will have no effect on the nodal displacements.

Remark 3.1. The necessary condition that the FE solution interpolates the exact solution at the nodes is that the Green's functions of the nodes lie in V_h. The Green's functions are piecewise homogeneous solutions. In equations such as

$$- EAu''(x) + c\,u(x) = p(x) \qquad EI\,w^{IV}(x) + c\,w(x) = p(x) \qquad (3.68)$$

the homogeneous solutions

$$u(x) = c_1\,e^{x\sqrt{c/EA}} + c_2\,e^{-x\sqrt{c/EA}} \qquad (3.69)$$

and

$$w(x) = e^{\beta\,x}(c_1\,\cos\beta\,x + c_2\,\sin\beta\,x) + e^{-\beta\,x}(c_3\,\cos\beta\,x + c_4\,\sin\beta\,x) \quad (3.70)$$

$$\beta = \sqrt[4]{\frac{c}{EI}} \qquad (3.71)$$

are not the typical shape functions of the FE space V_h. Hence, in these cases the FE solution does *not* interpolate the exact solution at the nodes. Theoretically it can also happen that V_h is too "smooth" i.e., for to generate a Green's function of the equation $-EA\,u'' = p$, we must allow for a discontinuous first-order derivative at the nodes. The FE space V_h of an Euler–Bernoulli beam would not do us the favor.

Hence, in some sense in FE analysis we must strike a balance between the regularity that is required by the energy and the non-regularity that is necessary in order to come close to the Green's functions. And the latter we achieve by letting $h \to 0$ because then we can model a nearly infinite slope, $1/h \to \infty$, with a nodal unit displacement $u = 1$.

3.4 Stiffness matrices

To calculate the stiffness matrix of the bar in Fig. 3.18, the differential equation that relates the axial load p to the axial displacement $u(x)$,

$$- EAu''(x) = p(x) \qquad (3.72)$$

and the general homogeneous solution of this equation

$$u_h(x) = a_0 + a_1 x \qquad (3.73)$$

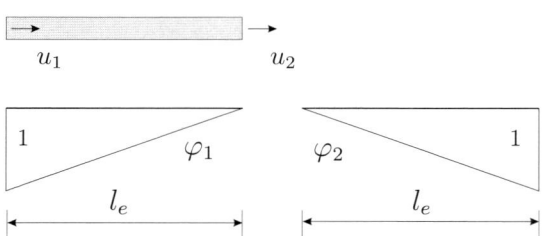

Fig. 3.18. Unit displacements of the bar. Horizontal displacements are plotted downwards

must both be known. By a proper choice of the coefficients a_0 and a_1, the unit displacements

$$\varphi_1(x) = \frac{1-x}{l} \qquad \varphi_1(0) = 1, \qquad \varphi_1(l) = 0,$$

$$\varphi_2(x) = \frac{x}{l} \qquad \varphi_2(0) = 0, \qquad \varphi_2(l) = 1 \tag{3.74}$$

are derived. Finally Green's first identity is formulated,

$$G(u, \hat{u}) = \underbrace{\int_0^l -EAu''\hat{u}\,dx + [N\hat{u}]_0^l}_{\delta W_e} - \underbrace{\int_0^l EAu'\,\hat{u}'dx}_{\delta W_i} = 0 \qquad N = EAu' \tag{3.75}$$

to provide the definition of the strain energy product δW_i, because the element k_{ij} of the stiffness matrix $\boldsymbol{K}$ is the strain energy product between the unit displacements, φ_i and φ_j

$$k_{ij} = \int_0^l EA\,\varphi_i'\varphi_j'\,dx = \int_0^l \frac{N_i N_j}{EA}\,dx, \tag{3.76}$$

so that

$$\boldsymbol{K} = \frac{EA}{l}\begin{bmatrix} 1 & -1 \\ -1 & 1 \end{bmatrix} \tag{3.77}$$

or $\boldsymbol{K}\,\boldsymbol{u} = \boldsymbol{f} + \boldsymbol{p}$, where the f_i are the end forces, and the p_i are the fixed-end forces $(\times -1)$ resulting from the distributed load

$$p_1 = \int_0^l p\,\varphi_1\,dx \qquad p_2 = \int_0^l p\,\varphi_2\,dx. \tag{3.78}$$

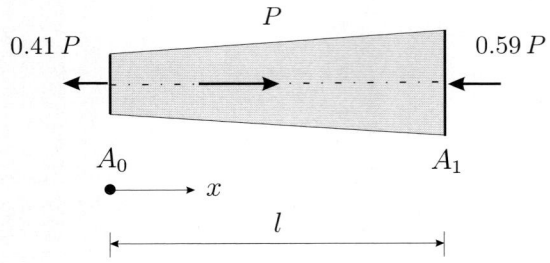

Fig. 3.19. Variable stiffness $EA = EA(x)$

Tapered beam

In a tapered beam with a cross-sectional area such as

$$A(x) = A_0 + A_1 x \,, \tag{3.79}$$

the differential equation for the longitudinal displacement is

$$-EA(x)u''(x) - EA'(x)u'(x) = p(x) \,. \tag{3.80}$$

With *Maple* or *Mathematica*, the homogeneous solution

$$u(x) = c_2 + c_1 \frac{\ln A(x)}{A_1} \tag{3.81}$$

can be found, and thereby the unit displacements

$$\varphi_1(x) = \frac{\ln A(x) - \ln A(l)}{\ln A(0) - \ln A(l)} \,, \qquad \varphi_2(x) = \frac{\ln A(0) - \ln A(x)}{\ln A(0) - \ln A(l)} \,. \tag{3.82}$$

Upon substituting these into the strain energy product

$$k_{ij} = \int_0^l EA(x)\, \varphi_i'\varphi_j' \, dx \,, \tag{3.83}$$

the stiffness matrix is obtained:

$$\boldsymbol{K} = k \begin{bmatrix} 1 & -1 \\ -1 & 1 \end{bmatrix} \qquad k = A_1\, E\, \frac{\ln A(l) - \ln A_0}{(\ln A_1 - \ln A_0)^2} \,. \tag{3.84}$$

According to the formula for the fixed-end forces

$$p_1 = \int_0^l p\, \varphi_1 \, dx \,, \qquad p_2 = \int_0^l p\, \varphi_2 \, dx \tag{3.85}$$

a single force P will generate the following actions at the fixed ends

$$p_1 = P \cdot \varphi_1(x_P) \,, \quad p_2 = P \cdot \varphi_2(x_P) \,, \quad x_P = \text{location of } P. \tag{3.86}$$

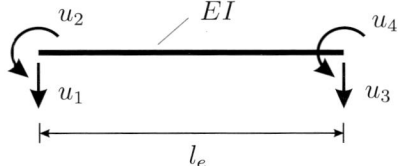

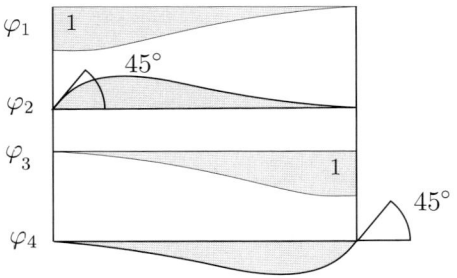

Fig. 3.20. Unit displacements of a beam

The sum of the support reactions p_i must equal $P = p_1 + p_2$, i.e., at each point x the sum of $\varphi_1(x)$ and $\varphi_2(x)$ must be 1

$$\varphi_1(x) + \varphi_2(x) = 1 \qquad (100\%). \tag{3.87}$$

If the cross-sectional area varies as

$$A(x) = A_0 + A_1 x = 1 + 1 \cdot x \qquad \text{length } l = 1 \tag{3.88}$$

and if a single force P acts at the center of the shaft (see Fig. 3.19), then because

$$\varphi_1(0.5) = 0.415 \qquad \varphi_2(0.5) = 0.585 \tag{3.89}$$

about 41% of P will pull at the left end and about 59% will press on the other end of the shaft.

Euler–Bernoulli beam

This technique for the derivation of stiffness matrices can be applied to all possible deformations in a beam, such as the *twist* $\varphi(x)$ of the axis, *shear deformations* $w_s(x)$, or simply the *deflection* $w(x)$ in which case the differential equation is

$$EIw^{IV}(x) = p(x). \tag{3.90}$$

Starting with the homogeneous solution

$$w_h(x) = a_0 + a_1 x + a_2 x^2 + a_3 x^3, \tag{3.91}$$

the four unit displacements (see Fig. 3.20) are found:

$$\varphi_1(x) = 1 - \frac{3x^2}{l^2} + \frac{2x^3}{l^3} \qquad \varphi_3(x) = \frac{3x^2}{l^2} - \frac{2x^3}{l^3}$$

$$\varphi_2(x) = -x + \frac{2x^2}{l} - \frac{x^3}{l^2} \qquad \varphi_4(x) = \frac{x^2}{l} - \frac{x^3}{l^2} \, .$$

(3.92)

Green's first identity is

$$G(w, \hat{w}) = \underbrace{\int_0^l EIw^{IV}\, \hat{w}dx + [V\hat{w} - M\hat{w}']_0^l}_{\delta W_e} - \underbrace{\int_0^l EIw''\hat{w}''dx}_{\delta W_i} = 0 \quad (3.93)$$

(3.94)

and the element k_{ij} of the stiffness matrix $\boldsymbol{K}$ is the *strain energy product* between the unit displacements φ_i and φ_j

$$k_{ij} = \int_0^l EI\, \varphi_i''\varphi_j''\, dx \, ,$$

(3.95)

so that

$$\boldsymbol{K} = \frac{EI}{l^3} \begin{bmatrix} 12 & -6l & -12 & -6l \\ -6l & 4l^2 & 6l & 2l^2 \\ -12 & 6l & 12 & 6l \\ -6l & 2l^2 & 6l & 4l^2 \end{bmatrix} \, .$$

(3.96)

The (negative) end-fixing forces p_i (= equivalent nodal forces) of a distributed load p are the scalar product of p and the unit displacements:

$$p_i = \int_0^l p\, \varphi_i\, dx \, .$$

(3.97)

Timoshenko beam

In the following it is assumed that the bending stiffness EI, the effective shear cross section A_s and the shear modulus G are constant. The deformations of the beam are described by the deflection w and the rotation θ (see Fig. 3.21). The constitutive equations are

strains	$\theta' - \kappa = 0$	$w' + \theta - \gamma = 0$	(3.98)
material law	$GA_s\gamma - V = 0$	$EI\kappa - M = 0$	(3.99)
equilibrium	$M' - V = 0$	$-V' = p$	(3.100)

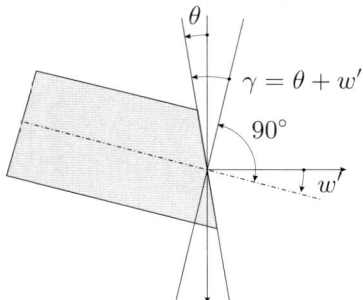

Fig. 3.21. Timoshenko beam

or

$$- EI\,\theta'' + GA_s\,(w' + \theta) = 0 \qquad (3.101)$$

$$-GA_s\,(w'' + \theta') = p\,. \qquad (3.102)$$

The latter system can be read as the application of an operator $-\boldsymbol{L}$ (minus because of second order) to the vector-valued function $\boldsymbol{u} = [w, \theta]^T$. Green's first identity for this operator $-\boldsymbol{L}$ is

$$G(\boldsymbol{u}, \hat{\boldsymbol{u}}) = \int_0^l -\boldsymbol{L}\,\boldsymbol{u} \boldsymbol{\cdot} \hat{\boldsymbol{u}}\,dx + \Big[V\,\hat{w} + M\,\hat{\theta}\Big]_0^l - a(\boldsymbol{u}, \hat{\boldsymbol{u}}) = 0 \qquad (3.103)$$

where

$$a(\boldsymbol{u}, \hat{\boldsymbol{u}}) = \int_0^l [V\,\hat{\gamma} + M\,\hat{\kappa}]\,dx$$

$$= \int_0^l [GA_s(w' + \theta)\,(\hat{w}' + \hat{\theta}) + EI\,\theta'\,\hat{\theta}']\,dx \qquad (3.104)$$

is the strain energy product. By some simple algebra, it follows that the homogeneous solutions of the system (3.101) and (3.102) must satisfy the equations

$$w^{IV} = 0 \qquad w' = -\theta + \frac{EI}{GA_s}\,\theta''\,. \qquad (3.105)$$

The homogeneous solutions of the first equation are

$$w(x) = c_1 + c_2\,\xi + c_3\,\xi^2 + c_4\,\xi^3\,, \qquad \xi = x/l \qquad (3.106)$$

and the matching function $\theta(x)$ is found to be

$$\theta(x) = c_1 \cdot 0 - c_2\,\frac{1}{l} - 2\,c_3\,\frac{\xi}{l} - c_4(\frac{\eta}{2\,l} + \frac{3}{l}\,\xi^2)\,, \qquad \eta = \frac{12}{l^2}\,\frac{EI}{GA_s}\,. \qquad (3.107)$$

By a proper choice of the constants c_i, the following nodal unit deflections for the two ends of a beam are found [198]:

$$\downarrow \qquad w_1(x) = \frac{1}{1+\eta} \left[1 - 3\,\xi^2 + 2\,\xi^3 + \eta\,(1-\xi) \right] \qquad (3.108)$$

$$\curvearrowright \qquad w_2(x) = \frac{l}{1+\eta} \left[-\xi + 2\,\xi^2 - \xi^3 - \frac{\eta}{2}\,(\xi - \xi^2) \right] \qquad (3.109)$$

$$\downarrow \qquad w_3(x) = \frac{1}{1+\eta} \left[3\,\xi^2 - 2\,\xi^3 + \eta\,\xi \right] \qquad (3.110)$$

$$\curvearrowright \qquad w_4(x) = \frac{l}{1+\eta} \left[\xi^2 - \xi^3 + \frac{\eta}{2}\,(\xi - \xi^2) \right]. \qquad (3.111)$$

The corresponding rotations are

$$\theta_1(x) = \frac{1}{1+\eta} \left[-\frac{6}{l}\,\xi\,(\xi - 1) \right] \qquad (3.112)$$

$$\theta_2(x) = \frac{1}{1+\eta} \left[1 - 4\xi + 3\,\xi^2 + (1-\xi)\,\eta \right] \qquad (3.113)$$

$$\theta_3(x) = \frac{1}{1+\eta} \left[-\frac{6}{l}\,\xi\,(1-\xi) \right] \qquad (3.114)$$

$$\theta_4(x) = \frac{1}{1+\eta} \left[-\xi\,(2 - 3\xi - \eta) \right]. \qquad (3.115)$$

The strain energy products, $k_{ij} = a(\boldsymbol{u}_i, \boldsymbol{u}_j)$, of the nodal unit deformations

$$\boldsymbol{u}_1 = \begin{bmatrix} w_1 \\ \theta_1 \end{bmatrix} \quad \boldsymbol{u}_2 = \begin{bmatrix} w_2 \\ \theta_2 \end{bmatrix} \quad \boldsymbol{u}_3 = \begin{bmatrix} w_3 \\ \theta_3 \end{bmatrix} \quad \boldsymbol{u}_4 = \begin{bmatrix} w_4 \\ \theta_4 \end{bmatrix} \qquad (3.116)$$

constitute the stiffness matrix

$$\boldsymbol{K} = \frac{EI}{l^3\,(1+\eta)} \begin{bmatrix} 12 & -6\,l & -12 & -6\,l \\ -6\,l & (4+\eta)\,l^2 & 6\,l & (2-\eta)\,l^2 \\ -12 & 6\,l & 12 & 6\,l \\ -6\,l & (2-\eta)\,l^2 & 6\,l & (4+\eta)\,l^2 \end{bmatrix}. \qquad (3.117)$$

The nodal degrees of freedom u_i have the same meaning as in an Euler–Bernoulli beam (see Fig. 3.20).

3.5 Approximations for stiffness matrices

A stiffness matrix is exact if the strain energy product $a(.,.)$ is exact, if the unit displacements φ_i are exact, and if the integration is done exactly:

$$a(\varphi_i, \varphi_j) = \int_0^l \frac{M_i\,M_j}{EI}\,dx\,. \qquad (3.118)$$

Green's first identity explains how the strain energy product is defined. It is more difficult to find the homogeneous solution of the differential equation. The unit displacements are based on this solution. Approximate stiffness matrices are usually based on approximate unit displacements, which are substituted into the correct strain energy product.

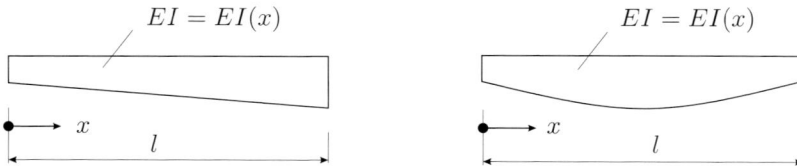

Fig. 3.22. Variable width

In a belly-shaped beam (Fig. 3.22) for example, the bending stiffness $EI(x)$ varies with x, so the beam equation becomes lengthy:

$$\underbrace{EI''(x)\,w''(x) + 2EI'(x)\,w'''(x)}_{additional\ terms} + EI(x)w^{IV}(x) = p(x)\,, \qquad (3.119)$$

while in the strain energy product little changes, only that $I(x)$ now depends on x:

$$a(w, \hat{w}) = \int_0^l \frac{M\,\hat{M}}{EI(x)}\,dx = \int_0^l EI(x)\,w''\,\hat{w}''\,dx\,. \qquad (3.120)$$

If the width h varies linearly,

$$EI(x) = E\,\frac{bh^3(x)}{12} \qquad h(x) = a_0 + a_1\,x \qquad (3.121)$$

it is possible to find the homogeneous solution of (3.119). But even *Maple* or *Mathematica* will have difficulty finding the homogeneous solutions of more complicated shapes.

Then the only way out is to substitute the unit displacements of the beam with a constant bending stiffness EI into the exact strain energy product and to work with this approximate matrix $\tilde{\boldsymbol{K}}$:

$$\tilde{k}_{ij} = \int_0^l EI(x)\,\varphi_i''(x)\,\varphi_j''(x)\,dx\,. \qquad (3.122)$$

Because the unit displacements are not exact, also the equivalent nodal forces (or negative end-fixing forces p_i) will not be consistent, i.e., they are only approximations of the real f_i:

$$\tilde{f}_i = \int_0^l p\,\varphi_i(x)\,dx\,. \qquad (3.123)$$

In the case of a beam on an elastic foundation (see Fig. 3.23),

$$EIw^{IV}(x) + c\,w(x) = p(x)\,, \qquad (3.124)$$

the exact unit displacements are well known, but program authors prefer to use the unit displacements of the standard beam element, because this

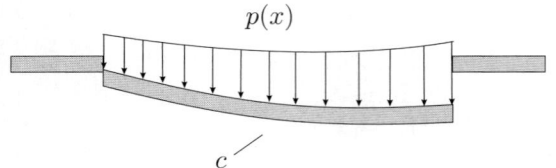

Fig. 3.23. Beam on an elastic foundation (Winkler)

facilitates program maintenance. That is, they calculate the elements of the stiffness matrix

$$k_{ij} = \int_0^l [\frac{M_i M_j}{EI} + c\,\varphi_i\,\varphi_j]\,dx \tag{3.125}$$

with the unit displacements $\varphi_i(x)$ of the standard beam element. This yields the approximation

$$\tilde{K} = \frac{EI}{l^3}\begin{bmatrix} 12 & -6\,l & -12 & -6\,l \\ -6\,l & 4\,l^2 & 6\,l & 2\,l^2 \\ -12 & 6\,l & 12 & 6\,l \\ -6\,l & 2\,l^2 & 6\,l & 4\,l^2 \end{bmatrix} + \frac{c}{420}\begin{bmatrix} 156\,l & -22\,l^2 & 54\,l & 13\,l^2 \\ -22\,l^2 & 4\,l^3 & -13\,l^2 & -3\,l^3 \\ 54\,l & 13\,l^2 & 156\,l & 22\,l^2 \\ 13\,l^2 & -3\,l^3 & 22\,l^2 & 4\,l^3 \end{bmatrix}. \tag{3.126}$$

Hence the FE program connects the end points of the deformed beam (end displacements u_i) with a curve w that is the sum of the four unit displacements of the standard beam,

$$w(x) = \sum_{i=1}^4 u_i\,\varphi_i(x)\,. \tag{3.127}$$

This elastic curve deviates from the exact shape because it is not a homogeneous solution of (3.124). Rather, a residual force appears,

$$EIw^{IV}(x) + c\,w(x) = c[u_1\varphi_1(x) + u_2\varphi_2(x) + u_3\varphi_3(x) + u_4\varphi_4(x)]\,, \tag{3.128}$$

which is just the distributed load p that must be applied to force the beam into the shape $w(x)$ (see Eq. (3.127)).

In second-order beam theory,

$$EIw^{IV}(x) + P\,w''(x) = p\,, \tag{3.129}$$

the procedure is virtually the same. If the unit displacements $\varphi_i(x)$ of the standard beam are substituted into the exact strain energy product

$$a(w, \hat{w}) = \int_0^l [EIw''\hat{w}'' - P\,w'\hat{w}']\,dx\,, \tag{3.130}$$

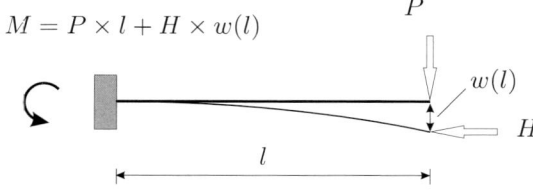

$$M = P \times l + H \times w(l)$$

Fig. 3.24. Second-order beam theory

the resulting approximate stiffness matrix

$$\tilde{K} = \frac{EI}{l^3} \begin{bmatrix} 12 & -6l & -12 & -6l \\ -6l & 4l^2 & 6l & 2l^2 \\ -12 & 6l & 12 & 6l \\ -6l & 2l^2 & 6l & 4l^2 \end{bmatrix} - \frac{P}{30\,l} \begin{bmatrix} 36 & -3\,l & -36 & -3\,l \\ -3\,l & 4l^2 & 3\,l & -l^2 \\ -36 & 3l & 36 & 3\,l \\ -3\,l & -l^2 & 3\,l & 4l^2 \end{bmatrix}, \quad (3.131)$$

is the sum of the first-order stiffness matrix and the so-called *geometric matrix*. This procedure amounts to a Taylor expansion of the exact stiffness matrix $K = K(P)$ at the point $P = 0$.

If in a Timoshenko beam element $[-1, 1]$ the deflection $w(\xi)$ and the rotation $\theta(\xi)$ are linear

$$\begin{bmatrix} w(\xi) \\ \theta(\xi) \end{bmatrix} = \frac{1}{2} \begin{bmatrix} 1-\xi & 0 & 1+\xi & 0 \\ 0 & 1-\xi & 0 & 1+\xi \end{bmatrix} \begin{bmatrix} w_1 \\ \theta_2 \\ w_3 \\ \theta_4 \end{bmatrix} \quad (3.132)$$

then the resulting stiffness matrix is only an approximation

$$K = \frac{EI}{l_e} \begin{bmatrix} 0 & 0 & 0 & 0 \\ 0 & 1 & 0 & -1 \\ 0 & 0 & 0 & 0 \\ 0 & -1 & 0 & 1 \end{bmatrix} + \frac{G\,A_s}{6\,l_e} \begin{bmatrix} 6 & 3\,l_e & 6 & 3\,l_e \\ 3\,l_e & 2\,l_e^2 & -3\,l_2 & l_e^2 \\ 6 & -3\,l_e & 6 & -3\,l_e \\ 3\,l_e & l_e^2 & -3\,l_2 & 2\,l_e^2 \end{bmatrix} \quad (3.133)$$

because the linear shape functions do not satisfy the differential equations (3.105).

Equilibrium ?

In first-order beam theory the load and the support reactions maintain equilibrium with regard to the undeformed structure and in second-order beam theory with regard to the deformed structure—correct? No. Second-order theory actually is a mixture of both; see Fig. 3.24. The lateral deformation is considered, but the longitudinal displacement of the axis is neglected.

Theoretically it is therefore not possible to check the equilibrium of a frame, that was analyzed with second-order beam theory, because first-order and second-order effects contribute to the movements of the joints. If we ignore

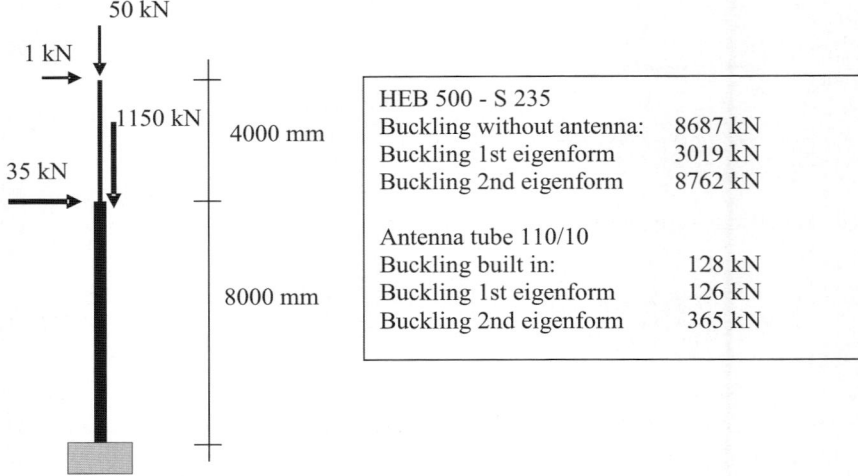

Fig. 3.25. Buckling

this and check the equilibrium conditions by employing the nodal displacements of the deformed structure then the check will fail. The reason that it does not fail in practice is that normally the displacements are so small, that the deviations are considered to be rounding errors.

Buckling length

Many design codes allow to establish the stability of a structure by determining the buckling loads of the individual members. The critical load P_{crit} of a frame member is expressed in the form

$$P_{crit} = \frac{\pi^2 EI}{(K \cdot l)^2} \tag{3.134}$$

where K is termed *effective length factor*. That is the load P_{crit} is expressed in terms of the critical load of an equivalent pin-ended frame member of length $s_K = K \cdot l$.

The standard FE-approach instead is a global approach based on 2nd order theory and usually also including possible imperfections of the structure. The result of the analysis are the eigenvalues λ of the structure based on the elastic stiffness matrix $\boldsymbol{K}$ and the geometric stiffness matrix $\boldsymbol{K}_G$

$$(\boldsymbol{K} + \lambda \, \boldsymbol{K}_G) \, \boldsymbol{u} = \boldsymbol{0} \qquad \boldsymbol{u} \neq \boldsymbol{0}. \tag{3.135}$$

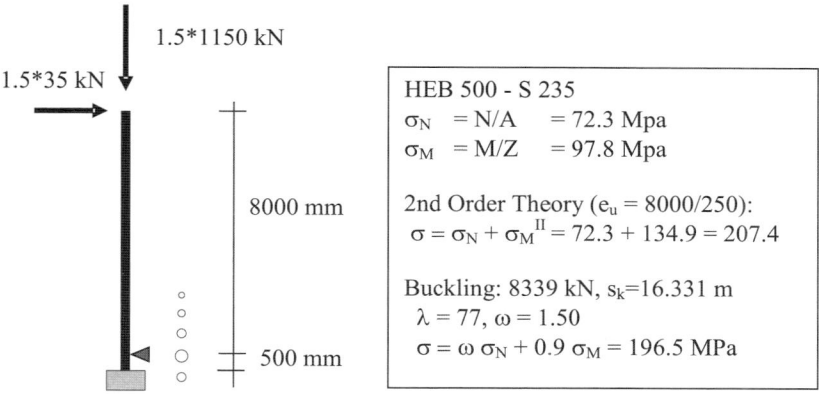

Fig. 3.26. Flag pole

However it is not always easy to determine the critical eigenvalues, sometimes even negative eigenvalues are observed which pose a problem for most of the algorithms, [215].

In the case of a multi-story building as in Fig. 3.25, with a slender antenna on top of the roof, we clearly see that it is not the smallest eigenvalue which is the critical eigenvalue but that we have to check a whole range of eigenvalues.

Another problem pose those structures where the length of the individual members can change. In the case of the flag pole in Fig. 3.26 the maximum bending moment occurs just above the horizontal support. The analysis of the upper part of the structure is standard, but in the lower part, the short beam, the buckling length is quite small, and it is not possible to find a correct stress in that part. If the horizontal support also carries vertical loads the normal force in the lower part is zero and then it is not possible to apply the buckling length approach to the design of this part.

This holds also true for the structure in Fig. 3.27 where

$$N_1 = F \qquad N_2 = N_3 = N_4 = N_5 = -S \qquad \varepsilon := l \sqrt{\frac{F}{EI}}. \qquad (3.136)$$

The work done by the two forces F and the moment M_a on acting through the displacements in Fig. 3.27 b (= influence function for M_a) must be zero

$$\delta W_e = (M_a + F \cdot l) \cdot 1 - S \cdot L \cdot (-0.5) \cdot (-0.5) \cdot 2 - S \cdot L \cdot 0.5 \cdot 0.5 \cdot 2$$
$$= M_a + F \cdot l - S \cdot L = 0. \qquad (3.137)$$

Given a horizontal force H, see Fig. 3.28, the bending moments are

$$M_a^I = H \cdot l \qquad M_a^{II} = M_a = H \cdot l \cdot \frac{\tanh \varepsilon}{\varepsilon} \le H \cdot l. \qquad (3.138)$$

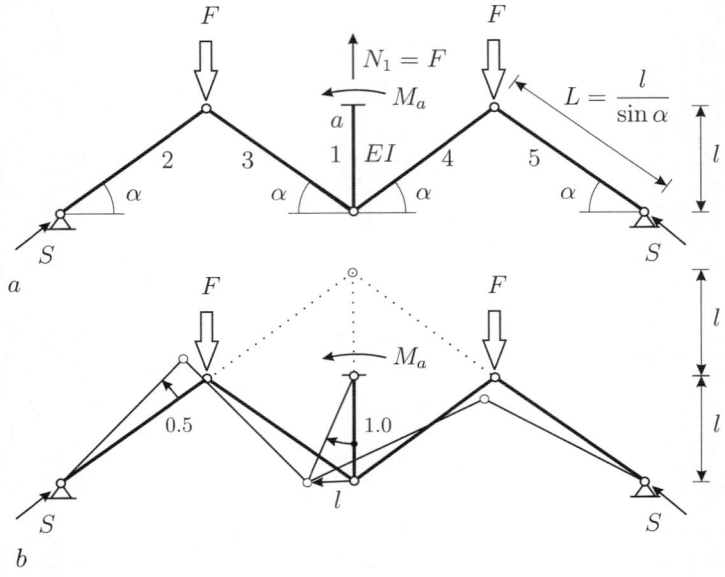

a

b

Fig. 3.27. What is the critical load? **a)** structure; **b)** influence function for M_a, displacements not to scale

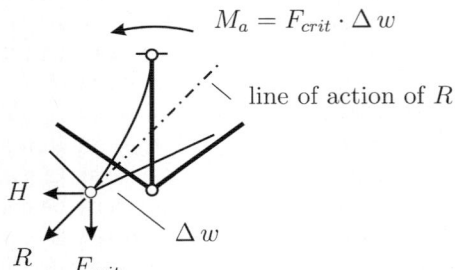

$$M_a = F_{crit} \cdot \Delta w$$

line of action of R

H

R F_{crit} Δw

Fig. 3.28. The line of action of R intersects the frame element only once

The difference between M_a^I and M_a^{II} is due to the force F

$$M_a^I - M_a^{II} = M_a \cdot \left(\frac{\varepsilon}{\tanh \varepsilon} - 1\right) = F \cdot l \cdot 1.0 \tag{3.139}$$

where we assumed that—as in the influence function—the tip of the element moves sideways by $\Delta w = l \cdot \tan \psi = l \cdot 1.0$ units. So that with

$$M_a = \frac{F \cdot l \cdot 1.0}{\varepsilon / \tanh \varepsilon - 1} \qquad S = -\frac{F}{2 \cdot \sin \alpha} \tag{3.140}$$

it follows

$$\frac{\tanh \varepsilon}{\varepsilon} = \cos (2 \cdot \alpha). \tag{3.141}$$

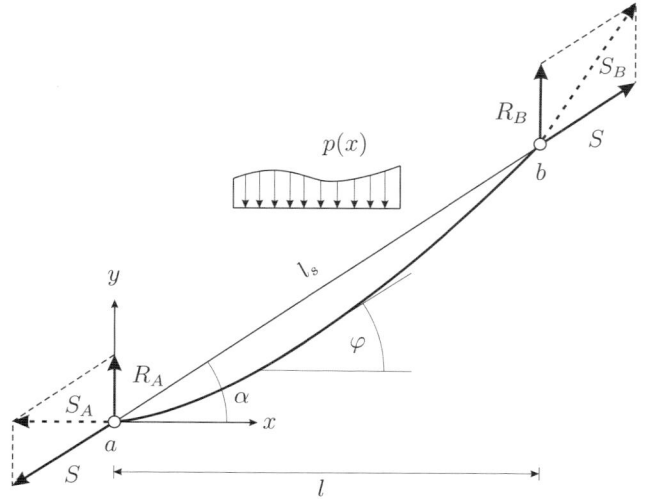

Fig. 3.29. Cable

Because of

$$\frac{\tanh \varepsilon}{\varepsilon} \leq 1 \qquad \Rightarrow \qquad \cos(2 \cdot \alpha) \leq 1 \qquad \Rightarrow \qquad \alpha \leq \frac{\pi}{4} \qquad (3.142)$$

the system is stable if $\alpha > \pi/4$ while for $\alpha \leq \pi/4$, the system will buckle if F exceeds the critical load $F_{crit} = F_{crit}(\alpha)$. But because the line of action of the resultant R intersects the buckled frame element only once (see Fig. 3.28) the buckling length approach is not applicable to this problem, [210].

3.6 Cables

In the first chapter a rope served us well to introduce the basic principles of the finite element technique. Here we want to add some more details to the analysis of ropes and cables.

The cable in Fig. 3.29 carries a vertical load $p(x)$ and is prestressed by a force S. If the angle of the chord is zero, $\alpha = 0$, the prestressing force S acts in horizontal direction, $S \cdot \cos \alpha = H = S \cdot 1$, as in chapter one.

In the following the bending moment $M(x)$ is the bending moment in a beam which carries the same load p and the forces R_A and R_B in Fig. 3.29 are identical with the support reactions of the beam.

Because the bending stiffness of the cable is neglected, $EI = 0$, the bending moment must be zero at any point x (let $p(x) = p$ a uniform load, though it can be any load)

$$\widehat{x} \sum M: \quad \underbrace{R_A \cdot x - \frac{p \cdot x^2}{2}}_{M(x)} - S \cdot \sin \alpha \cdot x + S \cdot \cos \alpha \cdot y = 0 \quad (3.143)$$

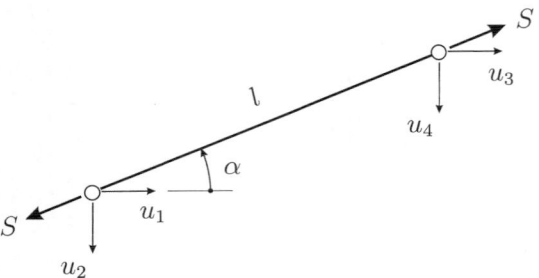

Fig. 3.30. Cable element

or

$$M(x) - S \cdot \sin \alpha \cdot x + S \cdot \cos \alpha \cdot y = 0 \qquad M(x) + H\,y = 0 \quad (3.144)$$

(the second equation applies to a horizontal cable) from which follows the equation for the shape of the cable

$$y(x) = \tan \alpha \cdot x - \frac{M(x)}{S \cdot \cos \alpha} \qquad y(x) = -\frac{M(x)}{H} . \qquad (3.145)$$

Note that the y-axis in Fig. 3.29 points upward. In this system the differential equation for the deflection has a positive sign

$$y(x) = -\frac{M(x)}{H} \quad \Rightarrow \quad H\,y''(x) = -M(x)'' = p(x) \qquad (3.146)$$

while if y points downward as in chapter one the sign is negative, $-H\,y''(x) = p(x)$.

The maximum tension in the rope is

$$S_B = \sqrt{(S\,\cos\alpha)^2 + (R_B + S\,\sin\alpha)^2} . \qquad (3.147)$$

With

$$y' = \tan\alpha - \frac{1}{S\,\cos\alpha} \cdot M' = \tan\alpha - \frac{V}{S \cdot \cos\alpha} \qquad (3.148)$$

the expression for the length s of the rope becomes

$$s = \int_0^l \sqrt{1 + (y')^2}\,dx = \int_0^l \sqrt{1 + (\tan\alpha - (\frac{V}{S \cdot \cos\alpha})^2)}\,dx . \quad (3.149)$$

The length s is equal to the length s_0 of the unstretched cable plus its elastic elongation

$$\Delta s = \int_0^s \varepsilon\,dx = \frac{S \cdot \cos\alpha}{EA} \int_0^l (1 + (y')^2)\,dx$$

$$= \frac{S \cdot \cos\alpha}{EA} \int_0^l (1 + (\tan\alpha - \frac{V}{S \cdot \cos\alpha})^2)\,dx \qquad (3.150)$$

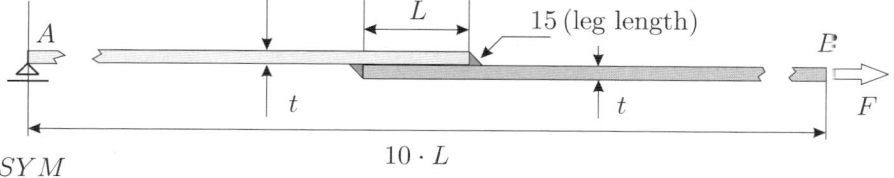

Fig. 3.31. Single lap joint

plus—eventually—thermal effects

$$s = s_0 + \Delta s + s_0 \, \alpha_T \, \Delta T \tag{3 151}$$

or

$$s = \int_0^l \sqrt{1 + (\tan \alpha - \frac{V}{S \cdot \cos \alpha})^2)} \, dx$$
$$= s_0 \, (1 + \alpha_T \, \Delta T) + \underbrace{\frac{S \cdot \cos \alpha}{EA} \int_0^l (1 + (\tan \alpha - \frac{V}{S \cdot \cos \alpha})^2) \, dx}_{\Delta s} \, .$$

This equation allows to determine for any type of load the prestressing force S and then with (3.145) the shape y of the cable.

A horizontal cable with cross section A assumes under gravity

$$g(x) = \frac{g}{\cos \varphi} \qquad g = \gamma \, A \, , \qquad \varphi = \arctan y' \tag{3.152}$$

the shape of a catenary (the origin, $x = 0, y = 0$, is the deepest point of the cable)

$$y(x) = \frac{H}{g} \left(\cosh \frac{x \cdot g}{H} - 1 \right) \tag{3.153}$$

or if we let $g(x) \simeq g$ as in a shallow cable the shape of a parabola (x and y as in Fig. 3.29)

$$y(x) = -\frac{g}{2 \, H} (l - x) \, x \, . \tag{3.154}$$

The two parts of the stiffness matrix of a cable element (2-D) (see Fig. 3.30) represent the longitudinal stiffness (EA/l) and the so-called geometric stiffness (S/l) of the cable, $c = \cos \alpha, s = \sin \alpha$,

$$\boldsymbol{K} = \frac{EA}{l} \begin{pmatrix} c^2 & -c \cdot s & -c^2 & c \cdot s \\ -c \cdot s & s^2 & c \cdot s & -s^2 \\ -c^2 & c \cdot s & c^2 & -c \cdot s \\ c \cdot s & -s^2 & -c \cdot s & s^2 \end{pmatrix} + \frac{S}{l} \begin{pmatrix} s^2 & c \cdot s & -s^2 & -c \cdot s \\ c \cdot s & c^2 & -c \cdot s & -c^2 \\ -s^2 & -c \cdot s & s^2 & c \cdot s \\ -c \cdot s & -c^2 & c \cdot s & c^2 \end{pmatrix} \, .$$
$$\tag{3.155}$$

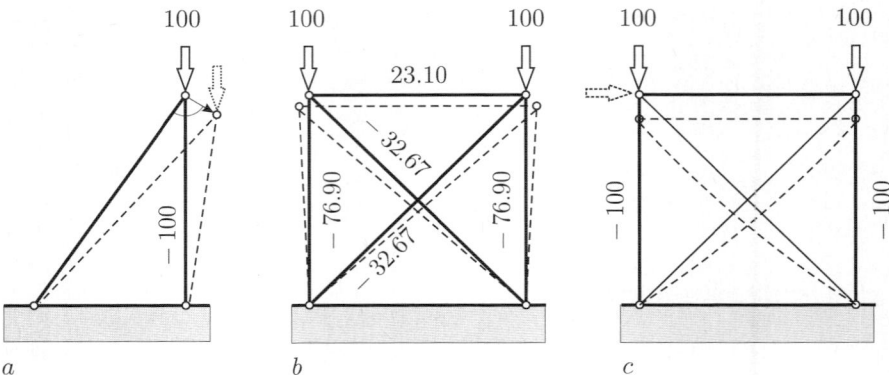

Fig. 3.32. Linear analysis: normal forces in **a)** a truss and **b)** a frame, **c)** a sway frame with cables

While a cable element is straight the real cable will sag and create its own prestress by its own weight. The effect can easily be described by a subdivision of the cable in different elements and an updated Lagrangian approach.

If we consider for example a horizontal cable with a length of 10.0 m, a cross section of 84 mm^2 and prestressed by a force of 1 kN the displacement of the straight prestressed rope under gravity load will be about 8 ‰ of the length. If we take into account the nonlinear effects of the displacements the tensile force will increase by a factor of 2 and the deformations and the eigen frequencies will change considerably, see Table 3.1.

Table 3.1. Linear and nonlinear analysis of a cable

	N	u	f_1	f_2	f_3
linear	1.0	83.36	1.928	3.809	5.596
nonlinear	1.9	43.95	2.374/3.468	4.690	6.890

Of special interest is the doubling of the first eigen value. Introduced by the deformation we have unsymmetric stiffness and different frequencies for a movement up and down. As far as we know these effects are not included automatically in finite element programs. So it is up to the user to detect such behaviour. There are many examples where those nonlinear effects (which are favorable in general) have not been included. Even for a simple problem as in Fig. 3.31 where a single lap joint was modeled with shell elements. The geometric non linear effects markedly reduced the eccentricity of the load and so the bending stress was reduced by a factor of 2.

Even if we stick to linear analysis the analysis of cable structures is anything but simple. Take for example a sway frame with bracings by two diagonal cables; see Fig. 3.32. The system is defined in general without prestress. In

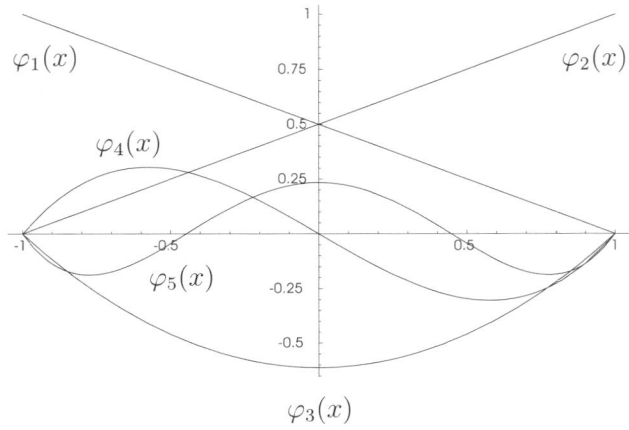

Fig. 3.33. Bar element extending from -1 to $+1$. The linear shape functions $\varphi_1(x), \varphi_2(x)$ and the hierarchical functions $\varphi_3(x), \varphi_4(x), \varphi_5(x)$

a linear analysis both elements will stay effective. The vertical payload will shorten the cables and both cables will thus carry compressive forces and consequently they will be deactivated in a nonlinear analysis. Then we might have a kinematic system, and it is only an additional horizontal force which will introduce a tensile stress in one of the cables and thus stabilize the whole structure. In the construction process the cables will be inserted at a time when some of the vertical loading has been already applied and for true cables a small prestress will be applied. To be able to cope with those problems a series of modifications and assumptions has to be made.

3.7 Hierarchical elements

In the p-method the element size is kept fixed while the degree p of the polynomial expansion is increased to order $2, 3, \ldots$. If the added shape functions are orthogonal in the sense of the strain energy product to the previous set of functions the new stiffness matrix is simply obtained by amending the previous matrix. Such elements are called hierarchical elements.

Consider a bar element $[-1, 1]$. To its two linear shape functions

$$\varphi_1(x) = \frac{1}{2}\,(1 - x) \qquad \varphi_2(x) = \frac{1}{2}\,(1 + x) \tag{3.156}$$

we add three shape functions which vanish at the end points of the element $[-1, +1]$ (see Fig. 3.33)

$$\varphi_3(x) = \frac{1}{\sqrt{6}}[-1 + \frac{1}{2}\,(-1 + 3\,x^2)] \qquad \varphi_4(x) = \frac{1}{\sqrt{10}}[-x + \frac{1}{2}\,(-3\,x + 5\,x^3)]$$

$$\varphi_5(x) = \frac{1}{\sqrt{14}}[\frac{1}{2}\,(1 - 3\,x^2) + \frac{1}{8}\,(3 - 30\,x^2 + 35\,x^4)]\,. \tag{3.157}$$

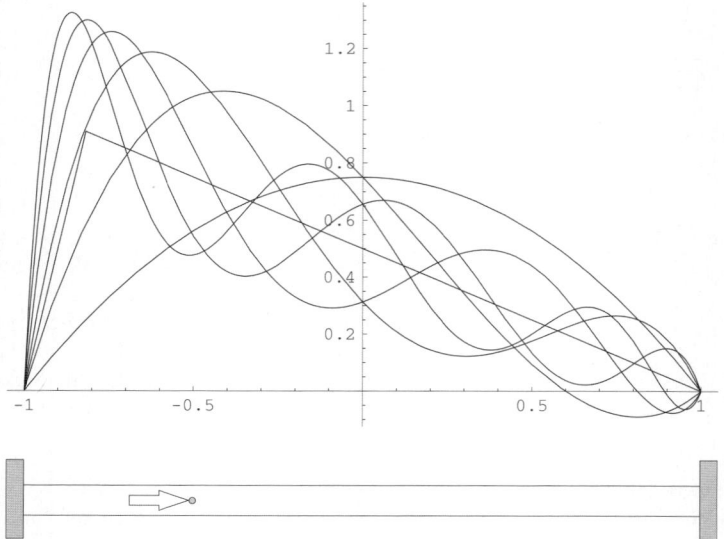

Fig. 3.34. Approximation of the Green's function for the support reaction at the left end with linear elements and hierarchical elements (increasing order of $p = 2 \ldots 10$)

These new functions are orthogonal to the first two functions φ_1 and φ_2

$$k_{1j} = a(\varphi_1, \varphi_j) = k_{2j} = a(\varphi_2, \varphi_j) = 0 \qquad j = 3, 4, 5 \qquad (3.158)$$

and they are also mutually orthogonal

$$k_{ij} = a(\varphi_i, \varphi_j) = \delta_{ij} \cdot EA \qquad i, j = 3, 4, 5 \qquad (3.159)$$

so that the amended matrix is simply, [26] p. 252,

$$\boldsymbol{K} = \frac{EA}{2} \begin{bmatrix} 1 & -1 & 0 & 0 & 0 \\ -1 & 1 & 0 & 0 & 0 \\ 0 & 0 & 2 & 0 & 0 \\ 0 & 0 & 0 & 2 & 0 \\ 0 & 0 & 0 & 0 & 2 \end{bmatrix}. \qquad (3.160)$$

For an application we consider a bar $[-1, +1]$ fixed at the left end and stretched by uniform forces $p = 1$

$$-EAu''(x) = 1 \qquad u(-1) = 0 \qquad N(1) = EA\,u'(1) = 0. \qquad (3.161)$$

Because the left end is fixed $u_1 = 0$ but the other four u_i are unknown. The equivalent nodal forces are

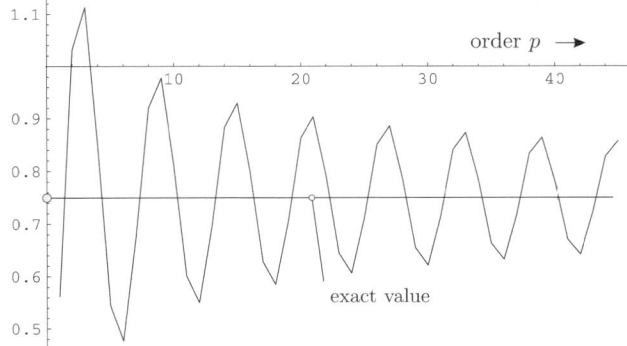

Fig. 3.35. In the p-method the support reaction oscillates considerably with the order p of the shape functions

$$f_2 = \int_{-1}^{+1} 1 \cdot \varphi_2(x)\,dx = 1 \qquad f_3 = \int_{-1}^{+1} 1 \cdot \varphi_3(x)\,dx = -\sqrt{\frac{2}{3}}$$

$$f_4 = \int_{-1}^{+1} 1 \cdot \varphi_4(x)\,dx = 0 \qquad f_5 = \int_{-1}^{+1} 1 \cdot \varphi_5(x)\,dx = 0. \qquad (3.162)$$

The solution of the system

$$\boldsymbol{K} = \frac{EA}{2} \begin{bmatrix} 1 & 0 & 0 & 0 \\ 0 & 2 & 0 & 0 \\ 0 & 0 & 2 & 0 \\ 0 & 0 & 0 & 2 \end{bmatrix} \begin{bmatrix} u_2 \\ u_3 \\ u_4 \\ u_5 \end{bmatrix} = \begin{bmatrix} f_2 \\ f_3 \\ f_4 \\ f_5 \end{bmatrix} \qquad (3.163)$$

is (we let $EA = 1$)

$$\boldsymbol{u} = \{2, -\sqrt{2/3}, 0, 0\}^T \qquad (3.164)$$

so that

$$u_h(x) = 2 \cdot \varphi_2(x) - \sqrt{\frac{2}{3}} \cdot \varphi_3(x) = \frac{3}{2} + x - \frac{x^2}{2} \qquad (3.165)$$

which is the exact solution.

Note that the sum of the equivalent nodal forces

$$\sum_{i=1}^{5} f_i = \int_{-1}^{+1} p \cdot \underbrace{(\varphi_1 + \varphi_2 + \varphi_3 + \varphi_4 + \varphi_5)}_{\neq 1}\,dx = 1.1835$$

$$\neq 2.0 = \int_{-1}^{+1} p\,dx = \int_{-1}^{+1} p \cdot \underbrace{(\varphi_1 + \varphi_2)}_{=1}\,dx = f_1 + f_2 \qquad (3.166)$$

is not the sum of the applied load (2.0) because the extended set of shape functions does not form a partition of unity

$$\sum_{i=1}^{5} \varphi_i(x) = 0.62 - 0.79\,x - 0.79\,x^2 + 0.79\,x^3 + 1.17\,x^4 \qquad (3.167)$$

—simply said the sum is not 1 at each point x. So in checking the global equilibrium condition in the p-method we must restrict the count to the original f_i.

Point loads and hierarchical elements

The p-method will improve the accuracy considerably if the exact solution is smooth but it can run into difficulties in the presence of point loads, [238] p. 196.

The model problem is a bar $[-1, +1]$ that is fixed at both ends and is subjected to a horizontal point load $P = 1$ at the quarter point $x = -0.5$. We study the support reaction at the left end of the bar, at $x = -1$. The exact Green's function of the support reaction is the straight line $G_1(x) = -0.5\,x + 0.5$, dropping from $+1$ at the left end to zero at the right end of the bar. The triangle in Fig. 3.34 is the best approximation with 10 linear elements. For our purposes it is perfect because the FE Green's function G_1^h is exact at $x = -0.5$ so that the linear model gives the correct answer, $N_h(0) = N(0) = 0.75$.

If we solve the same problem with the p-method — just one element but different orders p of shape functions (P_i = Lagrange polynomial)

$$\varphi_i(x) = \frac{P_{i+1}(x) - P_{i-1}(x)}{\sqrt{2}\sqrt{2(i+1) - 1}} \qquad i = 1, 2, 3, \ldots p \qquad (3.168)$$

then the support reaction oscillates considerably (see Fig. 3.35) because in the p-method the approximate Green's functions G_1^h tend to wobble; see Fig. 3.34. Note that the exact solution $u(x)$ does not lie in the trial space V_h of the p-method because all $\varphi_i(x)$ have continuous first-order derivatives and also the Green's function $G_1(x)$ is not contained in V_h—linear functions are excluded because of the boundary conditions—so we must expect an error in the support reaction.

If we would place a node where the single force is applied the solution would lie in V_h so the problem could have readily been resolved — in this case — but evidently care must be taken in the presence of point loads.

Remark 3.2. We add some details. The stiffness matrix in the foregoing problem is $EA \times I$ (the unit matrix), the equivalent nodal forces for G_1^h are $f_i = N(\varphi_i)(-1) = EA\,\varphi_i'(-1)$ so that the nodal values u_i of the Green's function G_1^h are the derivatives φ_i' of the shape functions at $x = -1$ and thus

$$G_1^h(x) = \sum_{i=1}^{p} \varphi_i'(-1)\,\varphi_i(x). \qquad (3.169)$$

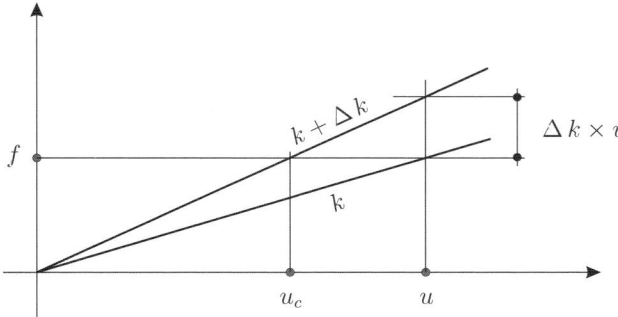

Fig. 3.36. Change in the stiffness of a spring

If the load p is uniformly distributed all is well. In the influence integral (G_1^h, p) only the integral of the first term of the series (3.169) is not zero and because the slope of φ_1 at $x = -1$ and the integral of φ_1 are well tuned the result

$$N_h(-1) = \int_{-1}^{+1} G_1^h(x)\, p\, dx = p\, \varphi_1'(-1) \int_{-1}^{+1} \varphi_1(x)dx$$
$$= p \cdot 0.816 \cdot 1.22474 = p \cdot 1.0 \tag{3.170}$$

is exact—the total load is $p \cdot 2.0$.

3.8 Sensitivity analysis

In sensitivity analysis we want to predict how changes in a structure affect the internal actions in a structure. The essential tool for this analysis are, as we want to show, the Green's functions.

If the stiffness of a member changes then the response of a structure will change too, as in a spring, $k\, u = f$; see Fig. 3.36. The response of the spring to the unit load $f = 1$ is $G = 1/k$ and an increase in the stiffness, $k + \Delta k$, changes the response to $G_c = 1/(k + \Delta k)$ so that for any load f the response before and after the change is

$$u = \frac{1}{k}\, f \qquad u_c = \frac{1}{k + \Delta k}\, f\,. \tag{3.171}$$

If we do a Taylor expansion of the updated "Green's function"

$$\frac{1}{k + \Delta k} = \frac{1}{k} - \frac{1}{k^2}\, \Delta k + \ldots \tag{3.172}$$

then we better understand how a structure reacts to such changes

$$u_c \simeq \left[\frac{1}{k} - \frac{1}{k}\frac{\Delta k}{k}\right] f = u - \frac{1}{k}\, \underbrace{\Delta k \times u}_{force}\,. \tag{3.173}$$

$$\Delta M \simeq \frac{\Delta EI}{EI} \int_0^l \frac{M\,M_2}{EI}\,dy = d(w, G_2) = 0.4\,\text{kNm}$$

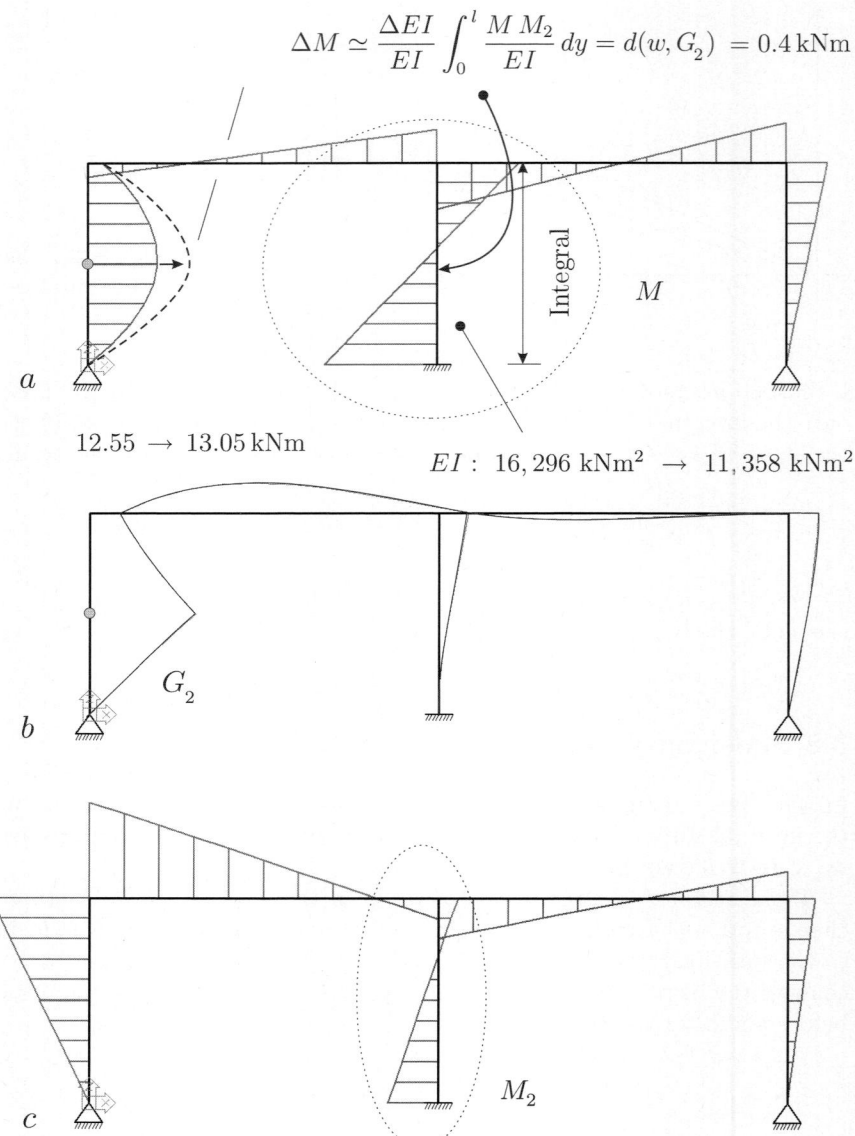

$12.55 \rightarrow 13.05\,\text{kNm}$

$EI:\ 16,296\ \text{kNm}^2 \rightarrow 11,358\ \text{kNm}^2$

Fig. 3.37. A drop in the bending stiffness in the central member will lead to an increase in the bending moment in the leftmost member **a)** original bending moment distribution M **b)** influence function G_2 for $M(x)$ **c)** bending moment M_2 of G_2

The stiffening of the spring, $k \to k + \Delta k$, makes that the original u produces too large a reaction $(k + \Delta k)\, u = f + \Delta k\, u$ in the spring and to annihilate this excess force the opposite movement $\Delta u = \Delta k\, u / k$ (approximately) must be superimposed on u.

In stiffness matrices the Taylor expansion has the following form, [75],

$$(\boldsymbol{K} + \boldsymbol{\Delta K})^{-1} = \boldsymbol{K}^{-1} - \boldsymbol{K}^{-1} \boldsymbol{\Delta K}\, \boldsymbol{K}^{-1} + \dots \qquad (3.174)$$

and the analogue between (3.173) and

$$\boldsymbol{u}_c = (\boldsymbol{K} + \boldsymbol{\Delta K})^{-1} \boldsymbol{f} \simeq \boldsymbol{u} - \boldsymbol{K}^{-1} \boldsymbol{\Delta K}\, \boldsymbol{u} \qquad (3.175)$$

is evident.

Let us apply some heuristic arguments to this equation

$$\boldsymbol{u}_c - \boldsymbol{u} = -\boldsymbol{K}^{-1} \boldsymbol{\Delta K}\, \boldsymbol{u} \qquad (3.176)$$

and let us consider a beam with a change $EI \to EI + \Delta EI$. If we simply give (3.176) an integral form then we obtain

$$w_c(x) - w(x) = - \underbrace{\int_0^l G_0(y,x)}_{\boldsymbol{K}^{-1}}\, \underbrace{\Delta EI\, \frac{d^4}{dy^4}}_{\boldsymbol{\Delta K}}\, \underbrace{\int_0^l G_0(y,z)\, p(z)\, dz}_{\boldsymbol{u}}\, dy \quad (3.177)$$

which even makes sense if the change $EI \to EI + \Delta EI$ is uniform because then

$$w_c(x) - w(x) = - \int_0^l G_0(y,x)\, \frac{\Delta EI}{EI} p(y)\, dy \qquad (3.178)$$

that is a negative change, $\Delta EI < 0$, is like an increase in the load that the beam must carry. And a local change obviously means a local increase in p.

Single values

To trace the change of single values $J(u)$ $(= u(x), \sigma(x), \dots)$ we can adopt the approach in Sect. 1.27 p. 113

$$J(u_c) - J(u) = -d(u, G_c) = -d(u_c, G) \simeq -d(u, G) \qquad (3.179)$$

where $d(.,.)$ is the symmetric term which we add to the strain energy product of the structure, $a(.,.) + d(.,.)$, to incorporate the change in the stiffness.

In the following we will consider some typical examples.

Cracks in a beam, $EI \rightarrow EI + \Delta EI$

If a single member, $[x_1, x_2]$, in a frame cracks, $EI \rightarrow EI + \Delta EI$, then the change in the bending moment $M(x)$ at a point x—which must not lie on the cracked element, it can be any point of the frame—is

$$M_c(x) - M(x) = -\int_{x_1}^{x_2} \Delta EI \, w_c'' \, G_2'' \, dy = -\int_{x_1}^{x_2} \frac{\Delta EI}{EI} \frac{M_c \, M_2}{EI} \, dy$$

$$(3.180)$$

where integration is done along the cracked member—*only*. The deflection w_c is the deflection of the cracked beam and the influence function G_2 (= deflection of the member under the action of the Dirac delta δ_2 at x) is from the uncracked frame. If the cracks are not too large we may approximate w_c by the original deflection w of the frame element. We tested this with the frame in Fig. 3.37 where the bending stiffness in the central member, $EI = 16,296$ kNm^2, dropped by nearly $1/3$ to $EI + \Delta EI = 11,358$ kNm^2. According to (3.180) the change of the bending moment M in the leftmost member should be approximately

$$\Delta M \simeq -\frac{\Delta EI}{EI} x \int_0^l \underbrace{\frac{M \, M_2}{EI}}_{ante} \, dy = 0.4 \, \text{kNm} \qquad (3.181)$$

while the true change is $\Delta M = 0.5$ kNm.

Rule #1 : If the stiffness changes in a part $[x_1, x_2]$ of a frame then the change in any quantity $\partial^i w$ (w, w', M, V) at any point x is

$$\partial^i w_c(x) - \partial^i w(x) \simeq -\int_{x_1}^{x_2} \frac{\Delta EI}{EI} \frac{M \, M_i}{EI} \, dx \qquad (3.182)$$

where M_i is the bending moment of the influence function G_i for $\partial^i w$ and M is the bending moment due to the load p.

Change in an elastic support, $k \rightarrow k + \Delta k$

If the stiffness of a spring changes, $k \rightarrow k + \Delta k$, then the strain energy product of a beam transforms as follows:

$$a(w, \hat{w}) + k \, w(l) \, \hat{w}(l) \quad \Rightarrow \quad a(w, \hat{w}) + (k + \Delta k) \, w(l) \, \hat{w}(l) \qquad (3.183)$$

so that in this case $d(w, \hat{w}) = \Delta k \, w(l) \, \hat{w}(l)$ and we have:

Rule #2 : If the stiffness of a spring changes then the change in any quantity $\partial^i w$ (w, w', M, V) at any point x is

$$\partial^i w_c(x) - \partial^i w(x) \simeq -\Delta k \cdot G_i(l, x) \cdot w(l) \qquad (3.184)$$

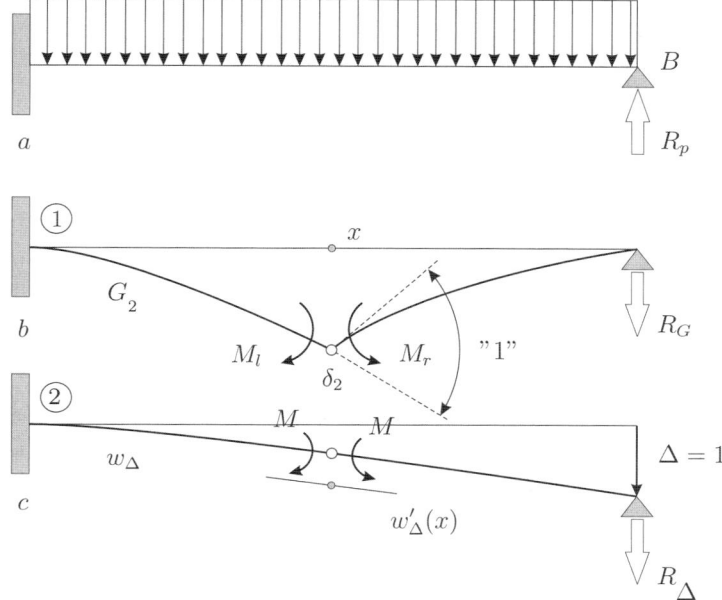

Fig. 3.38. Failure of support B and study of the changes in $M(x)$

where G_i is the movement in the spring due the action of the Dirac delta δ_i (influence function for $\partial^i w$) and $w(l)$ is the movement of the spring due to the load p.

Of course if the increase is too large, say if $\Delta k \to \infty$, then only the exact formula

$$J(w_c) - J(w) = -d(w, G_c) = -\Delta k \cdot G_c(l, x) \cdot w(l) = -R_G \cdot w(l)$$
$$(3.185)$$

can predict the change correctly. Note that $R_G = \infty \cdot 0$ (or very nearly), which is the support reaction in the spring due to the Dirac delta at x, tends to a value corresponding to a fixed support.

Loss of a support

How we argue if the support of a structure fails—here the support B of the beam in Fig. 3.38—may be illustrated by studying the consequences for the bending moment M at the mid-point x. The influence function for the bending moment $M(x)$ is the response of the beam to the action of a Dirac delta, δ_2 at x; see Fig. 3.38 b.

The support reaction R_G due to δ_2 equals the bending moment $M(x)$ at x if the support B settles by one unit length. This follows from Betti's theorem

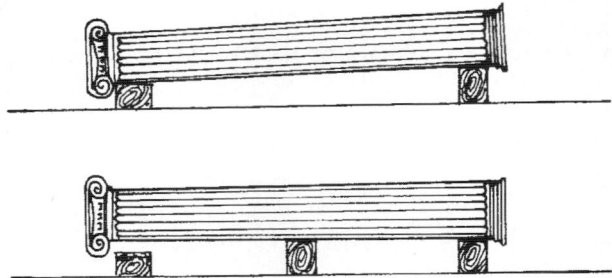

Fig. 3.39. The thoughtful engineer, [41]

(we ignore the actions on the left side at the fixed ends because they contribute no work)

$$W_{1,2} = \underbrace{M_l\, w'_\Delta(x_-) - M_r\, w'_\Delta(x_+)}_{=\,0} + R_G \cdot 1 = +R_G \cdot 1$$

$$= M(x) \cdot 1 + R_\Delta \cdot 0 = W_{2,1}\,. \tag{3.186}$$

The two moments M_l and M_r are of the same size but rotate in opposite directions and because the slope $w'_\Delta(x)$ is continuous at x the work done by the pair M_l and M_r is zero, so that indeed $M(x) = R_G$.

Now if the support B is removed in the load case p then to compensate for this loss the support reaction R_p must be applied in the opposite direction. This will cause the deflection $w(l) = R_p\, l^3/(3\,EI)$ and hence the change in the bending moment is

$$\Delta M = -R_G \cdot w(l) = \underbrace{-R_G \cdot R_p}_{ante} \cdot \frac{l^3}{3\,EI} \qquad \leftarrow \frac{1}{k_S}\,. \tag{3.187}$$

Rule #3 : If a support fails then the change in any quantity $\partial^i w$ (w, w', M, V) at any point x is

$$\partial^i w_c(x) - \partial^i w(x) = -R_G \cdot R_p \cdot \frac{1}{k_S} \tag{3.188}$$

where R_G is the support reaction due the action of the Dirac delta δ_i (influence function for $\partial^i w$), R_p is the support reaction due to the load p and k_S is the stiffness of the structure in the direction of the missing support.

Let us apply this result to the problem of the thoughtful engineer which Galileo mentions in his *discorsi*, [41]: Il mecanico places a heavy marble column on an additional support (see Fig. 3.39). But in so doing he unwillingly lifts the left end of the column from its support and so instead of lowering the bending moment by a factor of four as intended his maneuver effects only a change in the sign of the bending moment $M = M_{max}$.

Because if the weight of the column is $g = 10$ kN/m and its length $l = 8$ m then

$$\underbrace{M = \frac{10 \cdot 8^2}{8} = 80}_{2\,supp.} \qquad \underbrace{M = -\frac{10 \cdot 4^2}{8} = -20}_{3\,supp.} \qquad \underbrace{M = -\frac{1}{2} 10 \cdot 4^2 = -80}_{cantilever}$$

(3.189)

so nothing is gained.

In the transition from the 2-span beam to the cantilever the change of the bending moment is according to (3.188)

$$M_c(x) - M(x) = -R_G \cdot R_p \cdot \frac{1}{k_S} = -8,480 \cdot 15 \cdot \frac{1}{2,120} = -60 \,\text{kNm}$$

(3.190)

which agrees with (3.189). A point load $P = 1$ kN at the end of the cantilever beam effects the deflection 0.4708 mm and so $k_S = 1/0.4708 \cdot 10^3 = 2,120$ kN/m. The force $R_G = 8,480$ is the support reaction in the 2-span beam from the Dirac delta δ_2 ($\equiv \tan \varphi = 1$).

If a rigid support yields

If a rigid support turns soft ($\infty \to k_S$) the logic is the same. It is only that the end of the beam is now working against the bending stiffness of the beam ($3\,EI/l^3$) *and* the soft support with its relic stiffness k_S so that—the two springs are working in parallel—the resulting stiffness is

$$k = \frac{3\,EI}{l^3} + k_S$$

(3.191)

and consequently

$$\Delta M = -R_G \cdot w(l) = \underbrace{-R_G \cdot R_p}_{ante} \cdot \frac{1}{k}$$

(3.192)

where R_G and R_p have the same meaning as before. It is evident that this logic applies to all types of supports and to internal nodes as well. If a previously fixed support begins to rotate then (3.192) would have to be replaced by

$$\Delta M = -M_G \cdot \varphi(l) = \underbrace{-M_G \cdot M_p}_{ante} \cdot \frac{1}{k_\varphi}$$

(3.193)

where M_G is the fixed end reaction in the load case δ_i and M_p is the fixed end reaction in the load case p and k_φ is the rotational stiffness of the support.

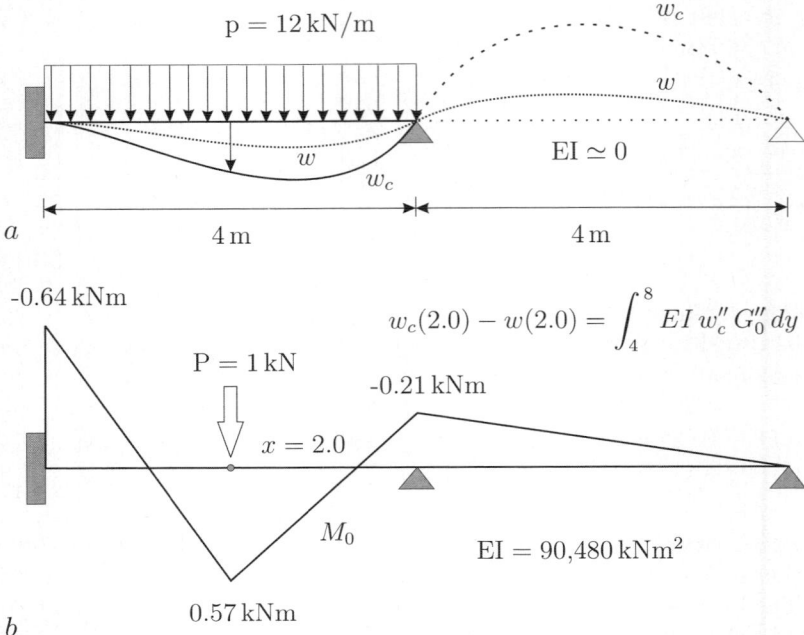

Fig. 3.40. Loss of a member—the energy needed to bend a spline into the shape w_c in the second span is indicative of the effect the loss will have on the structure **a)** deflection w and w_c **b)** bending moment M_0 of the Green's function G_0

Rule #4 : If a rigid support softens then the change in any quantity $\partial^i w$ (w, w', M, V) at any point x is

$$\partial^i w_c(x) - \partial^i w(x) = -R_G \cdot R_p \cdot \frac{1}{k} \tag{3.194}$$

where R_G is the support reaction due the action of the Dirac delta δ_i (influence function for $\partial^i w$), R_p is the support reaction due to the load p and k is the relic stiffness of the structure in the direction of the support.

Loss of a frame member

A complete loss of a frame member corresponds to $EI = 0$ or in terms of the equation $EI + \Delta EI = 0$ means $\Delta EI = -EI$ so that in the end

$$-d(w_c, G) = -\int_0^l \Delta EI\, w_c''\, G''\, dy = \int_0^l EI\, w_c''\, G''\, dy \tag{3.195}$$

where w_c is the shape of the member when it has lost all the stiffness; see Fig. 3.40 a. Its curvature $-w_c''$ is the quotient of the bending moment M_c in the member and the residual stiffness $EI + \Delta EI$

$$- w_c'' = \lim_{\Delta EI \to -EI} \frac{M_c}{EI + \Delta EI} \equiv \frac{0}{0}. \tag{3.196}$$

The deflection w_c in the frame member can be found as follows: remove the member and let the frame find its new equilibrium position. Next bridge the gap by a spline which attaches *seamlessly* to the two displaced nodes on both sides of the gap and which has the stiffness EI. This is correct, not $EI = 0$, because we need the bending moment $M_c = -EI\, w_c''$ in the second equation (3.195). Also the spline is not really attached to the nodes, rather prestressing forces at the ends of the spline keep the element in the shape w_c. If we would reattach the member and release the prestressing forces in the member then the structure immediately would snap back into its original shape.

The effect $EI = 0$ is cared for by releasing the two nodes that is by letting the structure find its equilibrium position without the member. After that we calculate how much the member must be prestressed to position its ends opposite to the displaced nodes of the structure.

● The energy needed for this maneuver is an indication of how important the element is for the structure that is how much $J(w)$ will change if the member fails.

The two-span beam in Fig. 3.40 loses its member in the second span. How much will this affect the deflection at the center of the first span? The rotation of the beam at the end of the first span is $w_c'(4.0) = -0.00018$ so that the spline w_c in the next span is the solution of the following problem

$$EI\, w_c^{IV} = 0 \qquad w_c(4.0) = w_c(8.0) = M_c(8.0) = 0 \qquad w_c'(4.0) = -0.00018\,. \tag{3.197}$$

The bending moment distribution of this spline is linear, $M_c(4.0) = -12.24$ kNm and $M_c(8.0) = 0$, so that according to (3.195)

$$w_c(2.0) - w(2.0) = \int_4^8 EI\, w_c''\, G_0''\, dy = \int_4^8 \frac{M_c\, M_0}{EI}\, dy$$
$$= \frac{1}{3}(-0.21) \cdot (-12.24) \cdot 4.0 \cdot \frac{1}{90,480} = 0.04\,\text{mm} \tag{3.198}$$

which agrees with the exact result

$$w_c(2.0) - w(2.0) = 0.18\,\text{mm} - 0.14\,\text{mm} = 0.04\,\text{mm}\,. \tag{3.199}$$

Of course (3.195) is purely theoretical because the spline w_c requires the calculation of the displacements and rotations of the two neighboring nodes *after* the member has been removed but then one can compare the two systems directly.

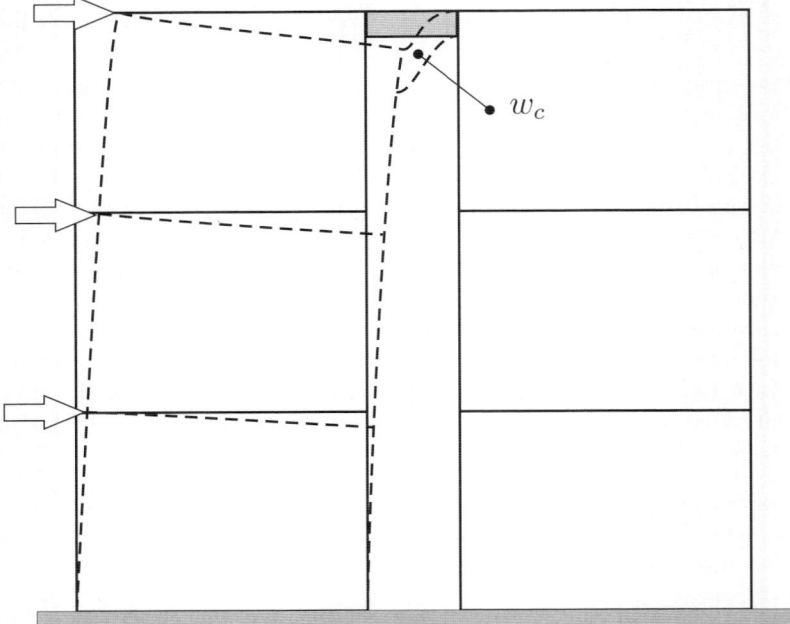

Fig. 3.41. Release of the connection between two frames. The energy needed to reconnect the spline w_c with the two released nodes is an indication of how much influence the stiff element exerts on the structure

The idea to substitute for $w_c \simeq w$ the original deflection would be in most cases too crude an approximation—here 0.02 mm instead of the exact 0.04 mm—but it could suffice to signal the trend into which direction things will be moving if the member fails.

Note that

$$|d(w_c, G)| \leq d(w_c, w_c) \cdot d(G, G) = \int_0^l EI\,(w_c'')^2\,dx \cdot \int_0^l EI\,(G'')^2\,dx$$

(3.200)

so that the product of the strain energy of the spline w_c and the Green's function G in the member is an upper bound for the change $J(w_c) - J(w)$. If either of these two terms is small the change will not be very pronounced.

On the other hand imagine a short bolt which ties two structures together and forces the structures to move in unison; see Fig. 3.41. Assume this bolt fails—technically a very stiff, very short frame element—then the released nodes will probably undergo large displacements and rotations and so the spline w_c which later must reconnect with the two released nodes will have to assume a serpent like shape and to stretch a long way. Consequently the

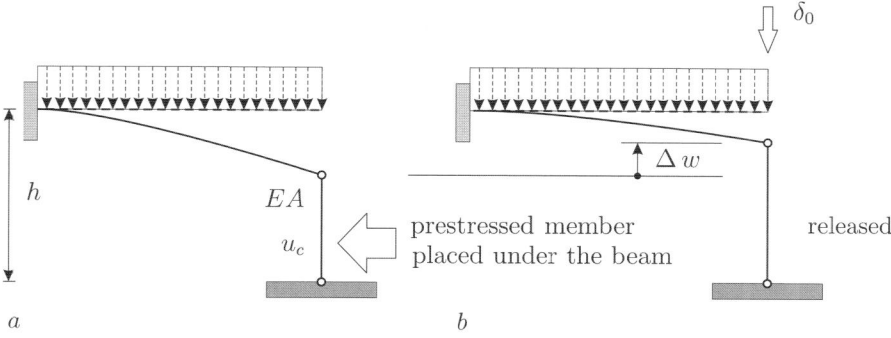

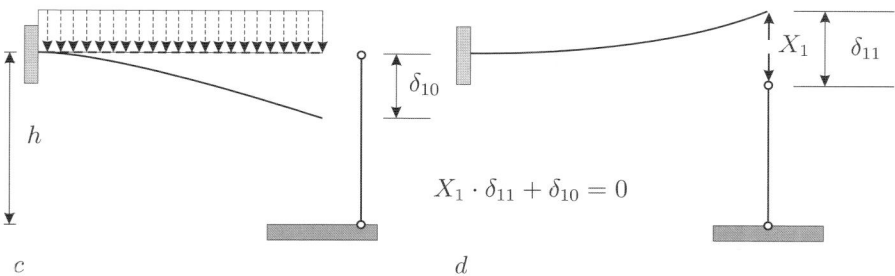

Fig. 3.42. A prestressed member **a)** is placed under the end of the cantilever beam and **b)** then released. This is the same logic as in the case of a loss of a member—only in the other direction. If the prestressing forces are released the element assumes its original shape **c)** and the gap δ_{10} must be closed **d)** by the redundant $X_1 = -\delta_{10}/\delta_{11}$.

strain energy in this very stiff spline will be quite large, that is the change (3.195) will no longer be negligible.

Rule #5 : If a member $[x_1, x_2]$ fails, $EI \to 0$, then the change in any quantity $\partial^i w$ (w, w', M, V) at any point x is

$$\partial^i w_c(x) - \partial^i w(x) = \int_{x_1}^{x_2} EI\, w_c''\, G_i''\, dy \qquad (3.201)$$

where w_c is the shape the member assumes if it is drained of all its stiffness and G_i is the influence function for $\partial^i w$.

Adding a member to a structure

All this can be applied in the opposite direction as well: if the end of a cantilever beam is placed on a vertical pin-jointed frame element (see Fig. 3.42)

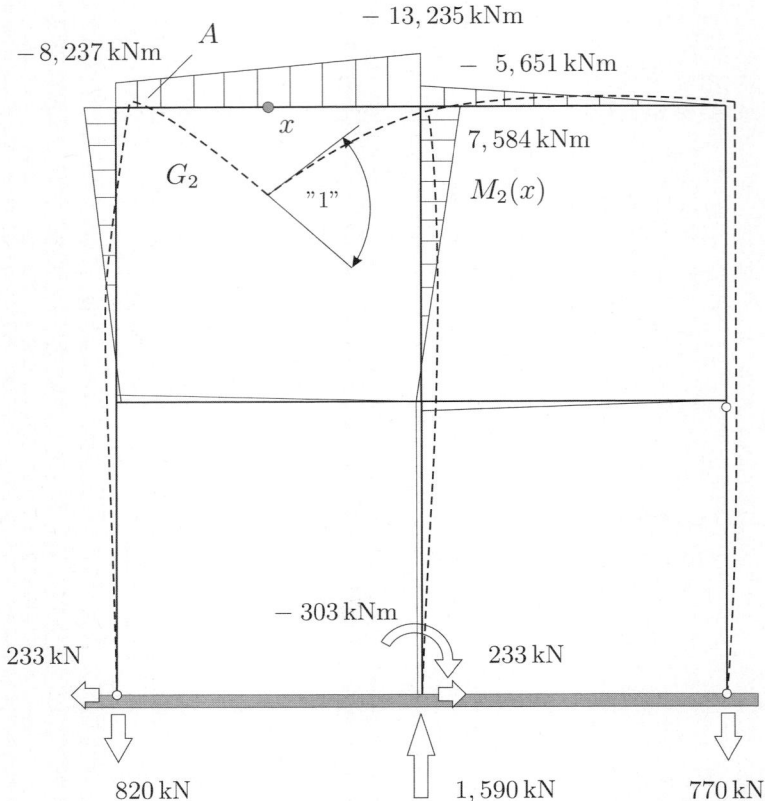

Fig. 3.43. Footprint of a Green's function or sensitivity map: which changes do influence $M(x)$ the most? $EI = 90,625$ kNm², members $0.5 \cdot 0.3 \cdot 6.0$ m

this can be done in two ways: (1) the length of the new frame element is $h - w(l)$ where h is the height of the cantilever beam before the load was applied or (2) the length of the new element is h. In the first case the beam would only benefit from the support when it carries additional loads. In the second case the element first must be compressed so that it fits under the beam—this would be the shape u_c (in axial direction). Next the end of the element is released so that it presses the cantilever beam upward. The change in any quantity of the beam would then be

$$J(e_w) = -\int_0^l EA\,G_i'\,u_c'\,dy = -\int_0^l \frac{N_i\,N_c}{EA}\,dy \qquad (3.202)$$

where N_i is the normal force in the frame element due to the Green's function for $\partial^i w$ and N_c is the prestressing force. Not surprisingly all quantities w_c, w_c', M_c, V_c would be just as large as if the additional support had been present from the start. (In some sense the technique to prestress an element—so that it fits into the statically determinate structure—and then to release it is the force method in disguise).

So we better understand the role of the splines w_c in Fig. 3.195 and Fig. 3.41. They act like batteries, they provide the energy to restore the original shape of the structure. By removing an element from the frame the strain energy of the structure increases which means that additional energy can be gained from the sagging load. To restore the original shape we must give this energy back to the structure, that is the load must be pushed up to its previous level. The energy needed for this maneuver comes from the prestressing forces in the splines.

The case of the missing member

If in the construction phase of a structure a member has been omitted unwillingly the defect can be amended by prestressing the member so that it fits into the deformed structure and by then releasing the prestressing force. The shape of the structure is then the same as if the member would have been present from the beginning.

We follow a similar procedure when we replace structural members, say, the rusty old pier of a bridge. When the hydraulic jacks on top of the new pier have lifted the bridge just enough to release the old pier the new pier is prestressed by precisely the right amount to replace the old pier without affecting the stress distribution in the bridge.

The important point is, of course, that the unstressed member has the correct length. So eventually the sequence of the single construction stages must be taken into account.

Summary

A change in any quantity $J(u) \to J(u) + \Delta J(u)$ means an increase or decrease in the Dirac energy. In the case of a support this means

$$\Delta J(u) \cdot 1 = \text{force} \cdot \text{displacement increment} . \qquad (3.203)$$

The force is the support reaction R_G due to the Dirac delta and the displacement is the incremental (additional) movement of the support induced by the change of the stiffness. So if the support of a node fails completely (or partially) and the load p presses the node downward by $w(x_i)$ *additional* units then the change is $R_G \cdot w(x_i)$.

In (3.187) the original movement is zero (rigid support). With the loss of the support the system becomes a cantilever beam which deflects at its end by R_p/k units; this is the deflection increment.

In (3.192) the original movement is zero too. But this time the support does not yield completely and so R_p must work against the stiffness of the node plus the relic stiffness of the support so that the deflection increment will be somewhat lower.

The sensitivities in a structure are encapsulated in the "footprints" of the Green's functions; see Fig. 3.43. In this figure are plotted the bending moment distribution (M_2) and the support reactions of the influence function for $M(x)$, which is the bending moment at the center of the upper left horizontal member. From this figure we can learn for example that if the central support yields by 1 mm in vertical direction then the change in $M(x)$ is $-1,590$ kN·0.001 m = - 1.59 kNm. Or a slip of 0.001 m in horizontal direction would effect a change of $233 \cdot 0.001$ m = 0.23 kNm. And a rotation tan $\varphi = 0.001$ of the support would change $M(x)$ by $-303 \cdot 0.001$ = -0.3 kNm.

But also the bending moments M_2 of the influence function tell a story. If at the upper left node, on the side A of the node, the concrete cracks—so that the fixed end begins to resemble a rotational spring—then for any differential degree φ of rotation between the horizontal member and the vertical member the change in $M(x)$ is $8,237$ kNm $\cdot$ tan φ.

Safety of structures

So the failure of a support is critical for the safety of a structure if (1) the support reaction R_p is large because then the incremental movement $w(l)$ will be large too (probably) *and* if (2) the support reaction R_G in the load case δ_i (= influence function for the internal actions $M(x)$ or $V(x)$ at x) is large, see (3.187).

Hence, to assess the safety of a structure we could adopt the following strategy: (1) Find the points x where the internal actions, $M(x), V(x)$, etc. attain their maximum values. (2) Calculate the influence functions for $M(x)$ and $V(x)$. (3) Study how changes in the stiffness of the structure would influence the distribution of the influence function and therewith the maximum values of $M(x)$ and $V(x)$. The same can be done with the support reactions.

Of course a seasoned structural engineer needs no computer to see where the weak points of a structure are but this is not the point here. Rather this technique eventually could allow us to assess the safety of a structure computationally.

4. Plane problems

We start with an elementary example to explain the FE technique in detail.

4.1 Simple example

The cantilever plate in Fig. 4.1 is subject to an edge load and subdivided into four *bilinear elements* of length l and width h.

Each of the four nodes of an element has two degrees of freedom u_i^e, so that the stiffness matrix $\boldsymbol{K}^e$ is of size 8×8. If Poisson's ratio is zero, $\nu = 0$, then the matrix is

$$
\begin{bmatrix}
\frac{l}{6h}+\frac{h}{3l} & \frac{l}{12h}-\frac{h}{3l} & -\frac{l}{12h}-\frac{h}{6l} & -\frac{l}{6h}+\frac{h}{6l} & \frac{1}{8} & -\frac{1}{8} & -\frac{1}{8} & \frac{1}{8} \\
 & \frac{l}{6h}+\frac{h}{3l} & -\frac{l}{6h}+\frac{h}{6l} & -\frac{l}{12h}-\frac{h}{6l} & \frac{1}{8} & -\frac{1}{8} & -\frac{1}{8} & \frac{1}{8} \\
 & & \frac{l}{6h}+\frac{h}{3l} & \frac{l}{12h}-\frac{h}{3l} & -\frac{1}{8} & \frac{1}{8} & \frac{1}{8} & -\frac{1}{8} \\
 & & & \frac{l}{6h}+\frac{h}{3l} & -\frac{1}{8} & \frac{1}{8} & \frac{1}{8} & -\frac{1}{8} \\
 & & & & \frac{l}{3h}+\frac{h}{6l} & \frac{l}{6h}-\frac{h}{6l} & -\frac{l}{6h}-\frac{h}{12l} & -\frac{l}{3h}+\frac{h}{12l} \\
 & & & & & \frac{l}{3h}+\frac{h}{6l} & -\frac{l}{3h}+\frac{h}{12l} & -\frac{l}{6h}-\frac{h}{12l} \\
 & sym. & & & & & \frac{l}{3h}+\frac{h}{6l} & \frac{l}{6h}-\frac{h}{6l} \\
 & & & & & & & \frac{l}{3h}+\frac{h}{6l}
\end{bmatrix} .
$$

$$\tag{4.1}$$

All values are to multiplied by $E \cdot t$, the product of the modulus of elasticity E and the thickness t of the plate.

In case the dimensions are $l = 2, h = 1$, the matrix $\boldsymbol{K}^e$ becomes very simple:

$$
\boldsymbol{K}^e = \frac{E\,t}{8}
\begin{bmatrix}
4 & 1 & 0 & -1 & -2 & -1 & -2 & 1 \\
1 & 6 & 1 & 2 & -1 & -3 & -1 & -5 \\
0 & 1 & 4 & -1 & -2 & -1 & -2 & 1 \\
-1 & 2 & -1 & 6 & 1 & -5 & 1 & -3 \\
-2 & -1 & -2 & 1 & 4 & 1 & 0 & -1 \\
-1 & -3 & -1 & -5 & 1 & 6 & 1 & 2 \\
-2 & -1 & -2 & 1 & 0 & 1 & 4 & -1 \\
1 & -5 & 1 & -3 & -1 & 2 & -1 & 6
\end{bmatrix} .
\tag{4.2}
$$

The product of the element matrix $\boldsymbol{K}^e$ and the nodal displacements $\boldsymbol{u}^e$ yields the equivalent nodal forces $\boldsymbol{f}^e$:

$$
\boldsymbol{K}^e \, \boldsymbol{u}^e = \boldsymbol{f}^e
\tag{4.3}
$$

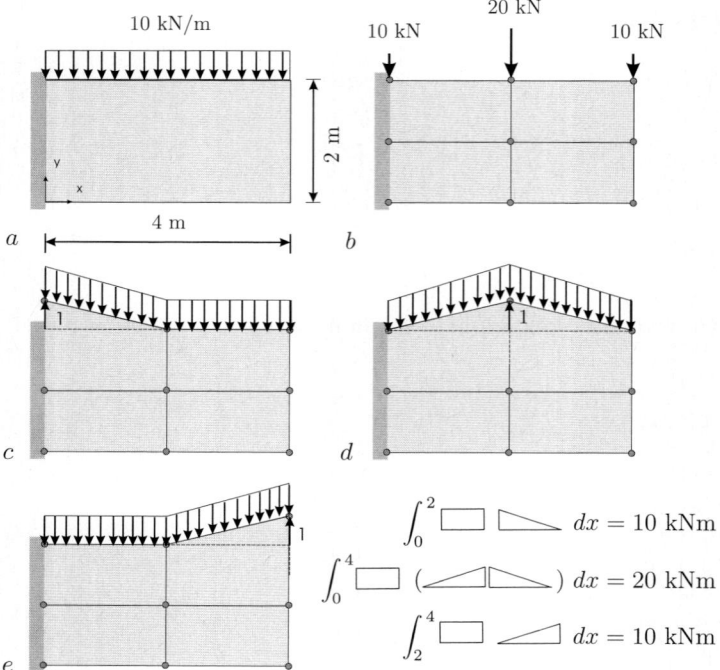

Fig. 4.1. Cantilever plate: **a)** system and load, **b)** equivalent nodal forces: these fictitious nodal forces are work-equivalent to the edge load with respect to the unit nodal displacements of the edge nodes as plotted in **c)**, **d)**, and **e)**. A unit force of 20 kN at the center node of the edge contributes the same work as the distributed load in panel **d)** on acting through the nodal unit displacement

or

$$\boldsymbol{K}^e\,\boldsymbol{u}^e = u_1 \begin{bmatrix} \boldsymbol{c}_1 \end{bmatrix} + u_2 \begin{bmatrix} \boldsymbol{c}_2 \end{bmatrix} + \ldots + u_8 \begin{bmatrix} \boldsymbol{c}_8 \end{bmatrix} = \boldsymbol{f}^e\,, \tag{4.4}$$

i.e.,

$$u_1 \begin{bmatrix} k_{11} \\ k_{21} \\ \ldots \\ k_{81} \end{bmatrix} + u_2 \begin{bmatrix} k_{12} \\ k_{22} \\ \ldots \\ k_{82} \end{bmatrix} + \ldots + u_8 \begin{bmatrix} k_{18} \\ k_{28} \\ \ldots \\ k_{88} \end{bmatrix} = \begin{bmatrix} f_1 \\ f_2 \\ \ldots \\ f_8 \end{bmatrix}\,. \tag{4.5}$$

Obviously the eight columns $\boldsymbol{c}_i$ of $\boldsymbol{K}$ are the equivalent nodal forces of the eight unit displacements $\boldsymbol{u}^e = \boldsymbol{e}_i$, $i = 1, 2, \ldots, 8$.

The nodal displacements of the individual elements and of the nodes of the plate are the same, so that if $\boldsymbol{u}_{(18)} = [u_1, u_2, \ldots, u_{18}]^T$ is the list of the nodal displacements and $\boldsymbol{u}^l_{(32)} = [\boldsymbol{u}^{(1)}, \boldsymbol{u}^{(2)}, \boldsymbol{u}^{(3)}, \boldsymbol{u}^{(4)}]^T$ the list of the element

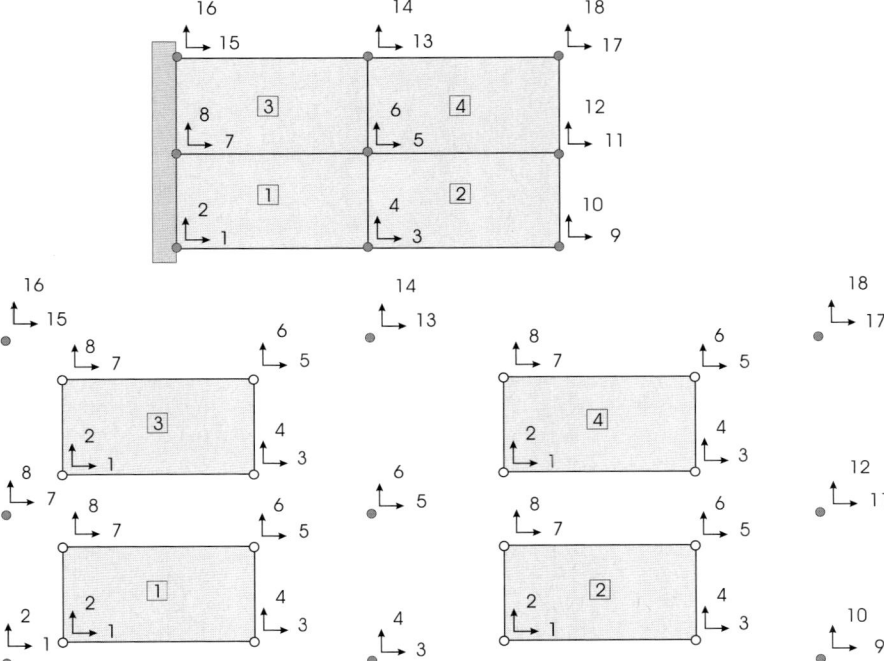

Fig. 4.2. Subdivision into elements and nodes. The dark circles are the nodes of the mesh and the bright circles are the nodes of the elements

nodal displacements, then there exists a *boolean matrix* $\boldsymbol{A}$ that maps the nodal displacements onto the nodal element displacements

$$\boldsymbol{u}^l_{(32)} = \boldsymbol{A}_{(32 \times 18)}\, \boldsymbol{u}_{(18)}\,. \tag{4.6}$$

The information in the matrix $\boldsymbol{A}$ is also provided by the following table

	1 2 3 4 5 6 7 8
Element 1	1 2 3 4 5 6 7 8
Element 2	3 4 9 10 11 12 5 6
Element 3	7 8 5 6 13 14 15 16
Element 4	5 6 11 12 17 18 13 14

$$\tag{4.7}$$

which shows for each element how the eight element degrees of freedom (top row) are associated with the global degrees of freedom; see Fig. 4.2.

In the reverse order, the equivalent nodal forces f_i at each node are balanced by the element nodal forces. Because $(\boldsymbol{A}\,\boldsymbol{u})^T \boldsymbol{K}\,\boldsymbol{A}\,\boldsymbol{u} = \boldsymbol{u}^T \boldsymbol{A}^T \boldsymbol{K}\,\boldsymbol{A}\,\boldsymbol{u}$, this equilibrium condition amounts to

$$\boldsymbol{f}_{(18)} = \boldsymbol{A}^T_{(18 \times 32)}\, \boldsymbol{f}^l_{(32)}\,. \tag{4.8}$$

If the element matrices are placed on the diagonal of a 32×32 matrix

$$\boldsymbol{K}^l_{(32\times32)} = \begin{bmatrix} \boldsymbol{K}^e_1 & 0 & 0 & 0 \\ 0 & \boldsymbol{K}^e_2 & 0 & 0 \\ 0 & 0 & \boldsymbol{K}^e_3 & 0 \\ 0 & 0 & 0 & \boldsymbol{K}^e_4 \end{bmatrix}, \tag{4.9}$$

the global stiffness matrix becomes

$$\boldsymbol{K}_{(18\times18)} = \boldsymbol{A}^T_{(18\times32)}\,\boldsymbol{K}^l_{(32\times32)}\,\boldsymbol{A}_{(32\times18)} \tag{4.10}$$

or

$$\boldsymbol{K} = \frac{Et}{8}
\begin{bmatrix}
4 & 1 & 0 & -1 & -2 & -1 & -2 & 1 & 0 & 0 & 0 & 0 & 0 & 0 & 0 & 0 & 0 & 0 \\
1 & 6 & 1 & 2 & -1 & -3 & -1 & -5 & 0 & 0 & 0 & 0 & 0 & 0 & 0 & 0 & 0 & 0 \\
0 & 1 & 8 & 0 & -4 & 0 & -2 & 1 & 0 & -1 & -2 & -1 & 0 & 0 & 0 & 0 & 0 & 0 \\
-1 & 2 & 0 & 12 & 0 & -10 & 1 & -3 & 1 & 2 & -1 & -3 & 0 & 0 & 0 & 0 & 0 & 0 \\
-2 & -1 & -4 & 0 & 16 & 0 & 0 & 0 & -2 & 1 & 0 & 0 & -4 & 0 & -2 & 1 & -2 & -1 \\
-1 & -3 & 0 & -10 & 0 & 24 & 0 & 4 & 1 & -3 & 0 & 4 & 0 & -10 & 1 & -3 & -1 & -3 \\
-2 & -1 & -2 & 1 & 0 & 0 & 8 & 0 & 0 & 0 & 0 & 0 & -2 & -1 & -2 & 1 & 0 & 0 \\
1 & -5 & 1 & -3 & 0 & 4 & 0 & 12 & 0 & 0 & 0 & 0 & -1 & -3 & -1 & -5 & 0 & 0 \\
0 & 0 & 0 & 1 & -2 & 1 & 0 & 0 & 4 & -1 & -2 & -1 & 0 & 0 & 0 & 0 & 0 & 0 \\
0 & 0 & -1 & 2 & 1 & -3 & 0 & 0 & -1 & 6 & 1 & -5 & 0 & 0 & 0 & 0 & 0 & 0 \\
0 & 0 & -2 & -1 & 0 & 0 & 0 & 0 & -2 & 1 & 8 & 0 & -2 & 1 & 0 & 0 & -2 & -1 \\
0 & 0 & -1 & -3 & 0 & 4 & 0 & 0 & -1 & -5 & 0 & 12 & 1 & -3 & 0 & 0 & 1 & -5 \\
0 & 0 & 0 & 0 & -4 & 0 & -2 & -1 & 0 & 0 & -2 & 1 & 8 & 0 & 0 & -1 & 0 & 1 \\
0 & 0 & 0 & 0 & 0 & -10 & -1 & -3 & 0 & 0 & 1 & -3 & 0 & 12 & 1 & 2 & -1 & 2 \\
0 & 0 & 0 & 0 & -2 & 1 & -2 & -1 & 0 & 0 & 0 & 0 & 0 & 1 & 4 & -1 & 0 & 0 \\
0 & 0 & 0 & 0 & 1 & -3 & 1 & -5 & 0 & 0 & 0 & 0 & -1 & 2 & -1 & 6 & 0 & 0 \\
0 & 0 & 0 & 0 & -2 & -1 & 0 & 0 & 0 & 0 & -2 & 1 & 0 & -1 & 0 & 0 & 4 & 1 \\
0 & 0 & 0 & 0 & -1 & -3 & 0 & 0 & 0 & 0 & -1 & -5 & 1 & 2 & 0 & 0 & 1 & 6
\end{bmatrix}.$$

$$\tag{4.11}$$

Of course the matrix multiplication (4.10) is never carried out in an FE program. Instead the entries k_{ij} are simply assembled by adding the corresponding stiffnesses of the element nodes, as would be done in a system of springs connected in parallel. This global stiffness matrix (18×18) embodies the interaction of the nodal displacements and the equivalent nodal forces of the plate:

$$\boldsymbol{K}\,\boldsymbol{u} = \boldsymbol{f} \qquad \text{or} \qquad \boldsymbol{A}^T_{(18\times32)}\,\boldsymbol{K}^l_{(32\times32)}\,\boldsymbol{A}_{(32\times18)}\,\boldsymbol{u}_{(18)} = \boldsymbol{f}_{(18)}. \tag{4.12}$$

Stream model

The nature of the assembled system of equations (4.12) is best understood in terms of a stream model, where it is assumed that each node possesses a certain potential u_i. Because the individual nodes have different potentials and the elements different physical properties, strains (and thus stresses) will

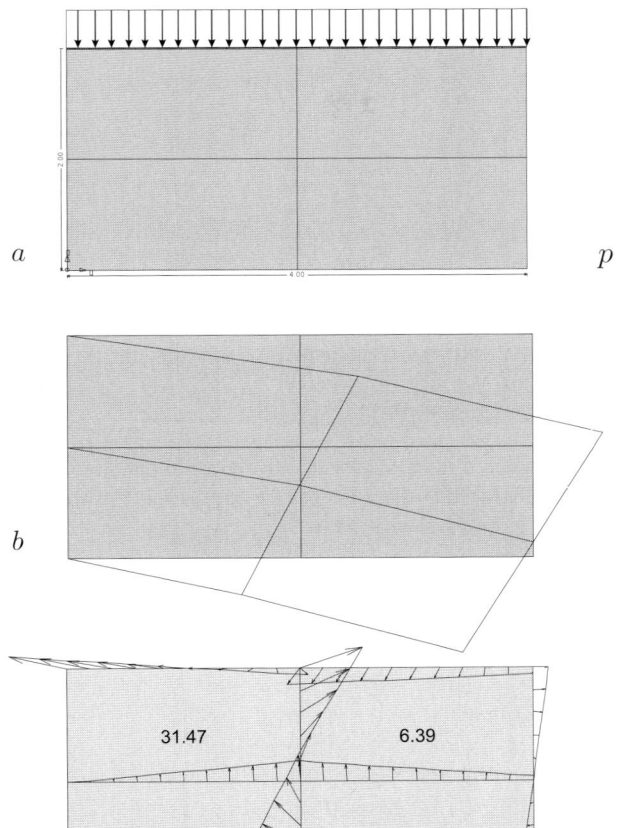

Fig. 4.3. FE analysis of a plate: **a)** system and original load case, **b)** deformed plate, **c)** load case p_h with resulting volume forces $p_\Omega = [(p_x^h, p_x^h) + (p_y^h, p_y^h)]^{1/2}$

develop between the nodes. These stresses flow back to the nodes in the form of element nodal forces, and these forces in turn are balanced by the external nodal forces f_i [232].

In the first step, $\boldsymbol{A}\,\boldsymbol{u}$ (Equ. (4.12) must be read from right to left), the nodal potentials u_i are distributed over the element nodes, $u_i \rightarrow u_i^e$. In each element the different nodal potentials generate stresses, resulting in element nodal forces $\boldsymbol{f}^e = \boldsymbol{K}^e\,\boldsymbol{u}^e$. In the second step all these element nodal forces are bundled at the nodes, $\boldsymbol{A}^T \boldsymbol{f}^l$ and are balanced by the external nodal forces $\boldsymbol{f}$, i.e., $\boldsymbol{A}^T \boldsymbol{K}^l \boldsymbol{A}\,\boldsymbol{u} = \boldsymbol{f}$, or simply $\boldsymbol{K}\,\boldsymbol{u} = \boldsymbol{f}$.

Equivalent nodal forces

To transform the edge load into equivalent nodal forces f_i the work is calculated which the edge load contributes on acting through the nodal unit displacements of the degrees of freedom (d.o.f.) $u_{16} = 1$, $u_{14} = 1$, and $u_{18} = 1$ on the upper edge:

$$f_{14} = \frac{1}{2} \cdot 1 \cdot (-10) \cdot 2 \cdot 2 = -20 \tag{4.13}$$

$$f_{16} = f_{18} = \frac{1}{2} \cdot 1 \cdot (-10) \cdot 2 = -10. \tag{4.14}$$

Note that u_{16} is also activated even though this d.o.f is fixed. Only this guarantees that the sum of the equivalent nodal forces is equal to the applied load. Hence, some part of the load flows directly to the support nodes and will not contribute any strains and stresses; see Fig. 4.1 b.

Because the support nodes are fixed, six degrees of freedom are zero:

$$u_1 = u_2 = u_7 = u_8 = u_{15} = u_{16} = 0, \tag{4.15}$$

so that 12 (out of 18) nodal displacements u_i are unknown. This is the degree of *kinematic indeterminacy* of the structure. The set of equations

$$\boldsymbol{K}_{(12 \times 12)}\, \boldsymbol{u}_{(12)} = \boldsymbol{f}_{(12)} \tag{4.16}$$

for these 12 nodal displacements u_i is obtained if in the global stiffness matrix (4.12) the rows and columns that belong to fixed degrees of freedom are eliminated:

$$\frac{Et}{8}
\begin{bmatrix}
8 & 0 & -4 & 0 & 0 & -1 & -2 & -1 & 0 & 0 & 0 & 0 \\
0 & 12 & 0 & -10 & 1 & 2 & -1 & -3 & 0 & 0 & 0 & 0 \\
-4 & 0 & 16 & 0 & -2 & 1 & 0 & 0 & -4 & 0 & -2 & -1 \\
0 & -10 & 0 & 24 & 1 & -3 & 0 & 4 & 0 & -10 & -1 & -3 \\
0 & 1 & -2 & 1 & 4 & -1 & -2 & -1 & 0 & 0 & 0 & 0 \\
-1 & 2 & 1 & -3 & -1 & 6 & 1 & -5 & 0 & 0 & 0 & 0 \\
-2 & -1 & 0 & 0 & -2 & 1 & 8 & 0 & -2 & 1 & -2 & -1 \\
-1 & -3 & 0 & 4 & -1 & -5 & 0 & 12 & 1 & -3 & 1 & -5 \\
0 & 0 & -4 & 0 & 0 & 0 & -2 & 1 & 8 & 0 & 0 & 1 \\
0 & 0 & 0 & -10 & 0 & 0 & 1 & -3 & 0 & 12 & -1 & 2 \\
0 & 0 & -2 & -1 & 0 & 0 & -2 & 1 & 0 & -1 & 4 & 1 \\
0 & 0 & -1 & -3 & 0 & 0 & -1 & -5 & 1 & 2 & 1 & 6
\end{bmatrix}
\begin{bmatrix}
u_3 \\ u_4 \\ u_5 \\ u_6 \\ u_9 \\ u_{10} \\ u_{11} \\ u_{12} \\ u_{13} \\ u_{14} \\ u_{17} \\ u_{18}
\end{bmatrix}
=
\begin{bmatrix}
0 \\ 0 \\ 0 \\ 0 \\ 0 \\ 0 \\ 0 \\ 0 \\ 0 \\ -20 \\ 0 \\ -10
\end{bmatrix}. \tag{4.17}$$

Results and interpretation

The shape of the deformed structure is displayed in Fig. 4.3 b, and the distribution of the bending stresses at the fixed edge is plotted in Fig. 4.4. In Table 4.1 the FE solution is compared to a BE solution and a beam solution. The plate was subdivided into 4, 8 and 32 elements, respectively. The material properties were $E = 29\,000$ MN/m^2, $t = 0.2$ m, and $\nu = 0.0$.

Table 4.1. Comparison of the deflection at the lower corner and the normal stresses at the fixed edge

Elements	Deflection (mm)	Compression (kN/m²)	Tensile stresses (kN/m²)
4	6.83E-02	−248	251
8	8.67E-02	−413	420
32	9.51E-02	−546	567
Beam	8.28E-02	−600	600
BE	9.86E-02	−828	1055

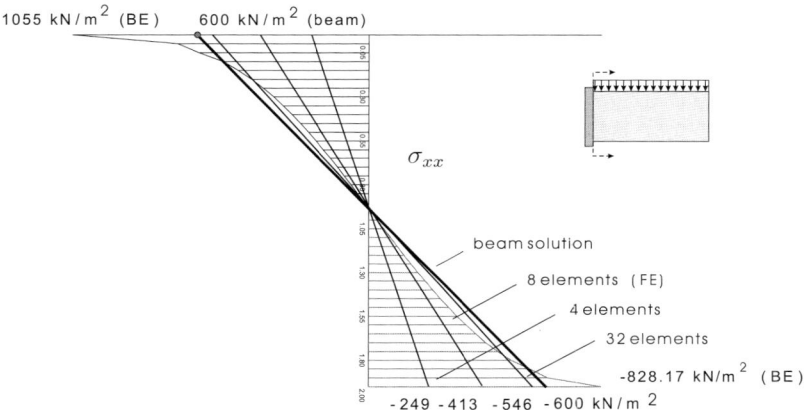

Fig. 4.4. Bending stresses

The relatively small vertical displacement at the lower right corner and the slow convergence of the bending stresses is an indication that bilinear elements have difficulties with bending-dominated problems. The stress distribution of the BE solution on the other hand deviates from the linear stress distribution of the beam theory, and the extreme values seem to tend to $\pm\infty$, which are obviously the exact bending stresses in the extreme fibers according to the theory of elasticity.

This simple problem is an indication that questions concerning the modeling are at least as important in FE analysis as questions concerning numerical details: What is to be calculated? What do we expect from the FE model? Is it the beam solution

$$\sigma_{xx} = \frac{M\,h}{2\,EI} = \frac{\pm 80 \cdot 2.0}{2 \cdot 2.9 \cdot 10^7 \cdot 0.1\bar{3}} = \pm\,600\,\text{kN/m}^2 \qquad (4.18)$$

or is it the stress concentration factor, or is it the size and location of the plastic zones?

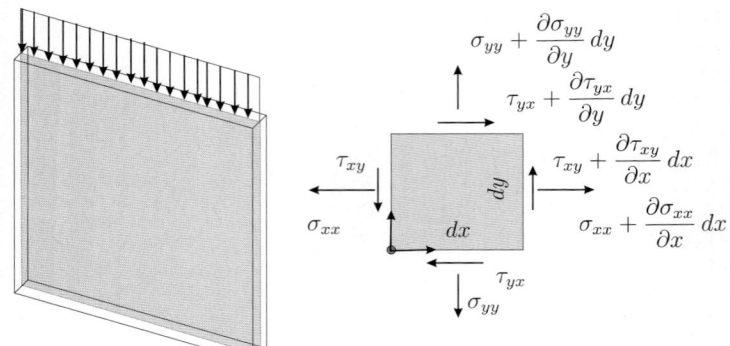

Fig. 4.5. Plate

4.2 Strains and stresses

The deformation of a plate is described by the displacement vector

$$\boldsymbol{u}(x,y) = \begin{bmatrix} u(x,y) \\ v(x,y) \end{bmatrix} \qquad \begin{matrix} \text{horizontal displacement} \\ \text{vertical displacement} \end{matrix} \qquad (4.19)$$

of the individual points. The stresses (see Fig. 4.5) are not proportional to the magnitude of the displacements, but to the change in the displacements per unit length, that is the gradient (strains) of the displacement field

$$\varepsilon_{xx} = \frac{\partial u}{\partial x} \qquad \varepsilon_{yy} = \frac{\partial v}{\partial y} \qquad \gamma_{xy} = \frac{\partial v}{\partial x} + \frac{\partial u}{\partial y} \qquad \varepsilon_{xy} = \frac{1}{2}\,\gamma_{xy}\,. \qquad (4.20)$$

In a state of *plane stress* (see Fig. 4.6), where $\sigma_{zz} = \tau_{yz} = \tau_{xz} = 0$,

$$\underbrace{\begin{bmatrix} \sigma_{xx} \\ \sigma_{yy} \\ \tau_{xy} \end{bmatrix}}_{\sigma} = \underbrace{\frac{E}{1-\nu^2} \begin{bmatrix} 1 & \nu & 0 \\ \nu & 1 & 0 \\ 0 & 0 & (1-\nu)/2 \end{bmatrix}}_{E} \underbrace{\begin{bmatrix} \varepsilon_{xx} \\ \varepsilon_{yy} \\ \gamma_{xy} \end{bmatrix}}_{\varepsilon} \qquad (4.21)$$

and in a state of *plane strain*,

$$\begin{bmatrix} \sigma_{xx} \\ \sigma_{yy} \\ \tau_{xy} \end{bmatrix} = \frac{E}{(1+\nu)(1-2\,\nu)} \begin{bmatrix} (1-\nu) & \nu & 0 \\ \nu & (1-\nu) & 0 \\ 0 & 0 & (1-2\,\nu)/2 \end{bmatrix} \begin{bmatrix} \varepsilon_{xx} \\ \varepsilon_{yy} \\ \gamma_{xy} \end{bmatrix}. \qquad (4.22)$$

To recover the strains from the stresses, the formula

$$\begin{bmatrix} \varepsilon_{xx} \\ \varepsilon_{yy} \\ \gamma_{xy} \end{bmatrix} = \begin{bmatrix} 1/E & -\nu/E & 0 \\ -\nu/E & 1/E & 0 \\ 0 & 0 & 1/G \end{bmatrix} \begin{bmatrix} \sigma_{xx} \\ \sigma_{yy} \\ \tau_{xy} \end{bmatrix} \qquad (4.23)$$

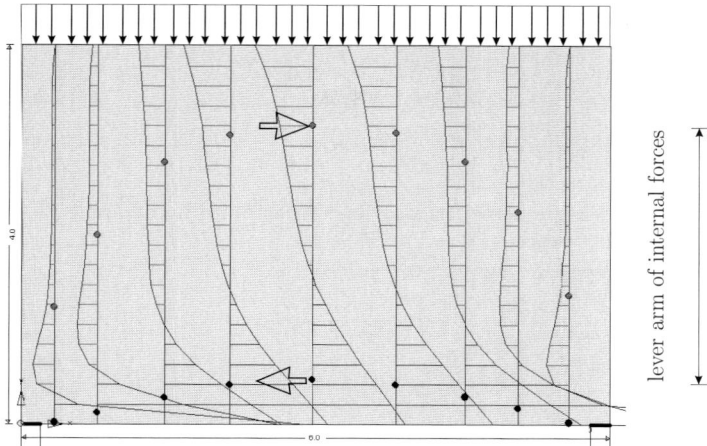

Fig. 4.6. Stress distribution in a wall. The distance between the stress resultants is proportional to the magnitude of the internal bending moment

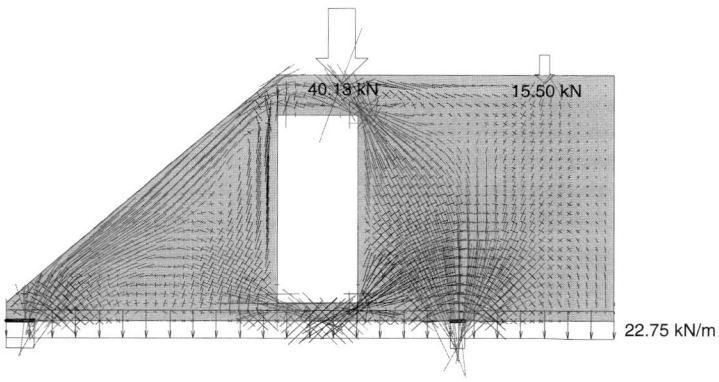

Fig. 4.7. Principal stresses in a plate

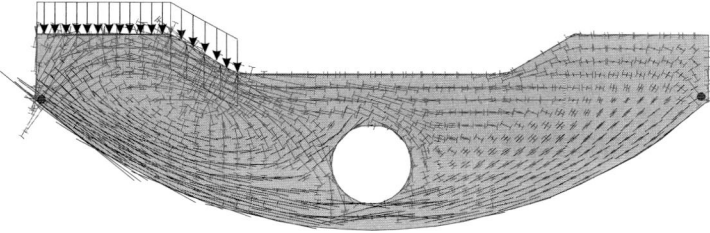

Fig. 4.8. At free edges the principal stresses always run parallel to the edge. This provides a visual check on the FE results

is used, where $G = 0.5\,E/(1 + \nu)$ is the shear modulus of the material.

In rubber-like materials where Poisson's ratio is close to 0.5, the stresses become infinite in a state of plane strain. Special efforts are necessary to deliver useful results with an FE program close to this point; see Sect. 4.17, p. 393.

Table 4.2. Critical angles for a plate; stresses become infinite if the angle of the boundary point exceeds these values.

Boundary conditions	angle	
fixed–fixed	180°	
fixed–roller	90°	
fixed–tangential	90°	
fixed–free	61.7°	$\nu = 0.29$ plane stress state
roller–roller	90°	
roller–tangential	45°	
roller–free	90°	
tangential–tangential	90°	
tangential–free	128.73°	
free–free	180°	

The angle

$$\tan 2\varphi = \frac{2\,\tau_{xy}}{\sigma_{xx} - \sigma_{yy}} \tag{4.24}$$

defines the orientation of the principal planes where the *principal stresses*

$$\sigma_{I,II} = \frac{\sigma_{xx} + \sigma_{yy}}{2} \pm \sqrt{\left[\frac{\sigma_{xx} - \sigma_{yy}}{2}\right]^2 + \tau_{xy}^2} \tag{4.25}$$

are acting. The shear stresses are zero in these planes. They attain their maximum values if the planes are rotated by 45°. The stress trajectories (see Fig. 4.7 and Fig. 4.8) provide a graphic description of the stress state.

If u_n and u_s denote the edge displacements in the normal and tangential direction and t_n and t_s the tractions in these directions, four combinations of support conditions are possible

$$
\begin{aligned}
u_n = u_s = 0 & \qquad \text{fixed edge} \\
u_n = 0\,, t_s = 0 & \qquad \text{roller support} \\
u_s = 0\,, t_n = 0 & \qquad \text{tangential support} \\
t_n = t_s = 0 & \qquad \text{free edge}
\end{aligned}
$$

The stress singularities at corner points depend on these boundary conditions and on the angle of the corner points; see Table 4.2 [206], [252].

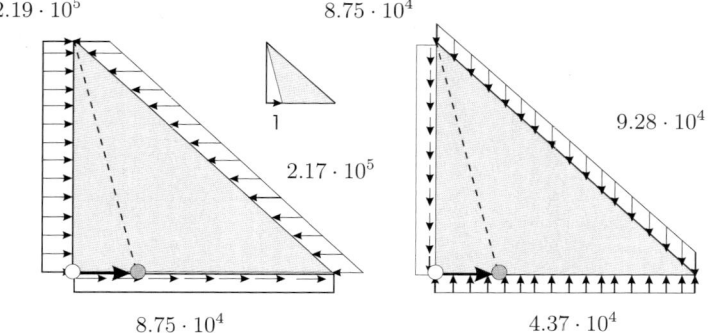

Fig. 4.9. CST element, $E = 2.1 \cdot 10^9$ kN/m^2, $\nu = 0.2$, $t = 0.1$ m. Displayed are the edge loads (kN/m) necessary to push the lower left node to the right while all other nodes are kept fixed

4.3 Shape functions

An element Ω_e with n nodes $\boldsymbol{x}_i$ is usually equipped with n local shape functions $\psi_i^e, i = 1, 2, \ldots, n$ with the property $\psi_i^e(\boldsymbol{x}_j) = \delta_{ij}$. By continuing these shape functions across interelement boundaries, *global* shape functions ψ_i are generated.

With these global shape functions, "monochrome" displacement fields can be generated which represent unit displacements fields of the nodes:

$$\boldsymbol{\varphi}_1 = \begin{bmatrix} \psi_1 \\ 0 \end{bmatrix} \qquad \boldsymbol{\varphi}_2 = \begin{bmatrix} 0 \\ \psi_1 \end{bmatrix} \qquad \begin{matrix} \leftarrow \\ \leftarrow \end{matrix} \quad \begin{matrix} \text{horizontal displacement} \\ \text{vertical displacement} \end{matrix} \qquad (4.26)$$

These are called monochrome because to portray a horizontal displacement of a node, no vertical component is needed, and vice versa, even though such monochrome displacement fields will also cause stresses in the other direction.

The FE displacement field is an expansion in terms of these $2n$ unit displacement fields $\boldsymbol{\varphi}_i$:

$$\boldsymbol{u}_h(x, y) = \sum_{i=1}^{2n} u_i \, \boldsymbol{\varphi}_i(x, y)$$

$$= u_1 \overbrace{\begin{bmatrix} \psi_1 \\ 0 \end{bmatrix}}^{node\ 1} + u_2 \begin{bmatrix} 0 \\ \psi_1 \end{bmatrix} + u_3 \overbrace{\begin{bmatrix} \psi_2 \\ 0 \end{bmatrix}}^{node\ 2} + u_4 \begin{bmatrix} 0 \\ \psi_2 \end{bmatrix} + \ldots \qquad (4.27)$$
$$\underbrace{}_{\boldsymbol{\varphi}_1 \rightarrow} \quad \underbrace{}_{\boldsymbol{\varphi}_2 \uparrow} \quad \underbrace{}_{\boldsymbol{\varphi}_3 \rightarrow} \quad \underbrace{}_{\boldsymbol{\varphi}_4 \uparrow}$$

The unit load case $\boldsymbol{p}_i$ which can be associated with a unit displacement field $\boldsymbol{\varphi}_i$ is simply the set of all forces necessary to force the plate into the shape $\boldsymbol{\varphi}_i$; see Fig. 4.9.

Furthermore the superposition of all these load cases $\boldsymbol{p}_i$ is the FE load case

$$\boldsymbol{p}_h = \sum_{i=1}^{2n} u_i \, \boldsymbol{p}_i \,, \tag{4.28}$$

which is tuned in such a way—by adjusting the nodal displacements u_i—that it is work-equivalent to the original load case $\boldsymbol{p}$ with respect to all unit displacement fields:

$$\delta W_e(\boldsymbol{p}, \boldsymbol{\varphi}_i) = \delta W_e(\boldsymbol{p}_h, \boldsymbol{\varphi}_i) \qquad \text{for all } \boldsymbol{\varphi}_i \,. \tag{4.29}$$

4.4 Plane elements

The unit displacement fields of the assembled structure should be able to represent rigid-body motions exactly, as well as constant strain and stress states, because otherwise the FE solution will not converge to the exact solution as the mesh size h tends to zero. Of course the unit displacement fields must be continuous across interelement boundaries, because no gap is allowed in the structure (C^0-elements). Depending on the order of the polynomial shape functions, the elements are called *linear, quadratic,* or *cubic elements*.

Preferably the shape functions should be *complete*, i.e., they should contain all terms $x^i \, y^j$ up to order $n = 1, 2$ or 3 (cubic elements). If not, they should at least contain the correlated terms $x \, y^2$ and $y \, x^2$, for example, in order that the element be *geometrically isotropic*—if the load is rotated by $90°$ then the stresses should follow—and it should be guaranteed that the quality of the interpolation is invariant with respect to rotation of the elements (*rotational invariance*).

CST element

The simplest element is a triangular element, with three nodes and linear shape functions (see Fig. 4.10)

$$\psi_1^e(x, y) = \frac{1}{2\,A_e} \left[(x_2 \, y_3 - x_3 \, y_2) + y_{23} \, x + x_{32} \, y \right] \tag{4.30}$$

$$\psi_2^e(x, y) = \frac{1}{2\,A_e} \left[(x_3 \, y_1 - x_1 \, y_3) + y_{31} \, x + x_{13} \, y \right] \tag{4.31}$$

$$\psi_3^e(x, y) = \frac{1}{2\,A_e} \left[(x_1 \, y_2 - x_2 \, y_1) + y_{12} \, x + x_{21} \, y \right] \tag{4.32}$$

where

$$x_{ij} = x_i - x_j \qquad y_{ij} = y_i - y_j \qquad 2\,A_e = x_{21} \, y_{31} - x_{31} \, y_{21} \tag{4.33}$$

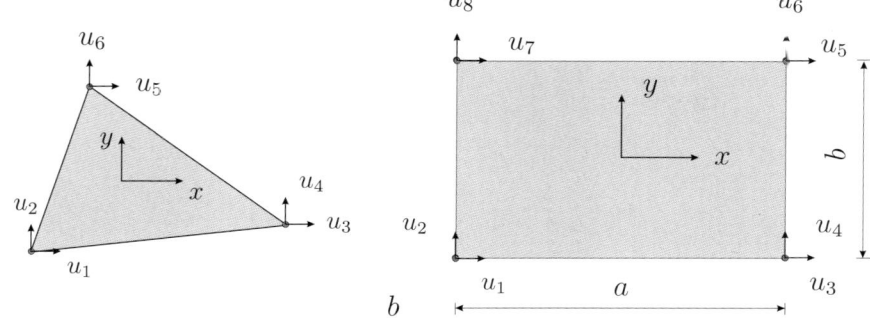

Fig. 4.10. Plane elements: **a)** CST element, **b)** bilinear element

so that the displacement field of the element is an expansion with respect to the six nodal displacements:

$$\boldsymbol{u}^e(x,y) = \begin{bmatrix} u^e \\ v^e \end{bmatrix} = \begin{bmatrix} \psi_1^e & 0 & \psi_2^e & 0 & \psi_3^e & 0 \\ 0 & \psi_1^e & 0 & \psi_2^e & 0 & \psi_3^e \end{bmatrix} \begin{bmatrix} u_1 \\ v_1 \\ u_2 \\ v_2 \\ u_3 \\ v_3 \end{bmatrix} \quad \text{or} \quad \boldsymbol{u}^e = \boldsymbol{\Psi}^e\,\boldsymbol{d}^e\,. \ (4.34)$$

As the name of the element implies, the strains are constant:

$$\begin{bmatrix} \varepsilon_{xx} \\ \varepsilon_{yy} \\ \gamma_{xy} \end{bmatrix} = \frac{1}{2\,A_e} \begin{bmatrix} y_{23} & 0 & y_{31} & 0 & y_{12} & 0 \\ 0 & x_{32} & 0 & x_{13} & 0 & x_{21} \\ x_{32} & y_{23} & x_{13} & y_{31} & x_{21} & y_{21} \end{bmatrix} \begin{bmatrix} u_1 \\ v_1 \\ u_2 \\ v_2 \\ u_3 \\ v_3 \end{bmatrix} \quad \text{or} \quad \boldsymbol{\varepsilon} = \boldsymbol{B}\,\boldsymbol{d}\,. \ (4.35)$$

The stress vector $\boldsymbol{\sigma} = [\sigma_{xx}, \sigma_{yy}, \tau_{xy}]^T$ is $\boldsymbol{\sigma} = \boldsymbol{E}\,\boldsymbol{\varepsilon} = \boldsymbol{E}\,\boldsymbol{B}\,\boldsymbol{d}$, where $\boldsymbol{E}$ is the matrix in (4.21) and the stiffness matrix of the element is

$$\boldsymbol{K}^e = t \int_{\Omega_e} \boldsymbol{B}^T\,\boldsymbol{E}\,\boldsymbol{B}\,d\Omega \qquad t = \text{ element thickness}\,.$$

The elements k_{ij}^e are the strain energy products between the unit displacement fields $\boldsymbol{\varphi}_i^e$ and $\boldsymbol{\varphi}_j^e$:

$$k_{ij}^e = a(\boldsymbol{\varphi}_i^e, \boldsymbol{\varphi}_j^e) = t \int_{\Omega_e} \boldsymbol{\sigma}^{(i)} \bullet \boldsymbol{\varepsilon}^{(j)}\,d\Omega$$

$$= t \int_{\Omega_e} \left[\sigma_{xx}^{(i)} \varepsilon_{xx}^{(j)} + \tau_{xy}^{(i)} \gamma_{xy}^{(j)} + \sigma_{yy}^{(i)} \varepsilon_{yy}^{(j)} \right] d\Omega\,. \qquad (4.36)$$

The CST element is the simplest plane element, but certainly not a very good one, as can be seen in Fig. 4.11.

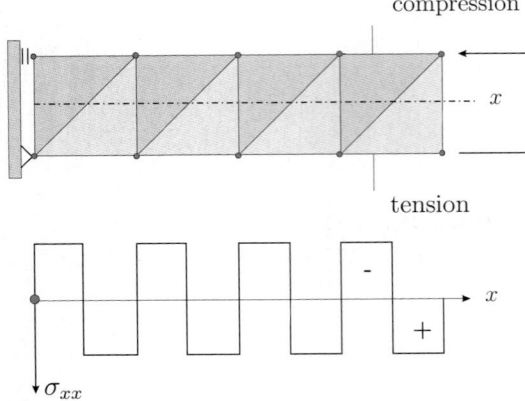

Fig. 4.11. Cantilever plate with end moment; analysis with CST elements. The normal stress along the axis of the plate oscillates

Bilinear elements

The simplest choice for a rectangular element are bilinear shape functions, i.e., shape functions that are products of linear polynomials, $(c_1 + c_2\,x)\,(d_1 + d_2\,y)$. If $2a$ denotes the length and $2b$ the width of the element, then the shape functions of the four nodes are (see Fig. 4.10)

$$\psi_1^e = \frac{1}{4\,a\,b}\,(a - 2x)(b - 2y) \qquad \psi_2^e = \frac{1}{4\,a\,b}\,(a + 2x)(b - 2y) \tag{4.37}$$

$$\psi_3^e = \frac{1}{4\,a\,b}\,(a + 2x)(b + 2y) \qquad \psi_4^e = \frac{1}{4\,a\,b}\,(a - 2x)(b + 2y)\,. \tag{4.38}$$

In such an element the strains and stresses vary linearly,

$$\varepsilon_{xx} = a_1 + a_3\,y\,, \qquad \varepsilon_{yy} = b_2 + b_3\,x\,, \quad \gamma_{xy} = (a_2 + b_1) + a_3\,x + b_3\,y\,, \tag{4.39}$$

but in the "wrong direction" if the x-axis is assumed to be the principal direction. In the case $\nu = 0$ the normal stresses are constant in the x-direction while they vary linearly in the y-direction. Only the shear stress varies in both directions (see Fig. 4.12).

This element is too stiff (Fig. 4.13), because the element cannot display any curvature. If two end moments M rotate the ends of a beam by an angle φ in a bilinear element with ratio a/b, the moments

$$M_{FE} = \frac{1}{1 + \nu}\left[\frac{1}{1 - \nu} + \frac{1}{2}\left(\frac{a}{b}\right)^2\right]M$$

are needed to achieve the same effect. It is easy to see that $M_{FE} > M$. If the length of the element and thus the ratio a/b increases, then the moments M_{FE} increase quadratically; in other words the element will lock.

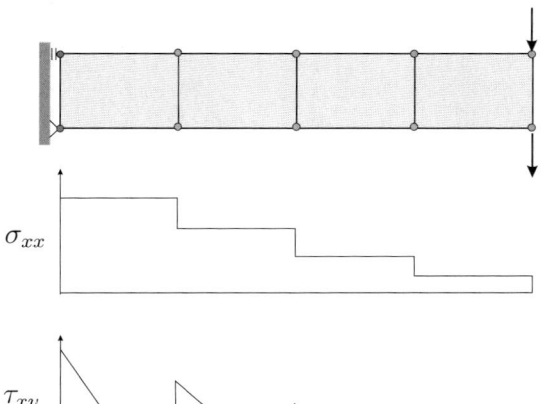

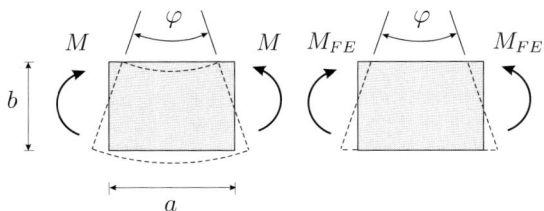

Fig. 4.12. Cantilever plate with end forces, analysis with bilinear elements, $\nu = 0$.

Fig. 4.13. The sides of a bilinear elements remain straight

LST element

A more flexible element is the linear strain triangle (LST element) which is based on quadratic polynomials,

$$u(x, y) = a_0 + a_1 x + a_2 y + a_3 x\, y + a_4\, x^2 + a_5\, y^2 \qquad (4.40)$$
$$v(x, y) = b_0 + b_1 x + b_2 y + b_3 x\, y + b_4\, x^2 + b_5\, y^2 \,, \qquad (4.41)$$

so that the strains ε_{xx} and ε_{yy} vary in *both* directions x and y

$$\varepsilon_{xx} = a_2 + 2\, a_4\, x + a_5\, y\,, \qquad \varepsilon_{yy} = b_2 + b_3\, x + 2\, b_5\, y\,, \qquad (4.42)$$
$$\gamma_{xy} = (a_2 + b_1) + (a_3 + 2\, b_4)\, x + (2\, a_5 + b_3)\, y\,. \qquad (4.43)$$

Bilinear + 2

Wilson [253] had the idea to enrich the bilinear element in each direction with two quadratic shape functions

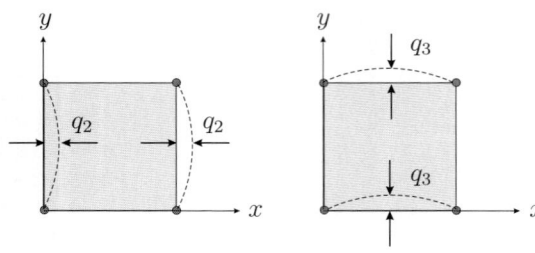

Fig. 4.14. Quadratic shape functions $u = (1 - \eta^2)\, q_2$ and $v = (1 - \xi^2)\, q_3$

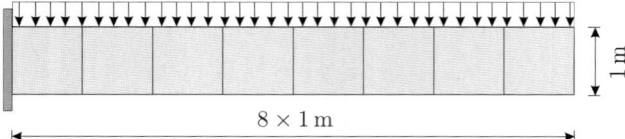

$$8 \times 1\,\mathrm{m}$$

Fig. 4.15. Cantilever beam with edge load

Table 4.3. Deflection v at the end of the cantilever plate, see Fig. 4.15, Q4 = bilinear element, Q4 + 2 = Wilson's element. The exact value is $v = 1.024$ cm

Mesh	Q4	Q4+2
1 x 8	0.715 cm	1.035 cm
2 x 16	0.939 cm	1.036 cm
4 x 32	1.010 cm	1.038 cm
8 x 80	1.021 cm	1.039 cm

$$u = \ldots + (1 - \xi^2)\, q_1 + (1 - \eta^2)\, q_2 \qquad \xi = x/a \qquad \eta = y/b \qquad (4.44)$$
$$v = \ldots + (1 - \xi^2)\, q_3 + (1 - \eta^2)\, q_4\,, \tag{4.45}$$

which can display constant curvatures (see Fig. 4.14), so that the element can curve upwards and sideways. Because no coordination exists between neighboring elements—the q_i are internal degrees of freedom which by *static condensation* (= Gaussian elimination) of the element matrix are later removed—the elements penetrate or gaps develop between them.

If the element size shrinks, the strains become nearly constant, i.e., the displacements are at most linear, and therefore the sides of the element might still rotate, but they will remain straight, hence the incompatible terms become superfluous ($q_i = 0$). This is probably also the reason why this element is so successful and why it is used in many commercial codes. If the implementation is done correctly, the elements are very stable elements [154].

The plate in Fig. 4.15 was analyzed with bilinear elements. Eight layers (rows) of elements, each comprising eighty elements, were necessary to come close to the exact deflection $w = 1.024$ cm at the end of the cantilever plate (see Table 4.3), while Wilson's element achieved the same result with just eight elements. Typical of a nonconforming element, the FE solution overestimates the deflection.

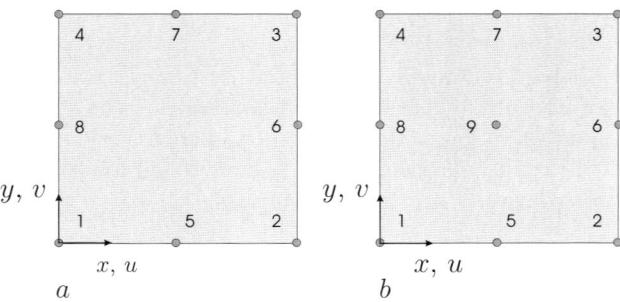

Fig. 4.16. Plane elements: **a)** square eight-node serendipity-element Q8, **b)** 9-node Lagrange element with bi-quadratic shape functions

Because the internal degrees of freedom q_i are later eliminated by static condensation, the associated equivalent nodal forces are set to zero. Before the condensation, the basic set of equations is

$$\begin{bmatrix} \boldsymbol{K}_{uu} & \boldsymbol{K}_{uq} \\ \boldsymbol{K}_{qu} & \boldsymbol{K}_{qq} \end{bmatrix} \begin{bmatrix} \boldsymbol{u}_{(8)} \\ \boldsymbol{q}_{(4)} \end{bmatrix} = \begin{bmatrix} \boldsymbol{f}_{(8)} \\ \boldsymbol{0}_{(4)} \end{bmatrix} \qquad (4.46)$$

and afterwards it separates into two parts

$$\boldsymbol{K}_{(8\times 8)}\boldsymbol{u}_{(8)} = \boldsymbol{f}_{(8)} \qquad \boldsymbol{q}_{(4)} = \boldsymbol{Q}_{(4\times 8)}\,\boldsymbol{u}_{(8)}\,, \qquad (4.47)$$

with

$$\boldsymbol{K} = \boldsymbol{K}_{uu} - \boldsymbol{K}_{uq}^T\,\boldsymbol{K}_{qq}^{-1}\,\boldsymbol{K}_{qu}\,, \qquad \boldsymbol{Q} = -\boldsymbol{K}_{qq}^{-1}\,\boldsymbol{K}_{uq}\,, \qquad (4.48)$$

where the q_i are the dependent degrees of freedom.

In the standard FE notation see Sect. 1.47, p. 221, the strain vector $\boldsymbol{\varepsilon}$ which results from a deformation $\boldsymbol{u}, \boldsymbol{q}$ is

$$\boldsymbol{\varepsilon} = \boldsymbol{B}_u\,\boldsymbol{u} + \boldsymbol{B}_q\,\boldsymbol{q}\,, \qquad (4.49)$$

and after condensation it becomes

$$\boldsymbol{\varepsilon} = \boldsymbol{B}\,\boldsymbol{u} \qquad \boldsymbol{B} := \boldsymbol{B}_u + \boldsymbol{B}_q\,\boldsymbol{Q}\,. \qquad (4.50)$$

Lagrange and serendipity elements

In *Lagrange elements* the shape functions are Lagrange polynomials. These elements are rectangular elements with edge nodes and internal nodes. *Serendipity elements* dispense with internal nodes, but they are therefore incomplete. The square eight-node element Q8 (Fig. 4.16 a) is such an element. The shape functions of this element are obtained if the quadratic polynomials of the LST element (4.40) and (4.41) are enriched with cubic terms:

$$u(x, y) = \ldots + a_6\,x^2\,y + a_7\,x\,y^2\,, \qquad (4.51)$$
$$v(x, y) = \ldots + b_6\,x^2\,y + b_7\,x\,y^2\,. \qquad (4.52)$$

Higher-order elements

In elements with higher-order polynomials (cubic, etc.) derivatives $u_{,x}$ or $v_{,y}$ appear as nodal degrees of freedom, which will cause difficulties if the stresses are discontinuous at a node. Also the coupling of such elements with other elements can lead to difficulties, because the displacements are no longer compatible at interelement boundaries.

Drilling degrees of freedom

The nodes of a plate have no rotational stiffness, which makes it difficult to couple a plate to a beam.

To add such a stiffness, one starts for example with an LST-element (triangular element with quadratic polynomials and additional nodes at the sides), and pretends that the element has rotational degrees of freedom at the corner points. Such rotational degrees of freedom would enable one to calculate the displacements at the mid-side nodes. Hence, it seems possible to sacrifice the degrees of freedom at the mid-side nodes to establish rotational degrees of freedom at the corner nodes instead. This is the basic idea, though some additional mathematical tricks are necessary to make this idea work ([69], [42], [5]).

The invaluable advantage of these elements is the incorporation of rotational degrees of freedom, which makes these elements suitable for modeling folded plates or shells. But because only one rotational degree of freedom is added at the corner nodes, while the mid-side nodes actually have two degrees of freedom, the possibilities of this modified element are somewhat restricted compared with either a fully isoparametric or nonconforming element.

4.5 The patch test

Irons proposed the patch test originally to check the convergence of nonconforming elements [125]. Although passing a patch test—theoretically at least—is neither necessary nor sufficient for the convergence of an FE solution [235], it is a very good test to check and compare elements.

The patch test is based on the observation that the stress distribution becomes more and more uniform the smaller the elements become. Therefore convergence can only be expected if an FE program can solve load cases with uniform stress states exactly.

In a wider sense, a patch test is simply a test to reproduce a certain stress distribution on a given mesh.

Wilson's improved bilinear element Q4 + 2 often yields better results than the original bilinear element Q4, even though it is a nonconforming element. To study the behavior of these two elements side by side, a cantilever plate was subjected to three standard load cases, producing

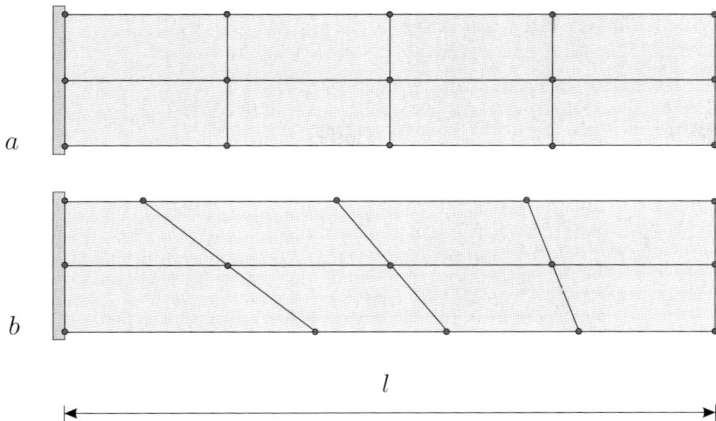

a

b

l

Fig. 4.17. These meshes should have no difficulty reproducing simple stress states: **a)** regular mesh, **b)** irregular mesh

- a constant moment
- linear bending moments = constant shear forces
- quadratic bending moments = linear shear forces

Each load case was solved on a relatively coarse mesh consisting of eight rectangular elements, and alternatively on a distorted mesh; see Fig. 4.17 b.

When the plate is stretched uniformly, this will produce a *constant normal force*, and this load case must be solved exactly. This is the original patch test. Both elements passed this test. But in the other three load cases, distinct deviations from the beam solution appeared; see Table 4.4. As expected the

Table 4.4. Normal stress σ (kN/m^2), R = regular mesh, I = irregular mesh, Q4 = bilinear element, Q4 + 2 = Wilson

Moment	Constant		Linear		Quadratic	
Mesh	$x = 0.0$	$x = 1/2$	$x = 0.0$	$x = 1/2$	$x = 0.0$	$x = 1/2$
Exact	1500	1500	1200	600	1200	300
R. Q4 + 2	1500	1500	1051	600	940	337
I. Q4 + 2	1322	1422	940	701	773	452
R. Q4	1072	1072	745	428	659	240
I. Q4	687	578	454	187	393	172

stresses at edge nodes were not as accurate as stresses at internal nodes, but nevertheless it is remarkable how much difficulty the bilinear element had in modeling the bending states on such a coarse mesh. In particular, the errors in the shear stresses were large; see Table 4.5. On the regular mesh

Table 4.5. Shear stress τ (kN/m^2)

Moment	Constant		Linear		Quadratic	
Mesh	$x = 0.0$	$x = 1/2$	$x = 0.0$	$x = 1/2$	$x = 0.0$	$x = 1/2$
Exact	0	0	50	50	100	50
R. Q4 + 2	0	0	50	50	87.5	50
I. Q4 + 2	58	28	65	80	130	73
R. Q4	438	0	364	8	376	8
I. Q4	502	220	380	294	366	11

a

b

Fig. 4.18. Displacements produced on the irregular mesh under constant horizontal volume forces: **a)** nonconforming element Q4+2 (Wilson), **b)** conforming element Q4 (bilinear)

the nonconforming Wilson element yielded the exact solution (the value 87.5 instead of 100 in the last column is due to the fact that some of the load is reduced directly into the support nodes). In the bilinear element the incorrect shear forces have nearly the same magnitude as the normal stresses. In the nonconforming solution they are a factor of 4 to 10 smaller.

The poor properties of the bilinear element also become apparent if a constant horizontal volume force is applied; see Fig. 4.18. Though the stresses of the two solutions, Q4 and Q4 + 2, are similar, the lateral displacements of the bilinear elements must cause concern. These displacements are due to a Poisson ratio $\nu > 0$. They are not that large, but they cause asymmetries—even in the stresses—and if the structure is statically indeterminate, we should be very careful.

4.6 Volume forces

Volume forces are transformed into equivalent nodal forces by letting the load act through the nodal unit displacements

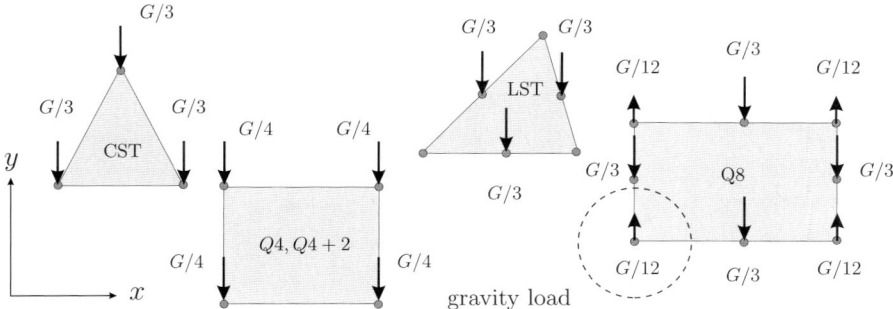

Fig. 4.19. Equivalent nodal forces in load case g as a fraction of the total dead weight $G = \gamma A_e\, t$ for different elements. Note that the corner forces of the *LST*-element are zero, and those of the $Q8$ element even have a negative sign

$$f_i = \int_\Omega \boldsymbol{p} \boldsymbol{\cdot} \boldsymbol{\varphi}_i \, d\Omega \,. \tag{4.53}$$

These equivalent nodal forces are in agreement with the elementary rules of structural mechanics. Constant volume forces as $\boldsymbol{g} = [0, \gamma]^T$ (gravity load) are distributed in equal parts onto the nodes of a CST element (see Fig. 4.19), while the corner nodes of a quadratic LST element are load-free, because the integrals of the shape functions are zero in this load case; see Fig. 4.19. Note that the corner forces of the $Q8$ element point upward, because the integrals of the pertinent shape functions are negative. Of course the sum of all equivalent nodal forces f_i is equal to the total weight $G \cdot 1$.

4.7 Supports

Fixed supports are perfectly rigid by nature. Such supports can be point supports or line supports. While point supports are ambiguous in nature, line supports are legitimate constraints, see Fig. 4.20 and Fig. 4.21.

At a roller support, the displacements normal to the support are zero, $\boldsymbol{u}^T \boldsymbol{n} = 0$. If the roller support is inclined, *kinematic constraints* couple the horizontal and vertical displacements. A slight misalignment of the supports can lead to dramatically different results. Plates are very sensitive to geometrical constraints or other such incompatibilities.

Care should also be taken not to constrain a structure unintentionally in the horizontal direction, because it could easily suggest a load bearing capacity that later fails to materialize because of nonexistent abutments.

Therefore a correct assessment of the support characteristics is very important. Only if the supports of the continuous beam in Fig. 4.22 are perfectly rigid will the support reactions agree with the ratios 0.36:1:1:0.36 that we expect in a continuous beam. If instead the four rigid supports are replaced by four 0.24 m × 0.24 m × 2.88 m columns with vertical stiffness

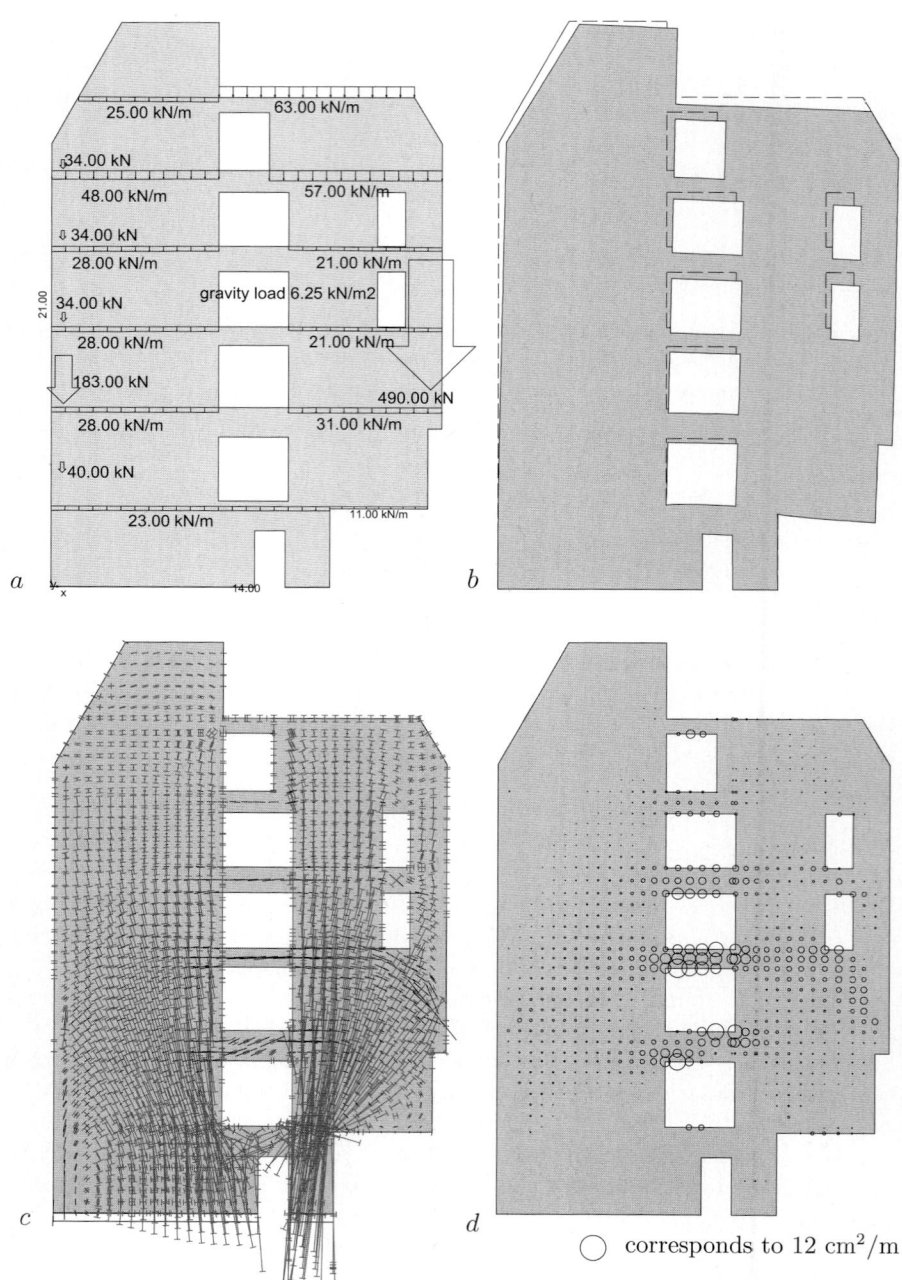

Fig. 4.20. Wall: **a)** loading, **b)** displacements, **c)** principal stresses, **d)** qualitativ representation of the horizontal reinforcement as-x cm²/m

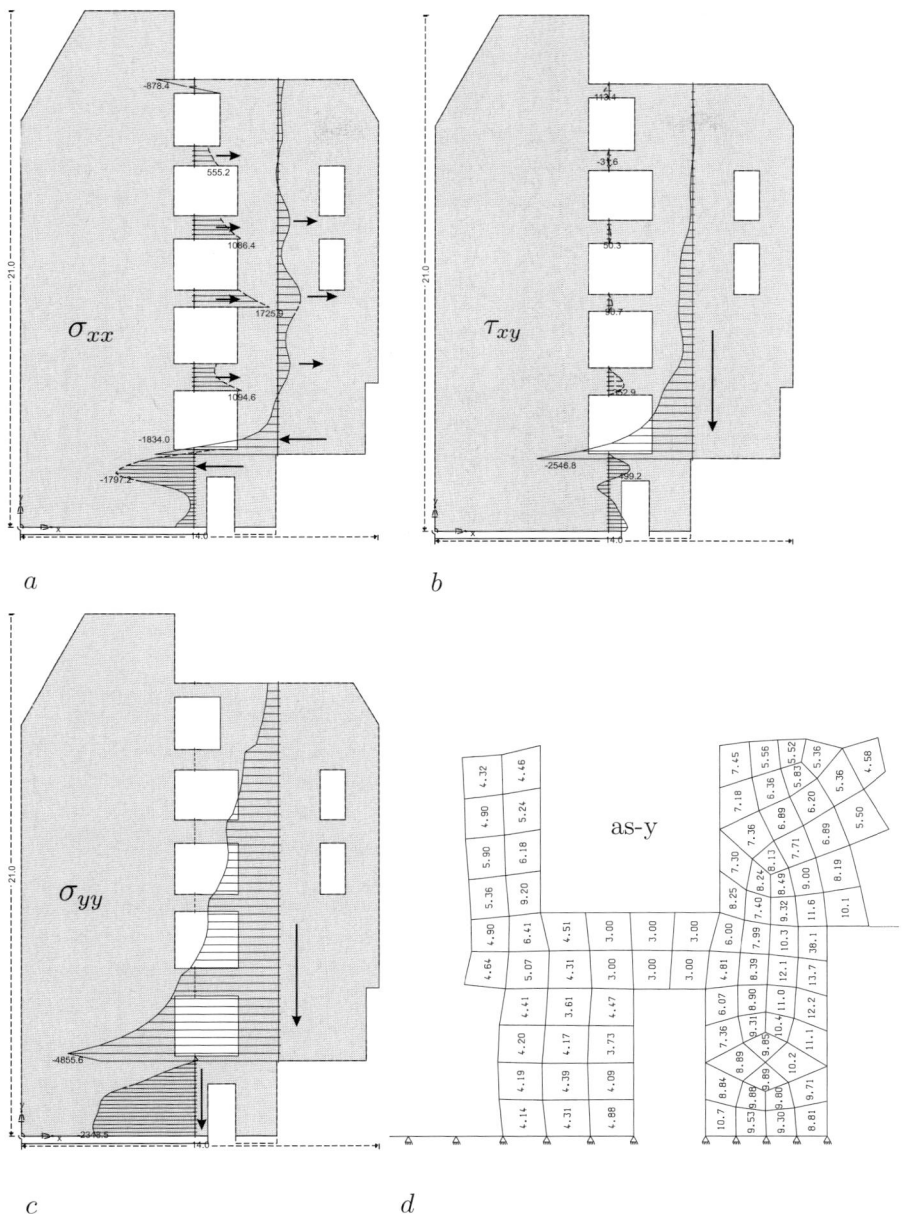

Fig. 4.21. Stresses and reinforcement: **a)** σ_{xx} **b)** τ_{xy} **c)** σ_{yy} **d)** part of the reinforcement as-y cm^2/m in the vertical direction

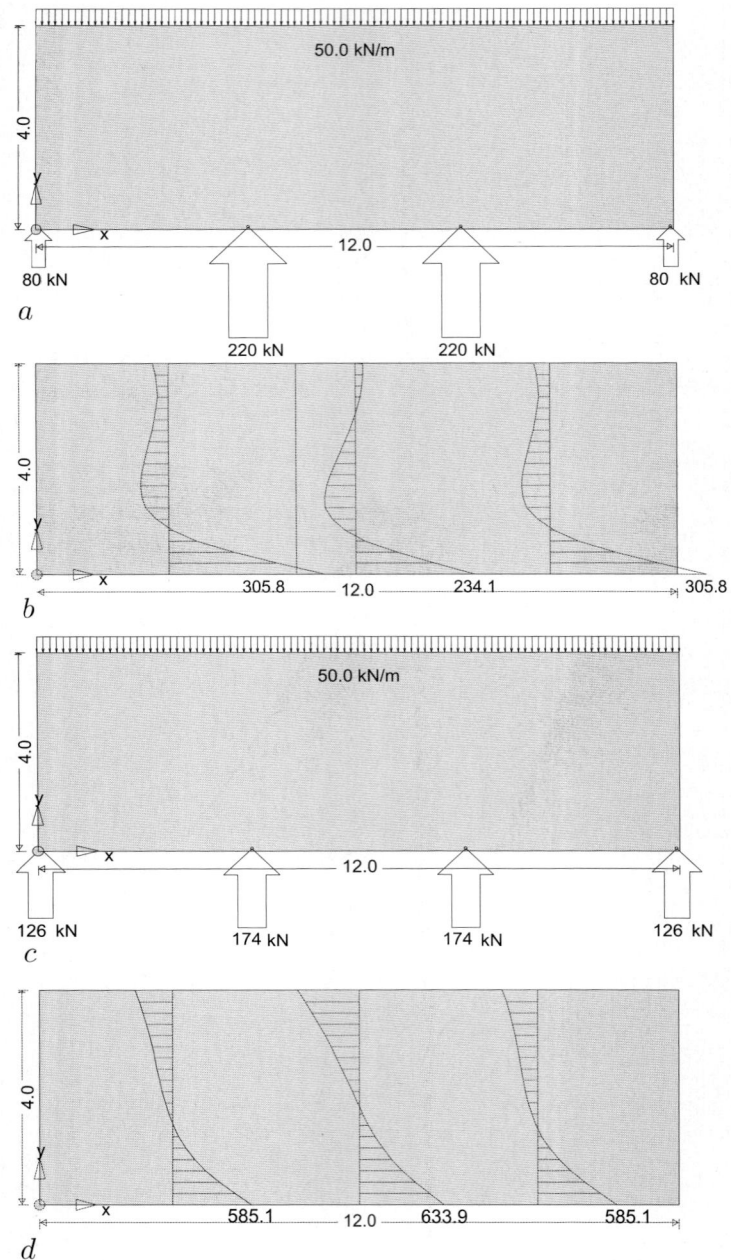

Fig. 4.22. Influence of the support stiffness: **a)** rigid support, **b)** stresses σ_{xx}, **c)** soft support (brickwork columns), **d)** stresses σ_{xx} kN/m^2

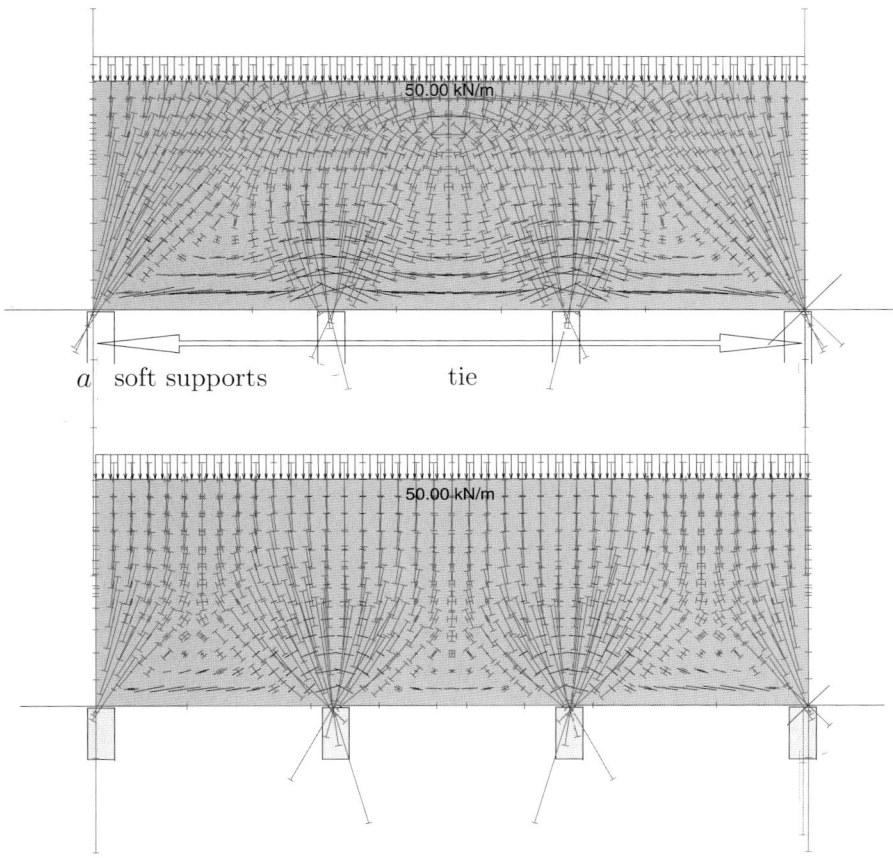

a soft supports   tie

b hard supports

Fig. 4.23. A shear wall: **a)** on soft supports the wall tends to behave like a long flat arch held taut by a tie rod; **b)** hard supports allow the plate to carry the load directly to the intermediate supports

$$k = \frac{E\,A}{h} = \frac{30\,000\text{MN/m}^2 \cdot 0.24 \cdot 0.24\,\text{m}^2}{2.88\,\text{m}} = 6.0 \cdot 10^5 \text{kN/m}\,, \quad (4.54)$$

the support reactions are much more evenly distributed, as indicated by the ratios 0.72:1:1:0.72.

The softer the supports, the closer the structural behavior of the plate to a long flat arch held taut by a tie rod; see Fig. 4.23.

Whenever possible, the support reactions should be compared with the support reactions of an equivalent beam model (see Fig. 4.24) and the reinforcement should be checked by working with approximate lever arms z [103]. In the following formulas d denotes the width of the wall, l the length of the span, and z_F and z_S the lever arms in the span and at the intermediate

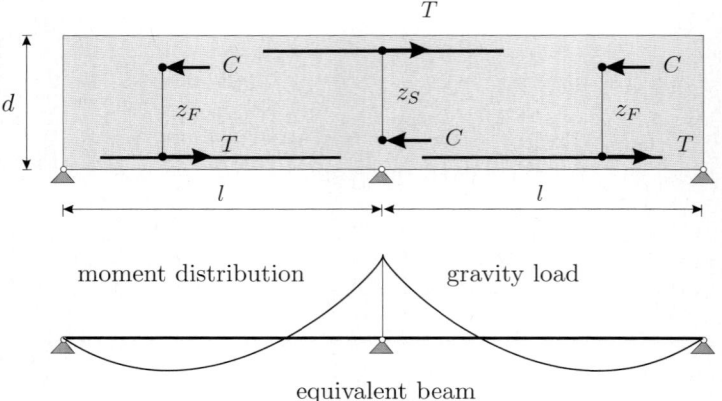

Fig. 4.24. Shear wall and equivalent beam

supports:

One span

$$z_F = 0.3 \cdot d \cdot (3 - d/l) \qquad\qquad 0.5 < d/l < 1.0$$
$$z_F = 0.6 \cdot l \qquad\qquad d/l \geq 1.0$$

Two-span beams and last span of a continuous beam

$$z_F = z_S = 0.5 \cdot d \cdot (1.9 - d/l) \qquad\qquad 0.4 < d/l < 1.0$$
$$z_F = z_S = 0.45 \cdot l \qquad\qquad d/l \geq 1.0$$

Interior spans of continuous beams

$$z_F = z_S = 0.5 \cdot d \cdot (1.8 - d/l) \qquad\qquad 0.3 < d/l < 1.0$$
$$z_F = z_S = 0.4 \cdot l \qquad\qquad d/l \geq 1.0$$

Cantilever beam with length l_k

$$z_S = 0.65 \cdot l_k + 0.10 \cdot d \qquad\qquad 1.0 < d/l_k < 2.0$$
$$z_F = 0.85 \cdot l_k \qquad\qquad d/l \geq 2.0$$

If M is the bending moment of an equivalent beam, the tensile force $T = M/z$ and the reinforcement are

$$A_s = \frac{T}{\beta_s/\gamma} = \frac{T \text{ kN}}{28.6 \text{ kN/cm}^2} \cdot \qquad\qquad (4.55)$$

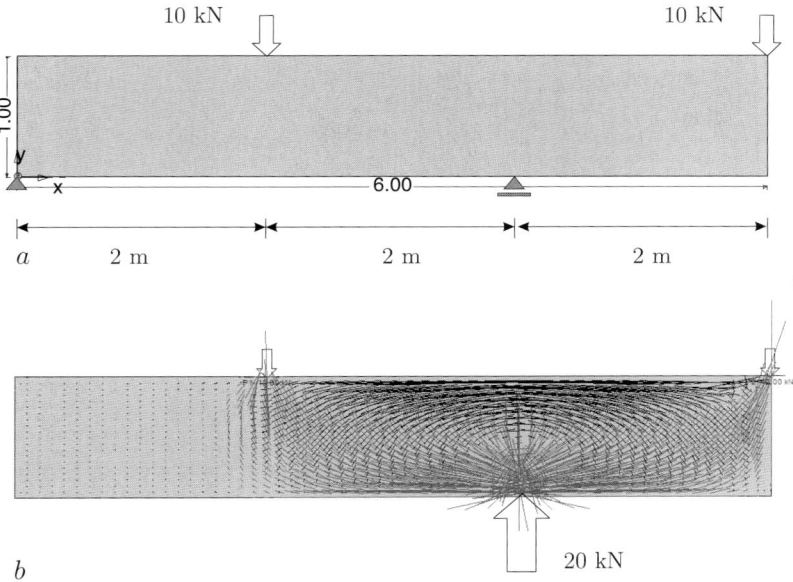

Fig. 4.25. Plate analysis of a beam with bilinear elements (Q4): **a)** system and load, where the point supports were modeled by keeping two nodes fixed, and the single forces were input as nodal forces; **b)** principal stresses, for which the support reactions agree with the beam theory

Here 28.6 kN/cm^2 is the allowable steel stress.

A check of the plate in Fig. 4.22 with these formulas shows a good agreement

	Span 1	Support	Span 2
Moment (kN/m^2)	64	-80	20
Lever arm z (m)	1.8	1.8	1.6
Tensile force T (kN)	35.6	44.4	12.5
A_s (cm^2)	1.24	1.55	0.43
A_s FE (cm^2)	1.35	1.60	0.60

Point supports

True point supports are not compatible with the theory of elasticity, because the exact support reaction would be zero. But if a node is kept fixed, it is not a point support (see Sect. 1.16, p. 55, and 1.24, p. 99). Instead the resulting support reaction agrees with the beam solution (see Fig. 4.25). Only the stresses signal that the solution is singular.

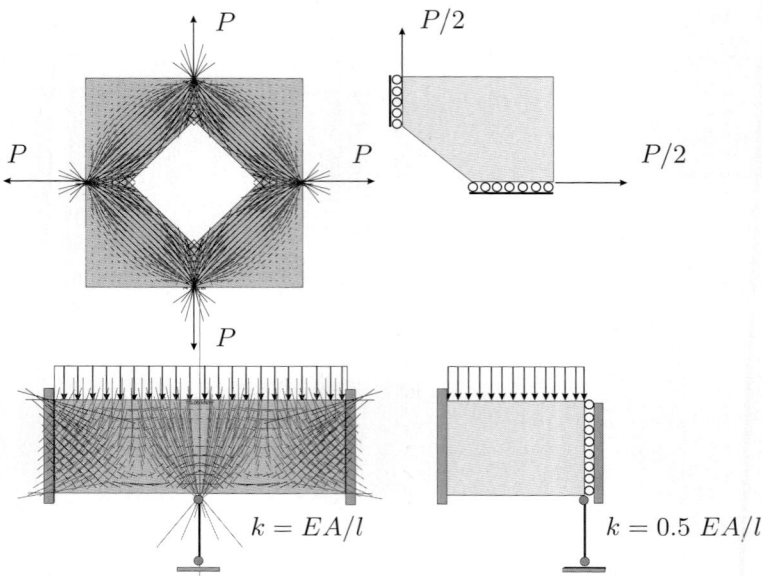

Fig. 4.26. Symmetric structures and symmetric loads

Symmetries

If the system and load are symmetric the FE analysis can be restricted to parts of a structure. Loads that happen to lie on the symmetry axis, are halved as is the stiffness of supports that lie on the axis (see Fig. 4.26).

Displaced point supports

The situation is the same as for point supports. According to the theory of elasticity, no force is necessary to displace a single point of a plate while an FE program produces the beam solution.

Tangential supports

A shear wall that is connected to floor plates experiences a deformation impediment in the tangential direction (see Fig. 4.27). Without this constraint (see Fig. 4.27 b and c; principal stresses and reinforcement), the structural behavior of the wall is similar to an arch held taut by a tie rod, while if the upper and lower floor plate constrain the wall, the distribution of the stresses in the wall is much more homogeneous and the reinforcement required is only about half as much as before.

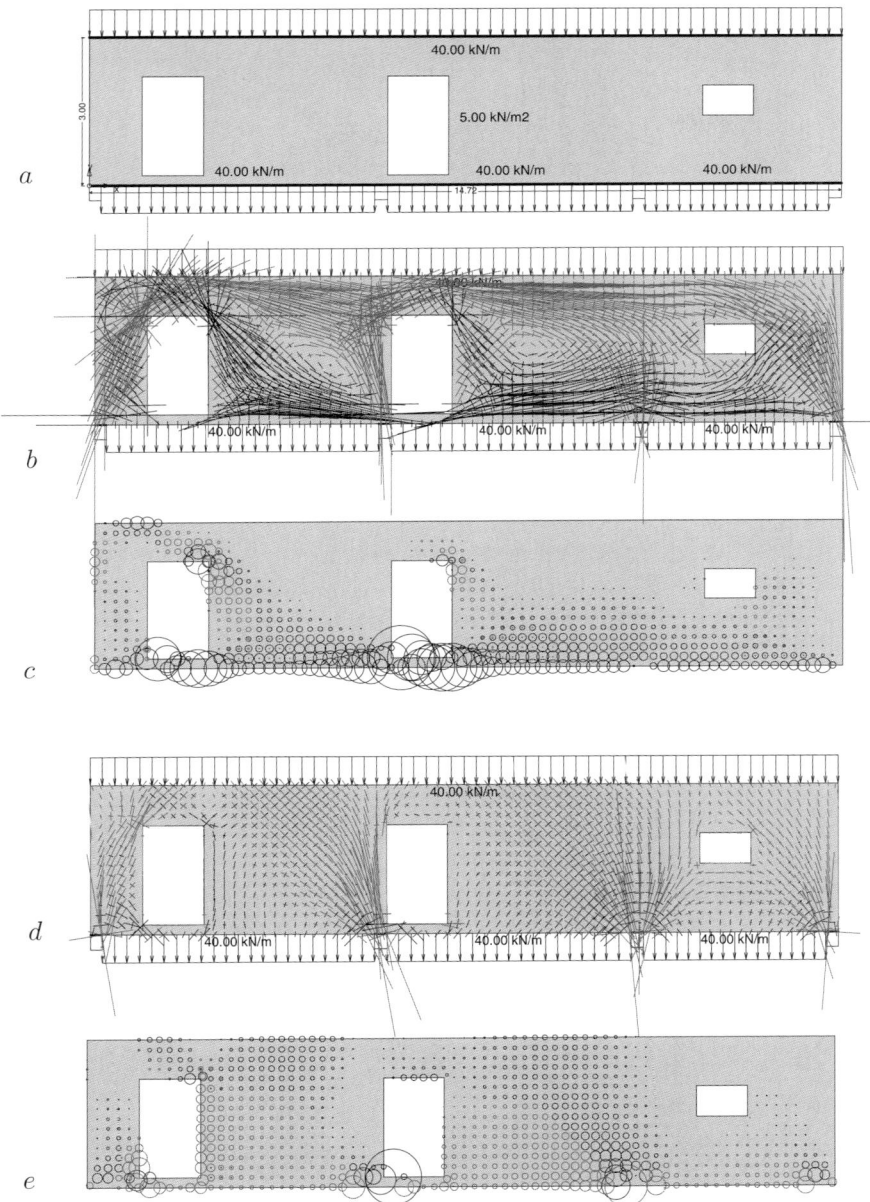

Fig. 4.27. Shear wall: **a)** system and load, **b)** principal stresses, **c)** reinforcement—only qualitatively—without tangential support, **d)** and **e)** with such a support at the upper and lower edge

truss model

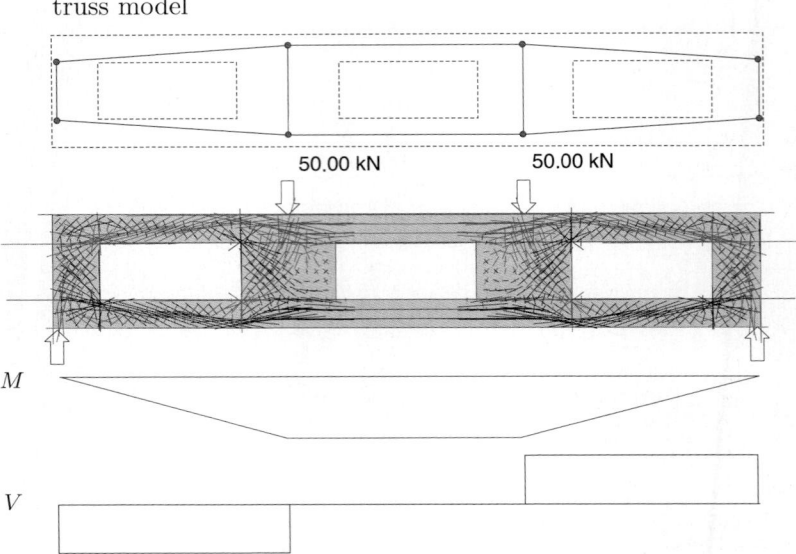

Fig. 4.28. The transfer of the shear force in the chords requires the compressive and tensile forces to be inclined. The shear forces lead to asymmetric bending moments in the chords

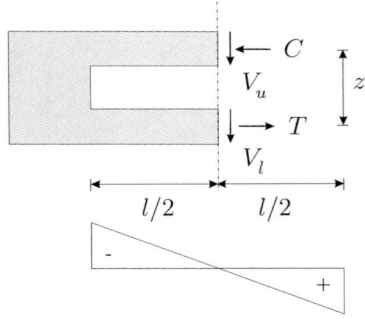

$M(x)$ within the chords

Fig. 4.29. Internal forces and bending moment distribution in the chords

Beam-like elements

Door or window lintels carry shear forces (see Fig. 4.28). The normal forces (compression and tension) in these elements are

$$C = T = \frac{M}{z} \qquad z = \text{vertical distance of the lintels}, \qquad (4.56)$$

and it can be assumed—if the concrete is not cracked—that the shear force V is evenly carried by both elements $V_u = V_l = 0.5\,V$, so that the bending moment in one element becomes (see Fig. 4.29)

$$M_{\text{element}} = 0.5 \, V \, \frac{l}{2} \qquad l = \text{length of the element}. \tag{4.57}$$

If the concrete in the lower element cracks, the stiffness will shrink, and it can then be assumed that all the shear force is carried by the upper beam which makes it necessary to provide hanger reinforcement to carry the shear force to the upper element.

To investigate how Wilson's element can represent such bending states, the column in Fig. 4.30 was analyzed. The height of the column is 3 m and its width is 50 cm. It is fixed at the base and it can slide horizontally at the top where a horizontal force $P = 20$ kN is applied. The structural system corresponds to the system in Fig. 4.30 c. As can be seen from the following table

Elements Width (m) × Height (m)	M (kN m)	u (mm)
0.500 × 0.600	25	0.74
0.250 × 0.250	27.6	0.76
0.125 × 0.150	28	0.76
exact	30	0.72

with elements of size 25 cm × 25 cm, two elements across the width of the column, good results are achieved. The error in the bending moments at the top is about 8%.

4.8 Nodal stresses and element stresses

In a bilinear element of length a and width b as in Fig. 4.10, the stresses are

$$\sigma_{xx}(x,y) = \frac{E}{a\,b\,(-1+\nu^2)} \cdot \Big[b\,(u_1 - u_3) + a\,\nu\,(u_2 - u_8) +$$

$$+ x\,\nu\,(-u_2 + u_4 - u_6 + u_8) + y\,(-u_1 + u_3 - u_5 + u_7) \Big] \tag{4.58}$$

$$\sigma_{yy}(x,y) = \frac{E}{a\,b\,(-1+\nu^2)} \cdot \Big[b\,\nu\,(u_1 - u_3) + a\,(u_2 - u_8) +$$

$$+ x\,(-u_2 + u_4 - u_6 + u_8) + y\,\nu\,(-u_1 + u_3 - u_5 + u_7) \Big] \tag{4.59}$$

$$\sigma_{xy}(x,y) = \frac{-E}{2\,a\,b\,(1+\nu)} \cdot \Big[b\,(u_2 - u_4) + a\,(u_1 - u_7) +$$

$$+ x\,(-u_1 + u_3 - u_5 + u_7) + y\,(-u_2 + u_4 - u_6 + u_8) \Big]. \tag{4.60}$$

The stresses in Wilson's element are

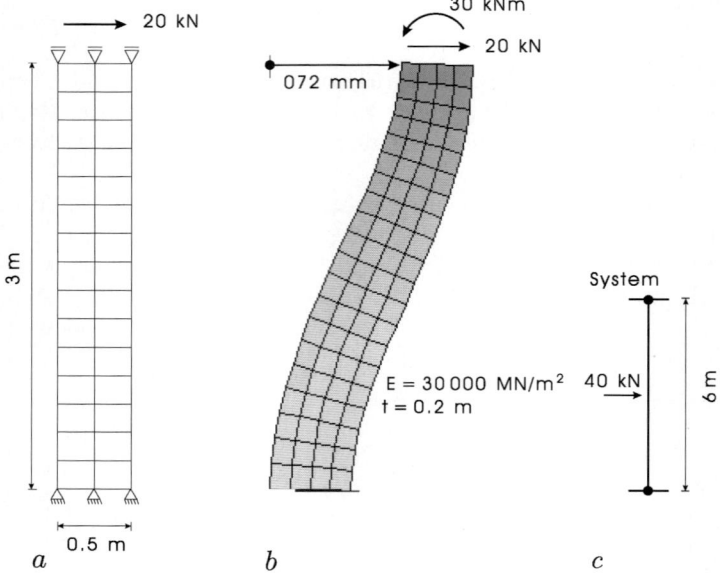

Fig. 4.30. Analysis of a column with Wilson's element

$$\sigma_{xx}(x,y) = -\frac{E}{a^2 b^2(-1+\nu^2)}\Big[8\,b^2 q_1\,x + a^2\nu\big(b(-4\,q_4 + u_2 - u_8) + 8\,q_4\,y\big)$$
$$+\,a\,b\big(b(-4\,q_1 + u_1 - u_3) + \nu\,x(-u_2 + u_4 - u_6 + u_8)\big)$$
$$+\,y(-u_1 + u_3 - u_5 + u_7)\Big] \tag{4.61}$$

$$\sigma_{yy}(x,y) = -\frac{E}{a^2 b^2(-1+\nu^2)}\Big[8\,b^2\nu q_1\,x + a^2\big(b(-4\,q_4 + u_2 - u_8) + 8\,q_4\,y\big)$$
$$+\,a\,b\big(b\nu(-4\,q_1 + u_1 - u_3) + x(-u_2 + u_4 - u_6 + u_8)\big)$$
$$+\,\nu y(-u_1 + u_3 - u_5 + u_7)\Big] \tag{4.62}$$

$$\sigma_{xy}(x,y) = -\frac{E}{2a^2 b^2(1+\nu)}\Big[-8\,b^2\nu q_3\,x + a^2\big(b(-4\,q_2 - u_1 + u_7) - 8\,q_2\,y\big)$$
$$+\,a\,b\big(b(4\,q_3 - u_2 + u_4) + x(u_1 - u_3 + u_5 - u_7)\big)$$
$$+\,y(u_2 - u_4 + u_6 - u_8)\Big]. \tag{4.63}$$

Remark 4.1. The origin of the coordinate system in these expressions is assumed to be the lower left corner of the element while in Fig. 4.10 it is the center of the element.

These equations can be simplified further. Because of (4.47),

$$q_1 = \frac{a\,\nu}{8\,b}\,[u_2 - u_4 + u_6 - u_8] \qquad q_2 = \frac{b}{8\,a}\,[u_2 - u_4 + u_6 - u_8] \quad (4.64)$$

$$q_3 = \frac{a\,\nu}{8\,b}\,[u_1 - u_3 + u_5 - u_7] \qquad q_4 = \frac{b\,\nu}{8\,a}\,[u_1 - u_3 + u_5 - u_7] \quad (4.65)$$

so that

$$\sigma_{xx}(x, y) = -\frac{E}{2\,a\,b\,(-1 + \nu^2)}\Big[b(2(u_1 - u_3) + \nu^2(-u_1 + u_3 - u_5 + u_7))$$
$$+ a\,\nu(u_2 + u_4 - u_6 - u_8) + 2(-1 + \nu^2)$$
$$(u_1 - u_3 + u_5 - u_7)\,y\Big] \qquad (4.66)$$

$$\sigma_{yy}(x, y) = -\frac{E}{2\,a\,b\,(-1 + \nu^2)}\Big[a(2(u_2 - u_8) + \nu^2(-u_2 + u_4 - u_6 + u_8))$$
$$+ b\,\nu(u_1 - u_3 - u_5 + u_7) + 2(-1 + \nu^2)$$
$$(u_2 - u_4 + u_6 - u_8)\,y\Big] \qquad (4.67)$$

$$\sigma_{xy}(x, y) = -\frac{E}{4\,a\,b(1 + \nu)}\Big[a(u_1 + u_3 - u_5 - u_7)$$
$$+ b(u_2 - u_4 - u_6 + u_8)\Big] \qquad (4.68)$$

and here it is seen that the horizontal stresses σ_{xx} only depend on y, the vertical stresses σ_{yy} only on x (in a bilinear element this is true only if $\nu = 0$), and σ_{xy} is constant (whereas it varies in a bilinear element). Therefore the volume forces in Wilson's element are zero (!):

$$- \sigma_{xx,x} - \sigma_{xy,y} = 0 \qquad (4.69)$$
$$- \sigma_{yx,x} - \sigma_{yy,y} = 0. \qquad (4.70)$$

Stress averaging

If the stress distribution is linear, the stresses are discontinuous at the element edges. This is straightened out by interpolating the stresses at the midpoints of the elements. Even in the presence of gross stress discontinuities, the results at the centers are often acceptable; see Sect. 1.22, p. 88. Similar behavior is shown at the Gauss points; see Sect. 1.25, p. 104.

According to the FE algorithm, the weighted average of the error in the stresses is zero on each patch Ω_i,

$$\int_{\Omega_i} (\boldsymbol{\sigma} - \boldsymbol{\sigma}_h) \bullet \varepsilon_i \, d\Omega = 0 \qquad i = 1, 2, \ldots . \qquad (4.71)$$

where $\varepsilon_i = [\varepsilon_{xx}^{(i)}, \varepsilon_{yy}^{(i)}, \gamma_{xy}^{(i)}]^T$ are the strains that belong to the unit displacement fields $\boldsymbol{\varphi}_i$, and the patch Ω_i is the support of the field (where the strains

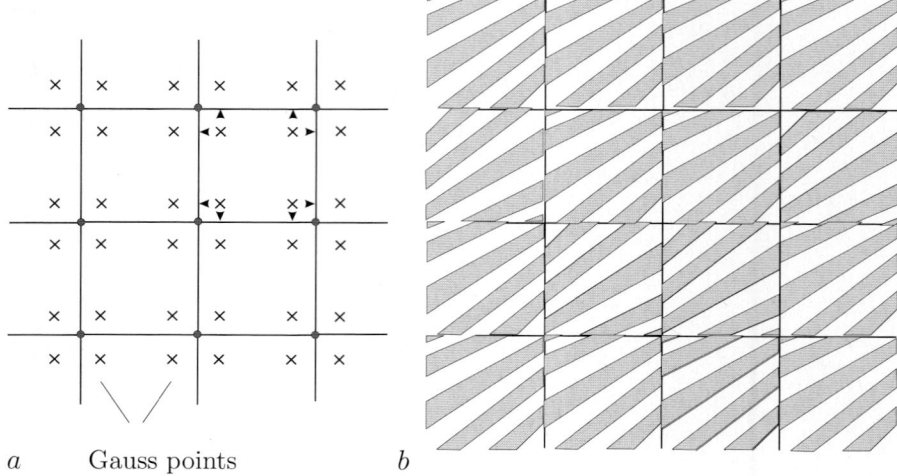

a Gauss points b

Fig. 4.31. Gauss points: **a)** the stresses at the Gauss points are extrapolated to the boundary, **b)** contour lines of the stresses σ_{xx}, the jumps are a measure of the quality of an FE mesh (Bathe)

are nonzero). Hence we may expect that the error in the stresses $\boldsymbol{\sigma} - \boldsymbol{\sigma}_h = [\sigma_{xx} - \sigma_{xx}^h, \sigma_{yy} - \sigma_{yy}^h, \tau_{xy} - \tau_{xy}^h]^T$ changes its sign on Ω_i repeatedly. Good candidates for these zeros are obviously the centers of the elements and the Gauss points; see Sect. 1.25, p. 104.

For the same reason the stresses at the edge of an element are the least reliable, and they are often replaced by values extrapolated from the Gauss points out to the edge; see Fig. 4.31. In the next step the solution is "improved" by averaging the stresses between the elements and at the nodes, because this is what a user wants to see. But one has to be careful.

In between two elements with different thickness (see Fig. 4.32 a), the normal force $N_n^{(1)} = \sigma_{nn}^{(1)} \cdot t_1 = \sigma_{nn}^{(2)} \cdot t_2 = N_n^{(2)}$ (orthogonal to the common edge) is the same, but the strains are discontinuous, $\varepsilon_{nn}^{(1)} \neq \varepsilon_{nn}^{(2)}$. If the thickness is the same but the modulus of elasticity changes, the stresses σ_{tt} parallel to the common edge are discontinuous. Because the strains ε_{tt} parallel to the edge are the same, if we let $\nu = 0$, then

$$\sigma_{tt}^{(1)} = E_1 \, \varepsilon_{tt} \neq E_2 \, \varepsilon_{tt} = \sigma_{tt}^{(2)} \, . \tag{4.72}$$

This means that eventually the reinforcement parallel to the edge is different (see Fig. 4.33). Initial stresses σ_{ij}^0 would complicate things even more. A good FE program will average the stresses only if the elements have the same elastic properties and the elements are load free, and no supports interfere with the stress distribution.

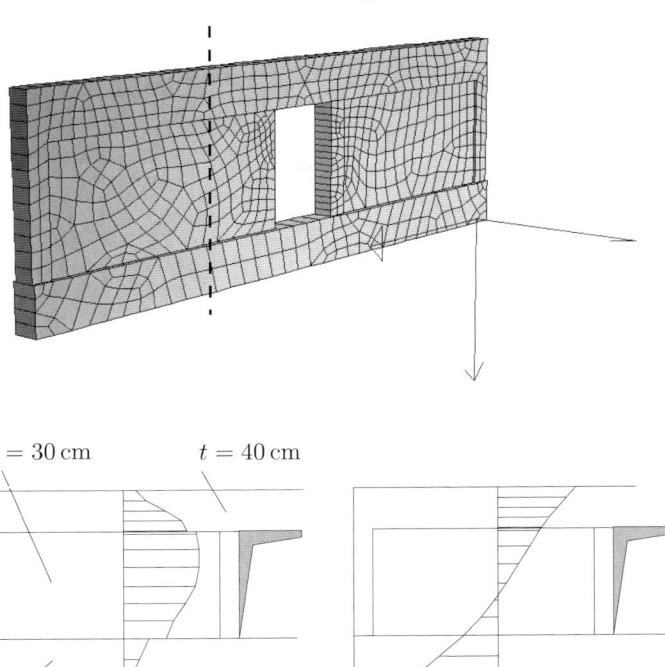

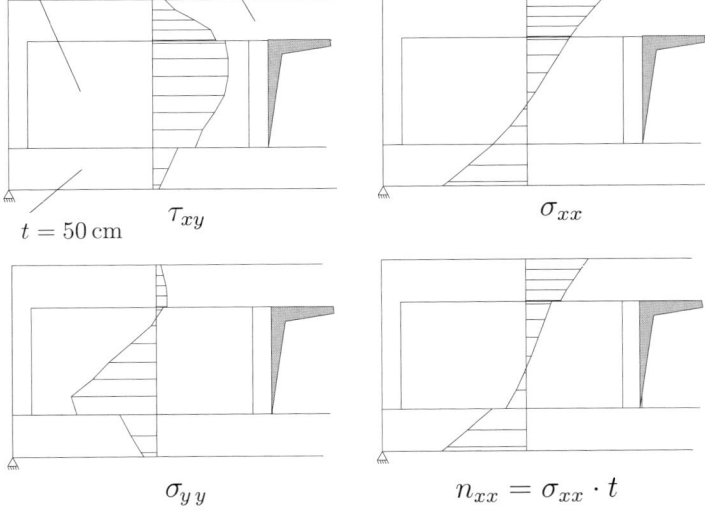

Fig. 4.32. Stress distribution in a vertical cross section of a shear wall with varying thickness under gravity load

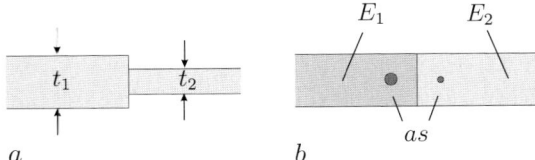

Fig. 4.33. a) The thickness of the plate changes; **b)** the modulus of elasticity changes, and therefore the stresses parallel to the interface are not the same

free edge

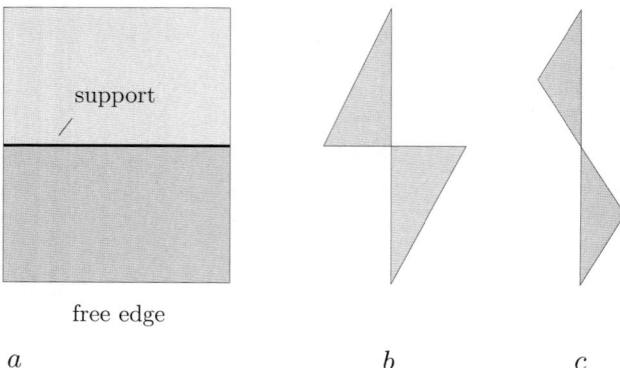

support

free edge

a   *b*   *c*

Fig. 4.34. The upper element presses on the support and the lower element pulls at the support: **a)** system, **b)** normal stresses in the element without averaging, **c)** with averaging (each time with two elements)

A small example will illustrate the implications a *smoothing process* can have when it ignores the structural context. If two elements—one placed upon the other—are fixed along their common boundary, the gravity load will produce compression in the upper element and tension in the lower element; see Fig. 4.34.

If the nodal stresses are averaged, the stresses at the fixed nodes will be zero. Hence if the system consists of just two elements, the stresses at all the nodes will be zero, and therefore—if the results are extrapolated into the interior—the stresses will be zero everywhere. If the two elements are split into two elements each, the stresses will still be halved. Only if very many elements are used will the averaging process leave no trace.

The misalignment of the contour lines of the raw stresses at the interelement boundaries offers a good visual control of the quality of a mesh (see Fig. 4.31). But very often these discontinuities are so strong that the user is irritated, and therefore program authors tend to display only the averaged stresses. Nevertheless stress discontinuities in the range 5 to 15% are in no way unusual, and not a warning sign. Even discrepancies of up to 40% can be tolerated if the design is based on the stresses at the midpoints of the elements.

Averaging the stresses at the edges and the nodes is the simplest way to improve the results. More sophisticated methods use an L_2 projection to improve the stresses; see Sect. 1.31, p. 147.

What is seen on the screen is often not the raw output; see Fig. 4.35. Hence to judge an FE program one must know which filters the program uses, how it displays the result, and what smoothing algorithms are employed.

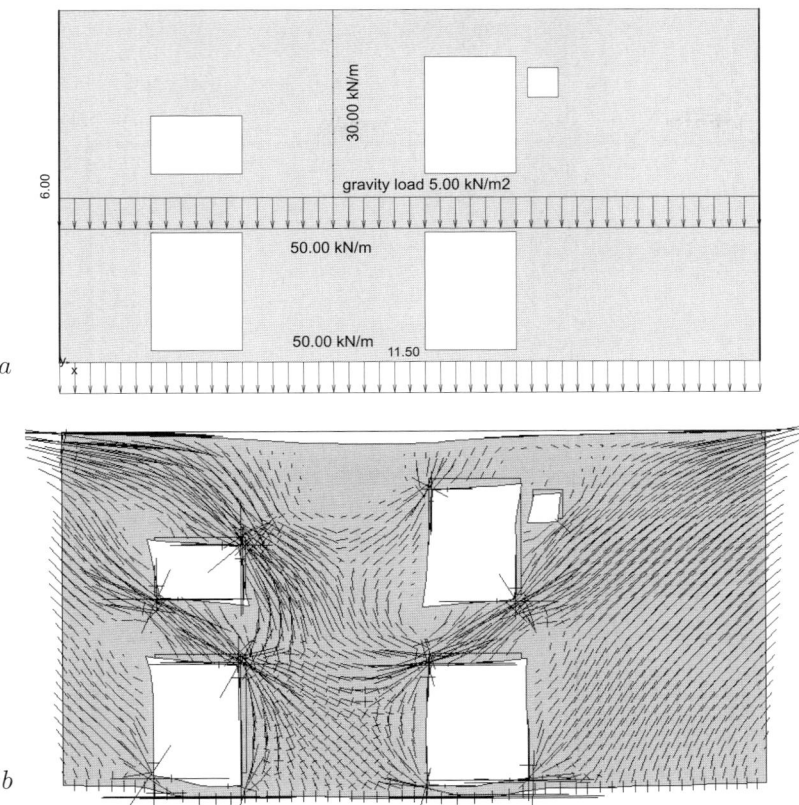

Fig. 4.35. Only the tensile stresses are displayed. The shear wall is fixed on the left- and right-hand side

4.9 Truss and frame models

If a truss is used to approximate a plate (similar to Fig. 4.36), then in each bar element the longitudinal displacement is a linear function

$$u_e(x) = u_1 \left(\frac{l-x}{l}\right) + u_2 \frac{x}{l} \tag{4.73}$$

and the deformed shape of the truss is found by minimizing the potential energy

$$\Pi(u) = \frac{1}{2} \sum_e \int_0^l \frac{N_e^2}{EA} \, dx - \sum_e \int_0^l p \, u_e \, dx \,. \tag{4.74}$$

In an FE analysis with plate elements, the expression for the potential energy is instead

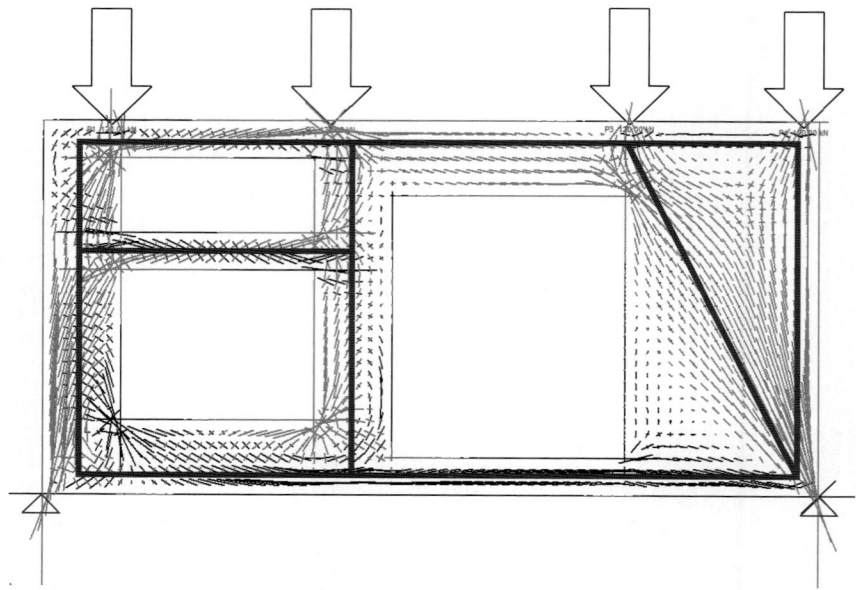

Fig. 4.36. Frame model of a shear wall

$$\Pi(\boldsymbol{u}) = \frac{1}{2}\int_\Omega \boldsymbol{S}\cdot\boldsymbol{E}\,d\Omega - \int_\Omega \boldsymbol{p}\cdot\boldsymbol{u}\,d\Omega\,. \qquad (4.75)$$

These two equations are not that different. If the plate Ω were subdivided in accordance with the direction of the principal forces into elements of constant width, and if all elements with stresses below a certain threshold limit were neglected, and if it were assumed that the stresses across the elements are constant, then the result would be Eq. (4.74). Hence truss or frame models of 2-D and 3-D continua are simplified FE models. Or stated otherwise, a truss model is not per se a better or more authentic engineering model than an FE model. It even has severe defects because it does not register the stress concentrations at the corner points, see Fig. 4.36. And if we ignore these stresses and other finer details what is an FE analysis of a plate good for?

But the true value of a truss model is that it enables us to visualize easily how the loads are carried by a structure. The more FE technology advances and the more massively parallel computing power is employed, the more a sound analysis tool is needed that is capable of filtering the essential and important information from the vast output produced by a computer program. Thus, it seems clear that we will soon see a revival of the old manual methods—but under a different name. Experienced structural engineers have always trusted these methods more than FE methods...

Table 4.6. Support reactions (kN) of the wall in Fig. 4.37, Q4 = bilinear elements, Q4+2 = Wilson's element

support	Q4	Q4+2	BE	beam
A	87	87	86	75
B	281	282	284	302
C	132	131	130	123
sum	500	500	500	500

4.10 Two-bay wall

The two-bay wall in Fig. 4.37, which carries a constant edge load of 50 kN/m on the upper edge, was analyzed with bilinear elements (Q4), Wilson's element (Q4 + 2), and boundary elements. As expected (see Table 4.6), the support reactions of the three solutions are nearly identical, but also the stresses in section $A - A$ do not deviate very much; see Fig. 4.38. This is a bit of a surprise, considering the seemingly large distance between the load case $\boldsymbol{p}_h$ (Q4) and the original load case $\boldsymbol{p}$; see Figure 4.37 a and b.

4.11 Multistory shear wall

The multistory shear wall in Fig. 4.39 can serve as an illustration for how the results of a larger FE analysis can be checked with simple models [32].

The thickness of the wall is 25 cm. The modulus of elasticity is 30 000 MN/m², and Poisson's ratio is $\nu = 0.0$. The load consists of gravity loads, floor loads, and wind loads; see Fig. 4.39. The FE analysis is based on Wilson's element (Q4 + 2).

As a first check, the support reactions were calculated for an equivalent two-span beam with equivalent stiffness

$$
\begin{aligned}
EI &= E \cdot \frac{h^3 \cdot b}{12} \\
&= \frac{30\,000\,\mathrm{MN/m^2} \cdot (13.06\,\mathrm{m})^3 \cdot 0.25\,\mathrm{m}}{12} = 1\,392\,225\,\mathrm{MNm^2} \quad (4.76)
\end{aligned}
$$

placed on elastic supports (see Fig. 4.40)

$$
c_w = \frac{EA}{h} = \frac{30\,000\,\mathrm{MN/m^2} \cdot 0.4\,\mathrm{m} \cdot 0.25\,\mathrm{m}}{4.74\,\mathrm{m}} = 633\,\mathrm{MN/m} \quad (4.77)
$$

and carrying a corresponding equivalent load. The results agree quite well with the FE/BE results; see Table 4.7.

The internal actions were checked in two sections (see Fig. 4.41), $x = 3$ m and $x = 10$ m; see Table 4.8 and Fig. 4.42. The normal force $N = N_x$ and the shear force $V = N_{xy}$ were calculated by integrating the stresses σ_{xx} and σ_{xy},

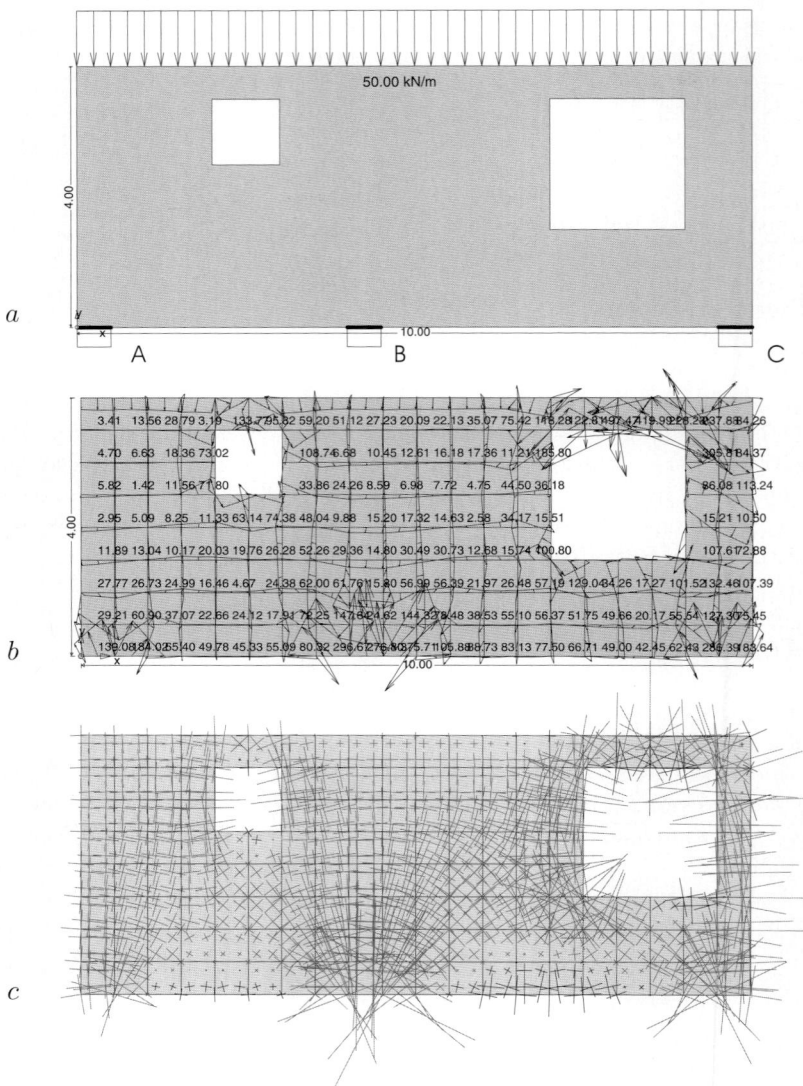

Fig. 4.37. Analysis of a shear wall with bilinear elements (Q4): **a)** the original load case p, **b)** the FE load case p_h consists of line loads and volume forces, **c)** principal stresses

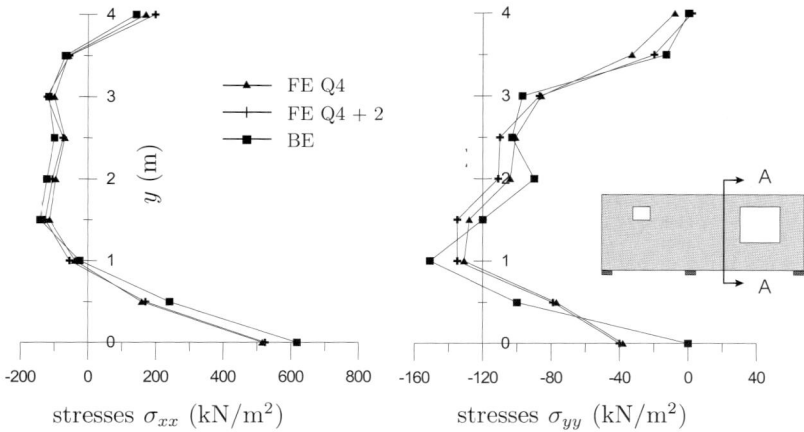

Fig. 4.38. Bending stresses σ_{xx} and shear stresses σ_{xy} in a vertical section; study of three different results

Table 4.7. Support reactions (kN) of the shear wall in Fig. 4.39

support	FE	beam	deviations %	BE
A_H	41	41	0.0	41
A_V	1493	1536	2.90	1498
B_V	1701	1589	6.60	1686
C_V	1503	1570	4.50	1511

Table 4.8. Comparison of the resultant stresses in the shear wall in Fig. 4.39

x = 3.0 m	FE	beam	deviations %
N (kN)	0	0	0
V (kN)	330	425	29
M (kN m)	4506	4898	8.7
x = 10.0 m	**FE**	**beam**	**deviations %**
N (kN)	0	0	0
V (kN)	−458	−590	28.8
M (kN m)	2461	2787	13.2

respectively. In the section $x = 10.0$, the compressive force C and the tensile force T were ± 584.6 kN. The distance between the two forces C and T was 4.21 m, so the moment is

$$M = 584.6\,\text{kN} \cdot 4.21\,\text{m} = 2461.1\,\text{kN m}, \qquad (4.78)$$

which agrees quite well with the beam moment of 2787 kN m.

A check of the lintels above the doors is not simple, because it is not clear what portion of the total shear force in the cross section is carried by

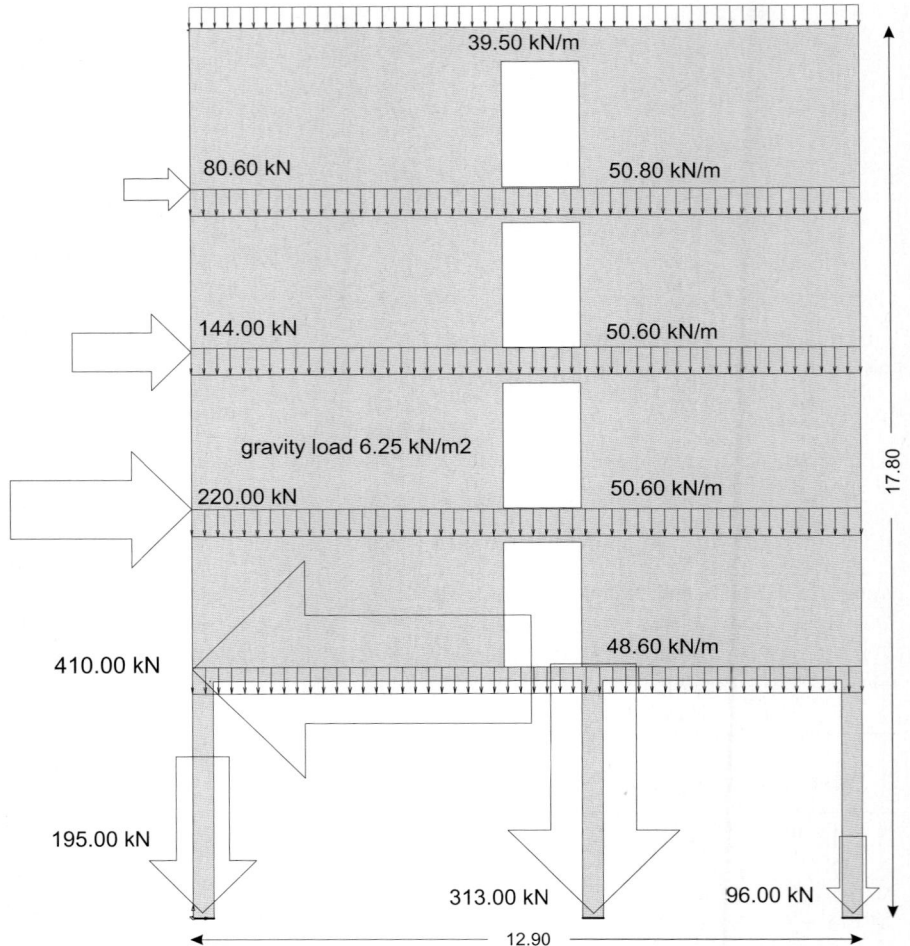

Fig. 4.39. Multistory shear wall

an individual lintel. With regard to the lintel above the ground floor, it was assumed that the concrete is cracked and that it therefore does not carry any shear force. If it is assumed that all stories carry the same load, the shear force grows linearly in the vertical direction. Hence the uppermost lintel on the fourth floor carries a shear force V_4, the next lintel a force $2 \cdot V_4$, etc., so that the total shear force across the height of the building is distributed as follows:

$$V = V_4 + V_3 + V_2 + V_1$$
$$= V_4 + 2 \cdot V_4 + 3 \cdot V_4 + 4 \cdot V_4 = 10 \cdot V_4 \,. \tag{4.79}$$

Similar considerations can be applied to check the compressive and tensile

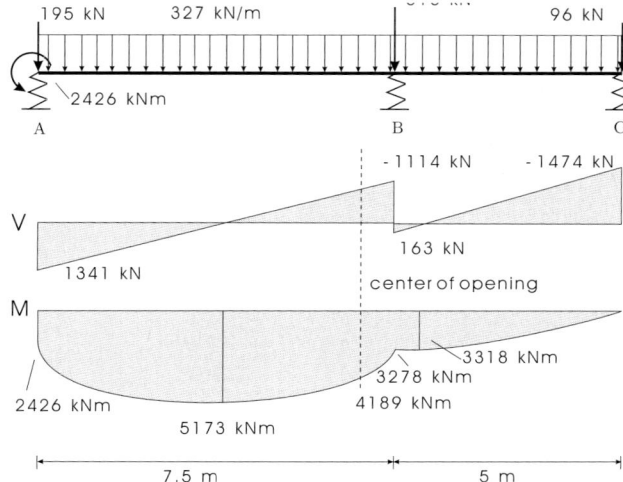

Fig. 4.40. Equivalent system

forces in the lintels. It is simply assumed that in the upper lintels—down to the second floor—only compressive forces act, which are neglected because they would reduce the reinforcement, while it is assumed that in the two lower lintels tensile forces are acting. The lever arm z is equal to the distance between the uppermost and lowermost lintel:

$$z = 13.06\,\mathrm{m} - \frac{1}{2}\,0.7\,\mathrm{m} - \frac{1}{2}\,0.26\,\mathrm{m} = 12.58\,\mathrm{m}\,. \tag{4.80}$$

The tensile force $T = M/z$ is split evenly into two parts, which are carried by the two lintels

$$T_1 = T_{\mathrm{floor}} = \frac{1}{2}\frac{M}{z}\,. \tag{4.81}$$

According to beam theory the moment M and the shear force V in the cross section are (see Fig. 4.40)

$$M = 4189.2\,\mathrm{kN\,m} \qquad V = -802.9\,\mathrm{kN}\,, \tag{4.82}$$

which yield the following forces:

$$V_4 = \frac{1}{10}\,(-802.9\,\mathrm{kN}) = -80.3\,\mathrm{kN} \tag{4.83}$$

$$V_3 = -160.6\,\mathrm{kN}\,, \quad V_2 = -240.9\,\mathrm{kN}\,, \quad V_1 = -321.2\,\mathrm{kN} \tag{4.84}$$

$$Z_{\mathrm{1st\ floor}} = Z_{\mathrm{ground\ floor}} = \frac{1}{2}\frac{4189.2\,\mathrm{kN\,m}}{12.58\,\mathrm{m}} = 166.5\,\mathrm{kN}\,. \tag{4.85}$$

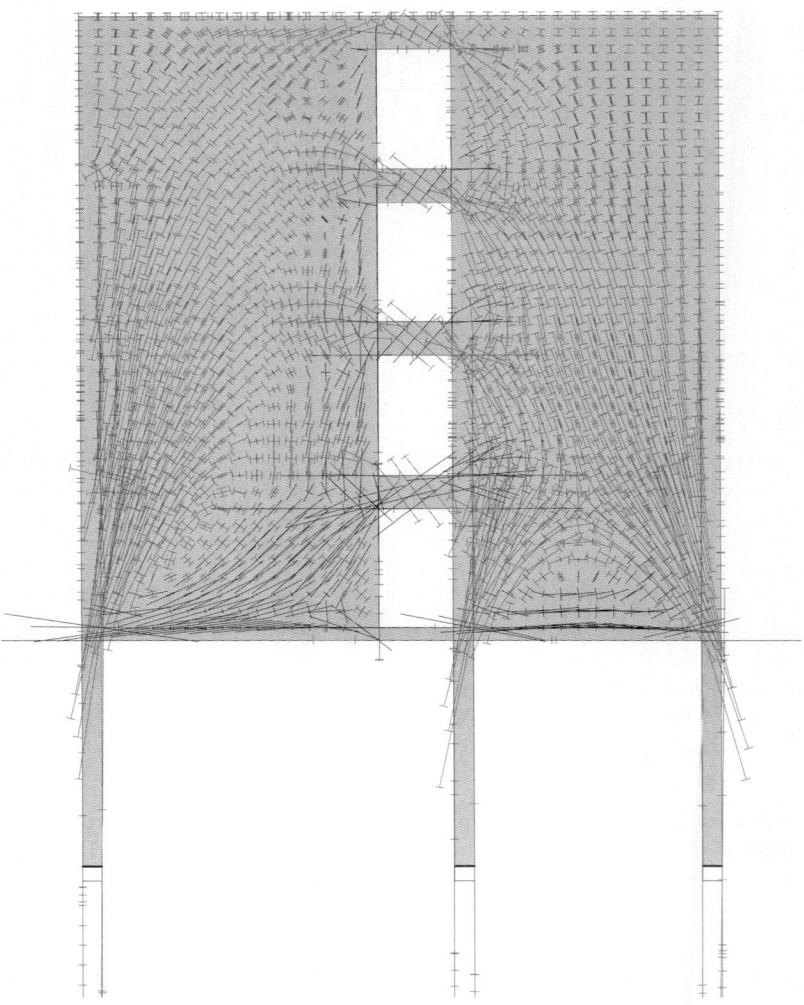

Fig. 4.41. Principal stresses

4.12 Shear wall with suspended load

Next a shear wall with a large opening is considered; see Fig. 4.43. The supports of the walls have a width of 25 cm. To be on the safe side the FE model has a span of 7.75 m and the plate is placed on two point supports.

The focus is on two details: the chord below the opening and the stress singularities at the corner points. In a beam with the same length and carrying the same distributed load the bending moment M at the center of the opening is 531 kN m, so it can be assumed that the compressive and tensile force in the upper and lower chord is approximately

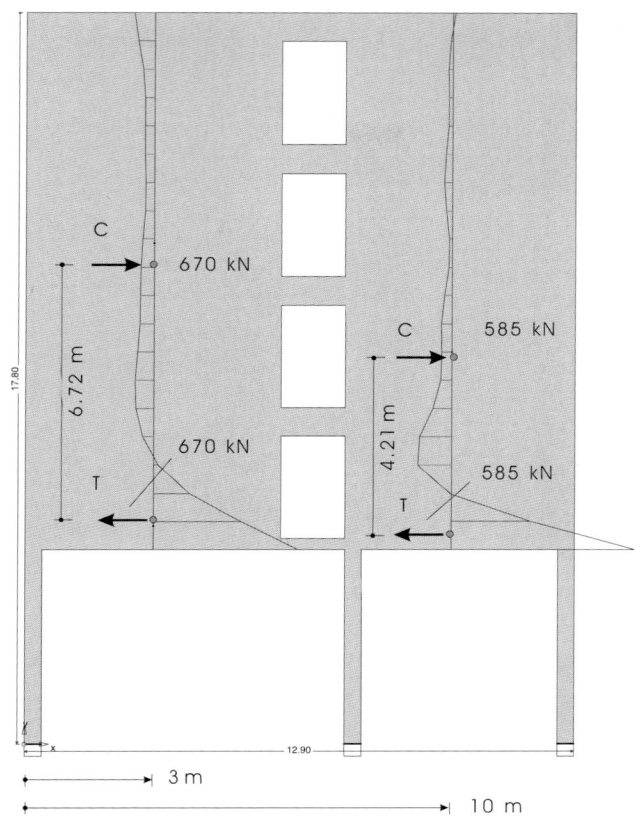

Fig. 4.42. Stresses σ_{xx} in two vertical sections, and resulting forces

$$C = T = \pm\frac{531}{3.51} = 151.3\,\text{kN} \quad z = 3.51\,\text{m} \quad \text{(distance of the chords)}.$$
(4.86)

According to beam theory the stress resultants in the lower chord are

$$M = \frac{(40 + 5 \cdot 0.33) \cdot 3.3^2}{8} = 56.7\,\text{kN m}, \qquad V = \frac{41.65 \cdot 3.3}{2} = 68.7\,\text{kN}.$$
(4.87)

Displayed in Fig. 4.44 are the values of the horizontal reinforcement as-x (cm^2/m). The total value is

$$A_s = \frac{6\,\text{cm}^2/\text{m} + 16.5\,\text{cm}^2/\text{m} \cdot 2 + 60.6\,\text{cm}^2/\text{m}}{2} \cdot 0.33\,\text{m} = 16.4\,\text{cm}^2,$$
(4.88)

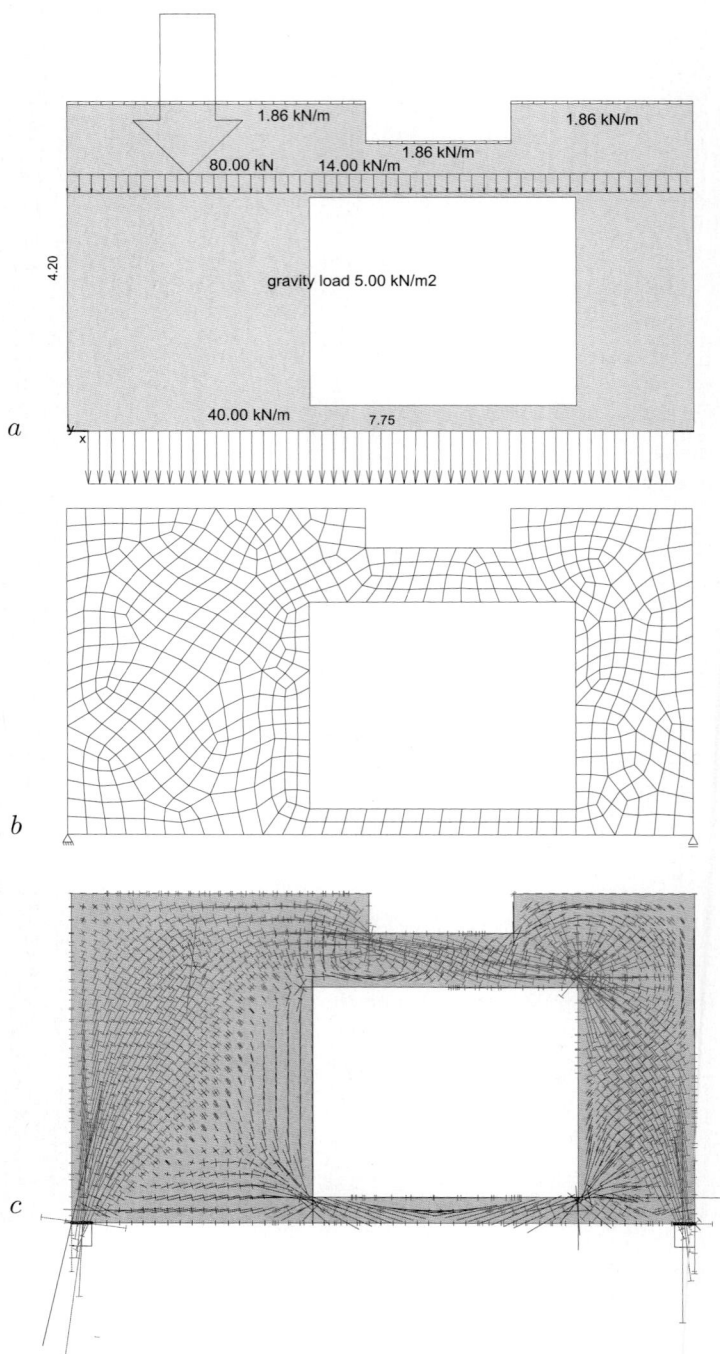

Fig. 4.43. Shear wall **a)** system and load, **b)** FE mesh, **c)** principal stresses

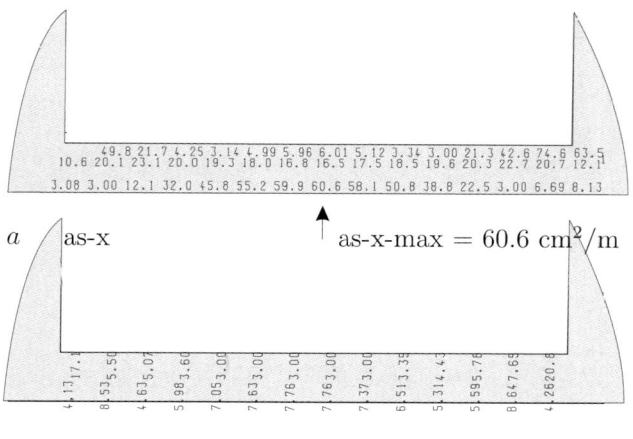

a   as-x   ↑   as-x-max $= 60.6$ cm^2/m

b   as-y

Fig. 4.44. Horizontal and vertical reinforcement as-x and as-y (cm^2/m) in the lower chord

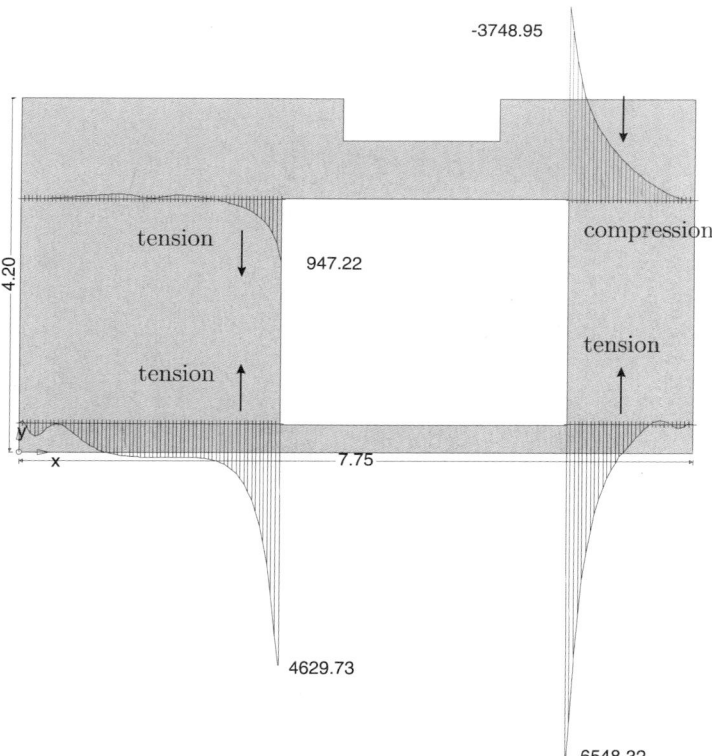

Fig. 4.45. Stresses σ_{xx} (kN/m^2) in horizontal sections

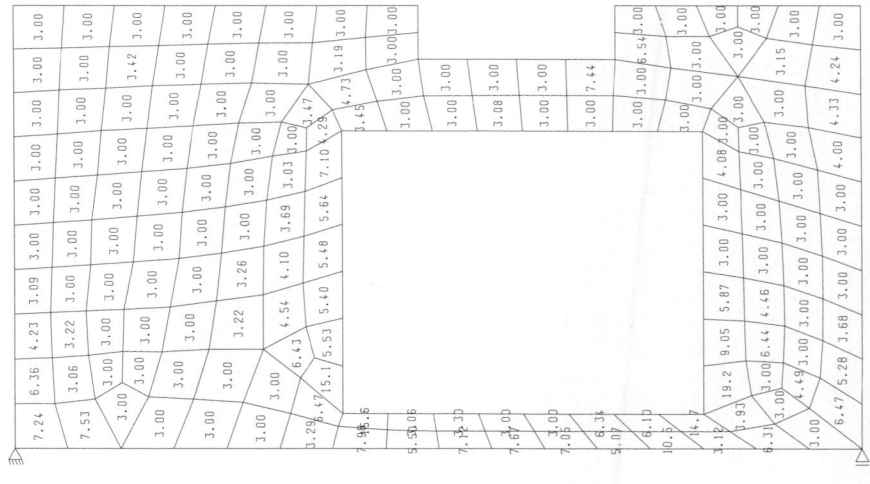

a

b

Fig. 4.46. Reinforcement **a)** as-x (cm²/m) and **b)** as-y (cm²/m), based on an average element length of 0.5 m

which exceeds the value obtained with beam analysis. Substituting the shear force of the beam solution for V the necessary shear reinforcement is[1]

---

[1] According to Eurocode

$$\tau_0 = \frac{V}{b \cdot z} = \frac{0.687 \, \text{MN}}{0.2 \, \text{m} \cdot 0.85 \cdot 0.28 \, \text{m}} = 1.44 \, \text{MN/m}^2 < \tau_{02} = 1.80 \, \text{MN/m}^2 \tag{4.89}$$

$$\tau = \frac{1.44^2}{1.80} = 1.15 \, \text{MN/m}^2 \qquad a_s = \frac{1.15 \cdot 20}{2.86} = 8 \, \text{cm}^2/\text{m} \, . \tag{4.90}$$

The integral of as-y across the width of the lower chord yields, according to Fig. 4.44

$$\int_0^{0.33} \text{as-y} \, dy = \frac{1}{2} \cdot \frac{4.13 + 2 \cdot 17.3 + 0}{2} \cdot 0.33 = 3.20 \, \text{cm}^2 \, . \tag{4.91}$$

The length of the elements—and therewith the distance of the nodes in x-direction—is 0.25 m and hence the shear reinforcement is $3.20 \cdot 4 = 12.8$ cm^2/m. This too exceeds the reinforcement in the equivalent beam. These checks may suffice.

The second topic is the singular stresses at the corner points of the opening; see Fig. 4.45. The hope to catch the singular stresses by refining the mesh is illusory. The more the elements are refined, the greater the stresses. But the integral of the stresses should be relatively stable; see Sect. 1.21, p. 86.

Along a horizontal line 50 cm long extending inwards from the lower left corner of the opening, the resultant stresses for three different element lengths 0.5 m, 0.25 m, and 0.20 m, are

$$N_x = 238 \, \text{kN} \qquad (0.50 \, \text{m}) \tag{4.92}$$
$$N_x = 246 \, \text{kN} \qquad (0.25 \, \text{m}) \tag{4.93}$$
$$N_x = 254 \, \text{kN} \qquad (0.20 \, \text{m}) \, . \tag{4.94}$$

The maximum value would require the reinforcement

$$A_s = \frac{254 \, \text{kN}}{28.6 \, \text{kN/cm}^2} = 8.9 \, \text{cm}^2 \tag{4.95}$$

where $28.6 \, \text{kN/cm}^2$ is the allowable steel stress.

In the same way, the vertical reinforcement can be checked. To this reinforcement must be added the hanger reinforcement for the suspended load.

Analogously the other corner points can be treated. Outside these critical regions the reinforcement calculated by the FE program is sufficient. In most parts of the shear wall only the minimum reinforcement is required; see Fig. 4.46.

4.13 Shear wall and horizontal load

The shear wall in Fig. 4.47 must sustain four different loadings. First gravity loads, then the live load of the floor plates. In the third load case (see

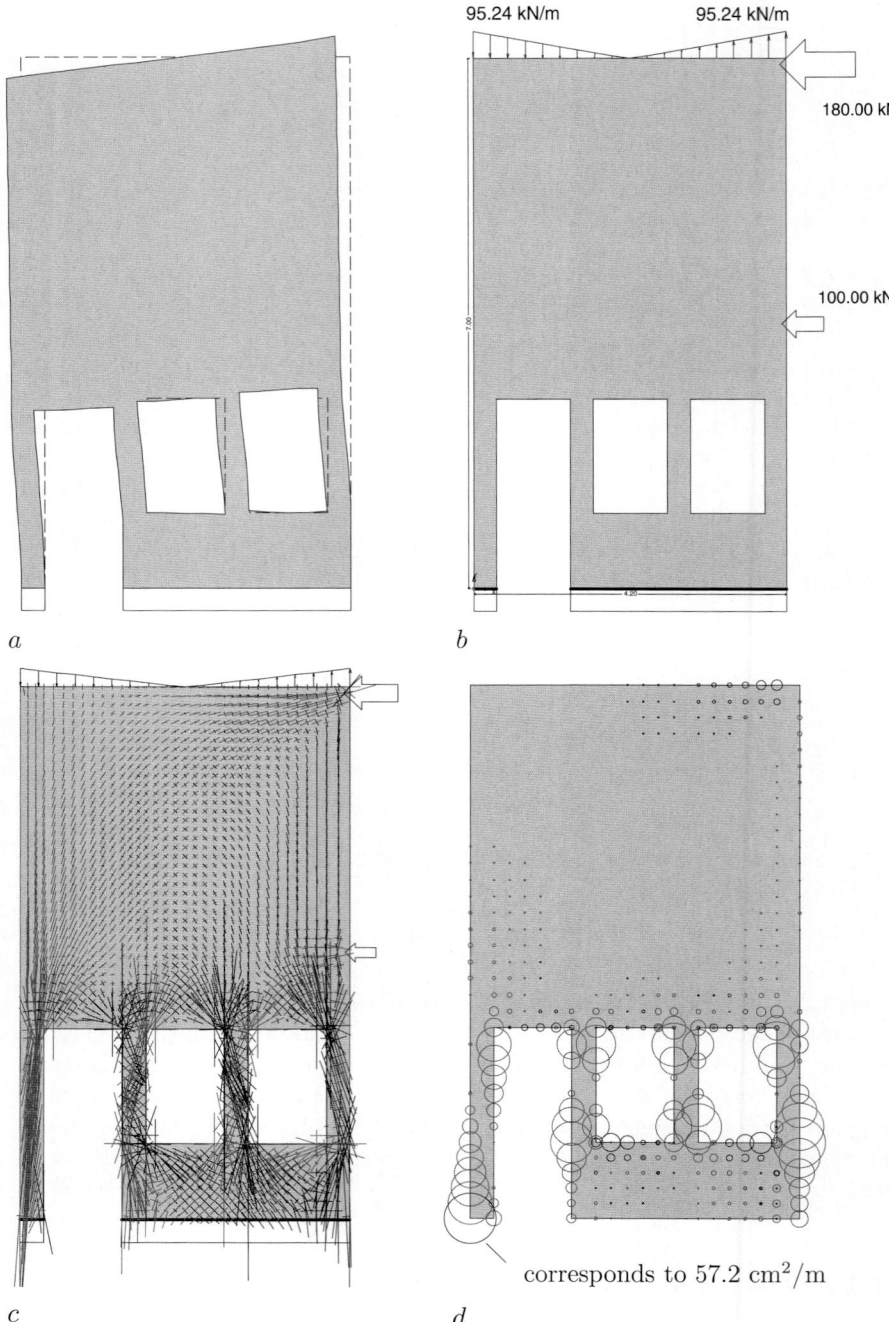

Fig. 4.47. Shear wall: **a)** displacements, **b)** load, **c)** principal stresses, and **d)** vertical reinforcement (designed for a total of 4 load cases)

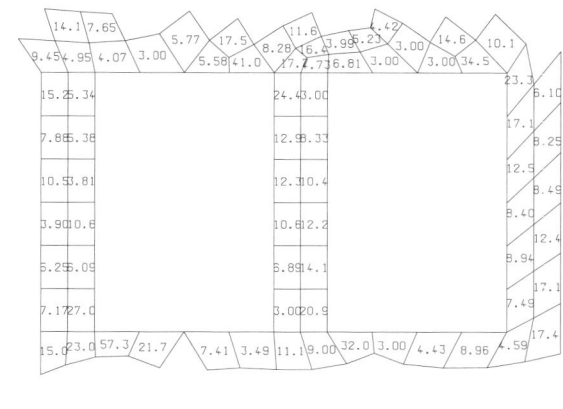

a

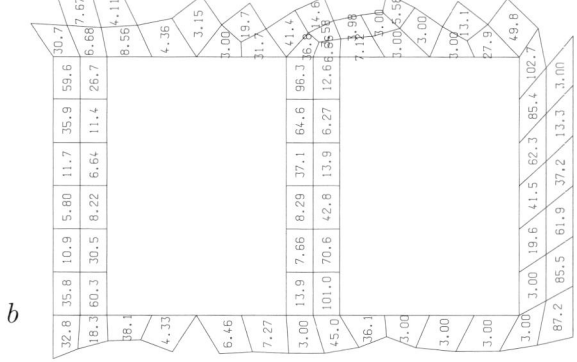

b

Fig. 4.48. Detail **a)** reinforcement as-x in cm^2/m and **b)** reinforcement as-y in cm^2/m in the columns

Fig. 4.47 b), the wall is subject to horizontal forces and to a moment $M = 280$ kN m on the upper edge, which is represented by an antisymmetric line load $p = \pm 95.24$ kN/m; see Fig. 4.47 b. The fourth load case is the same load case, except that the forces act in the opposite direction.

In the third load case the four columns must sustain a shear force $V = 180 + 100 = 280$ kN. If it is assumed that only the three short columns carry the load then each column carries a shear force V of 93.3 kN. The height of the columns is 1.5 m, so the bending moment becomes

$$M = 93.3\,\text{kN m} \cdot \frac{1.5\,\text{m}}{2} = 70\,\text{kN m} . \tag{4.96}$$

The moment resulting from the antisymmetric line load on the upper edge is split into two vertical forces:

$$C = T = \pm \frac{280\,\text{kN m}}{3.9\,\text{m}} = 72\,\text{kN} \qquad z = 3.9\,\text{m} . \tag{4.97}$$

Hence the design moment for the column with the heaviest load is

$$M_s = 70\,\text{kN m} - 72\,\text{kN} \cdot 0.1\,\text{m} = 62.8\,\text{kN m} , \tag{4.98}$$

and with[2]

$$k_h = \frac{25}{\sqrt{62.8/0.3}} = 1.73 \quad \rightarrow \quad k_s = 4.5 \tag{4.99}$$

we have

$$A_s = 4.5 \cdot \frac{62.8}{25} + \frac{72}{28.6} = 13.8 \,\mathrm{cm}^2 \,. \tag{4.100}$$

A check of Fig. 4.48 shows that the reinforcement

$$A_s = \frac{1}{2} \cdot (102.7 \,\mathrm{cm}^2/\mathrm{m} + 3.0 \mathrm{cm}^2/\mathrm{m}) \cdot 0.3 \,\mathrm{m} = 15.8 \,\mathrm{cm}^2 \tag{4.101}$$

is sufficient. The same holds for the shear reinforcement.

4.14 Equilibrium of resultant forces

In general the resultant stresses of the FE solution in any cross section do not balance the exterior load, because FE programs cannot generate the exact influence functions for the resultant stresses; see Sect. 1.21, p. 86.

The influence function for the horizontal force N_{yx} (the shear force) in section $A - A$ in Fig. 4.49 is a dislocation of the upper part of the structure by one unit of displacement to the right.

To produce this influence function the resultant shear stress $\sigma_{xy}^{(i)}$ of the unit displacement fields φ_i in the section $A - A$ must be applied as equivalent nodal forces:

$$f_i = \int_{A-A} \sigma_{xy}^{(i)} \, dx \,. \tag{4.102}$$

The resulting shape (see Fig. 4.49 c) is not the exact influence function, because the horizontal displacement at the upper left corner is 2.09 units of displacements instead of the exact 1.0. A simple test confirms this result: when a point load P is applied at the upper left corner, the stress resultant in section $A - A$ is exactly

$$\int_{A-A} \sigma_{yx}^h \, dx = 2.09 \, P \,. \tag{4.103}$$

This is a large error.

Next we let the section $A-A$ pass right through the *center* of the elements and—to our surprise—now the upper part of the wall moves sideways by exactly one unit of displacement, i.e., the resultant shear force N_{xy} in the section $A - A$ is exactly $1.0\,P$.

---

[2] According to Eurocode

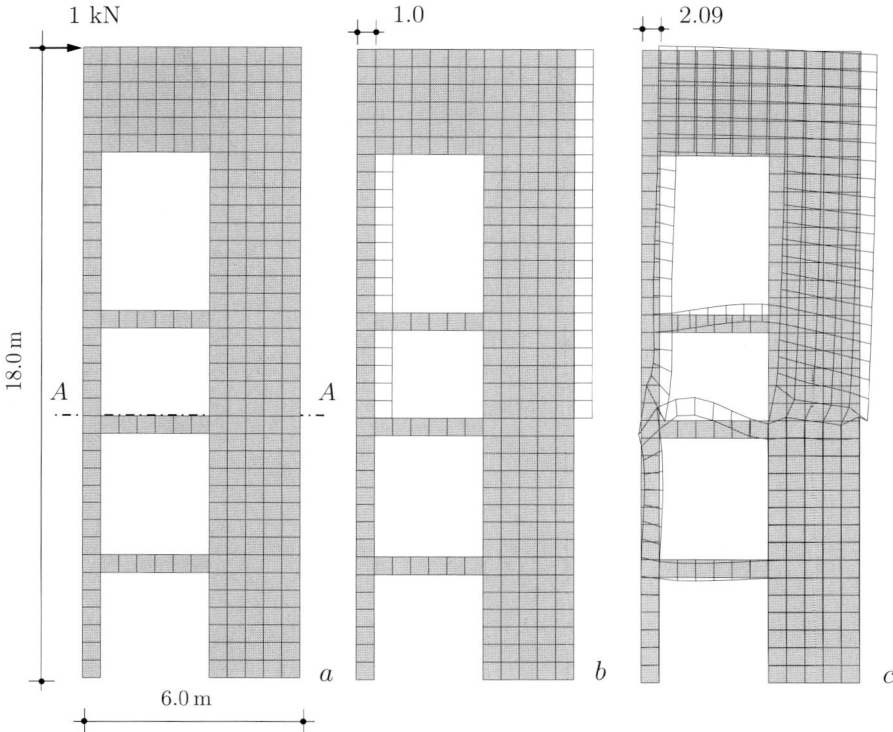

Fig. 4.49. Shear wall: **a)** section $A - A$, **b)** exact influence function for N_{yx} in section $A - A$, **c)** FE approximation

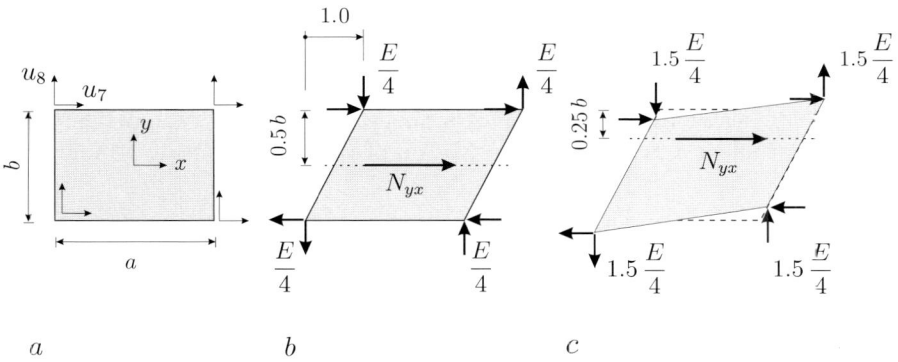

Fig. 4.50. Single bilinear element: **a)** degrees of freedom; equivalent nodal forces for **b)** the influence function for N_{yx} at the center, and **c)** at the upper quarter point, all horizontal forces are the same, $f_i = E\,a/4\,b$

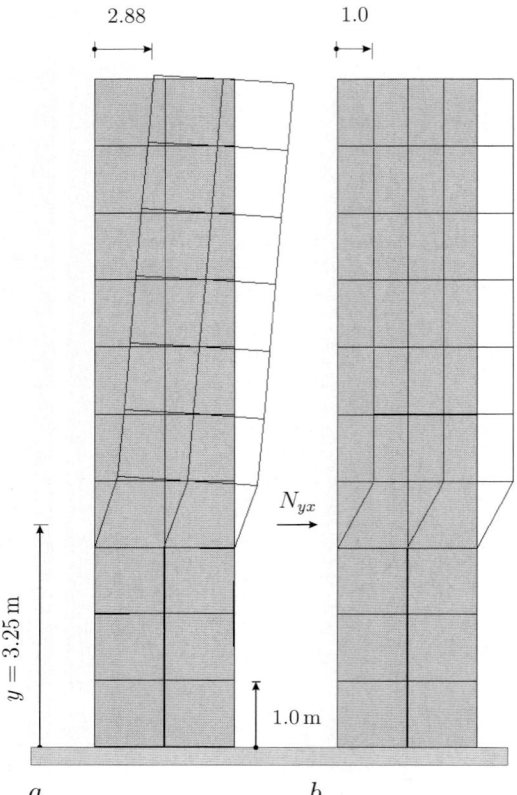

2.88

1.0

N_{yx}

$y = 3.25\,\mathrm{m}$

1.0 m

a

b

Fig. 4.51. Influence function for N_{xy} at $y = 3.25$: **a)** bilinear element, **b)** Wilson's element. Lateral deformations not to scale

To understand this phenomenon it is best to study a single bilinear element (see Fig. 4.50), and to begin with the shear force N_{yx} in a horizontal section which passes through the center of the element. The equivalent nodal forces that will generate the (approximate) influence function for N_{yx} are the same forces that effect a pure shear deformation of the element (see Fig. 4.50 a),

$$u(x,y) = \frac{y + 0.5\,b}{b} \qquad v(x,y) = 0 \qquad \varepsilon_{xy} = \frac{1}{2\,b} \qquad \varepsilon_{xx} = \varepsilon_{yy} = 0. \quad (4.104)$$

Note that at the upper edge $u(x, y = 0.5\,b) = 1$. The proof is given below.

Next it is assumed that the line intersects the element at a distance of $0.25\,b$ units from the upper edge; see Fig. 4.50 b. The equivalent horizontal nodal forces do not change, but the vertical forces do, and they effect a rotation which is responsible for the large error, 2.09 versus 1.0, that was observed previously at the upper left corner of the shear wall.

To formalize these observations let

$$f_i^{kj} = \int_0^l \sigma_{kj}(\boldsymbol{\varphi}_i)ds \qquad (= N_{kj}(\boldsymbol{\varphi}_i)) \qquad (4.105)$$

denote the resultant stresses of the unit displacement fields. Then the following condition can be formulated.

Shape condition An element satisfies the shape condition in a particular cross section if the shape generated by the equivalent nodal forces f_i^{kj} is in agreement with the equilibrium conditions.

That is, under the action of the f_i^{yx} (N_{yx}), the upper edge should slide sidewards, $u = 1$, and the f_i^{yy} (N_{yy} = normal force) should lift the upper edge by one unit, $v = 1$, etc. If the shape condition is satisfied, the resultant stress in the cross section maintains equilibrium with an edge load.

It follows that the shape condition is satisfied if the forces that effect a unit displacement of the element edge can be recovered in the pertinent cross section, which is just the same as saying that $\sum H = 0, \sum V = 0$.

While the bilinear element satisfies the shape condition for N_{yx} only at the center, Wilson's element satisfies the shape condition for N_{xy} in *any* horizontal section; see Fig. 4.51. Regardless of where the cut passes through Wilson's element, the f_i^{xy} are always the same, and because the f_i^{xy} at the center yield the correct shape, any other section has this property as well.

It is evident that the influence functions for the normal forces N_{xx} and N_{yy} in an element satisfy the shape condition. The associated equivalent nodal forces stretch the element in the horizontal or vertical direction by one unit of displacement regardless of where the element is intersected. More difficult is the situation—as seen—for the shear force N_{xy} while with regard to rotations ($\sum M = 0$) the situation is hopeless, because the influence functions (see Fig. 1.60 p. 89) definitely do not lie in V_h.

Proof

We show that the equivalent nodal forces that effect the shear deformation (see Fig. 4.50 b) are the same equivalent nodal forces f_i^{xy} that provide the influence function for N_{xy} at the center. The unit displacement field $\varphi_7(\boldsymbol{x}) = [u, v]^T$ of the upper left node has the components

$$u = \frac{(0.5\,a - x)(0.5\,b + y)}{a\,b} \qquad v(x, y) = 0 \qquad (4.106)$$

$$\varepsilon_{xy} = \frac{1}{2}u_{,y} = \frac{0.5\,a - x}{2\,a\,b} \qquad \varepsilon_{xx} = \ldots \qquad \varepsilon_{yy} = \ldots \qquad (4.107)$$

so that with the displacement field $\boldsymbol{u}$ of (4.104), the equivalent nodal force f_7 becomes

$$f_7 = a(\boldsymbol{u}, \varphi_7) = \int_\Omega 2\,\varepsilon_{xy}\,\sigma_{xy}\,d\Omega = 2\,E \int_\Omega \frac{1}{2\,b}\,\frac{(0.5\,a - x)}{2\,a\,b}\,d\Omega = \frac{a}{4\,b}\,E$$

$$(4.108)$$

which is the same as the resultant shear stress of the displacement field $\varphi_7(\boldsymbol{x})$:

$$f_7 = \int_{-a/2}^{+a/2} \sigma_{xy}\, dx = E \int_{-a/2}^{+a/2} \frac{(0.5\,a - x)}{2\,a\,b}\, dx = \frac{a}{4\,b}\, E = N_{xy}\,. \quad (4.109)$$

Equivalent nodal forces for influence functions

In principle, it is easy to arrange for an FE program code to display the FE influence functions on the screen. This can provide a better understanding of the behavior of the discretized structure. The implementation is simple because the equivalent nodal forces can be calculated on an element-per-element basis, as explained in the following.

First let us recall how the influence function $\boldsymbol{G}_1$ for, say, the stress σ_{xx} at the center $\boldsymbol{x}_c$ of an element is calculated. The equivalent nodal forces f_i are the stresses $\sigma_{xx}^{(i)}(\boldsymbol{x}_c)$ of the nodal unit displacement fields φ_i at the center. This is consequence of

$$a(\boldsymbol{G}_1[\boldsymbol{x}_c], \varphi_i) = (\delta_1[\boldsymbol{x}_c], \varphi_i) = \sigma_{xx}^{(i)}(\boldsymbol{x}_c)\,. \quad (4.110)$$

If these equivalent nodal forces f_i are applied at the four nodes of a bilinear element, the result is a displacement field $G_1^h[\boldsymbol{x}_c]$ that simulates a point dislocation $u(\boldsymbol{x}_c^+) - u(\boldsymbol{x}_c^-) = 1$ in the horizontal direction at $\boldsymbol{x}_c$.

To obtain the FE influence functions for the resultant stresses

$$N_x = \int_0^b \sigma_{xx}\, dy \qquad N_{xy} = \int_0^b \sigma_{xy}\, dy \qquad (4.111)$$

$$N_{yx} = \int_0^a \sigma_{yx}\, dx \qquad N_y = \int_0^a \sigma_{yy}\, dx \qquad (4.112)$$

in a vertical or horizontal section of an element, the integrals of these stresses must be applied as equivalent nodal forces f_i. In a horizontal section passing through the point (x, y) of a bilinear element, we have [96]

$$\int_0^a \sigma_{yy}\, dx = \frac{E}{2\,b\,(-1 + \nu^2)} \cdot \Big[2\,b\,\nu\,(u_1 - u_3) +$$

$$+ a\,(u_2 + u_4 - u_6 - u_8) + 2\,y\,\nu\,(-u_1 + u_3 - u_5 + u_7) \Big] \quad (4.113)$$

$$\int_0^a \sigma_{xy}\, dx = \frac{E}{4\,b\,(1 + \nu)} \cdot \Big[a\,(-u_1 - u_3 + u_5 + u_7) +$$

$$- 2\,(b\,(u_2 - u_4) + y\,(-u_2 + u_4 - u_6 + u_8)) \Big] \quad (4.114)$$

and in a vertical section

$$\int_0^b \sigma_{xx}\, dy = \frac{E}{2\,a\,(-1+\nu^2)} \cdot \Big[b\,(u_1 - u_3 - u_5 + u_7) +$$

$$+ 2\,\nu\,(a\,(u_2 - u_8) + x\,(-u_2 + u_4 - u_6 + u_8)) \Big] \qquad (4.115)$$

$$\int_0^b \sigma_{xy}\, dy = \frac{E}{4\,a\,(1+\nu)} \cdot \Big[-2\,a\,(u_1 - u_7) +$$

$$+ b\,(-u_2 + u_4 + u_6 - u_8) + 2\,x\,(u_1 - u_3 + u_5 - u_7) \Big]. \qquad (4.116)$$

Here a denotes the length of the element, b is the width of the element, and the origin of the coordinate system lies in the lower left corner. To obtain the nodal force f_1^e for the influence function for N_y in a horizontal section passing through the point (x, y), for example, let $u_1 = 1$ in (4.113) and let all other $u_i = 0$.

4.15 Adaptive mesh refinement

The plate in Fig. 4.52 was analyzed with bilinear elements, with an attempt to improve the results by adaptively refining the mesh in three steps. At the start the mesh consisted of 160 elements, and the final mesh consisted of 1231 elements. At that stage in the analysis, the energy error was about the same in most of the elements.

The thickness of the plate was $t = 0.25$ m, and the material properties were $E = 3.4 \cdot 10^4$ MN/m^2, $\nu = 0.167$.

The energy norm (squared) of the FE solution is the strain energy product between the stresses and the strains,

$$||\boldsymbol{u}_h||_E^2 = \int_\Omega \boldsymbol{\sigma}_h \bullet \varepsilon_h\, d\Omega, \qquad (4.117)$$

and the energy norm squared of the error $\boldsymbol{e} = \boldsymbol{u} - \boldsymbol{u}_h$ is

$$||\boldsymbol{e}_h||_E^2 = \int_\Omega (\boldsymbol{\sigma} - \boldsymbol{\sigma}_h) \bullet (\varepsilon - \varepsilon_h)\, d\Omega. \qquad (4.118)$$

Because the exact solution $\boldsymbol{u}$ is unknown the energy norm is replaced—as discussed in Sect. 1.31, p. 147—by the following estimate [132],

$$||\boldsymbol{e}_h||_E^2 \le \eta^2 = \sum_i \eta_i^2 = \sum_i \frac{0.4^2\,h^2}{\lambda + 5\,\mu}\,||\boldsymbol{r}_i||_0^2 + \frac{1.2^2\,h}{\lambda + 5\,\mu}\,||\boldsymbol{j}_i||_0^2 \qquad (4.119)$$

where $\boldsymbol{r}_i = \boldsymbol{p} - \boldsymbol{p}_h$ are the residual forces within an element Ω_i, and where $\boldsymbol{j}_i$ are the jumps of the traction vector $\times$ 0.5 on the edges of the element Ω_i:

$$||\boldsymbol{r}_i||_0^2 = \int_{\Omega_i} (\boldsymbol{p} - \boldsymbol{p}_h)^2\, d\Omega \qquad ||\boldsymbol{j}_i||_0^2 = 0.5 \int_{\Gamma_i} \boldsymbol{t}_\Delta^2\, ds. \qquad (4.120)$$

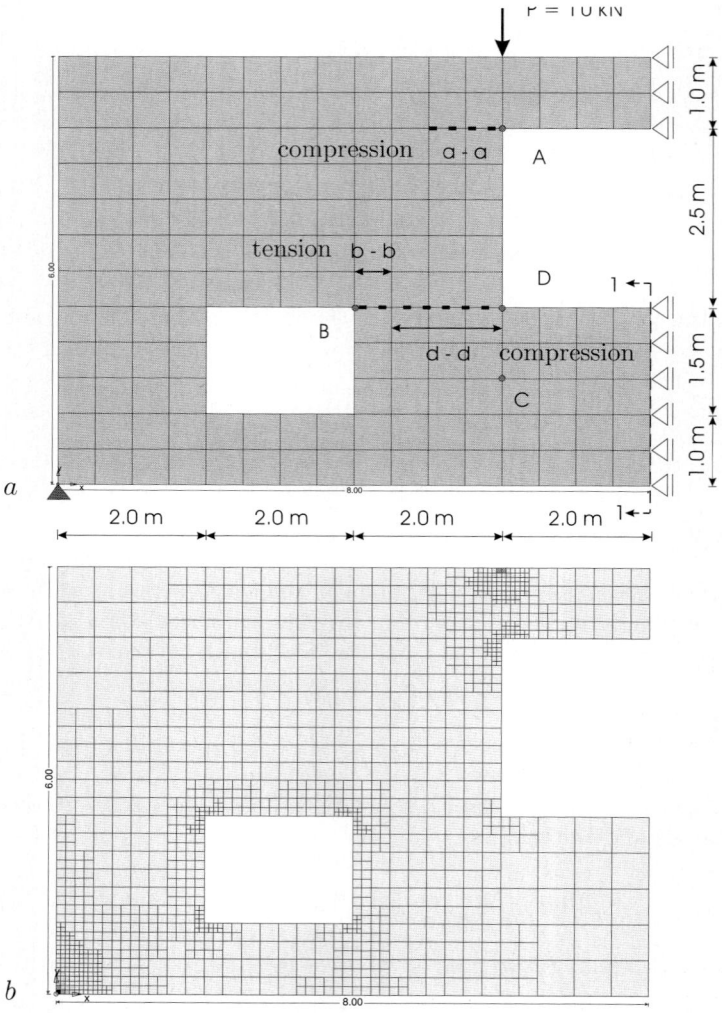

Fig. 4.52. Adaptive refinement of a shear wall: **a)** original mesh, **b)** final mesh

At the start of the adaptive refinement the relative error is

$$\eta^{rel} := \frac{\eta}{\sqrt{||\boldsymbol{u}_h||_E^2 + \eta^2}} = \sqrt{\frac{1.78}{25 + 1.78}} \cdot 100\% = 25.78\% \qquad (4.121)$$

and then it slowly decreases as can be seen in Table 4.9.

Theoretically the mesh should be refined further near the supports and the base of the single force—if this makes sense. Close to these critical points, the grey is a very dark grey.

Table 4.9. Adaptive refinement

Step	$\|\boldsymbol{u}_h\|_E^2$	$\|\boldsymbol{e}_h\|_E^2$	η^{rel}	Nodes	Elements	d.o.f.
0	2.50E+06	1.78E+05	25.78%	199	160	388
1	2.77E+06	1.59E+05	23.30%	407	328	802
2	3.01E+06	1.49E+05	21.68%	821	685	1630
3	3.25E+06	1.43E+05	20.53%	1442	1231	2872

The form of the resulting mesh essentially depends on the percentage of elements that are refined. Usually only the first 20, 30 or 50% of the elements that exceed the critical value are refined. The lower the percentage the better the refinement process will concentrate on the truly singular points. Here only the first 50% were refined. The relative error η^{rel} dropped from about 25% to 20.53% which is not a great gain.

In a second analysis only the first 30% were refined and at the end the estimated relative error was $\eta = 20.87\%$. This level was reached with 781 elements, while the 50% strategy required 1,231 elements—a larger effort for practically the same result. This observation indicates that the two *hot spots*, the point support and the single force, dominate the error.

The corner points of the openings, where the stresses oscillated, were also critical points (see Table 4.10), while the results at the more backward interior point C, for example, were stable. The refinement of the mesh hardly had any effect on the stresses at that point. The same holds for the section $1 - 1$; see Table 4.12.

Table 4.10. Stresses (kN/m^2) at selected nodes at the beginning (0) and after three refinements (3) and BE stresses

Nodes	A	B	C	D
$\sigma_{xx}^{(0)}$	−39.87	32.26	16.23	7.43
$\sigma_{xx}^{(3)}$	−106.96	90.44	16.76	9.36
σ_{xx}^{BE}	−76.814	68.94	16.41	11.84
$\sigma_{yy}^{(0)}$	−54.33	22.88	−1.66	−12.32
$\sigma_{yy}^{(3)}$	−127.28	79.52	−1.24	−15.39
σ_{yy}^{BE}	−99.95	56.96	−0.91	−16.71

To confirm that stress resultants are more stable than isolated stresses, the stresses were integrated along three sections; see Table 4.11. These resultant forces also vary, but they are more stable than the stresses at the isolated points.

Hence the presence of singular points does not necessarily imply that the whole solution is worthless. Rather the results show that in regions where the

Table 4.11. Integral of the stresses σ_{yy} (kN/m) along three sections

Cross section	a-a	b-b	d-d
Step 0	−26.80	5.12	−11.63
Step 3	−32.1	9.85	−12.64
BE	−29.32	6.92	−13.48

solution is smooth, the stresses are stable, although this doesn't necessarily mean that they are accurate! It is difficult to judge how large the influence of the corner singularities is, that is how large the pollution error is, but we think that in structural analysis—considering the error margins we are accustomed to—pollution is a secondary effect. In standard situations, modeling errors will probably have a more negative impact on the accuracy of, say, the support reactions or the bending moments than an unresolved singularity on the boundary.

Table 4.12. Stress distribution (kN/m^2) in section 1-1

Level	$y = 2.5$	$y = 2.0$	$y = 1.5$	$y = 1.0$	$y = 0.5$	$y = 0.0$
$\sigma_{xx}^{(0)}$	7.86	13.09	18.11	23.43	28.96	34.44
$\sigma_{xx}^{(3)}$	8.72	13.67	18.38	23.36	28.54	33.64
$\sigma_{xx}^{(BE)}$	8.9	13.62	18.18	22.97	27.91	32.69

4.16 Plane problems in soil mechanics

It is easier to work in soil mechanics with plane models than with full 3-D models. In this context there are three topics to be discussed.

Self-equilibrated stress states and primary load cases

In soil mechanics particular attention must be paid to the various construction stages, because in nonlinear analysis it must be possible to define a primary stress state to assess the stress history correctly. Elements are removed or their stiffness is reduced, if for example, an injection is washed out or a certain segment of the soil is defrosted. Then the vanishing stresses generate loads. To handle these stresses in a program it must be possible to identify the stresses with certain loads, which are so tuned that the sum of these loads cancels in the undisturbed zones, while they produce true loads at the edge of the disturbed zone.

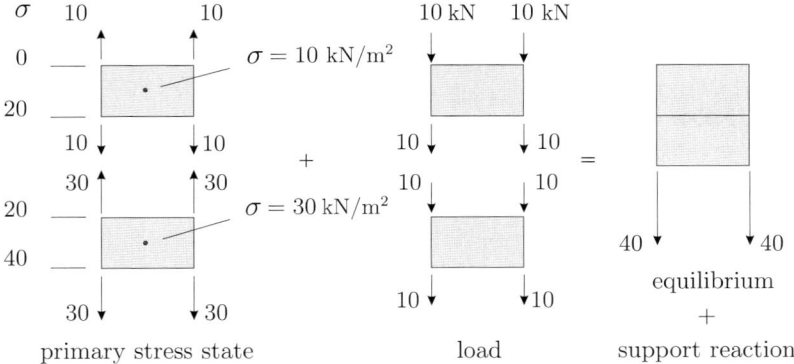

Fig. 4.53. Primary stress state and additional load

The nodal forces of the primary state are determined using the principle of virtual displacements. Let us assume that as in Fig. 4.53, the stresses due to the gravity loads are zero at the upper edge, 20 kN/m² at the interface between the two elements, and 40 kN/m² at the lower edge. At the centers of the elements the stresses are 10 kN/m² and 30 kN/m², respectively. Hence in the primary state there results a pair of opposing forces: at the upper edge a force of 10 kN and at the lower edge one of 30 kN; the length of the elements is assumed to be 1.0 m. If the gravitational forces of 10 kN are added to all element nodes, the resultant forces vanish at the upper nodes, while at the bottom the nodal forces equal the total load.

If the gravity load had not been applied, the loads obtained would have pointed upward and been of the same magnitude as the external load, hence the displacements and stresses in the primary stress state would be zero.

There is a peculiar effect with regard to the horizontal stresses. Because these stresses do not enter the equilibrium conditions directly they can only be recovered if the vertical stresses are multiplied by $\nu/(1-\nu)$. But very often the so-called *lateral pressure ratio* of the soil will not agree with Poisson's ratio because of geologic preloads and possible plastifications. Even after a complete removal of the vertical load the soil will not be stress free.

Settlements

A strange effect also exists with regard to the calculation of the settlements under a footing: *the more the mesh extends in all directions the greater the settlements.*

This effect has to do with the behavior of the natural logarithm, $\ln r$. When a single surface load $P = 1$ is applied at the edge of the elastic half-plane, the deflection on the left- and right-hand side of the point load essentially resembles $\ln r$, that is, at the base of the source point $r = 0$ the deflection is infinite, $w = -\infty$. (The y-axis points upward, the load P points downward.)

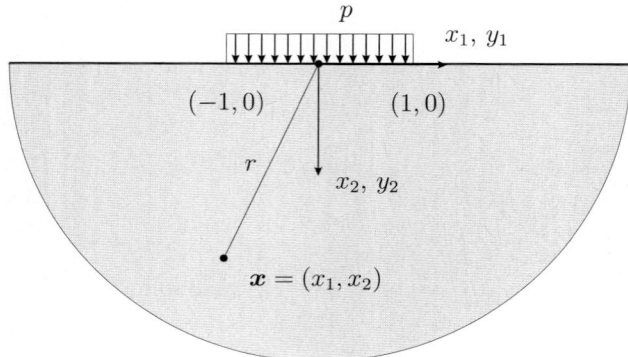

Fig. 4.54. Distributed load at the edge of the elastic half-plane

The strange thing is that there is also a singularity at the other end: moving away from the source point, the absolute value of the deflection $|w|$ will first decrease, and then after a while it will *increase* again, and will tend to $+\infty$ as $r \to \infty$.

Let us assume that a load $p(\boldsymbol{x}) = p(x_1, x_2)$ acts at the edge ($x_2 = 0$) of the half-plane in the interval $-1 \leq x_1 \leq +1$ (see Fig. 4.54). The vertical displacement of the soil at a point $\boldsymbol{x} = (x_1, x_2)$ is, if we simplify somewhat and concentrate on the essentials,

$$v(\boldsymbol{x}) = \int_{-1}^{+1} \frac{1}{2\,\pi} \ln r \, p(\boldsymbol{y}) \, ds_{\boldsymbol{y}} \, . \tag{4.122}$$

From the standpoint of a mole that burrows into the soil, the edge load at the surface more and more resembles a point force, so that at a large enough distance r from the surface the deflection becomes

$$v(\boldsymbol{x}) = \frac{1}{2\,\pi} \ln r \int_{-1}^{+1} p(y_1, 0) \, dy_1 = \frac{1}{2\,\pi} \ln r \, P \, ,$$

where P is the resultant force of the edge load.

This means that in any system of loads that are not self-equilibrated, any increase in the radius of the mesh will increase the displacements, and if the radius tends to infinity, so will the displacements.

For this tendency to prevail to the very end the lateral parts of the mesh must not be chopped off, because otherwise the sheet piling ($u = 0$) might act as an abutment, and the load would be carried by an arch which develops beneath the footing. Therefore an unimpeded expansion of the soil must be possible.

Various remedies overcome this problem: firstly one could simply set a limit on the mesh size and set the vertical displacement to zero at the bottom of the mesh, $v = 0$. Secondly one could modify the load to make it self-equilibrated. Thirdly one could let the modulus of elasticity increase in the vertical direction. This would lead to finite displacements.

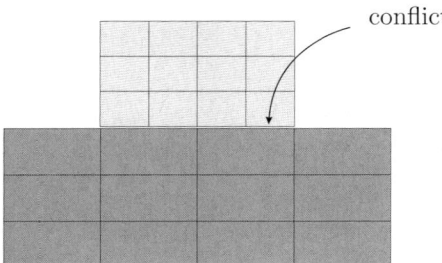
conflict

Fig. 4.55. Building bricks

Discontinuities

A foundation slab and the soil have different material properties. This alone may cause trouble. Putting a building brick on top of a second brick (see Fig. 4.55) looks harmless, but at the interface between the two bricks, stress peaks and even singularities may be observed. The magnitude of the stress peaks depends on the stiffness ratio of the two bricks. The extreme values are marked by an infinitely flexible slab and a perfectly rigid slab.

In the first case—if the modulus of elasticity of the lower brick is distinctively greater—the pressure on the lower brick will be nearly constant and the singularity will be hardly noticeable. Outside the loaded region, the vertical stress abruptly drops to zero, but this discontinuity soon gives way to a continuous stress field near the interface.

But the stress discontinuity at the interface between the two blocks cannot be modeled, because the condition that the two blocks have one edge in common enforces a constant strain in both elements over the whole length, which leads to a conflict; see Fig. 4.55.

In the second case (rigid punch on a half-space) the soil pressure becomes infinite at the edges of the punch. To see these stress peaks, the mesh must be sufficiently refined. The problem was analyzed with a coarse and a fine *irregular* mesh. Because of the minor irregularities in the layout of the mesh (Wilson's element), the stresses on the left- and right-hand side were slightly different, even though the system and the load were symmetric. Asymmetries in the displacements were not noticeable. Different ratios of the modulus of elasticity were tested. Table 4.13 lists the vertical stresses at the outer contact nodes under a total load of 100 MPa.

The results are remarkable. Even though no inferior elements were used, the bandwidth of the results is surprising. In the case of a soft slab, the stress at the extreme node of the interface is 50 MPa, which is simply the average value of 0 and 100. The more the mesh is refined, the greater the stresses become. Even on the finest mesh, there are noteworthy deviations between the results for the two corner nodes on the left- and right-hand side. But it would make no sense to put more effort into the analysis in order to drive the stresses towards infinity. To assess the crack sensitivity, the results of a medium-sized mesh are in general sufficient.

Table 4.13. Vertical stresses left/right (MPa) at the extreme nodes in the contact area for different ratios of $\eta = E_{upper}/E_{lower}$

elements	$\eta = 0.01$	$\eta = 0.1$	$\eta = 1.0$	$\eta = 10$	$\eta = 100$
521	44 / 55	55 / 67	102 / 107	138 / 133	143 / 136
884	39 / 72	56 / 94	152 / 179	226 / 216	222 / 208
2308	52 / 61	88 / 103	303 / 341	404 / 459	321 / 398

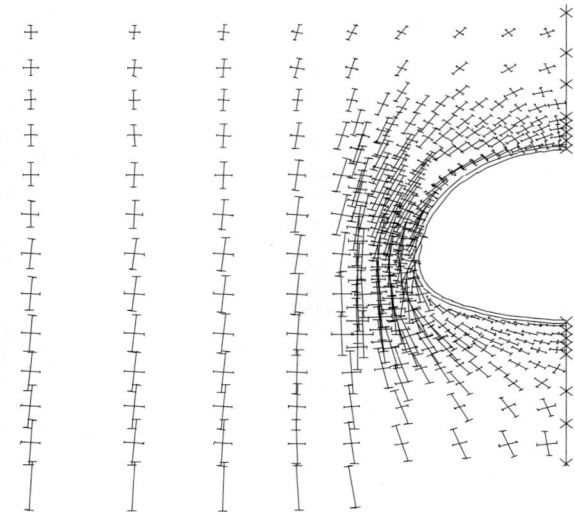

Fig. 4.56. The principal stresses flow around the tunnel

Tunnels

Shallow tunnels are designed to carry the full weight of the soil. But with increasing depth—beginning at a distance of about a full diameter, and depending on the quality of the soil material or the rock—the stresses are redirected around the tunnel (see Fig. 4.56), so that the load is carried by the rock, and the tunnel shell simply provides shelter for the traffic passing through the tunnel.

Modeling the soil with a 3-D mesh and simulating the complete construction process by successively deactivating the tunnel elements and activating new shell elements (tunnel lining) brought into the tunnel results (even in the case of a short tunnel) in a very large system of equations. After each step, this system must be rearranged and solved iteratively, due to the nonlinear effects. Therefore a conventional 3-D analysis will be restricted to short sections of the tunnel at the very front of the excavation, near crossings, and crevices.

Thus, the stress analysis of tunnels nowadays is mostly based on plane models, where by suitable modifications one also tries to take into account the redistribution of the internal forces in the direction of the tunnel axis. The various construction stages and the sequence of excavations complicate

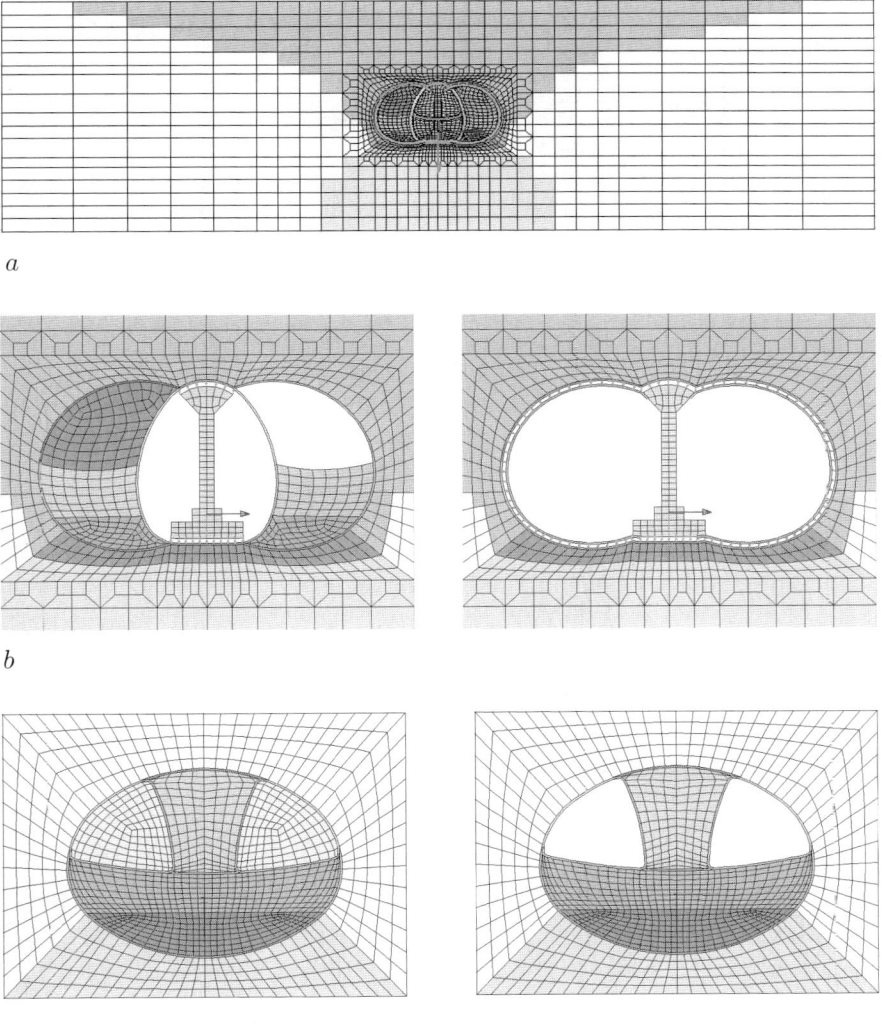

a

b

c

Fig. 4.57. 2-D analysis: **a)** global mesh, **b)** two excavation steps, **c)** cross section

the numerical analysis even of the plane model, because (see Fig. 4.57) the front of the tunnel is divided into a variety of different cross sections, and the nonlinearity of the strains and stresses in the rock near the tunnel front must be considered.

In an FE model, this sequence of events is normally modeled by an iterative procedure, where a new step in the analysis starts with the stress state of the

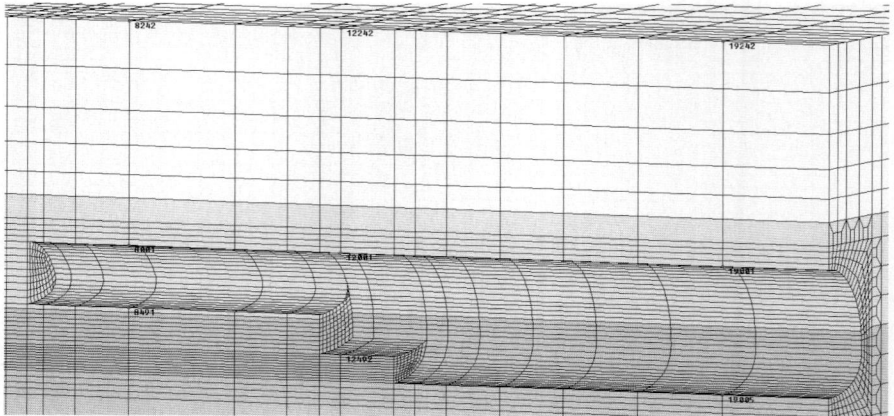

Fig. 4.58. 3D-model of the excavation

previous steps, and where the stiffness of the elements is reduced according to the construction process by a suitable empirical factor $0 < \alpha < 1$.

The alternative to a plane model is a 3-D model with a so-called excavation by driving, where the excavation sequence is modeled by a special 3-D mesh (see Fig. 4.58). Imagine a sequence of, say, 20 tunnel "slices" (cross sections with a thickness of about 1 m) that wander as a package through the rock. Because the mesh is moving with the excavation—from the point of view of the mesh, the rock is moving—the mesh can be kept relatively small. The iteration proceeds as follows:

1. First the primary stress state in the untouched rock is calculated. Next the sequence of excavations follows and the complete tunnel lining is activated. The first step in this sequence yields unrealistic results.
2. By iteratively rearranging the results the stresses in the rock are moved towards the tunnel opening.
3. This modified stress state then serves as the starting point for the next calculation. This reanalysis is repeated until the modifications become negligible.

In this manner, the actual front of analysis advances through the rock in concert with the excavation. The unrealistic results of the first step move towards the tunnel opening and leave the rock zone.

The advantage in comparison with a conventional 2D-analysis is that both structural behavior in the longitudinal direction and the relaxation of the rock are accounted for by the analysis.

4.17 Incompressible material

The constitutive model in linear elasticity depends on two parameters, most often the modulus of elasticity E and Poisson's ratio ν. In soil mechanics the *bulk modulus* and the *shear modulus* are preferred instead:

$$K = \frac{E}{3(1-2\nu)}, \qquad G = \frac{E}{2(1+\nu)}. \tag{4.123}$$

The stresses are split into deviatoric shear stresses and a uniform pressure $\boldsymbol{\sigma} = (G\,\boldsymbol{E}_G + K\,\boldsymbol{E}_K)\,\varepsilon$, or

$$
\begin{bmatrix} \sigma_{xx} \\ \sigma_{yy} \\ \sigma_{zz} \\ \sigma_{xy} \\ \sigma_{xz} \\ \sigma_{yz} \end{bmatrix}
= \left\{ G \begin{bmatrix} 4/3 & -2/3 & -2/3 & 0 & 0 & 0 \\ -2/3 & 4/3 & -2/3 & 0 & 0 & 0 \\ -2/3 & -2/3 & 4/3 & 0 & 0 & 0 \\ 0 & 0 & 0 & 1 & 0 & 0 \\ 0 & 0 & 0 & 0 & 1 & 0 \\ 0 & 0 & 0 & 0 & 0 & 1 \end{bmatrix}
+ K \begin{bmatrix} 1 & 1 & 1 & 0 & 0 & 0 \\ 1 & 1 & 1 & 0 & 0 & 0 \\ 1 & 1 & 1 & 0 & 0 & 0 \\ 0 & 0 & 0 & 0 & 0 & 0 \\ 0 & 0 & 0 & 0 & 0 & 0 \\ 0 & 0 & 0 & 0 & 0 & 0 \end{bmatrix} \right\}
\begin{bmatrix} \varepsilon_{xx} \\ \varepsilon_{yy} \\ \varepsilon_{zz} \\ \varepsilon_{xy} \\ \varepsilon_{xz} \\ \varepsilon_{yz} \end{bmatrix}.
\tag{4.124}
$$

Near $\nu = 0.5$, the bulk modulus exceeds the modulus of elasticity and the shear modulus by a large margin. Theoretically the bulk modulus K becomes infinite (see Fig. 4.60). This is why the material is said to be incompressible even though water, for example, retains a bulk modulus of about 2,000 MPa. Close to $\nu = 0.5$, special techniques are necessary to avoid locking, as for example by using so-called enhanced strain elements, or by introducing three-field mixed formulations [258].

In a Lagrangian approach of the dynamic analysis of nearly incompressible fluids the rotational derivatives of the deformations have to be suppressed with a penalty function to avoid spurious modes (see Fig. 4.59) because the mass matrix has still modes which are suppressed in the stiffness matrix.

4.18 Mixed methods

In FE methods we distinguish between displacement-based approaches and mixed methods. In displacement methods, the structural behavior is solely governed by the displacements, so approximate displacements suffice. The classical FE methods are displacement-based methods. But these technologies are limited in scope, in that incompressible material or simply Kirchhoff plates and shell structures pose serious problems [26]. Mixed methods can overcome these difficulties.

In mixed methods *separate approximations* are chosen for the displacements and the stresses, so that for example a bar element has four degrees of freedom: the two end displacements u_1, u_2 and the stresses σ_1, σ_2 at the end cross sections; see Fig. 4.61.

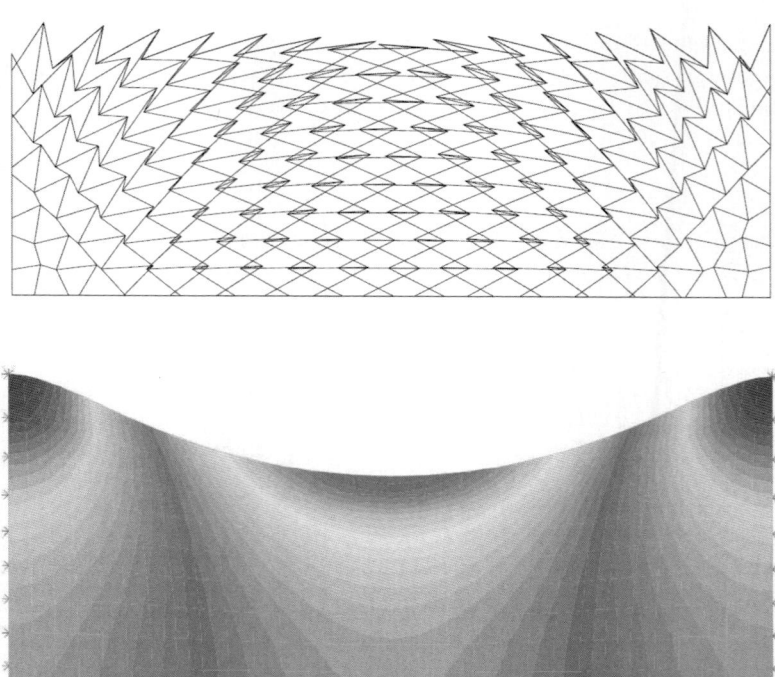

Fig. 4.59. Spurious displacement field in a water tank and regular field

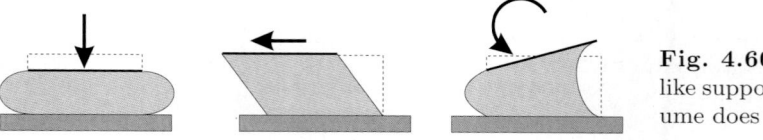

Fig. 4.60. Rubber-like support – the volume does not change

On first glance this approach might look strange, but the mixed approach is in good agreement with the nature of structural mechanics because structural mechanics is "mixed", the displacements u_i, the strains ε_{ij}, and the stresses σ_{ij} in a plate are coupled by a system of first-order differential equations:

$$\begin{aligned}
\boldsymbol{E}(\boldsymbol{u}) - \boldsymbol{E} &= \boldsymbol{E}^+ \\
\boldsymbol{C}[\boldsymbol{E}] - \boldsymbol{S} &= \boldsymbol{S}^+ \\
-\mathrm{div}\,\boldsymbol{S} &= \boldsymbol{p}
\end{aligned} \tag{4.125}$$

where

$$\boldsymbol{E}(\boldsymbol{u}) = \frac{1}{2}(\nabla \boldsymbol{u} + \nabla \boldsymbol{u}^T) = \frac{1}{2}\begin{bmatrix} 2\,u_{1,1} & u_{1,2} + u_{2,1} \\ u_{2,1} + u_{1,2} & 2\,u_{2,2} \end{bmatrix} \tag{4.126}$$

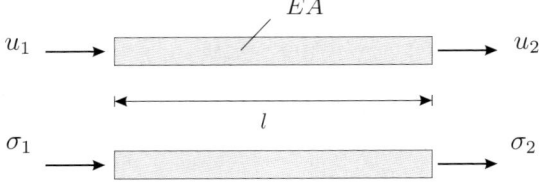

Fig. **4.61.** A bar element with independent functions for displacements and stresses

and $C[\ ]$ is the elasticity tensor

$$C[\boldsymbol{E}] = 2\,\mu\,\boldsymbol{E} + \lambda\,(\mathrm{tr}\ \boldsymbol{E})\,\boldsymbol{I}\,. \qquad (4.127)$$

Here $\boldsymbol{I}$ is the unit matrix, $\lambda = 2\,\mu\,\nu/(1-2\nu)$, and $\mathrm{tr}\ \boldsymbol{E} = \varepsilon_{11} + \varepsilon_{22}$ is the trace of $\boldsymbol{E}$. The terms on the right-hand side in (4.125) are initial strains ε_{ij}^{+} and stresses σ_{ij}^{+}.

If we call $\mathcal{S} = \{\boldsymbol{u}, \boldsymbol{E}, \boldsymbol{S}\}$ an *elastic state*, (4.125) can be read as the application of a differential operator $\boldsymbol{A}$ to the elastic state $\mathcal{S}$.

When the same operator $\boldsymbol{A}$ is applied to the FE solution $\mathcal{S}_h = \{\boldsymbol{u}_h, \boldsymbol{E}_h, \boldsymbol{S}_h\}$, the result is

$$\begin{aligned}
\boldsymbol{E}(\boldsymbol{u}_h) - \boldsymbol{E}_h &= \boldsymbol{E}_h^{+}\\
C[\boldsymbol{E}_h] - \boldsymbol{S}_h &= \boldsymbol{S}_h^{+}\\
-\mathrm{div}\ \boldsymbol{S}_h &= \boldsymbol{p}_h\,.
\end{aligned} \qquad (4.128)$$

The operator $\boldsymbol{A}$ satisfies the identity

$$G(\mathcal{S}, \hat{\mathcal{S}}) = \underbrace{\langle \boldsymbol{A}(\mathcal{S}), \hat{\mathcal{S}} \rangle + \int_{\Gamma} \boldsymbol{S}\,\boldsymbol{n} \boldsymbol{\cdot} \hat{\boldsymbol{u}}\,ds}_{\delta W_e} - \underbrace{a(\mathcal{S}, \hat{\mathcal{S}})}_{\delta W_i} = 0 \qquad (4.129)$$

where the angular brackets denote the scalar product of $\boldsymbol{A}(\mathcal{S})$ and the virtual elastic state $\hat{\mathcal{S}}$

$$\begin{aligned}
\langle \boldsymbol{A}(\mathcal{S}), \hat{\mathcal{S}} \rangle :&= \int_0^l (\boldsymbol{E}(\boldsymbol{u}) - \boldsymbol{E}) \boldsymbol{\cdot} \hat{\boldsymbol{S}}\,d\Omega + \int_{\Omega} (C[\boldsymbol{E}] - \boldsymbol{S}) \boldsymbol{\cdot} \hat{\boldsymbol{E}}\,d\Omega\\
&+ \int_{\Omega} -\mathrm{div}\ \boldsymbol{S} \boldsymbol{\cdot} \hat{\boldsymbol{u}}\,d\Omega
\end{aligned} \qquad (4.130)$$

and the strain energy product is

$$\begin{aligned}
a(\mathcal{S}, \hat{\mathcal{S}}) = &\int_{\Omega} (\boldsymbol{E}(\boldsymbol{u}) - \boldsymbol{E}) \boldsymbol{\cdot} \hat{\boldsymbol{S}}\,d\Omega + \int_{\Omega} C[\boldsymbol{E}] \boldsymbol{\cdot} \hat{\boldsymbol{E}}\,d\Omega\\
&+ \int_{\Omega} \boldsymbol{S} \boldsymbol{\cdot} (\boldsymbol{E}(\hat{\boldsymbol{u}}) - \hat{\boldsymbol{E}})\,d\Omega\,.
\end{aligned} \qquad (4.131)$$

If the initial stresses are zero, $\sigma_{ij}^{+} = 0$, the strains ε_{ij} are determined by the stresses

$$\boldsymbol{E} = \boldsymbol{C}^{-1}[\boldsymbol{S}]\,. \tag{4.132}$$

and the states $\mathcal{S}$ contain only two independent quantities—the displacements u_i and the stresses σ_{ij}:

$$\mathcal{S} = \{\boldsymbol{u}, \boldsymbol{C}^{-1}[\boldsymbol{S}], \boldsymbol{S}\}\,. \tag{4.133}$$

This is the starting point for the so-called *Hellinger–Reissner principle*. Mixed methods are mostly based on this principle.

The FE state

$$\mathcal{S}_h = s_1\,\mathcal{S}_1 + s_2\,\mathcal{S}_2 + \ldots = \sum_i s_i\,\mathcal{S}_i \tag{4.134}$$

consists of a series of "unit states", to which weights s_i are attached. It is best to split the unit states into u-states $\mathcal{S}_u$ and σ-states $\mathcal{S}_\sigma$:

$$\mathcal{S}_u = \{\boldsymbol{\varphi}_i, \boldsymbol{0}, \boldsymbol{0}\}\,, \qquad \mathcal{S}_\sigma = \{\boldsymbol{0}, \boldsymbol{C}^{-1}[\boldsymbol{S}_i], \boldsymbol{S}_i\}\,. \tag{4.135}$$

The u-states only represent displacements, while the σ-states only represent strains + stresses, so the FE solution looks like

$$\mathcal{S}_h = \sum_{i=1}^n u_i\,\{\boldsymbol{\varphi}_i, \boldsymbol{0}, \boldsymbol{0}\} + \sum_{i=1}^m \sigma_i\,\{\boldsymbol{0}, \boldsymbol{C}^{-1}[\boldsymbol{S}_i], \boldsymbol{S}_i\}\,, \tag{4.136}$$

where the $\boldsymbol{\varphi}_i(\boldsymbol{x})$ are the unit displacements and the $\boldsymbol{S}_i(\boldsymbol{x})$ the stress states.

A unit state like $\mathcal{S}_u = \{\boldsymbol{\varphi}_i, \boldsymbol{0}, \boldsymbol{0}\}$ is stress-free if there exists initial strains $\boldsymbol{E}_i^+$ that annihilate the strains induced by the displacement field $\boldsymbol{\varphi}_i$, i.e., if $\boldsymbol{E} = \boldsymbol{E}(\boldsymbol{\varphi}_i) + \boldsymbol{E}_i^+ = \boldsymbol{0}$. Hence, the $\mathcal{S}_u$ can be considered the solutions of problems $\boldsymbol{A}(\mathcal{S}) = [-\boldsymbol{E}(\boldsymbol{\varphi}_i), \boldsymbol{0}, \boldsymbol{0}]^T$.

Similar things can be said about the states $\mathcal{S}_\sigma$. The strains induced by the stresses are

$$\boldsymbol{E} = \boldsymbol{C}^{-1}[\boldsymbol{S}_i] + \boldsymbol{C}^{-1}[\boldsymbol{S}^+]\,. \tag{4.137}$$

Normally these strains induce displacements. But if $\boldsymbol{S}^+ = -\boldsymbol{S}_i$ the strains are zero and therefore a plate must not undergo any compensating movement, $\boldsymbol{u} = \boldsymbol{0}$, i.e. the $\mathcal{S}_\sigma$ solve problems like $\boldsymbol{A}(\mathcal{S}) = [\boldsymbol{0}, -\boldsymbol{S}_i^+, -\operatorname{div}\boldsymbol{S}_i]^T$.

The strain energy product associated with the *Hellinger–Reissner principle* is

$$a(\mathcal{S}, \hat{\mathcal{S}}) = \int_\Omega [\boldsymbol{C}^{-1}[\boldsymbol{S}] \bullet \hat{\boldsymbol{S}} + \boldsymbol{E}(\boldsymbol{u}) \bullet \hat{\boldsymbol{S}} + \boldsymbol{S} \bullet \boldsymbol{E}(\hat{\boldsymbol{u}})]\, d\Omega\,, \tag{4.138}$$

hence the stiffness matrix $k_{ij} = a(\mathcal{S}_i, \mathcal{S}_j)$ is

$$\begin{bmatrix} \boldsymbol{0}_{(n\times n)} & \boldsymbol{A}_{(n\times m)} \\ \boldsymbol{A}^T_{(m\times n)} & \boldsymbol{B}_{(m\times m)} \end{bmatrix} = \begin{bmatrix} \boldsymbol{u}_{(n)} \\ \boldsymbol{\sigma}_{(m)} \end{bmatrix} = \begin{bmatrix} \boldsymbol{p}_{(n)} \\ \boldsymbol{0}_{(m)} \end{bmatrix}\,, \tag{4.139}$$

where

$$a_{ij} = \int_\Omega \boldsymbol{S}_j \cdot \boldsymbol{E}(\boldsymbol{\varphi}_i)\, d\Omega \quad b_{ij} = \int_\Omega \boldsymbol{C}^{-1}[\boldsymbol{S}_i] \cdot \boldsymbol{S}_j\, d\Omega \qquad (4.140)$$

and

$$p_i = \int_\Omega \boldsymbol{p} \cdot \boldsymbol{\varphi}_i\, d\Omega\,. \qquad (4.141)$$

The first set of equations corresponds to the equilibrium condition $-\mathrm{div}\,\boldsymbol{S} = \boldsymbol{p}$. It is the task of the stress states to balance the external load $\boldsymbol{p}$:

$$\boldsymbol{A}\,\boldsymbol{\sigma} = \boldsymbol{p}\,. \qquad (4.142)$$

The second set of equations

$$\boldsymbol{A}^T\,\boldsymbol{u} + \boldsymbol{B}\,\boldsymbol{\sigma} = \boldsymbol{0} \qquad (4.143)$$

corresponds to the *compatibility condition* $\boldsymbol{E}(\boldsymbol{u}) - \boldsymbol{E} = \boldsymbol{0}$. As long as there are no initial strains this is an internal affair between the displacements and stresses, which is why the zero vector appears on the right-hand side.

Of course the compatibility condition is only satisfied in the L_2-sense, not in a strict pointwise sense, because in the second set of equations it is only required that the error be orthogonal to the test functions $\boldsymbol{S}_j$, the stress states:

$$\int_\Omega (\boldsymbol{E}(\boldsymbol{u}_h) - \boldsymbol{E}) \cdot \boldsymbol{S}_j\, d\Omega = 0 \qquad j = 1, 2, \ldots m\,. \qquad (4.144)$$

Hence the difference between $\boldsymbol{E}(\boldsymbol{u}_h)$ and the strains $\boldsymbol{E}_h$—which come from the stresses $\boldsymbol{E}_h = \boldsymbol{C}^{-1}[\boldsymbol{S}]$—is zero only in the weighted L_2 sense. The residual can be interpreted as initial strain $\boldsymbol{E}_h^+$:

$$\boldsymbol{E}(\boldsymbol{u}_h) - \boldsymbol{E} = \boldsymbol{E}_h^+\,. \qquad (4.145)$$

Hence in mixed methods (based on the Hellinger–Reissner principle) the substitute load case that is work-equivalent to the original load case $[\boldsymbol{0}, \boldsymbol{0}, \boldsymbol{p}]^T$ solved by the FE program consists of substitute loads $\boldsymbol{p}_h$, and includes initial strains in each element, $[\boldsymbol{E}_h^+, \boldsymbol{0}, \boldsymbol{p}_h]$.

Now the model has u- and σ-degrees of freedom and the next question then is how many to choose of each (see Fig. 4.62):

$$n = \text{number of displacement degrees of freedom}$$
$$m = \text{number of stress degrees of freedom}\,.$$

Linear algebra provides the answer. For the system $\boldsymbol{A}\boldsymbol{\sigma} = \boldsymbol{p}$ to have a solution, the right-hand side, the vector $\boldsymbol{p}$, must be orthogonal to all solutions $\boldsymbol{u}$ of the adjoint system

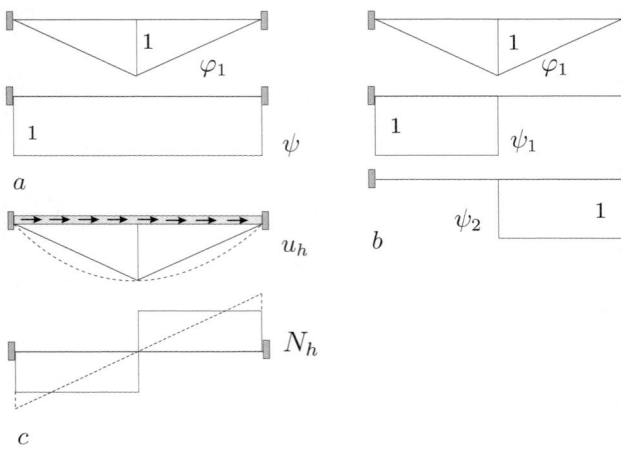

Fig. 4.62. a) Not enough stress functions, the *Babuška-Brezzi-condition* is violated, b) more stress functions than displacement functions, the condition is satisfied, c) FE solution

$$\boldsymbol{A}^T \boldsymbol{u} = \boldsymbol{0} \, . \tag{4.146}$$

The vectors $\boldsymbol{u}$ that solve this homogeneous system of equations form the so-called *kernel* of the matrix $\boldsymbol{A}^T$, and two vectors are orthogonal if their scalar product is zero, $\boldsymbol{p}^T \boldsymbol{u} = 0$.

If the system (4.146) is regular, i.e., if the columns of $\boldsymbol{A}^T$ are linearly independent, the problem $\boldsymbol{A}^T \boldsymbol{u} = \boldsymbol{0}$ has only the trivial solution $\boldsymbol{u} = \boldsymbol{0}$, and the right-hand side $\boldsymbol{p}$ is then orthogonal to the kernel of $\boldsymbol{A}^T$. One need not worry that there are load cases that the FE program cannot solve.

The columns are linearly independent if and only if

$$\boldsymbol{A}^T \boldsymbol{u} = \boldsymbol{0} \qquad \Longrightarrow \qquad \boldsymbol{u} = \boldsymbol{0} \tag{4.147}$$

i.e., if the fact that the scalar product between all the strains

$$\boldsymbol{E} = \sum_{i=1}^{n} u_i \, \boldsymbol{E}(\boldsymbol{\varphi}_i) \tag{4.148}$$

and all the stresses $\boldsymbol{S}_i(\boldsymbol{x})$ is zero,

$$\sum_{i=1}^{n} u_i \int_{\Omega} \boldsymbol{E}(\boldsymbol{\varphi}_i) \bullet \boldsymbol{S}_j \, d\Omega = 0 \qquad j = 1, 2, \ldots m \, , \tag{4.149}$$

implies that the nodal displacements u_i are zero.

Obviously this can only happen if there are more stress states $\boldsymbol{S}_i$ than displacement states $\boldsymbol{\varphi}_i$, because otherwise one could easily construct, say, a 12-term displacement field that is so balanced that the scalar product with, say, all seven stress states is zero, even though the displacement field—or rather its strains—is not.

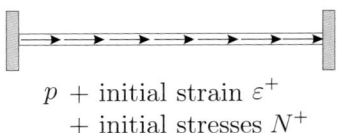

p + initial strain ε^+
+ initial stresses N^+

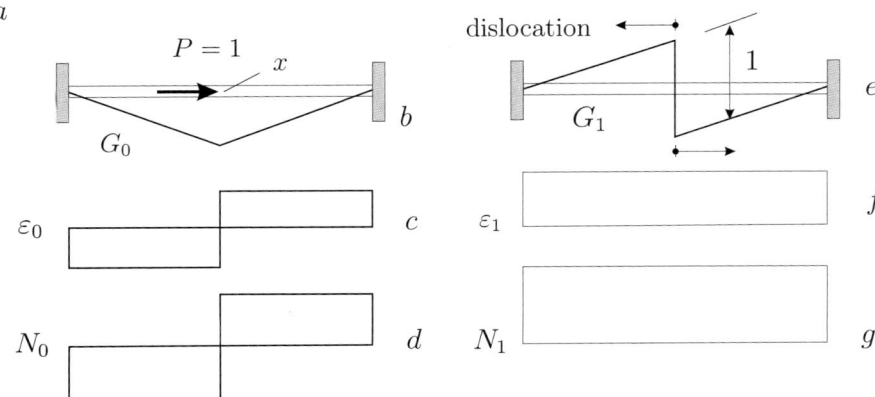

Fig. 4.63. Elastic bar **a)** system and load, **b)** Green's function G_0 **c)** strain ε_0

Hence the referee must always have more degrees of freedom than that which he is to check, i.e., there must always be more stress degree of freedoms than displacement degrees of freedom.

The columns of the matrix $\boldsymbol{A}^T$ are linearly independent if and only if there exists a constant β such that

$$0 < \beta\,|\boldsymbol{u}| < \inf_{\boldsymbol{\delta}} \sup_{\boldsymbol{\sigma}} \frac{\boldsymbol{\sigma}^T \boldsymbol{A}^T \boldsymbol{u}}{|\boldsymbol{\sigma}|}. \tag{4.150}$$

This is the *Babuška–Brezzi condition.*

The lower bound β ensures that there is always a large enough distance from zero. When the elements shrink in size, i.e., if the vectors $\boldsymbol{\delta}$ and $\boldsymbol{\sigma}$ grow in length, the angle between the two vectors will definitely be greater than zero.

4.19 Influence functions for mixed formulations

To simplify the formulation, let us switch to a 1-D problem, an elastic bar. The governing equations

$$u' - \varepsilon = \varepsilon^+$$
$$EA\,\varepsilon - N = N^+ \tag{4.151}$$
$$-N' = p$$

have the same structure as before, $\mathcal{S} = \{u, \varepsilon, N\}$,

$$G(\mathcal{S}, \hat{\mathcal{S}}) = \langle \boldsymbol{A}(\mathcal{S}), \hat{\mathcal{S}} \rangle + [N\,\hat{u}]_0^l - a(\mathcal{S}, \hat{\mathcal{S}}) = 0 \qquad (4.152)$$

and

$$a(\mathcal{S}, \hat{\mathcal{S}}) = \int_0^l (u' - \varepsilon)\,\hat{N}\,dx + \int_0^l EA\,\varepsilon\,\hat{\varepsilon}\,dx + \int_0^l N\,(\hat{u}' - \hat{\varepsilon})\,dx\,. \quad (4.153)$$

First we need Betti's theorem

$$B(\mathcal{S}, \hat{\mathcal{S}}) = \langle \boldsymbol{A}(\mathcal{S}), \hat{\mathcal{S}} \rangle + [N\,\hat{u}]_0^l - [u\,\hat{N}]_0^l - \langle \mathcal{S}, \boldsymbol{A}(\hat{\mathcal{S}}) \rangle = 0\,. \quad (4.154)$$

Next let $\mathcal{S} = \{u, \varepsilon, N\}$ the solution of $\boldsymbol{A}(\mathcal{S}) = [\varepsilon^+, N^+, p]^T$ (see Fig. 4.63 a), and let $\hat{\mathcal{S}} = \mathcal{S}_0 = \{G_0, \varepsilon_0, N_0\}$ the point solution (see Fig. 4.63 b) where

$$\varepsilon_0 = \frac{d}{dy}\,G_0(y, x) \qquad N_0 = EA\,\varepsilon_0\,. \qquad (4.155)$$

Because the point solution lacks the necessary regularity, we formulate Green's second identity (4.154) on the punctured interval $I_\varepsilon := [0, x - \varepsilon] \cup [x + \varepsilon, l]$ and take the limit. Note that on I_ε the right-hand side of the Green's function $\mathcal{S}_0$ is zero, $\boldsymbol{A}(\mathcal{S}_0) = [0, 0, 0]^T$, so that

$$\lim_{\varepsilon \to 0} B(\mathcal{S}, \mathcal{S}_0)_{I_\varepsilon} = \langle \boldsymbol{A}(\mathcal{S}), \mathcal{S}_0 \rangle + [N\,G_0]_0^l - [u\,N_0]_0^l - u(x) = 0\,, \quad (4.156)$$

where the single term $u(x)$ is the limit of

$$\lim_{\varepsilon \to 0} \{N_0(x - \varepsilon)\,u(x - \varepsilon) - N_0(x + \varepsilon)\,u(x + \varepsilon)\} = 1 \cdot u(x)\,. \quad (4.157)$$

If we take the boundary conditions into account, it follows

$$u(x) = \langle \boldsymbol{A}(\mathcal{S}), \mathcal{S}_0 \rangle = \int_0^l [N_0\,\varepsilon^+ + \varepsilon_0\,N^+ + G_0\,p]\,dx\,, \qquad (4.158)$$

and in the same way we obtain the influence function for the normal force

$$N(x) = \langle \boldsymbol{A}(\mathcal{S}), \mathcal{S}_1 \rangle = \int_0^l [N_1\,\varepsilon^+ + \varepsilon_1\,N^+ + G_1\,p]\,dx\,. \qquad (4.159)$$

It is not possible to derive an influence function for the strain ε, because there are only two boundary operators in the "boundary integral" $[N\,\hat{u}]$—the identity and the normal force operator, $N = EA\,du/dx$; see Sect. 7.6, p. 529. But of course N and ε are—up to the factor EA—the same. This problem and its solution ($\varepsilon \simeq N$) has its origin in the nature of the system (4.151): first we define the strains (one differential operator), then the stresses (no differential operator only the "elasticity tensor" is applied!), and finally we define what equilibrium means (one differential operator).

4.20 Error analysis

Let $\wp$ denote the right-hand side $A(\mathcal{S}) = \wp$ and let $V_h = \{\mathcal{S}_i\}$ the trial space. Regardless of whether we use the *Hu-Washizu principle* or the *Hellinger–Reissner principle* the FE solution $\mathcal{S}_h$ is characterized by the property

$$a(\mathcal{S}_h, \mathcal{S}_i) = \langle \wp, \mathcal{S}_i \rangle \qquad \mathcal{S}_i \in V_h \tag{4.160}$$

where $a(.,.)$ represents a symmetric bilinear form and $\langle .,. \rangle$ a linear form. To make statements about the existence and uniqueness of an FE solution we would need to see that these forms are continuous and coercive. We simply assume that this is the case.

Most of the variational properties which we are used to attribute to FE solutions, see Sect. 7.12, p. 568, can be carried over to mixed problems, because to a large extent they are simply based on Green's first identity and some simple algebra. Hence it is evident that also the FE solution of mixed problems satisfies the Galerkin orthogonality

$$a(\mathcal{S} - \mathcal{S}_h, \mathcal{S}_i) = 0 \qquad \mathcal{S}_i \in V_h \,, \tag{4.161}$$

and also Tottenham's equation holds for the solution of mixed problems, i.e.,

$$u_h(x) = \int_0^l [N_0^h \, \varepsilon^+ + \varepsilon_0^h \, N^+ + G_0^h \, p] \, dx \,. \qquad (\bullet \ \bullet) \tag{4.162}$$

The proof is done as in the case of the original equation (1.210) on p. 64.

Recall (1.228) on p. 69, where we stated that

$$u_h(x) = p_h(G_0) = \underbrace{p_h(G_0^h)}_{\bullet} = \underbrace{p(G_0^h)}_{\bullet \ \bullet} = (\delta_0, u_h) = (\delta_0^h, u_h) = (\delta_0^h, u)$$

$$\tag{4.163}$$

and therefore we have for example as well

$$u_h(x) = \int_0^l [N_0^h \, \varepsilon_h^+ + \varepsilon_0^h \, N_h^+ + G_0^h \, p_h] \, dx \,. \qquad (\bullet) \tag{4.164}$$

Of course also the basic formula for goal-oriented recovery techniques holds as well

$$|e(x)| = |a(\mathcal{S}_0 - \mathcal{S}_0^h, \mathcal{S} - \mathcal{S}_h)| \leq ||\mathcal{S}_0 - \mathcal{S}_0^h||_E \, ||\mathcal{S} - \mathcal{S}_h||_E \tag{4.165}$$

where $||\mathcal{S}||_E^2 = a(\mathcal{S}, \mathcal{S})$.

4.21 Nonlinear problems

In the triple $\{u, E, S\}$ we let now E the Green-Lagrangian strain tensor and S the second Piola-Kirchhoff stress tensor, and we assume the material to be hyperelastic, i.e., there exists a strain-energy function $\boldsymbol{W}$ such that $\boldsymbol{S} = \partial \boldsymbol{W}/\partial \boldsymbol{E}$. Given volume forces $\boldsymbol{p}$ the elastic state $\mathcal{S} = \{\boldsymbol{u}, \boldsymbol{E}, \boldsymbol{S}\}$ satisfies at every point $\boldsymbol{x}$ of the undeformed body the system

$$\boldsymbol{E}(\boldsymbol{u}) - \boldsymbol{E} = 0 \qquad \frac{1}{2}\left(u_{i,j} + u_{j,i} + u_{k,i}\, u_{k,j}\right) - \varepsilon_{ij} = 0$$

$$\boldsymbol{W}'(\boldsymbol{E}) - \boldsymbol{S} = 0 \qquad \frac{\partial W}{\partial \varepsilon_{ij}} - \sigma_{ij} = 0 \qquad (4.166)$$

$$-\mathrm{div}(\boldsymbol{S} + \nabla\boldsymbol{u}\,\boldsymbol{S}) = \boldsymbol{p} \qquad -(\sigma_{ij} + u_{i,k}\,\sigma_{kj})_{,j} = p_i$$

and satisfies displacement boundary conditions $\boldsymbol{u} = \bar{\boldsymbol{u}}$ on a part Γ_D of the boundary and stress boundary conditions $\boldsymbol{t}(\boldsymbol{S}, \boldsymbol{u}) = \bar{\boldsymbol{t}}$ on the complementary part Γ_N where

$$\boldsymbol{t}(\boldsymbol{S}, \boldsymbol{u}) := (\boldsymbol{S} + \nabla\boldsymbol{u}\,\boldsymbol{S})\,\boldsymbol{n} \qquad (4.167)$$

is the traction vector at a boundary point with outward normal vector $\boldsymbol{n}$.

With symmetric stress tensors $\boldsymbol{S}$ we have the identity

$$\int_{\Omega} -\mathrm{div}(\boldsymbol{S} + \nabla\boldsymbol{u}\,\boldsymbol{S}) \bullet \hat{\boldsymbol{u}}\, d\Omega$$

$$= -\int_{\Gamma} \boldsymbol{t}(\boldsymbol{S}, \boldsymbol{u}) \bullet \hat{\boldsymbol{u}}\, ds + \int_{\Omega} \boldsymbol{E}_{\boldsymbol{u}}(\hat{\boldsymbol{u}}) \bullet \boldsymbol{S}\, d\Omega \qquad (4.168)$$

where

$$\boldsymbol{E}_{\boldsymbol{u}}(\hat{\boldsymbol{u}}) := \frac{1}{2}\left(\nabla\hat{\boldsymbol{u}} + \nabla\hat{\boldsymbol{u}}^T + \nabla\boldsymbol{u}^T\,\nabla\hat{\boldsymbol{u}} + \nabla\hat{\boldsymbol{u}}^T\,\nabla\boldsymbol{u}\right) \qquad (4.169)$$

is the Gateaux derivative of the matrix $\boldsymbol{E}(\boldsymbol{u})$

$$\frac{d}{d\varepsilon}[\boldsymbol{E}(\boldsymbol{u} + \varepsilon\,\hat{\boldsymbol{u}})]_{|\varepsilon=0} = \boldsymbol{E}_{\boldsymbol{u}}(\hat{\boldsymbol{u}}). \qquad (4.170)$$

Collecting terms we can formulate Green's first identity of the operator $\boldsymbol{A}(\mathcal{S})$, that is the system (4.166)

$$G(\mathcal{S}, \hat{\mathcal{S}}) = \underbrace{\langle \boldsymbol{A}(\mathcal{S}), \hat{\mathcal{S}} \rangle + \int_{\Gamma} \boldsymbol{t}(\boldsymbol{S}, \boldsymbol{u}) \bullet \hat{\boldsymbol{u}}\, ds}_{\delta W_e} - \underbrace{a(\mathcal{S}, \hat{\mathcal{S}})}_{\delta W_i} = 0 \qquad (4.171)$$

where $\langle \boldsymbol{A}(\mathcal{S}), \hat{\mathcal{S}} \rangle$ is similar to (4.130) and where

$$a(\mathcal{S}, \hat{\mathcal{S}}) = \int_{\Omega} (\boldsymbol{E}(\boldsymbol{u}) - \boldsymbol{E}) \bullet \hat{\boldsymbol{S}}\, d\Omega$$

$$+ \int_{\Omega} (\boldsymbol{W}'(\boldsymbol{E}) - \boldsymbol{S}) \bullet \hat{\boldsymbol{E}}\, d\Omega + \int_{\Omega} \boldsymbol{E}_{\boldsymbol{u}}(\hat{\boldsymbol{u}}) \bullet \boldsymbol{S}\, d\Omega. \qquad (4.172)$$

The identity (4.171) is the basis of many variational principles in nonlinear mechanics and can be formulated in the same way also for beams and slabs, [115].

In the case of a pure displacement formulation $\mathcal{S} = \{\boldsymbol{u}, \boldsymbol{E}(\boldsymbol{u}), \boldsymbol{W}'(\boldsymbol{E}(\boldsymbol{u}))\}$ and it is $\hat{\boldsymbol{u}} = \boldsymbol{0}$ on Γ_D, so that (4.171) reduces to

$$G(\boldsymbol{u}, \hat{\boldsymbol{u}}) = \int_\Omega \boldsymbol{p} \cdot \hat{\boldsymbol{u}} \, d\Omega + \int_{\Gamma_N} \bar{\boldsymbol{t}} \cdot \hat{\boldsymbol{u}} \, ds - \int_\Omega \boldsymbol{E}_{\boldsymbol{u}}(\hat{\boldsymbol{u}}) \cdot \boldsymbol{S} \, d\Omega = 0, \quad (4.173)$$

where $\boldsymbol{S} = \boldsymbol{W}'(\boldsymbol{E}(\boldsymbol{u}))$.

Next let $\boldsymbol{u}_h = \sum_j^n u_j \, \boldsymbol{\varphi}_j(\boldsymbol{x})$ the FE solution and let $\hat{\boldsymbol{u}} = \boldsymbol{\varphi}_i$ a virtual displacement then

$$\underbrace{\int_\Omega \boldsymbol{E}_{\boldsymbol{u}_h}(\boldsymbol{\varphi}_i) \cdot \boldsymbol{W}'(\boldsymbol{E}(\boldsymbol{u}_h)) \, d\Omega}_{k_i} = \underbrace{\int_\Omega \boldsymbol{p} \cdot \boldsymbol{\varphi}_i \, d\Omega + \int_{\Gamma_N} \bar{\boldsymbol{t}} \cdot \boldsymbol{\varphi}_i \, ds}_{f_i} \quad (4.174)$$

or

$$\boldsymbol{k}(\boldsymbol{u}) = \boldsymbol{f} \quad (4.175)$$

where $\boldsymbol{u}$ is the vector of nodal coordinates.

Linearization

For computational purposes a linearization of (4.175) is necessary. Let $\boldsymbol{p}_\Delta$ and $\bar{\boldsymbol{t}}_\Delta$ be load increments, and let $\boldsymbol{u} + \boldsymbol{u}_\Delta$ be the displacement field corresponding to $\boldsymbol{p} + \boldsymbol{p}_\Delta$ and $\bar{\boldsymbol{t}} + \bar{\boldsymbol{t}}_\Delta$ then

$$G(\boldsymbol{u} + \boldsymbol{u}_\Delta, \hat{\boldsymbol{u}}) = \int_\Omega (\boldsymbol{p} + \boldsymbol{p}_\Delta) \cdot \hat{\boldsymbol{u}} \, d\Omega + \int_{\Gamma_N} (\bar{\boldsymbol{t}} + \bar{\boldsymbol{t}}_\Delta) \cdot \hat{\boldsymbol{u}} \, ds$$
$$- a(\boldsymbol{u} + \boldsymbol{u}_\Delta, \hat{\boldsymbol{u}}) = 0, \quad (4.176)$$

where

$$a(\boldsymbol{u} + \boldsymbol{u}_\Delta, \hat{\boldsymbol{u}}) := \int_\Omega \boldsymbol{E}_{\boldsymbol{u}+\boldsymbol{u}_\Delta}(\hat{\boldsymbol{u}}) \cdot \boldsymbol{W}'(\boldsymbol{E}(\boldsymbol{u} + \boldsymbol{u}_\Delta)) \, d\Omega. \quad (4.177)$$

The Gateaux derivative of the strain energy product

$$a(\boldsymbol{u}, \hat{\boldsymbol{u}}) := \int_\Omega \boldsymbol{E}_{\boldsymbol{u}}(\hat{\boldsymbol{u}}) \cdot \boldsymbol{W}'(\boldsymbol{E}(\boldsymbol{u})) \, d\Omega \quad (4.178)$$

with respect to a displacement increment $\boldsymbol{u}_\Delta$ is

$$a_T(\boldsymbol{u}, \boldsymbol{u}_\Delta, \hat{\boldsymbol{u}}) := \left[\frac{d}{d\varepsilon} a(\boldsymbol{u} + \varepsilon \, \boldsymbol{u}_\Delta, \hat{\boldsymbol{u}}) \right]_{\varepsilon=0}$$
$$= \int_\Omega [\nabla \boldsymbol{u}_\Delta \, \boldsymbol{W}'(\boldsymbol{E}(\boldsymbol{u})) \cdot \nabla \hat{\boldsymbol{u}} + \boldsymbol{E}_{\boldsymbol{u}}(\hat{\boldsymbol{u}}) \cdot \boldsymbol{C}[\boldsymbol{E}_{\boldsymbol{u}}(\boldsymbol{u}_\Delta)]] \, d\Omega, \quad (4.179)$$

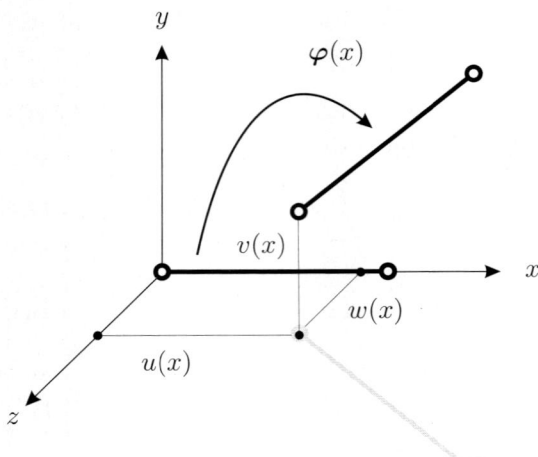

Fig. 4.64. Truss element

where the tensor

$$C = \frac{\partial^2 W}{\partial E \, \partial E} = \frac{\partial}{\partial E} S \qquad (4.180)$$

is the second derivative of W, which is to be evaluated at u. Note that $a_T(u, u_\Delta, \hat{u})$ is linear in the second and third argument, u_Δ and $\hat{u}$.

We then let

$$a(u + u_\Delta, \hat{u}) \simeq a(u, \hat{u}) + a_T(u, u_\Delta, \hat{u}), \qquad (4.181)$$

so that (4.175) becomes

$$K_T(u) \, u_\Delta = f - k(u), \qquad (4.182)$$

where now u_Δ is the vector of nodal displacements of the field u_Δ and K_T is the tangential stiffness matrix:

$$(K_T)_{ij} = a_T(u, \varphi_j, \varphi_i). \qquad (4.183)$$

A truss element

We consider a truss element, that extends along the x-axis; see Fig. 4.64. The deformation of the element is

$$\varphi(x, y, z) = (x + u(x)) \, e_1 + (y + v(x)) \, e_2 + (z + w(x)) \, e_3, \qquad (4.184)$$

where $u(x) = [u(x), v(x), w(x)]^T$ is the displacement vector. By definition the deformation gradient is

$$\boldsymbol{F} = \nabla\boldsymbol{\varphi} = \begin{bmatrix} 1 + u' & 0 & 0 \\ v' & 1 & 0 \\ w' & 0 & 1 \end{bmatrix} = \boldsymbol{I} + \nabla\boldsymbol{u}\,, \tag{4.185}$$

and the Green-Lagrangian strain tensor

$$\boldsymbol{E}(\boldsymbol{u}) = \frac{1}{2}\,(\boldsymbol{F}^T\,\boldsymbol{F} - \boldsymbol{I}) = \begin{bmatrix} \varepsilon_x & \frac{1}{2}\,v' & \frac{1}{2}\,w' \\ \frac{1}{2}\,v' & 0 & 0 \\ \frac{1}{2}\,w' & 0 & 0 \end{bmatrix}, \tag{4.186}$$

where

$$\varepsilon(\boldsymbol{u}) := \varepsilon_x(\boldsymbol{u}) = \frac{1}{2}\,((1+u')^2 + (v')^2 + (w')^2 - 1)$$

$$= u' + \frac{1}{2}\,((u')^2 + (v')^2 + (w')^2) \qquad (\ )' = \frac{d}{dx}\,. \tag{4.187}$$

Green's first identity reads, if all functions are functions of x only,

$$G(\boldsymbol{u}, \hat{\boldsymbol{u}}) = \int_0^l -\mathrm{div}\,(\boldsymbol{S} + \nabla\boldsymbol{u}\,\boldsymbol{S}) \bullet \hat{\boldsymbol{u}}\,A\,dx + [(\boldsymbol{S} + \nabla\boldsymbol{u}\,\boldsymbol{S})\,\boldsymbol{e}_1 \bullet \hat{\boldsymbol{u}}\,A]_0^l$$

$$- \int_0^l \boldsymbol{E}_{\boldsymbol{u}}(\hat{\boldsymbol{u}}) \bullet \boldsymbol{S}\,A\,dx = 0\,. \tag{4.188}$$

We then let simply $\sigma = \sigma_{xx} = E\,\varepsilon$, where E is Young's modulus, and we let all other $\sigma_{ij} = 0$, so that with $N = A(\sigma + u'\,\sigma)$

$$G(\boldsymbol{u}, \hat{\boldsymbol{u}}) = \int_0^l -N'\,\hat{u}\,dx + [N\,\hat{u}]_0^l \underbrace{- \int_0^l \varepsilon_{\boldsymbol{u}}(\hat{\boldsymbol{u}})\,\sigma\,A\,dx}_{a(\boldsymbol{u}, \hat{\boldsymbol{u}})} = 0\,, \tag{4.189}$$

where

$$\varepsilon_{\boldsymbol{u}}(\hat{\boldsymbol{u}}) = (1 + u')\,\hat{u}' + v'\,\hat{v}' + w'\,\hat{w}'\,. \tag{4.190}$$

The Gateaux derivative of the strain energy product $a(\boldsymbol{u}, \hat{\boldsymbol{u}})$ is

$$a_T(\boldsymbol{u}, \boldsymbol{u}_\Delta, \hat{\boldsymbol{u}}) := \left[\frac{d}{d\varepsilon}\,a(\boldsymbol{u} + \varepsilon\boldsymbol{u}_\Delta, \hat{\boldsymbol{u}})\right]_{\varepsilon=0}$$

$$= \int_0^l [\varepsilon_{\boldsymbol{u}_\Delta}(\hat{\boldsymbol{u}})\,\sigma + \varepsilon_{\boldsymbol{u}}(\hat{\boldsymbol{u}})\,\sigma]\,A\,dx\,, \tag{4.191}$$

where

$$\varepsilon_{\boldsymbol{u}_\Delta}(\hat{\boldsymbol{u}}) = \left[\frac{d}{d\varepsilon}\,\varepsilon_{\boldsymbol{u}+\varepsilon\,\boldsymbol{u}_\Delta}(\hat{\boldsymbol{u}})\right]_{\varepsilon=0} = u'_\Delta\hat{u}' + v'_\Delta\,\hat{v}' + w'_\Delta\,\hat{w}' \tag{4.192}$$

and

$$\sigma = \left[\frac{d}{d\varepsilon} \sigma(\boldsymbol{u} + \varepsilon \, \boldsymbol{u}_\Delta) \right]_{\varepsilon=0} = E \, \varepsilon_{\boldsymbol{u}}(\boldsymbol{u}_\Delta) \, . \tag{4.193}$$

The elements of the tangential stiffness matrix are therefore

$$(K_T(\boldsymbol{u}))_{ij} = a_T(\boldsymbol{u}, \boldsymbol{\varphi}_i, \boldsymbol{\varphi}_j) \, , \tag{4.194}$$

where the vector-valued functions $\boldsymbol{\varphi}_i(\xi)$ are the nodal unit displacements (three at each node).

With linear shape functions on an element Ω_e with length l_e

$$u_e(x) = \sum_{i=1}^{2} u_i^e \, \varphi_i^e(\xi) \, , \; v_e(x) = \sum_{i=1}^{2} v_i^e \, \varphi_i^e(\xi) \, , \; w_e(x) = \sum_{i=1}^{2} w_i^e \, \varphi_i^e(\xi) \, , \tag{4.195}$$

we obtain for the nodal vector $\boldsymbol{u}^e = [u_1, v_1, w_1, u_2, v_2, w_2]^T$ the 6×6 tangential stiffness matrix [255]

$$\boldsymbol{K}_T^e(\boldsymbol{u}) = \begin{bmatrix} (\boldsymbol{A}_1 + \boldsymbol{A}_2) & -(\boldsymbol{A}_1 + \boldsymbol{A}_2) \\ -(\boldsymbol{A}_1 + \boldsymbol{A}_2) & (\boldsymbol{A}_1 + \boldsymbol{A}_2) \end{bmatrix} \, , \tag{4.196}$$

where

$$\boldsymbol{A}_1 = \frac{E \, A}{l_e} \begin{bmatrix} (1 + u_e')^2 & (1 + u_e') \, v_e' & (1 + u_e') \, w_e' \\ (1 + u_e') \, v_e' & (v_e')^2 & v_e' \, w_e' \\ (1 + u_e') \, w_e' & v_e' \, w_e' & (w_e')^2 \end{bmatrix} \tag{4.197}$$

and

$$\boldsymbol{A}_2 = \frac{\sigma \, A}{l_e} \begin{bmatrix} 1 & 0 & 0 \\ 0 & 1 & 0 \\ 0 & 0 & 1 \end{bmatrix} \, , \tag{4.198}$$

so that after the assemblage

$$\boldsymbol{K}_T(\boldsymbol{u}) \, \boldsymbol{u}_\Delta = \boldsymbol{f} - \boldsymbol{k}(\boldsymbol{u}) \, . \tag{4.199}$$

For an extension of these concepts to 3-D beam problems see [104].

Plane problem

Let $\boldsymbol{X}$ denote the initial coordinates of a point and $\boldsymbol{x}$ the coordinates of the current configuration (see Fig. 4.65)

$$\boldsymbol{x} = \boldsymbol{X} + \boldsymbol{u} \, , \quad \text{or} \quad x_i = X_i + u_i \, , \tag{4.200}$$

where $\boldsymbol{u}$ is the displacement vector.

The deformation gradient is

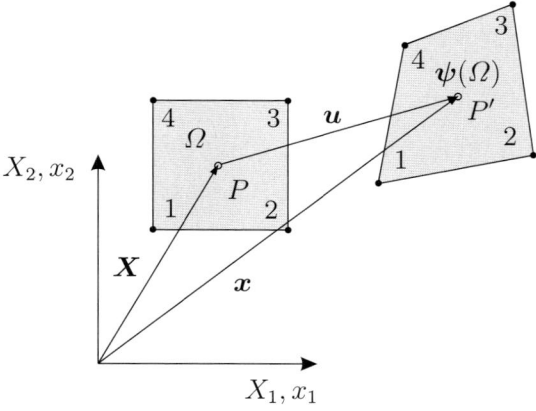

Fig. 4.65. Deformations of a plane element

$$F = \frac{\partial \boldsymbol{x}}{\partial \boldsymbol{X}} = \boldsymbol{I} + \nabla \boldsymbol{u} = \begin{bmatrix} 1 + u_{1,1} & u_{1,2} & 0 \\ u_{2,1} & 1 + u_{2,2} & 0 \\ 0 & 0 & 1 \end{bmatrix} , \qquad (4.201)$$

so that the Green-Lagrangian strain tensor

$$\boldsymbol{E} = \frac{1}{2}(\boldsymbol{F}^T \boldsymbol{F} - \boldsymbol{I}) = \begin{bmatrix} E_{11} & E_{12} \\ \text{sym.} & E_{22} \end{bmatrix} \qquad (4.202)$$

has the components

$$E_{11} = u_{1,1} + \frac{1}{2} \left(u_{1,1}^2 + u_{2,1}^2 \right)$$

$$E_{22} = u_{2,2} + \frac{1}{2} \left(u_{2,2}^2 + u_{1,2}^2 \right) \qquad (4.203)$$

$$E_{12} = \frac{1}{2} \left(u_{1,2} + u_{2,1} \right) + \frac{1}{2} \left(u_{1,1}\, u_{1,2} + u_{2,2}\, u_{2,1} \right) .$$

The FE displacement field can be written in two ways

$$\boldsymbol{u}_h = \sum_{i=1}^{\text{DOFS}} u_i\, \boldsymbol{\varphi}_i(\boldsymbol{x}) \equiv \sum_i u_i \begin{bmatrix} \cdot \\ \cdot \end{bmatrix} , \quad \boldsymbol{u}_h = \sum_{i=1}^{\text{NODES}} \boldsymbol{u}_i\, \psi_i(\boldsymbol{x}) \equiv \sum_i \begin{bmatrix} \cdot \\ \cdot \end{bmatrix} \psi_i ,$$

$$(4.204)$$

where the sum extends either over the degrees of freedom u_i or the nodes of the structure. The $\boldsymbol{\varphi}_i(\boldsymbol{x})$ are the nodal unit displacement fields associated with the u_i, see (4.27) on p. 337, while the ψ_i in the second formula are the scalar-valued shape functions of the nodes ($\psi_i(\boldsymbol{x}_j) = \delta_{ij}$) and the vectors $\boldsymbol{u}_i = [u_1^{(i)}, u_2^{(i)}]^T$ are the nodal displacements at node i. Here, we use the second notation so that, for example, the virtual exterior work of the volume forces $\boldsymbol{p}$ becomes

$$\delta W_e = \int_\Omega \boldsymbol{p} \cdot \boldsymbol{u}_h \, d\Omega = \sum_{i=1}^n \int_\Omega \boldsymbol{p} \cdot \boldsymbol{u}_i \, \psi_i \, d\Omega = \sum_{i=1}^n \boldsymbol{u}_i^T \int_\Omega \boldsymbol{p} \, \psi_i \, d\Omega = \sum_{i=1}^n \boldsymbol{u}_i^T \boldsymbol{f}_i \, .$$
(4.205)

The gradient of the FE displacement field is then

$$\nabla \boldsymbol{u}_h = \sum_{i=1}^n \boldsymbol{u}_i \otimes \nabla \psi_i$$
(4.206)

and the deformation gradient,

$$\boldsymbol{F} = \boldsymbol{I} + \nabla \boldsymbol{u}_h = \boldsymbol{I} + \sum_{i=1}^n \boldsymbol{u}_i \otimes \nabla \psi_i \, .$$
(4.207)

The gradient of the virtual displacement field $\hat{\boldsymbol{u}}_h = \sum_i \hat{\boldsymbol{u}}_i \, \psi_i$ follows (4.206) so that Gateaux derivative becomes (we drop the subscript h on $\boldsymbol{u}_h$ and $\hat{\boldsymbol{u}}_h$ for a moment)

$$\begin{aligned} \boldsymbol{E}_{\boldsymbol{u}}(\hat{\boldsymbol{u}}) :&= \frac{1}{2} \left(\nabla \hat{\boldsymbol{u}} + \nabla \hat{\boldsymbol{u}}^T + \nabla \boldsymbol{u}^T \nabla \hat{\boldsymbol{u}} + \nabla \hat{\boldsymbol{u}}^T \nabla \boldsymbol{u} \right) \\ &= \frac{1}{2} \left(\boldsymbol{F}^T \nabla \hat{\boldsymbol{u}} + \nabla \hat{\boldsymbol{u}}^T \boldsymbol{F} \right) \\ &= \frac{1}{2} \sum_{i=1}^n [\boldsymbol{F}^T (\hat{\boldsymbol{u}}_i \otimes \nabla \psi_i) + (\nabla \psi_i \otimes \hat{\boldsymbol{u}}_i) \boldsymbol{F}] \, . \end{aligned}$$
(4.208)

As in the linear theory where $\boldsymbol{E} \cdot \boldsymbol{S} = \boldsymbol{\varepsilon} \cdot \boldsymbol{\sigma}$ the symmetry of $\boldsymbol{E}_{\boldsymbol{u}}(\hat{\boldsymbol{u}})$ and $\boldsymbol{S}$ is motivation to introduce a "Gateaux strain vector"

$$\boldsymbol{\varepsilon}_{\boldsymbol{u}}(\hat{\boldsymbol{u}}) := \begin{bmatrix} (\boldsymbol{E}_{\boldsymbol{u}}(\hat{\boldsymbol{u}}))_{11} \\ (\boldsymbol{E}_{\boldsymbol{u}}(\hat{\boldsymbol{u}}))_{22} \\ 2\,(\boldsymbol{E}_{\boldsymbol{u}}(\hat{\boldsymbol{u}}))_{12} \end{bmatrix} = \sum_{i=1}^n \boldsymbol{B}_{L_i} \, \hat{\boldsymbol{u}}_i \, ,$$
(4.209)

where

$$\boldsymbol{B}_{L_i} = \begin{bmatrix} F_{11}\,\psi_{i,1} & F_{21}\,\psi_{i,1} \\ F_{12}\,\psi_{i,2} & F_{22}\,\psi_{i,2} \\ F_{11}\,\psi_{i,2} + F_{12}\,\psi_{i,1} & F_{21}\,\psi_{i,2} + F_{22}\,\psi_{i,1} \end{bmatrix} \, .$$
(4.210)

Given a St. Venant type material

$$\boldsymbol{S} = \boldsymbol{C}[\boldsymbol{E}] = 2\mu \boldsymbol{E} + \lambda (\operatorname{tr} \boldsymbol{E})\, \boldsymbol{I}$$
(4.211)

the stress vector of the second Piola-Kirchhoff stress tensor is:

$$\boldsymbol{\sigma} = \begin{bmatrix} S_{11} \\ S_{22} \\ S_{12} \end{bmatrix} = \begin{bmatrix} \lambda + 2\mu & \lambda & 0 \\ \lambda & \lambda + 2\mu & 0 \\ 0 & 0 & \mu \end{bmatrix} \begin{bmatrix} E_{11} \\ E_{22} \\ 2E_{12} \end{bmatrix} \, ,$$
(4.212)

where

$$\mu = \frac{E}{2(1+\nu)}, \qquad \lambda = \frac{E\nu}{(1+\nu)(1-2\nu)}. \qquad (4.213)$$

Hence the weak form of the equilibrium conditions is

$$G(\boldsymbol{u}_h, \hat{\boldsymbol{u}}_h) = \sum_{i=1}^{n} \hat{\boldsymbol{u}}_i^T [\underbrace{\int_\Omega \boldsymbol{B}_{L_i}^T \boldsymbol{\cdot} \boldsymbol{\sigma} \, d\Omega}_{\boldsymbol{k}_i} - \underbrace{\int_\Omega \boldsymbol{p} \, \psi_i \, d\Omega - \int_{\Gamma_N} \bar{\boldsymbol{t}} \, \psi_i \, ds}_{\boldsymbol{f}_i}] = 0$$

$$(4.214)$$

or

$$\boldsymbol{k}(\boldsymbol{u}) - \boldsymbol{f} = \boldsymbol{0}. \qquad (4.215)$$

As on p. 403 we linearize this equation with the help of a Gateaux derivative of the strain energy product in the direction of $\boldsymbol{u}_\Delta$

$$a_T(\boldsymbol{u}, \boldsymbol{u}_\Delta, \hat{\boldsymbol{u}}) = \int_\Omega [\nabla \boldsymbol{u}_\Delta \, \boldsymbol{S} \boldsymbol{\cdot} \nabla \hat{\boldsymbol{u}} + \boldsymbol{E}_{\boldsymbol{u}}(\hat{\boldsymbol{u}}) \boldsymbol{\cdot} \boldsymbol{C}[\boldsymbol{E}_{\boldsymbol{u}}(\boldsymbol{u}_\Delta)]] \, d\Omega. \quad (4.216)$$

The discretization of the first term in (4.216) yields with

$$\nabla \boldsymbol{u}_{\Delta_h} = \sum_{j=1}^{n} \boldsymbol{u}_{\Delta j} \otimes \nabla \psi_j, \qquad \nabla \hat{\boldsymbol{u}}_h = \sum_{i=1}^{n} \hat{\boldsymbol{u}}_i \otimes \nabla \psi_i \qquad (4.217)$$

the so-called initial stress stiffness matrix

$$\int_\Omega \nabla \boldsymbol{u}_\Delta \, \boldsymbol{S} \boldsymbol{\cdot} \nabla \hat{\boldsymbol{u}} \, d\Omega = \sum_{i=1}^{n} \sum_{j=1}^{n} \hat{\boldsymbol{u}}_i^T \int_\Omega G_{ij} \, \boldsymbol{I} \, d\Omega \, \boldsymbol{u}_{\Delta j} \qquad (4.218)$$

where

$$G_{ij} := \nabla^T \psi_i \, \boldsymbol{S} \, \nabla \psi_j = [\psi_{i,1} \ \psi_{i,2}] \begin{bmatrix} S_{11} & S_{12} \\ S_{21} & S_{22} \end{bmatrix} \begin{bmatrix} \psi_{j,1} \\ \psi_{j,2} \end{bmatrix}. \qquad (4.219)$$

With

$$\boldsymbol{E}_{\boldsymbol{u}_h}(\boldsymbol{u}_\Delta) = \frac{1}{2} \sum_{j=1}^{n} [\boldsymbol{F}^T (\boldsymbol{u}_{\Delta j} \otimes \nabla \psi_j) + (\nabla \psi_j \otimes \boldsymbol{u}_{\Delta j})\boldsymbol{F}] = \sum_{j=1}^{n} \boldsymbol{B}_{L_j} \boldsymbol{u}_{\Delta j}$$

$$(4.220)$$

the second term in (4.216) becomes:

$$\int_\Omega \boldsymbol{E}_{\boldsymbol{u}}(\hat{\boldsymbol{u}}) \boldsymbol{\cdot} \boldsymbol{C}[\boldsymbol{E}_{\boldsymbol{u}}(\boldsymbol{u}_\Delta)] \, d\Omega = \sum_{i=1}^{n} \sum_{j=1}^{n} \hat{\boldsymbol{u}}_i^T \int_\Omega \boldsymbol{B}_{L_i}^T \boldsymbol{C} \boldsymbol{B}_{L_j} \, d\Omega \, \boldsymbol{u}_{\Delta j}. \quad (4.221)$$

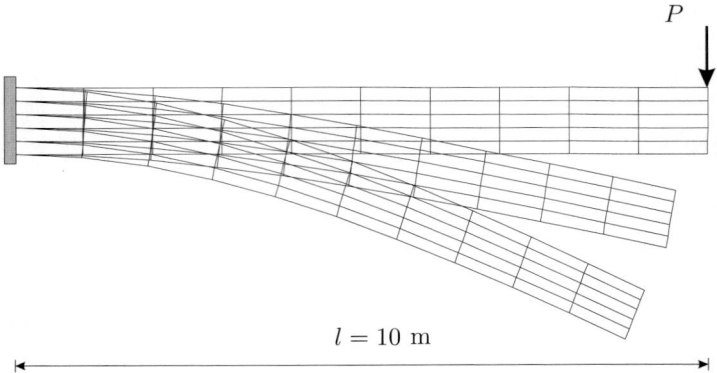

$l = 10$ m

Fig. 4.66. Nonlinear analysis of a cantilever plate that carries an equivalent nodal force; large displacements, small strains

Adding all terms we obtain

$$a_T(\boldsymbol{u}, \boldsymbol{u}_\Delta, \hat{\boldsymbol{u}}) = \sum_{i=1}^{n} \sum_{j=1}^{n} \hat{\boldsymbol{u}}_i^T \, \boldsymbol{K}_{T_{ij}} \, \boldsymbol{u}_{\Delta j} \tag{4.222}$$

where

$$\boldsymbol{K}_{T_{ij}} := \int_\Omega \left[G_{ij} \, \boldsymbol{I} + \boldsymbol{B}_{L_i}^T \boldsymbol{C} \boldsymbol{B}_{L_j} \right] d\Omega \,. \tag{4.223}$$

In the case of a bilinear element the (2×2) sub matrices $\boldsymbol{K}_{T_{ij}}$ of the tangential stiffness matrix of a single element are arranged as follows:

$$\boldsymbol{K}_T^e = \begin{bmatrix} \boldsymbol{K}_{T_{11}} & \boldsymbol{K}_{T_{12}} & \boldsymbol{K}_{T_{13}} & \boldsymbol{K}_{T_{14}} \\ \boldsymbol{K}_{T_{21}} & \boldsymbol{K}_{T_{22}} & \boldsymbol{K}_{T_{23}} & \boldsymbol{K}_{T_{24}} \\ \boldsymbol{K}_{T_{31}} & \boldsymbol{K}_{T_{32}} & \boldsymbol{K}_{T_{33}} & \boldsymbol{K}_{T_{34}} \\ \boldsymbol{K}_{T_{41}} & \boldsymbol{K}_{T_{42}} & \boldsymbol{K}_{T_{43}} & \boldsymbol{K}_{T_{44}} \end{bmatrix} . \tag{4.224}$$

After the assemblage, the resulting system of equations

$$\boldsymbol{K}_T(\boldsymbol{u}) \, \boldsymbol{u}_\Delta = \boldsymbol{f} - \boldsymbol{k}(\boldsymbol{u}) \,, \tag{4.225}$$

is solved with the *Newton-Raphson* method.

A cantilever plate

A cantilever plate of length $l = 10$ m, width $h = 1$ m and thickness $t = 1$ m, is loaded at its end with a vertical point force $P = 50,000$ kN; see Fig. 4.66 and 4.68. The parameter of the St. Venant type material are $E = 10^7$ kN/m^2 and $\nu = 0$. The analysis was done with plane bilinear elements. The vertical

Table 4.14. Deflection w (m) at the foot of the point load

element	linear	nonlinear
mixed shell elements	19.950	7.533
bilinear plate elements	13.413	7.307

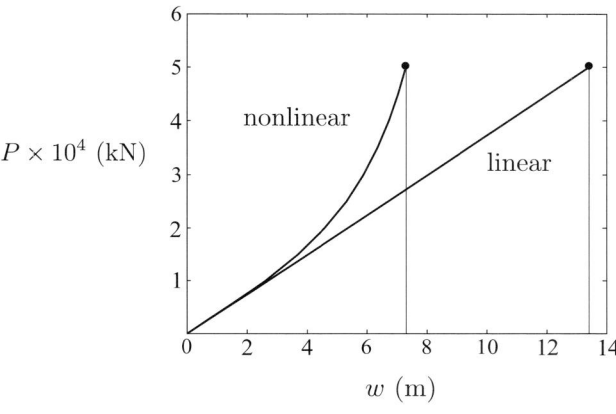

Fig. 4.67. Deflection at the foot of the point load

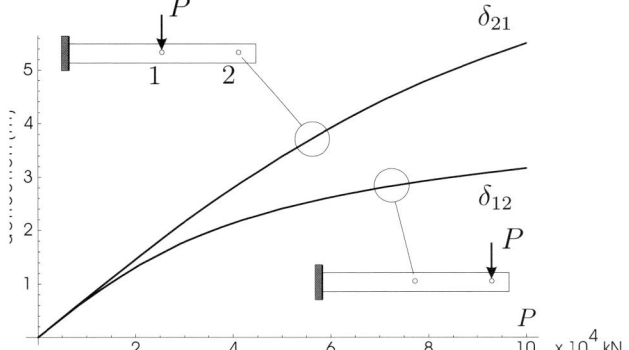

Fig. 4.68. Maxwell's theorem is no longer true

deflection w at the base of the point load is plotted in Fig. 4.67. For small values of P the linear and the nonlinear results coincide. But if the load P increases the increase in the deflection slows down, i.e., in linear analysis the deflections are overestimated! This was confirmed in independent tests with ADINA. The reference solution in Table 4.66 is based on a mixed four-node shell element [45]. The nonlinear results agree quite well, while in the linear case—one is tempted to say: as usual—the deviations are larger. Obviously the bilinear element fares better in nonlinear problems than in linear problems!

Goal-oriented refinement

In nonlinear problems the dual problem for the generalized Green's function z is formulated at the equilibrium point u, see Sect. 7.5, p. 526,

$$z \in V \qquad a_T^*(u; z, v) = J'(u; v) \qquad \forall\, v \in V\,. \tag{4.226}$$

and a_T^* corresponds to the tangential form a_T of the original problem at the equilibrium point, see (7.181) p. 528, so that z_h is the solution of

$$K_T(u_h)\, z = j \tag{4.227}$$

where $K_T(u_h)$ is the tangential stiffness matrix at u_h and the components

$$j_k = J'(u_h; \varphi_k)\,. \tag{4.228}$$

of j are the equivalent nodal forces. While in linear problems the equivalent nodal forces j_k are simply the values $J(\varphi_k)$ of the different shape functions we now must evaluate the Gateaux derivative (with respect to u) of the functional $J(.)$ for each single φ_k.

Let us assume we calculate the stress $\sigma_{ij}(x)$ at a point x, that is

$$J(u) = \sigma_{ij}(u)(x)\,. \tag{4.229}$$

The Gateaux derivative of this functional is

$$J'(u; v) := \left[\frac{d}{d\varepsilon}\, J(u + \varepsilon\, v)\right]_{\varepsilon=0} = \left[\frac{d}{d\varepsilon}\, \sigma_{ij}(u + \varepsilon\, v)\right]_{\varepsilon=0} \tag{4.230}$$

or with $S = C[E(u)]$,

$$\left[\frac{d}{d\varepsilon}\, S(u + \varepsilon\, v)\right]_{\varepsilon=0} = C\left[\frac{d}{d\varepsilon}\, E(u + \varepsilon\, v)\right]_{\varepsilon=0} = C[E_u(v)] =: \hat{S} \tag{4.231}$$

where $E_u(v)$ is the Gateaux derivative of the *Green-Lagrangian* strain tensor, see (4.208). Hence the equivalent nodal forces

$$j_k = J'(u_h; \varphi_k) = \hat{\sigma}_{ij}(\varphi_k)(x) \tag{4.232}$$

are the component $\hat{\sigma}_{ij}$ of the "tangential" stress tensor $C[E_{u_h}(\varphi_k)]$. These are the stress increments at the actual equilibrium point u_h resulting from the displacement field φ_k.

For a second example we consider the integral of the stresses along a cross section A-A

$$J(u) = \int_{A-A} \sigma_{ij}(u)\, ds\,. \tag{4.233}$$

The Gateaux derivative of this functional is

$$J'(\boldsymbol{u}; \boldsymbol{v}) = \int_{A-A} \hat{\sigma}_{ij}(\boldsymbol{v}) \, ds \qquad (4.234)$$

so that the equivalent nodal forces are the integrals of the stresses $\hat{\sigma}_{ij}$ resulting from the unit displacement fields $\boldsymbol{\varphi}_k$

$$j_k = J'(\boldsymbol{u}_h; \boldsymbol{\varphi}_k) = \int_{A-A} \hat{\sigma}_{ij}(\boldsymbol{\varphi}_k) \, ds \qquad (4.235)$$

The aim of goal oriented methods is to improve the accuracy of the output functional $J(u)$ by minimizing the energy error in the associated generalized Green's functional and in the solution u itself.

Now how do we proceed? We apply the first load increment $\boldsymbol{p}_1$ (let $\boldsymbol{p} = \boldsymbol{p}_1 + \boldsymbol{p}_2 + \dots$) and we find the equilibrium point $\boldsymbol{u}_1$ and the generalized Green's function $\boldsymbol{z}_1$. Then the mesh is adaptively refined so that both errors (in $\boldsymbol{u}_1$ as well as in $\boldsymbol{z}_1$) are below a certain threshold value. This completes the cycle and we apply the next increment $\boldsymbol{p}_2$ and repeat the process, etc. At the end we have a mesh which is optimal for the output value $J(\boldsymbol{u})$.

As in the linear case the final values of the stresses $\sigma_{ij}(\boldsymbol{x})$ are computed directly by differentiating the FE solution. Actually, we have no other choice. Unlike linear problems where

$$J(\boldsymbol{u}) = \int_{\Omega} \boldsymbol{p} \cdot \boldsymbol{z} \, d\Omega \qquad (4.236)$$

no such formula exists in nonlinear problems. And also such formulations as

$$J(\boldsymbol{u}) \simeq \int_{\Omega} \boldsymbol{p}_1 \cdot \boldsymbol{z}_1 \, d\Omega + \int_{\Omega} \boldsymbol{p}_2 \cdot \boldsymbol{z}_2 \, d\Omega + \int_{\Omega} \boldsymbol{p}_3 \cdot \boldsymbol{z}_3 \, d\Omega + \dots \qquad (4.237)$$

or

$$J(\boldsymbol{u}) \simeq \int_{\Omega} \boldsymbol{p} \cdot \boldsymbol{z}_n \, d\Omega \qquad \boldsymbol{z}_n = \text{final Green's function} \qquad (4.238)$$

lead to nowhere.

The results in Fig. 4.69 illustrate nicely how the actions of the Green's functions automatically (!) follow the movement of the structure—a cantilever plate to which a nodal force is applied. Depicted is the orientation of the nodal forces of the FE Green's functions for the two functionals

$$J(\boldsymbol{u}) = \sigma_{ij}(\boldsymbol{u})(\boldsymbol{x}) \qquad J(\boldsymbol{u}) = \int_{A-A} \sigma_{ij}(\boldsymbol{u}) \, ds \qquad (4.239)$$

at the final stage. For not to complicate the drawings the meshes were not refined. Just a normal nonlinear iterative analysis was performed for each of the two solutions $\boldsymbol{u}$ and $\boldsymbol{z}$.

original problem

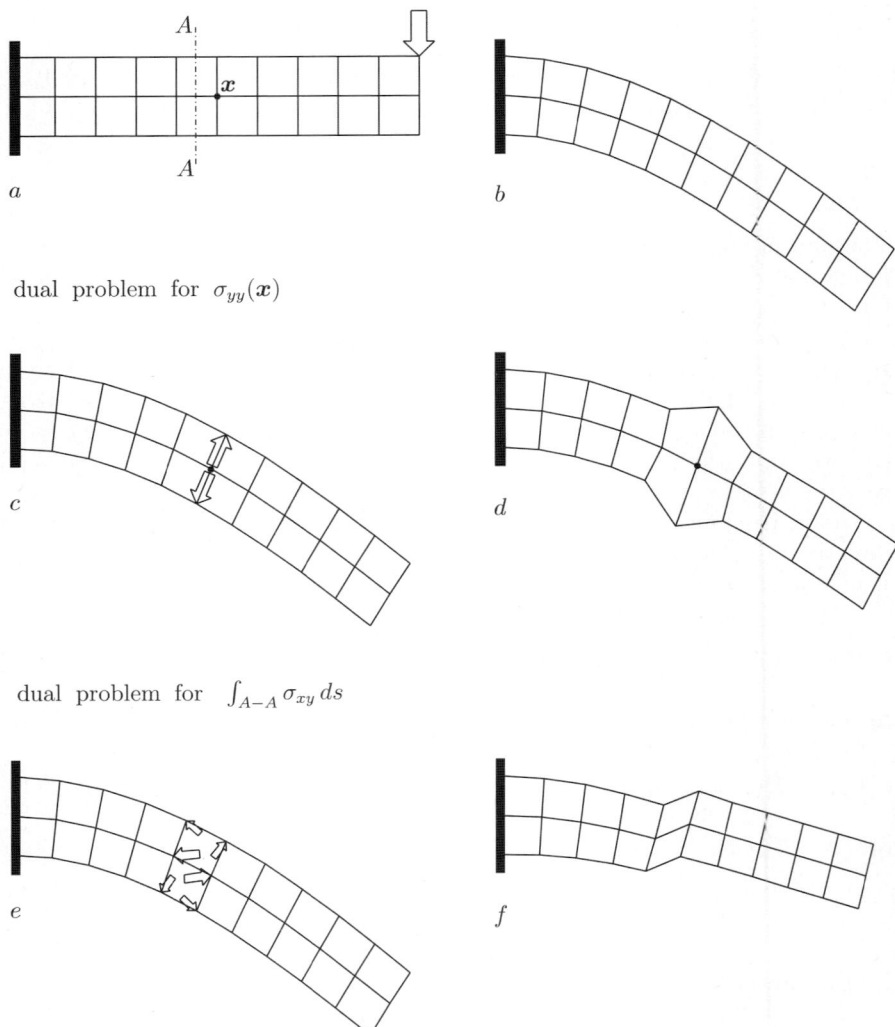

Fig. 4.69. Nonlinear analysis of a cantilever plate **a)** original load **b)** deformed structure **c)** equivalent nodal forces of the dual problem for $\sigma_{yy}(\boldsymbol{x})$ at the equilibrium point **d)** FE approximation of the dual solution for $\sigma_{yy}(\boldsymbol{x})$ **e)** equivalent nodal forces for $\int_A \sigma_{xy}\, ds$ in cross-section A-A **f)** FE approximation of the dual problem for $\int_A \sigma_{xy}\, ds$, [162]

5. Slabs

The bending theory of plates can be viewed as an extension of beam theory. In an Euler–Bernoulli-beam, $EIw^{IV} = p$, shear deformations are neglected, i.e., a straight line that is initially normal to the neutral axis remains so after the load is applied. In a Timoshenko beam, the line instead rotates by an angle γ; see Fig. 5.1 b.

The extension of the Euler–Bernoulli-beam to plate theory is the Kirchhoff plate, $K \Delta \Delta w = p$, and the extension of the Timoshenko beam is the Reissner–Mindlin plate. The first finite elements developed for plate bending problems were based on the Kirchhoff plate theory. But the problem that the shape functions must be C^1 and must be easy to be implement at the same time soon favored Reissner–Mindlin plate elements, where C^0 suffices for the shape functions, see Fig. 5.2.

Normally slabs are relatively thin with negligible shear deformations, so that good Reissner–Mindlin plate elements tend to produce the same results

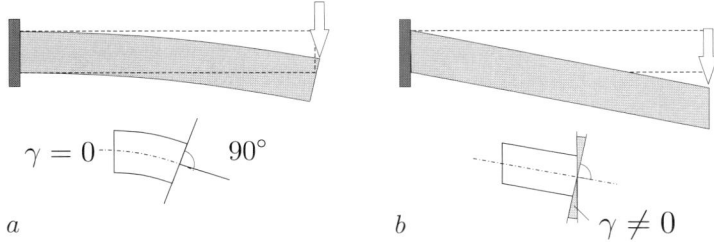

Fig. 5.1. Cantilever beams: **a)** Euler–Bernoulli beam **b)** Timoshenko beam

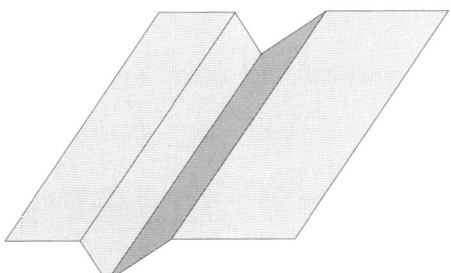

Fig. 5.2. A Kirchhoff plate cannot be folded like sheet metal

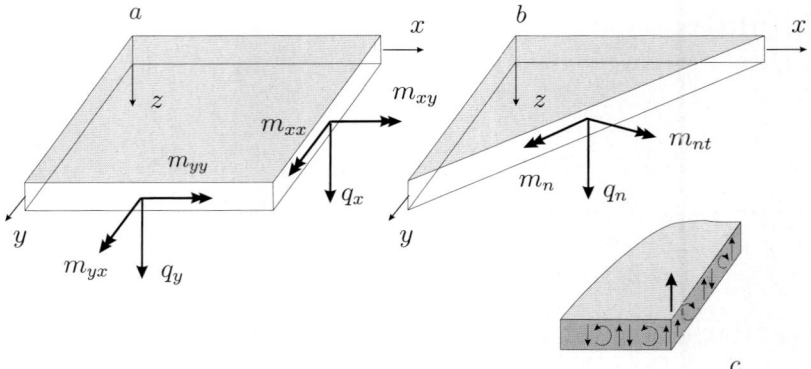

Fig. 5.3. Resultant stresses in a slab

as Kirchhoff plate elements, with the exception possible of zones close to the boundary. The switch to the Reissner–Mindlin plate theory in FE programming therefore probably remained unnoticed in the engineering community, even though the most popular plate elements, like the DKT element and the Bathe–Dvorkin element are not genuine Reissner–Mindlin plate elements but rather an ingenious mixture of Kirchhoff and Reissner–Mindlin plate theories.

5.1 Kirchhoff plates

The deflection w, the three curvature terms κ_{ij} and the three bending moments m_{ij} are coupled by a system of seven differential equations, which in indicial notation reads

$$\kappa_{ij} - w_{,ij} = 0\,, \qquad (3 \text{ eqs.}),$$
$$K\{(1-\nu)\kappa_{ij} + \nu\kappa_{kk}\delta_{ij}\} + m_{ij} = 0\,, \qquad (3 \text{ eqs.}), \qquad (5.1)$$
$$-m_{ij,ji} = p\,, \qquad (1 \text{ eq.})\,.$$

The constant

$$K = \frac{Eh^3}{12(1-\nu^2)}\,, \qquad h = \text{slab thickness} \qquad (5.2)$$

is the plate stiffness and ν is Poisson's ratio.

This system is equivalent to the biharmonic differential equation

$$K(w_{,xxxx} + 2\,w_{,xxyy} + w_{,yyyy}) = K\Delta\Delta w = p \qquad (5.3)$$

for the deflection surface $w(x,y)$ of the slab.

The similarity of (5.3) to the beam equation $EIw^{IV} = p$ is obvious. As in a beam, the bending moments in a slab are proportional to the curvature of the deflection surface (see Fig. 5.3)

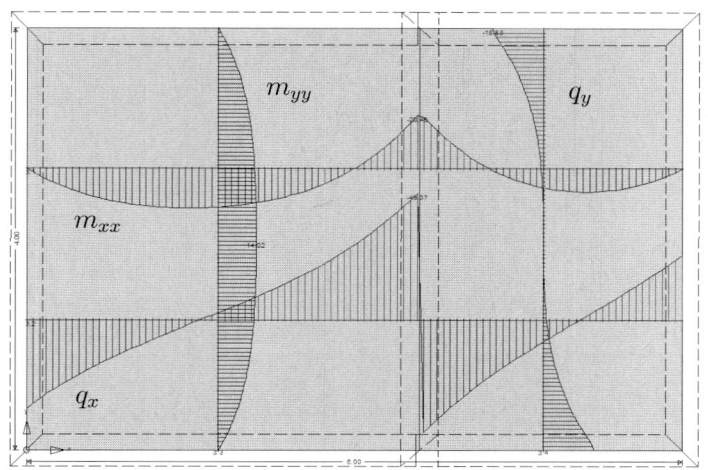

Fig. 5.4. Bending moments and shear forces

$$m_{xx} = -K(w_{,xx} + \nu\, w_{,yy})\,, \quad m_{yy} = -K(w_{,yy} + \nu\, w_{,xx})\,,$$
$$m_{xy} = -(1-\nu)K w_{,xy}\,,$$

and the shear forces are proportional to the third derivatives:

$$q_x = -K(w_{,xxx} + w_{,yyx})\,, \qquad q_y = -K(w_{,xxy} + w_{,yyy})\,.$$

The bending moments m_{xx} are accounted for by reinforcement in the x-direction and the bending moments m_{yy} by reinforcement in the y-direction. All resultant stresses are resultant stresses per unit length; see Fig. 5.4.

The extreme values of the normal curvature κ_{11} and κ_{22} at a given point on a surface are called the principal curvatures, and they occur in the direction

$$\tan 2\varphi = \frac{2\, m_{yy}}{m_{xx} - m_{yy}}\,. \tag{5.4}$$

The principal curvatures determine the maximum and minimum bending of a slab at any given point (see Fig. 5.5):

$$m_{I,II} = \frac{m_{xx} + m_{yy}}{2} \pm \sqrt{\left[\frac{m_{xx} - m_{yy}}{2}\right]^2 + m_{xy}^2}\,. \tag{5.5}$$

In an arbitrary direction the resultant stresses are, in indicial notation,

$$m_{nn} = m_{ij}\, n_i\, n_j\,, \qquad m_{nt} = m_{ij}\, n_i\, t_j\,, \qquad q_n = q_i\, n_i\,,$$

where $\boldsymbol{n} = [n_x, n_y]^T$ is the normal vector and $\boldsymbol{t} = [t_x, t_y]^T = [-n_y, n_x]^T$ is the tangent vector.

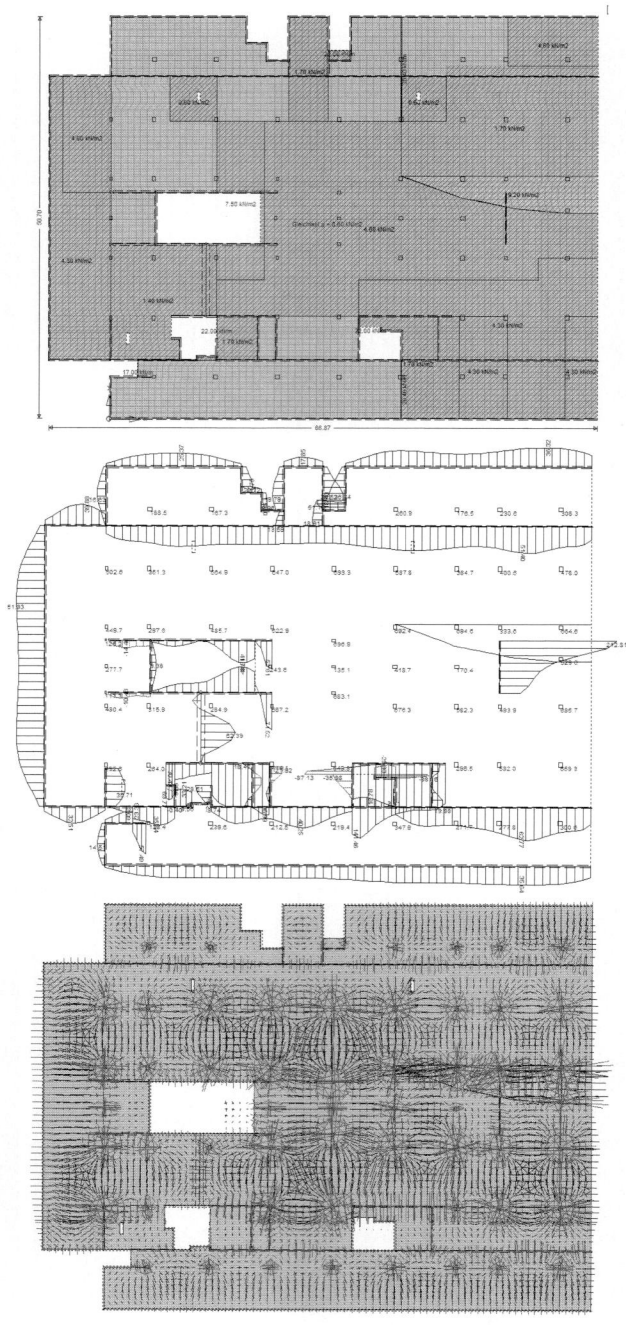

Fig. 5.5. Slab: **a)** plan view, **b)** support reactions, **c)** principal moments

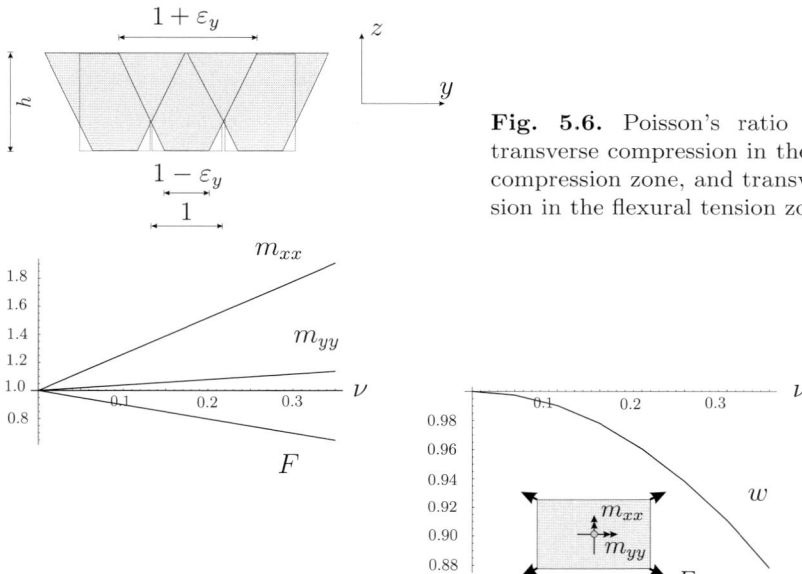

Fig. 5.6. Poisson's ratio leads to transverse compression in the flexural compression zone, and transverse tension in the flexural tension zone

Fig. 5.7. Influence of Poisson's ratio ν on the bending moments, the deflection at midspan and the corner forces F of a hinged slab. Plotted are the ratios with respect to $\nu = 0$. The bending moments increase if ν increases while the corner force F and the deflection w decrease

Poisson's ratio

Poisson's ratio ν ensures that the concrete widens in the compression zone and that it narrows in the tension zone; see Fig. 5.6.

The larger ν gets, the more the bending moments increase, while the deflection and also the corner forces F decrease; see Fig. 5.7. The deflection w becomes smaller because the slab stiffness K increases; see Eq. (5.2).

In uncracked concrete ν has a value of about 0.2. Many tables are based on $\nu = 0.0$, and the bending moments for $\nu \neq 0$ are obtained with

$$m_{xx}(\nu) = m_{xx}(0) + \nu \, m_{yy}(0), \qquad m_{yy}(\nu) = m_{yy}(0) + \nu \, m_{xx}(0). \quad (5.6)$$

The effects of an incorrect Poisson's ratio ν may well exceed the approximation error, as the following example of a quadratic slab with clamped edges shows. The correct value of Poisson's ratio is assumed to be $\nu = 0$. Plotted in Fig. 5.8 is the error in the FE bending moment $m_{xx} = m_{yy}$ at the center of the plate, as a function of the number of elements and ν. Obviously the error due to deviations in Poisson's ratio is larger than the approximation error. If the analysis were based on $\nu = 0.3$, the best FE result would overestimate the bending stresses for $\nu = 0$ by about 30%. Even at $\nu = 0.1$, which is relatively

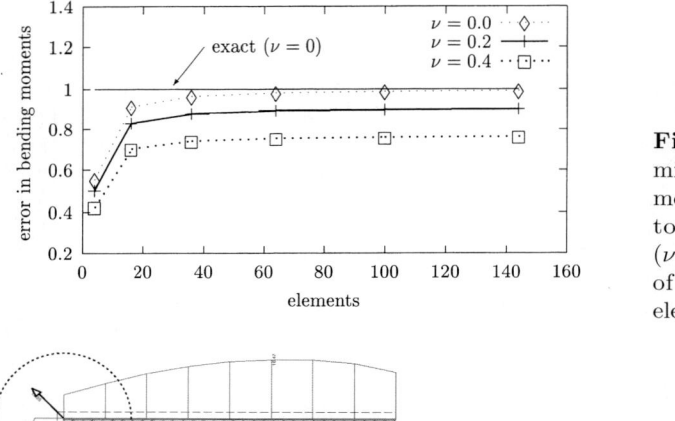

Fig. 5.8. Error in the midspan FE bending moment as compared to a series solution ($\nu = 0$), as a function of the number of elements

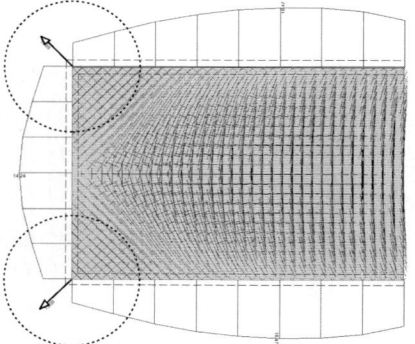
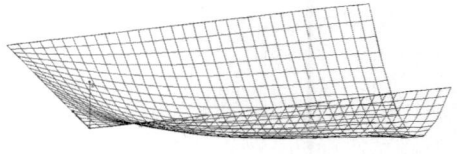

Fig. 5.9. Slab with edge anchorage

close to the true value $\nu = 0$ the error on a fine mesh with 144 elements is still relatively high, 10.96%, while on a mesh with 16 elements but with the correct value $\nu = 0$ the approximation error is only 9.65%. These results suggest that a correct assessment of the elastic constants and parameters is very important for accurate analysis.

Equilibrium

The *Kirchhoff shear* v_n rather than the shear forces q_n, maintains the equilibrium with the applied load. The Kirchhoff shear v_n is the shear force q_n plus the derivative of the twisting moment m_{nt} with regard to the arc length s on the edge,

$$v_n = q_n + \frac{dm_{nt}}{ds} , \tag{5.7}$$

so that along a vertical ($|$) or a horizontal ($-$) edge, respectively,

$$v_x = q_x + \frac{dm_{xy}}{dy} \qquad v_y = q_y + \frac{dm_{xy}}{dx} . \tag{5.8}$$

The additional force

$$\frac{dm_{nt}}{ds} \qquad \left(\frac{\text{kN m/m}}{\text{m}} = \text{kN/m}\right) \tag{5.9}$$

is easily understood if the twisting moment is split into pairs of opposing forces; see Fig. 5.3.

Because the equivalent nodal forces are work-equivalent to the resultant stresses along the edge of an element, the nodal forces represent the action of the Kirchhoff shear, and not of the shear forces q_n.

The shear force q_n and the twisting moments m_{nt} are not zero along a free edge, but they are so tuned that the change in m_{nt} is balanced by q_n, so that the total effect vanishes

$$v_n = q_n + \frac{dm_{nt}}{ds} = 0\,. \tag{5.10}$$

On a clamped edge the twisting moment is zero, $m_{nt} = 0$, so the shear force q_n coincides with the Kirchhoff shear v_n. For other support conditions the two are not the same, $q_n \neq v_n$, but the difference is usually not very large.

Corner force

The jump in the twisting moment at a corner point

$$F := m_{nt}^+ - m_{nt}^- \tag{5.11}$$

can be identified with a corner force F; see Fig. 5.3. If the corner is not held down, no physical reaction is possible and the slab will tend to move away from the support. This is the source of the "corner lifting" phenomenon that can be observed on a laterally loaded square slab with simply supported edges that do not prevent lifting; see Fig. 5.9. This effect does not appear if the edges meeting at the corner point are clamped, $m_{nt} = 0$.

5.2 The displacement model

In an ideal slab model the unit deflections $\varphi_i(x, y)$ of the nodes describe what happens to the plate if a node deflects, $w = 1$, or if it rotates by $45°$ about the y- or x-axis, $w,_x = 1$ and $w,_y = 1$.

These unit deflections form the basis of the trial space V_h, and the FE solution is an expansion in terms of these $3n$ unit deflections,

$$w_h(x, y) = \sum_{i=1}^{3n} u_i\, \varphi_i(x, y)\,, \tag{5.12}$$

where the nodal degrees of freedom u_i (in this sequence: $u_1 = w, u_2 = w,_x, u_3 = w,_y$ etc.) denote a deflection and two rotations.

A unit load case p_i can be associated with each of these unit deflections and so, as before, the FE load case can be considered a superposition of these $3n$ unit load cases,

$$p_h(x, y) = \sum_{i=1}^{3n} u_i\, p_i\,,\qquad(5.13)$$

where the weights u_i are chosen in such a way that the FE load case p_h is work-equivalent to the original load case in terms of the $3n$ unit deflections

$$\delta W_e(p_h, \varphi_i) = \delta W_e(p, \varphi_i)\qquad i = 1, 2, \ldots 3n\,.\qquad(5.14)$$

Conforming shape functions are C^1 but not C^2, therefore the curvatures on opposite sides of the interelement boundaries are different. In a beam, such abrupt changes in the curvature would be attributed to the action of nodal moments

$$- EI\,\kappa(x_i^-) + EI\,\kappa(x_i^+) = M(x_i^-) - M(x_i^+) = M(x_i)\,,\qquad(5.15)$$

while here these discontinuities are attributed to the action of line moments (kN m/m), and jumps in the Kirchhoff shear to the action of line loads (kN/m), so that the attributes of a typical FE load case p_h (as in Fig. 5.10) are

- line loads—along the interelement edges (kN/m)
- line moments—along the interelement edges (kN m/m)
- surface loads—on each element (kN/m^2)
- (eventually) forces P_i—at the nodes (kN)

The forces P_i result if the corner forces F^e of the individual elements are added at the nodes.

The slab in Fig. 5.10 gives the impression of such a load case p_h. The original loading consists of a uniformly distributed surface load $p = 6.5$ kN/m^2. The element is a conforming rectangular element with 16 degrees of freedom which is based on the shape functions of a beam; see Eq. (5.17).

5.3 Elements

The natural choice for a plate element would be a triangular element with degrees of freedom $w, w_{,x}, w_{,y}$ at each of the three corner nodes, and a cubic polynomial to interpolate w between. This would result in a linear variation of the bending moments and constant shear forces. But because a complete cubic polynomial has 10 terms instead of the $3 \times 3 = 9$ terms, the element would be nonconforming, that is, the first derivatives would not be continuous across interelement boundaries.

A conforming rectangular element can be derived from the unit deflections $\varphi_i(x)$ of a beam (see Fig. 5.11)

a)

6.50	6.50	6.50	6.50	6.50	6.50	6.50	6.50	6.50	6.50
6.50	6.50	6.50	6.50	6.50	6.50	6.50	6.50	6.50	6.50
6.50	6.50					6.50	6.50	6.50	6.50
6.50	6.50					6.50	6.50	6.50	6.50
6.50	6.50					6.50	6.50	6.50	6.50
6.50	6.50	6.50	6.50	6.50	6.50	6.50	6.50	6.50	6.50
6.50	6.50	6.50	6.50	6.50	6.50	6.50	6.50	6.50	6.50
6.50	6.50	6.50	6.50	6.50	6.50	6.50	6.50	6.50	6.50
6.50	6.50	6.50	6.50	6.50	6.50	6.50	6.50	6.50	6.50
6.50	6.50	6.50	6.50	6.50	6.50	6.50	6.50	6.50	6.50

b)

3.98	3.35	-2.55	0.53	0.10	-3.91	2.45	3.68	1.61	1.63
3.54	42.36	-29.62	-3.47	-4.77	-37.85	43.34	6.44	2.79	1.27
-2.21	-29.15					-27.23	-0.28	1.71	0.82
0.51	-7.29					-9.53	-1.17	0.51	0.50
-4.18	-38.20					-55.25	-5.75	-0.02	0.30
1.99	39.95	-22.53	-3.74	-6.62	-48.01	62.59	6.19	1.84	0.60
3.28	7.03	1.17	-1.22	-1.91	-2.45	8.62	7.96	3.62	1.09
1.57	3.13	2.47	1.39	1.02	1.58	3.59	4.58	3.40	1.27
1.36	2.45	2.14	1.65	1.42	1.61	2.22	2.83	2.62	1.43
1.58	1.25	0.90	0.69	0.60	0.64	0.81	1.06	1.38	1.62

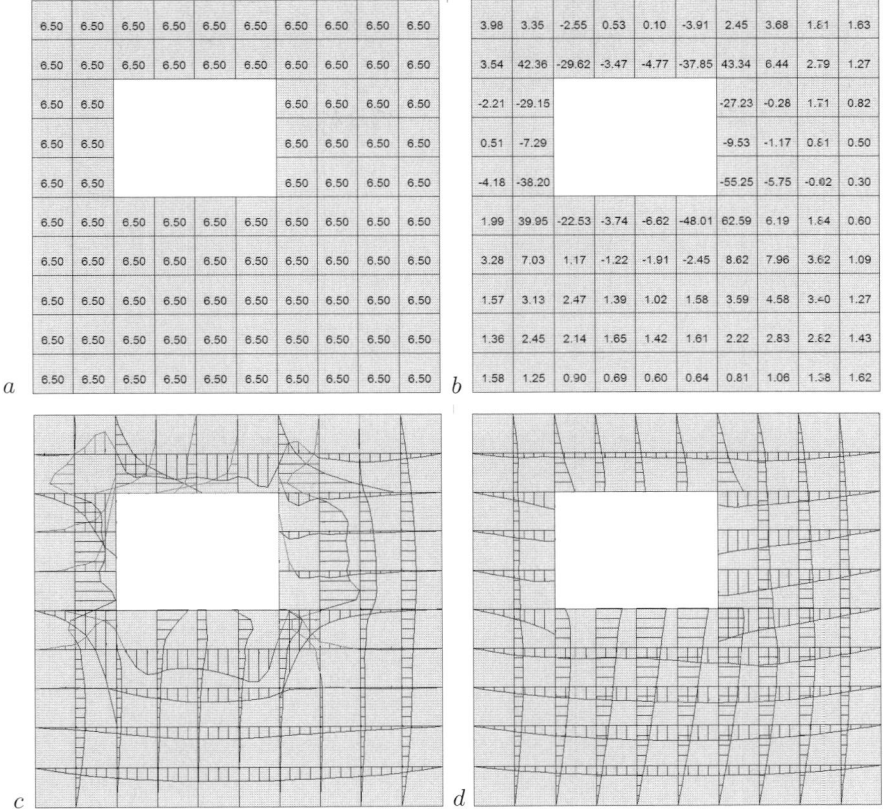

c)   **d)**

Fig. 5.10. Slab: **a)** uniform load $p = 6.5\ \mathrm{kN/m^2}$ and the equivalent FE load case p_h, **b)** element loads, **c)** vertical forces, and **d)** moments along interelement boundaries

$$\varphi_1(x) = 1 - \frac{3x^2}{l^2} + \frac{2x^3}{l^3} \qquad\qquad \varphi_3(x) = \frac{3x^2}{l^2} - \frac{2x^3}{l^3}$$

$$\varphi_2(x) = -x + \frac{2x^2}{l} - \frac{x^3}{l^2} \qquad\qquad \varphi_4(x) = \frac{x^2}{l} - \frac{x^3}{l^2}\,, \tag{5.16}$$

using a product approach

$$\varphi^e_{..}(x,y) = \varphi_i(x)\,\varphi_j(y) \qquad i,j = 1,2,3,4\,, \tag{5.17}$$

so that each node has four degrees of freedom $w, w_{,x}, w_{,y}, w_{,xy}$. Unfortunately this approach is limited to rectangular slabs. Also the degree of freedom $w_{,xy}$ is not easily accounted for if the element is coupled to other elements.

A truly isoparametric conforming quadrilateral element also requires that the mapping of the master element onto the single elements be C^1. Because

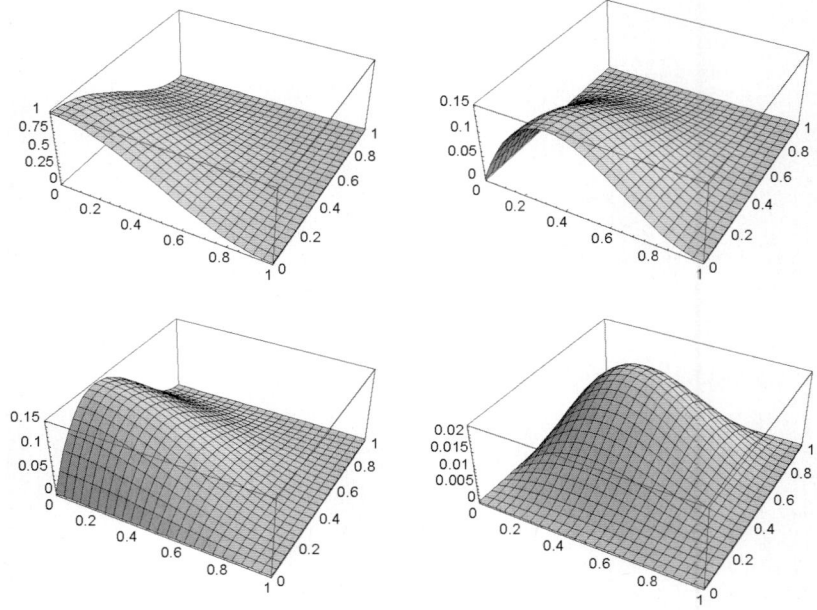

Fig. 5.11. Shape functions based on Hermite polynomials

of the chain rule

$$w_{,x} = w_{,\xi}\ \xi_{,x} + w_{,\eta}\ \eta_{,x} \tag{5.18}$$

the derivatives $\xi_{,x}$ and $\eta_{,x}$, etc., of the mapping functions $x(\xi,\eta)$ and $y(\xi,\eta)$ and the inverse functions $\xi(x,y)$ and $\eta(x,y)$ must be continuous. This means that the coordinate lines $\xi = const.$ and $\eta = const.$ cannot change direction abruptly (no kinks!) upon crossing the interelement boundaries. It is very costly to establish this property numerically.

Theoretically the simplest method of deriving a conforming triangular element is to choose a complete fifth-order polynomial [215] with the following six degrees of freedom at the vertices,

$$w,\ w_{,x}\ ,\ w_{,y}\ ,\ w_{,xx}\ ,\ w_{,xy}\ ,\ w_{,yy} \tag{5.19}$$

and the normal derivatives at the mid-side nodes as additional degrees of freedom. This conforming element has 21 degrees of freedom.

Stiffness matrices

The strain energy product of two deflection surfaces w and $\hat{w}$ is the scalar product of the bending moment tensor $\boldsymbol{M} = [m_{ij}]$ of w and the curvature tensor $\boldsymbol{K} = [\kappa_{ij}]$ of $\hat{w}$,

$$a(w, \hat{w}) = \int_{\Omega} \boldsymbol{M} \bullet \hat{\boldsymbol{K}} \, d\Omega = \int_{\Omega} m_{ij} \, \hat{\kappa}_{ij} \, d\Omega \qquad (5.20)$$

$$= \int_{\Omega} [m_{xx} \hat{\kappa}_{xx} + 2 \, m_{xy} \hat{\kappa}_{xy} + m_{yy} \hat{\kappa}_{yy}] \, d\Omega \,, \qquad (5.21)$$

or in equivalent notation the scalar product of the vector $\boldsymbol{m} = [m_{xx}, m_{yy}, m_{xy}]^T$ and the vector $\hat{\boldsymbol{\kappa}} = [\kappa_{xx}, \kappa_{yy}, 2 \, \kappa_{xy}]^T$,

$$a(w, \hat{w}) = \int_{\Omega} \boldsymbol{m} \bullet \hat{\boldsymbol{\kappa}} \, d\Omega = \int_{\Omega} m_i \, \hat{\kappa}_i \, d\Omega$$

$$= \int_{\Omega} [m_{xx} \hat{\kappa}_{xx} + 2 \, m_{xy} \hat{\kappa}_{xy} + m_{yy} \hat{\kappa}_{yy}] \, d\Omega \,, \qquad (5.22)$$

where

$$\underbrace{\begin{bmatrix} m_{xx} \\ m_{yy} \\ m_{xy} \end{bmatrix}}_{\boldsymbol{m}} = \underbrace{\begin{bmatrix} K & \nu \, K & 0 \\ \nu \, K & K & 0 \\ 0 & 0 & (1-\nu) \, K/2 \end{bmatrix}}_{\boldsymbol{D}} \underbrace{\begin{bmatrix} \kappa_{xx} \\ \kappa_{yy} \\ 2 \, \kappa_{xy} \end{bmatrix}}_{\boldsymbol{\kappa}} . \qquad (5.23)$$

Hence if $\boldsymbol{m}_i$ and $\boldsymbol{\kappa}_i$ denote the "bending moment" and "curvature" vector of the element shape function φ_i^e, an individual element of the stiffness matrix is

$$k_{ij}^e = \int_{\Omega_e} \boldsymbol{m}_i \bullet \boldsymbol{\kappa}_j \, d\Omega = \int_{\Omega_e} \boldsymbol{m}_i^T \boldsymbol{\kappa}_j \, d\Omega = \int_{\Omega_e} (\boldsymbol{D} \, \boldsymbol{\kappa}_i)^T \boldsymbol{\kappa}_j \, d\Omega$$

$$= \int_{\Omega_e} \boldsymbol{\kappa}_i^T \boldsymbol{D} \, \boldsymbol{\kappa}_j \, d\Omega \qquad (5.24)$$

and the stiffness matrix of an element with n degrees of freedom is

$$\boldsymbol{K}_{(n \times n)}^e = \int_{\Omega_e} \boldsymbol{B}_{(n \times 3)}^T \boldsymbol{D}_{(3 \times 3)} \, \boldsymbol{B}_{(3 \times n)} \, d\Omega \,, \qquad (5.25)$$

where column i of matrix $\boldsymbol{B}$ contains the curvatures $\kappa_{xx}, \kappa_{yy}, 2 \, \kappa_{xy}$ of the deflection surface φ_i^e.

5.4 Hybrid elements

By a special technique triangular elements can be developed which have only the three degrees of freedom $w, w_{,x}, w_y$ at the three nodes, but which yield better results than simple nonconforming displacement-based elements. The best known elements of this type are the triangular HSM element (hybrid-stress-model), [29], and the DKT element. The latter is a modified Reissner–Mindlin plate element which will be discussed later. First we discuss the HSM element.

The starting point is the principle of minimum complementary energy. According to this principle, the moment tensor $\boldsymbol{M} = [m_{ij}]$ of the exact solution minimizes the complementary energy of the slab,

$$\Pi_c(\boldsymbol{M}) = -\frac{1}{2} \int_\Omega \boldsymbol{M} \bullet \boldsymbol{C}^{-1}[\boldsymbol{M}] \, d\Omega = -\frac{1}{2} \int_\Omega m_{ij} \, \kappa_{ij} \, d\Omega \qquad (5.26)$$

on the set V_c, which is the set of all bending moment tensors $\boldsymbol{M}$ that satisfy the equilibrium condition

$$- \operatorname{div}^2 \boldsymbol{M} = p \qquad \text{or} \qquad - m_{ij,ji} = p \qquad (5.27)$$

and if necessary, static boundary conditions such as

$$m_{ij} \, n_i \, n_j = \bar{m}_n \,, \qquad \frac{d}{ds} m_{ij} \, n_i \, t_j + m_{ij,i} \, n_j = \bar{v}_n \qquad \text{on } \Gamma_N \,. \quad (5.28)$$

Here $\boldsymbol{n} = [n_x, n_y]^T$ and $\boldsymbol{t} = [t_x, t_y]^T$ are the normal- and tangent vectors at the edge of the slab. The boundary conditions (5.28) mean that along a portion Γ_N of the edge, the moments $\bar{m}_n$ and forces $\bar{v}_n$ are prescribed. Other combinations of static boundary conditions are possible as well. Geometric boundary conditions like $w = 0$ (hinged support) or $w = w,_n = 0$ (clamped edge) are of no concern for the purpose of this principle.

To construct a subset V_h^c of the space V^c, a moment tensor $\boldsymbol{M}_p$ is chosen which satisfies the equilibrium condition (5.27) and the static boundary conditions (5.28), and this tensor $\boldsymbol{M}_p$ is paired with a string of homogeneous tensors $\boldsymbol{M}_i$

$$V_h^c \equiv \boldsymbol{M}_p \oplus \sum_i \sigma_i \, \boldsymbol{M}_i \qquad - \operatorname{div}^2 \boldsymbol{M}_i = 0 \,, \qquad (5.29)$$

where *homogeneous* also implies that the tensors $\boldsymbol{M}_i$ satisfy the boundary conditions (5.28) in a homogeneous form, i.e., $\bar{v}_n = \bar{m}_n = 0$ on Γ_N.

This is the same solution technique as in the force method, where

$$M = M_0 + X_i \, M_i \qquad - M_0'' = p \qquad - M_i'' = 0 \,. \qquad (5.30)$$

The tensor $\boldsymbol{M}_p$ corresponds to the bending moment M_0 of the primary state and the σ_i are the redundants X_i.

If the bending moments m_{kj} in the tensors $\boldsymbol{M}_i$ are first-order polynomials, the equilibrium condition $-\operatorname{div}^2 \boldsymbol{M}_i = 0$ is satisfied in each element, but the bending moments are discontinuous at the interelement boundaries, which violates the definition of V_c.

To overcome this obstacle the continuity requirement for the bending moments is added to the functional Π_c, using Lagrange multipliers:

$$\Pi_c(\boldsymbol{M}, w) = -\frac{1}{2} \int_\Omega \boldsymbol{M} \bullet \boldsymbol{C}^{-1}[\boldsymbol{M}] \, d\Omega$$

$$+ \sum_i \int_{\Gamma_i} [(m_n^+ - m_n^-) \, w,_n + (v_n^+ - v_n^-) \, w] \, ds + \sum_k F_k \, w(\boldsymbol{x}_k) \,, \quad (5.31)$$

where the Γ_i are the interelement boundaries and the superscripts $+$ and $-$ denote the force terms on the left- and right-hand side of the interelement boundary. Each F_k is the sum of the corner forces (discontinuity of m_{nt}) of the single elements at the nodes $\boldsymbol{x}_k$. The two Lagrange multipliers are the deflection w and the rotation $w_{,n}$ at the interelement boundary Γ_i.

The corner forces F_c appear in Green's first identity if the boundary integral of $m_{nt}\,\hat{w}_{,s}$ is integrated by parts

$$\int_{\partial\Omega_e} m_{ij}\,n_i\,t_j\partial_t\hat{w}\,ds = -\sum_c F_c\,\hat{w}(\boldsymbol{x}_c) - \int_{\partial\Omega_e} \frac{d}{ds}(m_{ij}\,n_i\,t_j)\,\hat{w}\,ds\,. \quad (5.32)$$

Here $\partial\Omega_e$ is the edge of the element and the $\boldsymbol{x}_c$ are the corner points of the element. If this procedure is reversed, the corner forces can be brought under the integral sign, and because the normal vectors on two adjacent element edges point in opposite directions, the sum of $(v_n^+ - v_n^-)\,w$, etc., over the interelement boundaries Γ_i can be written as a sum over the element boundaries Γ_e, so that the following equation is equivalent to Eq. (5.31)

$$\Pi_c(\boldsymbol{M}, w) = -\frac{1}{2}\sum_e \int_{\Omega_e} \boldsymbol{M}\cdot\boldsymbol{C}^{-1}[\boldsymbol{M}]\,d\Omega$$

$$+ \sum_e \int_{\partial\Omega_e} [w\,q_n - w_{,n}\,m_n - w_{,s}\,m_{nt}]\,ds\,. \quad (5.33)$$

The FE program constructs the tensors $\boldsymbol{M}_i$ by linearly interpolating the bending moments m_{ij}. This guarantees that the equilibrium condition

$$-\operatorname{div}^2 \boldsymbol{M} = \operatorname{div}\,(\operatorname{div}\,\boldsymbol{M}) = m_{ij,ji}$$

$$= m_{11,11} + m_{12,12} + m_{21,21} + m_{22,22} = 0 \quad (5.34)$$

is satisfied in each element. The deflection w at the edge is interpolated with cubic polynomials, and therefore the complementary energy of an element becomes

$$\Pi_c = -\frac{1}{2}\boldsymbol{\sigma}^T\boldsymbol{B}\,\boldsymbol{\sigma} + \boldsymbol{\sigma}^T\boldsymbol{C}\,\boldsymbol{w}\,, \quad (5.35)$$

with $\boldsymbol{\sigma}$ and $\boldsymbol{w}$ as the nine nodal variables of the bending moments, and the nodal deformations $w^i, w_{,x}^i, w_{,y}^i$. The condition that the first variation with respect to the parameters σ_i of the bending moments must vanish,

$$\frac{\partial\Pi_c}{\partial\boldsymbol{\sigma}} = -\boldsymbol{B}\,\boldsymbol{\sigma} + \boldsymbol{C}\,\boldsymbol{w} = \boldsymbol{0}\,, \quad (5.36)$$

implies $\boldsymbol{\sigma} = \boldsymbol{B}^{-1}\boldsymbol{C}\,\boldsymbol{w}$, and therefore the stiffness matrix can be expressed in terms of the nodal values of the deflection w alone:

$$\boldsymbol{K}_{(9\times9)} = \boldsymbol{C}^T\boldsymbol{B}^{-1}\boldsymbol{C}\,. \quad (5.37)$$

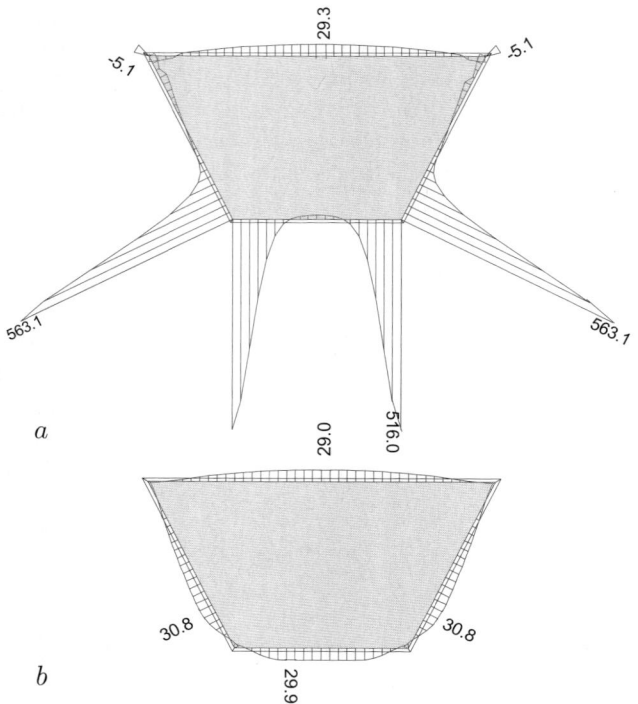

Fig. 5.12. Support reactions of a hinged trapezoidal slab subject to a uniform load. **a)** At an obtuse corner point the support reactions become singular. **b)** These effects vanish if the rotations $w,_x$ and $w,_y$ are set free at these points

By this detour a stiffness matrix in terms of the natural nodal degrees of freedom of a triangular plate element is derived. The whole technique is very similar to the derivation of a DKT element, and as in the case of a DKT element, a consistent approach for the calculation of the equivalent nodal forces is not defined. Theoretically we should proceed as in the force method, where the terms on the right-hand side (called δ_{i0} in the force method and not f_i) are the scalar product between the bending moment M_0 and the bending moments M_i,

$$\delta_{i0} = \int_0^l \frac{M_0 \, M_i}{EI} \, dx \, , \qquad (5.38)$$

that is, the scalar product of the bending moment tensor $\boldsymbol{M}_p$ of the particular solution and the tensors $\boldsymbol{M}_i$ should define the equivalent nodal forces:

$$f_i = \int_\Omega \boldsymbol{M}_p \cdot \boldsymbol{M}_i \, d\Omega \, . \qquad (5.39)$$

But instead, to achieve the "same effect" a) the deflection w that originally only lives on the interelement boundaries is continued by a choice of appro-

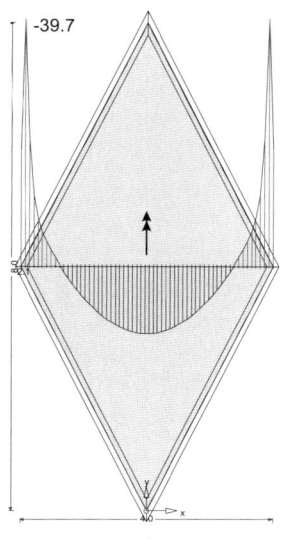

-39.7

18.8

moments m_{xx}

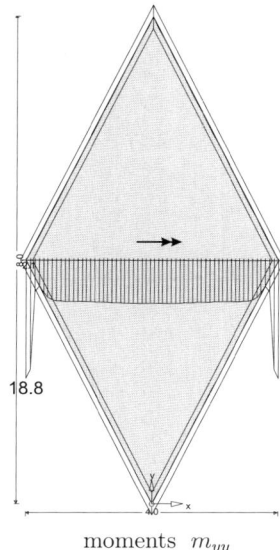

moments m_{yy}

Fig. 5.13. Hinged slab. At the corner points the bending moments become singular, $m_{xx} \to -\infty$ and $m_{yy} \to +\infty$

priate global shape functions $w_i(\boldsymbol{x})$ across the whole slab, and b) the scalar product of the load and these shape functions is calculated

$$\tilde{f}_i = \int_\Omega p\, w_i\, d\Omega\,. \tag{5.40}$$

In practice this means that the principle of minimum complementary energy is only used to derive the stiffness matrix, and that we later switch back to the principle of minimum potential energy.

5.5 Singularities of a Kirchhoff plate

The handicap of a Kirchhoff plate is its lesser inner flexibility. Unlike a Reissner–Mindlin plate, in which cross-sectional planes can rotate independently of the position of the mid-surface, in a Kirchhoff plate the rotations are wedded to the rotations of the mid-surface.

A Reissner–Mindlin plate can lie flat on the ground, giving no notice that the cross-sectional planes in the interior tilt to the left; see Fig. 5.19, p. 434. Or at a clamped edge the slab can perform a feat which is impossible for a Kirchhoff plate: it can fold like sheet metal, and descend steeply.

This (relatively) inflexible behavior of a Kirchhoff plate can lead to problems at corner points (see Table 5.1), as for example at angular points of a hinged slab (see Fig. 5.12), because the gradient $\nabla w = [w_{,x}\,, w_{,y}\,]^T$ vanishes at a hinged corner point. This is a consequence of the fact that the derivatives in the direction of the two hinged edges (tangent vectors $\boldsymbol{t}_R$ and $\boldsymbol{t}_L$) are zero:

Table 5.1. Corner singularities of a Kirchhoff plate, [165]

Support conditions	Bending moments	Kirchhoff shear
clamped–clamped	180°	126°
clamped–hinged	129°	90°
clamped–free	95°	52°
hinged–hinged	90°	60°
hinged–free	90°	51°
free–free	180°	78°

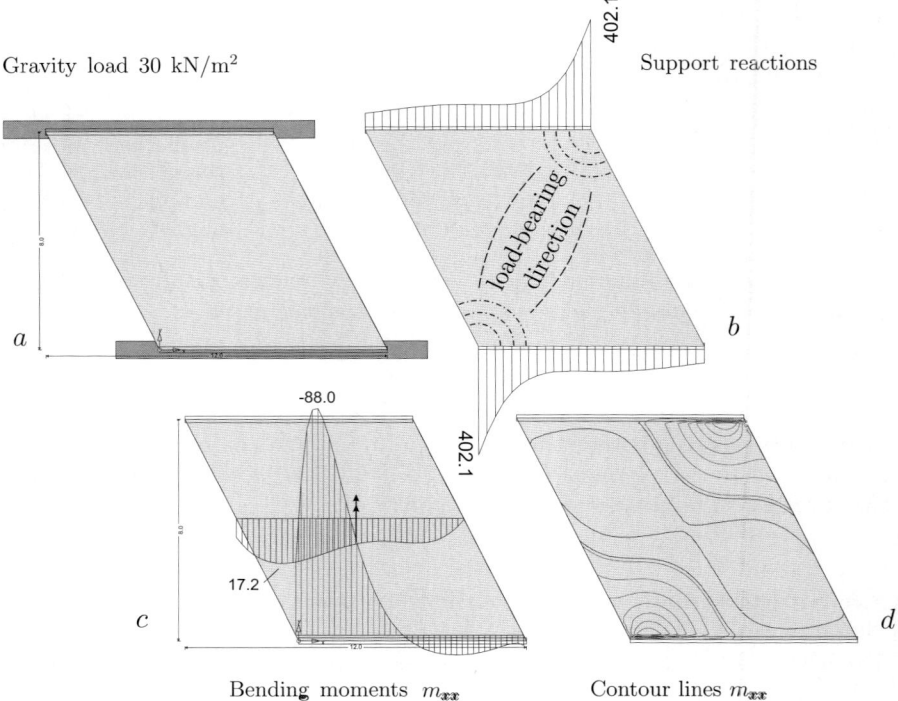

Fig. 5.14. At the obtuse-angled corner points, the support reactions and bending moments m_{xx} become infinite

$$\nabla w \cdot \boldsymbol{t}_R = 0 \qquad \nabla w \cdot \boldsymbol{t}_R = 0 \qquad \Rightarrow \qquad \nabla w = \boldsymbol{0} \,. \tag{5.41}$$

At such points the plate is clamped:

$$w = w_{,x} = w_{,y} = 0 \,. \tag{5.42}$$

The singularity vanishes immediately if the rotations $w_{,x}$ and $w_{,y}$ are set free.

In a hinged plate with a rhombic shape, a strange singularity is observed at the wide-angled corner points. The bending moment m_{xx} tends to $-\infty$ and the bending moment m_{yy} to $+\infty$; see Fig. 5.13. Again by releasing the rotations $w_{,x}$ and $w_{,y}$ the singularity disappears.

A skew bridge mainly carries the load from its lower wide-angled corner to the upper wide-angled corner—this is the shortest path between the two supports. Unfortunately the bending moments and the support reaction (Kirchhoff shear) become singular precisely at these corner points; see Fig. 5.14.

If the lower edge of the bridge coincides with the x-axis (because $w = 0$) the rotations $w_{,x}$ in the tangential direction are zero. In the terminology of Reissner–Mindlin plates this would be a *hard support*, while it would be considered a *soft support* if the rotations $w_{,x}$ were released. In a Kirchhoff plate hinged supports are normally modeled as hard supports, $w = w_{,x} = 0$, but it eventually helps to release the rotations near critical points.

Whenever possible the flexibility of the supports should be taken into account, because this helps to avoid stress peaks.

5.6 Reissner–Mindlin plates

A Reissner–Mindlin plate forms a kink if it is loaded with line loads, as illustrated by analogy with the Timoshenko beam in Fig. 5.16 b. The shearing strain $\gamma = w/0.5\,l$ activates the shear stress $\tau = G\gamma$, which keeps the balance with the applied load, $P/2 = \tau A = GA\gamma$. If the load is evenly distributed, the shearing strain varies linearly and the result is a well-rounded deflection curve. Because kinks are a natural feature of a Reissner–Mindlin plate C^0-elements are sufficient for such plates; see Fig. 5.17.

The deformations of a Reissner–Mindlin plate are described by the *deflection* and *rotations* of the planes $x = $ constant and $y = $ constant (see Fig. 5.18):

$$w, \quad \theta_x, \quad \theta_y \,. \tag{5.43}$$

In a Kirchhoff plate the rotations θ_x and θ_y are not independent quantities, because the planes maintain their position with respect to the midsurface (which deflects) and the planes rotate by the same angle by which the deflection surface w rotates: $\theta_x = -w_{,x}$ and $\theta_y = -w_{,y}$. The expressions

$$\gamma_x = w_{,x} + \theta_x \qquad \gamma_y = w_{,y} + \theta_y \tag{5.44}$$

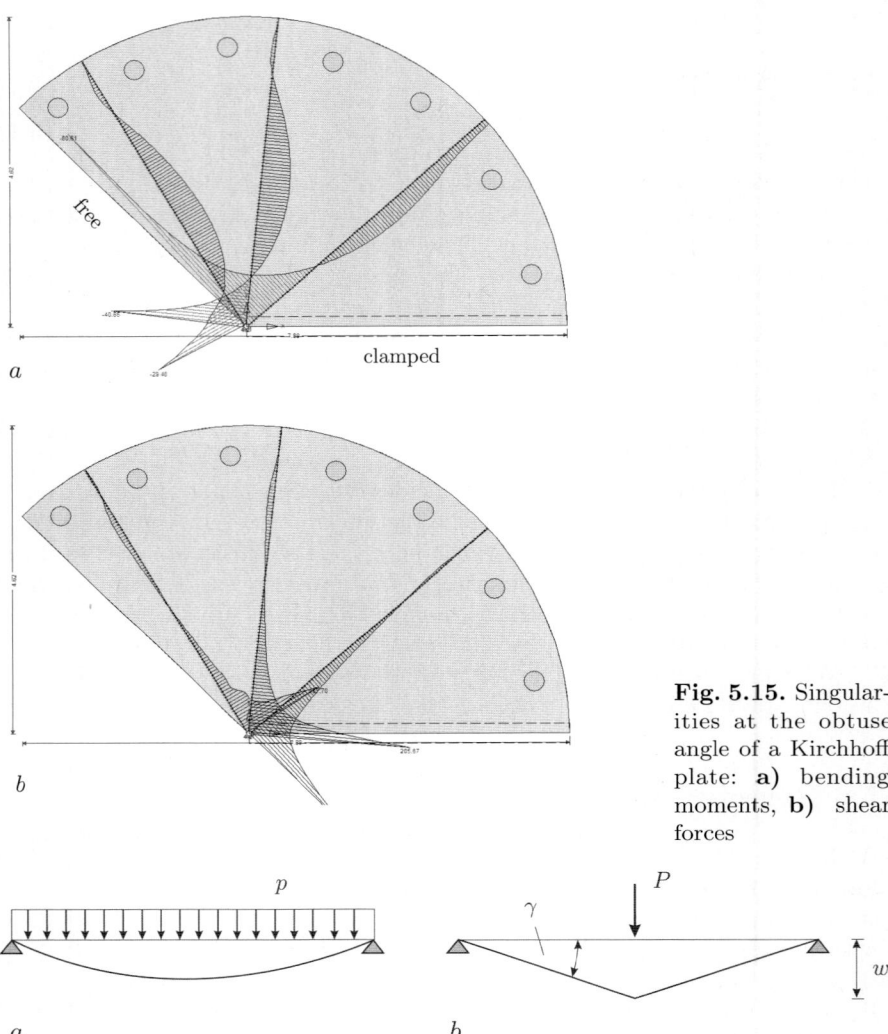

Fig. 5.15. Singularities at the obtuse angle of a Kirchhoff plate: **a)** bending moments, **b)** shear forces

Fig. 5.16. Timoshenko beam **a)** subjected to a distributed load and **b)** a single force

are the shearing strains. In a Kirchhoff plate the shearing strains are zero.

The system of differential equations for w, θ_x, θ_y in indicial notation is

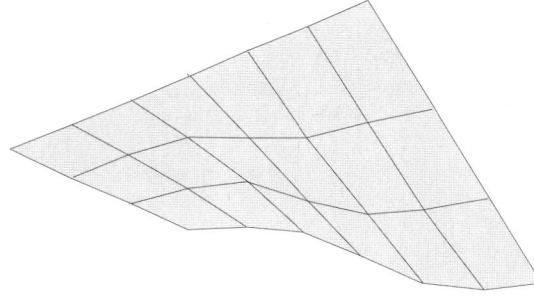

Fig. 5.17. A Reissner–Mindlin plate can have kinks

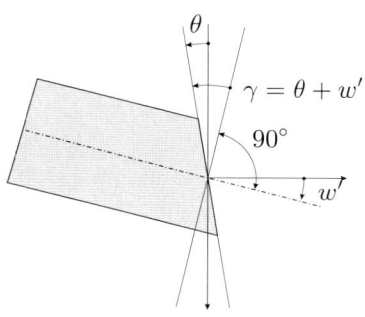

Fig. 5.18. The shearing strain γ

$$K(1-\nu)\{-(\frac{1}{2}(\theta_{\alpha,\beta}+\theta_{\beta,\alpha})+\frac{\nu}{1-\nu}\theta_{\gamma,\gamma}\,\delta_{\alpha\beta}),_{\beta}$$

$$+\bar{\lambda}^2(\theta_\alpha+w,_\alpha)\}=\frac{\nu}{1-\nu}\frac{1}{\bar{\lambda}^2}\,p,_\alpha\quad\alpha=1,2\ (5.45)$$

$$-\frac{1}{2}K(1-\nu)\bar{\lambda}^2(\theta_\alpha+w,_\alpha),_\alpha=p$$

where

$$K=\frac{Eh^3}{12(1-\nu^2)}\,,\qquad\bar{\lambda}^2=\frac{10}{h^2}\,,\qquad h=\text{slab thickness}\,.\qquad(5.46)$$

The terms on the right-hand side are the gradient $\nabla p=[p,_x,p,_y]^T$ of the vertical load and the vertical load p itself.

In the same way as the displacement components of an elastic solid form a vector $\boldsymbol{u}$ the displacement components of a Reissner–Mindlin plate can be assembled into a vector

$$\boldsymbol{u}(x,y)=[w(x,y),\theta_x(x,y),\theta_y(x,y)]^T\,.\qquad(5.47)$$

This is not a true displacement vector because $\boldsymbol{x}+\boldsymbol{u}$ is not the position of the point $\boldsymbol{x}$ after the deformation; the coordinates of the new position $\boldsymbol{x}'$ are instead

$$x'=\theta_x\,z\qquad y'=\theta_y\,z\qquad z'=w\,.\qquad(5.48)$$

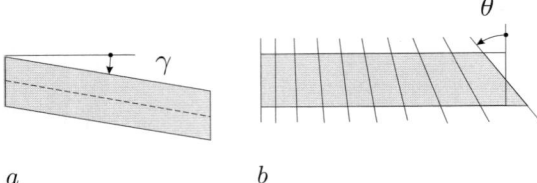

Fig. 5.19. Possible movements of a Mindlin type beam element. **a)** Type 1: $w = ax$, $\theta = 0$; **b)** Type 2: $w = 0$, $\theta = ax$

The bending moment depends only on the rotations of the planes,

$$m_{xx} = K\left(\theta_{x,x} + \nu\,\theta_{y,y}\right), \qquad m_{yy} = K\left(\theta_{y,y} + \nu\,\theta_{x,x}\right),$$
$$m_{xy} = (1 - \nu)\, K\left(\theta_{x,y} + \theta_{y,x}\right), \tag{5.49}$$

while the shear forces also depend on the gradient of the deflection w:

$$q_x = K\,\frac{1-\nu}{2}\,\bar{\lambda}^2\left(\theta_x + w_{,x}\right), \qquad q_y = K\,\frac{1-\nu}{2}\,\bar{\lambda}^2\left(\theta_y + w_{,y}\right). \tag{5.50}$$

Unlike a Kirchhoff plate, a Reissner–Mindlin plate does not sustain the attack of a single force. It shares this property with a wall beam or simply with 2-D and 3-D elastic solids.

But otherwise the structural analysis of a Reissner–Mindlin plate is not different from a Kirchhoff plate, because as long as the thickness to length ratio is small, the shear deformations are small, and then it hardly matters (apart perhaps from the results in a zone close to the boundary) which of the two slab models is used; see Sect. 5.20, p. 480.

Because the differential system of equations (5.45) is of second order, the symmetric strain energy

$$a(\boldsymbol{u}, \hat{\boldsymbol{u}}) = \int_{\Omega} [m_{xx}\,\hat{\theta}_{x,x} + m_{xy}\,\hat{\theta}_{x,y} + m_{yx}\,\hat{\theta}_{y,x} + m_{yy}\,\hat{\theta}_{y,y}$$
$$+ q_x\left(\hat{\theta}_x + \hat{w}_{,x}\right) + q_y\left(\hat{\theta}_y + \hat{w}_{,y}\right)]\,d\Omega$$

contains only first-order derivatives.

Formally the solution of a Reissner–Mindlin plate is a vector-valued function (see Eq. (5.47)), hence the FE solution becomes

$$\boldsymbol{u}_h = u_1 \underbrace{\begin{bmatrix} \psi_1 \\ 0 \\ 0 \end{bmatrix} + u_2 \begin{bmatrix} 0 \\ \psi_2 \\ 0 \end{bmatrix} + u_3 \begin{bmatrix} 0 \\ 0 \\ \psi_3 \end{bmatrix}}_{\text{node 1}}$$

$$\underbrace{+ u_4 \begin{bmatrix} \psi_4 \\ 0 \\ 0 \end{bmatrix} + u_5 \begin{bmatrix} 0 \\ \psi_5 \\ 0 \end{bmatrix} + u_6 \begin{bmatrix} 0 \\ 0 \\ \psi_6 \end{bmatrix}}_{\text{node 2}} \dots, \tag{5.51}$$

Kirchhoff plate

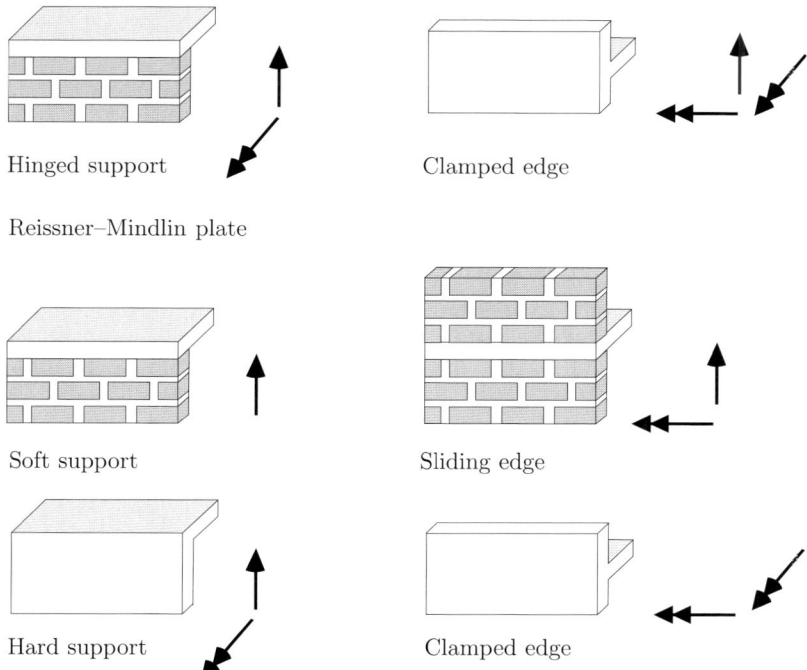

Fig. 5.20. Boundary conditions

where the three degrees of freedom u_i and the three associated shape functions $\psi_i(x, y)$ at each node are the deflection and the rotations in x-direction and y-direction, respectively, so that the basic pattern which repeats at each node is a sequence of three special unit deformations:

$$
\begin{bmatrix} w \\ 0 \\ 0 \end{bmatrix} \quad \begin{bmatrix} 0 \\ \theta_x \\ 0 \end{bmatrix} \quad \begin{bmatrix} 0 \\ 0 \\ \theta_y \end{bmatrix} . \tag{5.52}
$$

First the element inclines (see Fig. 5.19) but the planes remain vertical, because $\theta_x = \theta_y = 0$ means that the shearing strains $\gamma_x = w,_x$ and $\gamma_y = w,_y$ counterbalance the rotations of the plate mid-surface. In the second and third deformation the element remains flat, $w = 0$, but the planes rotate; see Fig. 5.19.

5.7 Singularities of a Reissner–Mindlin plate

Slab models are usually developed by starting with a 3-D elastic continuum and then simplifying the kinematics. Depending on the assumptions made in this process, the result is either a Kirchhoff plate or a Reissner–Mindlin plate.

The loss of accuracy in the transition from a 3-D model to a 2-D model is often felt the most near the boundary where the models differ most. With regard to the Reissner–Mindlin plate we speak of a *boundary layer effect* [9], [238]. To some extent this effect seems to be due to the fact that we cannot put an elastic solid on a line support such as a hinged support (see Sect. 1.16, p. 55), and the Reissner–Mindlin plate—not surprisingly—seems to suffer more from such adverse effects.

In a Reissner–Mindlin plate we have a choice of two support conditions for a hinged edge: the *hard support*, $w = w_{,t} = 0$, which corresponds to the hinged support of a Kirchhoff plate, and the *soft support*, $w = 0$, where the slope $w_{,t} = w_{,x}\, t_x + w_{,y}\, t_y$ in the tangential direction is released, $w_{,t} \neq 0$; see Fig. 5.20.

As in a Kirchhoff plate, the boundary conditions and the angle of the boundary points determine when and where singularities will develop; see Table 5.2 [205].

Table 5.2. Corner singularities of a Reissner–Mindlin plate

Boundary conditions	Bending moment	Shear force
clamped–clamped	180°	180°
sliding edge–sliding edge	90°	180°
hard support–hard support	90°	180°
soft support–soft support	180°	180°
free-free	180°	180°
clamped–sliding edge	90°	180°
clamped–hard support	90°	180°
clamped–soft support	$\approx 61.70°\ (\nu = 0.29)$	180°
clamped–free	$\approx 61.70°\ (\nu = 0.29)$	90°
sliding edge–hard support	45°	90°
sliding edge–soft support	90°	180°
sliding edge–free	90°	90°
hard support–soft support	$\approx 128.73°$	180°
hard support–free	$\approx 128.73°$	90°
soft support–free	180°	90°

Shear locking

The main advantages of a Reissner–Mindlin plate are essentially the relaxed continuity requirements for the shape functions, and its inner "richness" of kinematic variables. On the other hand *shear locking* can become a problem. The transition from a Reissner–Mindlin plate model—relatively thick slabs (foundation slabs)—to a Kirchhoff plate model, which is the standard model for thin slabs, can cause problems.

Clearly the Reissner–Mindlin plate theory subsumes the Kirchhoff slab theory, because the transition from the former to the latter model is simply achieved by setting the shearing strains to zero:

$$\gamma_x = w_{,x} + \theta_x = 0 \qquad \gamma_y = w_{,y} + \theta_y = 0 \,. \tag{5.53}$$

Because in normal slabs shear deformations are negligible, a Reissner–Mindlin plate should behave like a Kirchhoff plate. But this does not happen. The more the slab thickness h shrinks, the more a Reissner–Mindlin plate tends to stiffen, until the slab ultimately seems to freeze.

This is primarily a problem of the finite elements. If the equations could be solved exactly, then if h tends to zero the Reissner–Mindlin results should tend (in the sense of the strain energy [238], p. 263) to the results of a Kirchhoff plate.

Shear locking is best explained by studying the example of a Timoshenko beam, $\boldsymbol{u} = [w, \theta]^T$, see Fig. 5.21. The strain energy product is

$$a(\boldsymbol{u}, \hat{\boldsymbol{u}}) = \int_0^l [EI \, \theta' \, \hat{\theta}' + GA_s \, (w' + \theta) \, (\hat{w}' + \hat{\theta})] \, dx \,, \tag{5.54}$$

so that with appropriate unit displacements (2 at each node)

$$\underbrace{\boldsymbol{\varphi}_1(x) = \begin{bmatrix} w_1 \\ 0 \end{bmatrix} \quad \boldsymbol{\varphi}_2(x) = \begin{bmatrix} 0 \\ \theta_2 \end{bmatrix}}_{\text{node 1}} \quad \underbrace{\boldsymbol{\varphi}_3(x) = \begin{bmatrix} w_3 \\ 0 \end{bmatrix} \quad \boldsymbol{\varphi}_4(x) = \begin{bmatrix} 0 \\ \theta_4 \end{bmatrix}}_{\text{node 2}} \dots$$

$$\tag{5.55}$$

for example linear functions

$$w_i(x) = \frac{l - x}{l} \qquad \theta_j(x) = \frac{x}{l} \tag{5.56}$$

a result like

$$(\boldsymbol{K}_B + \boldsymbol{K}_S)\boldsymbol{u} = \boldsymbol{f} \tag{5.57}$$

is obtained, where the entries in the matrix $\boldsymbol{K}_B$ come from the bending terms, and the entries in $\boldsymbol{K}_S$ from the shear terms:

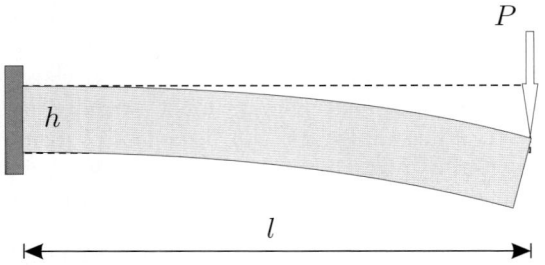

Fig. 5.21. Cantilever beam

$$k_{ij}^B = \int_0^l EI\,\theta_i'\,\theta_j'\,dx \qquad k_{ij}^S = \int_0^l GA_s\,(w_i' + \theta_i)\,(w_j' + \theta_j)\,dx\,. \quad (5.58)$$

If the cantilever beam in Fig. 5.21 is modeled with just one element, the end deflection is [70],

$$w(l) = \frac{12(h/l)^2 + 20}{12(h/l)^2 + 5} \cdot 1.2\,\frac{Pl}{GA_s} \qquad (5.59)$$

where A_s is the equivalent shear cross-sectional area. For a short beam, $l \ll 1$, the first fraction is approximately 1 and the end deflection is identical to the shear deformation

$$w(l) = 1.2\,\frac{Pl}{GA_s}\,. \qquad (5.60)$$

If $l \gg h$, i.e., if the length l is much greater than the width h of the beam, the first fraction is about $20/5$ and the end deflection will be much too small:

$$w(l) = 4 \cdot 1.2\,\frac{Pl}{GA_s} = 9.6\,\frac{Pl}{E\,bh} \qquad (G = 0.5\,E,\ \nu = 0) \qquad (5.61)$$

compared with the exact value (let $l/h = 8$)

$$w(l) = \frac{Pl^3}{3\,EI} = 4\,\frac{Pl}{E\,bh}\left(\frac{l}{h}\right)^2 = 256\,\frac{Pl}{E\,bh}\,. \qquad (5.62)$$

This is shearlocking.

The reason for this stiffening effect is the different sensitivity of the bending stiffness EI and the shear stiffness GA_s with respect to the width h of the beam:

$$EI = \frac{b\,h^3}{12} \qquad GA_s \simeq G\,b\,h \qquad \text{(rectangular cross section)}\,. \qquad (5.63)$$

If the width h—and thus the shear deformations $\gamma = w' + \theta$—tend to zero, the bending stiffness decreases much faster than the shear stiffness. As in an equation such as

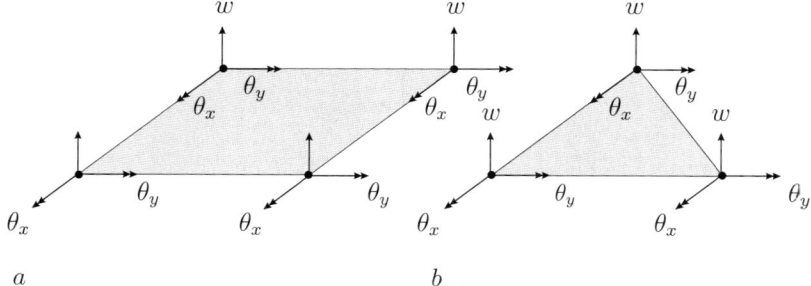

Fig. 5.22. Plate elements: **a)** Bathe–Dvorkin element **b)** DKT element. At the nodes of a DKT element, the shearing strain is zero, $\theta_{xi} = -\partial w/\partial x_i$, $\theta_{yi} = -\partial w/\partial y_i$

$$(1 + 10^5)\, u = 10 \qquad (\text{solution } u = 9.9999 \cdot 10^{-5} \approx 0)\,, \qquad (5.64)$$

where the large second term enforces a nearly zero solution, the matrix $\boldsymbol{K}_S$—because of the rapidly increasing influence of GA_s—tends to dominate (5.57). If (5.57) could be solved exactly, the increasing influence of GA_s would be canceled by the opposing tendency of $w' + \theta = \gamma \mapsto 0$.

All this holds for slabs as well: the transition from a Reissner–Mindlin plate model (average to large slab thickness) to a Kirchhoff plate model (small thickness) does not succeed numerically.

A whole catalog of measures has been proposed to circumvent shearlocking. Reduced integration is the best-known remedy. Although a better and simpler approach is to raise the order of the polynomials. This holds in similar situations as well, where internal constraints force an element to sacrifice all the degrees of freedom to satisfy the constraints, so that nothing is left to describe the movements of the element [70].

5.8 Reissner–Mindlin elements

A multitude of plate elements are based on the Reissner–Mindlin plate model. Only three such elements, the Bathe–Dvorkin element, the DKT element and the DST element, will be discussed here because they are probably the most popular.

Bathe–Dvorkin element

It seems that this element (see Fig. 5.22 a) was first developed by Hughes and Tezduyar [122], and later extended by Bathe and Dvorkin [27] to shells.

The element is an isoparametric four-node element with bilinear functions for the deflection w and the rotations θ_x and θ_y. According to the equations

$$\gamma_x = w,_x + \theta_x \qquad \gamma_y = w,_y + \theta_y\,, \qquad (5.65)$$

the polynomial shape function for w should have a degree one order higher than the rotations. To begin, one chooses for w a ninth-order Lagrange polynomial—4 corner nodes + 4 nodes at the mid-sides + 1 node at the center of the element—and a matching bilinear polynomial for the rotations. The idea is to calculate the shearing strains γ_x and γ_y independently of the deflection at the center and deflections at the mid-side nodes. Hence the stiffness matrix only depends on the deflections w at the four corner points, therefore bilinear functions for w suffice. This simplification is based on the observation that the shearing strains parallel to the sides of the element at the mid-side nodes are independent of the deflections at the mid-side nodes and at the center of the element.

The main advantage of this element is the easy transition from thick slabs to thin slabs, so that the element is universally applicable. The bending moment m_{xx} in the principal direction—which is assumed here to be the x-axis—is constant, and the bending moment m_{yy} is linear. As in the Wilson element, quadratic terms can be added, so that the bending moments also vary linearly in the x-direction. The shear forces q_x and q_y are of course constant.

DKT element

The DKT element (see Fig. 5.22 b) can be considered a modified Kirchhoff plate element or a modified Reissner–Mindlin plate element [236].

The derivations starts with a Reissner–Mindlin plate element and the assumption is that the shearing strains in the element are zero [29]. Hence the strain energy product is simply the scalar product of the bending moments and the curvature terms:

$$a(\boldsymbol{u}, \boldsymbol{u}) = \int_\Omega [m_{xx}\, \theta_{x,x} + m_{xy}\, \theta_{x,y} + m_{yx}\, \theta_{y,x} + m_{yy}\, \theta_{y,y}]\, d\Omega\,. \quad (5.66)$$

But the element shape functions satisfy the condition $\gamma_x = \gamma_y = 0$ only at discrete points, namely the corner points of the triangular element and the mid-side nodes of the element, which is why this element is called a *discrete Kirchhoff triangle*.

For the rotations, linear functions are chosen:

$$\theta_x = \sum_{i=1}^3 \theta_{xi}\, \varphi_i(\boldsymbol{x}) \qquad \theta_y = \sum_{i=1}^3 \theta_{yi}\, \varphi_i(\boldsymbol{x})\,. \quad (5.67)$$

The deflection w is instead only defined along the edge, and interpolated by *Hermite polynomials*. Next one assumes: a) linear rotations θ_n ($=$ slope) on each side, b) zero shearing strains at the corner points and at the mid-side nodes. In particular the latter assumption, $\theta_\alpha = w_{,\alpha}$, makes it possible to couple the rotations to the deflection w, and it is thereby possible to reduce the model to the 3×3 degrees of freedom $w_i, w_{,xi}, w_{,yi}$ at the three corner points, so that the result is a triangular plate element with the nine "natural" degrees of freedom. Because there are no shape functions for the deflection w inside the element, a calculation of consistent equivalent nodal forces is not possible. In other words, it is up to the user how to distribute the load over the nodes. To calculate shear forces additional assumptions must be introduced.

The DKT element is very popular, because with minimal effort (C^0-functions suffice) a triangular element with the nodal degrees of freedom $w, w_{,x}, w_{,y}$ is obtained. But of course this element also is nonconforming [51].

DST element

The DST element is closely related to the DKT element [31]. Unlike the DKT element, the shearing strains γ_i are not zero at the nodes. The starting point is a weak formulation of the Reissner–Mindlin equations using a Hellinger–Reissner functional with $w, \varphi_x, \varphi_y, \gamma_x, \gamma_y, q_x, q_y$ as independent variables. By a corresponding weak coupling of the terms (L_2 orthogonality), a triangular element can be derived in which each of the three nodes has degrees of freedom w, φ_x, φ_y. The name *discrete* is justified: as in a DKT element the shearing strains γ_x and γ_y are coupled at only three points (the mid-side nodes) to the other degrees of freedom w, φ_x, φ_y.

5.9 Supports

Standard support conditions are

$$\begin{array}{ll} \text{hinged:} & w = 0, m_n = 0 \\ \text{clamped:} & w = 0, w_n = 0 \\ \text{free:} & m_n = v_n = 0 \end{array}$$

Hinged or clamped supports are often idealized as being completely rigid. But the correct assessment of the stiffness of a load-bearing wall or an edge beam is important, because the distribution of the support reactions strongly depends on the stiffness of the supports; see Fig. 5.23. In Fig. 5.24 a point load is applied at the end of the load-bearing wall. If the support were really rigid, the applied load would cause no stresses in the slab.

The more flexible the supports, the more "beautiful" the results, because the slab has a chance to circumvent constraints that might otherwise lead to singularities; see Fig. 5.25.

It seems that intermittent supports which typically occur at doors and window openings (see Fig. 5.26) can be modeled as continuous supports as long as $l/h \le 7$, where $l = $ length of the opening, $h = $ slab thickness. The effect of a sleeping beam on the structural behavior is often overrated. The increase in stiffness due to additional reinforcement is too little to be noticeable.

The vertical stiffness of a load-bearing wall with modulus of elasticity E is

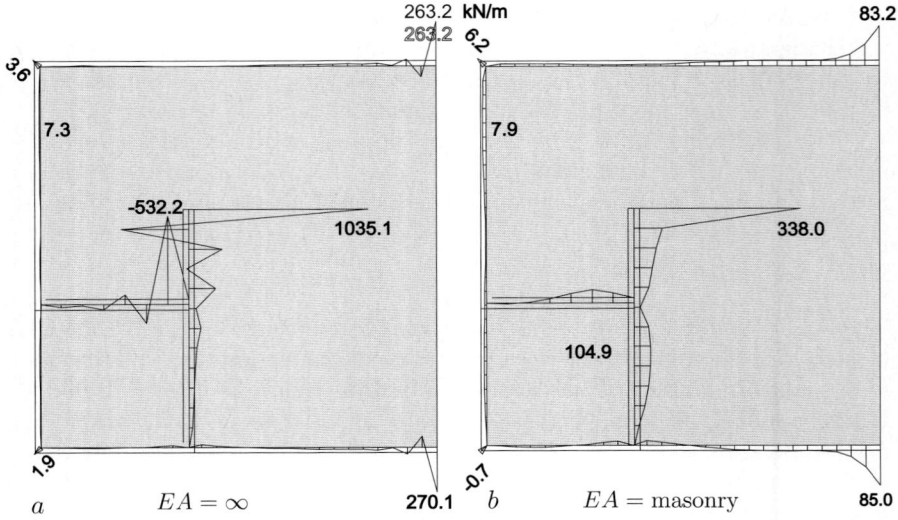

Fig. 5.23. Support reactions of a slab (8m × 8 m) under gravity load $g = 9$ kN/m^2 having a free edge on the right-hand side. **a)** Rigid supports, **b)** soft support (brickwork)

$$k = \frac{E\,d}{h} \quad (\text{kN/m}^2)\,, \tag{5.68}$$

where d is the thickness of the wall and h is its height. This coefficient k times the displacement w of the wall yields the support reaction (kN/m). In the same sense,

$$k = \frac{E\,A}{h} \quad (\text{kN/m}) \tag{5.69}$$

is the stiffness of a column with cross-sectional area A, height h, and modulus of elasticity E.

The rotational stiffness c_φ of a wall is the bending moment (kN m/m) that effects a rotation of 45° of the upper edge. The rotational stiffness of the head of a column depends on the support conditions at the bottom of the column:

$$k_\varphi = \frac{3\,EI}{h} \qquad \text{hinged support} \tag{5.70}$$

$$k_\varphi = \frac{4\,EI}{h} \qquad \text{clamped support}\,. \tag{5.71}$$

It is obvious that if a column forms a rigid joint with the slab, the support reaction will increase, because the influence function for the support reaction will widen.

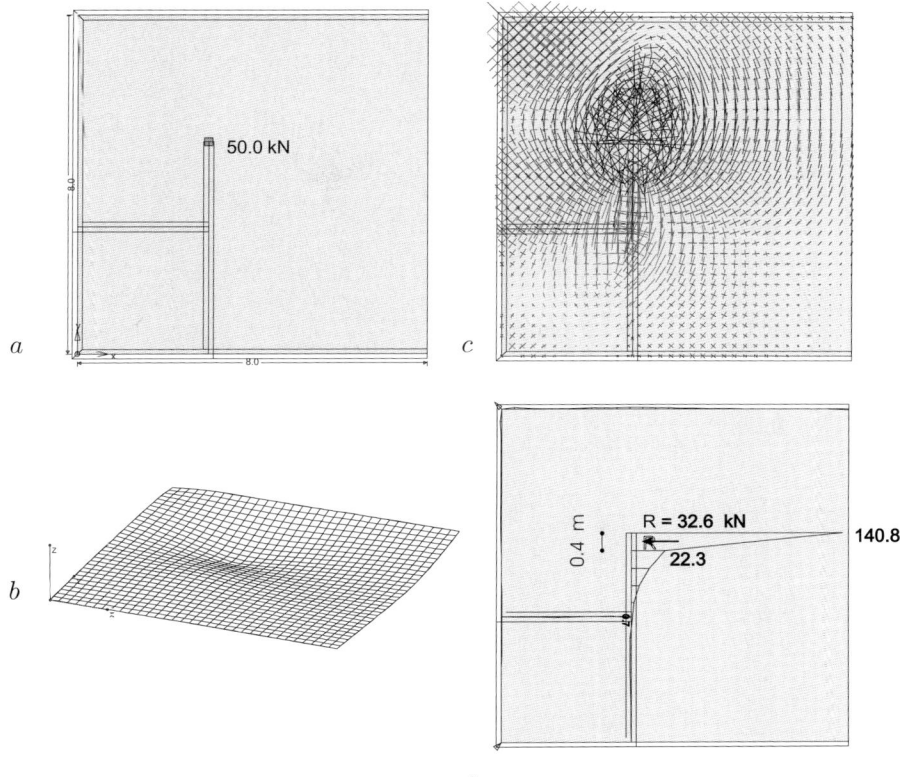

Fig. 5.24. Slab on a system of brickwork walls. **a)** System and single force applied at the end of an interior wall; **b)** deflection surface of the slab; **c)** principal moments; **d)** support reactions and assumed punching shear

5.10 Columns

Reliable estimation of the magnitude of the bending moments near columns and similar point supports is one of the main problems in FE analysis.

Near a point support the bending moments can be split into smooth polynomial (p) parts and singular parts (s),

$$m_{ij}(\boldsymbol{x}) = m^p_{ij}(\boldsymbol{x}) + P\,m^s_{ij}(\boldsymbol{y}_c, \boldsymbol{x}) \quad \boldsymbol{y}_c = \text{center of the column} \quad (5.72)$$

where P is the support reaction. The singular parts (for simplicity it is assumed that $\nu = 0$ and polar coordinates $\boldsymbol{x} \to r, \varphi$ are used)

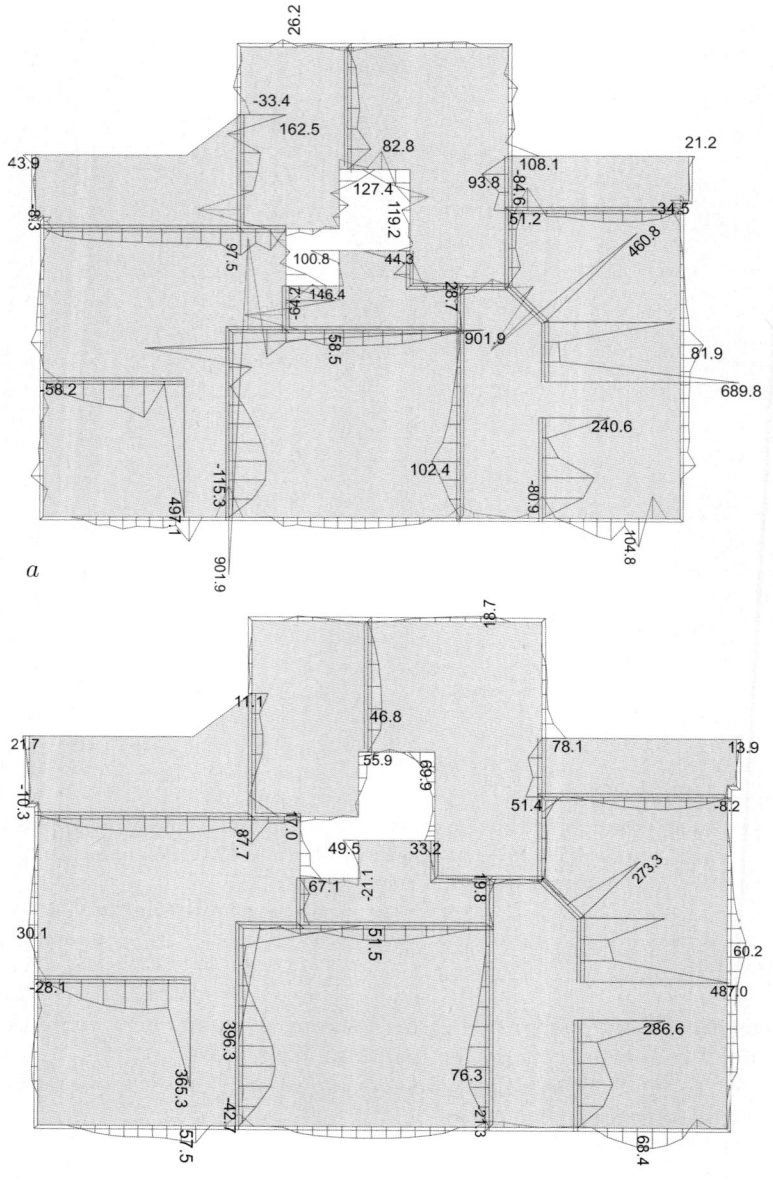

Fig. 5.25. Support reactions (kN/m) **a)** produced by rigid supports, $EA = \infty$, and **b)** by masonry walls

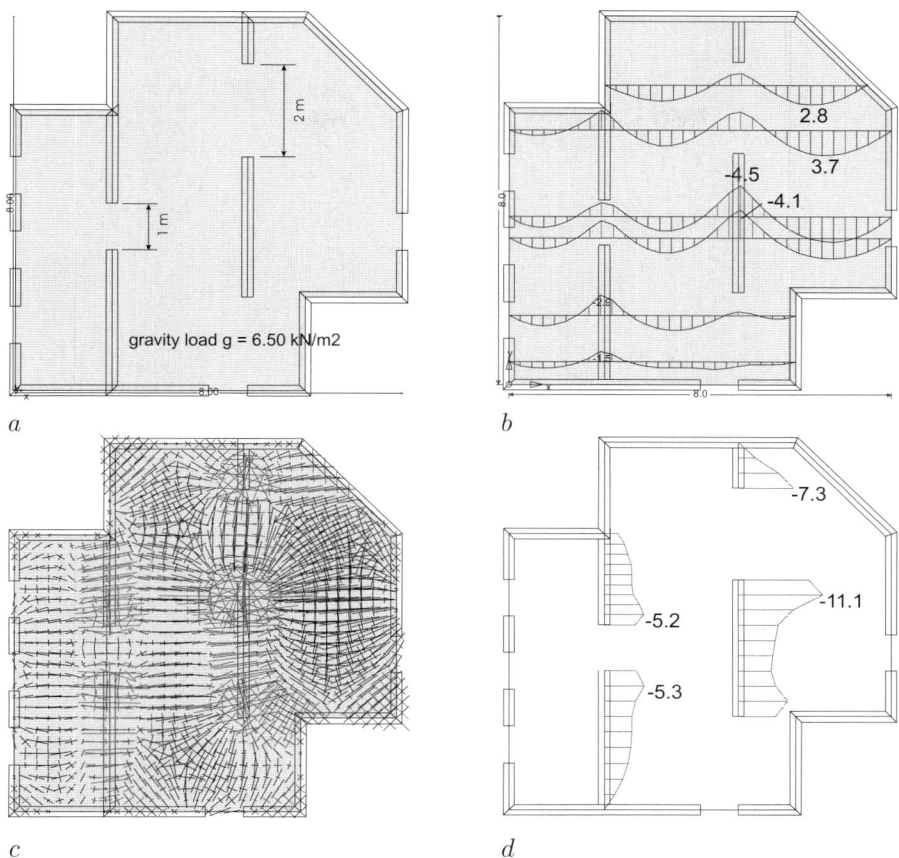

Fig. 5.26. Slab with intermittent support, slab thickness $h = 0.2$ m. **a)** System and gravity load g; **b)** bending moments m_{xx} kNm/m; **c)** principal moments; **d)** bending moments in the slab along the interior walls

$$m^s_{xx}(r, \varphi) = -\frac{1}{8\,\pi}[(3 + 2\ln r)\cos^2\varphi + (1 + 2\ln r)\sin^2\varphi]\,, \qquad (5.73)$$

$$m^s_{yy}(r, \varphi) = -\frac{1}{8\,\pi}[(3 + 2\ln r)\sin^2\varphi + (1 + 2\ln r)\cos^2\varphi]\,, \qquad (5.74)$$

$$m^s_{xy}(r, \varphi) = -\frac{1}{8\,\pi}[(4 + 4\ln r)\sin\varphi\,\cos\varphi]\,, \qquad (5.75)$$

are independent of the shape and size of the slab, and of the support conditions [116]. Only the smooth parts differ.

These singular parts would also be dominant if the single force P were replaced by the bearing pressure $p = P/\Omega_c$ at the head of the column,

$$m_{ij}(\boldsymbol{x}) = m^p_{ij}(\boldsymbol{x}) + \int_{\Omega_c} m^s_{ij}(\boldsymbol{y}, \boldsymbol{x})\,p\,d\Omega_{\boldsymbol{y}}\,, \qquad (5.76)$$

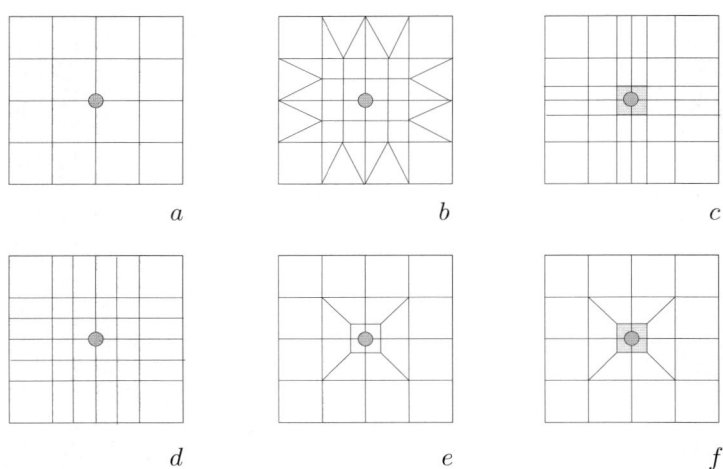

Fig. 5.27. Modeling of a point support

because not too far from the column, this integral would be identical to the influence of the single force P

$$\int_{\Omega_c} m_{ij}^s(\boldsymbol{y}, \boldsymbol{x}) \, p \, d\Omega_{\boldsymbol{y}} \cong P \, m_{ij}^s(\boldsymbol{y}_c, \boldsymbol{x}) \,. \tag{5.77}$$

The problem of the FE method is that polynomial shape functions are not very good at approximating the singular functions m_{ij}^s.

The best strategy would be to add the singular functions to the FE code. Otherwise the mesh must be refined (see Fig. 5.27), or techniques such as

- spreading the point support onto multiple nodes (multiple nodes model)
- multiple nodes + elastic support
- multiple nodes and plate stiffness $K = \infty$ near the column
- a single rigid plate element ($K = \infty$) which can rotate freely placed on an elastic pinned support

may be employed to ease the burden for an FE program to approximate singular bending moments with polynomial shape functions; see Fig. 5.28.

Recommendations

To keep a reasonable ratio between effort and accuracy we make the following recommendations:

- Columns should always be modeled with their natural stiffness.
- If no special coupling elements are used, as in Fig. 5.31 d, the element size should decrease gradually towards the column center.

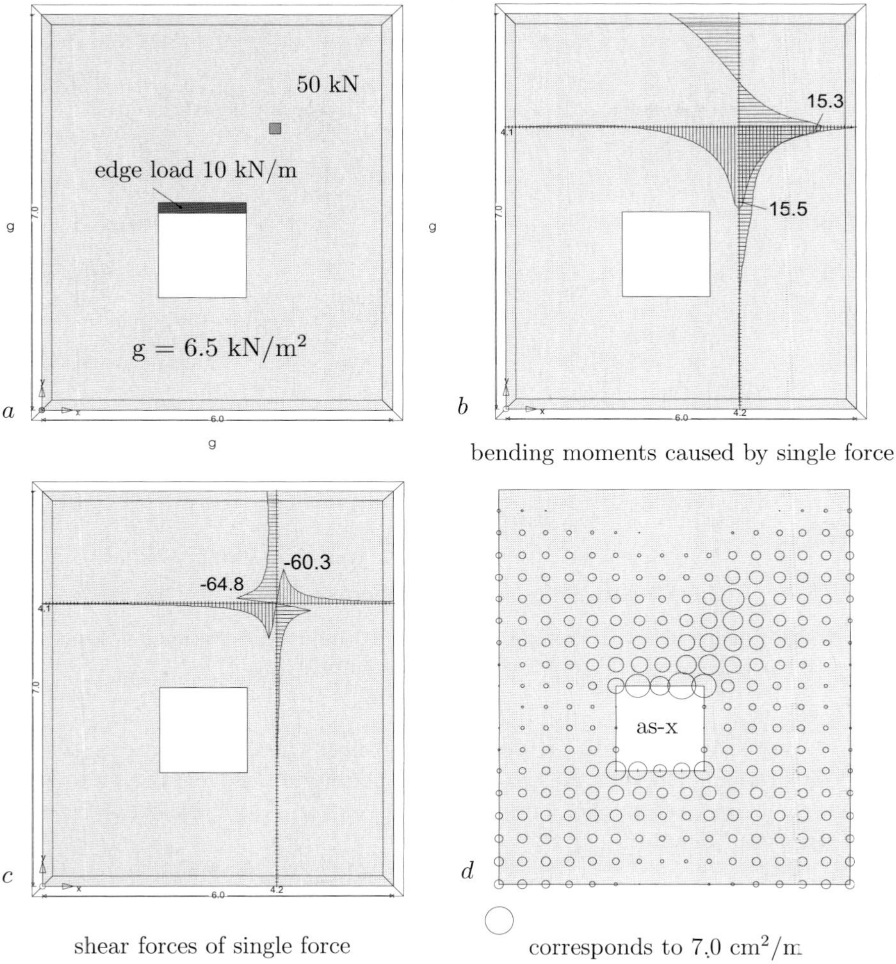

50 kN

edge load 10 kN/m

$g = 6.5 \text{ kN/m}^2$

a

g

15.3

15.5

b   bending moments caused by single force

-60.3

-64.8

c

as-x

d

shear forces of single force

corresponds to 7.0 cm²/m

Fig. 5.28. Slab with opening: **a)** gravity load $g = 6.5$ kN/m², single force 50 kN + edge load 10 kN/m, **b)** bending moments (kNm/m) caused by the single force, **c)** shear forces (kN/m) caused by the single force in two sections, **d)** reinforcement as-x at the bottom (qualitative)

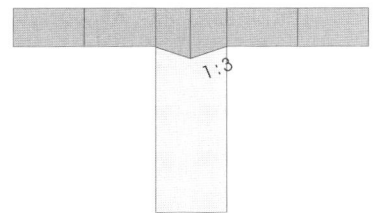

1:3

Fig. 5.29. Widening of the elements above the column head

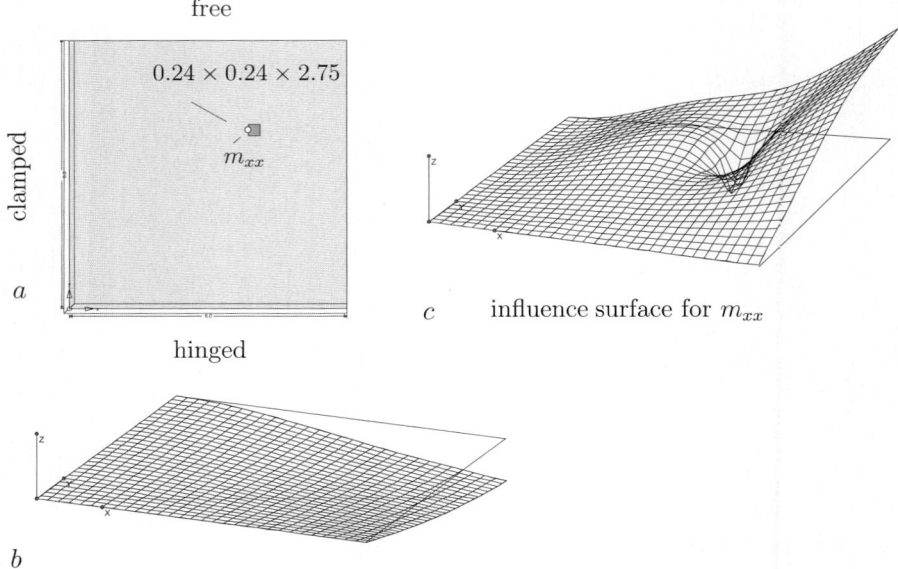

free

$0.24 \times 0.24 \times 2.75$

m_{xx}

clamped

hinged

a

c   influence surface for m_{xx}

b

influence surface for column support reaction

Fig. 5.30. Square slab with column: **a)** system; **b)** influence function for the support reaction; **c)** influence function for the bending moment m_{xx}

- It is a good idea to subdivide the region near the column into a patch of four elements, so that the midpoint of the patch is the column, and to let the thickness of the elements increase from the edge of the patch to the center; see Fig. 5.29.
- If possible the center of the elements should coincide with the vertices of the column cross section.
- It is sufficient if the center node is supported. A multinode model does not increase accuracy. In particular a multinode model can easily lead to unilateral rotational constraints, or simulate a rigid joint.
- In no case should a rigid multi-node model be used.

Other strategies and refinement techniques are possible. The important point is that something should be done to alleviate for an FE program the task to approximate the influence functions for the bending moments at the columns (see Fig. 5.30 c), because these are the functions an FE program must approximate if it is to calculate the bending moments.

Things are different with regard to the support reaction in the column. The influence function for the support reaction (see Fig. 5.30 b) is much smoother and more regular. If the focus were only on the column reactions, no mesh refinement near point supports would be necessary.

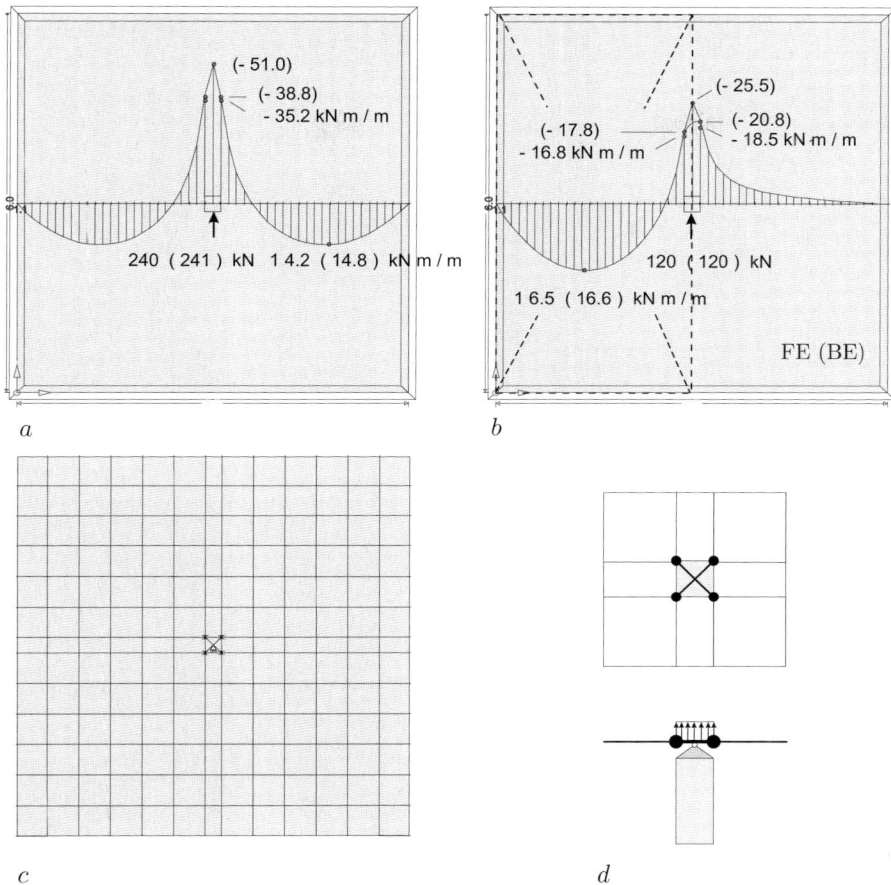

Fig. 5.31. Hinged square slab with interior column 0.25 m × 0.25 m, length $l = 2.75$ m. **a)** Uniform load; **b)** traffic load on one side; **c)** FE mesh; **d)** rigid plate element

Boundary elements

In a BE program the correct singularities (influence functions, Eq. (5.75)), are built into the code. The column reaction P is so determined that the deflection of the slab and the compression of the column at *the center of the column* are the same, and later—when the bending moments are calculated— the single force P is replaced by the load bearing pressure $p = P/\Omega_c$ to avoid the bending moments becoming singular; see Fig. 5.31.

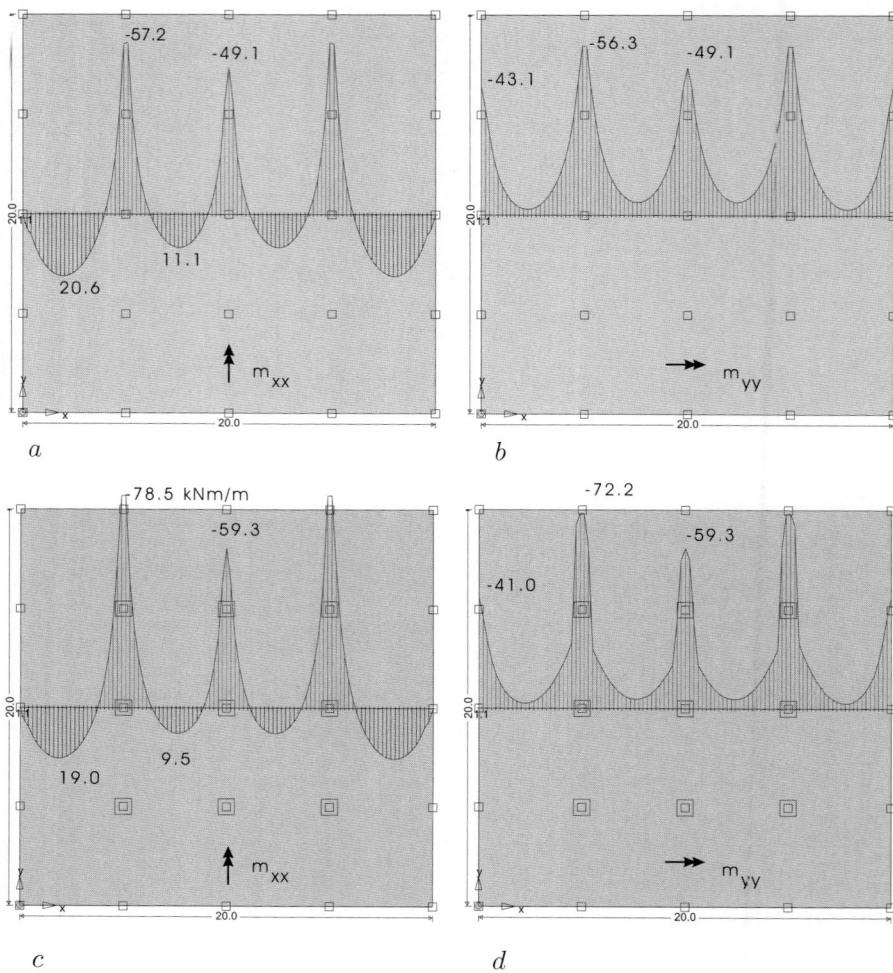

Fig. 5.32. Slab, $h = 0.4$ m, columns $0.4 \times 0.4 \times 2.45$ and columns with drop panels $0.2 \times 0.2 \times 0.2$. **a)** Bending moments m_{xx}, **b)** bending moments m_{yy}, **c)** bending moments m_{xx}, **d)** bending moments m_{yy}

Example

As mentioned earlier one possibility for modeling the column–slab interaction is to use a rigid slab element, $K = \infty$, which sits on top of the column on a pinned support so that it can rotate freely.

A hinged square slab 6 m × 6 m, supported at its center by a single column 25 cm × 25 cm with a length of 2.75 m was analyzed with such an element. The modulus of elasticity was 30 000 MN/m², the slab thickness was $h = 0.2$ m, and Poisson's ratio $\nu = 0.2$. In the first load case the loading consisted of a constant surface load $p = 20$ kN/m^2, and in the second load case the same load was applied only to the left portion of the slab. The agreement with a BE solution is very good (see Fig. 5.31 and the following table).

Uniform load	Rigid		Elastic		
	FE	BE	FE	BE	
m_{xx} column	−37	−43.6	−35.2	−38.8	kN m
m_{xx} span	14.2	14.4	14.8	14.9	kN m
Support reaction	253	257	240	241	kN
One-sided load			Elastic		
			FE	BE	
m_{xx} column, left			−16.8	−17.8	kN m
m_{xx} column, right			−18.5	−20.8	kN m
m_{xx} span			16.5	16.6	kN m
Support reaction			120	120	kN

Rigid means stiff supports, $EA = \infty$, and elastic that the stiffness of the column was $EA = 6.82 \cdot 10^5$ kN/m and the walls had a stiffness of $2.62 \cdot 10^6$ kN/m^2, corresponding to a modulus of elasticity of 30 000 MN/m^2, a wall thickness of 24 cm, and a height of 2.75 m.

Drop panels and column capitals

Drop panels and column capitals ensure that the bending moments migrate to the column capitals, as can be seen in Fig. 5.32. The discontinuity in the slab thickness ensures that the bending moments in a cross section perpendicular to the discontinuity jump (see Fig. 5.39).

5.11 Shear forces

Shear forces are the least reliable quantities in FE analysis. They easily oscillate and tend to exhibit erratic behavior; see Fig. 5.33.

In a Kirchhoff plate model the shear forces are the third-order derivatives of the unit deflections,

$$q_x = -K(w_{,xxx} + w_{,yyx}), \qquad q_y = -K(w_{,xxy} + w_{,yyy}) \qquad (5.78)$$

while in a mixed model they are the first-order derivatives of the bending moments,

$$q_x = m_{xx,x} + m_{xy,y} \qquad q_y = m_{yy,y} + m_{yx,x}, \qquad (5.79)$$

and they are therefore often constant because in mixed methods mainly linear functions are used to approximate the bending momente m_{xx}, m_{xy}, m_{yy}.

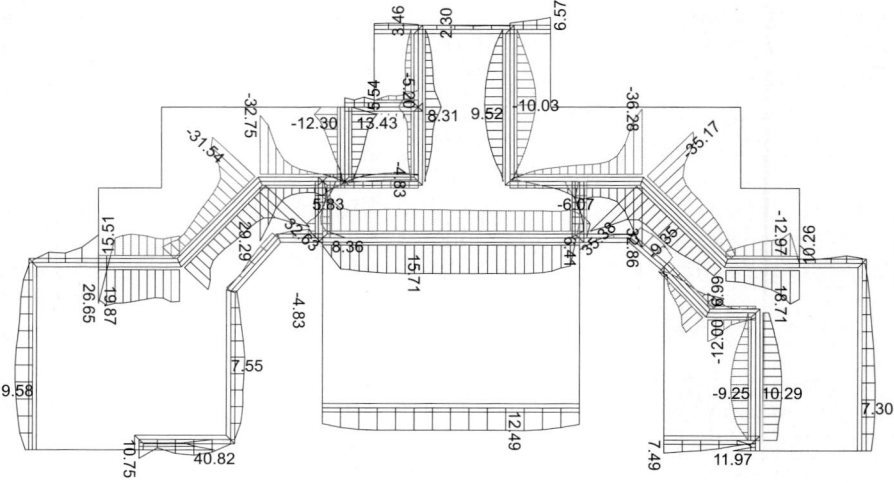

Fig. 5.33. Distribution of shear forces in a slab near the supports

In a Reissner–Mindlin model the shear forces are proportional to the shearing strains γ_x and γ_y, and thus proportional to the rotations θ_x, θ_y, and the derivatives of w:

$$q_x = K\,\frac{1-\nu}{2}\,\bar\lambda^2\,(\theta_x + w_{,x}) \qquad q_y = K\,\frac{1-\nu}{2}\,\bar\lambda^2\,(\theta_y + w_{,y}). \qquad (5.80)$$

In slabs, no shear reinforcement is necessary if the shear stresses remain below some threshold limits like $\tau \le 0.5$ MN/m^2; see Fig. 5.34. Only at certain critical points the shear stresses exceed these limits. But even then it is questionable whether it is really necessary to provide shear reinforcement, because while the numbers indicate a trend, the magnitude of the numbers itself is dubious.

In Fig. 5.35 the distribution of the shear forces in a horizontal (q_x) and a vertical (q_y) cross section in front of a wall is plotted. While the shear force q_x exhibits normal variability the shear force q_y grows exponentially to a peak value of 104 kN/m. At such points it is more appropriate to calculate an equivalent punching strain, as in the case of the slab in Fig. 5.36. Nowadays this is done routinely by most FE programs; see Fig. 5.37.

5.12 Variable thickness

If a slab has a smooth surface but the thickness varies the midsurfaces of the single panels will lie at differing levels; see Fig. 5.38 a. To accurately model such a plate would require elements for which such a shift of the midsurface

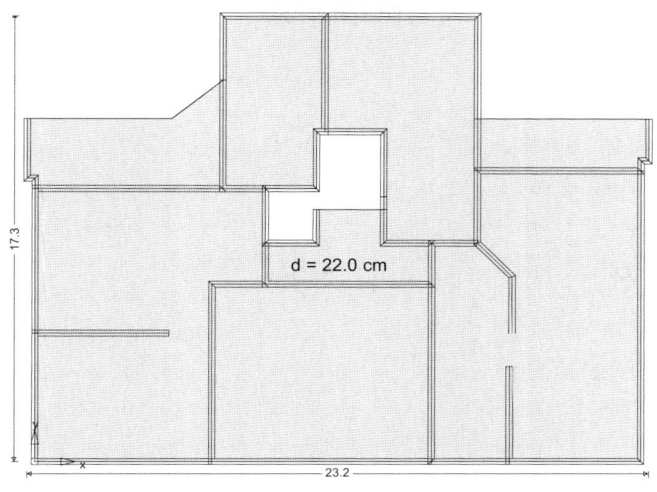

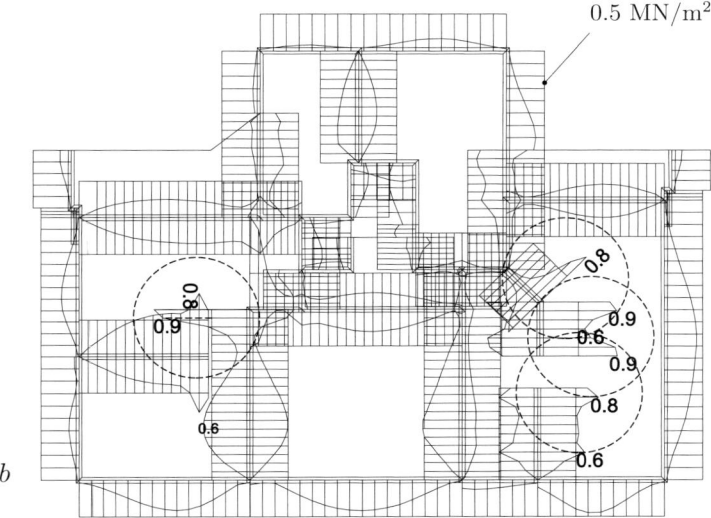

Fig. 5.34. Slab: **a)** system, **b)** shear stresses at the supports usually remain below the threshold values for shear reinforcement, here 0.5 MN/m^2

is possible. Conventional plate elements model such a slab with a uniform midsurface; see Fig. 5.38 e.

Variations in the thickness of the slab will produce jumps in the internal actions; see Fig. 5.38. At the interface between two such zones the bending moment m_{xx} and the curvature $\kappa_{yy} = w,_{yy}$ must be the same,

$$m_{xx}^L = -K^L(w^L,_{xx} + \nu\, w,_{yy}) = -K^R(w^R,_{xx} + \nu\, w,_{yy}) = m_{xx}^R , \quad (5.81)$$

while the bending moment m_{yy} will be discontinuous.

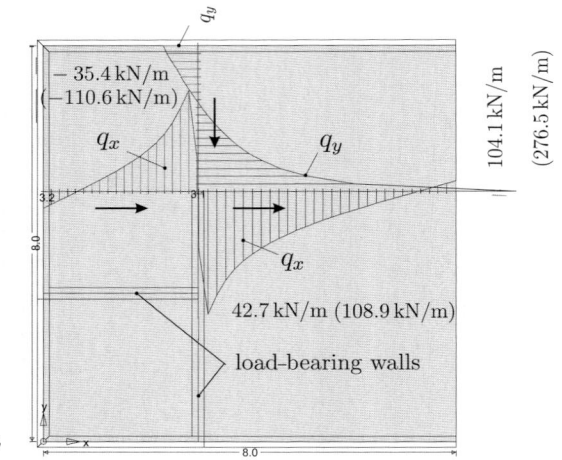

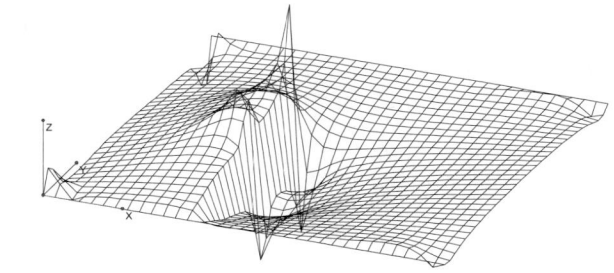

Fig. 5.35. Slab on masonry walls: **a)** shear forces q_x and q_y in a horizontal and a vertical section, respectively. The values in brackets are the results for rigid supports, $EA = \infty$; **b)** 3-D view of q_x

If Poisson's ratio is assumed to be zero, $\nu = 0$, then the ratio of these two bending moments becomes approximately

$$\frac{m_{yy}^L}{m_{yy}^R} = \frac{K^L}{K^R}\frac{(w,_{yy}+\nu\, w^L,_{xx})}{(w,_{yy}+\nu\, w^R,_{xx})} \simeq \frac{K^L}{K^R} = \frac{h_L^3}{h_R^3} = \frac{0.2^3}{0.4^3} = \frac{1}{8}\,. \qquad (5.82)$$

Hence if the thickness h doubles, then because of the h^3 the bending moment increases by a factor of eight.

At column capitals or drop panels, the bending moments peak at an earlier stage, and they stay at that level for a longer time; see Fig. 5.39.

In the slab in Fig. 5.40 the singularity in the support reactions is very pronounced and mainly due to the rather large change in the thickness of the slab from 0.25 m to 0.60 m. Such situations are not uncommon in the analysis of slabs and then elaborate mathematical theories will not help very much—rather a sound engineering judgement must cope with such problematic results.

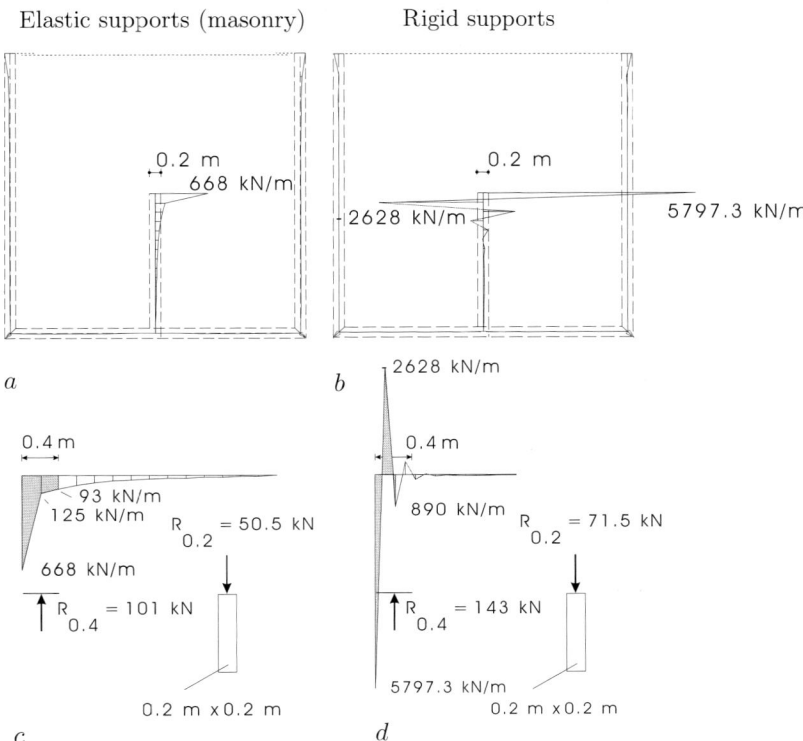

Fig. 5.36. Support reactions and equivalent punching shear for assumed columns 0.4×0.4 and 0.2×0.2 respectively at the end of a wall

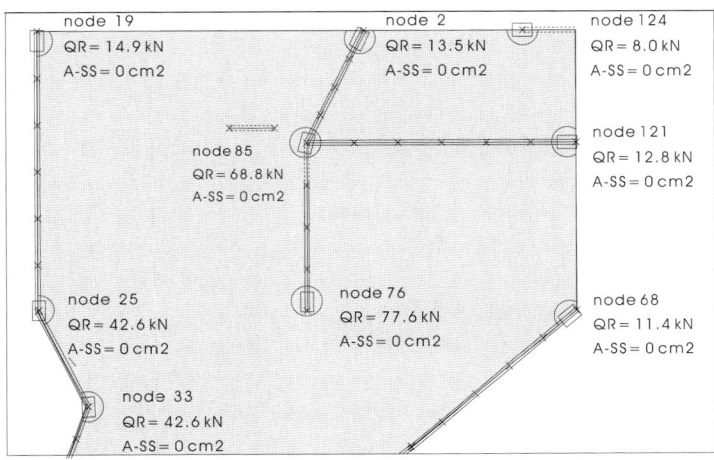

Fig. 5.37. Nowadays punching shear checks are done routinely by FE programs at the end points of load-bearing walls

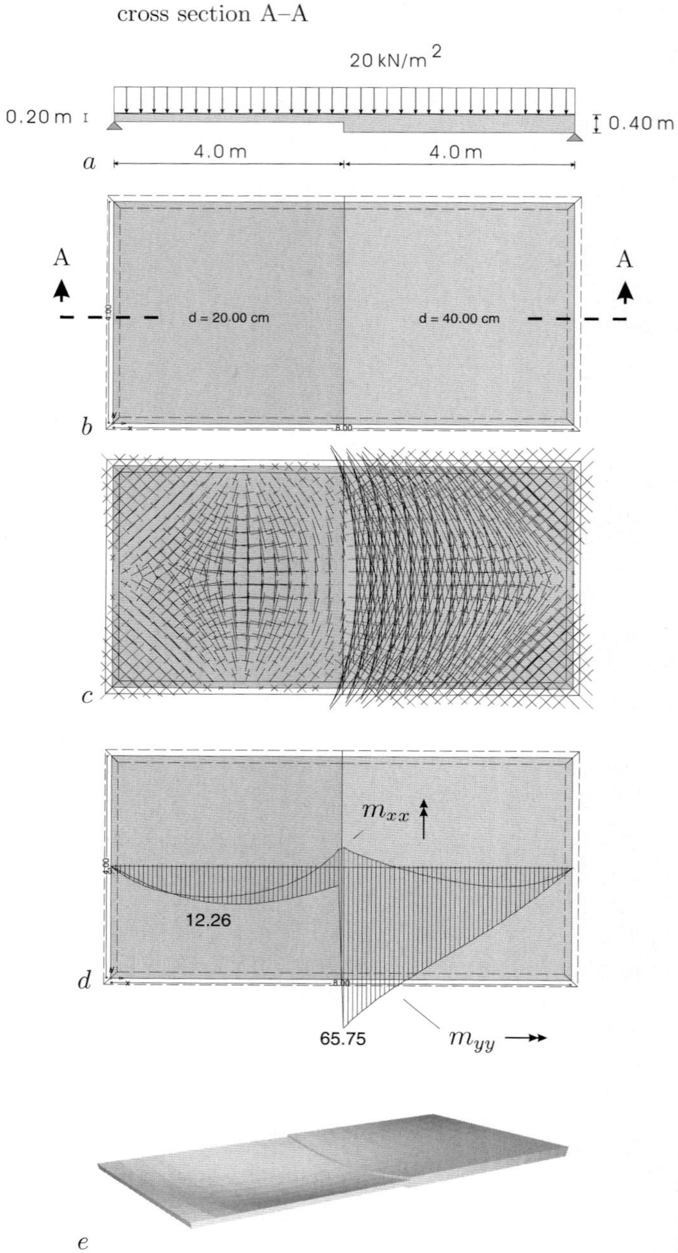

cross section A–A

20 kN/m^2

0.20 m

0.40 m

4.0 m   4.0 m

a

A   A

d = 20.00 cm   d = 40.00 cm

b

c

m_{xx}

12.26

d

65.75 m_{yy} →→

e

Fig. 5.38. Hinged slab: **a)** cross section; **b)** system; **c)** principal moments; **d)** bending moments m_{xx} and m_{yy}; **e)** 3D-view

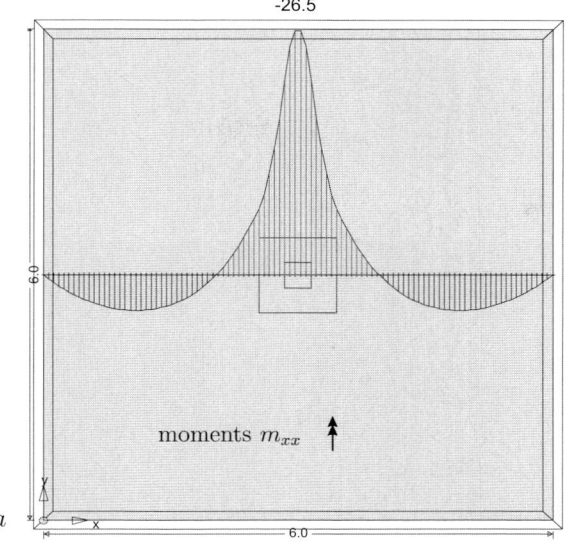

-26.5

moments m_{xx}

6.0

6.0

a

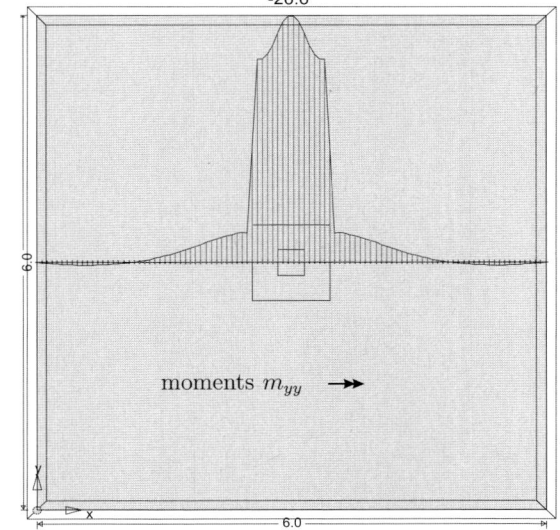

-26.6

moments m_{yy}

6.0

6.0

b

Fig. 5.39. Interior column of a hinged slab with a drop panel: **a)** bending moments m_{xx}, **b)** bending moments m_{yy} in a horizontal section. In a vertical section m_{yy} would be continuous and m_{xx} would be discontinuous

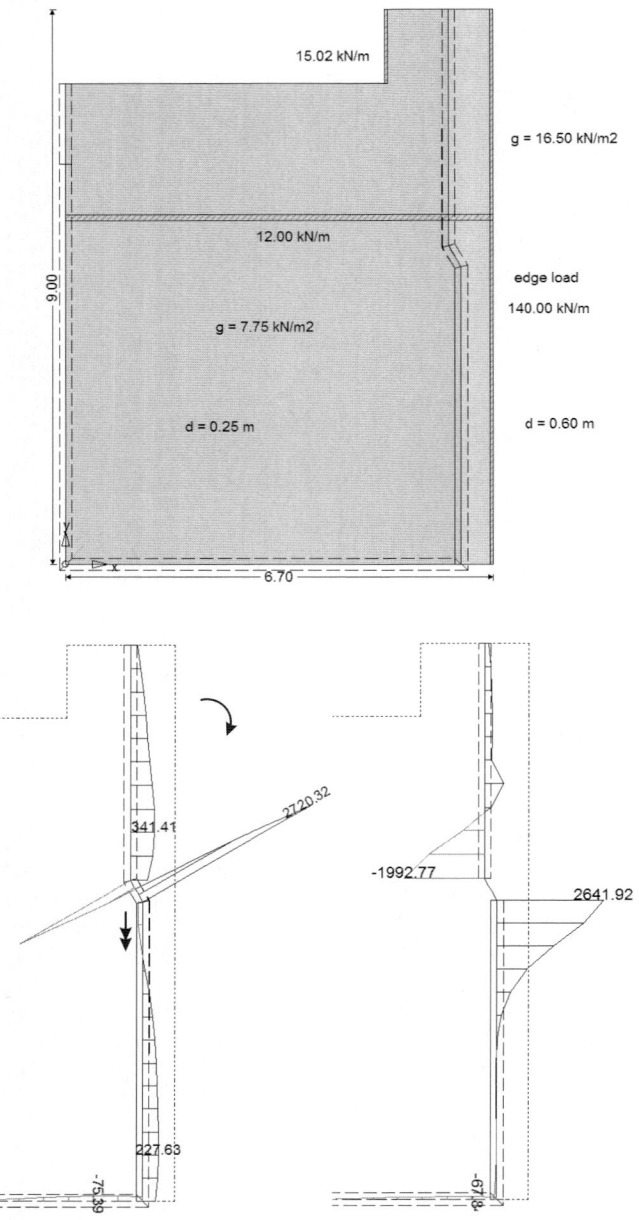

Fig. 5.40. The edge load rotates the slab downward but the slab is stabilized by the torque built up by the support reactions: **a)** slab and loading **b)** support reactions of the continuous support **c)** support reactions with an intermission in the supporting wall

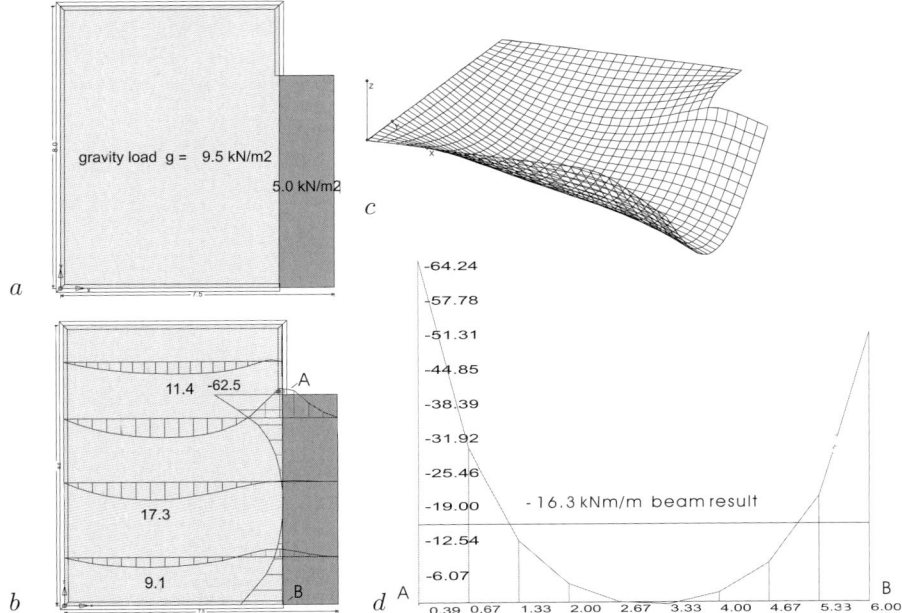

Fig. 5.41. Slab with an attached balcony load case $g + p$ (balcony): **a)** system, **b)** bending moments m_{xx} in various sections, **c)** 3-D view of the deflection surface, **d)** bending moment m_{xx} in the transition zone between slab and balcony, slab solution and beam solution

5.13 Beam models

One must be careful when FE results are checked with beam models. The gravity load of the slab in Fig. 5.41 is $g = 9.5$ kN/m^2 and the live load on the balcony (overhang) $p = 5$ kN/m^2. According to beam theory, the bending moment in the balcony should be $m_{xx} = -(g+p)\,l^2/2 = -14.5 \cdot 1.5^2/2 = -16.3$ kN m/m, while the FE result is $m_{xx} \simeq 0.0$ kN m/m at the middle of the transition zone between the slab and the balcony.

A 3-D view of the deflection surface (see Fig. 5.41 c) explains why it makes no sense to compare the two bending moments. The two solutions happen to agree only at the quarter points; see Fig. 5.41 d.

At the upper and lower end of the transition zone, the bending moments m_{xx} of the FE solution considerably exceed the beam results. This makes sense because the integral of the bending moments in the vertical direction must be identical to the bending moment in a cantilever beam with the same vertical extension b,

$$M = \int_0^b m_{xx}\,dy = \frac{(g+p)\,b\,l^2}{2}\,, \qquad (5.83)$$

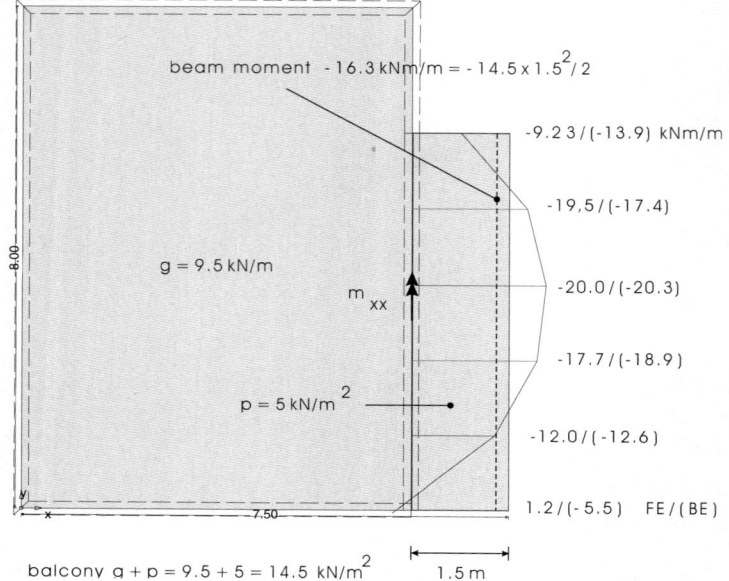

Fig. 5.42. Bending moments in the transition zone between the slab and the balcony
when the balcony is supported in the transition zone by a rigid wall

and these two sides will only match if the bending moment m_{xx} of the slab
increases towards the end points.

If a rigid support is placed under the balcony in the transition zone (see
Fig. 5.42), the bending moments show the opposite tendency: they increase
towards the middle and decrease towards the end points!

5.14 Wheel loads

Let us assume that the contact area of a tire is 40×20 cm and that the load
it carries is 50 kN. If the slab thickness is 60 cm then at the midsurface the
load carrying area has widened to 100×80 cm; see Fig. 5.43.

Figure 5.44 gives an impression of how an FE program resolves six such
wheel loads representing a heavy truck traversing a hinged rectangular slab
into a series of line loads that push and pull at the slab; see Fig. 5.44. The
bending moments (see Fig. 5.45) instead give the impression that the slab
carries the original wheel loads.

The elements are conforming square elements based on the shape functions
in (5.17). The element size is 50 cm which means—if the load distribution in
the lesser direction of the tire is neglected—that the equivalent nodal force at
the node directly under the tire is 25 kN and that the two neighboring nodes
carry a load of 12.5 kN.

50 kN

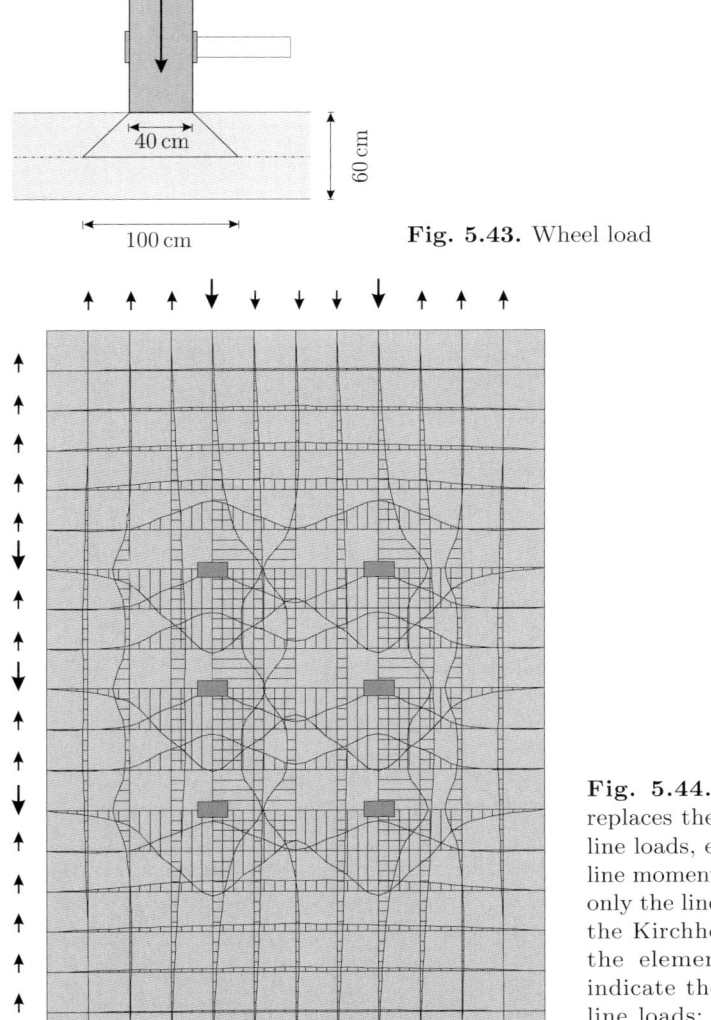

40 cm

60 cm

100 cm

Fig. 5.43. Wheel load

Fig. 5.44. An FE program replaces the wheel loads with line loads, element loads, and line moments. Shown here are only the line loads = jumps in the Kirchhoff shear between the elements. The arrows indicate the direction of the line loads: ↓ = compressive forces, ↑ = tensile forces

5.15 Circular slabs

As noted earlier, zero deflection ($w = 0$) at the corner point of a hinged slab implies that the gradient of the deflection surface is also zero, $\nabla w = \mathbf{0}$. Hence, if the edge of a hinged circular slab is approximated by an n-sided polygon, then at $n + 1$ points on the boundary the slab is no longer able to rotate, $w,_x = w,_y = 0$ (see Fig. 5.46). The strange thing is that the first

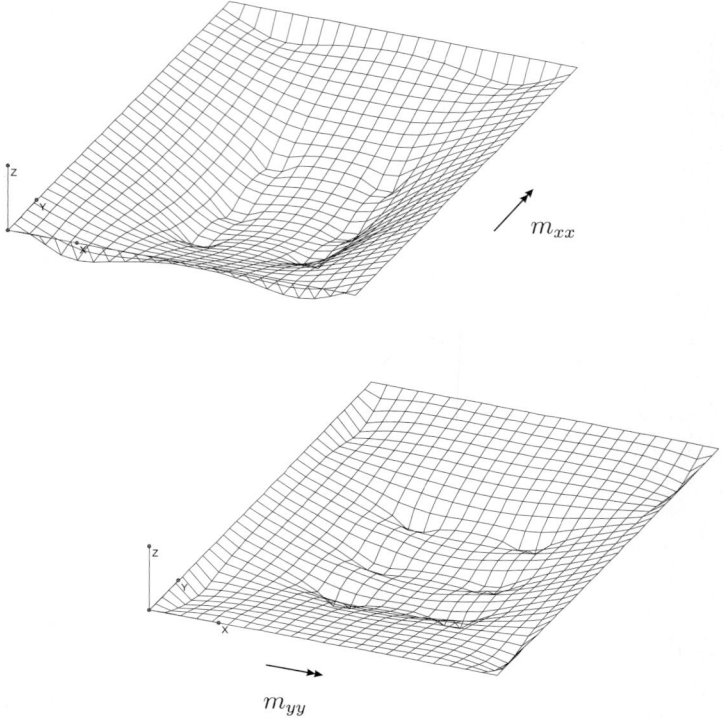

Fig. 5.45. The bending moments generated by the truck

the subdivision becomes, the more the edge will appear to be clamped, and therefore the error in the approximate solution increases, Babuškas paradoxon [17].

The problem can be ameliorated by switching to a *soft support* $w = 0$, i.e., when the rotations are set free, $w_{,x} \neq 0, w_{,y} \neq 0$. Hence, a Reissner–Mindlin plate should have less trouble with such slabs, but see [9].

5.16 T beams

Hardly any problem attracts so much attention as the modeling of T beams; see Fig. 5.47. Basically the interaction between a T beam and a slab is a complex three-dimensional problem. But even though computers are becoming more and more powerful, true 3-D solutions are much too complicated and simply require too much effort. Engineers always tried to handle the complex situation with simplified models, by working with an effective beam width or other approximate models. The choice of the model depends heavily on the accuracy to be achieved. Therefore, a multitude of possible models exist, with differing degrees of accuracy:

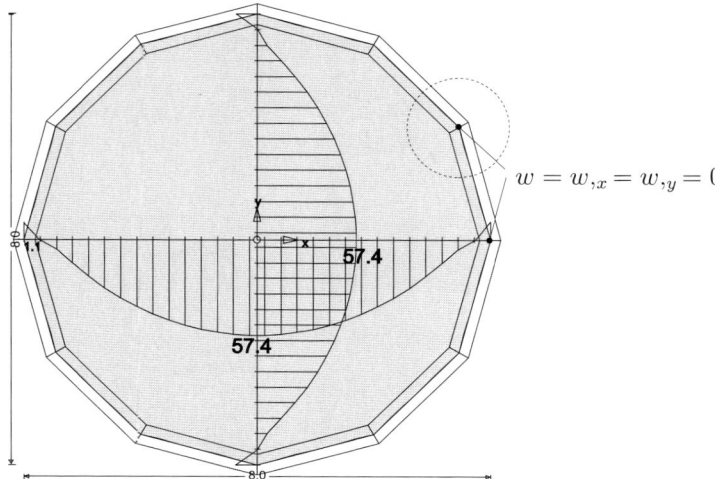

$$w = w_{,x} = w_{,y} = 0$$

Fig. 5.46. Hinged circular slab, bending moments (kNm/m)

- The slab and the beam are modeled as folded plates (also called the shell model).
- The slab is modeled as a folded plate, and the T beam as an eccentric beam or plate.
- The slab is modeled as a slab, and the T beam as an eccentric beam (with a normal force).
- The slab is modeled as a slab, and the axis of the T beam lies in the midsurface of the slab.

The fact that some engineers model T beams as rigid supports ($EI = \infty$) suggests just how wide the range of possible models is. This approach may be sufficient to assess the limit load of a beam, but it does not suffice to produce an affordable solution, or to accurately predict the displacements of the structure.

The first two models differ in how they treat the web. In the first model the web is made up of plate elements, i.e., the distribution of the normal stresses in the web is nonlinear. In the second model, the classical linear stress distribution of beam theory prevails.

Both models are capable of predicting the distribution of the normal stresses in the slab very accurately because the effective width is a result of the analysis. On the screen it can be seen how the effective width b_m increases in the span and how it shrinks near the supports.

In all other cases the models simulate the coupling between a separate T beam and a slab (see Fig. 5.48), where the slab is either treated as a folded plate (m_{ij}, q_i, n_{ij}) or "simply" as a slab (m_{ij}, q_i).

The coupling in the finite element sense means that the movements of the beam and the movements of the slab are synchronized at the nodes, and

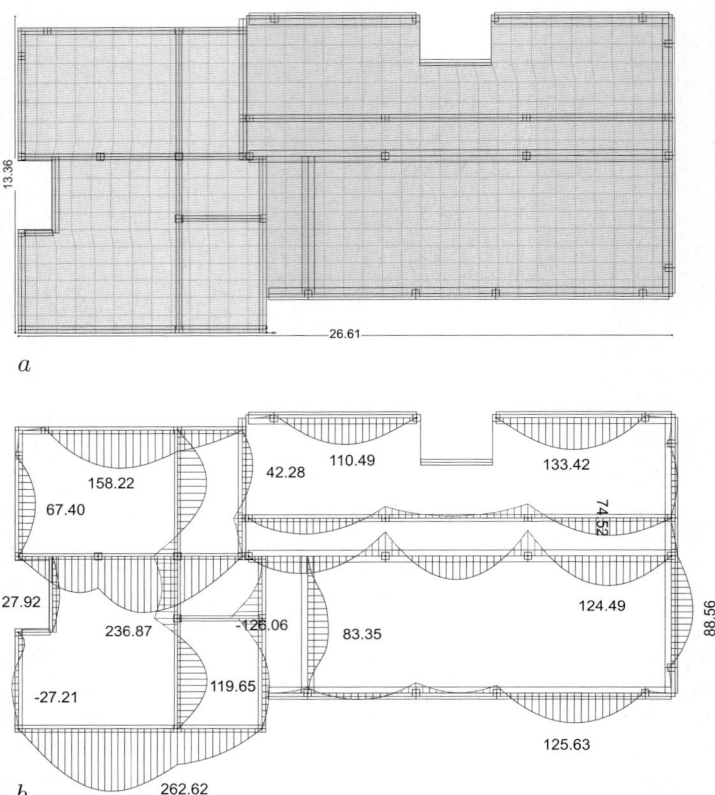

Fig. 5.47. Slab with T beams: **a)** system, **b)** bending moments in the T beams (kNm)

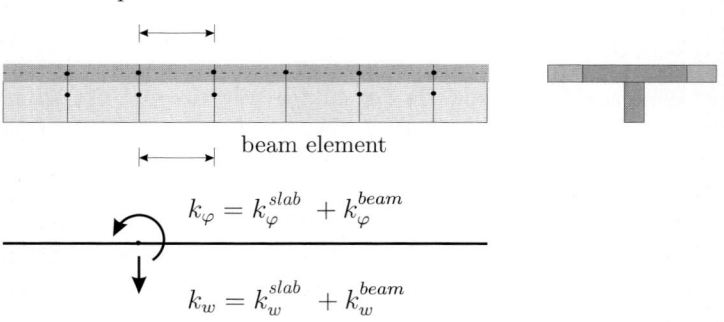

Fig. 5.48. The nodes of the slab lie atop the nodes of the T beam

that the work done on both sides of the joints is the same for any virtual displacement (energy coupling).

If u, w, and φ (= rotation) denote the corresponding degrees of freedom of the slab (S) and beam (B), then

$$w_B = w_S \qquad u_B = u_S + \varphi_S \, e \qquad \varphi_B = \varphi_S \,, \tag{5.84}$$

or if it is assumed that $EA = \infty$ in the slab, then the longitudinal displacements u_B are simply $u_B = \varphi_P \cdot e$, where e is the distance of the neutral axis of the beam from the midsurface of the slab. That is, the slab and the beam are imagined to be connected by a rigid bar of length e, so that rotations in the slab lead to longitudinal displacements in the beam.

Depending on whether the eccentricity e is considered or not, we speak of a centric or eccentric beam model. The models distinguish how the normal forces resulting from this eccentricity are introduced into the analysis.

As a result of the coupling of various structural elements, a whole series of incompatibilities and errors arise. Recall that at the interface of two structural elements, the stresses are no longer pointwise opposite (see Sect. 1.34, p. 177). Only the *virtual work* of the resultant stresses (plus the volume forces, etc.) is the same. Furthermore the beam and slab deflect differently, because the deflection curve w of the beam will in general not be the same as the deflection surface $w(x, y)$ of the slab. Often a Reissner–Mindlin plate that exhibits shear deformations is coupled to an Euler–Bernoulli beam that knows no such deformations. Any idea of a pointwise match between stresses on the two sides of the interface is misleading.

What is more, we will also see an error in the transfer of the shear stresses, because the part of the longitudinal displacements resulting from the eccentricity is normally of quadratic type, while the displacements due to the normal force are most often only linear. This error decreases quadratically, but it requires that the span be subdivided into several elements, and makes its presence felt by a stepwise distribution of the normal force.

Even if the stiffnesses are added only at the nodes, it is helpful to think in terms of the flexural rigidity of the whole system. The flexural rigidity of the slab increases if the corresponding terms of the beam elements are added:

$$k_w = b_m \cdot \frac{Eh^3}{12(1 - \mu^2)} + EI + EA \cdot e^2 \,. \tag{5.85}$$

This is also seen if the modified stiffness matrix of the beam is studied:

$$K = \begin{bmatrix} 12EI/l^3 & -6EI/l^2 & -12EI/l^3 & 6EI/l^2 \\ . & 4EI/l + EA/l \cdot e^2 & 6EI/l^2 & 2EI/l - EA/l \cdot e^2 \\ . & . & 12EI/l^3 & 6EI/l^2 \\ \text{sym.} & . & . & 4EI/l + EA/l \cdot e^2 \end{bmatrix} \tag{5.86}$$

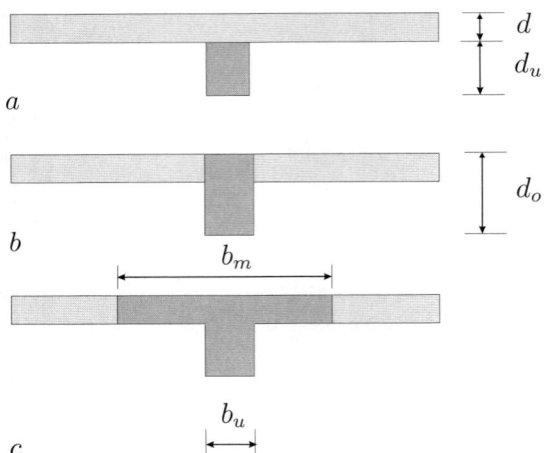

a

b

c

Fig. 5.49. Position of the T beam with respect to the slab

Next most important is—besides the choice of the eccentricity e—which moment of inertia I is attributed to the T beam. i.e., which cross-sectional parts we apportion to the T beam; see Fig. 5.49.

If the T beam reaches the upper surface of the slab (Fig. 5.49 b) then

$$A_B = b_U\, d_o \qquad e = \frac{d_o - d}{2} \qquad I_B = \frac{b_U\, d_0^3}{12} - \frac{b_U\, d^3}{12}. \tag{5.87}$$

If it is assumed that the stiffness of the T beam corresponds to the cross-sectional values, then

$$A_B = b_U\, d_U + b_m\, d \qquad e = \frac{b_U\, d_U}{b_U\, d_U + b_m\, d}\, \frac{d_0}{2} \tag{5.88}$$

$$I_B = \frac{b_U\, d_U^3}{12} - \frac{b_U\, d_U\, b_m\, d}{b_U\, d_U + b_m\, d}\, \frac{d_0^2}{4}. \tag{5.89}$$

If only the cross-sectional area of the web is attributed to the T beam (see Fig. 5.49 a) then

$$I_B = \frac{b_U\, d_U^3}{12} + b_U\, d_U\, \frac{d_0^2}{4}. \tag{5.90}$$

Before any discussion about what the right choice might be we should pause and recall how "virtual" the coupling is, because neither are the resultant stresses between the beam and the slab the same, nor are the displacements the same. In other words, the only chance we have is to think in terms of equivalent nodal forces. But then the effects that an error in the coupling has on the results is no longer so serious and studies have shown that the simplest strategies often yield the best results.

Therefore, not too much effort should be put into the modeling of the coupling of T beams and slabs. The trouble taken by the engineer is not

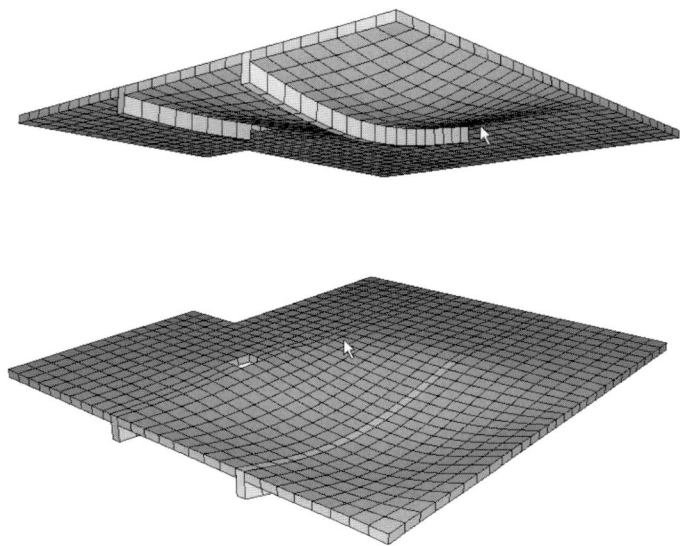

Fig. 5.50. Deformations of slab and T beam

honored by the FE program. Whether the contributions of the parallel axis theorem are considered or not, whether cross-sectional areas are considered twice, and whether the neutral axis runs in the midsurface or below the slab is only important insofar as the flexural rigidity EI of the T beam and thus its rotational and lateral flexibility is modeled more accurately; see Fig. 5.50.

The best strategy emerges if the sum of the individual stiffnesses corresponds to the actual stiffness. By choosing a certain effective width from which the joint stiffness of the combined system (slab + T beam) follows—if the slab stiffness is subtracted—and an eccentricity e is chosen, the stiffness of the T beam can be calculated. Normally the effective width has little influence on the results; a choice of $l_0/3$ is sufficient in most cases, [139].

In the BE method, a T beam is simply a beam (with the same bending stiffness as the actual T beam) that supports the slab, and as in the force method, the support reactions are determined such that the deflection of the slab is the same as that of the T beam.

A study of various FE models—$A =$ T beam as centric beam, $B =$ T beam as a centric beam, $b_m = \infty$, $C =$ T beam as eccentric beam, $D =$ T beam as rigid support—has shown that the results of the various models are not much different.

Model	M in the beam	m_{yy} (kN m/m)	m_{xx} (kN m/m)	f (mm)
FE A	481	4.7	-30.0	1.8
FE B	493	4.5	-31.2	1.5
FE C	490	4.3	-30.9	1.6
FE D	$-$	0.0	-36.4	0.0
BE	485	5.3	-31.4	1.7

The slab was a two-span slab with the T beam running down the middle of the plate in the vertical direction (y-direction). The values f are the maximum deflections of the T beam, M is the maximum bending moment in the T beam, and m_{yy} and m_{xx} are the bending moments in the slab.

If we consider how much uncertainty we must cope with in the design of reinforced concrete, the choice of the model does not seem that important. But in a commercial program, the limiting values of the design variables must also yield reasonable results, and in that respect a systematic disregard of certain effects, for example the axial displacements of the slab or the postulate that the design value of the stiffness corresponds to the actual stiffness of the coupled system, can lead to deviations in the results that cannot be neglected in the extreme cases.

Recommendations

The best results are obtained if the T beam is modeled as an eccentric beam—with an increase in the stiffness corresponding to the eccentricity e—and if axial displacements in the slab are taken into account, because then the effective width is a result of elasticity theory, i.e., it is automatically determined by the FE program. In most cases however, it will be sufficient to neglect axial displacements in the slab and work with an approximate effective width. In that case one should choose the equivalent moment of inertia $\tilde{I}$ in such a way that the sum of the flexural rigidities is equal to the flexural rigidity of the full T beam:

$$EI_{tot} = b_m \cdot \frac{E \cdot d^3}{12(1 - \nu^2)} + \frac{b_0 \cdot d_u^{\,3}}{12} + E \cdot b_m \cdot d \cdot e_p^2 + E \cdot b_0 \cdot d_u \cdot e_b^2 \,,$$

$$\tag{5.91}$$

$$EI_{tot} = b_m \cdot \frac{E \cdot d^3}{12(1 - \nu^2)} + E\tilde{I} \,. \tag{5.92}$$

Here e_p and e_b denote the distances of the slab midsurface and neutral axis of the beam from the center of gravity. This implies that $\tilde{I}$ is the moment of inertia of the total cross section of the T beam minus the stiffness of the slab itself.

In any case we recommend that every engineer test his model with regard to the limits of the design variables.

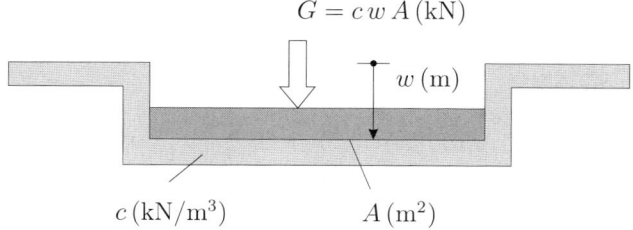

$$G = c\,w\,A\,(\mathrm{kN})$$

$$w\,(\mathrm{m})$$

$$c\,(\mathrm{kN/m^3}) \qquad A\,(\mathrm{m^2})$$

Fig. 5.51. Winkler model and gravity load

Resultant stresses

Once the nodal displacements are determined, the resultant stresses m_{ij}, q_i, n_{ij} in the slab and M_B, V_B, N_B in the beam, can be calculated. The next problem is that many programs calculate the resultant stresses separately for the slab and the beam, because no other strategy exists. If one tries to avoid this dilemma by modifying the stiffnesses, nothing is gained in the end; rather, the confusion increases. The result are designs in which the reinforcement lies in the compression zone of the slab, or where it is simply ignored, which leads either to unsafe or to wasteful designs.

A design can only be considered correct if the reinforcement is determined for the complete cross section, and if the fact that the neutral axis lies below the midsurface of the slab is taken into account:

$$M = M_{beam} + N_{beam} \cdot e_b + \int_{slab} (m_{yy} + n_{xx} \cdot e_p)\, dz\,, \qquad (5.93)$$

$$V = V_{beam} + \int_{slab} q_z\, dz\,. \qquad (5.94)$$

With these resultant stresses, the correct reinforcement for the T beam can be designed.

5.17 Foundation slabs

Winkler model

In the Winkler model it is assumed that the soil acts like a system of isolated springs that move independently and exert a force $c\,w$ on the underside of the slab. This leads to the differential equations

$$EIw^{IV} + c\,w = p \qquad \text{beam} \qquad (5.95)$$

$$K\,\Delta\Delta\,w + c\,w = p \qquad \text{slab}\,. \qquad (5.96)$$

The strain energy products associated with these two equations are

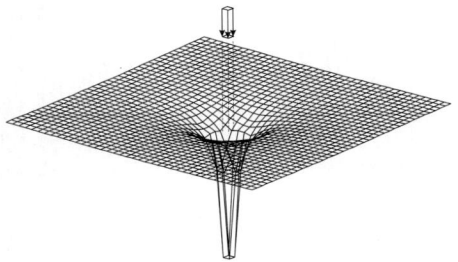

Fig. 5.52. Concentrated forces acting on the surface of the half-space

$$a(w, \hat{w}) = \int_0^l \frac{M \, \hat{M}}{EI} \, dx + c \int_0^l w \, \hat{w} \, dx \qquad (5.97)$$

$$a(w, \hat{w}) = \int_\Omega \boldsymbol{m} \cdot \hat{\boldsymbol{\kappa}} \, d\Omega + c \int_\Omega w \, \hat{w} \, d\Omega \,. \qquad (5.98)$$

Hence to the stiffness matrix $\boldsymbol{K}$ of a beam or plate element must only be added the mass matrix $\boldsymbol{M}$,

$$m_{ij} = \int_\Omega \varphi_i \, \varphi_j \, d\Omega \,, \qquad (5.99)$$

so that

$$(\boldsymbol{K} + c \, \boldsymbol{M}) \, \boldsymbol{u} = \boldsymbol{f} \,. \qquad (5.100)$$

A consequence of this simple spring model is that under gravity load G, the slab deflection $w = G/c$ is constant and the bending moments are zero; see Fig. 5.51. Because the individual springs move independently, the soil outside the loaded region simply retains its original level; see Fig. 5.51.

Inasmuch as this model does not properly replicate even the most basic aspects of actual soil behavior, it is no surprise that the modulus of subgrade reaction c is not a genuine physical quantity, but an artificial (albeit convenient) tool to simplify the design of foundation slabs. Theoretically the coefficient not only depends on the soil, but also on the extensions of the foundation slab. And in principle it is also not a constant, because otherwise it would not be possible to recover the rapidly increasing soil pressure p near the edge of a rigid punch from the formula $p(\boldsymbol{x}) = c(\boldsymbol{x}) \, w$; see Fig. 1.123 on p. 175.

Various strategies have been proposed to circumvent these defects. Mostly these techniques modify the modulus of subgrade reaction c iteratively and locally in such a way that the shape of the deformed slab and the soil are approximately the same; see Fig. 5.53.

Half-space model

In a linear elastic isotropic half-space, the displacement field $\boldsymbol{u}(\boldsymbol{x})$ is the solution of the system

$$L_{ij}\, u_j := -\,\mu\, u_{i,jj} - \frac{\mu}{1-2\,\nu}\, u_{j,ji} = p_i \qquad (5.101)$$

where the p_i are volume forces and μ and ν are elastic constants.

The displacement field due to a concentrated normal force P on the edge of the elastic half-space was found by Boussinesq [49]. The vertical displacement is (see Fig. 5.52)

$$u_B := u_3(\boldsymbol{x}, \boldsymbol{y}) = \frac{1+\nu}{2\pi\, E}\left[\frac{(x_3 - y_3)^2}{r^3} + 2\,\frac{1-\nu}{r}\right] P \qquad (5.102)$$

and the stresses are

$$\sigma_{ij} = -\,\frac{P}{2\,\pi\, r}\,\frac{r_{,i}\, r_{,j}\, r_{,2}}{r} \qquad r_{,i} = \frac{\partial r}{\partial y_i} = \frac{y_i - x_i}{r}\,. \qquad (5.103)$$

Note that the stresses σ_{ij} are independent of the modulus of elasticity E. How this solution can be extended to layered soils was explained in Sect. 2.2, p. 256.

Next let us consider the coupled problem of a foundation slab and the soil. On the free surface the tractions must be zero, $\boldsymbol{t} = \boldsymbol{0}$, while at the interface between the slab and the soil the deflection must be the same, and the vertical stresses must have opposite signs,

$$w = w_S \qquad t_3 - p_S = 0 \qquad (5.104)$$

where p_S is the soil pressure, which also appears on the right-hand side of the plate equation with a negative sign because it opposes the load p coming from the building:

$$K\,\Delta\,\Delta\,w = p - p_S\,. \qquad (5.105)$$

If the soil pressure $p(\boldsymbol{y})$ is expanded in terms of nodal shape functions $\psi_i(\boldsymbol{y})$ (hat functions)

$$p(\boldsymbol{y}) = \sum_i \psi_i(\boldsymbol{y})p_i\,, \qquad (5.106)$$

the soil deflection at a point $\boldsymbol{x}$ is

$$w_S(\boldsymbol{x}) = \sum_i \int_\Omega u_B(\boldsymbol{x}, \boldsymbol{y})\, \psi_i(\boldsymbol{y})\, d\Omega_{\boldsymbol{y}}\, p_i = \sum_i \eta_i(\boldsymbol{x})\, p_i \qquad (5.107)$$

and the stress in the soil is

$$\sigma_{zz}(\boldsymbol{x}) = \sum_i \int_\Omega \frac{3}{2\pi}\,\frac{(x_3 - y_3)^3}{r^3}\,\psi_i(\boldsymbol{y})\, d\Omega_{\boldsymbol{y}}\, p_i = \sum_i \theta_i(\boldsymbol{x})\, p_i\,. \qquad (5.108)$$

Hence the coupled problem leads to the system

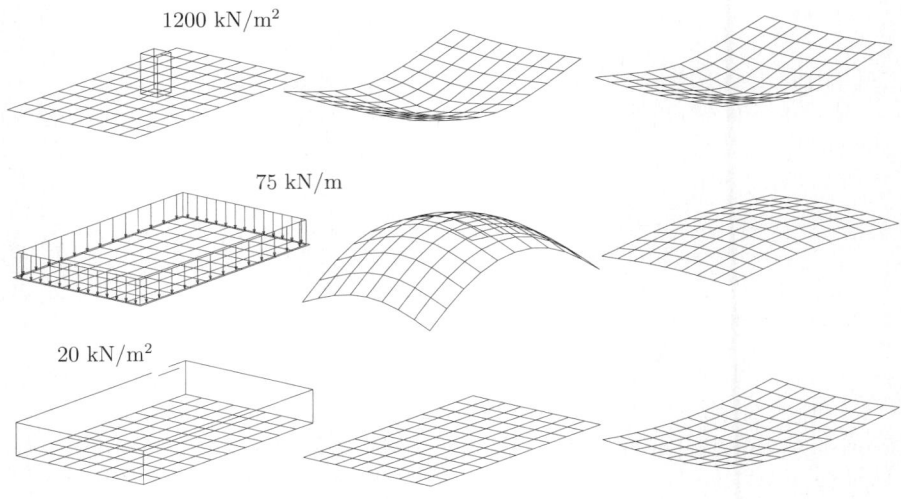

1200 kN/m²

75 kN/m

20 kN/m²

Winkler model   Half-space model

Fig. 5.53. Deformation patterns

$$\begin{bmatrix} \boldsymbol{K}_{(nn)} & \boldsymbol{L}_{(nm)} \\ \boldsymbol{I}^w_{(mn)} & -\boldsymbol{J}_{(mm)} \end{bmatrix} \begin{bmatrix} \boldsymbol{u}_{(n)} \\ \boldsymbol{p}_{(m)} \end{bmatrix} = \begin{bmatrix} \boldsymbol{f}_{(n)} \\ \boldsymbol{0}_{(m)} \end{bmatrix} \qquad \begin{array}{l} \text{FE with soil pressure} \\ w_{\text{slab}} - w_{\text{soil}} = 0 \end{array} \qquad (5.109)$$

where

$$k_{ij} = a(\varphi_i, \varphi_j) \qquad l_{ij} = \int_\Omega \psi_i \, \varphi_j \, d\Omega \qquad J_{ij} = \eta_j(\boldsymbol{x}_i). \qquad (5.110)$$

The matrix $\boldsymbol{I}^w$ results if the $n \times n$ unit matrix $\boldsymbol{I}$ is reduced to an $m \times m$ matrix by deleting certain rows and columns. Because only the deflections at the soil interface are equated, the rotational degrees of freedom in the vector $\boldsymbol{u}$ are meaningless. So that if the underlined rows

$$1, \underline{2}, \underline{3}, 4, \underline{5}, \underline{6}, 7, \underline{8}, \underline{9}, \dots \qquad (5.111)$$

in the unit matrix are deleted, the result is the matrix $\boldsymbol{I}^w$.

With the equivalent nodal forces $-\boldsymbol{L}\,\boldsymbol{p}$ of the soil pressure the extended set of equations becomes $\boldsymbol{K}\boldsymbol{u} = \boldsymbol{f} - \boldsymbol{L}\,\boldsymbol{p}$, or $\boldsymbol{K}\,\boldsymbol{u} + \boldsymbol{L}\,\boldsymbol{p} = \boldsymbol{f}$.

The Boussinesq solution is based on the linear theory of elasticity, so E is Young's modulus, while in soil mechanics a one-dimensional modulus E_s is used instead, which corresponds to consolidation or oedometer testing, and is therefore also called the *constrained modulus*. The two moduli are related via elasticity theory:

$$E = \frac{(1 + \nu)(1 - 2\,\nu)}{(1 - \nu)} E_s\,, \qquad (5.112)$$

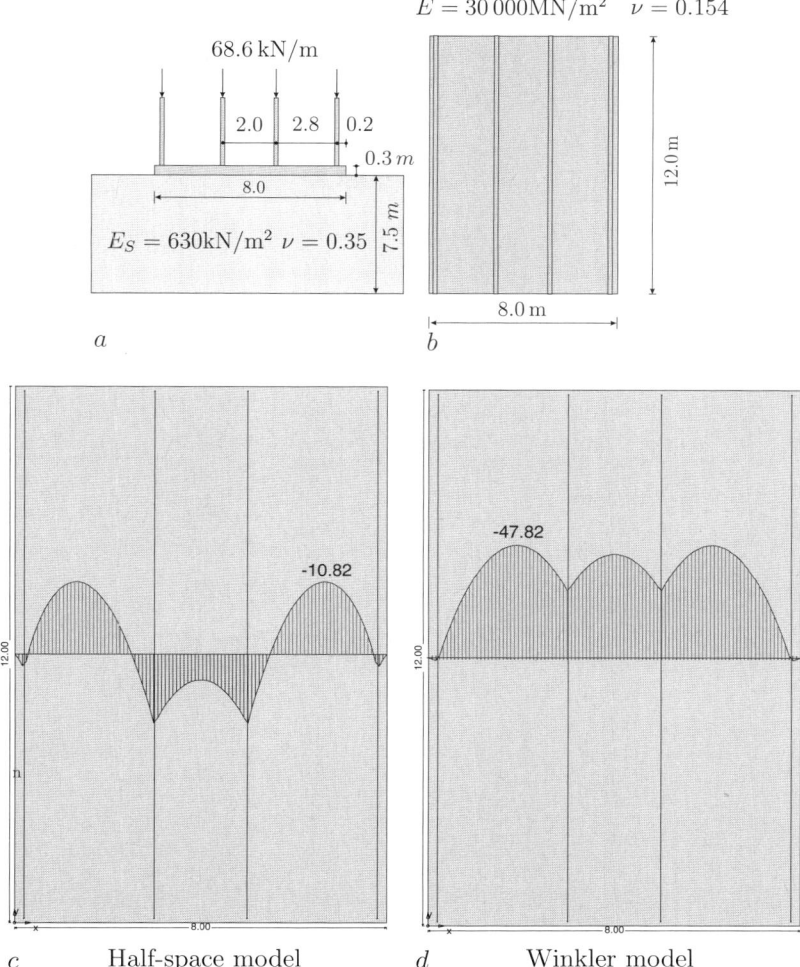

Fig. 5.54. Foundation slab: **a)** model; **b)** cross section; **c)** bending moments m_{xx} in the half-space model; **d)** Winkler model

where ν is the (drained) Poisson's ratio. For $\nu = 0$ this becomes $E = E_s$ and for $\nu = 0.2$ Young's modulus E is 90 percent of E_s so there is not much difference between the two.

Comparisons of the Winkler model and half-space model

Ultimately there is little agreement between results based on the Winkler model and the half-space model, as in the case of the slab in Fig. 5.54 or the slab in Fig. 5.55, which is stiffened by a ring of shear walls.

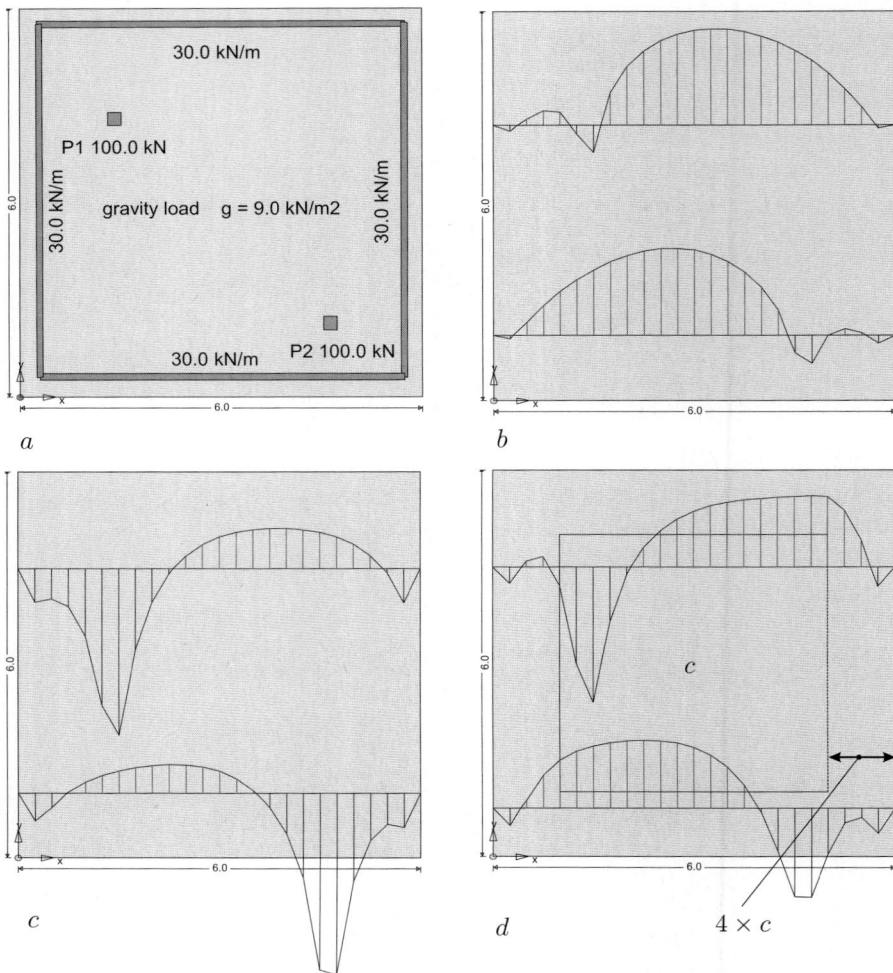

Fig. 5.55. Modified Winkler model: **a)** plan view and load; **b)** bending moments m_{xx} in two sections, constant modulus of subgrade reaction: **c)** half-space model; **d)** in a 1 m strip the modulus of subgrade reaction was raised by a factor of 4

In the half-space model the soil pressure increases towards the edge of the slab, while in the Winkler model the deflection increases near the edge; see Fig. 5.55 b. In a first analysis, the modulus of subgrade reaction was assumed to be constant, $c = 10\,000$ kN/m^3 (Fig. 5.55 b), and in a second analysis the modulus was increased by a factor of 4 in a 1 m strip along the edge. In this second improved model the agreement of the bending moments m_{xx} with the half-space model ($E_S = 50\,000$ kN/m^2) (Fig. 5.55 c and d) is much better.

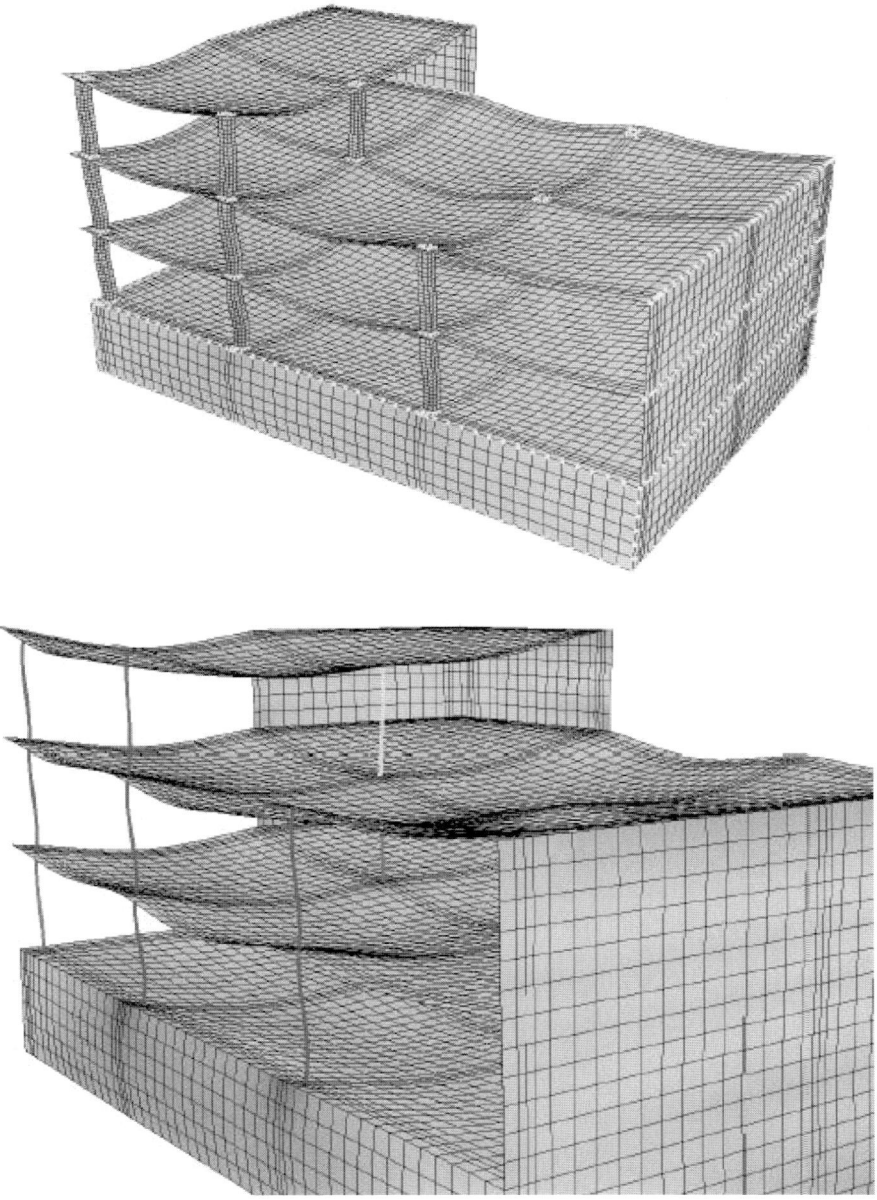

Fig. 5.56. 3-D model of a multi-story building on an elastic foundation: **a)** global view of the deformed model (the deformation are of course not to scale); **b)** partial view

3-D models

A 3-D analysis of a multi-story building itself is a complex undertaking. Even for an experienced structural engineer it is not easy to verify the results of a 3-D FE analysis with independent models. Too many effects combine—the modeling of the girders, the floor plates, shear walls, columns, etc. usually has a much larger influence on the accuracy than questions of mathematical rigor: 'If the mechanical model is false high-precision arithmetic will not save the analysis'.

Given the large deviations between the Winkler model and the half-space model it should be no surprise then that the choice of the foundation model has a great impact on practically all results in a 3-D model of a building, see Fig. 5.56. This shows clearly in numerical studies, [94]. Placing a sophisticated 3-D model of a building on a simple Winkler foundation cannot be recommended.

5.18 Direct design method

The direct design method is a simplified method for the design of concrete slabs that is applicable if the slab consists of (not too overlong) rectangular panels and meets other nominal requirements; see for example ACI 318R-95. Although the method is restricted to evenly distributed loads, it allows an easy check of FE calculations; see Fig. 5.57.

Other techniques are grid-framework methods and the elastic frame method. The simplest technique is the slab strip method, in which the slabs are assumed to carry the load in only one direction (one-way slabs).

The slab in Fig. 5.58 was analyzed a) with finite elements, b) the direct design method, and c) as a system of one-way slabs. The original design called for a slab with a thickness of 14 cm. In the "manual method" (one-way slabs), the T beams had to be analyzed separately. To meet the limitations of the direct design method the slab thickness was changed to 18 cm. In the following table the total weight of the reinforcement resulting from the different design procedures are listed.

Area = 171.53 m²	Upper kg	Lower kg	Total kg	kg/m²	Difference kg
FE h = 14 cm	378.1	678.0	1056.1	6.16	+20.9
FE h = 18 cm	350.1	664.7	1035.2	6.03	0
Direct design method	557.5	688.1	1245.6	7.26	+210.4
Uniaxial	599.9	662.2	1262.1	7.36	+226.9

m_{xx}/m_{yy}

11.9/5.5 direct design method
12.0/7.1 FE 1
11.7/7.1 FE 2
11.5/6.9 BE

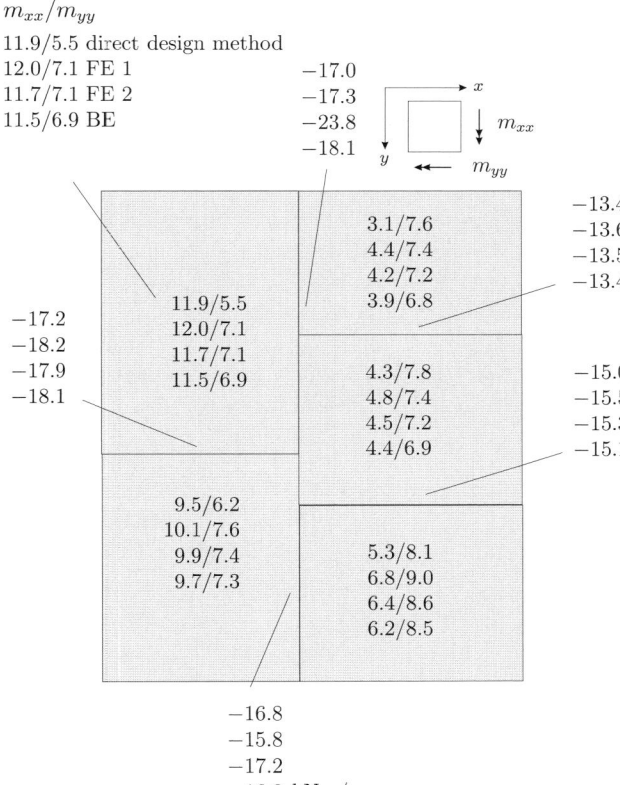

Fig. 5.57. Comparison between direct design method, finite elements and boundary elements

5.19 Point supports

Slabs on a regular grid of point supports are often modeled by subdividing the slab into *column strips* and *middle strips*, so that the bending moments at characteristic points

$$m_{ss} = \text{support moments in the column strip}$$
$$m_{sf} = \text{negative support moments in the middle strip}$$
$$m_{fg} = \text{midspan bending moment in the column strip}$$
$$m_{ff} = \text{midspan bending moment in the middle strip}$$

can be calculated with plate strip methods. A comparative study of the results of a plate strip method and an FE analysis (see Figure 5.59) shows good agreement between the two methods. Influence functions (see Fig. 5.60) can be a great help in understanding the structural behavior of such slabs.

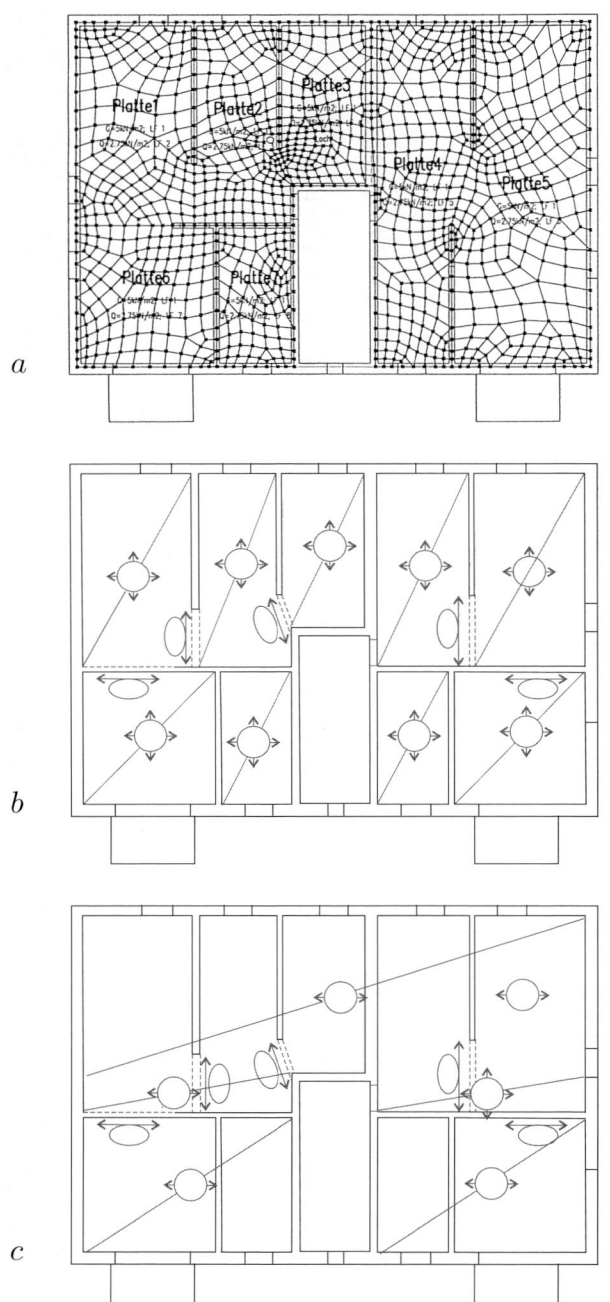

Fig. 5.58. Three design methods: **a)** finite elements, **b)** direct design method, and **c)** one-way slabs

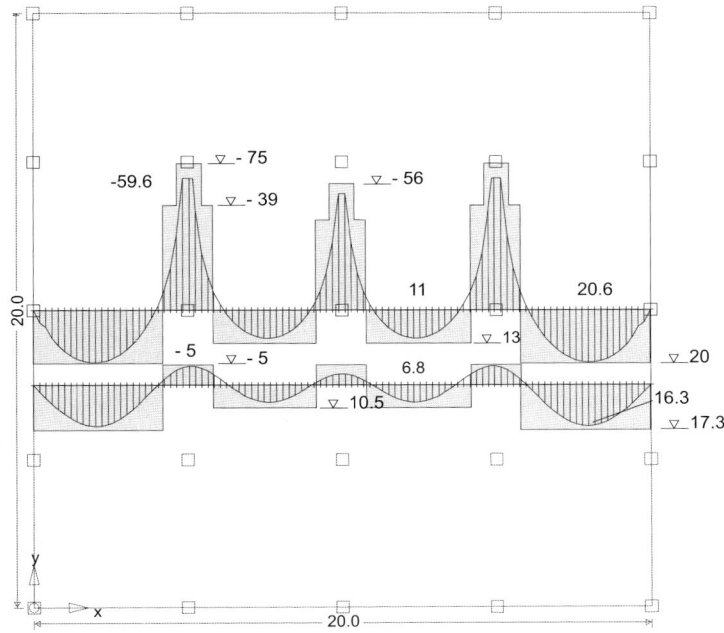

Fig. 5.59. Slab on point supports, gravity load, comparison with a plate strip method

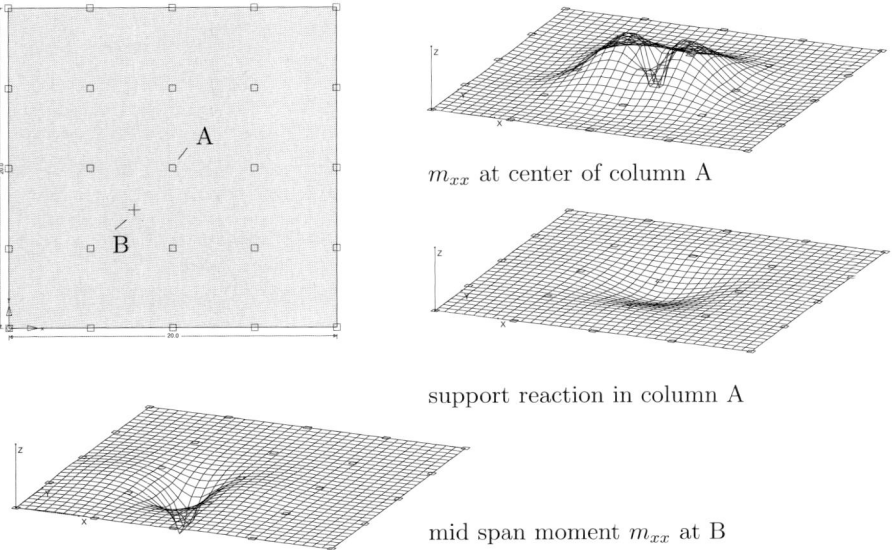

m_{xx} at center of column A

support reaction in column A

mid span moment m_{xx} at B

Fig. 5.60. Influence functions

5.20 Study

The slab in Fig. 5.61 a was analyzed with two different FE programs (SOFiSTiK and Nemetschek) and a BE program [24].

The agreement between the results is very good; see Figs. 5.62, 5.63, 5.64. Both FE programs used Reissner–Mindlin elements, although the elements themselves were not identical. The BE solution is based on Kirchhoff's plate theory. That the Kirchhoff results agree so well with the Reissner–Mindlin results demonstrates how small the difference is between the two plate theories, as long as slabs of normal thickness are considered. Reissner–Mindlin plate elements essentially behave like Kirchhoff plate elements if shear deformations are negligible. The quality of the shear forces is also surprisingly good; see Fig. 5.65.

5.21 Sensitivity analysis

The ideas which we developed in Section 1.27 can also be applied to slabs. If the stiffness k_s of a line support Γ—typically a wall—changes, $k_s \to k_s + \Delta k_s$, then the increment in the Dirac energy is the integral

$$
\begin{aligned}
- d(G_i^c, w) &= - \int_\Gamma \Delta k_s \, G_i^c(\boldsymbol{y}, \boldsymbol{x}) \, w(\boldsymbol{y}) \, ds_{\boldsymbol{y}} \\
&\simeq - \int_\Gamma \Delta k_s \, G_i(\boldsymbol{y}, \boldsymbol{x}) \, w(\boldsymbol{y}) \, ds_{\boldsymbol{y}}
\end{aligned}
\tag{5.113}
$$

or if we use a one-point quadrature rule

$$
- d(G_i^c, w) \simeq -\Delta k_s \cdot \frac{R_G}{k_s} \cdot \frac{R_p}{k_s} \cdot \frac{1}{l_\Gamma}
\tag{5.114}
$$

that is if we replace the distributed forces by their resultants, $R_G = (G_i, k_s)$ and $R_p = (w, k_s)$ along the wall $[0, l_\Gamma]$.

We applied this idea to the slab in Fig. 5.66 where each single value next to a wall signals by how much the bending moment $m_{yy}(\boldsymbol{x})$ at the mid point of the central slab will change if the stiffness of this wall drops by 50 %. The changes are very small probably because with $k_s \to 0.5\,k_s$ the load is transferred to the neighboring walls which—in this model—are assumed to retain their original stiffness.

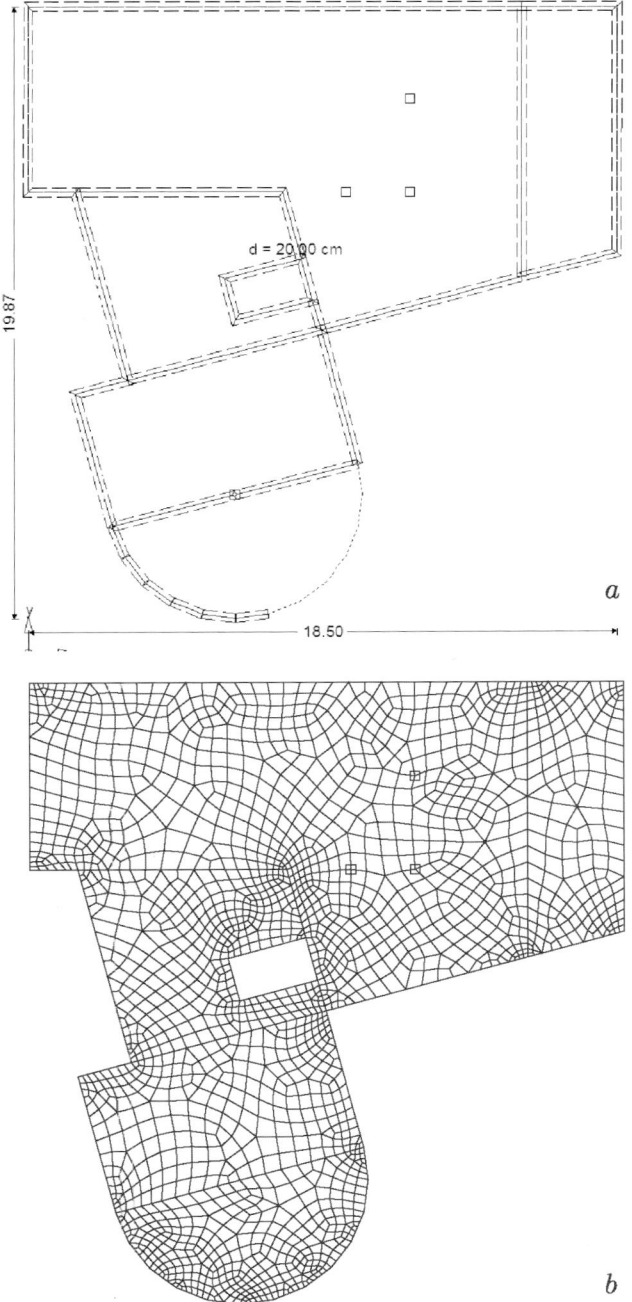

Fig. 5.61. Slab: **a)** system, **b)** FE SOFiSTiK mesh

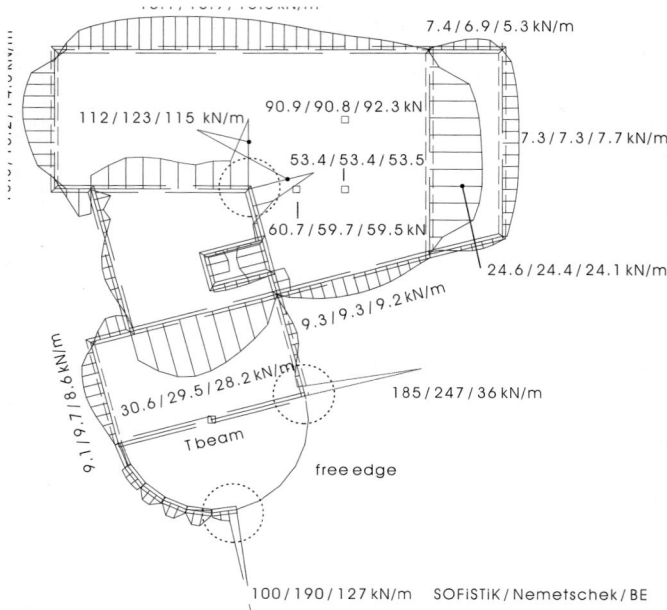

Fig. 5.62. Gravity load, support reactions

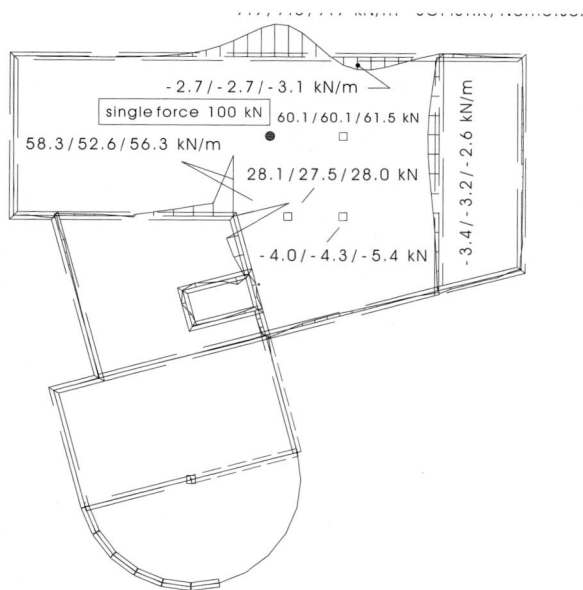

Fig. 5.63. Support reactions when a single force $P = 100$ kN is applied

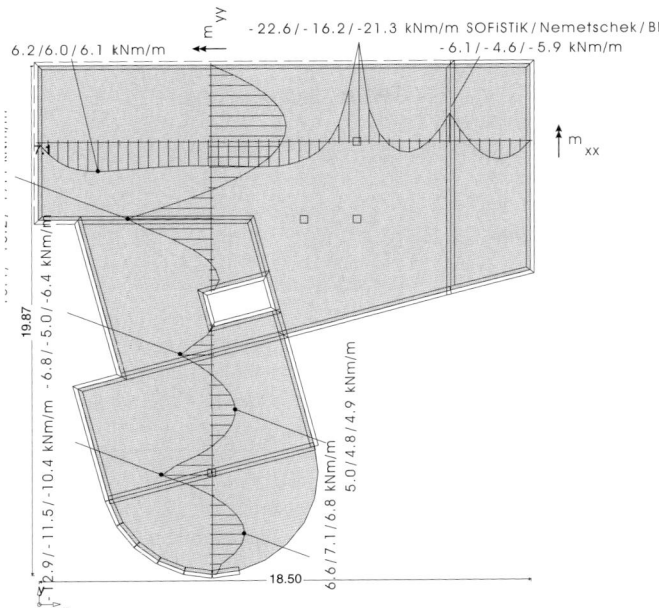

Fig. 5.64. Gravity load: bending moments m_{xx} and m_{yy} in two sections

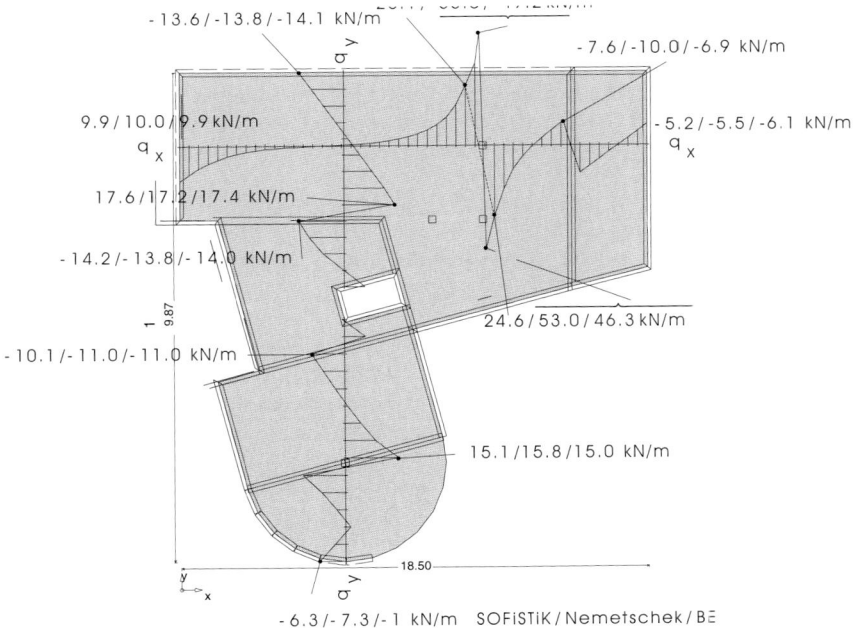

Fig. 5.65. Gravity load: shear forces q_x and q_y in two sections

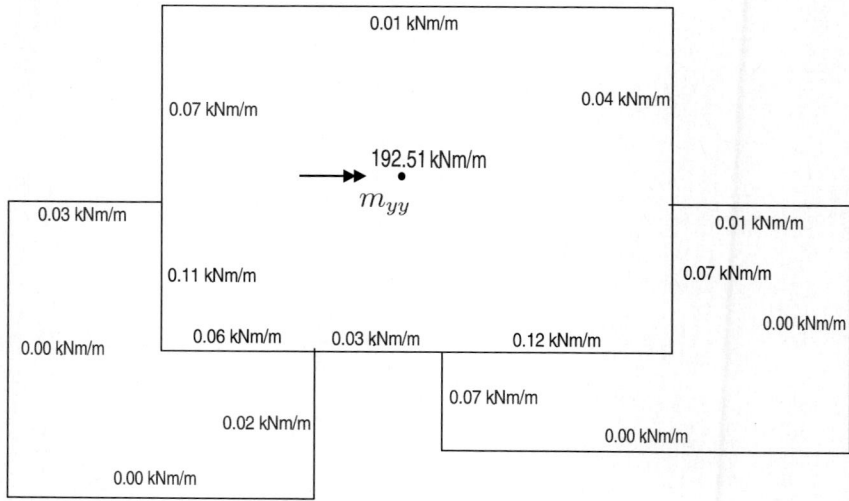

Fig. 5.66. Sensitivity analysis: the single numbers indicate how much m_{yy} will change if the stiffness of the walls change by 50 %

6. Shells

Shell elements are the most sophisticated elements because they must represent membrane and bending stresses equally well, and they must also model the coupling between these two stress states due to the curvature of the element; see Fig. 6.1. The topic is so complex that not all aspects of FE shell analysis can be discussed in this chapter. Instead we concentrate on the typical features.

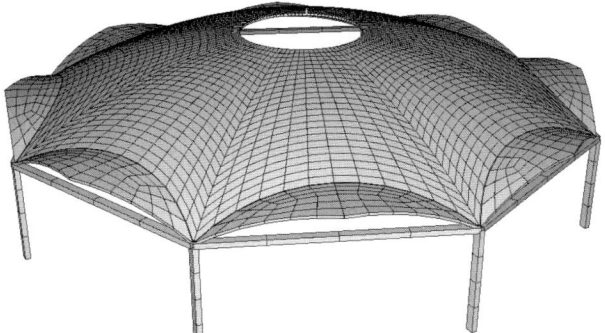

Fig. 6.1. Shell roof

6.1 Shell equations

The midsurface of the shell is represented by the position vector

$$\boldsymbol{x}(\theta_1, \theta_2) = [\boldsymbol{x}_1(\theta_1, \theta_2), x_2(\theta_1, \theta_2), x_3(\theta_1, \theta_2)]^T, \qquad (6.1)$$

which depends on the two parameters θ_1 and θ_2. If either of these is kept fixed, the position vector traces out parameter curves $\theta_i = c$ on the shell midsurface; see Fig. 6.2. The two tangent vectors

$$\boldsymbol{a}_1 = \frac{\partial \boldsymbol{x}}{\partial \theta_1}, \qquad \boldsymbol{a}_2 = \frac{\partial \boldsymbol{x}}{\partial \theta_2} \qquad (6.2)$$

and the associated normal vector

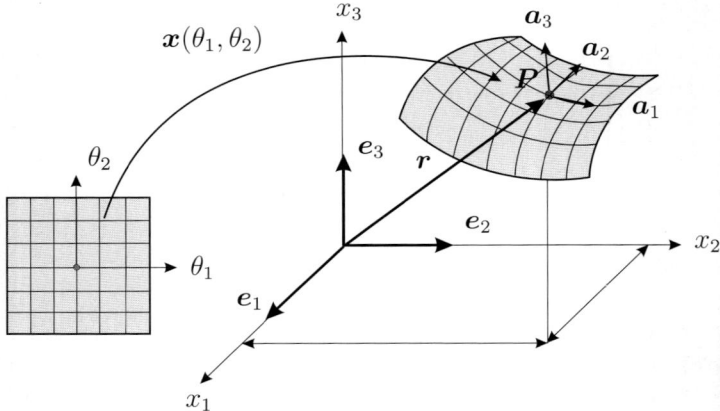

Fig. 6.2. Representation of the shell midsurface by a function $\boldsymbol{x}(\theta_1, \theta_2)$

$$\boldsymbol{a}_3 = \frac{\boldsymbol{a}_1 \times \boldsymbol{a}_2}{|\boldsymbol{a}_1 \times \boldsymbol{a}_2|} \qquad (6.3)$$

form the basis vectors of a system of curvilinear coordinates. The symmetric tensor

$$a_{ik} = \boldsymbol{a}_i \bullet \boldsymbol{a}_k \qquad \begin{bmatrix} a_{11} & a_{12} \\ a_{21} & a_{22} \end{bmatrix} = \begin{bmatrix} E & F \\ F & G \end{bmatrix} \qquad (6.4)$$

is called the *first fundamental form* of the surface, and the symmetric tensor

$$b_{\alpha\beta} = \frac{\partial \boldsymbol{a}_\alpha}{\partial \theta_\beta} \bullet \boldsymbol{a}_3 = \boldsymbol{a}_{\alpha,\beta} \bullet \boldsymbol{a}_3 \qquad (6.5)$$

is the *second fundamental form* of the surface (curvature tensor).

The basic property of a shell, as can be seen from the following equations, which are based on Koiters shell model,

$$-(\bar{n}^{\alpha\beta} - b_\lambda^\beta \bar{m}^{\lambda\alpha})|_\alpha + b_\alpha^\beta \bar{m}^{\lambda\alpha}|_\lambda = p^\beta \qquad \beta = 1, 2$$

$$-b_{\alpha\beta}(\bar{n}^{\alpha\beta} - b_\lambda^\beta \bar{m}^{\lambda\alpha}) - \bar{m}^{\alpha\beta}|_{\alpha\beta} = p^3, \qquad (6.6)$$

is that membrane and bending effects are coupled. Without the curvature terms $b_{\alpha\beta}$ and $b_\alpha^\beta = b_{\beta\rho} a^{\rho\alpha}$, the system would be decoupled. The displacements of the midsurface would be the solutions of a system of second-order differential equations, and the deflection w—in this model—would be the solution of the biharmonic equation $(\bar{m}^{\alpha\beta}|_{\alpha\beta} = p$ or $K \Delta\Delta w = p)$. Other shell models adopt the Reissner–Mindlin theory for the lateral deflection, but basically many aspects of 2-D elasticity theory can be carried over to shell theory—in particular, what was said about about concentrated forces, point supports, and infinite strain energy.

The strain energy product

$$a(\boldsymbol{u}, \hat{\boldsymbol{u}}) = \int_S \left[\bar{n}^{\alpha\beta}(\boldsymbol{u})\, \gamma_{\alpha\beta}(\hat{\boldsymbol{u}}) + \bar{m}^{\alpha\beta}(\boldsymbol{u})\, \rho_{\alpha\beta}(\hat{\boldsymbol{u}}) \right] ds \qquad (6.7)$$

contains strains

$$\gamma_{\alpha\beta} = \gamma_{\beta\alpha} = \frac{1}{2} \left(u_\alpha|_\beta + u_\beta|_\alpha \right) - b_{\alpha\beta}\, u_3 \qquad (6.8)$$

and curvature terms

$$\rho_{\alpha\beta} = \rho_{\beta\alpha} = - \left[u_3|_{\alpha\beta} - b_\alpha^\lambda\, b_{\lambda\beta}\, u_3 + b_\alpha^\lambda\, u_\lambda|_\beta + b_\beta^\lambda\, u_\lambda|_\alpha + b_\beta^\lambda|_\alpha\, u_\lambda \right] \quad (6.9)$$

that are multiplied by the conjugate resultant stresses

$$\bar{n}^{\alpha\beta} = t\, C^{\alpha\beta\lambda\delta}\, \gamma_{\lambda\delta} \qquad \bar{m}^{\alpha\beta} = \frac{t^3}{12}\, C^{\alpha\beta\lambda\delta}\, \rho_{\lambda\delta}\,, \qquad (6.10)$$

where t is the shell thickness. The elasticity tensor

$$C^{\alpha\beta\lambda\delta} = C^{\lambda\delta\alpha\beta} = \mu \left[a^{\alpha\lambda}\, a^{\beta\delta} + a^{\alpha\delta}\, a^{\beta\lambda} + \frac{2\,\nu}{1-\nu}\, a^{\alpha\beta}\, a^{\lambda\delta} \right] \qquad (6.11)$$

depends like the strain and curvature terms on the metric tensor $a^{ik} = \boldsymbol{a}^i \bullet \boldsymbol{a}^k$ of the shell midsurface.

What is different in shell theory is that in general the geometry of the shell midsurface must be approximated as well.

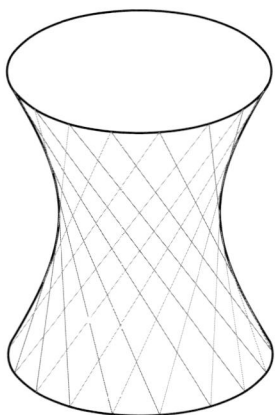

Fig. 6.3. The membrane stress state of a hyperboloid is governed by a system of hyperbolic differential equations

Membrane stresses

In a *membrane stress state* the load is carried solely by normal forces and plane shear forces:

$$n_{xx} = \int_{-t/2}^{t/2} \sigma_{xx} \, dz \qquad n_{yy} = \int_{-t/2}^{t/2} \sigma_{yy} \, dz \qquad n_{xy} = \int_{-t/2}^{t/2} \sigma_{xy} \, dz \, . \, (6.12)$$

Whereas in structural mechanics the differential equations are normally of elliptic type, the type of the equations that govern membrane stress states depend on the curvature of the shell. At each point of the shell midsurface, two orthogonal directions exist with respect to which the curvature $\kappa = 1/R$ attains its maximum (κ_1) and minimum (κ_2) value, respectively. The *Gaussian curvature*

$$K = \det b_\alpha^\beta = \frac{1}{\kappa_1 \, \kappa_2} \tag{6.13}$$

determines the type of the differential equations that relates the displacements to the load.

In cooling towers the Gaussian curvature is negative, $K < 0$ (see Fig. 6.3) and therefore the differential equations are of hyperbolic type. In cylindrical shells the Gaussian curvature is zero ($K = 0$), so that the equations are of parabolic type and only in a sphere where $K > 0$ are the equations of elliptic type; see Table 6.1. The problem is that St. Venant's principle holds only

Table 6.1. Gauss curvature and type of differential equations

Gauss curvature	type of equations	example
positive	elliptic	sphere
zero	parabolic	cylindrical shell
negative	hyperbolic	cooling tower

for systems of elliptic equations, i.e., in a cooling tower, local disturbances at the lower edge of the shell propagate along a generator straight up to the rim of the shell. (In reality, cooling towers are not pure hyperbolic shells.) In bending-dominated problems the situation is different, because the equations are of elliptic type [189] and St. Venant's principle applies.

6.2 Shells of revolution

In an axisymmetric shell and in a stress state with rotational symmetry only displacements normal to the meridian w, and in the tangential direction u will develop; see Fig. 6.4. Therefore a subdivision of the meridian (= the generator) into beam-like straight or curved elements suffices.

The relations between the arc length s on such an element and the characteristic quantities of a shell of revolution are (see Fig. 6.4)

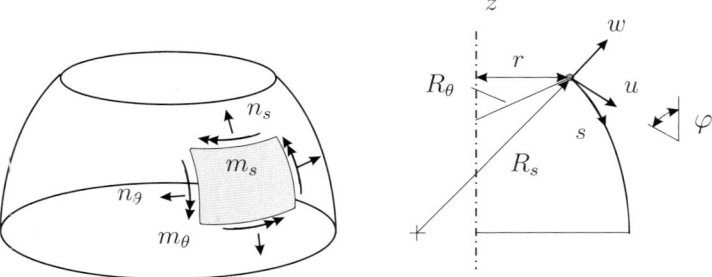

Fig. 6.4. Shell of revolution

$$R_\vartheta = \frac{r}{\cos\varphi}, \qquad R_s = -\frac{ds}{d\varphi}, \qquad \sin\varphi = \frac{dr}{ds}, \qquad \cos\varphi = -\frac{dz}{ds}. \; (6\;14)$$

Here R_ϑ and R_s are the principal radii of curvature. If straight elements are used, as in a frustum or cylindrical shell the radius of curvature in the plane of the meridian is infinite, $R_s = \infty$. The strains are

$$\varepsilon_s = \frac{du}{ds} + \frac{w}{R_s} \qquad\qquad \varepsilon_\vartheta = \frac{u\sin\varphi + w\cos\varphi}{r} \qquad\qquad (6.15)$$

$$\kappa_s = \frac{d}{ds}\left(\frac{u}{R}\right) - \frac{d^2 w}{ds^2} \qquad \kappa_\vartheta = \frac{\sin\varphi}{r}\left(\frac{u}{R_s} - \frac{dw}{ds}\right), \qquad (6.16)$$

where ε_s and ε_ϑ are the strains of the midsurface in the direction of a meridian (arc length s) or in the tangential direction (ϑ) and κ_s and κ_ϑ are the curvatures.

The strain energy product of an element is

$$a(u,u) = \varepsilon^T \int_0^l \begin{bmatrix} \boldsymbol{D}_M & \boldsymbol{0} \\ \boldsymbol{0} & \boldsymbol{D}_K \end{bmatrix} 2\,\pi\,r\,ds\,\varepsilon \qquad (6.17)$$

where with $C = E\,t/(1-\nu^2)$, $D = E\,t^3/(12(1-\nu^2))$,

$$\boldsymbol{D}_M = C\begin{bmatrix} 1 & \nu \\ \nu & 1 \end{bmatrix} \qquad \boldsymbol{D}_K = D\begin{bmatrix} 1 & \nu \\ \nu & 1 \end{bmatrix} \qquad \varepsilon = \begin{bmatrix} \varepsilon_s \\ \varepsilon_\vartheta \\ \kappa_s \\ \kappa_\vartheta \end{bmatrix}, \qquad (6.18)$$

and the resultant stresses are

$$\begin{bmatrix} n_s \\ n_\vartheta \end{bmatrix} = \frac{E\,t}{1-\nu^2}\begin{bmatrix} 1 & \nu \\ \nu & 1 \end{bmatrix}\begin{bmatrix} \varepsilon_s \\ \varepsilon_\vartheta \end{bmatrix} \qquad (6.19)$$

$$\begin{bmatrix} m_s \\ m_\vartheta \end{bmatrix} = \frac{E\,t^3}{12\,(1-\nu^2)}\begin{bmatrix} 1 & \nu \\ \nu & 1 \end{bmatrix}\begin{bmatrix} \kappa_s \\ \kappa_\vartheta \end{bmatrix}. \qquad (6.20)$$

In the sense of isoparametric elements, the element is interpreted as the C^1 map of a *master element* $-1 \le \xi \le +1$ on which four cubic shape functions corresponding to the two nodes $\xi_1 = -1, \xi_2 = +1$ and $\xi_0 = \xi_i\,\xi$, are defined:

$$\varphi_i^{(1)}(\xi) = \frac{1}{4}(\xi_0\,\xi^2 - 3\,\xi_0 + 2) \qquad \varphi_i^{(2)}(\xi) = \frac{1}{4}(1 - \xi_0)^2\,(1 + \xi_0)\,. \quad (6.21)$$

These make it possible to interpolate the shape of the element, i.e., the functions r and z,

$$r(\xi) = \sum_{i=1}^{2}\left[\varphi_i^{(1)}(\xi)\,r(\xi_i) + \varphi_i^{(2)}(\xi)\frac{dr}{d\xi}(\xi_i)\right] \qquad (6.22)$$

(as well as $z(\xi)$), and the displacements u and w of the meridian in a C^1 continuous fashion. The degrees of freedom at the element nodes are the displacements and the first-order derivative with respect to the arc length s

$$\boldsymbol{u}_e = [u_i, w_i, u_i', w_i']^T\,. \qquad (6.23)$$

The C^1 continuity of the displacement u is unusual, and must be dropped at nodes where the thickness of the shell changes, because then the strains ε_s are discontinuous [258].

6.3 Volume elements and degenerate shell elements

If shells are approximated by volume elements, the number of degrees of freedom easily becomes very large, and the large differences in the membrane and bending stiffnesses make the element sensitive to rounding errors.

A better strategy is to design special degenerate shell elements (see Fig. 6.5) by modifying volume elements. Because these shell elements inherit their properties from 3-D elements, they are of Reissner–Mindlin type, and are also called Mindlin shell elements.

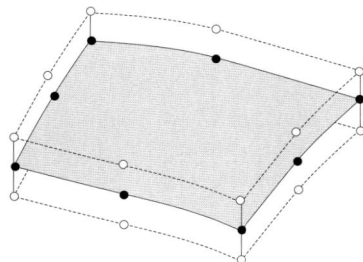

Fig. 6.5. Degenerate shell element, reduction of a volume element with 20 nodes to a shell element with 8 nodes

The reduction is essentially done by mapping all terms to the shell midsurface while maintaining contact with points outside by means of a vector $\boldsymbol{v}_3 \simeq \boldsymbol{n}$:

$$\boldsymbol{x}(\xi, \eta, \zeta) = \sum_i \boldsymbol{x}_i\,\varphi_i(\xi, \eta) + \sum_i \varphi_i(\xi, \eta)\,\frac{\zeta}{2}\,\boldsymbol{v}_{3i}\,. \qquad (6.24)$$

The first sum is an expansion in terms of the intrinsic shell coordinates ξ, η of the nodes, and the second sum is the part that extends beyond the midsurface.

In the same sense the displacement field of the shell is developed by starting at the midsurface ($\zeta = 0$)

$$\boldsymbol{u}(\xi, \eta) = \sum_i \boldsymbol{u}_i \, \varphi_i(\xi, \eta) + \sum_i \varphi_i(\xi, \eta) \, \frac{\zeta \, t_i}{2} \, [v_{1i} \, \alpha_i - v_{2i} \, \beta_i] \,, \qquad (6.25)$$

and letting the second part translate the rotations α_i and β_i (axes $\boldsymbol{v}_{1i}$ and $\boldsymbol{v}_{2i}$ in the tangential plane) into displacements at levels $\zeta \, t_i/2$ above the midsurface.

Next one can derive a stiffness matrix for a shell element by letting $\sigma_{33} = 0$:

$$\boldsymbol{K}^e = \int_{-1}^{+1} \int_{-1}^{+1} \int_{-1}^{+1} \boldsymbol{B}^T \, \boldsymbol{E} \, \boldsymbol{B} \, \det \boldsymbol{J} \, d\xi \, d\eta \, d\zeta \,. \qquad (6.26)$$

Here too one must be careful, because as $t \to 0$ shear-locking might set in, and if the element is curved, then so might membrane locking. But there is a whole catalog of countermeasures with which to improve the situation [26].

6.4 Circular arches

The problem of shear locking in shell elements is best explained by studying the modeling of arches with finite elements.

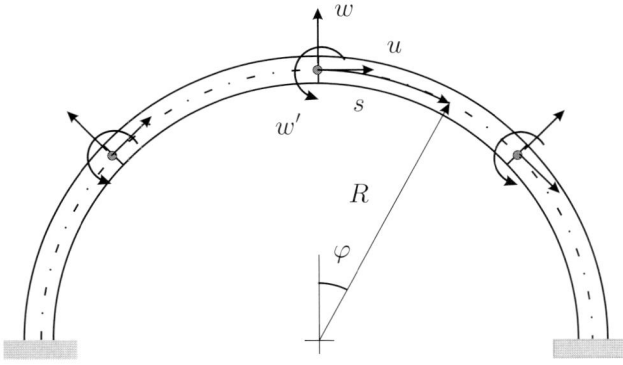

Fig. 6.6. Circular arch

The displacement of a point on the neutral axis can be split into a tangential movement u and a movement w orthogonal to the axis; see Fig. 6.6. To first order, the strains in a fiber at a distance z from the axis are [70]

$$\varepsilon_s = \varepsilon_m + z \, \kappa \,, \qquad \text{where} \qquad \varepsilon_m = u_{,s} + \frac{w}{R} \qquad \kappa = \frac{u_{,s}}{R} - w_{,ss} \,. \quad (6.27)$$

Taking the integral of the strain energy density $E\,\varepsilon_s^2$ across the arch thickness t, the strain energy product becomes

$$a(\boldsymbol{u},\boldsymbol{u}) = \int_0^l EA\,\varepsilon_m^2\,ds + \int_0^l EI\,\kappa^2\,ds \qquad \boldsymbol{u} = \{u,w\}\,, \qquad (6.28)$$

where E is the modulus of elasticity, $A = b\,t$ is the cross-sectional area and $I = b\,t^3/12$ is the moment of inertia of the cross section of the arch.

In the case of rigid-body motion, the strains and curvatures are zero, $\varepsilon_m = \kappa = 0$, so that

$$u = b_1\,\cos\varphi + b_2\,\cos\sin\varphi + b_3\,, \quad w = b_1\,\sin\varphi - b_2\,\cos\varphi\,, \quad \varphi = \frac{s}{R}\,.$$
$$(6.29)$$

The constants b_1 and b_2 represent displacements in two orthogonal directions and b_3 is a rotation about the center of the circular arch. Upon such a rotation of the arch $w = 0$, and all points move in a tangential direction, $u = b_3$.

In a thin arch the strain ε_m of the neutral axis is essentially zero, and all variations in u are mainly attributable to the deflection w:

$$\varepsilon_m = 0 \qquad \rightarrow \qquad u,_s + \frac{w}{R} = 0\,. \qquad (6.30)$$

Employing linear polynomials for u, and cubic for w,

$$u = a_0 + a_1\,s \qquad w = b_0 + b_1\,s + b_2\,s^2 + b_3\,s^3\,, \qquad (6.31)$$

we have

$$\varepsilon_m = (a_1 + \frac{b_0}{R}) + \frac{b_1}{R}\,s + \frac{b_2}{R}\,s^2 + \frac{b_3}{R}\,s^3\,. \qquad (6.32)$$

Next, if the thickness t of the arch tends to zero, then so must the strains in the neutral axis of the arch, $\varepsilon_m = 0$,

$$a_1 + \frac{b_0}{R} = b_1 = b_2 = b_3 = 0\,, \qquad (6.33)$$

implying that the flexibility of the arch must tend to zero, because only the term $w = b_0$ is left to model the deflections. All derivatives of this deflection curve are zero ($w,_s = w,_{ss} = w,_{sss} = 0$). As $t \to 0$ the element stiffens. This is the so-called *membrane locking*. The term EA tends to dominate the term EI in the strain energy product and any attempt to achieve $\varepsilon_m = 0$ by increasing the ratio $EA/EI \to \infty$ ensures that the deflection becomes much too small.

These effects can be minimized by *reduced integration*. The curvature terms in (6.28) are integrated with a two-point formula, but for the membrane part only a one-point formula is used, which means that the integrand is only evaluated at the center point, $s = 0$. At this point

$$a_1 + \frac{b_0}{R} = 0\,, \qquad (6.34)$$

and therefore only one degree of freedom must be sacrificed to comply with the constraint.

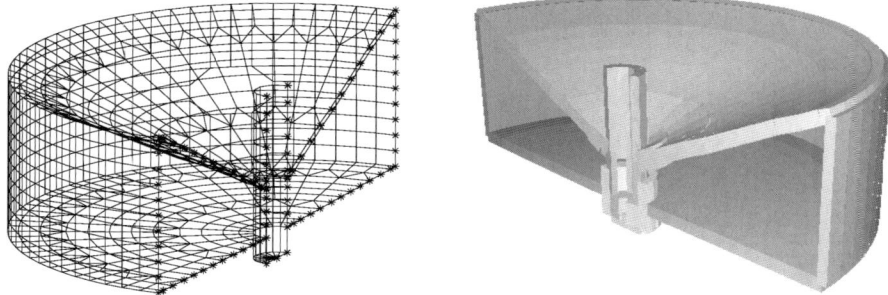

Fig. 6.7. Analysis of a water tank with flat plate elements

6.5 Flat elements

Flat elements are plate elements in which membrane and bending action are both included. The coupling between these two actions occurs at the nodes, and is due to the varying orientation of the elements.

Most shells are probably analyzed with such flat plate elements (see Fig. 6.7, 6.12 and 6.13), because the modeling is easy and the accuracy in most cases is sufficient. It is guaranteed that flat elements can represent rigid body motions and because the membrane and bending stresses within an element are decoupled it easy to understand and control the behavior of such elements.

The first idea is to use triangular elements; see Fig. 6.8. If each node of the triangle has three degrees of displacement and three degrees of rotation, then such an element has 18 degrees of freedom,

$$ \boldsymbol{u} = [\boldsymbol{u}_i, \boldsymbol{v}_i, \boldsymbol{\vartheta}_{zi}, \boldsymbol{w}_i, \boldsymbol{\vartheta}_{xi}, \boldsymbol{\vartheta}_{yi}]^T \,, \qquad (6.35) $$

where the individual vectors $\boldsymbol{u}_i = [u_1, u_2, u_3]^T$ and $\boldsymbol{\vartheta}_{zi} = [\vartheta_{z1}, \vartheta_{z2}, \vartheta_{z3}]^T$, etc. are respectively the displacements and rotations of the nodes.

Let the matrix $\boldsymbol{K}^M$ be the associated 9×9 stiffness matrix accounting for the membrane stresses of the element. For simplicity, a DKT element is chosen for the bending stresses. If $\boldsymbol{K}^B$ denotes the associated stiffness matrix, then membrane and bending stresses are indeed decoupled:

$$ \boldsymbol{K}^e \, \boldsymbol{u} = \begin{bmatrix} \boldsymbol{K}^M_{(9 \times 9)} & \boldsymbol{0}_{(9 \times 9)} \\ \boldsymbol{0}_{(9 \times 9)} & \boldsymbol{K}^B_{(9 \times 9)} \end{bmatrix} \begin{bmatrix} \boldsymbol{u}_i \\ \boldsymbol{v}_i \\ \boldsymbol{\vartheta}_i \\ \boldsymbol{w}_i \\ \boldsymbol{\vartheta}_{xi} \\ \boldsymbol{\vartheta}_{yi} \end{bmatrix} = \boldsymbol{f} \,. \qquad (6.36) $$

Only if the midsurfaces of the neighboring elements do not lie in the same plane will the two stress states (in general) become coupled.

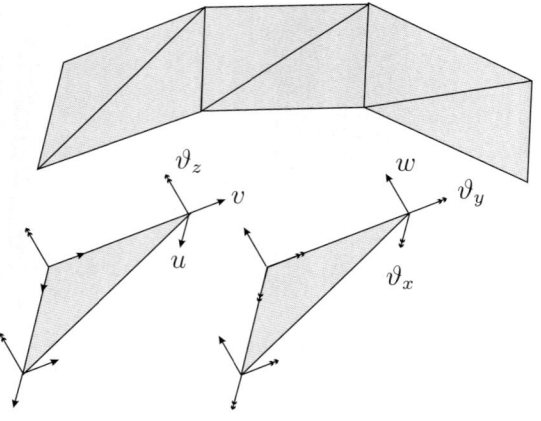

Fig. 6.8. Flat elements

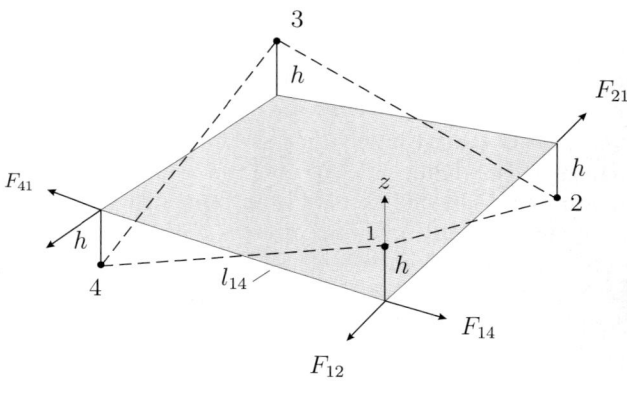

Fig. 6.9. An originally flat element where the four nodes no longer lie in a plane [160]

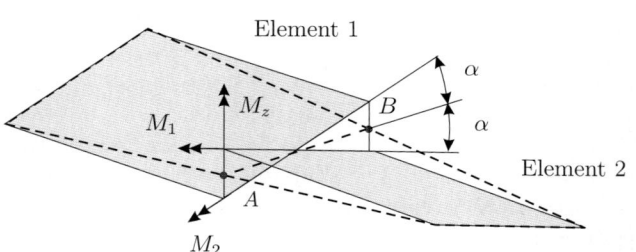

Fig. 6.10. Twisted beam problem [160]

If the *CST element* is used for membrane stresses, there are no rotational degrees of freedom about the vertical axes,

$$
\boldsymbol{K}^M \boldsymbol{u} = \left[\begin{array}{cc} \boldsymbol{K}^{CST}_{(6\times6)} & \boldsymbol{0}_{(6\times3)} \\ \boldsymbol{0}_{(3\times6)} & \boldsymbol{0}_{(3\times3)} \end{array} \right] \left[\begin{array}{c} \boldsymbol{u}_i \\ \boldsymbol{v}_i \\ \boldsymbol{\vartheta}_{zi} \end{array} \right] , \tag{6.37}
$$

which means that a "flat" node causes the global stiffness matrix to become singular. To avoid such unconstrained modes, artificial rotational degrees of freedom are built in:

$$\alpha\,E\,V \begin{bmatrix} 1.0 & -0.5 & -0.5 \\ -0.5 & 1.0 & -0.5 \\ -0.5 & -0.5 & 1.0 \end{bmatrix} \begin{bmatrix} \vartheta_{z1} \\ \vartheta_{z2} \\ \vartheta_{z3} \end{bmatrix} = \begin{bmatrix} M_{z1} \\ M_{z2} \\ M_{z3} \end{bmatrix}. \tag{6.38}$$

Here E and V are the modulus of elasticity and the volume of the element, and α is a scaling factor (< 0.5) [258]. In other words the null matrix on the diagonal in (6.37) is replaced by this matrix. One easily recognizes that rigid-body motions such as $\vartheta_{z1} = \vartheta_{z2} = \vartheta_{z3}$ will not give rise to couples at the nodes.

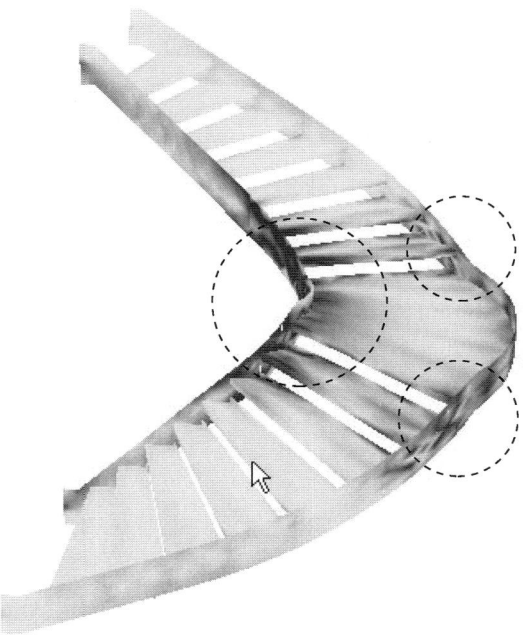

Fig. 6.11. Analysis of a curved staircase with flat elements: view of the deformed structure under gravity load

A good choice for flat elements is a combination of the Wilson Q4+2 element and a four-node Reissner–Mindlin element. But one must be careful: while the nodes of a triangular element always lie in a plane this is not guaranteed in quadrilaterals. Therefore one must modify the stiffness matrix of the element to account for the chance that the nodes do not lie in a plane; see Fig. 6.9. One idea is to write

$$\hat{K} = S^T K S, \tag{6.39}$$

where the matrix S represents the coupling between the degrees of freedom of the displaced nodes and the nodes in the plane of the element,

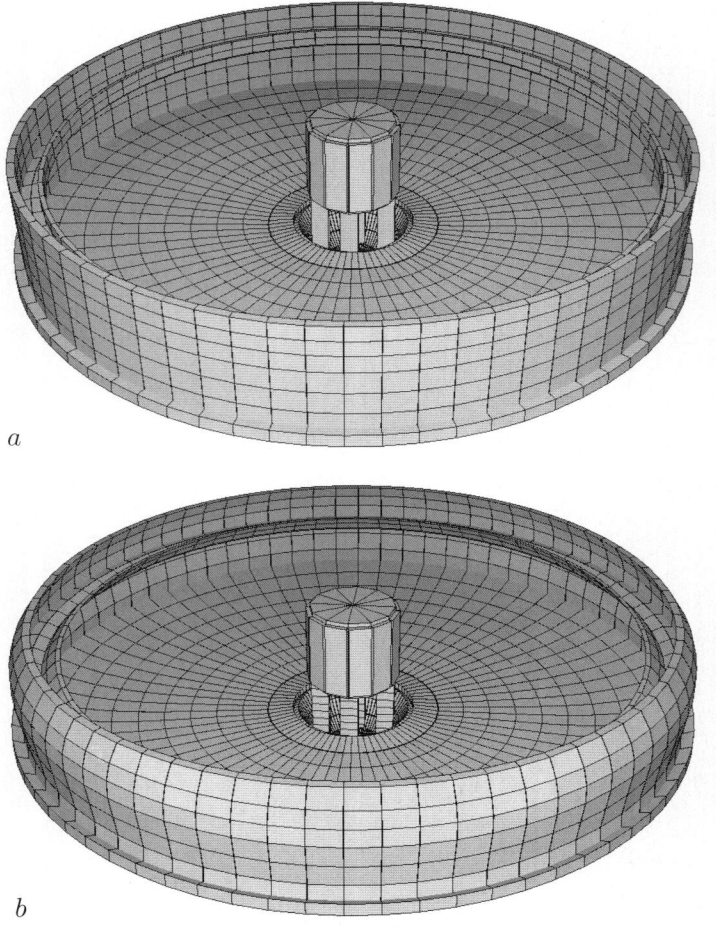

Fig. 6.12. Clarifier, analysis with flat elements: **a)** system, **b)** deformations due to nonuniform temperature changes

$$\boldsymbol{u} = \boldsymbol{S}\,\hat{\boldsymbol{u}}\,,\tag{6.40}$$

and the transposed matrix $\boldsymbol{S}^T$ transforms the equivalent nodal forces $\boldsymbol{f}$ of the element into the equivalent nodal forces $\hat{\boldsymbol{f}}$ of the displaced nodes:

$$\hat{\boldsymbol{f}} = \boldsymbol{S}^T\,\boldsymbol{f}\,,\qquad \boldsymbol{f}_i = [N_x^{(i)}, N_y^{(i)}, P_z^{(i)}, M_x^{(i)}, M_y^{(i)}, 0]^T\,.\tag{6.41}$$

Here it is assumed that the displaced nodes are connected to the element via small rigid bars of length h, and that the equilibrium conditions can thus be employed to formulate a relation between the equivalent nodal forces $\boldsymbol{f}$ and $\hat{\boldsymbol{f}}$, i.e., to establish the matrix $\boldsymbol{S}^T$.

The lever h will give rise to moments

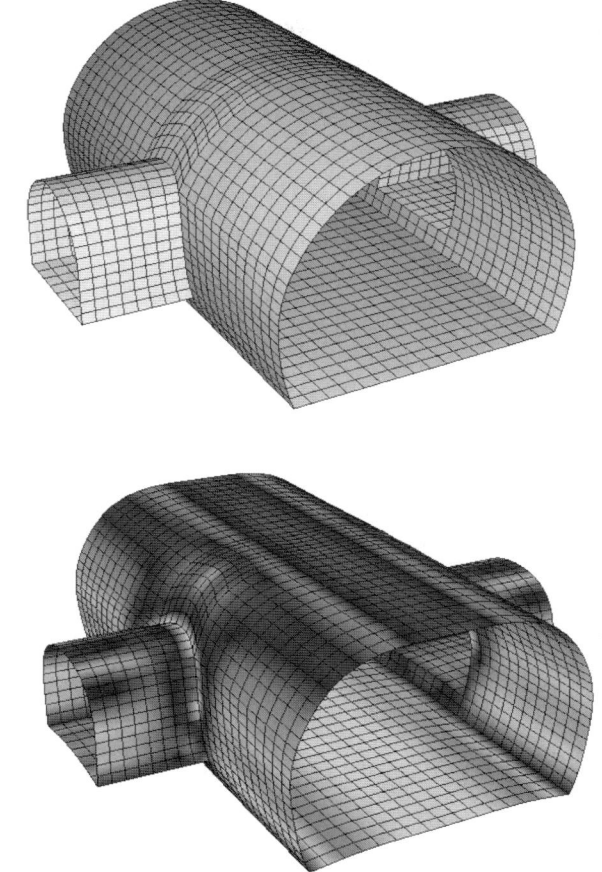

a

b

Fig. 6.13. Intersection of two tubes, analysis with flat elements: **a)** system, **b)** deformations

$$\hat{M}_x = \pm h\, N_y\,, \qquad \hat{M}_y = \pm h\, N_x \tag{6.42}$$

at the nodes. But as explained in [160], this technique is not to be recommended, because the nodal moments tend to disturb the pure membrane stress state. Therefore it is more appropriate to account for the moments due to the displaced nodes by means of vertical couples. In this way, for example, the moment due to the two forces F_{14} and F_{41} (see Fig. 6.9) is accounted for by two opposing forces:

$$F_{1z} = -F_{4z} = \frac{h}{l_{14}}(F_{14} - F_{41})\,. \tag{6.43}$$

If the twist of a plate strip is to be modeled (twisted beam problem), then by transferring the moment to the nodes, a moment M_z will also be generated about the vertical axis, which because there is no corresponding stiffness will lead to a failure of the model. To avoid this behavior, one can balance the z-component of the moment M_1 (see Fig. 6.10) with a pair of vertical forces:

$$F_A = -F_B = \frac{\sin \alpha}{l_{AB}} M_1 \,. \tag{6.44}$$

Flat elements model a shell as a faceted surface, and these models are more sensitive to singularities than a curved shell. This can be seen in the model of a steel staircase in Fig. 6.11, where under gravity load, stress singularities develop at those points where the curvature of the structure is large; see Fig. 6.11.

6.6 Membranes

Tents or similar space-like membranes can be analyzed with special flat elements by combining the structural behavior of a prestressed membrane with a rigid cloth.

If it is assumed that the horizontal prestressing force H in the membrane is uniform in all directions, the deflection of the membrane satisfies the differential equation

$$- H \left(w_{,xx} + w_{,yy} \right) = p \qquad p = \text{wind pressure} \,. \tag{6.45}$$

As expected, this is the extension to 2-D problems of the one-dimensional equation $-H \, w'' = p$ of a taut rope. If the model is extended and it is assumed that the prestressing force in the x-direction H_x is different from the force H_y in the y-direction, then it seems reasonable to modify the differential equation as follows

$$- H_x \, w_{,xx} - H_y \, w_{,yy} = p \,. \tag{6.46}$$

Green's first identity for this equation is

$$G(w, \hat{w}) = \int_\Omega \left(-H_x \, w_{,xx} - H_y \, w_{,yy} \right) \hat{w} \, d\Omega \tag{6.47}$$

$$+ \int_\Gamma \left(H_x \, w_{,x} \, n_x + H_y \, w_{,y} \, n_y \right) \hat{w} \, ds - a(w, \hat{w}) = 0 \,, \tag{6.48}$$

where the strain energy product is

$$a(w, \hat{w}) = \int_\Omega \left(H_x \, w_{,x} \, \hat{w}_{,x} + H_y \, w_{,y} \, \hat{w}_{,y} \right) d\Omega \,. \tag{6.49}$$

To understand how the analysis proceeds, let us consider a bar element.

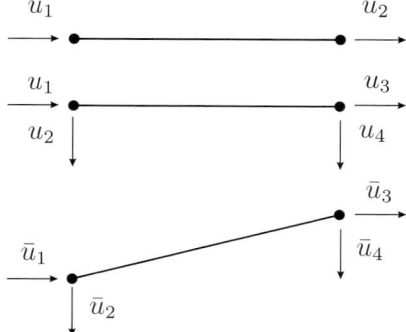

Fig. 6.14. Rotation of a bar element

Originally the stiffness matrix of the bar element is a 2×2-matrix, which is enlarged to a 4×4-matrix to account for possible rotations of the element (see Fig. 6.14):

$$\frac{EA}{l_e} \begin{bmatrix} 1 & -1 \\ -1 & 1 \end{bmatrix} \begin{bmatrix} u_1 \\ u_2 \end{bmatrix} \quad \Rightarrow \quad \frac{EA}{l_e} \begin{bmatrix} 1 & 0 & -1 & 0 \\ 0 & 0 & 0 & 0 \\ -1 & 0 & 1 & 0 \\ 0 & 0 & 0 & 0 \end{bmatrix} \begin{bmatrix} u_1 \\ u_2 \\ u_3 \\ u_4 \end{bmatrix}. \quad (6.50)$$

Assuming that the bar element is stabilized by a horizontal force H, we obtain

$$\boldsymbol{Ku} = \left\{ \frac{EA}{l_e} \begin{bmatrix} 1 & 0 & -1 & 0 \\ 0 & 0 & 0 & 0 \\ -1 & 0 & 1 & 0 \\ 0 & 0 & 0 & 0 \end{bmatrix} + \frac{H}{l_e} \begin{bmatrix} 0 & 0 & 0 & 0 \\ 0 & 1 & 0 & -1 \\ 0 & 0 & 0 & 0 \\ 0 & -1 & 0 & 1 \end{bmatrix} \right\} \begin{bmatrix} u_1 \\ u_2 \\ u_3 \\ u_4 \end{bmatrix} = \boldsymbol{f}. \quad (6.51)$$

The horizontal force H adds vertical stiffness to the bar, because the force H tends to pull the bar straight.

The structure of this matrix resembles the stiffness matrix of a beam in second-order beam theory. The first part is the linear stiffness matrix, and the second is the so-called geometric stiffness matrix, which is just the stiffness matrix of the rope,

$$\frac{H}{l_e} \begin{bmatrix} 1 & -1 \\ -1 & 1 \end{bmatrix} \begin{bmatrix} u_1 \\ u_2 \end{bmatrix} = \begin{bmatrix} f_1 \\ f_2 \end{bmatrix}, \quad (6.52)$$

extended to 4×4. Hence the stiffness matrix of a membrane element consists of the stiffness matrix $\boldsymbol{K}^S$ of the cloth (orthotropic material) and a membrane matrix $\boldsymbol{K}^M$:

$$\boldsymbol{K} = \boldsymbol{K}^S + \boldsymbol{K}^M. \quad (6.53)$$

An appropriate choice for the membrane part is the rectangular Q4+2 element, and to these four bilinear shape functions are added the unit deflections

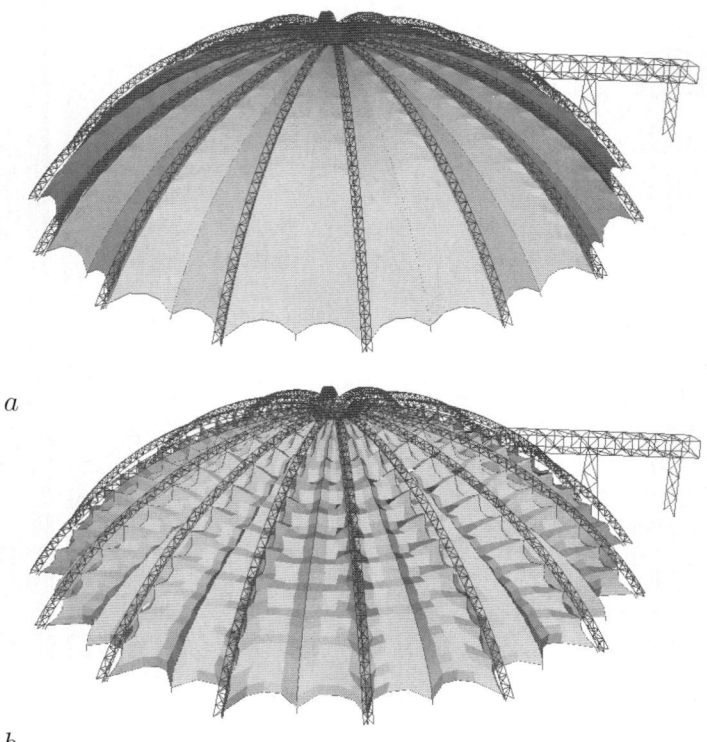

a

b

Fig. 6.15. Roofing of sand boxes with a prestressed membrane: **a)** system,
b) deformations due to gravity load + live load + snow

of the four nodes, so that the membrane matrix $\boldsymbol{K}^M$ contains the strain energy
found, then the stresses are calculated; see Fig. 6.15.

$$k_{ij}^M = \int_\Omega (H_x\, \varphi_{i,x}\ \varphi_{j,x} + H_y\, \varphi_{i,y}\ \varphi_{j,y})\, d\Omega \,. \tag{6.54}$$

Hence, a prestressed membrane is analyzed in two steps: first the shape is
found, then the stresses are calculated.

In the first step only the geometric matrix due to the prestressing forces is
activated. In other words, in this step the extensional stiffness is assumed to be
zero. To prevent the nodes from swimming on the surface of the membrane, as
if on a soap film, the nodes are stabilized in the tangential direction by small
springs. In the second step the stresses within the membrane are calculated.

Figure 6.16 a illustrates finding the shape of a tightly stressed rope. The
deflection of nodes 1 and 4 is given: $u_1 = 1.0$ m, $u_4 = 1.4$ m. The unknowns are
the associated nodal forces f_1, f_4 and displacements u_2, u_3 of the free nodes,
so that the system for the unknowns is

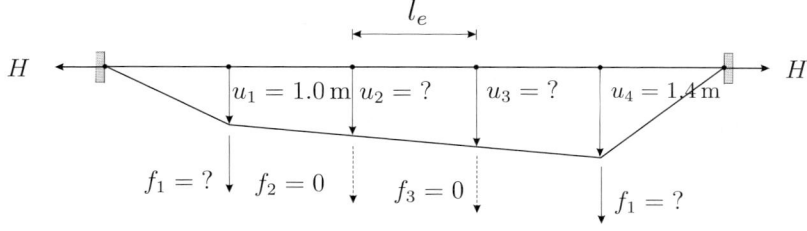

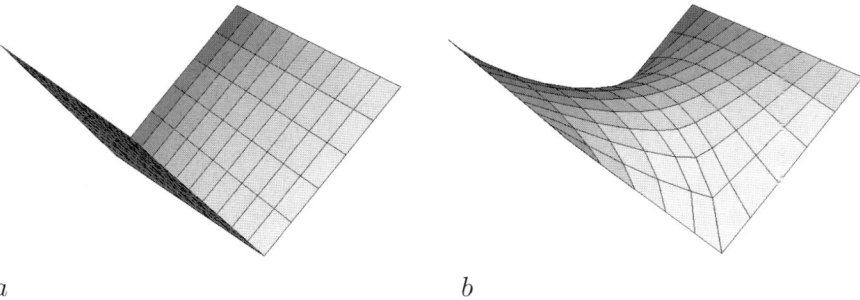

Fig. 6.16. Determining the shape **a)** of a taut rope, and **b)** of a membrane

Fig. 6.17. Determining the shape of a membrane supported along its edge: **a)** starting position, **b)** final shape

$$\frac{H}{l_e} \begin{bmatrix} 2 & -1 & 0 & 0 \\ -1 & 2 & -1 & 0 \\ 0 & -1 & 2 & -1 \\ 0 & 0 & -1 & 2 \end{bmatrix} = \begin{bmatrix} 1.0 \\ u_2 \\ u_3 \\ 1.4 \end{bmatrix} = \begin{bmatrix} f_1 \\ 0 \\ 0 \\ f_4 \end{bmatrix}. \tag{6.55}$$

The starting point for finding the shape of a membrane can either be a 3-D system (see Fig. 6.17) or a plane system. If the analysis starts with a 3-D shape, the individual elements are initially flat. If it starts with a plane

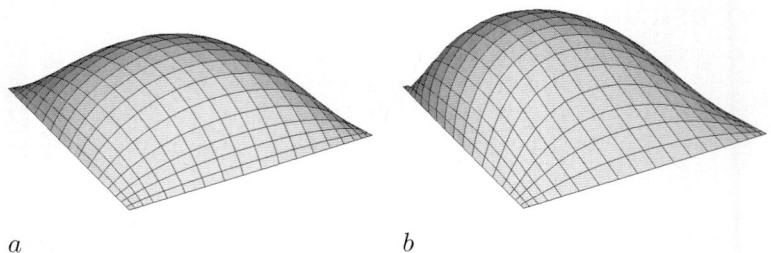

a   *b*

Fig. 6.18. Wind load: **a)** undeformed system, **b)** deformations due to wind load; (the maximum deflection at the center of the membrane was about 30 cm)

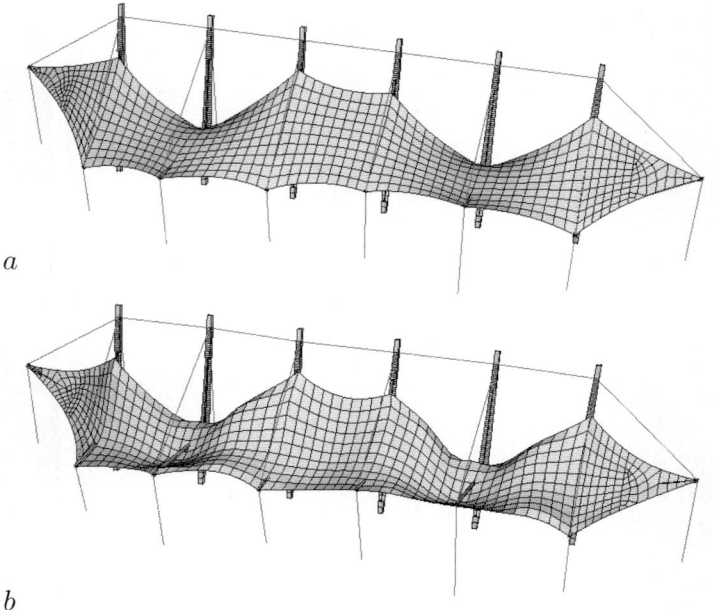

a

b

Fig. 6.19. Wind load **a)** undeformed system **b)** deformations under wind load, the maximum deflection at the center of the membrane was about 30 cm

system as in Fig. 6.16 b, the edge of the membrane is moored at the supports and pulled upwards.

Once the shape has been found, the stresses resulting from wind loads and snow must be determined. While snow is a simple load case, the wind load depends on the height, position and orientation of the individual elements; see Figs. 6.18 and 6.19. Under the action of high wind pressure, the magnitude of the tensile stresses from the prestressing forces may not be great enough and wrinkles will develop in the membrane. Because the full wind load does not converge in one step, the wind load must be applied in single steps instead.

7. Theoretical details

7.1 Scalar product

Small boldface letters denote vectors and capital bold letters matrices,

$$\boldsymbol{u} = \begin{bmatrix} u_x \\ u_y \end{bmatrix} \qquad \boldsymbol{E} = \begin{bmatrix} \varepsilon_{xx} & \varepsilon_{xy} \\ \varepsilon_{yx} & \varepsilon_{yy} \end{bmatrix} \qquad \boldsymbol{S} = \begin{bmatrix} \sigma_{xx} & \sigma_{xy} \\ \sigma_{yx} & \sigma_{yy} \end{bmatrix} \tag{7.1}$$

with the exception of $\boldsymbol{G}_0$ which denotes the vector-valued Green's function (displacement field) of a plate.

The gradient of a scalar-valued function u is a vector, and the gradient of a vector-valued function $\boldsymbol{u} = [u_1, u_2]^T$ is a matrix,

$$\nabla u = \begin{bmatrix} u_{,1} \\ u_{,2} \end{bmatrix} \qquad \nabla \boldsymbol{u} = \begin{bmatrix} u_{1,1} & u_{1,2} \\ u_{2,1} & u_{2,2} \end{bmatrix} \qquad u_{i,j} := \frac{\partial u_i}{\partial x_j} \tag{7.2}$$

while the operator div does the opposite. The divergence of a matrix-valued function is a vector-valued function, and the divergence of a vector-valued function $\boldsymbol{q} = [q_1, q_2]^T$ is a scalar-valued function:

$$\operatorname{div} \boldsymbol{S} = \begin{bmatrix} \sigma_{11,1} + \sigma_{12,2} \\ \sigma_{21,1} + \sigma_{22,2} \end{bmatrix} \qquad \operatorname{div} \boldsymbol{q} = q_{1,1} + q_{2,2} \ . \tag{7.3}$$

The following identity that relates these two operators

$$\int_\Omega \operatorname{div} \boldsymbol{S} \cdot \hat{\boldsymbol{u}} \, d\Omega = \int_\Gamma \boldsymbol{S} \boldsymbol{n} \cdot \hat{\boldsymbol{u}} \, ds - \int_\Omega \boldsymbol{S} \cdot \nabla \hat{\boldsymbol{u}} \, d\Omega \tag{7.4}$$

is fundamental for structural mechanics. Note that if $\boldsymbol{S} = \boldsymbol{S}^T$, then

$$\int_\Omega -\operatorname{div} \boldsymbol{S} \cdot \hat{\boldsymbol{u}} \, d\Omega + \int_\Gamma \boldsymbol{S} \boldsymbol{n} \cdot \hat{\boldsymbol{u}} \, ds = \int_\Omega \boldsymbol{S} \cdot \nabla \hat{\boldsymbol{u}} \, d\Omega$$

$$= \int_\Omega \boldsymbol{S} \cdot \frac{1}{2} (\nabla \hat{\boldsymbol{u}} + \nabla \hat{\boldsymbol{u}}^T) \, d\Omega, \tag{7.5}$$

which is just the statement that $\delta W_e = \delta W_i$ if $\hat{\boldsymbol{u}}$ is considered to be a virtual displacement field.

Vector fields $\boldsymbol{u}$ obey the same rule,

$$\int_{\Omega} \operatorname{div} \boldsymbol{u}\, \hat{u}\, d\Omega = \int_{\Gamma} (\boldsymbol{u} \bullet \boldsymbol{n})\, \hat{u}\, ds - \int_{\Omega} \boldsymbol{u} \bullet \nabla \hat{u}\, d\Omega \,, \tag{7.6}$$

and in 1-D problems div $= ()'$ and $\nabla = ()'$ are the same:

$$\int_0^l u'\, \hat{u}\, dx = [u\, \hat{u}]_0^l - \int_0^l u\, \hat{u}'\, dx \,. \tag{7.7}$$

By default all vectors are column vectors, and a dot indicates the scalar product of two vectors:

$$\boldsymbol{f} \bullet \boldsymbol{u} = f_x u_x + f_y u_y \,. \tag{7.8}$$

Occasionally the notation $\boldsymbol{f} \bullet \boldsymbol{u} = \boldsymbol{f}^T \boldsymbol{u}$ is also used. The dot also denotes the scalar product of the strain and stress tensor, as in

$$\begin{aligned} W_i &= \frac{1}{2} \int_{\Omega} \boldsymbol{E} \bullet \boldsymbol{S}\, d\Omega \\ &= \frac{1}{2} \int_{\Omega} \underbrace{[\varepsilon_{xx}\, \sigma_{xx} + \varepsilon_{xy}\, \sigma_{xy} + \varepsilon_{yx}\, \sigma_{yx} + \varepsilon_{yy}\, \sigma_{yy}]}_{\text{scalar product}}\, d\Omega \,. \end{aligned} \tag{7.9}$$

Other notations used in the literature for the scalar product of matrices are

$$\boldsymbol{E} \bullet \boldsymbol{S} = \operatorname{tr}(\boldsymbol{E} \otimes \boldsymbol{S}) = \boldsymbol{E} : \boldsymbol{S} \qquad (\operatorname{tr} = \text{trace}) \,. \tag{7.10}$$

where $\boldsymbol{E} \otimes \boldsymbol{S}$ is the *direct product* of the two tensors $\boldsymbol{E}$ and $\boldsymbol{S}$. The direct product of two vectors is a matrix

$$\boldsymbol{f} \otimes \boldsymbol{u} = \begin{bmatrix} f_x \\ f_y \end{bmatrix} \otimes \begin{bmatrix} u_x \\ u_y \end{bmatrix} = \begin{bmatrix} f_x \cdot u_x & f_x \cdot u_y \\ f_y \cdot u_x & f_y \cdot u_y \end{bmatrix} = \boldsymbol{A} \tag{7.11}$$

where $a_{ij} = f_i \cdot u_j$.

The scalar product of a strain and stress vector

$$\boldsymbol{\varepsilon} = \begin{bmatrix} \varepsilon_{xx} \\ \varepsilon_{yy} \\ \gamma_{xy} \end{bmatrix} \qquad \boldsymbol{\sigma} = \begin{bmatrix} \sigma_{xx} \\ \sigma_{yy} \\ \tau_{xy} \end{bmatrix} \qquad \gamma_{xy} = 2\,\varepsilon_{xy} \,, \qquad \tau_{xy} = \sigma_{xy} \,, \tag{7.12}$$

is—because of the factor 2 in $\gamma_{xy} = 2\,\varepsilon_{xy}$—the same as the scalar product of the tensors; $\boldsymbol{E} \bullet \boldsymbol{S} = \boldsymbol{\varepsilon} \bullet \boldsymbol{\sigma}$.

The scalar product

$$a(\boldsymbol{u}, \hat{\boldsymbol{u}}) = \int_{\Omega} \boldsymbol{S} \bullet \hat{\boldsymbol{E}}\, d\Omega = \int_{\Omega} \boldsymbol{C}[\boldsymbol{E}(\boldsymbol{u})] \bullet \boldsymbol{E}(\hat{\boldsymbol{u}})\, d\Omega \tag{7.13}$$

is called the *strain energy product* between two displacement fields. It is a *bilinear form*, because for any numbers c_i, d_i,

$$a(c_1 \boldsymbol{u}_1 + c_2 \boldsymbol{u}_2, d_1 \hat{\boldsymbol{u}}_1 + d_2 \hat{\boldsymbol{u}}_2) = \sum_{i,j=1}^{2} c_i \, d_j \, a(\boldsymbol{u}_i, \hat{\boldsymbol{u}}_j) \,. \tag{7.14}$$

The scalar product between the vector $\boldsymbol{u}$ and the vector $\boldsymbol{f}$ is the projection of the vector $\boldsymbol{u}$ onto the vector $\boldsymbol{f}$

$$\boldsymbol{u} \bullet \boldsymbol{f} = |\boldsymbol{u}| \, |\boldsymbol{f}| \cos \varphi \,. \tag{7.15}$$

Because the projection of $\boldsymbol{u}$ onto $\boldsymbol{f}$ should be the same as the projection of $\boldsymbol{f}$ onto $\boldsymbol{u}$ we expect the scalar product to be symmetric, $\cos(\varphi) = \cos(-\varphi)$, which the scalar product (7.13) is. According to Green's first identity—here in an abbreviated symbolic notation

$$G(\boldsymbol{u}, \hat{\boldsymbol{u}}) = \boldsymbol{p}(\hat{\boldsymbol{u}}) - a(\boldsymbol{u}, \hat{\boldsymbol{u}}) = 0 \,, \tag{7.16}$$

the strain energy product between $\boldsymbol{u}$ and $\hat{\boldsymbol{u}}$ is equivalent to the work done by the load $\boldsymbol{p}$ acting through $\hat{\boldsymbol{u}}$ and because of the symmetry of the scalar product this can also be expressed as

$$\boldsymbol{p}(\hat{\boldsymbol{u}}) = a(\boldsymbol{u}, \hat{\boldsymbol{u}}) = a(\hat{\boldsymbol{u}}, \boldsymbol{u}) = \hat{\boldsymbol{p}}(\boldsymbol{u}) \tag{7.17}$$

which is Betti's theorem.

The integral

$$\int_0^l p(x) \, w(x) \, dx =: (p, w) \tag{7.18}$$

is called the L_2 *scalar product* of p and w. The notations

$$(\boldsymbol{p}, \boldsymbol{u}) = \int_\Omega \boldsymbol{p} \bullet \boldsymbol{u} \, d\Omega = \int_\Omega [p_x u_x + p_y u_y + p_z u_z] \, d\Omega \tag{7.19}$$

and

$$\begin{aligned}
(\boldsymbol{S}, \boldsymbol{E}) &= \int_\Omega \boldsymbol{S} \bullet \boldsymbol{E} \, d\Omega \\
&= \int_\Omega [\sigma_{xx} \, \varepsilon_{xx} + \sigma_{xy} \, \varepsilon_{xy} + \sigma_{yx} \, \varepsilon_{yx} + \sigma_{yy} \, \varepsilon_{yy}] \, d\Omega
\end{aligned} \tag{7.20}$$

are extensions of this concept to vector-valued and matrix-valued functions, respectively. The expression

$$\|f\|_0 := (f, f)^{1/2} = \left[\int_0^l f(x)^2 \, dx \right]^{1/2} \tag{7.21}$$

is the L_2-*norm* of the function $f(x)$. The space of all functions defined on $(0, l)$ with a finite L_2-norm, $\|f\|_0 < \infty$, is called $L_2(0, l)$. Note that the function

$f(x) = 1/\sqrt{x}$ can be integrated but its L_2-norm is infinite because of the square in (7.21)

$$\int_0^1 \frac{1}{\sqrt{x}}\, dx = 2 \qquad \int_0^1 \frac{1}{x}\, dx = \infty. \tag{7.22}$$

On the other hand if two functions f and g lie in L_2, then the scalar product of f and g exists, it is bounded

$$||f||_0 < \infty, \ ||g||_0 < \infty \quad \Rightarrow \quad \int_0^l f\, g\, dx < \infty. \tag{7.23}$$

Note that $||f||_0 = ||g||_0$ does not imply that $||f - g||_0 = 0$. In the Euclidean norm, for example, all unit vectors $\boldsymbol{e}_i$ have the same length, $||\boldsymbol{e}_i|| = 1$ but of course their tips do not touch, so that $||\boldsymbol{e}_1 - \boldsymbol{e}_3|| \neq 0$.

Hence, if the FE solution seems to converge, because the variations in the strain energy $a(\boldsymbol{u}_h, \boldsymbol{u}_h) = \boldsymbol{f}^T \boldsymbol{u}$ come to a halt, then (theoretically at least) this does not imply that two consecutive solutions are the "same":

$$||\boldsymbol{u}_{h^{(1)}}||_E \sim ||\boldsymbol{u}_{h^{(2)}}||_E \quad \not\Rightarrow \quad ||\boldsymbol{u}_{h^{(1)}} - \boldsymbol{u}_{h^{(2)}}||_E \ll 1. \tag{7.24}$$

The inequality

$$\left| \int_0^l f\, g\, dx \right| \leq \left[\int_0^l f^2 dx \right]^{1/2} \left[\int_0^l g^2 dx \right]^{1/2} \tag{7.25}$$

or

$$|(f, g)| \leq ||f||_0 \, ||g||_0 \tag{7.26}$$

is known as *Cauchy-Schwarz inequality*.

The extension of the space $L_2(\Omega)$ to higher derivatives constitutes the Sobolev spaces. Imagine that we form a one-dimensional array that contains the function u and all its derivatives up to the order m, for example

$$\boldsymbol{u}^{(1)} := [u, u_{,x}, u_{,y}]^T \qquad m = 1. \tag{7.27}$$

The Sobolev space $H^m(\Omega)$ then consists of all functions u for which the L_2 scalar product of these vectors is bounded,

$$||u||_m^2 = \int_\Omega \boldsymbol{u}^{(m)} \bullet \boldsymbol{u}^{(m)}\, d\Omega := \int_\Omega [u\, u + u_{,x}\, u_{,x} + \ldots]\, d\Omega < \infty \tag{7.28}$$

i.e., u and all its derivatives up to order m are square integrable (they lie in $L_2(\Omega)$):

$$||u||_1^2 = \int_\Omega \boldsymbol{u}^{(1)} \bullet \boldsymbol{u}^{(1)}\, d\Omega = \int_\Omega [u\, u + u_{,x}\, u_{,x} + u_{,y}\, u_{,y}]\, d\Omega < \infty. \tag{7.29}$$

The space $H^m(\Omega)$ can also be seen as the completion of $C^\infty(\Omega)$ in the norm $||.||_m$, and the space $H_0^m(\Omega) \subset H^m(\Omega)$ is the completion of $C_0^\infty(\Omega)$ (= the functions in $C^\infty(\Omega)$ which vanish near the boundary).

On $H^2(\Omega)$ the scalar product of two functions is defined as

$$(u,v)_{H^2} = \int_\Omega \boldsymbol{u}^{(2)} \cdot \boldsymbol{v}^{(2)} \, d\Omega = \int_\Omega [u\,v + u_{,x}\,v_{,x} + u_{,y}\,v_{,y}$$
$$+ u_{,xx}\,v_{,xx} + u_{,xy}\,v_{,xy} + u_{,yx}\,v_{,yx} + u_{,yy}\,v_{,yy}]\,d\Omega \qquad (7.30)$$

and the norm is

$$||u||_2 = \sqrt{(u,u)_{H^2}}\,. \qquad (7.31)$$

The extension of these concepts to other spaces $H^m(\Omega)$ is obvious.

An expression such as

$$|u|_2 := \left[\int_\Gamma (u_{,xx}^2 + u_{,xy}^2 + u_{,yx}^2 + u_{,yy}^2)\,d\Omega \right]^{1/2} \qquad (7.32)$$

would be called a *semi-norm*, because $|u|_2 = 0$ with $u = a + b\,x + c\,y$ does not imply that $u = 0$.

In abstract terms the FE displacement field $\boldsymbol{u}_h$ is the solution of the variational problem

$$a(\boldsymbol{u}_h, \boldsymbol{v}) = \boldsymbol{p}(\boldsymbol{v}) \qquad \text{for all } \boldsymbol{v} \in V_h \subset V\,, \qquad (7.33)$$

where V is a Hilbert space usually endowed with a Sobolev norm $||.||_m$, and $\boldsymbol{p}(\boldsymbol{v})$ is a continuous linear functional.

An important property of the strain energy product is that it establishes an equivalent norm on V,

$$c_1\,||\boldsymbol{u}||_m \le \sqrt{a(\boldsymbol{u},\boldsymbol{u})} \le c_2\,||\boldsymbol{u}||_m \qquad (7.34)$$

where c_1 and c_2 are independent of $\boldsymbol{u}$. Formally this so-called *energy norm*

$$||\boldsymbol{u}||_E := \sqrt{a(\boldsymbol{u},\boldsymbol{u})} = (\boldsymbol{S},\boldsymbol{E})^{1/2} = \left[\int_\Omega \boldsymbol{S} \cdot \boldsymbol{E}\,d\Omega \right]^{1/2} \qquad (7.35)$$

is a only a semi-norm. To actually be a norm on V, the space V must not allow rigid-body motions (that is, enough supports must be provided), because otherwise the energy norm cannot separate the elements of V. This property guarantees that if the norm of $\boldsymbol{u} - \hat{\boldsymbol{u}}$ is zero, then $\boldsymbol{u} = \hat{\boldsymbol{u}}$:

$$||\boldsymbol{u} - \hat{\boldsymbol{u}}||_E = 0 \qquad \Rightarrow \qquad \boldsymbol{u} = \hat{\boldsymbol{u}}\,. \qquad (7.36)$$

In this book the same letter p is used for the loads that constitute the load case p and the load case p itself. In an abstract sense, any load case p constitutes a *functional* $p(\varphi_i)$ on V_h,

$$p(\varphi_i) := \int_\Omega p\,\varphi_i\,d\Omega = (p, \varphi_i) \qquad (7.37)$$

where it is understood that the functional may contain additional terms, as in

$$p(\varphi_i) := \int_\Omega p\varphi_i\,d\Omega + \int_\Gamma t\,\varphi_i\,ds + P\,\varphi_i(\boldsymbol{x})\,, \qquad (7.38)$$

if edge loads, t, and point loads, P, are also present but the simplest form is (7.37).

7.2 Green's identities

'The principle of virtual displacements is nothing else than integration by parts' and so we start this section with repeating the rules for integration by parts before we formulate Green's identities which are based on integration by parts. These identities encapsulate the basic principles of mechanics and play a fundamental role in finite element analysis.

Integration by parts

Let u and $\hat{u}$ be two functions with continuous first derivatives in the interval $(0, l)$ then

$$\int_0^l u'\,\hat{u}\,dx = [u\,\hat{u}]_0^l - \int_0^l u\,\hat{u}'\,dx \qquad (7.39)$$

and in higher dimensions with functions u and $\hat{u}$ from $C^1(\Omega)$,

$$\int_\Omega u_{,x_i}\,\hat{u}\,d\Omega = \int_\Gamma u\,n_i\,\hat{u}\,ds - \int_\Omega u\,\hat{u}_{,x_i}\,d\Omega\,, \qquad (7.40)$$

where n_i is the i-th component of the normal vector $\boldsymbol{n}$ on the edge Γ of the domain Ω.

For example let $\hat{u} = 1$ and $u' = \varepsilon$ the strain in a rod then

$$\int_0^l \varepsilon\,dx = u(l) - u(0)\,. \qquad (7.41)$$

In a plate where $\varepsilon_{xx} = u_{x,x}$ the same statement is

$$\int_\Omega \varepsilon_{xx}\,d\Omega = \int_\Gamma u_x\,n_x\,ds\,. \qquad (7.42)$$

If for example $\Omega = a \times b$ is a rectangle with $n_x = \pm 1$ on the vertical edges Γ_L and Γ_R and $n_x = 0$ on the horizontal edges then the result resembles the 1-D result

$$\int_\Omega \varepsilon_{xx}\,d\Omega = \int_{\Gamma_R} u_x\,ds - \int_{\Gamma_L} u_x\,ds\,. \qquad (7.43)$$

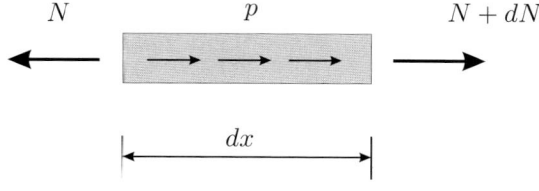

N p $N + dN$

dx

Fig. 7.1. Static equilibrium of a bar element $dN + p\,dx = 0$

Bars

The equilibrium condition $\sum H = 0$ for a bar element dx (see Figure 7.1) leads to the differential equation

$$- EA\,u'' = p\,,\tag{7.44}$$

where $u(x)$ is the longitudinal displacement, p is the applied load, and EA is the (constant) stiffness of the bar. If the left-hand side of the differential equation is multiplied by a virtual displacement δu and we integrate by parts

$$\int_0^l -EAu''\,\delta u\,dx = [-EA\,u'\,\delta u]_0^l - \int_0^l -EAu'\,\delta u'\,dx\,,\tag{7.45}$$

the result is Green's first identity:

$$G(u, \delta u) = \underbrace{\int_0^l -EAu''\,\delta u\,dx}_{start} + \underbrace{[N\,\delta u]_0^l - \int_0^l EA\,u'\,\delta u'\,dx}_{transformed\ terms}$$

$$= \underbrace{\int_0^l -EAu''\,\delta u\,dx + [N\,\delta u]_0^l}_{\delta W_e} - \underbrace{\int_0^l EA\,u'\,\delta u'\,dx}_{\delta W_i} = 0\,.\tag{7.46}$$

The terms in brackets

$$[N\,\delta u]_0^l = N(l)\,\delta u(l) - N(0)\,\delta u(0)\tag{7.47}$$

are the virtual external work done by the normal forces $N = EA\,u'$ at the ends of the bar.

The expression $B(u, \hat{u}) = G(u, \hat{u}) - G(\hat{u}, u) = 0 - 0 = 0$ is Green's second identity,

$$B(u, \hat{u}) = \underbrace{\int_0^l -EA\,u''\,\hat{u}\,dx + [N\,\hat{u}]_0^l}_{W_{1,2}} - \underbrace{[u\,\hat{N}]_0^l - \int_0^l u\,(-EA\,\hat{u}'')\,dx}_{W_{2,1}} = 0$$

$$\tag{7.48}$$

and it formulates Betti's theorem.

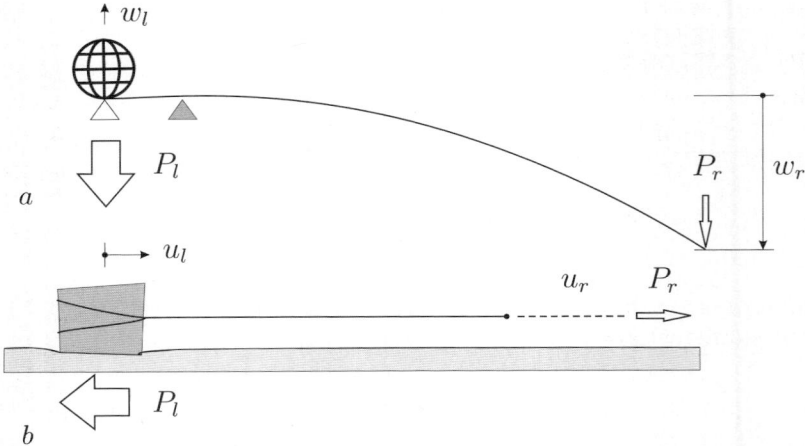

a

b

Fig. 7.2. Archimedes' dilemma: all effort is consumed by the strain energy **a)** the Earth will not move one iota **b)** the rubber band will stretch and stretch and ...

Green's first identity basically is of the form

$$G(u, \hat{u}) = p(\hat{u}) - a(u, \hat{u}) = \delta W_e(u, \hat{u}) - \delta W_i(u, \hat{u}) = 0 \tag{7.49}$$

or if it is solved for the strain energy product

$$a(u, \hat{u}) = p(\hat{u}) \,. \tag{7.50}$$

Note that $a(u, \hat{u})$ is the first-order derivative of the quadratic form

$$F(u) = \frac{1}{2}\, a(u, u) \,, \tag{7.51}$$

that is

$$F'(u) := \left[\frac{d}{d\varepsilon} \frac{1}{2} a(u + \varepsilon \hat{u}, u + \varepsilon \hat{u}) \right]_{\varepsilon=0} = a(u, \hat{u}) \,. \tag{7.52}$$

Beam

The differential equation of a beam with constant bending stiffness EI is

$$EI\, w^{IV}(x) = p(x) \,. \tag{7.53}$$

The bending moment is $M(x) = -EI\, w''(x)$ and the shear force is $V(x) = -EI\, w'''(x)$. Green's first identity for the beam equation is

$$G(w, \hat{w}) = \underbrace{\int_0^l EI\, w^{IV}\, \hat{w}\, dx + [V\, \hat{w} - M\, \hat{w}']_0^l}_{\delta W_e} - \underbrace{\int_0^l \frac{M\, \hat{M}}{EI}\, dx}_{\delta W_i} = 0 \,, \tag{7.54}$$

and $B(w, \hat{w}) = G(w, \hat{w}) - G(\hat{w}, w) = 0$ is Betti's theorem.

An application—Archimedes' dilemma

A place to stand does not suffice to move the Earth. Archimedes needs also a lever with $EI = \infty$; see Fig. 7.2 a. Otherwise the lever will only bend. Because of Green's first identity $G(w, w) = W_e - W_i = 0$ the exterior work is at any moment equal to the strain energy in the beam

$$W_e = P_r \cdot w_r - P_l \cdot w_l = a(w, w) = \int_0^l \frac{M^2}{EI} \, dx = W_i \qquad (7.55)$$

or

$$P_r \cdot w_r = P_l \cdot w_l + a(w, w) \,. \qquad (7.56)$$

So that all of Archimedes' effort, $P_r \cdot w_r$, will be consumed by the internal energy $a(w, w)$ and very little—effectively nothing—remains to lift the Earth.

The same happens if you try to pull a heavy weight across the wet sand on the beach (see Fig. 7.2 b)

$$W_e = \underbrace{P_r \cdot u_r}_{your\ effort} - P_l \cdot u_l = a(u, u) = \int_0^l \frac{N^2}{EA} \, dx = W_i \,. \qquad (7.57)$$

The rubber band (EA) will stretch and stretch and stretch, $u_r \to 1, 2, 3, \ldots$, that is $a(u, u)$ will increase but the weight will hardly move, $u_l \simeq 0$.

Poisson equation

Green's first identity for the differential equation $-\Delta u = p$ is

$$G(u, \hat{u}) = \underbrace{\int_\Omega -\Delta u \, \hat{u} \, d\Omega + \int_\Gamma \frac{\partial u}{\partial n} \, \hat{u} \, ds}_{\delta W_e} - \underbrace{\int_\Omega \nabla u \bullet \nabla \hat{u} \, d\Omega}_{\delta W_i} = 0 \,, \qquad (7.58)$$

where

$$a(u, \hat{u}) = \int_\Omega \nabla u \bullet \nabla \hat{u} \, d\Omega = \int_\Omega (u_{,x} \, \hat{u}_{,x} + u_{,y} \, \hat{u}_{,y}) \, d\Omega \qquad (7.59)$$

is the strain energy product.

Kirchhoff plate

The differential equation of the Kirchhoff plate is the biharmonic equation

$$K \, \Delta\Delta \, w = p \qquad K = \frac{E \, h^3}{12 \, (1 - \nu^2)} \,. \qquad (7.60)$$

The curvature tensor $\boldsymbol{K} = [\kappa_{ij}]$ has elements $\kappa_{ij} = w,_{ij}$, and the bending moment tensor is $\boldsymbol{M} = K\left\{(1-\nu)\,\boldsymbol{K} + \nu(\mathrm{tr}\,\boldsymbol{K})\,\boldsymbol{I}\right\}$, or

$$m_{xx} = -K\,(w,_{xx} + \nu\,w,_{yy}), \qquad m_{yy} = -K\,(w,_{yy} + \nu\,w,_{xx}), \quad (7.61)$$

$$m_{xy} = -(1-\nu)\,K\,w,_{xy}. \tag{7.62}$$

The shear forces are

$$q_x = -K\,(w,_{xxx} + w,_{yyx}) \qquad q_y = -K\,(w,_{xxy} + w,_{yyy}), \tag{7.63}$$

and the resultant stresses on the boundary are, in indicial notation,

$$m_n = m_{ij}\,n_i\,n_j \qquad m_{nt} = m_{ij}\,n_i\,t_j \tag{7.64}$$

$$q_n = q_i\,n_i \qquad v_n = \frac{d}{ds}\,m_{nt} + q_n. \tag{7.65}$$

Green's first identity is

$$G(w,\hat{w}) = \underbrace{\int_\Omega K\,\Delta\Delta w\,\hat{w}\,d\Omega + \int_\Gamma \left[v_n\,\hat{w} - m_n\,\hat{w},_n\right]ds + \sum_i F_i\,\hat{w}(\boldsymbol{x}_i)}_{\delta W_e}$$

$$- \underbrace{a(w,\hat{w})}_{\delta W_i} = 0, \tag{7.66}$$

where the strain energy product is the expression

$$a(w,\hat{w}) = \int_\Omega \left[w,_{xx}\,(\hat{w},_{xx} + \nu\,\hat{w},_{yy}) + 2(1-\nu)\,w,_{xy}\,\hat{w},_{xy}\right.$$

$$\left. + w,_{yy}\,(\hat{w},_{yy} + \nu\,\hat{w},_{xx})\right]d\Omega = \int_\Omega \boldsymbol{M} \bullet \hat{\boldsymbol{K}}\,d\Omega \tag{7.67}$$

and the F_i are the corner forces resulting from the jumps in the twisting moment m_{nt}:

$$F_i := F(w)(\boldsymbol{x}_i) = m_{nt}(\boldsymbol{x}_i^+) - m_{nt}(\boldsymbol{x}_i^-). \tag{7.68}$$

Reissner–Mindlin plate

The terms of a Reissner–Mindlin plate are the rotations $\boldsymbol{\varphi} = [\varphi_x, \varphi_y]^T$, the deflection w, the shearing strains $\boldsymbol{\gamma} = [\gamma_x, \gamma_y]^T$, the curvature tensor $\boldsymbol{K}$, the bending moment tensor $\boldsymbol{M}$, and the shear forces $\boldsymbol{q} = [q_x, q_y]^T$ which are governed by the equations

strains:	$\boldsymbol{K}(\boldsymbol{\varphi}) - \boldsymbol{K} = 0$	$\boldsymbol{\varphi} + \nabla w - \boldsymbol{\gamma} = 0$	(7.69)
material law:	$C[\boldsymbol{K}] - \boldsymbol{M} = 0$	$a\,\boldsymbol{\gamma} - \boldsymbol{q} = 0$	(7.70)
equilibrium:	$-\mathrm{div}\,\boldsymbol{M} + \boldsymbol{q} = b\,\nabla p$	$-\mathrm{div}\,\boldsymbol{q} = p$	(7.71)

where

$$\boldsymbol{K}(\boldsymbol{\varphi}) = \frac{1}{2}\left(\nabla\boldsymbol{\varphi} + \nabla\boldsymbol{\varphi}^{T}\right) = \frac{1}{2}\begin{bmatrix} 2\,\varphi_{x,x} & \varphi_{x,y} + \varphi_{y,x} \\ \varphi_{y,x} + \varphi_{x,y} & 2\,\varphi_{y,y} \end{bmatrix} \qquad (7.72)$$

$$\boldsymbol{C}[\boldsymbol{K}] = K\,(1-\nu)\,\boldsymbol{E} + \nu\,K\,(\mathrm{tr}\,\boldsymbol{K})\,\boldsymbol{I} \qquad (7.73)$$

and

$$K = \frac{E\,h^3}{12\,(1-\nu^2)}\,, \qquad a = K\,\frac{1-\nu}{2}\,\bar{\lambda}^2\,, \qquad b = \frac{\nu}{1-\nu}\,\frac{1}{\bar{\lambda}^2}\,, \qquad \bar{\lambda}^2 = \frac{10}{h^2}\,, \quad (7.74)$$

where h is the plate thickness. These equations are equivalent to the system

$$-\,\mathrm{div}\,\boldsymbol{C}[\boldsymbol{K}(\boldsymbol{\varphi})] + a(\boldsymbol{\varphi} + \nabla\,w) = b\,\nabla p \qquad (7.75)$$

$$-\mathrm{div}\,(a\,(\boldsymbol{\varphi} + \nabla\,w)) = p \qquad (7.76)$$

or in indicial notation

$$-\,m_{\alpha\,\beta,\beta} + a\,q_\alpha = b\,p_{,\alpha}\,, \qquad \alpha = 1,2 \qquad (7.77)$$

$$-a\,q_{\beta,\beta} = p \qquad (7.78)$$

where

$$m_{\alpha\,\beta} = K(1-\nu)\frac{1}{2}(\varphi_{\alpha,\beta} + \varphi_{\beta,\alpha}) + \nu\,K\,\varphi_{\gamma,\gamma}\,\delta_{\alpha\,\beta} \qquad (7.79)$$

$$q_\alpha = \varphi_\alpha + w_{,\alpha}\,. \qquad (7.80)$$

If this system is interpreted as the application of an operator $-\boldsymbol{L}$ to the vector-valued function $\boldsymbol{u} = [\varphi_x, \varphi_y, w]^T$ we have the identity

$$G(\boldsymbol{u}, \hat{\boldsymbol{u}}) = \int_{\Omega} -\boldsymbol{L}\,\boldsymbol{u}\boldsymbol{\cdot}\hat{\boldsymbol{u}}\,d\Omega + \int_{\Gamma}[\boldsymbol{M}\,\boldsymbol{n}\boldsymbol{\cdot}\hat{\boldsymbol{\varphi}} + \boldsymbol{q}\boldsymbol{\cdot}\boldsymbol{n}\,\hat{w}]\,ds - a(\boldsymbol{u}, \hat{\boldsymbol{u}}) = 0$$

$$(7.81)$$

where

$$a(\boldsymbol{u}, \hat{\boldsymbol{u}}) = \int_{\Omega}[\boldsymbol{M}\boldsymbol{\cdot}\hat{\boldsymbol{K}} + \boldsymbol{q}\boldsymbol{\cdot}\hat{\boldsymbol{\gamma}}]\,d\Omega\,. \qquad (7.82)$$

Linear elasticity

The governing equation is

$$\boldsymbol{L}\,\boldsymbol{u} := -\left[\mu\,\boldsymbol{\Delta} + \frac{\mu}{1-2\,\nu}\nabla\,\mathrm{div}\right]\boldsymbol{u} = \boldsymbol{p}\,, \qquad (7.83)$$

or in tensor notation

$$G(u,\hat{u}) = 0 \quad G(u,\hat{u}) = 0 \quad G(u,\hat{u}) = 0$$

$$-\mu\, u_{i,jj} - \frac{\mu}{1-2\nu}\, u_{j,ji} = p_i \qquad i = 1,2 \tag{7.84}$$

which is equivalent to

$$-\sigma_{ij,j} = p_i \qquad i = 1,2\,. \tag{7.85}$$

Green's first identity is

$$G(\boldsymbol{u},\hat{\boldsymbol{u}}) = \int_{\Omega} -\boldsymbol{L}\,\boldsymbol{u}\cdot\hat{\boldsymbol{u}}\,d\Omega + \int_{\Gamma} \boldsymbol{\tau}(\boldsymbol{u})\cdot\hat{\boldsymbol{u}}\,d\Omega - a(\boldsymbol{u},\hat{\boldsymbol{u}}) = 0\,, \tag{7.86}$$

where $\boldsymbol{\tau}(\boldsymbol{u}) = \boldsymbol{S}\,\boldsymbol{n}$ is the traction vector on the boundary and

$$a(\boldsymbol{u},\hat{\boldsymbol{u}}) = \int_{\Omega} [\sigma_{xx}\,\hat{\varepsilon}_{xx} + 2\,\sigma_{xy}\,\hat{\varepsilon}_{xy} + \sigma_{yy}\,\hat{\varepsilon}_{yy}]\,d\Omega = \int_{\Omega} \boldsymbol{S}\cdot\hat{\boldsymbol{E}}\,d\Omega \tag{7.87}$$

is the strain energy product. For more identities see [115].

Regularity

Because the identities are based on integration by parts, the functions u and $\hat{u}$ must be sufficiently regular. If that is not the case, the interval $(0,l)$ or the domain Ω can be subdivided into as many intervals or partitions as necessary:

$$G(u,\hat{u})_{(0,l)} = G(u,\hat{u})_{(0,l_1)} + G(u,\hat{u})_{(l_1,l_2)} + \ldots + G(u,\hat{u})_{(l_n,l)} = 0\,. \tag{7.88}$$

Typically the partitions are the individual elements; see Fig. 7.3.

Green's first identity and stiffness matrices

Substituting two nodal unit displacements (not necessarily the actual displacements but "any" displacements φ_i) into Green's first identity for a beam yields

$$G(\varphi_i,\varphi_j) = \underbrace{\int_0^l EI\varphi_i^{IV}\,\varphi_j\,dx + \left[V_i\,\varphi_j - M_i\,\varphi_j'\right]_0^l}_{p_{\,ij}} - \underbrace{\int_0^l EI\varphi_i''\,\varphi_j''dx}_{k_{\,ij}} = 0$$

$$\tag{7.89}$$

which means that

$$\delta W_e(p_i, \varphi_j) = p_{ij} = k_{ij} = \delta W_i(\varphi_i, \varphi_j) \tag{7.90}$$

or that the strain energy product (virtual internal energy) k_{ij} between two such nodal unit displacements φ_i and φ_j is equal to the virtual external work p_{ij} done by the unit load case p_i via the virtual displacements φ_j.

The load case p_i simply consists of all forces that produce the shape φ_i, i.e., the distributed load $EI\,\varphi_i^{IV}$, the shear forces $V_i(0), V_i(l)$, and the moments $M_i(0), M_i(l)$ at the ends of the beam. The double subscripted term p_{ij} is the virtual external work $\delta W_e(p_i, \varphi_j)$ corresponding to the load case p_i and the virtual displacement φ_j.

With $u_h = \sum_j u_j \varphi_j$ and the n-fold identity

$$G(u_h, \varphi_i) = 0 \qquad i = 1, 2, \ldots n \tag{7.91}$$

this is equivalent to

$$\boldsymbol{P}\,\boldsymbol{u} - \boldsymbol{K}\,\boldsymbol{u} = \boldsymbol{0} \qquad \text{or} \qquad \boldsymbol{f}_h - \boldsymbol{K}\,\boldsymbol{u} = \boldsymbol{0} \tag{7.92}$$

where $\boldsymbol{f}_h := \boldsymbol{P}\,\boldsymbol{u}$.

Strain energy = nodal forces × nodal displacements

It should be obvious by now that the strain energy in a single element

$$a(\boldsymbol{u}_h, \boldsymbol{u}_h)_{\Omega_e} = \int_{\Omega_e} \sigma_{ij} \cdot \varepsilon_{ij}\, d\Omega = \int_{\Omega_e} \boldsymbol{p}_h \bullet \boldsymbol{u}_h\, d\Omega + \int_{\Gamma_e} \boldsymbol{t}_h \bullet \boldsymbol{u}_h\, ds = \boldsymbol{f}_e^T\,\boldsymbol{u}_e \tag{7.93}$$

is the same as the scalar product between the equivalent nodal forces of that element

$$f_i^e = \int_{\Omega_e} \boldsymbol{p}_h \bullet \boldsymbol{\varphi}_i^e\, d\Omega + \int_{\Gamma_e} \boldsymbol{t}_h \bullet \boldsymbol{\varphi}_i^e\, ds \tag{7.94}$$

—the $\boldsymbol{t}_h$ are the tractions on the edge of the element—and the vector $\boldsymbol{u}_e$ of nodal displacements. Summing the contributions from all elements we obtain the well known formula for the strain energy stored in a structure

$$a(\boldsymbol{u}_h, \boldsymbol{u}_h) = \boldsymbol{u}^T\,\boldsymbol{K}\,\boldsymbol{u} = \boldsymbol{f}^T\,\boldsymbol{u}\,. \tag{7.95}$$

Green's first identity and projections

The FE solution u_h is the projection of the exact solution u onto V_h

$$u_h \in V_h : \qquad a(u - u_h, \varphi_i) = 0 \qquad \varphi_i \in V_h \tag{7.96}$$

or

$$u_h \in V_h : \qquad a(u_h, \varphi_i) = a(u, \varphi_i) = \delta W_e(u, \varphi_i) = f_i \qquad \varphi_i \in V_h \quad (7.97)$$

where $\delta W_e(u, \varphi_i)$ is short for

$$G(u, \varphi_i) = \underbrace{\int_0^l -EA\, u''\, \varphi_i\, dx + [N\, \varphi_i]_0^l}_{\delta W_e(u, \varphi_i)} -a(u, \varphi_i) = 0 \qquad (7.98)$$

so that Green's first identity allows to replace the term $a(u, \varphi_i)$ by an expression of external virtual work. This is the vector $\boldsymbol{f}$.

7.3 Green's functions

To solve the equation

$$3 \cdot x = 12 \qquad \Rightarrow \qquad x = \frac{1}{3} \cdot 12 \qquad (7.99)$$

the right-hand side is multiplied by the "Green's function" $x_G = 1/3$, which is the solution of $3 \cdot x = 1$.

The Green's functions are the solutions of the adjoint equations. Consider for example the system $\boldsymbol{K}\,\boldsymbol{u} = \boldsymbol{f}$ and the identity

$$B(\boldsymbol{u}, \hat{\boldsymbol{u}}) = \hat{\boldsymbol{u}}^T\,\boldsymbol{K}\,\boldsymbol{u} - \boldsymbol{u}^T\,\boldsymbol{K}^T\,\hat{\boldsymbol{u}} = 0\,, \qquad (7.100)$$

where $\boldsymbol{K}^T$ is the adjoint (= transpose) of the matrix $\boldsymbol{K}$. Clearly if $\boldsymbol{g}_i$ is a solution of $\boldsymbol{K}^T\,\boldsymbol{g}_i = \boldsymbol{e}_i$ then $u_i = \boldsymbol{g}_i^T\,\boldsymbol{f}$.

In linear structural mechanics the equations are self-adjoint (or symmetric $\boldsymbol{K} = \boldsymbol{K}^T$) so that the Green's functions are the solutions of the same equations, $EI\, G_0^{IV} = \delta_0$, as in the original problem, $EI\, w^{IV} = p$.

The complement of the Green's function is Green's second identity (Betti's theorem), which in the case of the Laplacian reads

$$B(u, \hat{u}) = \int_\Omega -\Delta u\, \hat{u}\, d\Omega + \int_\Gamma \frac{\partial u}{\partial n}\, \hat{u}\, ds - \int_\Gamma u\, \frac{\partial \hat{u}}{\partial n}\, ds - \int_\Omega u\, (-\Delta \hat{u})\, d\Omega = 0\,.$$
$$(7.101)$$

From this equation we can see what boundary conditions must be imposed on the Green's functions; see Fig. 7.4. In a Dirichlet problem

$$-\Delta u = p\,, \qquad u = g \text{ on } \Gamma \qquad (7.102)$$

things are easy:

$$-\Delta G_0 = \delta_0\,, \qquad G_0 = 0 \text{ on } \Gamma \qquad (7.103)$$

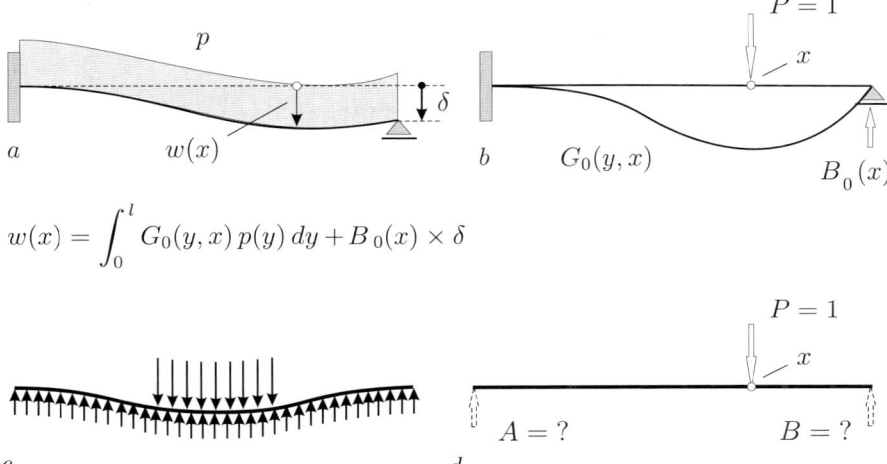

$$w(x) = \int_0^l G_0(y,x)\,p(y)\,dy + B_0(x) \times \delta$$

Fig. 7.4. Influence function for a beam: **a)** when a support is displaced, **b)** Green's function, **c)** theoretically no Green's function exists for a beam with no supports

and so

$$u = \int_\Omega G_0\,p\,d\Omega_{\boldsymbol{y}} - \int_\Gamma \frac{\partial G_0}{\partial n}\,g\,ds\,. \tag{7.104}$$

In a mixed problem

$$-\Delta u = p\,, \quad u = g \text{ on } \Gamma_D\,, \quad \frac{\partial u}{\partial n} = t \quad \text{on } \Gamma_N \tag{7.105}$$

we require that

$$-\Delta G_0 = \delta_0\,, \quad G_0 = 0 \quad \text{on } \Gamma_D\,, \quad \frac{\partial G_0}{\partial n} = 0 \quad \text{on } \Gamma_N \tag{7.106}$$

and so

$$u = \int_\Omega G_0\,p\,d\Omega_{\boldsymbol{y}} - \int_{\Gamma_D} \frac{\partial G_0}{\partial n}\,g\,ds + \int_{\Gamma_N} G_0\,t\,ds\,. \tag{7.107}$$

The support conditions of the beam in Fig. 7.4 a are of such a mixed type, because geometric, $w(0) = w'(0) = w(l)$, as well as static boundary conditions, $M(l) = 0$, are prescribed. Hence if the Green's function of the beam in Fig. 7.4 a solves the boundary value problem

$$EI\,G_0^{IV} = \delta_0(y-x)\,, \quad G_0(0) = G_0(l) = G_0'(0) = -EI\,G_0''(l) = 0 \tag{7.108}$$

then Betti's theorem yields

$$B(w, G_0) = \int_0^l p\, G_0\, dy + [V\, G_0 - M\, G_0']_0^l - [V_0\, w - M_0\, w']_0^l$$

$$- \int_0^l w\, \delta_0\, dy = \int_0^l p\, G_0\, dy + B_0(x)\, \delta - w(x) = 0 \quad (7.109)$$

where $-V_0(l)\, w(l) = B_0(x)\, \delta$ with the sign convention in Fig. 7.4.

But influence functions for the solution of Neumann problems do not exist, because one cannot place a force δ_0 on Ω and require at the same time that all the tractions on the boundary vanish

$$- \Delta G_0 = \delta_0 \qquad \frac{\partial G_0}{\partial n} = 0 \qquad \text{on } \Gamma \qquad ? \qquad (7.110)$$

The reason is that the solution of a Neumann problem is only unique up to a constant u_c as for example in the case of the beam in Fig. 7.4 c. that gives the impression of a beam on an elastic foundation, $EIw^{IV} + c\,w = p$, but it is a standard beam EIw^{IV}. It is only that the sum of the distributed load on both sides of the beam happens to be the same, so that no supports are necessary. Of course Green's functions for beams on an elastic foundation exist.

Naturally all these problems go away if the solution is made unique by specifying single values of the solution as $w(0) = w(l) = 0$, or in terms of structural mechanics, by adding supports to a structure.

Elastic supports

To be complete let us also discuss the case that the structure rests on an elastic support. Imagine that the hinged support in Fig. 7.4 is replaced by a spring with stiffness k. The decisive term in Betti's theorem (7.109) is the work term

$$V_0(l)\, w(l) = k\, G_0(l, x)\, w(l) = G_0(l, x)\, V_p = \frac{1}{k}\, V_0\, V_p \qquad (7.111)$$

which encapsulates the interaction between the compression $G_0(l, x)$ of the spring due to the point load $P = 1$ and the support reaction V_p in the load case p or—vice versa—the interaction between the support reaction V_0 due to $P = 1$ and the compression $w(l)$ of the spring in the load case p, so that the influence function becomes

$$w(x) = \int_0^l G_0(y, x)\, p(y)\, dy + V_0(l)\, w(l) \qquad (7.112)$$

which for the special case $w(l) = \delta$ is identical with (7.109). Note that $V_0(l)\, w(l)$ comes from the "boundary integral" [...]. Such edge contributions always appear in the influence functions if the structure rests on soft supports.

7.4 Generalized Green's functions

In an abstract sense the right-hand side of the weak formulation: *find a function $u \in V$ such that*

$$a(u, v) = p(v) \qquad \text{for all } v \in V \tag{7.113}$$

is a functional on V. Hence we are tempted to associate with any functional $J(v)$ an element z of V in the sense that

$$a(z, v) = J(v) \qquad \text{for all } v \in V. \tag{7.114}$$

We have switched names, $u \to z$ ($=$ generalized Green's function) and $p(v) \to J(v)$, to adopt the notation used in the literature.

Because of Green's first identity—here for a bar

$$G(v, z) = \underbrace{\int_0^l -EA\,v''\,z\,dx + [N\,z]_0^l}_{p(z)} - \underbrace{\int_0^l EA\,v'\,z'\,dx}_{a(v,z)} = 0 \tag{7.115}$$

the strain energy product can be replaced by $p(z)$ so that

$$p(z) = a(z, v) = J(v) \qquad v \in V. \tag{7.116}$$

That is, the value of the functional $J(v)$ is identical to $p(z)$ ($=$ work done by the applied load $p = \{-EA\,v'', N(l), N(0)\}$ on acting through z where p is the load case that belongs to v. If, for example, v is the solution of the problem

$$-EA\,v'' = p \qquad N(l) = P \qquad v'(0) = 0 \tag{7.117}$$

and z is the Green's function then

$$p(z) := \int_0^l p\,z\,dx + P \cdot z(l). \tag{7.118}$$

Note that (7.116) is simply Betti's theorem with the symmetric strain energy product in between.

An engineer would say that z is the solution of the load case J or rather δ_0 as in the case $J(u) = (\delta_0, u)$ and z is the Green's function G_0

$$p(G_0) = a(G_0, v) = J(v) = (\delta_0, v) \qquad v \in V. \tag{7.119}$$

The advantage of this abstract approach (7.114) is, that we can associate with any functional—not just the Dirac deltas

$$J(v) = \int_{\Omega_e} v\,d\Omega \qquad J(v) = \int_{\Omega} \delta_0\,v\,d\Omega \qquad J(v) = \int_0^l \sigma_{yy}\,dx \tag{7.120}$$

a Green's function z in the sense that $J(u) = p(z)$.

To make the mechanism more transparent, let L be a linear operator, let L^* be the adjoint operator, let u be the solution of $Lu = p$ and assume, that j is some functional, which by pairing it with u yields a result $J(u) = (u, j)$. Let z be the solution of the adjoint problem $L^*z = j$ then

$$J(u) = (u, j) = (u, L^*z) = (L u, z) = (p, z). \tag{7.121}$$

In the case $j = \delta_0$ and $z = G_0$, for example, we have

$$u(\boldsymbol{x}) = (u, \delta_0) = (u, L^*G_0) = (L u, G_0) = (p, G_0). \tag{7.122}$$

Because of Tottenham's equation (1.210), p. 64, we know that the FE program evaluates $J(u_h)$ by substituting for z the approximate generalized Green's function z_h

$$J(u_h) = p(z_h). \tag{7.123}$$

Hence, the more accurate z_h the more accurate $J(u_h)$. The strategy of the *goal oriented recovery* or simply *duality technique* can then be summarized as follows:

- Say you are interested in some point value or integral value of the solution.
- Interpret the value as the result of a functional applied to the solution

$$J(u) = (\delta_0, u) \qquad \text{point value} \tag{7.124}$$

$$J(u) = \int_0^l \sigma_{yy}\, dx \qquad \text{integral value} \tag{7.125}$$

which implies that there is a generalized Green's function z such that $J(u) = p(z)$.

- Formulate two weak boundary value problems: one for the original solution u and one for the generalized Green's function z and determine the corresponding FE solutions

$$a(u_h, \varphi_i) = p(\varphi_i) \qquad \varphi_i \in V_h, \tag{7.126}$$

$$a(z_h, \varphi_i) = J(\varphi_i) \qquad \varphi_i \in V_h. \tag{7.127}$$

- Calculate on each element Ω_e error indicators $\eta_e^{(p)}$ and $\eta_e^{(z)}$ for the two problems, multiply the two indicators $\eta_e = \eta_e^{(p)} \cdot \eta_e^{(z)}$ and refine the mesh where $\eta_e \geq TOL$ (some tolerance); see Fig. 7.5.
- By following this procedure the FE result $J(u_h)$ is automatically improved.
- Note that it is *not* necessary to actually calculate $p(z_h)$

$$J(u_h) = p(z_h), \tag{7.128}$$

because according to Tottenham's equation a direct evaluation of the FE solution yields the same result

$$J(u_h) = (\delta_0^h, u) = u_h(x) \qquad J(u_h) = \int_0^l \sigma_{yy}^h\, dx. \tag{7.129}$$

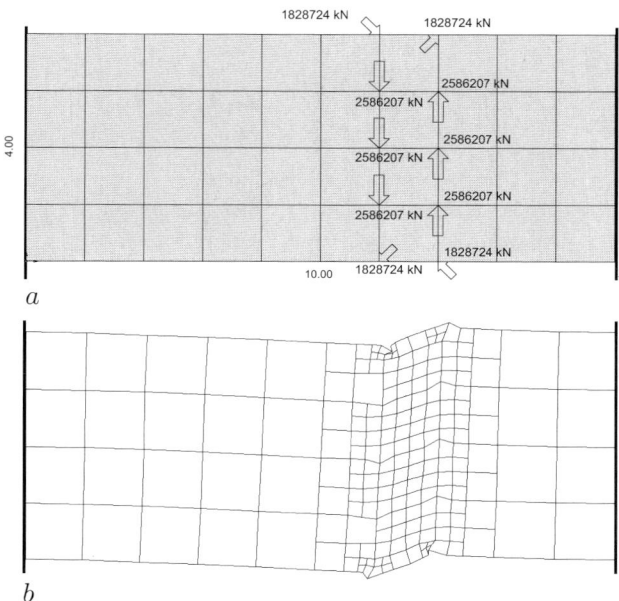

1828724 kN

1828724 kN

2586207 kN

2586207 kN

2586207 kN

2586207 kN

2586207 kN

2586207 kN

1828724 kN

1828724 kN

4.00

10.00

a

b

Fig. 7.5. Influence function for N_{xy}: **a)** initial mesh, **b)** adaptively refined mesh

If we can associate with each functional a generalized Green's function, the basic FE statement

$$a(u, v) = p(v) \qquad v \in V \tag{7.130}$$

implies that the equilibrium position u of a structure is the generalized Green's function of the functional $p(u) = (p, u)$ where p is the applied load. That is, the whole concept of a generalized Green's function is simply an application of the *Riesz' representation theorem*: for each linear (bounded) functional $J(v)$ there is an element $z \in V$ such that $a(z, v) = J(v)$ for each v.

The engineer's version would go like this: for each load p there is a strong solution $L\,u = p$, where L is the differential equation. In FE methods the load p becomes a functional $p(u)$ and the strong solution becomes a weak solution, $a(u, v) = p(v)$. Hence for each functional $p(v)$ there is a weak solution u. Now we extend this approach to just any (linear and continuous) functional $J(v)$, which not necessarily must be associated with a load case p, and we claim, that for any such functional there is a weak solution z such that $a(z, v) = J(v)$ for every $v \in V$. Finally we apply integration by parts, so that the virtual strain energy $a(z, u)$ becomes virtual external work $a(z, u) = (p, z)$, and so $J(u) = (p, z)$ where $L\,u = p$.

Essentially it is again Betti's theorem which allows to proceed from the symmetric middle term $a(z, u)$ in either direction

$$(p, z) \quad \overset{\leftarrow}{=} \quad a(z, u) \quad \overset{\rightarrow}{=} \quad (J, u) = J(u), \tag{7.131}$$

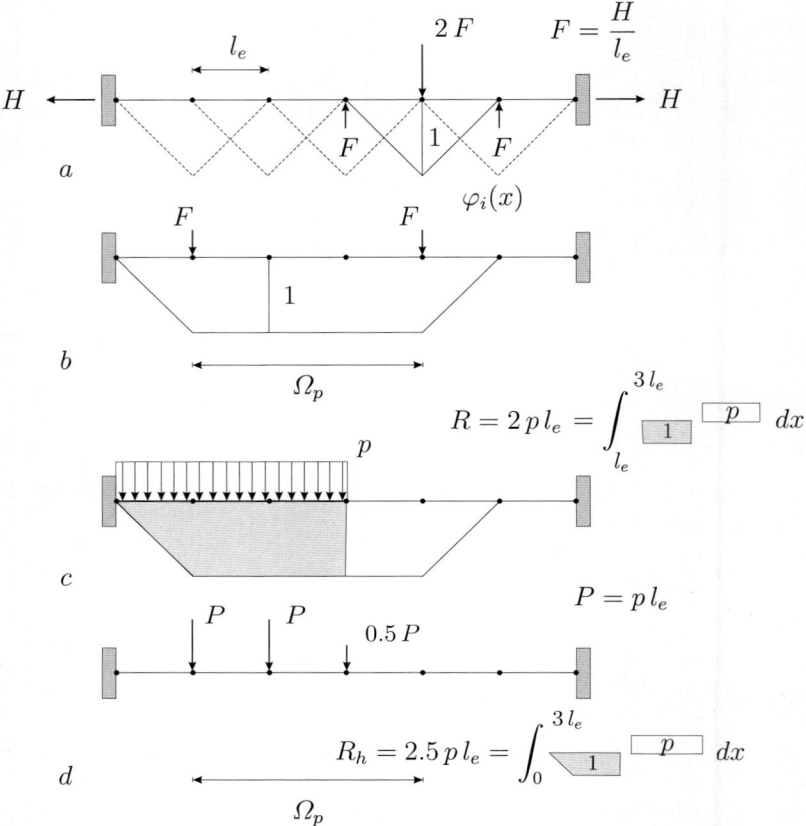

Fig. 7.6. Taut rope: **a)** unit load case p_i, **b)** equivalent nodal forces that generate the pseudo rigid-body motion $w_h = 1$, **c)** a distributed load, **d)** equivalent nodal forces

where $J = L z$ is the "load" (as for example $J = \delta_0$) in the functional $J(u)$.

That is $(J, u) = (\delta_0, u)$ with $J = L z$ is the "load form" of the functional and $J(u) = u(\boldsymbol{x})$ is the abstract form. The abstract form is that version where we directly evaluate the solution u while the load forms—actually there are two

$$J(u) = (\delta_0, u) = (L z, u) = (z, L^* u) = (z, p),\qquad(7.132)$$

are those versions where we evaluate $J(u)$ indirectly by forming e.g. the scalar product between z and p.

Examples

Consider the equation $-\boldsymbol{L}\boldsymbol{u} = \boldsymbol{p}$ of 2-D elasticity, see (7.83) p. 513. The functional which provides the sum of the horizontal forces, $\sum H$, in a patch Ω_p is $J(\boldsymbol{u}) = a(\boldsymbol{u}, \boldsymbol{e}_1)_{\Omega_p}$, because

$$G(\boldsymbol{u}, \boldsymbol{e}_1)_{\Omega_p} = \boldsymbol{p}(\boldsymbol{e}_1)_{\Omega_p} - a(\boldsymbol{u}, \boldsymbol{e}_1)_{\Omega_p}$$

$$= \underbrace{\int_{\Omega_p} \boldsymbol{p} \cdot \boldsymbol{e}_1 \, d\Omega + \int_{\Gamma_p} \boldsymbol{t} \cdot \boldsymbol{e}_1 \, ds}_{\sum H} - a(\boldsymbol{u}, \boldsymbol{e}_1)_{\Omega_p} = 0 , \quad (7.133)$$

so that the generalized Green's functions $\boldsymbol{z}$ is the solution of the variational problem

$$a(\boldsymbol{z}, \boldsymbol{v}) = J(\boldsymbol{v}) = a(\boldsymbol{v}, \boldsymbol{e}_1)_{\Omega_p} \qquad \boldsymbol{v} \in V . \qquad (7.134)$$

Clearly the solution is the step function $\boldsymbol{z} = \boldsymbol{e}_1$ (inside Ω_p and $\boldsymbol{z} = \boldsymbol{0}$ outside of Ω_p). The FE approximation $\boldsymbol{z}_h$ solves the variational problem

$$a(\boldsymbol{z}_h, \boldsymbol{\varphi}_i) = a(\boldsymbol{\varphi}_i, \boldsymbol{e}_1)_{\Omega_p} \qquad \boldsymbol{\varphi}_i \in V_h . \qquad (7.135)$$

It has the shape of the "ramp" in Fig. 1.133 p. 185, i.e., $\boldsymbol{z}_h = \boldsymbol{e}_1$ inside Ω_p and then it slowly—and not abruptly—drops to zero outside Ω_p.

The equivalent nodal forces $f_i = a(\boldsymbol{\varphi}_i, \boldsymbol{e}_1)_{\Omega_p}$ are zero if the support of the nodal unit displacement $\boldsymbol{\varphi}_i$ is contained in Ω_p, because the external loads which constitute the associated unit load case $\boldsymbol{p}_i$ are self-equilibrated, $f_i = a(\boldsymbol{\varphi}_i, \boldsymbol{e}_1) = \boldsymbol{p}(\boldsymbol{e}_1) = 0$. This is best seen in the case of a taut rope; see Fig. 7.6. The sum of the three nodal forces, that constitute the typical unit load case is zero, and therefore $f_i = -F + 2F - F = 0$. Only at the edge nodes of the patch Ω_p the balance is disturbed, because the nodal forces, that happen to lie outside of Ω_p are not taken into account, so that $f_i = -F + 2F = F$; see Fig. 7.6 b.

In the same sense the functional

$$J(\boldsymbol{v}) = a(\boldsymbol{v}, \boldsymbol{u}_{rot})_{\Omega_p} \qquad \boldsymbol{u}_{rot} = \tan \varphi \, \boldsymbol{e}_3 \times \boldsymbol{x} \qquad (7.136)$$

yields the resulting moment for the right portion Ω_p of the plate in Fig. 1.60, p. 89, for a given displacement field $\boldsymbol{v}$. Here it is assumed, that the point $\boldsymbol{x} = \boldsymbol{0}$ is the point about which the right portion is rotated by an angle φ. The FE solution $\boldsymbol{z}_h \in V_h$ of the variational equation

$$a(\boldsymbol{z}_h, \boldsymbol{\varphi}_i) = a(\boldsymbol{\varphi}_i, \boldsymbol{u}_{rot})_{\Omega_p} \qquad \boldsymbol{\varphi}_i \in V_h \qquad (7.137)$$

is not the true solution $\boldsymbol{z} = \tan \varphi \, \boldsymbol{e}_3 \times \boldsymbol{x}$, and this is why the resulting moment is wrong

$$M_h = 1.\bar{6} \, \mathrm{k\,Nm} = \int_{\Omega_p} \boldsymbol{z}_h \cdot \boldsymbol{p} \, d\Omega \neq \int_{\Omega} \boldsymbol{z} \cdot \boldsymbol{p} \, d\Omega = 2.5 \, \mathrm{k\,Nm} = M . \, (7.138)$$

Error analysis

So any point value or integral value is obtained by applying a linear functional $J(u)$ to the solution

$$J(u) = u(\boldsymbol{x}) \qquad J(u) = \sigma_{xx}(\boldsymbol{x}) \qquad J(u) = \int_0^l \sigma_{yy}\, dx\,, \qquad (7.139)$$

and the values we see on the computer screen are obtained by applying the same functionals to the FE solution u_h. Hence in linear problems the error of an FE solution is simply $J(e) = J(u - u_h)$ and this error is—as we will show in the following—"symmetric".

The Galerkin orthogonality implies

$$\begin{aligned} J(e) &= a(e, z) = a(e, z - z_h) = a(u - u_h, z - z_h) \\ &= a(u, z - z_h) = p(z - z_h)\,, \end{aligned} \qquad (7.140)$$

so that, for example, in the case $J(v) = (\delta_0, v)$

$$J(e) = u(\boldsymbol{x}) - u_h(\boldsymbol{x}) = \int_\Omega p\left(G_0 - G_0^h\right) d\Omega_{\boldsymbol{y}}\,. \qquad (7.141)$$

Now we have as well

$$J(e) = a(e, z) = (\delta_0, u - u_h) \qquad (7.142)$$

so that

$$J(e) = \frac{1}{2} p(z - z_h) + \frac{1}{2}\left(\delta_0, u - u_h\right)\,, \qquad (7.143)$$

that is we can look at the error $J(e)$ in both ways, either we attribute it to the error in the generalized Green's function $z - z_h$ or to the error in the FE solution $u - u_h$; see Fig. 7.13, p. 547. According to Betti's theorem both errors are the same. In nonlinear problems the two errors are different, see Sect. 7.5, p. 526.

Error bounds

Next let us derive error bounds. To this aim we consider the equation $-\boldsymbol{L}\boldsymbol{u} = \boldsymbol{p}$ of 2-D elasticity with boundary conditions $\boldsymbol{u} = \boldsymbol{0}$. Let $\boldsymbol{e} = \boldsymbol{u} - \boldsymbol{u}_h$ the error of the FE solution $\boldsymbol{u}_h$ then

$$G(\boldsymbol{e}, \boldsymbol{\varphi}_k) := \sum_i G(\boldsymbol{e}, \boldsymbol{\varphi}_k)_{\Omega_i} = \sum_i \{\int_{\Omega_i} \boldsymbol{r}_i \cdot \boldsymbol{\varphi}_k\, d\Omega + \int_{\Gamma_i} \boldsymbol{j}_i \cdot \boldsymbol{\varphi}_k\, ds$$
$$-a(\boldsymbol{e}, \boldsymbol{\varphi}_k)_{\Omega_i}\} = 0 \qquad \boldsymbol{\varphi}_k \in V_h \qquad (7.144)$$

where on each element Ω_i

$$\boldsymbol{r}_i := \boldsymbol{p} + \boldsymbol{L}\boldsymbol{u}_h \qquad \boldsymbol{j}_i := \frac{1}{2}\,[\boldsymbol{t}^+ + \boldsymbol{t}^-] \qquad (7.145)$$

are the element residuals and the evenly split jumps of the tractions between the elements.

Next let $J(\boldsymbol{v})$ be any functional and $\boldsymbol{z} \in V$ the associated generalized Green's function, the value of $J(\boldsymbol{e})$ is just the virtual work of the right-hand side of $\boldsymbol{e}$ acting through $\boldsymbol{z}$ or

$$J(\boldsymbol{e}) = \sum_i \{(\boldsymbol{r}_i, \boldsymbol{z})_{\Omega_i} + (\boldsymbol{j}_i, \boldsymbol{z})_{\Gamma_i}\} = \sum_i \{(\boldsymbol{r}_i, \boldsymbol{z} - \boldsymbol{z}_h)_{\Omega_i} + (\boldsymbol{j}_i, \boldsymbol{z} - \boldsymbol{z}_h)_{\Gamma_i}\}$$
$$(7.146)$$

where in the last equation we made use of the Galerkin orthogonality. Note also that because of $\boldsymbol{z} = \boldsymbol{z}_h = \boldsymbol{0}$ on Γ the boundary integrals are zero on each portion $\Gamma_i \subset \Gamma$.

By applying the Cauchy-Schwarz inequality if follows [22]

$$|J(\boldsymbol{e})| \le \eta_\omega := \sum_i \eta_i^{(p)}\,\eta_i^{(z)}\,, \qquad (7.147)$$

where with the diameter h_i of the single elements Ω_i

$$\eta_i^{(p)} := (\|\boldsymbol{r}_i\|_{0,\Omega_i}^2 + \frac{1}{h_i}\,\|\boldsymbol{j}_i\|_{0,\Gamma_i}^2)^{1/2} \qquad (7.148)$$

$$\eta_i^{(z)} := (\|\boldsymbol{z} - \boldsymbol{z}_h\|_{0,\Omega_i}^2 + h_i\|\boldsymbol{z} - \boldsymbol{z}_h\|_{0,\Gamma_i}^2)^{1/2}\,. \qquad (7.149)$$

Equation (7.147) is the basis of the duality approach or the goal-oriented techniques, see Sect. 1.31, p. 156.

The two special functionals

$$J(\boldsymbol{v}) := \frac{a(\boldsymbol{v}, \boldsymbol{e})}{\|\boldsymbol{e}\|_E} \;\to\; J(\boldsymbol{e}) = \frac{a(\boldsymbol{e}, \boldsymbol{e})}{\|\boldsymbol{e}\|_E} = \frac{\|\boldsymbol{e}\|_E^2}{\|\boldsymbol{e}\|_E} = \|\boldsymbol{e}\|_E \qquad (7.150)$$

$$J(\boldsymbol{v}) := \frac{(\boldsymbol{v}, \boldsymbol{e})}{\|\boldsymbol{e}\|_0} \;\to\; J(\boldsymbol{e}) = \frac{(\boldsymbol{e}, \boldsymbol{e})}{\|\boldsymbol{e}\|_0} = \frac{\|\boldsymbol{e}\|_0^2}{\|\boldsymbol{e}\|_0} = \|\boldsymbol{e}\|_0 \qquad (7.151)$$

coincide at $\boldsymbol{v} = \boldsymbol{e}$ with $\|\boldsymbol{e}\|_E$ and $\|\boldsymbol{e}\|_0$ respectively and consequently the *dual weighted residual* error estimate (7.147) can be applied to these global norms as well.

In a classical L_2-error estimate appears a constant c_I, that reflects the interpolation properties of the space V_h and a global stability constant c_S

$$\|\boldsymbol{e}\|_0 \le \eta_{L_2} := c_I\,c_S\left(\sum_i h_i^4\,(\eta_i^{(p)})^2\right)^{1/2} \qquad c_S := \|a(\boldsymbol{z}, \boldsymbol{z})\|_{E,\Omega}\,, \quad (7.152)$$

which represents the global energy norm of the generalized Green's function, while in the duality approach

$$||e||_0 \leq \eta_{L_2}^{\omega} := c_I \sum_i h_i^2 \, \eta_i^{(p)} \, \eta_i^{(z)} \qquad \eta_i^{(z)} := ||a(z,z)||_{E,\Omega_i} \qquad (7.153)$$

the weights $\eta_i^{(z)}$ reflect the local character of the generalized Green's function and "it can be beneficial to keep the dual weights within the error estimator rather than condensing them into just one global stability constant" [22].

7.5 Nonlinear problems

In a constrained minimization problem

$$J(u) \to \min \qquad A(u) = 0 \;\; \text{constraints} \qquad (7.154)$$

the Lagrangian functional

$$L(u,\lambda) = J(u) + <\lambda, A(u)> \qquad <.,.> \;\; \text{"duality pairing"} \qquad (7.155)$$

is stationary at the minimum point u_0, see e.g. [202], i.e.,

$$\delta J(u_0) + \lambda \, d(u_0) = 0 \,. \qquad (7.156)$$

This technique is now adopted[1] to estimate the error $J(u) - J(u_h)$ in nonlinear problems by introducing a "dual" variable z.

As a model problem we choose the linear Poisson equation on $V = \{u \in H_1(\Omega) \,|\, u = 0 \;\text{on}\; \Gamma\}$

$$A(u) := -\Delta u - p = 0 \,. \qquad (7.157)$$

We let

$$A(u)(\psi) := a(u,\psi) - (p,\psi) \qquad a(u,\psi) := (\nabla u, \nabla \psi) \qquad (7.158)$$

the weak form and we intend to evaluate the solution at a point x so that

$$J(u) = (\delta_0, u) = u(x) \,. \qquad (7.159)$$

The Gateaux derivatives of these functionals are

$$J'(u)(\varphi) := \left[\frac{d}{d\varepsilon} J(u + \varepsilon \, \varphi)\right]_{\varepsilon=0} = (\delta_0, \varphi) = J(\varphi) \qquad (7.160)$$

and

$$A'(u)(\varphi, z) := \left[\frac{d}{d\varepsilon} A(u + \varepsilon\varphi)(z)\right]_{\varepsilon=0} = a(\varphi, z) \,. \qquad (7.161)$$

---

[1] The following is based on Sect. 6.1 in [22]. Added in proof: a good summary can also be found in Ern A, Guermond J-L (2004) Theory and Practice of Finite Elements, Springer-Verlag

Next we define the Lagrangian functional

$$L(u, z) = J(u) - A(u)(z) \qquad (7.162)$$

and we seek for a stationary point $\{u, z\} \in V \times V$ of $L(.,.)$, i.e.,

$$L'(u, z)(\varphi, \psi) = \left\{ \begin{matrix} J'(u_h)(\varphi_h) - A'(u_h)(\varphi_h, z_h) \\ -A(u)(\psi) \end{matrix} \right\} = 0 \qquad (7.163)$$
$$\text{for all } \{\varphi_h, \psi_h\} \in V \times V.$$

Evidently in the linear case these two equations are identical to the standard approach

$$J(\varphi_h) - a(\varphi_h, z_h) = 0 \qquad \text{for all } \varphi_h \in V_h \qquad \to z_h \qquad (7.164)$$
$$a(u_h, \psi_h) - p(\psi_h) = 0 \qquad \text{for all } \psi_h \in V_h \qquad \to u_h. \qquad (7.165)$$

Under appropriate assumptions we have the error representation

$$J(u) - J(u_h) = \frac{1}{2}\rho(u_h)(z - z_h) + \frac{1}{2}\rho^*(u_h, z_h)(u - u_h) + R_h^{(3)} \qquad (7.166)$$

where

$$\rho(u_h)(z - z_h) := -A(u_h)(z - z_h) \qquad (7.167)$$
$$\rho_h^*(u_h, z_h)(u - u_h) := J'(u_h) - A'(u_h)(u - u_h, z_h) \qquad (7.168)$$

and where the remainder term $R_h^{(3)}$ is cubic in the "primal" and "dual" errors $e := u - u_h$ and $e^* := z - z_h$ and involves second and third order Gateaux derivatives of $J(.)$ and $A(.)$.

In the case of the linear model problem (7.157) we have

$$- A(u_h)(z - z_h) = -a(u_h, z - z_h) + p(z - z_h) = -a(u_h, z) + p(z)$$
$$= -p_h(z) + p(z) = -u_h(\boldsymbol{x}) + u(\boldsymbol{x}) \qquad (7.169)$$

and

$$J'(u - u_h) - A'(u_h)(u - u_h, z_h) = (\delta_0, u - u_h) - a(u - u_h, z_h)$$
$$= (\delta_0, u) - a(u, z_h) = (\delta_0, u) - (\delta_0^h, u) = u(\boldsymbol{x}) - u_h(\boldsymbol{x}) \qquad (7.170)$$

and of course $R^{(3)}$ is zero in this case so that indeed

$$J(u) - J(u_h) = \frac{1}{2}(u(\boldsymbol{x}) - u_h(\boldsymbol{x})) + \frac{1}{2}(u(\boldsymbol{x}) - u_h(\boldsymbol{x})). \qquad (7.171)$$

In the case of a nonlinear equation such as

$$A(u) := -\Delta u - u^3 - p = 0 \qquad (7.172)$$

the weak form is

$$A(u)(\psi) := (\nabla u, \nabla \psi) - (u^3, \psi) - (p, \psi) \tag{7.173}$$

and

$$A'(u)(\varphi, z) := (\nabla \varphi, \nabla z) - (3u^2 \varphi, z) \tag{7.174}$$

so that the generalized Green's function z is the solution of

$$(\nabla \varphi, \nabla z) - (3u^2 \varphi, z) = J'(\varphi) \qquad \text{for all } \varphi \in V \tag{7.175}$$

and the FE approximation z_h solves the variational problem

$$a_T(\boldsymbol{u}, \boldsymbol{z}_h, \boldsymbol{\varphi}_h) = J'(\boldsymbol{\varphi}_h) \qquad \text{for all } \boldsymbol{\varphi}_h \in V_h \tag{7.176}$$

where we have written $a_T(., ., .)$ for the Gateaux derivative, i.e., the left-hand side of (7.175). If J is linear then $J' = J$ so that

$$\boldsymbol{K}_T(\boldsymbol{u})\, \boldsymbol{z} = \boldsymbol{j} \qquad j_i = J(\boldsymbol{\varphi}_i) \tag{7.177}$$

where $\boldsymbol{K}_T$ is the tangential stiffness matrix and $\boldsymbol{z}$ is the vector of nodal values of the field $\boldsymbol{z}$.

A more pedestrian, engineering approach would go like this: let $a(\boldsymbol{u}, \boldsymbol{v}) = p(\boldsymbol{v})$ the nonlinear equation and let $\boldsymbol{u} = \boldsymbol{u}_h + \boldsymbol{e}$ then

$$a(\boldsymbol{u}_h + \boldsymbol{e}, \boldsymbol{v}) = p(\boldsymbol{v}) \qquad \text{for all } \boldsymbol{v} \in V \tag{7.178}$$

or if we do a "Taylor expansion"

$$a(\boldsymbol{u}_h, \boldsymbol{v}) + a_T(\boldsymbol{u}_h; \boldsymbol{e}, \boldsymbol{v}) + \ldots = p(\boldsymbol{v}) \tag{7.179}$$

and neglect the higher order terms $(\ldots)$

$$\begin{aligned}
a_T(\boldsymbol{u}_h; \boldsymbol{e}, \boldsymbol{v}) &= p(\boldsymbol{v}) - a(\boldsymbol{u}_h, \boldsymbol{v}) = p(\boldsymbol{v}) - p_h(\boldsymbol{v}) \\
&= \sum_i \left\{ \int_{\Omega_i} \boldsymbol{r} \cdot \boldsymbol{v}\, d\Omega + \int_{\Gamma_i} \boldsymbol{j} \cdot \boldsymbol{v}\, ds \right\} := r(\boldsymbol{v})
\end{aligned} \tag{7.180}$$

where $\boldsymbol{r}$ and $\boldsymbol{j}$ are defined as in (1.425) p. 150.

Next, let $a_T^*(\boldsymbol{u}_h; \boldsymbol{e}, \boldsymbol{v})$ the dual bilinear form defined by switching the last two arguments in a_T

$$a_T^*(\boldsymbol{u}_h; \boldsymbol{e}, \boldsymbol{v}) := a_T(\boldsymbol{u}_h; \boldsymbol{v}, \boldsymbol{e}) \,. \tag{7.181}$$

If a_T is symmetric in the last two arguments—as in hyperelasticity—then $a_T^* = a_T$. Now let $J(\boldsymbol{v})$ a linear functional on V and $\boldsymbol{z}$ the solution of

$$a_T^*(\boldsymbol{u}_h; \boldsymbol{z}, \boldsymbol{v}) = J(\boldsymbol{v}) \tag{7.182}$$

then

$$J(e) = a_T^*(\boldsymbol{u}_h; \boldsymbol{z}, \boldsymbol{e}) = a_T(\boldsymbol{u}_h; \boldsymbol{e}, \boldsymbol{z}) = r(\boldsymbol{z}).\tag{7.183}$$

If $\boldsymbol{z}_h$ is the FE solution of (7.182), we can invoke the Galerkin orthogonality

$$0 = a_T^*(\boldsymbol{u}_h; \boldsymbol{z} - z_h, \boldsymbol{v}) = a_T(\boldsymbol{u}_h; \boldsymbol{e}, \boldsymbol{z} - z_h) = r(\boldsymbol{z} - z_h)\tag{7.184}$$

and so we arrive at

$$J(e) = \sum_i \left\{ \int_{\Omega_i} \boldsymbol{r} \bullet (\boldsymbol{z} - z_h)\, d\Omega + \int_{\Gamma_i} \boldsymbol{j} \bullet (\boldsymbol{z} - z_h)\, ds \right\}.\tag{7.185}$$

Remark 7.1. In the mathematical literature the Gateaux derivative a_T is often replaced by a *secant form*

$$a_S(\boldsymbol{u}, \boldsymbol{u}_h; \boldsymbol{e}, \boldsymbol{v}) := \int_0^1 a_T(\boldsymbol{u}_h + s\,\boldsymbol{e}; \boldsymbol{e}, \boldsymbol{v})\, ds,\tag{7.186}$$

which can be interpreted as the average Fréchet derivative of $a(\boldsymbol{u}, \boldsymbol{v})$. In this case the error $\boldsymbol{e} = \boldsymbol{u} - \boldsymbol{u}_h$ is the solution of the linear variational problem

$$a_S(\boldsymbol{u}, \boldsymbol{u}_h; \boldsymbol{e}, \boldsymbol{v}) = a(\boldsymbol{u}, \boldsymbol{v}) - a(\boldsymbol{u}_h, \boldsymbol{v}) = r(\boldsymbol{v}) \quad \boldsymbol{v} \in V.\tag{7.187}$$

Often this approach leads to identical formulation—because the exact solution $\boldsymbol{u}$ is unknown and therefore compromises must be made—though in specific circumstances this formulation can be advantageous [153].

7.6 The derivation of influence functions

Influence functions or influence lines are based on Betti's theorem, $B(w_1, w_2) = 0$. Because the idea behind influence functions is central for the understanding of the distribution of the internal actions and also the support reactions in a structure we start with a short repetition of classical structural analysis.

Influence function for $V(x)$

To obtain the influence function for the shear force $V(x)$ we introduce a shear hinge at x and to keep the balance with the applied load the prior internal actions $V_i(x)$ and $V_r(x)$ now act as external forces. In a second load case two opposite forces spread the two faces of the shear hinge by one unit length, $G_3(x_-, x) - G_3(x_+, x) = 1$, apart. According to Betti's theorem the reciprocal external work of the two systems must be the same

$$W_{1,2} = -V(x) \cdot 1 + \int_0^l G_3(y, x)\, p(y)\, dy = 0 = W_{2,1}\tag{7.188}$$

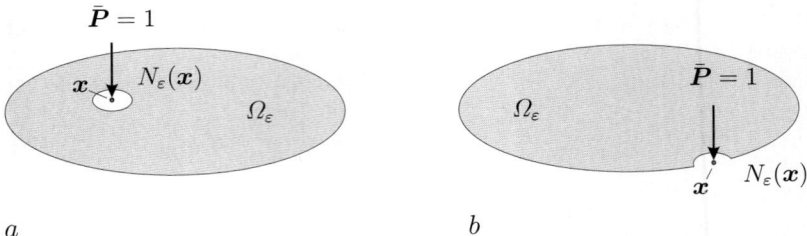

a

b

Fig. 7.7. The punctured domain: **a)** a point inside Ω, **b)** a point on the edge

or

$$V(x) = \int_0^l G_3(y, x) \, p(y) \, dy \, . \tag{7.189}$$

This is the standard procedure for influence functions. The work $W_{1,2} = 0$ because the two opposite forces that spread the shear hinge apart are of the same size so when they act through the same deflection $w(x)$ their effort is nil.

To see how influence functions in 2-D are derived, we consider the solution u of the Poisson equation

$$- \Delta u = p \quad \text{on } \Omega = \text{unit disk}\,, \quad u = 0 \quad \text{on } \Gamma = \text{unit circle}\,, \tag{7.190}$$

which can be identified with the deflection of a circular prestressed membrane which bears a pressure p.

The Green's function for the deflection at the center $x = 0$ of the unit disk

$$G_0(y, x) = -\frac{1}{2\pi} \ln r \tag{7.191}$$

is a homogeneous solution of the Laplace equation at all points $y \neq x$,

$$- \Delta G_0(y, x) = 0 \quad y \neq x \, . \tag{7.192}$$

Next a small circular hole N_ε with radius ε is punched in the unit disk. The center of the hole is the center of the disk, $x = 0$, and Γ_{N_ε} is the edge of the hole. At a point y on this circle the normal vector $n = n(y)$ points to the center $x = 0$, and it has components $n_1 = -\cos \varphi$ and $n_2 = -\sin \varphi$ if polar coordinates (r, φ) centered at $x = 0$ are used. For all points y on the circle, the distance $r = |y - x|$ from the center x is the same. The gradient $\nabla_y r$ of the distance r points in the direction opposite the normal vector n, because this is the direction into which the point y must be pushed if the distance from $x = 0$ is to increase at the fastest rate possible. Hence

$$-\frac{1}{2\pi} \frac{\partial}{\partial n} \ln r = -\frac{1}{2\pi} \frac{1}{\varepsilon} \nabla_y r \cdot n = \frac{1}{2\pi} \frac{1}{\varepsilon} n \cdot n = \frac{1}{2\pi} \frac{1}{\varepsilon} \, . \tag{7.193}$$

Next Betti's theorem is formulated on the domain $\Omega_\varepsilon = \Omega - N_\varepsilon(\boldsymbol{x})$. On the outer edge Γ both solutions are zero, $u = G_0 = 0$, so that:

$$B(G_0, u)_{\Omega_\varepsilon} = \int_{\Omega_\varepsilon} -\Delta G_0 \, u \, d\Omega_{\boldsymbol{y}} - \int_{\Gamma_{N_\varepsilon}} \frac{1}{2\pi} \frac{\partial}{\partial n} \ln r \, u \, ds_{\boldsymbol{y}}$$

$$+ \int_{\Gamma_{N_\varepsilon}} \frac{\partial u}{\partial n} G_0 \, ds_{\boldsymbol{y}} - \int_{\Omega_\varepsilon} G_0 \, p \, d\Omega_{\boldsymbol{y}}$$

$$\text{(7.194)}$$

The first integral is zero because $-\Delta G_0 = 0$ in Ω_ε and the third integral is of order $O(\varepsilon)$

$$\frac{\partial u}{\partial n} \cdot \ln \varepsilon \cdot \varepsilon \cdot d\varphi = O(1) \cdot \ln \varepsilon \cdot \varepsilon = O(\varepsilon) \qquad \text{(7.195)}$$

so that

$$B(G_0, u)_{\Omega_\varepsilon} = \int_{\Gamma_{N_\varepsilon}} \frac{1}{2\pi} \frac{\partial}{\partial n} \ln r \, u \, ds_{\boldsymbol{y}} + O(\varepsilon) - \int_{\Omega_\varepsilon} G_0 \, p \, d\Omega_{\boldsymbol{y}}$$

$$= \int_0^{2\pi} \frac{1}{2\pi\varepsilon} u(\underbrace{\boldsymbol{x} + \varepsilon \, \boldsymbol{n}(\varphi)}_{\boldsymbol{y}(\varphi)}) \, \varepsilon \, d\varphi + O(\varepsilon) - \int_{\Omega_\varepsilon} G_0 \, p \, d\Omega_{\boldsymbol{y}}$$

$$\text{(7.196)}$$

which in the limit becomes

$$\lim_{\varepsilon \to 0} B(G_0, u)_{\Omega_\varepsilon} = u(\boldsymbol{x}) - \int_\Omega G_0(\boldsymbol{y}, \boldsymbol{x}) \, p(\boldsymbol{y}) \, d\Omega_{\boldsymbol{y}} . \qquad \text{(7.197)}$$

The important observation is that in this approach $u(\boldsymbol{x})$ is not the limit of a domain integral, as the notation

$$- \Delta G_0(\boldsymbol{y}, \boldsymbol{x}) = \delta_0(\boldsymbol{y} - \boldsymbol{x}) \qquad \text{(7.198)}$$

seems to suggest, and Betti's theorem evidently seems to confirm

$$B(G_0, u) = \int_\Omega -\Delta G_0 \, u \, d\Omega - \int_\Omega p \, G_0 \, d\Omega_{\boldsymbol{y}}$$

$$= \int_\Omega \delta_0(\boldsymbol{y} - \boldsymbol{x}) \, u(\boldsymbol{y}) \, d\Omega_{\boldsymbol{y}} - \int_\Omega G_0(\boldsymbol{y}, \boldsymbol{x}) \, p(\boldsymbol{y}) \, d\Omega_{\boldsymbol{y}}$$

$$= u(\boldsymbol{x}) - \int_\Omega G_0(\boldsymbol{y}, \boldsymbol{x}) \, p(\boldsymbol{y}) \, d\Omega_{\boldsymbol{y}} . \qquad \text{(7.199)}$$

But juggling with Dirac's delta is not mathematics, rather it is an application of symbolic algebra. The Dirac delta is a handy symbol to express the algebraic properties of the Green's function, but the real properties can only be clarified by doing mathematics as in (7.194).

So all the single terms, the point values $\partial^0 u \equiv u(\boldsymbol{x}), \partial^1 u \equiv \sigma_{xx}(\boldsymbol{x})$, in the influence functions are the limits of certain boundary integrals over the edge of the hole

$$\lim_{\varepsilon \to 0} \int_{\Gamma_{N_\varepsilon}} \partial^i u \, \partial^j \hat{u} \, ds = 1 \cdot \partial^i u(\boldsymbol{x}) \qquad i + j = 2m - 1 \qquad (7.200)$$

where the conjugate kernel $\partial^j \hat{u}$ that makes $\partial^i u(\boldsymbol{x})$ emerge has the property

$$\lim_{\varepsilon \to 0} \int_{\Gamma_{N_\varepsilon}} \partial^j \hat{u} \, ds = 1 \,. \qquad (7.201)$$

In 1-D the neighborhood Γ_{N_ε} consists of two points, in 2-D it is a circle

$$\lim_{\varepsilon \to 0} \int_{\Gamma_{N_\varepsilon}} V_n \, ds = 1 \qquad V_n = \text{Kirchhoff shear} \qquad (7.202)$$

and in 3-D it is a sphere — two points become a circle become a sphere.

Hence one cannot monitor a dislocation in a slab by checking, say, the deflection only at two points, $w(\boldsymbol{x} + \boldsymbol{\varepsilon}) - w(\boldsymbol{x} - \boldsymbol{\varepsilon}) \neq 1$. Rather one must complete a full circle to sense a bend δ_2 or a dislocation δ_3 in a slab

$$\delta_2 : \lim_{\varepsilon \to 0} \int_{\Gamma_{N_\varepsilon}} \frac{\partial G_2}{\partial n} \, ds = 1 \qquad \delta_3 : \lim_{\varepsilon \to 0} \int_{\Gamma_{N_\varepsilon}} G_3 \, ds = 1 \,. \qquad (7.203)$$

Remark 7.2. For a detailed analysis of the derivation of the hyper-singular influence function for the slope in a slab (Kirchhoff plate, biharmonic equation)

$$\frac{\partial w}{\partial n} = \lim_{\varepsilon \to 0} B(G_1, w) = 0 \qquad (7.204)$$

see [116] p. 375.

Influence functions for crack tip singularities - don't exist

If there were an influence function for the singular stresses at a crack tip

$$\sigma_{yy}(\boldsymbol{x}) = \int_\Omega \boldsymbol{G}_1^{yy}(\boldsymbol{y}, \boldsymbol{x}) \bullet \boldsymbol{p}(\boldsymbol{y}) \, d\Omega_{\boldsymbol{y}} = \infty \,, \qquad (7.205)$$

then because the load $\boldsymbol{p}$ is finite, the stress could only become infinite if the kernel $\boldsymbol{G}_1^{yy}$ is infinite, but infinite displacements $|\boldsymbol{G}_1^{yy}| = \infty$ (each kernel is a displacement field) in almost any patch Ω_p of a plate make no sense, as they would tear the plate apart.

Hence there cannot exist an influence function for the stresses at a crack tip, only for the stress intensity factor [47].

The mathematical reason is the following: To derive an influence function for $\sigma_{yy}(\boldsymbol{x})$ at a point $\boldsymbol{x}$, first a small circular region of the point $\boldsymbol{x}$ must be excluded, and only then are we allowed to let the radius ε tend to zero. In standard situations, a single term $\sigma_{yy}(\boldsymbol{x}) \cdot 1$ will be recovered in the limit, but if the stress field is infinite at $\boldsymbol{x}$, the effect of the infinite stress is canceled in Green's second identity (Betti's theorem) by a second singular term of the opposite sign, so that in the limit—which is guaranteed to be zero,

$$\lim_{\varepsilon \to 0} B(\boldsymbol{u}, \boldsymbol{G}_1^{yy})_{\Omega_\varepsilon} = \infty - \infty + \text{finite terms} = 0 \qquad (7.206)$$

we are left with meaningless terms, "the ashes", which convey no real information.

Hence each influence function is the limit of an integral identity. Concepts as *Cauchy principal value* or Hadamard's *partie fini integral* are just other names for the limit

$$\lim_{\varepsilon \to 0} B(\boldsymbol{u}, \boldsymbol{G}_1^{yy})_{\Omega_\varepsilon} = 0 . \qquad (7.207)$$

By this process singular or hypersingular integrals are *automatically* regularized, because the critical terms drop out.

In Sect. 1.20, p. 80, we argued that in the presence of stress singularities it is more reasonable to work with resultant stresses. Now we can be more precise. If the stress σ_{yy} becomes singular at the crack tip but if the integral

$$N_y = \int_0^l \sigma_{yy} \, dx < \infty \qquad (7.208)$$

across the cut is bounded, there exists an influence function for the resultant stress N_y—in the sense of (7.207)—and the FE program has a chance to approximate this influence function.

Equivalent nodal forces

The equivalent nodal forces f_i^G for the numerical Green's functions are the displacements, the stresses, etc. of the shape functions at the point $\boldsymbol{x}$, see Sect. 1.19 p. 69,

$$f_i^G = \varphi_i(\boldsymbol{x}) \qquad f_i^G = \sigma_{xx}(\varphi_i)(\boldsymbol{x}) \qquad f_i^G(\boldsymbol{x}) = q_x(\varphi_i)(\boldsymbol{x}) . \qquad (7.209)$$

But the *dimension* of each f_i^G is force × displacements

$$f_i^G = \int_\Omega \delta_0(\boldsymbol{y} - \boldsymbol{x}) \, \varphi_i(\boldsymbol{y}) \, d\Omega_{\boldsymbol{y}} \equiv \text{kN} \times \text{m} \qquad (7.210)$$

To extract the stress from a field we apply a dislocation and calculate the work done by the stress on acting through the dislocation, $[\text{kN/m}^2] \times [\text{m}]$, and to extract a bending moment we apply a dimensionless unit rotation, so that the work done is $M \times w' = [\text{kNm}] \times [\ ]$, etc.. (A unit slope means $dw/dx = 1 \times [\text{m}]/[\text{m}] = 1 \times [\ ]$ or $\tan \varphi = 1$).

This result is in agreement with the fact that the influence functions are energy expressions

$$u(\boldsymbol{x})\,[\text{m}] \times 1\,[\text{kN}] = \dots \qquad \sigma_{xx}(\boldsymbol{x})\,[\text{kN/m}^2] \times 1\,[\text{m}] = \dots \qquad (7.211)$$

So when we calculate a point value by summing over the nodes

$$u_h(\boldsymbol{x})\,[\text{m}] \times 1\,[\text{kN}] = \sum_i f_i^G\,[\text{kNm}] \cdot u_i \qquad (7.212)$$

we must divide by the unit, here 1 kN, which extracts the displacement or the stress, etc., from the field $\boldsymbol{u}$ via (7.210), so that

$$u_h(\boldsymbol{x})\,[\text{m}] = \frac{1}{\text{kN}} \sum_i f_i^G\,[\text{kNm}] \cdot u_i = \sum_i \varphi_i(\boldsymbol{x})\,[\text{m}] \cdot u_i\,. \qquad (7.213)$$

Recall that the f_i^G of the Green's function for $u_h(\boldsymbol{x})$ are the nodal values of the shape functions φ_i at the point $\boldsymbol{x}$.

The nodal displacements u_i play the role of weights, that is pure numbers. The dimension [m] of the displacement $u(\boldsymbol{x}) = \sum_i u_i\,\varphi_i(\boldsymbol{x}) = \sum_i u_i \times (\varphi_i(\boldsymbol{x})[\text{m}])$ is attached, so to speak, to the φ_i and has already been consumed in the definition of the f_i^G in (7.210).

The net result is that nothing needs to be done: the equivalent nodal forces f_i^G of the Green's functions are the displacement, stresses, etc. of the shape functions at $\boldsymbol{x}$ —times the physical dimension of the Dirac delta, [m], [kN], etc., but because we later divide again by these terms we may ignore them from the start—and so when we multiply the f_i^G with the nodal values u_i of the FE solution we obtain—quite naturally—the pertinent values of the FE solution at the point $\boldsymbol{x}$

$$\sigma_{xx}^h(\boldsymbol{x}) = \sum_i f_i^G \cdot u_i = \sum_i \sigma_{xx}(\boldsymbol{\varphi}_i)(\boldsymbol{x}) \cdot u_i\,. \qquad (7.214)$$

The same can be expressed as

$$\sigma_{xx}^h(\boldsymbol{x}) = \sum_i f_i \cdot u_i^G \qquad (7.215)$$

where the u_i^G are the nodal values of the Green's function and the f_i are the equivalent nodal forces of the load case p, see Sect. 1.19 p. 79.

The first formula corresponds to

$$\sigma_{xx}^h(\boldsymbol{x}) = \int_\Omega \delta_1(\boldsymbol{y} - \boldsymbol{x})\,\boldsymbol{u}_h(\boldsymbol{y})\,d\Omega_y = \sum_i \underbrace{\int_\Omega \delta_1(\boldsymbol{y} - \boldsymbol{x})\,\varphi_i(\boldsymbol{y})\,d\Omega_y}_{\sigma_{xx}(\boldsymbol{\varphi}_i)(\boldsymbol{x})} \cdot u_i$$

$$(7.216)$$

and the second formula corresponds to

$$\sigma_{xx}^h(\boldsymbol{x}) = \int_\Omega G_1^h(\boldsymbol{y}, \boldsymbol{x})\, p(\boldsymbol{y})\, d\Omega_{\boldsymbol{y}} = \sum_i \underbrace{G_1^h(\boldsymbol{y}_i, \boldsymbol{x})}_{u_i^G} \cdot \underbrace{\int_\Omega \varphi_i(\boldsymbol{y})\, p(\boldsymbol{y})\, d\Omega_{\boldsymbol{y}}}_{f_i}$$

$$(7.217)$$

7.7 Weak form of influence functions

With Mohr's integral we can calculate the deflection of a beam

$$w(x) = \int_0^l \frac{M_0\, M}{EI}\, dy \qquad M(x) \neq \int_0^l \frac{M_2\, M}{EI}\, dy = 0 \qquad (7.218)$$

but not the bending moment $M(x)$. But if we do finite elements then everything fits perfectly

$$M_h(x) = \int_0^l \frac{M_2^h\, M_h}{EI}\, dy\,. \qquad (7.219)$$

This surprising result is the topic of this section.

The classical influence functions are "strong" formulations, are based on Betti's theorem or Green's second identity. Here we study "weak" formulations which are based on the principle of virtual forces/displacements or else, on Green's first identity. By weak we mean that the output value

$$J(w) = a(G_i, w) \qquad \equiv \qquad w(x) = \int_0^l \frac{M_0\, M}{EI}\, dy \qquad (7.220)$$

is calculated by forming the strain-energy product between the Green's function G_i and the solution w; an engineer would say: with Mohr's integral. This is only possible if $J(w)$ is a displacement or a deflection or rotation (Euler-Bernoulli beams and Kirchhoff) because the strain energy product of higher order Green's functions (for stresses and alike) is zero $a(G_i, w) = 0$ and so $J(w) = a(G_i, w)$ makes no sense.

The strange thing is that if we approximate the Green's function $z\ (= G_i)$ with finite elements by solving the problem

$$z_h \in V_h : \qquad a(z_h, \varphi_i) = J(\varphi_i) \qquad \varphi_i \in V_h \qquad (7.221)$$

then the FE solution z_h is a reasonable approximation of z and we have

$$J(w_h) = \int_0^l z_h(y, x)\, p(y)\, dy = a(z_h, w_h)\,. \qquad (7.222)$$

This is just Tottenham's equation (1.210), p. 64. In simpler terms the Green's function z_h just does what it is supposed to do

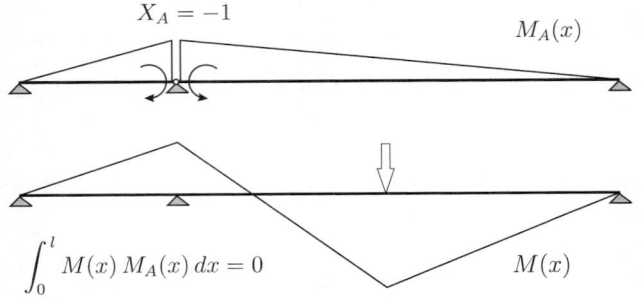

$X_A = -1$

$M_A(x)$

$\int_0^l M(x)\, M_A(x)\, dx = 0$

$M(x)$

Fig. 7.8. The strain energy product $a(M_A, M) = 0$ is zero

$$J(w_h) = \sum_i J(\varphi_i)\, u_i = \sum_i a(z_h, \varphi_i)\, u_i = a(z_h, w_h) \qquad (7.223)$$

because if z_h solves (7.221) and $w_h = \sum_i u_i\, \varphi_i$ then (7.223) follows naturally.

So in FE analysis we can use both formulas, Betti and $a(z_h, w_h)$, regardless of what we calculate. When z is exact then only Betti will do for force terms.

The solve this riddle let us compare the three variants of influence functions: the exact formulations

$$\partial^i w = \int_0^l G_i(y, x)\, p(y)\, dy = a(G_i, w) = \int_0^l \delta_i(y - x)\, w(y) dy \quad (7.224)$$

$$(1) \qquad\qquad\qquad (2) \qquad\qquad (3)$$

and the approximate formulations

$$\partial^i w_h = \int_0^l G_i^h(y, x)\, p(y)\, dy = a(G_i^h, w) = \int_0^l \delta_i^h(y - x)\, w(y) dy \quad (7.225)$$

$$(1h) \qquad\qquad\qquad (2h) \qquad\qquad (3h)$$

where $\partial^i w$ is any of the four values w, w', M, V.

Let us start at the end: the formula (3) with the Dirac delta is not an integral in the ordinary sense which can be looked up in an integral table because the Dirac delta is not a proper function and so (3) is a symbol for the point value $\partial^i w(x)$.

But the formula (3h) is *not* a symbol. It is an integral—though in a somewhat abbreviated notation. To see this compare Fig. 1.45 a and 1.45 b on page 67 where the horizontal displacement $u_x(\boldsymbol{x})$ and $\boldsymbol{u}_x^h(\boldsymbol{x})$ of a plate is calculated with two Dirac deltas $\boldsymbol{\delta}_0$ and $\boldsymbol{\delta}_0^h$ respectively.

The point load $\boldsymbol{\delta}_0$ in Fig. 1.45 a and its action, the integral

$$u_x(\boldsymbol{x}) = \int_\Omega \boldsymbol{\delta}_0(\boldsymbol{y} - \boldsymbol{x}) \bullet \boldsymbol{u}(\boldsymbol{y})\, d\Omega_{\boldsymbol{y}}\,, \qquad (7.226)$$

must be interpreted symbolically. But the action of the approximate Dirac delta $\boldsymbol{\delta}_0^h$, the integral

$$u_x^h(\boldsymbol{x}) = \int_\Omega \boldsymbol{\delta}_0^h(\boldsymbol{y} - \boldsymbol{x}) \cdot \boldsymbol{u}(\boldsymbol{y}) \, d\Omega_{\boldsymbol{y}} = \sum_e \int_{\Omega_e} \boldsymbol{p}_{0,e}^h \cdot \boldsymbol{u} \, d\Omega + \sum_k \int_{\Gamma_k} \boldsymbol{j}_0^h \cdot \boldsymbol{u} \, ds$$

$$(7.227)$$

can be calculated. It is the work done by the true displacement field $\boldsymbol{u}$ on acting through the residual volume forces $\boldsymbol{p}_{0,e}^h$ and jumps $\boldsymbol{j}_0^h$ in the stress vectors along the interelement boundaries—these forces try to imitate the Dirac delta δ_0. The forces $\boldsymbol{j}_0^h$ are the shaded triangles in Fig. 1.45 b. The volume forces $\boldsymbol{p}_{0,e}^h$ are given only as element resultants $r_e = ((p_x^h)^2 + (p_y^h)^2)_{\Omega_e}^{1/2}$.

With regard to the second equation, (2), there holds:

$$a(G_i, w) = \begin{cases} \partial^i w & \partial^i \, w = \text{displacement} \\ 0 & \partial^i \, w = \text{force term} \end{cases} \qquad (7.228)$$

that is force terms, $M(x)$ and $V(x)$, cannot be calculated with Mohr's integral or in more general terms: stresses σ_{ij} are out of reach for any weak formulation

$$\sigma_{ij} \neq a(\boldsymbol{G}_1, \boldsymbol{u}) = 0. \qquad (7.229)$$

This is familiar from beam analysis: the scalar product between the bending moment M in a beam and the bending moment M_A of a redundant X_A ($\equiv G_2$) (see Fig. 7.8) is zero which simply means that the slope w' of the deflection is continuous at the support.

So to summarize the results: in FE analysis—where we operate with substitute Green's functions G_i^h—all three equations $(1h), (2h)$ and $(3h)$ are valid formulations. If the Green's function G_i is exact (3) is not computable, it is a symbol, (2) is only applicable if $\partial^i u$ is a displacement, and only (1) will do in all cases.

Remark 7.3. We trust that the reader is now familiar with the technique and so we can quickly study the full range of possible weak formulations for influence functions. Let $-u'' = p$ with boundary values $u(0) = u(l) = 0$ then we obtain by formulating Green's first identity and taking the limit

$$\lim_{\varepsilon \to 0} G(G_i, u)_{\Omega_\varepsilon} = 0 \qquad (7.230)$$

the results

$$a(G_0, u) = \int_{\Omega_\varepsilon} -G_0'' \, u \, dy + [G_0' \, u]_{\Omega_\varepsilon} = u(x) \qquad (7.231)$$

$$a(u, G_0) = \int_{\Omega_\varepsilon} -u'' \, G_0 \, dy + [u' \, G_0]_{\Omega_\varepsilon} = \int_0^l p \, G_0 \, dy \qquad (7.232)$$

$$a(u, G_1) = \int_{\Omega_\varepsilon} -u'' \, G_1 \, dy + [u' \, G_1]_{\Omega_\varepsilon} = \int_0^l p \, G_1 \, dy - u'(x) \qquad (7.233)$$

$$a(G_1, u) = \int_{\Omega_\varepsilon} -G_1'' \, u \, dy + [G_1' \, u]_{\Omega_\varepsilon} = 0 + \lim_{\varepsilon \to 0} [\ldots] \qquad (7.234)$$

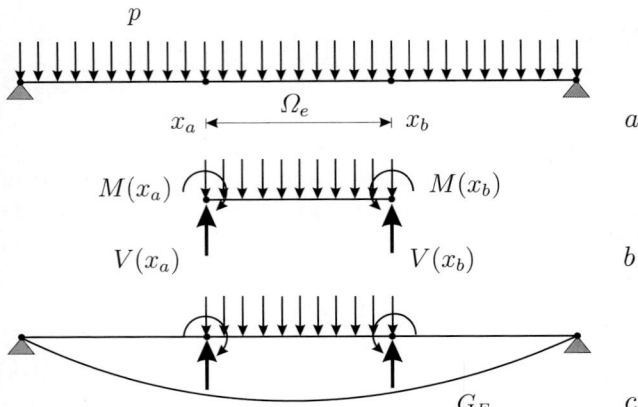

Fig. 7.9. The Green's function G_E for the strain energy in the element Ω_e

where the notation means

$$\lim_{\varepsilon \to 0} a(.,.)_{\Omega_\varepsilon} = \lim_{\varepsilon \to 0} \ldots = \text{result} \tag{7.235}$$

and

$$\Omega_\varepsilon = [0, x - \varepsilon] \cup [x + \varepsilon, l]. \tag{7.236}$$

Because (7.233) is Betti, $u'(x) = (G_1, p)$, we conclude that $a(u, G_1) = 0$ and therefore the limit of the jump terms in the last equation

$$\lim_{\varepsilon \to 0} \left\{ -G'_1(x + \varepsilon)\, u(x + \varepsilon) + G'_1(x - \varepsilon)\, u(x - \varepsilon) \right\} = 0 \tag{7.237}$$

must be zero as well. The two opposite horizontal forces G'_l and G'_r press the cut by one unit apart—in the force method this pair would be a pair $X_1 = 1$—and in the force method $a(G_1, u) = 0$ would be a test that u is continuous at x: if both sides of the cut move by the same amount $u(x)$ then the work done by the two opposite but equal forces, the pair G'_l, G'_r or $1, 1$ respectively, is zero.

So in this light $a(w, G_i) = 0$ *must* be zero if G_i is the Green's function for a force term—N, M or V—because with $a(w, G_i)$ we test whether u, w' or w is continuous at x. A value $a(w, G_i) \neq 0$ would signal a discontinuity at x.

Because $a(G_0, u) = a(u, G_0)$ also the first two results are the same and the combined result is Betti for $u(x)$. Note that $G''_0 = 0$ in Ω_e and that the limit of $[G'_0, u]$ (replace G_1 in (7.237) by G_0) is $u(x)$ because the normal force G'_0 jumps by one unit at x.

In textbooks $a(G_0, u) =$ the principle of virtual forces and $a(u, G_0) =$ the principle of virtual displacements. So both principles, $a(u, G_i) = a(G_i, u)$, return zero for the strain energy product if $\delta u = G_i$ is a Green's function for a force term while lower order Green's functions. $a(u, G_i) = a(G_i, u) = \partial^i u$, return displacements

$$a(G_i, u) = \begin{cases} J(u) & J(u) = \text{displacement} \\ 0 & J(u) = \text{force term} . \end{cases} \qquad (7.238)$$

7.8 Influence functions for other quantities

The reach of Green's second identity $B(u, \hat{u}) = 0$ extends well beyond the calculation of point values.

• To derive an influence function for *the strain energy in a single element* Ω_e we argue as follows: According to Green's first identity the strain energy in a single element $\Omega_e = [x_a, x_b]$ of the beam in Fig. 7.9 a is

$$a(w, w)_{\Omega_e} = \int_{x_a}^{x_b} EI\,(w'')^2\,dx = \int_{x_a}^{x_b} p\,w\,dx + [V\,w - M\,w']_{x_a}^{x_b}$$
$$= : p(w)_{\Omega_e} + [R, w]_{\Omega_e} \qquad (7.239)$$

where in the abbreviated notation R is short for the internal actions V and M at the ends of the element Ω_e which balance the applied load p. Next let G_E the deflection of the beam if the distributed load p acts on Ω_e alone and when the forces R are applied; see Fig. 7.9 c. Evidently then

$$a(w, w)_{\Omega_e} = \underbrace{p(w)_{\Omega_e} + [R, w]_{\Omega_e}}_{W_{1,2}} = a(G_E, w) = \underbrace{\int_0^l G_E\,p\,dx}_{W_{2,1}} . \qquad (7.240)$$

Hence the solution G_E is the Green's function for the strain energy in the element Ω_e. The Green's function for the strain energy in the whole beam is the solution w itself because $a(w, w) = (p, w)$. Note that G_E is load case dependent.
The implication for the FE solution is that

$$a(w_h, w_h)_{\Omega_e} = \int_0^l G_E^h\,p\,dx \qquad (7.241)$$

where G_E^h is the FE solution of the auxiliary load case p_{Ω_e}, R. This result is based on integration by parts and (7.408)

$$a(w_h, w_h)_{\Omega_e} = (p_h, w_h) + [R_h, w_h] = \int_0^l G_E^h\,p_h\,dx$$
$$= \int_0^l G_E^h\,p\,dx . \qquad (7.242)$$

- To derive an influence function for the L_2-*norm* squared of a displacement field

$$||\boldsymbol{u}||_0^2 = \int_\Omega \boldsymbol{u} \cdot \boldsymbol{u} \, d\Omega \qquad (7.243)$$

we apply the displacement field $\boldsymbol{u}$ as volume forces. Let $\boldsymbol{u}_u$ the solution of this problem, that is

$$a(\boldsymbol{u}_u, \boldsymbol{v}) = (\boldsymbol{u}, \boldsymbol{v}) \qquad \boldsymbol{v} \in V \,, \qquad (7.244)$$

and with $\boldsymbol{u} \in V$ it follows

$$W_{1,2} = \int_\Omega \boldsymbol{u}_u \cdot \boldsymbol{p} \, d\Omega = a(\boldsymbol{u}_u, \boldsymbol{u}) = (\boldsymbol{u}, \boldsymbol{u}) = W_{2,1} \,, \qquad (7.245)$$

where $\boldsymbol{p}$ is the right-hand side (the volume forces) which belongs to the original displacement field $\boldsymbol{u}$. In this particular case the technique is also known as the *Aubin–Nitsche trick*.
The extension of (7.245) to the FE solution is evident

$$\int_\Omega \boldsymbol{u}_u^h \cdot \boldsymbol{p} \, d\Omega = (\boldsymbol{u}_h, \boldsymbol{u}_h) \,. \qquad (7.246)$$

Here $\boldsymbol{u}_u^h$ is the FE solution if the volume forces $\boldsymbol{u}_h$ are applied.
- The influence function for the *integral value of the displacement* in a bar under the action of a distributed load p is the solution of $-EA\,G_I'' = 1$ because

$$W_{1,2} = \int_0^l 1 \times u \, dx = \int_0^l G_I \times p \, dx = W_{2,1} \,. \qquad (7.247)$$

That is G_I is a *quadratic function*. Now this is interesting. The influence function for the sum of the horizontal forces is $G_\Sigma = 1$, that is a *constant function*. The influence function for the integral value of the stress $\sigma_x = N/A$ is a *linear function* because two opposite forces ± 1 pull at the ends of the bar to generate the influence function G_σ.

$$\text{const} \qquad \Longrightarrow \qquad \sum H = 0 = \sum H^h = N_h(l) - N_h(0) + \int_0^l p_h \, dx$$

$$\text{linear} \qquad \Longrightarrow \qquad \int_0^l \sigma_x(x) \, dx = \int_0^l \sigma_x^h(x) \, dx$$

$$\text{quadratic} \qquad \Longrightarrow \qquad \int_0^l u(x) \, dx = \int_0^l u_h(x) \, dx \,.$$

That is, if the element shape functions can model constant displacements then the equilibrium condition is satisfied. If they can model linear displacements then the integral values of the stresses coincide, $(\sigma_x, 1) =$

$(\sigma_x^h, 1)$, and if they even can solve the equation $-EA\,u'' = 1$ exactly then it is guaranteed that the integral value of the displacement is the same

$$\int_0^l u\,dx = \int_0^l u_h\,dx\,.$$

(7.248)

(Note that by dividing with the length l we could establish the same results for the average values). The higher the degree of the shape functions the higher the moments that will agree $(-EA\,\varphi_i'' = x^k)$

$$\int_0^l u\,x^k\,dx = \int_0^l u_h\,x^k\,dx$$

(7.249)

where $k = p - 2$ and $p \geq 2$ is the degree of the polynomial shape functions.

Hence it seems that for any quantity we are interested in, there is an influence function and the important point is that the FE program replaces the exact Green's functions in these formulas by approximate solutions or as we say by shifted Green's functions.

7.9 Shifted Green's functions

In more abstract terms most properties of an FE solution are based on a *Shifted Green's function theorem*, that we want to formulate in the following.

The model boundary value problem is the Poisson equation:

$$-\Delta u = p \quad \text{in } \Omega \qquad u = 0 \quad \text{on } \Gamma\,.$$

(7.250)

The associated identities are

$$G(u, \hat{u}) = \int_\Omega -\Delta u\,\hat{u}\,d\Omega + \int_\Gamma \frac{\partial u}{\partial n}\,\hat{u}\,ds - \int_\Omega \nabla u \bullet \nabla \hat{u}\,d\Omega = 0$$

(7.251)

and

$$B(u, \hat{u}) = \int_\Omega -\Delta u\,\hat{u}\,d\Omega - \int_\Gamma \frac{\partial u}{\partial n}\,\hat{u}\,ds - \int_\Gamma u\,\frac{\partial \hat{u}}{\partial n}\,ds$$
$$- \int_\Omega u\,(-\Delta \hat{u})\,d\Omega = 0\,.$$

(7.252)

In the following $G(u/p, \hat{u}) = 0$ denotes the formulation of Green's first identity if in $G(u, \hat{u})$ the term $-\Delta u$ is replaced by p and the trace of u (= boundary value) on Γ by 0, i.e. in a first step for the left-hand side the data on the right-hand side of the boundary value problem (7.250) are substituted—wherever possible—and in a second step the remaining slots are filled with the function u or its derivatives, where u is the argument to the left of the slash in u/p. In

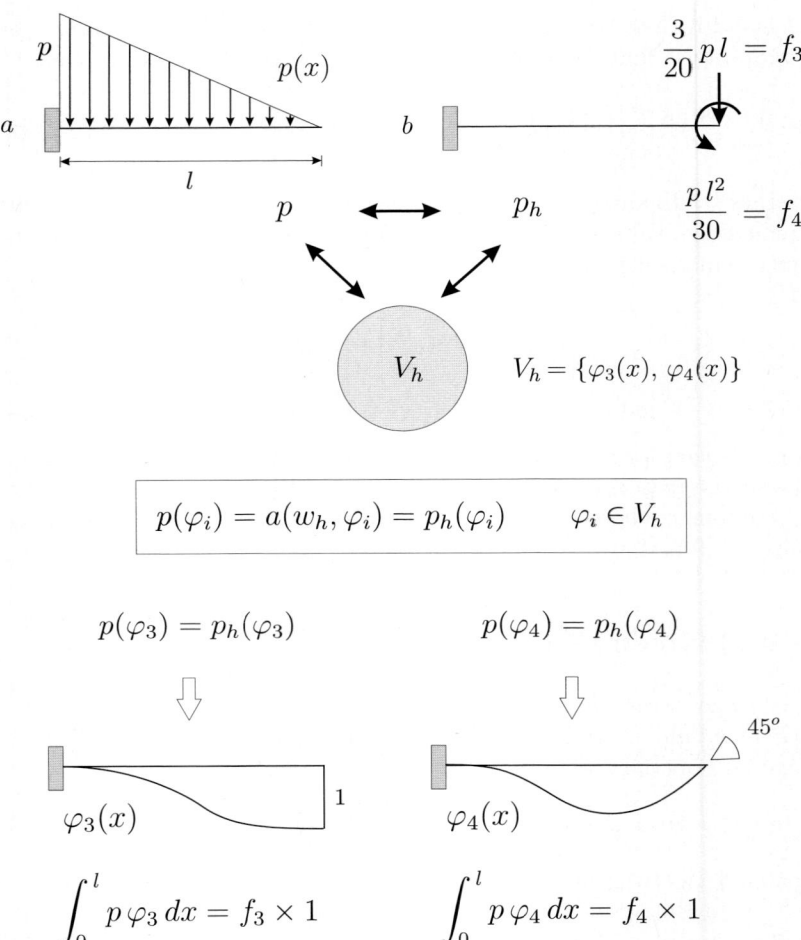

$$V_h = \{\varphi_3(x),\ \varphi_4(x)\}$$

$$p(\varphi_i) = a(w_h, \varphi_i) = p_h(\varphi_i) \qquad \varphi_i \in V_h$$

$$p(\varphi_3) = p_h(\varphi_3) \qquad\qquad p(\varphi_4) = p_h(\varphi_4)$$

$$\int_0^l p\,\varphi_3\,dx = f_3 \times 1 \qquad\qquad \int_0^l p\,\varphi_4\,dx = f_4 \times 1$$

Fig. 7.10. The essence of the FE method—the substitute load p_h is work equivalent to p with respect to all virtual displacements $\varphi_i \in V_h$: **a)** original load case, **b)** equivalent load case

the same sense a formulation like $B(u/p, \hat{u}/\hat{p}) = 0$ is understood where $\hat{u}$ is the solution of a problem

$$-\Delta\hat{u} = \hat{p} \quad \text{in } \Omega \qquad \hat{u} = 0 \quad \text{on } \Gamma. \tag{7.253}$$

With these substitutions, the identities become instances of the principle of virtual displacements:

$$G(u/p, \delta u) = \underbrace{\int_\Omega p\,\delta u\,d\Omega}_{\delta W_e(p,\delta u)} - \underbrace{\int_\Omega \nabla u \cdot \nabla \delta u\,d\Omega}_{\delta W_i(u,\delta u)} = 0, \tag{7.254}$$

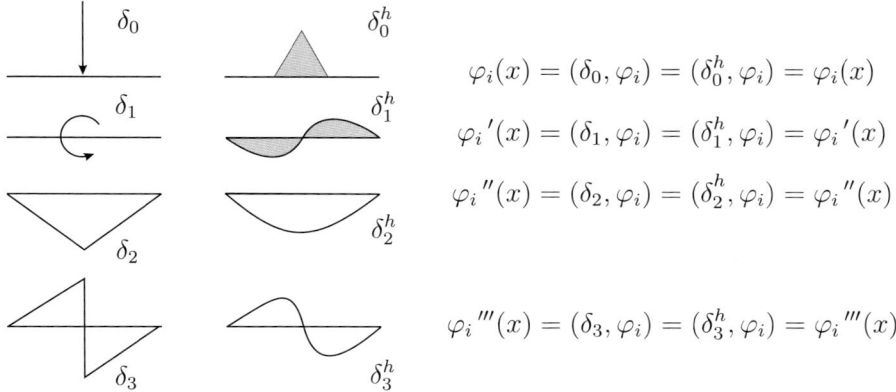

$$\varphi_i(x) = (\delta_0, \varphi_i) = (\delta_0^h, \varphi_i) = \varphi_i(x)$$

$$\varphi_i{}'(x) = (\delta_1, \varphi_i) = (\delta_1^h, \varphi_i) = \varphi_i{}'(x)$$

$$\varphi_i{}''(x) = (\delta_2, \varphi_i) = (\delta_2^h, \varphi_i) = \varphi_i{}''(x)$$

$$\varphi_i{}'''(x) = (\delta_3, \varphi_i) = (\delta_3^h, \varphi_i) = \varphi_i{}'''(x)$$

Fig. 7.11. On V_h the approximate Dirac deltas δ_0^h are perfect replacements or proxies for the original Dirac deltas δ_i

and Betti's theorem:

$$B(u/p, \hat{u}/\hat{p}) = \int_\Omega p\,\hat{u}\,d\Omega - \int_\Omega u\,\hat{p}\,d\Omega = 0\,. \qquad (7.255)$$

If u is the exact solution of (7.250), the expression $G(u/p, \delta u)$ is the same as $G(u, \delta u)$. Things will become interesting if the exact data are mixed with the FE solution u_h as in $G(u_h/p, \varphi_i)$. This is what is done in FE methods.

Principle of virtual displacements

The essential property of the exact solution u is that it satisfies the principle of virtual displacements (7.254) for any (sufficiently regular) virtual displacement δu.

The characteristic property of the FE solution u_h is that it satisfies (7.254) only with regard to the trial functions $\varphi_i \in V_h$ (see Fig. 7.10):

$$G(u_h/p, \varphi_i) = \int_\Omega p\,\varphi_i\,d\Omega - \int_\Omega \nabla u_h \bullet \nabla \varphi_i\,d\Omega = 0 \qquad \varphi_i \in V_h \quad (7.256)$$

but not for arbitrary admissible virtual displacements δu (admissible means $\delta u = 0$ on Γ)

$$G(u_h/p, \delta u) = \int_\Omega p\,\delta u\,d\Omega - \int_\Omega \nabla u_h \bullet \nabla \delta u\,d\Omega \neq 0\,, \qquad (7.257)$$

because $-\Delta u_h \neq p$.

Equation (7.256) essentially means that $G(u/p, \varphi_i) = 0$ and $G(u_h/p, \varphi_i) = 0$ are the "same". In the expression $G(u/p, \varphi_i) = 0$ the exact solution u can be replaced by the FE solution u_h.

An immediate consequence of the equivalence between p and its FE counterpart p_h on V_h is that the approximate Dirac deltas δ_0^h are perfect substitutes for the exact deltas on V_h; see Fig. 7.11.

Betti's theorem

By similar reasoning Betti's theorem can be extended to FE solutions. Given the solutions u_1 and u_2 of two model problems

$$- \Delta u_1 = p_1 \quad u_1 = 0 \quad \text{on } \Gamma , \qquad - \Delta u_2 = p_2 \quad u_2 = 0 \quad \text{on } \Gamma \quad (7.258)$$

it follows that

$$B(u_1, u_2) = 0 \quad \text{or} \quad \int_\Omega p_1 \, u_2 \, d\Omega = \int_\Omega p_2 \, u_1 \, d\Omega , \qquad (7.259)$$

and the message is that in the last equation the exact solutions u_1 and u_2 can be replaced by their FE counterparts u_1^h and u_2^h,

$$\int_\Omega p_1 \, u_2^h \, d\Omega = \int_\Omega p_2 \, u_1^h \, d\Omega , \qquad (7.260)$$

which means that $B(u_1/p_1, u_2/p_2) = 0$ and $B(u_1^h/p_1, u_2^h/p_2) = 0$ are on V_h the "same". The proof rests on the Equivalence Theorem (see Eq. (7.402)),

$$\int_\Omega p_1 \, u_2^h \, d\Omega = \int_\Omega p_1^h \, u_2^h \, d\Omega \qquad \int_\Omega p_2 \, u_1^h \, d\Omega = \int_\Omega p_2^h \, u_1^h \, d\Omega \quad (7.261)$$

and Betti's theorem,

$$B(u_1^h, u_2^h) = \int_\Omega p_1^h \, u_2^h \, d\Omega - \int_\Omega p_2^h \, u_1^h \, d\Omega = W_{1,2} - W_{2,1} = 0 . \quad (7.262)$$

Hence (7.260) means that the FE solutions can serve on V_h as "proxies" for the exact solutions; see Fig. 7.12.

Principle of virtual forces

Here the sequence of functions is reversed[2]. The auxiliary state $\hat{u}$ and the virtual forces $\hat{p}$ (the right-hand side of $\hat{u}$) comes first

$$G(\hat{u}/\hat{p}, u) = \int_\Omega \hat{p} \, u \, d\Omega + \int_\Gamma \frac{\partial \hat{u}}{\partial n} u \, ds - \int_\Omega \nabla \hat{u} \bullet \nabla u \, d\Omega = 0 \quad (7.263)$$

and the rule is now the following: (i) if $\hat{u}_h \in V_h$ is the FE solution of a load case $\hat{p}$, and if (ii) the second argument u lies in V_h (therefore we write u_h

---

[2] We mention this principle only to be complete. The result essentially is contained in the previous formulations

Betti's Theorem

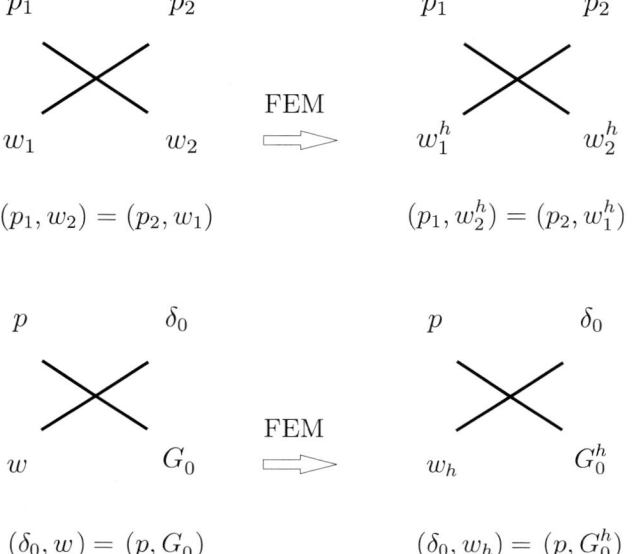

$(p_1, w_2) = (p_2, w_1)$ $(p_1, w_2^h) = (p_2, w_1^h)$

$(\delta_0, w) = (p, G_0)$ $(\delta_0, w_h) = (p, G_0^h)$

Fig. 7.12. Betti's theorem and its extension to FE solutions. Tottenham's equation is the most prominent application of this extension

instead of u), then for the virtual forces $\hat{p}$ may be substituted the FE forces $\hat{p}_h$

$$G(\hat{u}/\hat{p}_h, u_h) = \int_\Omega \hat{p}_h \, u_h \, d\Omega + \int_\Gamma \frac{\partial \hat{u}_h}{\partial n} \, u_h \, ds - \int_\Omega \nabla \hat{u}_h \cdot \nabla u_h \, d\Omega = 0 \, .$$
(7.264)

To prove this rule, we calculate the deflection $\varphi_i \in V_h$ of a beam by applying a "virtual force" $P = 1$ at a point x:

$$G(G_0/\delta_0, \varphi_i) = \int_0^l \delta_0 \, \varphi_i \, dy - \int_0^l \frac{M_0 \, M_i}{EI} \, dy = 0 \, .$$
(7.265)

Next recall that

$$a(G_0 - G_0^h, \varphi_i) = 0$$
(7.266)

and that

$$a(G_0, \varphi_i) = (\delta_0, \varphi_i) \qquad a(G_0^h, \varphi_i) = (\delta_0^h, \varphi_i) \, .$$
(7.267)

Hence it follows that

$$G(G_0/\delta_0^h, \varphi_i) = \int_0^l \delta_0^h \, \varphi_i \, dy - \int_0^l \frac{M_0 \, M_i}{EI} \, dy = 0 \,, \qquad (7.268)$$

which means that

$$\int_0^l \delta_0 \, w_h \, dy = \int_0^l \delta_0^h \, w_h \, dy \qquad w_h \in V_h \,. \qquad (7.269)$$

Hence any point value $w_h(x)$ of a function $w_h \in V_h$ is equal to the work done by the approximate load $\delta_0^h(y - x)$ acting through $w_h(y)$, or stated otherwise, on V_h the kernel δ_0^h is a perfect replacement for the kernel δ_0. This is a truly remarkable result, which of course also holds for the higher Dirac deltas.

Note that the notation $G(G_0/\delta_0^h, u_h)$ implies that in the extension of the principle of virtual forces the FE load case $\hat{p}$ is substituted for $\hat{p}$ but that G_0 is left untouched! This is just the reverse of the previous substitutions.

Summary

The practical importance of these three extensions, (7.256), (7.260), and (7.264) is that probably *all post-processing in mechanics is applied duality, is based on Green's first or second identity*:

$$G(u/p, \hat{u}) = 0 \qquad \text{(principle of virtual displacements)} \qquad (7.270)$$
$$G(\hat{u}/\hat{p}, u) = 0 \qquad \text{(principle of virtual forces)} \qquad (7.271)$$
$$B(u/p, \hat{u}/\hat{p}) = 0 \qquad \text{(Betti's theorem)} \,. \qquad (7.272)$$

Thus to extract information from the solution for u, the exact solution is substituted (and therewith the right-hand sides p and 0, etc.), and the place of $\hat{u}$ is taken by appropriate auxiliary functions. The function $\hat{u}$ can be a rigid body motion $\hat{u} = 1$ so that

$$G(u, 1) = \int_\Omega -\Delta u \cdot 1 \, d\Omega + \int_\Gamma \frac{\partial u}{\partial n} \cdot 1 \, ds = 0 \,, \qquad (7.273)$$

provides the sum of the vertical forces ($\hat{u} = 1$ would then be called a generalized Green's function) or it can be a genuine Green's function if, say, the stress $\sigma(\boldsymbol{x}) = \nabla u \bullet \boldsymbol{n}$ in the membrane at a specific point in a particular direction $(\boldsymbol{n})$

$$\sigma(\boldsymbol{x}) = \int_\Omega G_1(\boldsymbol{y}, \boldsymbol{x}) \, p(\boldsymbol{y}) \, d\Omega_{\boldsymbol{y}} \qquad (7.274)$$

is to be calculated. This equation is identical to $B(u, G_1[\boldsymbol{x}]) = 0$. Similarly, the unit-dummy-load method of structural mechanics, which is used to calculate the deflection of a beam at a specific point x

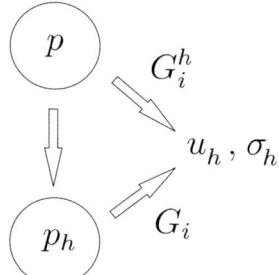

Fig. 7.13. The two approaches to FE analysis

$$w(x) = \int_0^l \frac{M(y)\, M_2(y, x)}{EI}\, dy \qquad M_2 = -EI \frac{d^2}{dy^2}\, G_2(y, x)\,, \quad (7.275)$$

is identical to $G(G_2[x], w) = 0$ (principle of virtual forces).

Hence the (original) Green's function G_i and the generalized Green's function z allow us to extract information from the exact solution via Green's identities.

If we simply speak of Green's function—and drop the artificial distinction between original Green's functions and generalized Green's functions—we can formulate the following theorem:

Shifted Green's function theorem: *the FE solution satisfies all identities or tests with regard to the projections of the Green's functions*

$$G(u_h/p, G_i^h) = 0 \quad G(G_i/\delta_i^h, u_h) = 0 \quad B(u_h/p, G_i^h/\delta_i) = 0\,, \quad (7.276)$$

where the projections $G_i^h \in V_h$ are the FE approximations of the Green's functions.

To appreciate this theorem the reader must understand the importance of the Green's identities and Green's functions for structural mechanics. When we say that the support reactions maintain the equilibrium with the applied load then this actually means that

$$G(u/p, \hat{u}) = 0 \qquad \hat{u} = a + b\, x = \text{rigid-body motion}\,. \qquad (7.277)$$

Any property that we are used to attribute to the exact solution as the satisfaction of the equilibrium conditions or the fact, that the deflection of a cantilever beam carrying a point load P is $w(l) = P\, l^3/(3\, EI)$, is a result that can be reproduced by substituting for u the exact solution and for $\hat{u}$ an appropriate Green's function into Green's identities. And all what we do in FE analysis is that we create a *shadow world* V_h where anything which is true in the real world V is true as well *if only we consequently substitute for the exact Green's function the projections G_i^h.*

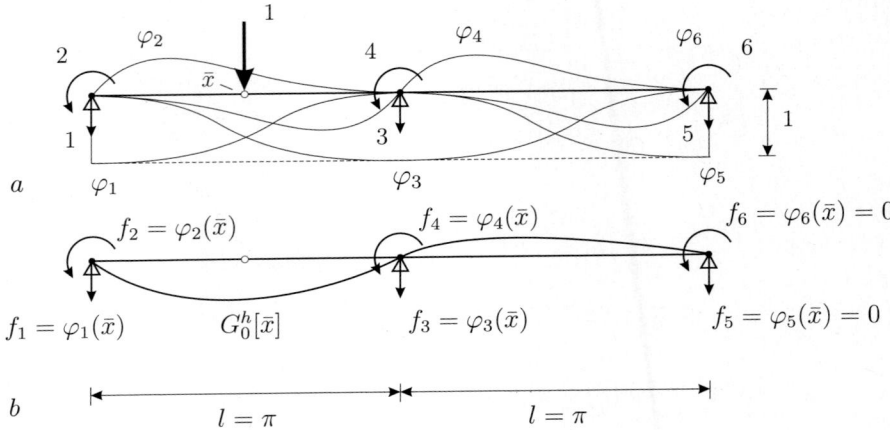

Fig. 7.14. Four load cases

That is with any mesh (or trial space V_h) we can associate a shift-operator which maps the exact Green's functions $G_i[\boldsymbol{x}]$ onto functionals $G_i^h[\boldsymbol{x}]$ in the dual space $H'_m(\Omega)$ and the "only" problem with the FE method is that the algorithm considers these shifted Green's functions to be the real Green's functions.

Just as a change in the elastic parameters or a change in the stiffness of a support effects a shift in the Green's functions so an FE mesh produces a shift of the Green's functions. This seems to be the whole point of FE analysis.

Two approaches

From the standpoint of a reviewing engineer the FE method basically can be classified in two ways. Both are depicted in Fig. 7.13.

• In the first approach the FE solution is identified with the solution of an equivalent loadcase p_h and the displacements and the stresses

$$u_h(x) = \int_0^l G_0(y,x)\, p_h\, dy \qquad \sigma_h(x) = \int_0^l G_1(y,x)\, p_h\, dy \quad (7.278)$$

are the scalar product between the *exact* Green's functions and the equivalent load.

• In the second approach the same displacements and stresses

$$u_h(x) = \int_0^l G_0^h(y,x)\, p\, dy \qquad \sigma_h(x) = \int_0^l G_1^h(y,x)\, p\, dy \quad (7.279)$$

are the scalar product between the *approximate* Green's functions (the shifted Green's functions) and the original load.

In the first approach the load p is replaced by a work-equivalent load p_h and in the second approach the Green's functions G_i are replaced by equivalent Green's functions G_i^h, where equivalent means that on P_h (= the set of all FE load cases) the two coincide; $(G_i, p_j) = (G_i^h, p_j)$ for every unit load case p_j. Note that a change in the Green's function is the same as a change in the governing equation.

Proxies

As mentioned earlier, the kernel δ_0^h is a perfect substitute on V_h for the kernel δ_0, namely

$$w_h(x) = (\delta_0, w_h) = (\delta_0^h, w_h) = (\delta_0^h, \sum_i \varphi_i u_i)$$

$$= \sum_i (\delta_0, \varphi_i) u_i = \sum_i \varphi_i(x) u_i . \qquad (7.280)$$

The inner workings of this result are best understood by studying a two-span beam, which is subdivided into two elements; see Fig. 7.14 a. The influence function for the deflection at the center $\bar{x}$ of the first span is the deflection curve $G_0[\bar{x}]$ if a single force $P = 1$ is applied at $\bar{x}$. To solve this load case on V_h, the deflections of the nodal unit displacements at the point $\bar{x}$ must be applied as equivalent nodal forces f_i; see Fig. 7.14 b. This strange rule—deflections become equivalent nodal forces—is easily understood if this substitute Dirac delta

$$\delta_0^h(\bar{x} - y) = \{f_1, f_2, f_3, f_4, f_5, f_6\}$$
$$= \{\varphi_1(\bar{x}), \varphi_2(\bar{x}), \varphi_3(\bar{x}), \varphi_4(\bar{x}), 0, 0\} \qquad (7.281)$$

is applied to a function $w_h \in V_h$. Then indeed the value of w_h at $\bar{x}$ is recovered,

$$\int_0^{2l} \delta_0^h(\bar{x} - y) w_h(y) \, dy$$
$$= f_1 w_h(0) - f_2 w_h'(0) + f_3 w_h(l) - f_4 w_h'(l) + 0 w_h(2l) + 0 w_h'(2l)$$
$$= -\varphi_2(\bar{x}) w_h'(0) - \varphi_4(\bar{x}) w_h'(l) = w_h(\bar{x}) , \qquad (7.282)$$

(a positive f_2 contributes negative work upon acting through a positive rotation $w_h'(0)$), because w_h lies in V_h and therefore

$$w_h(\bar{x}) = [-\varphi_2(x) w_h'(0) - \varphi_4(x) w_h'(l) - \varphi_6(x) w_h'(2l)]_{x=\bar{x}}$$
$$= [-\varphi_2(\bar{x}) w_h'(0) - \varphi_4(\bar{x}) w_h'(l)] . \qquad (7.283)$$

On the larger space V this substitute Dirac delta δ_0^h will not work. Consider for example the function $w(x) = \sin x$. The result

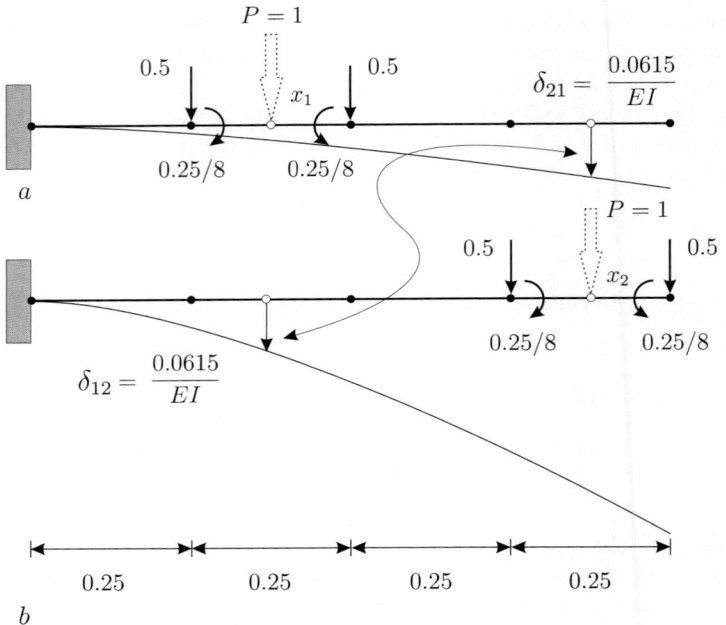

Fig. 7.15. Maxwell's theorem can be extended to the substitute Dirac deltas δ_0^h

$$\int_0^{2l} \delta_0^h(\bar{x} - y) \sin y \, dy = -\varphi_2(\bar{x}) \cos(0) - \varphi_4(\bar{x}) \cos(\pi)$$

$$= -(-\frac{\pi}{8}) \cdot 1 - \frac{\pi}{8} \cdot (-1) = \frac{\pi}{4} = 0.785 \tag{7.284}$$

does not fit, because $\sin \bar{x} = \sin(\pi/2) = 1$. Note that in the first span

$$\varphi_2(x) = -x + 2\frac{x^2}{\pi} - \frac{x^3}{\pi^2} \qquad \varphi_2(\bar{x}) = -\pi/8 \tag{7.285}$$

$$\varphi_4(x) = \frac{x^2}{\pi} - \frac{x^3}{\pi^2} \qquad \varphi_4(\bar{x}) = +\pi/8 \, . \tag{7.286}$$

Maxwell's theorem

Maxwell's theorem, $\delta_{ij} = \delta_{ji}$, seems to make no sense in FE analysis, because we cannot study the effects of true point loads with an FE program since the program replaces any point load by work-equivalent surface loads and line loads. But because these substitute loads δ_0^h are (on V_h) a perfect substitute for the original Dirac deltas, it follows that

$$w_2^h(x_1) = (\delta_0^{(1,h)}, w_2^h) = (\delta_0^{(2,h)}, w_1^h) = w_1^h(x_2) \, . \tag{7.287}$$

Here $\delta_0^{1,h}$ is the assemblage of loads that simulate the action of $P = 1$ at x_1 (see Fig. 7.15 a; simply the nodal forces f_i in this 1-D problem), and $\delta_0^{2,h}$ has the equivalent meaning (see Fig. 7.15 b).

Of course linear algebra provides the same result: let

$$\boldsymbol{K}\,\boldsymbol{u}_i = \boldsymbol{e}_i \qquad \boldsymbol{K}\,\boldsymbol{u}_j = \boldsymbol{e}_j \qquad (7.288)$$

then

$$\delta_{ij} = \boldsymbol{u}_j^T\,\boldsymbol{K}\,\boldsymbol{u}_i = \boldsymbol{u}_i^T\,\boldsymbol{K}\,\boldsymbol{u}_j = \delta_{ji}\,. \qquad (7.289)$$

And also the extension to arbitrary other pairs of Dirac deltas (that's what the point loads are after all) is evident.

Classical Maxwell

$$\delta_{1,2}^0 = \int_\Omega G_0(\boldsymbol{y}, \boldsymbol{x}_1)\,\delta_0(\boldsymbol{y} - \boldsymbol{x}_2)\,d\Omega_{\boldsymbol{y}} = \int_\Omega G_0(\boldsymbol{y}, \boldsymbol{x}_2)\,\delta_0(\boldsymbol{y} - \boldsymbol{x}_1)\,d\Omega_{\boldsymbol{y}} = \delta_{2,1}^0$$

$$(7.290)$$

or, to keep it short,

$$\delta_{1,2}^0 = (G_0[\boldsymbol{x}_1], \delta_0[\boldsymbol{x}_2]) = (G_0[\boldsymbol{x}_2], \delta_0[\boldsymbol{x}_1]) = \delta_{2,1}^0 \qquad (7.291)$$

—the superscript 0 stands for displacement—is simply Betti (L is the self-adjoint differential operator)

$$(G_0[\boldsymbol{x}_1], \delta_0[\boldsymbol{x}_2]) = (G_0[\boldsymbol{x}_1], L\,G_0[\boldsymbol{x}_2]) = (L\,G_0[\boldsymbol{x}_1], G_0[\boldsymbol{x}_2]) = (\delta_0[\boldsymbol{x}_1], G_0[\boldsymbol{x}_2])\,.$$

$$(7.292)$$

The extension to arbitrary pairs $\{i, j\}$ of Green's functions amounts to

$$\delta_{1,2}^i = (G_i[\boldsymbol{x}_1], \delta_j[\boldsymbol{x}_2]) = (G_j[\boldsymbol{x}_2], \delta_i[\boldsymbol{x}_1]) = \delta_{2,1}^j\,. \qquad (7.293)$$

Let for example $i = 3$ and $j = 2$ then the equation

$$\delta_{1,2}^3 = (G_3[\boldsymbol{x}_1], \delta_2[\boldsymbol{x}_2]) = (G_2[\boldsymbol{x}_2], \delta_3[\boldsymbol{x}_1]) = \delta_{2,1}^2 \qquad (7.294)$$

means that the shear force at the point $\boldsymbol{x}_1$ of the influence function for the bending moment at the point $\boldsymbol{x}_2$ in a slab is the same as the bending moment at the point $\boldsymbol{x}_2$ of the influence function for the shear force at the point $\boldsymbol{x}_1$.

Equ. (7.294) must be read as: $G_3[\boldsymbol{x}_1]$ picks from the "load" δ_2 the shear force (3) of the field $G_2[\boldsymbol{x}_2]$ ($= L\,\delta_2$) at point $\boldsymbol{x}_1$, etc..

The extension of this result to finite elements is more or less obvious (we skip the details)

$$\delta_{1,2}^{i,h} = (G_i^h[\boldsymbol{x}_1], \delta_j^h[\boldsymbol{x}_2]) = (G_j^h[\boldsymbol{x}_2], \delta_i^h[\boldsymbol{x}_1]) = \delta_{2,1}^{j,h}\,. \qquad (7.295)$$

Are all results of point loads adjoint?

Consider the following question: in load case 1 a single force P_1 acts at a point $\boldsymbol{x}_1$ and in load case 2 a force P_2 at a point $\boldsymbol{x}_2$. Are the stresses caused by P_1 at the foot of P_2, say $\sigma_{xx}^{(1)}(\boldsymbol{x}_2)$, the same as the stress $\sigma_{xx}^{(2)}(\boldsymbol{x}_1)$ caused by P_2 at the foot of P_1? No, this is not true

$$\sigma_{xx}^{(1)}(\boldsymbol{x}_2) \neq \sigma_{xx}^{(2)}(\boldsymbol{x}_1)\,. \tag{7.296}$$

To see this let $\boldsymbol{K}\,\boldsymbol{u}_1 = \boldsymbol{f}_1$ and $\boldsymbol{K}\,\boldsymbol{u}_2 = \boldsymbol{f}_2$ two FE solutions then the symmetry of $\boldsymbol{K}$ implies that

$$\boldsymbol{u}_2^T\,\boldsymbol{f}_1 = \boldsymbol{u}_1^T\,\boldsymbol{f}_2 \tag{7.297}$$

which is Betti. If the vectors $\boldsymbol{f}$ are the equivalent nodal forces of Green's functions then we have Maxwell

$$(\boldsymbol{u}_2^G)^T\,\boldsymbol{f}_1^G = (\boldsymbol{u}_1^G)^T\,\boldsymbol{f}_2^G\,. \tag{7.298}$$

So Betti is "two" and Maxwell is "two" but in (7.296) we operate with four states

$$\sigma_{xx}^{(1)}(\boldsymbol{x}_2) = (\boldsymbol{u}^{(P_1)})^T\,\boldsymbol{f}^{G_2^\sigma} \neq (\boldsymbol{u}^{(P_2)})^T\,\boldsymbol{f}^{G_1^\sigma} = \sigma_{xx}^{(2)}(\boldsymbol{x}_1) \tag{7.299}$$

and so we cannot use the symmetry of $\boldsymbol{K}$ to switch sides. Why, after all, should the value of a functional $J_1(\boldsymbol{u}_2)$ at $\boldsymbol{u}_2$ be the same as the value of a second functional $J_2(\boldsymbol{u}_1)$ at $\boldsymbol{u}_1$? The numbering scheme, 1,2, alone is no proof.

7.10 The dual space

Recall *Sobolev's Embedding Theorem*, see p. 46, which states that:
    If Ω is a bounded domain in $\mathbb{R}^n$ with a smooth boundary, and if $2\,m > n$, then

$$H^{i+m}(\Omega) \subset C^i(\bar{\Omega}) \tag{7.300}$$

and there exist constants $c_i < \infty$ such that for all $u \in H^{i+m}(\Omega)$

$$||u||_{C^i(\bar{\Omega})} \leq c_i\,||u||_{H^{i+m}(\Omega)}\,. \tag{7.301}$$

This seems to be an abstract theorem with no immediate consequences for structural mechanics—besides of course clarifying our ideas about point loads—(see Sect. 1.14, p. 44). But we wish to comment on some interesting consequences of this theorem.
    Recall that the norm $||u||_{C^i(\bar{\Omega})}$ of a function is the maximum absolute value of u and its derivatives up to order i on $\bar{\Omega}$. Hence if two deflection surfaces

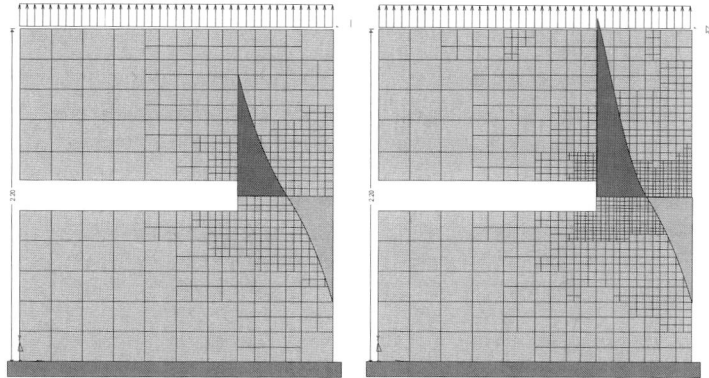

Fig. 7.16. If two solutions are close in the sense that $a(\boldsymbol{u} - \hat{\boldsymbol{u}}, \boldsymbol{u} - \hat{\boldsymbol{u}}) \ll 1$, the maximum stresses must not be the same

w_1 and w_2 of a Kirchhoff plate are close in the sense of the Sobolev space $H^2(\Omega)$—if their strain energy is about the same—the maximum deflections of the two surfaces must also be nearly the same:

$$\|w_1 - w_2\|_2 \ll 1 \qquad \Rightarrow \qquad \max w_1 \sim \max w_2 . \qquad (7.302)$$

This follows from (7.301)—using a somewhat symbolic notation in the last step,

$$\max |w_1 - w_2| = \|w_1 - w_2\|_{C^0(\bar{\Omega})} \le c_0 \, \|w_1 - w_2\|_2 \ll c_0 \cdot 1 , \quad (7.303)$$

if we assume that the constant c_0 is not too pessimistic, i.e., too large.

For displacement fields $\boldsymbol{u} = [u_x, u_y]^T$ of plates—which we typically associate with the space $\boldsymbol{H}^1(\Omega) = H^1(\Omega) \times H^1(\Omega)$—this is not necessarily true

$$\|\boldsymbol{u}_1 - \boldsymbol{u}_2\|_1 \ll 1 \qquad \not\Rightarrow \qquad \max |\boldsymbol{u}_1| \sim \max |\boldsymbol{u}_2| \qquad (7.304)$$

because the inequality $2\,m = 2 \cdot 1 > 2 = n$ is not true.

To study the consequences of this theorem more systematically, we need to introduce the concept of the dual space of a Sobolev space $H^m(\Omega)$. The dual space is defined as the set of all continuous linear functionals $p(.)$ on $H^m(\Omega)$ as for example

$$p(w) := \int_\Omega p \, w \, d\Omega . \qquad (7.305)$$

In the following the focus is on the Sobolev space $H^2(\Omega)$, endowed with the norm

$$\|w\|_2 := \left[\int_\Omega (w^2 + w_{,x}^2 + w_{,y}^2 + w_{,xx}^2 + w_{,xy}^2 + w_{,yx}^2 + w_{,yy}^2) \, d\Omega \right]^{1/2} , \ (7.306)$$

which is the energy space of Kirchhoff plates.

For a very special reason we focus initially on the subspace $H_0^2(\Omega) \subset H^2(\Omega)$ of this space. All the functions $w \in H^2(\Omega)$ with vanishing deflection and slope on the boundary, $w = \partial w/\partial n = 0$, constitute this subspace $H_0^2(\Omega)$. If a plate is clamped, the deflections w lie in $H_0^2(\Omega)$. The nice feature of $H_0^2(\Omega)$ is, that its dual can be identified with the Sobolev space $H^{-2}(\Omega)^3$ [202].

To understand this "negative" space, recall that the regularity of the functions in H^m increases with the index m. The opposite is true with regard to the negative spaces H^{-m}: the more negative, the worse the regularity. Such functions "borrow" their regularity from their brethren in H^m.

Take for example the Green's function $G_0(y, x)$ of a taut rope (prestressed with a force $H = 1$) with its triangular shape. This function has no second derivative at the source point y, the foot of the point load, so that the integral

$$\int_0^l G_0''(y, x)\, \delta w(x)\, dx \qquad (\ )'' = \frac{d^2}{dx^2} \qquad (7.307)$$

makes no sense. The point load $\delta_0 = G_0''$ does not lie in $L_2(0, l) = H^0(0, l)$, because the integral of G_0'' squared does not exist:

$$\int_0^l [G_0''(y, x)]^2\, dx = \infty\,. \qquad (7.308)$$

Hence the point load must lie in a weaker space, in some negative Sobolev space $H^{-m}(0, l)$, namely $H^{-2}(0, l)$.

To understand this choice, note that if the virtual displacement δw lies in $H^2(0, l)$, then integration by parts can be applied twice, and because G_0 and δw have zero boundary values (the rope is fixed at its ends), the result is

$$\int_0^l G_0''(y, x)\, \delta w(x)\, dx = [G_0'\, \delta w]_0^l - \int_0^l G_0'(y, x)\, \delta w'(x)\, dy$$

$$= [G_0'\, \delta w - G_0\, \delta w']_0^l + \int_0^l G_0(y, x)\, \delta w''(x)\, dx$$

$$= \int_0^l G_0(y, x)\, \delta w''(x)\, dx \qquad (7.309)$$

i.e., the work done by the point load acting through δw can be expressed in terms of the work done by the distributed load $\delta w''$ acting through G_0.

The point load $\delta_0 = G_0''$ is called the *generalized second derivative* of G_0 and because this technique can always be applied if $\delta w \in H^2(0, l)$ it is said that the point load δ_0 lies in $H^{-2}(0, l)$.

From an engineering point of view, the concept of generalized derivatives is an application of Betti's theorem $W_{1,2} = W_{2,1}$. "If $W_{1,2}$ seems to make no sense or cannot be calculated, then try $W_{2,1}$!".

---

[3] We spare the reader a definition of negative Sobolev spaces, because the essence of such spaces hopefully will become clear in the following discussion.

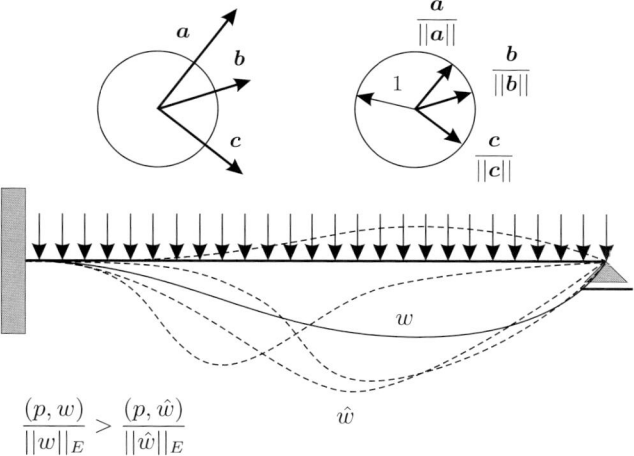

$$\frac{(p, w)}{||w||_E} > \frac{(p, \hat{w})}{||\hat{w}||_E} \qquad \hat{w}$$

Fig. 7.17. In the unit ball $B_1 = \{w \,|\, ||w||_E = 1\}$ the normalized exact solution $w/||w||_E$ gets "the most mileage" out of p, i.e. the virtual work exceeds that of any other normalized virtual deflection $\hat{w}/||\hat{w}||_E$.

Continuous functionals

If a functional p is continuous, there exists a constant c such that

$$|p(\hat{w})| \le c\,||\hat{w}||_E\,. \tag{7.310}$$

The lowest bound c—divide the equation by $||\hat{w}||_E$—is defined as the norm of the functional p

$$||p\,||_{-E} := \sup_{\substack{\hat{w} \in V \\ \hat{w} \ne 0}} \frac{|p(\hat{w})|}{||\hat{w}||_E} = \sup_{\substack{\hat{w} \in V \\ ||\hat{w}||_E = 1}} |p(\hat{w})|\,. \tag{7.311}$$

If w is the solution of the load case p, then

$$\frac{|p(\hat{w})|}{||\hat{w}||_E} = \frac{|a(w, \hat{w})|}{||\hat{w}||_E} \le \frac{||w||_E\,||\hat{w}||_E}{||\hat{w}||_E} = ||w||_E \tag{7.312}$$

i.e., $||w||_E$ is an upper bound and because of

$$\frac{|p(w)|}{||w||_E} = \frac{|a(w, w)|}{||w||_E} = ||w||_E \tag{7.313}$$

it is also the lowest upper bound. Hence the norm $||p\,||_{-E}$ of a load case p is just the norm of the solution

$$||p\,||_{-E} = ||w||_E = \int_0^l \frac{M^2}{EI}\,dx \qquad \text{(in a beam)}\,. \tag{7.314}$$

This means that the exact solution w is that deflection in V which gets "the most mileage" out of p in the sense of (7.311); see Fig. 7.17.

It seems intuitively clear that any surface load or line load p represents a continuous functional

$$|p(\delta w)| < c \, ||\delta w||_2 \, . \tag{7.315}$$

(We switch again to Sobolev norms).

But does this also hold true for point loads acting on a Kirchhoff plate? The answer is yes. The Dirac delta δ_0—a point load of magnitude $P = 1$—belongs to $H^{-2}(\Omega)$. The reason is that the functions $w \in H^2(\Omega)$ lie also in $C(\Omega)$ and because the embedding $H^2(\Omega) \subset C(\Omega)$ is—according to Sobolev's Embedding Theorem continuous, $(m - i > n/2$ or $2 - 0 > 1)$—it follows

$$\max_{\boldsymbol{x} \in \Omega} |w(\boldsymbol{x})| \leq c \, ||w||_2 \tag{7.316}$$

where the constant c does not depend on w. Thus any function $w \in H^2(\Omega)$ is guaranteed to have a bounded value $w(\boldsymbol{x})$ at *every* point $\boldsymbol{x} \in \Omega$, and therefore an expression such as

$$\delta_0(w) := \int_{\Omega} \delta_0(\boldsymbol{y} - \boldsymbol{x}) \, w(\boldsymbol{y}) \, d\Omega_{\boldsymbol{y}} = w(\boldsymbol{x}) \tag{7.317}$$

makes sense, and is a continuous functional on $H^2(\Omega)$

$$|\delta_0(w)| = |w(\boldsymbol{x})| \leq c \, ||w||_2 \, . \tag{7.318}$$

That is if $||w||_2 \to 0$ then also $|\delta_0(w)| \to 0$. So if the strain energy of a slab is zero ($||w||_E = a(w,w)^{1/2}$ and $||w||_2$ are equivalent norms) then $w \equiv 0$ and no single point is allowed to break ranks while in a plate (2-D elasticity) this is possible: the influence function for the point support is zero—the plate does not move, $||\boldsymbol{u}||_E = 0$—but one single point leaves the plate and travels downward by one unit length, see Fig. 1.73 p. 103. Hence the conclusion is

- A slab with finite strain energy, $w \in H^2(\Omega)$, is smooth, i.e., the deflection is continuous—no sudden jumps—and the maximum value of w is bounded.

But it is not guaranteed that *all* functions $w \in H^2(\Omega)$ have a well-defined slope at *all* the points $\boldsymbol{x} \in \Omega$, because the embedding of $H^2(\Omega)$ into $C^1(\Omega)$ is not continuous, because the inequality $2 - 1 > 1$ is not true. Hence a single moment $M = 1$ (Dirac delta δ_1) is not a continuous functional on $H^2(\Omega)$.

But the embedding of $H^3(\Omega)$ into $C^1(\Omega)$ is continuous, so $\delta_1 \in H^{-3}(\Omega)$, and δ_2 lies in H^{-4} and δ_3 lies in H^{-5}:

H^{-5}	H^{-4}	H^{-3}	H^{-2}	H^{-1}	$H^0 = L_2(\Omega)$	H^1	H^2	H^3	H^4	H^5
δ_3	δ_2	δ_1	δ_0				C^0	C^1	C^2	C^3

$$w(x) = \frac{M\,l\,x}{6\,EI}\left(\frac{x^2}{l^2} - 0.25\right) \in H^2(0,l)\,, \notin H^3(0,l) \;\Rightarrow w \in C^1, w \notin C^2$$

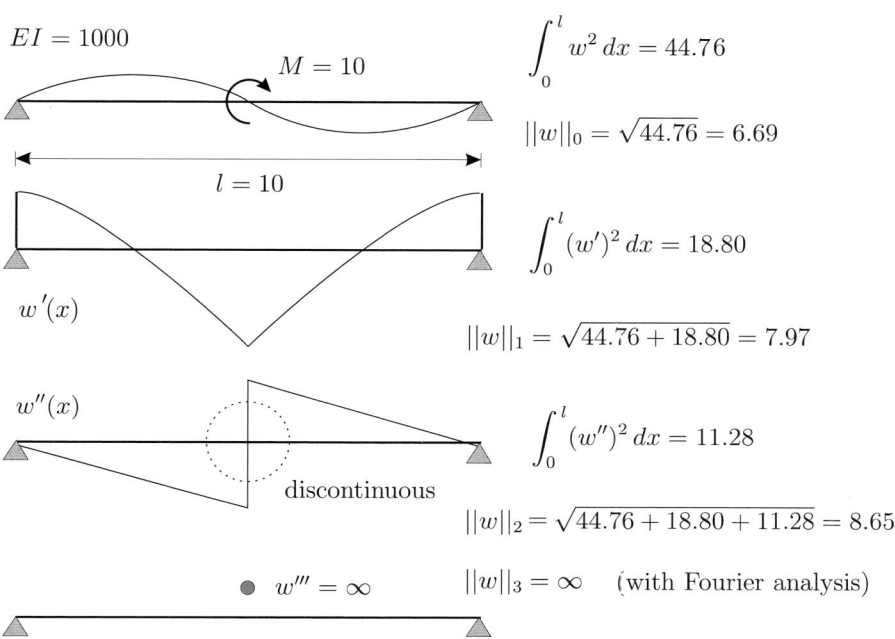

$EI = 1000$

$$\int_0^l w^2\,dx = 44.76$$

$M = 10$

$$\|w\|_0 = \sqrt{44.76} = 6.69$$

$l = 10$

$$\int_0^l (w')^2\,dx = 18.80$$

$w'(x)$

$$\|w\|_1 = \sqrt{44.76 + 18.80} = 7.97$$

$w''(x)$

$$\int_0^l (w'')^2\,dx = 11.28$$

discontinuous

$$\|w\|_2 = \sqrt{44.76 + 18.80 + 11.28} = 8.65$$

$w''' = \infty$

$\|w\|_3 = \infty$ (with Fourier analysis)

Fig. 7.18. The deflection curve w caused by the moment M lies in $H^2(0,l)$ but not in $H^3(0,l)$, otherwise the bending moment would have to be continuous, because $H^3(0,l) \subset C^2(0,l), 3-2 > 1/2$

From a theoretical point of view the only load cases p that are admissible lie in the dual of the energy space $H^m(\Omega)$, i.e., there must exist a bound c such that

$$|p(w)| \le c\,\|w\|_m\,. \tag{7.319}$$

Because the constant c is just the norm of the solution of the load case p, i.e., $c = \|w\|_m$ and because the norm $\|w\|_m$ and the energy norm $\|w\|_E = a(w,w)^{1/2}$ are equivalent, the constant c is proportional to $\|w\|_E$.

Unbounded point functionals

Normal structural loads do lie in the dual space. The solution can be approximated by minimizing the distance in the energy. This even holds true—at least for the classical point loads $P = 1$ and $M = 1$—in beams; see Fig. 7.18. But in higher dimensions point loads (*point functionals, Dirac deltas*) are critical.

To prescribe the displacement $u_1(\boldsymbol{x})$ at a particular point $\boldsymbol{x}$ of a plate makes no sense, because the embedding of the energy space $\boldsymbol{H}^1(\Omega) := H^1(\Omega) \times H^1(\Omega)$ of the displacement fields $\boldsymbol{u} = [u_x, u_y]^T$ into $\boldsymbol{C}^0(\Omega) := C^0(\Omega) \times C^0(\Omega)$ is not continuous.

The displacement field $\boldsymbol{u}(\boldsymbol{x}) = [\ln(\ln(1/r)), 0]^T$ for example has finite energy, in any circular domain $\Omega_{1-\varepsilon} = \{(r, \varphi) | 0 \leq r \leq 1 - \varepsilon, 0 \leq \varphi \leq 2\pi\}$ but the horizontal displacement $u_x = \ln(\ln(1/r))$ is infinite at $r = 0$.

Hence the Dirac delta $\boldsymbol{\delta}_0$ does not lie in the dual of $\boldsymbol{H}^1(\Omega)$, because there is no constant c such that *for all* $\boldsymbol{u} \in \boldsymbol{H}^1(\Omega)$

$$|(\boldsymbol{\delta}_0^x, \boldsymbol{u})| = |u_x(\boldsymbol{x})| \leq c \, ||\boldsymbol{u}||_1 \, . \tag{7.320}$$

Rather we are lead to conclude that $c = \infty$. This even makes sense, because a point load will generate a stress field with infinite strain energy and so $c = ||\boldsymbol{u}||_1 \cong a(\boldsymbol{u}, \boldsymbol{u})^{1/2} = \infty$.

Even more critical are the stresses (the first derivatives), because a result such as

$$|(\boldsymbol{\delta}_1^{xx}, \boldsymbol{u})| = |\sigma_{xx}(\boldsymbol{x})| \leq c_1 \, ||\boldsymbol{u}||_1 \tag{7.321}$$

would require that the embedding of the energy space $\boldsymbol{H}^1(\Omega)$ into $\boldsymbol{C}^1(\Omega)$ is continuous—which is not true in 2-D and 3-D elasticity. Hence the point functional $\boldsymbol{\delta}_1^{xx}$ which extracts the stress σ_{xx} at a particular point (post processing!) is not a continuous functional on the energy space $\boldsymbol{H}^1(\Omega)$. If we pick an arbitrary point $\boldsymbol{x} \in \Omega$, there is no *global* bound on the stress, say, $\sigma_{xx}(\boldsymbol{x})$ at this point, i.e, which is a bound for the stress $\sigma_{xx}(\boldsymbol{x})$ of *all* displacement fields $\boldsymbol{u}$ in $\boldsymbol{H}^1(\Omega)$ in the sense of (7.321). A displacement field $\boldsymbol{u}$ can have a bounded strain energy, $a(\boldsymbol{u}, \boldsymbol{u}) < \infty$, (the norms $||\boldsymbol{u}||_1$ and $||\boldsymbol{u}||_E$ are equivalent) but the stresses may become infinite at some points inside Ω. This is no contradiction.

- *Hence, if we calculate the stress at a point, we apply an unbounded point functional, even though we think we only evaluate the polynomial function which represents the stress distribution.*

Also note that if two displacement fields have nearly zero distance in the metric of the Sobolev space $\boldsymbol{H}^1(\Omega)$, $||\boldsymbol{u} - \hat{\boldsymbol{u}}||_1 \ll 1$, it is not guaranteed that the maximum stresses are about the same; see Fig. 7.16, p. 553.

Riesz' representation theorem

We have mentioned Riesz' representation theorem before, but it deserves more than a place in a footnote because this theorem is central to Green's functions and finite elements.

Extracting information from a structure means—in an abstract sense—to apply a functional $J(u)$ to the solution

$$J(u) = u(x) \qquad J(u) = \sigma_{xx}(x) \qquad J(u) = \int_\Omega u \, d\Omega \qquad \text{etc.} \quad (7.322)$$

According to Riesz' representation theorem for each linear, bounded functional $J()$ there is an element $z \in V$ such that

$$a(z, u) = J(u) \,. \qquad (7.323)$$

The function z is of course the (generalized) Green's function G.

In structural mechanics most functionals are unbounded

$$J(u) = u(x) \qquad J(u) = \sigma(x) \qquad J(u) = u_{,x}(x) \qquad \ldots \quad (7.324)$$

that is it cannot be guaranteed that the functional is less than the energy of u times a *global* constant (not depending on the single u)

$$|J(u)| \le c \, ||u||_E \qquad (7.325)$$

(a displacement field $\boldsymbol{u}$ can be infinite at one point $\boldsymbol{x}$, that is $J(\boldsymbol{u}) = u_x(\boldsymbol{x}) = \infty$, but the energy is finite, $||\boldsymbol{u}||_E < \infty$) and also the strain energy of the Green's functions $G_i \, (= z)$ is infinite

$$a(G_i, G_i) = ||G_i||_E^2 = \infty \qquad (7.326)$$

so that—theoretically at least—Riesz' representation theorem is not applicable. But we know that if we replace point values, i.e. point functionals $J(u) = u(x)$, by average values

$$u(x) \qquad \rightarrow \qquad \bar{u}(x) = \frac{1}{|\Omega_\varepsilon|} \int_{\Omega_\varepsilon} u \, d\Omega \qquad (7.327)$$

then the functionals $J()$ are bounded and the corresponding generalized Green's function $||G_i||_E < \infty$ have finite energy. So we may assume that Riesz theorem is "very nearly" applicable to our problems.

Now to each mesh belongs a test and trial space $V_h \subset V$ and an abstract operator $\mathcal{P}$ which maps each functional $J()$ onto a functional[4] $J_h()$

$$\mathcal{P}: \qquad J() \in V' \Rightarrow J_h() \in V_h' \qquad (7.328)$$

such that (see Fig. 1.45 p. 67)

$$J(v) = J_h(v) \qquad \text{for each } v \in V_h \,. \qquad (7.329)$$

To the mapping $J() \to J_h()$ corresponds a mapping

$$V \ni G \quad \rightarrow \quad G_h \in V_h \qquad (7.330)$$

---

[4] V' and V_h' are the duals of V and V_h that is the set of all functionals defined on V and V_h resp.

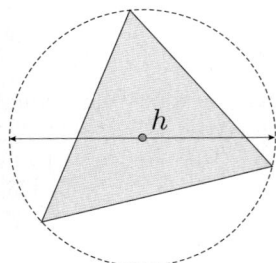

Fig. 7.19. The element size h

of the Riesz element.

The interesting aspect of this operator $\mathcal{P}$ is that it allows to characterize FE solutions via the functionals $J_h()$: namely in FE analysis we choose $u_h \in V_h$ in such a way that for all $J()$

$$J(u_h) = a(G, u_h) = a(G_h, u_h) = (p, G_h) = J_h(u) \qquad (7.331)$$

or, to keep it short,

$$J(u_h) = J_h(u) . \qquad (7.332)$$

Surprisingly this statement is equivalent to

$$a(u_h, v) = (p, v) \qquad v \in V_h . \qquad (7.333)$$

For a proof of (7.329) and (7.332) see (1.228) p. 69.

Note that all this happens automatically. A mesh is a space V_h, is an operator $\mathcal{P}$, is an assemblage of Green's functions G_i^h and all this before even one single load case has been solved on this mesh.

7.11 Some concepts of error analysis

Asymptotic error estimates

These estimates tell what kind of convergence rate we can expect if the mesh size h tends to zero; see Fig. 7.19. The Taylor series of a function

$$u(x) = \underbrace{u(0) + u'(0)\, x + \ldots + u^{(n)}(0)\, \frac{x^n}{n!}}_{n\text{-th degree polynomial}} + \underbrace{u^{(n+1)}(\xi)\frac{x^{(n+1)}}{(n+1)!}}_{\text{remainder}} \qquad (7.334)$$

consists of an n-th degree polynomial and a remainder term, which is essentially the derivative $u^{(n+1)}$ at an unknown point ξ between 0 and x.

Let us apply this series to the exact solution $u(x)$ in an element $[x_i, x_{i+1}]$ of length $h = x_{i+1} - x_i$. If the Taylor series is truncated after the first derivative, then at a point $x_i \leq x \leq x_{i+1}$ we have

$$u(x) = u(x_i) + u'(x_i)\, x + u''(\xi)\frac{x^2}{2} \qquad x_i \le \xi \le x_{i+1}\,. \qquad (7.335)$$

Next let us assume that the FE solution consists of a string of first-degree polynomials (hat functions), and the exact solution is interpolated at the nodes. Then the error $e_I(x) = u(x) - u_I(x)$ of the interpolating function $u_I(x)$ is

$$e_I(x) = u(x_i) + u'(x_i)\, x + u''(\xi)\frac{x^2}{2} - u_I(x_i) - u_I'(x_i)\, x$$

$$= e'(x_i)\, x + u''(\xi)\frac{x^2}{2} \qquad \text{because of } u_I(x_i) = u(x_i)\,. \qquad (7.336)$$

Because the error at the other end of the element is zero as well, $u_I(x_{i+1}) = u(x_{i+1})$, the error $e_I(x)$ must have its maximum at some point s in between, and because $e_I'' = u'' - u_I'' = u''$ it follows that

$$e_I'(x) = \int_s^x u''(z)\, dz \le \int_{x_i}^{x_{i+1}} |u''(z)|\, dz \le h \max_{x_i \le z \le x_{i+1}} |u''(z)|\,. \qquad (7.337)$$

If (7.337) and (7.336) are combined, then we have the estimate

$$|e_I(x)| \le h^2 \max_{x_i \le \xi \le x_{i+1}} |u''(\xi)|\,. \qquad (7.338)$$

This can be generalized: if the shape functions can represent any polynomial up to degree k exactly (completeness!) and if the derivatives of the shape functions are uniformly bounded,[5] then in plate problems the error in the displacements is of order $O(h^{k+1})$ and the error in the stresses of order $O(h^k)$, and the constant factor in the error bound (see (7.338)) depends on the derivatives of order $k + 1$ of the solution $u(x)$.

If quadratic shape functions are used in plate problems, $k = 2$, then the error in the displacements is of order $O(h^3)$ and the error in the stresses is of order $O(h^2)$. In beam or in plate bending analysis, complete cubics ($k = 3$) would yield $O(h^4)$ for the error in the deflection, $O(h^2)$ for the error in the moments, and $O(h)$ for the error in the shear forces. Each order of differentiation reduces the order of the convergence by 1.

All this holds of course only for smooth solutions. In the presence of singularities, or if the solution is simply not that smooth enough the convergence rate is lower, because the Taylor series terminates with the last regular derivative of u.

If for example the distributed load abruptly drops to zero as in Fig. 7.20 a the third derivative u''' is a delta function and the Taylor series of $u(x)$ is truncated with the remainder $u''(\xi)$

$$u(x) = u(0) + u'(0)\, x + u''(\xi)\frac{x^2}{2} \qquad \leftarrow \qquad \text{forced stop}\,, \qquad (7.339)$$

---

[5] [230] p. 137

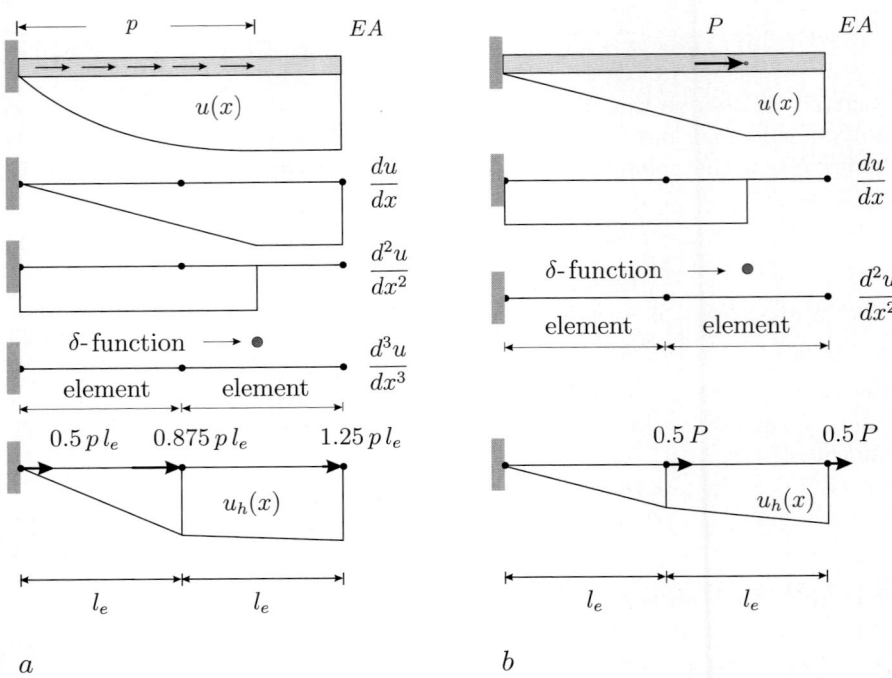

a   b

Fig. 7.20. An unfavorable arrangement of the elements reduces the order of the convergence

and the estimate does not extend beyond $|e_I| \leq h^2 \max |u''|$. Even quadratic shape functions, which theoretically allow a rate $O(h^3)$, do not fare better than $O(h^2)$.

The presence of a single force ensures that the second derivative u'' is already a delta function and the Taylor series is terminated with the remainder $u'(\xi)$:

$$u(x) = u(0) + u'(\xi)\, x\,, \tag{7.340}$$

which means that it is not possible to do better than $O(h)$.

Hence, if the elements are of order k, and if under normal circumstances the bound on the error in the interpolating function is

$$|e_I| \leq h^{k+1} \max |u^{(k+1)}|\,, \tag{7.341}$$

then with each derivative that is not smooth, one order in the convergence rate is lost.

These examples illustrate how important it is to arrange the elements in such a way that single forces are applied at the nodes or that load discontinuities occur at interelement boundaries. Of course the same applies to discontinuities in the modulus of elasticity or, say, the cross section A of a bar, though all this is evident.

In an energy method such as FE analysis the pointwise interpolation error $e_I(x) = u(x) - u_I(x)$ is not the typical error studied. Rather, it is the L_2-measure

$$||e_I||_m = \left[\int_0^l [e_I^2 + (e_I')^2 + (e_I'')^2 + \ldots + (e_I^{(m)})^2] \, dx \right]^{1/2}, \quad (7.342)$$

that is, the error e is measured in terms of *Sobolev norms* of order m, where m typically has the same order as the energy:

$$2\,m = \begin{cases} 2 & \text{bar, plates, Reissner–Mindlin plates} \\ 4 & \text{Euler–Bernoulli beams, Kirchhoff plates} \end{cases} \quad (7.343)$$

From (7.338) follows the estimate for the interpolation error:

$$||e_I||_m = c\,h^{k+1-m}\,||u||_{k+1}. \quad (7.344)$$

The highest derivatives in the Sobolev norm $||.||_m$ are of the same order as the highest derivative in the strain energy $a(.,.)$. This is one of the reasons why under regular circumstances the energy norm $||u||_E = \sqrt{a(u,u)}$ and the Sobolev norm $||.||_m$ are equivalent

$$c_1\,||u||_m \le \sqrt{a(u,u)} \le c_2\,||u||_m \qquad \text{with constants } c_1 \text{ and } c_2. \quad (7.345)$$

In an Euler–Bernoulli beam, which is governed by the equation $EI\,w^{IV} = p$, the Sobolev norm

$$||w||_2 = \left[\int_0^l (w^2 + (w')^2 + (w'')^2) \, dx \right]^{1/2} \quad (7.346)$$

is, if the beam cannot perform any rigid-body motions, equivalent to the energy norm

$$||w||_E := a(w,w)^{(1/2)} = \left[\int_0^l \frac{M^2}{EI} \, dx \right]^{1/2} \quad (7.347)$$

i.e., there exist constants c_1 and c_2 such that

$$c_1\,||w||_2 \le ||w||_E \le c_2\,||w||_2. \quad (7.348)$$

This is a remarkable fact, because the energy norm only measures the second-order derivatives. The equivalence implies that if the bending moments and therefore $||w||_E$ are zero in a beam, then w and w' also are zero (in the L_2 sense).

The index $k+1$ in the norm $||u||_{k+1}$ on the right-hand side of Eq. (7.344) comes from the remainder $u^{k+1}(\xi)$ of the Taylor series. For other indices r, estimates such as

$$||e_I||_m = c\,h^\alpha\,||u||_r \qquad \alpha = \min(k+1-m, r-m) \qquad (7.349)$$

are obtained where either the degree of the element is decisive: the greater the value of k, the more the Taylor series can be expanded and powers of h gained. On the other hand the regularity of the solution (steering the index r) might not allow one to go that far, and the expansion must stop. The weakest term in the chain determines the possible convergence rate.

Up to now we only talked about the interpolation error, not the error in the FE solution. But thanks to *Céa's lemma*,

$$||u-u_h||_m \le c\,\inf_{v_h \in V_h} ||u-v_h||_m \qquad (7.350)$$

this is only a small additional step and we obtain, see (7.349), the basic error estimate for the error $e = u - u_h$ of the FE solution

$$||e||_m \le c\,h^\alpha\,||u||_r \qquad \alpha = \min(k+1-m, r-m)\,. \qquad (7.351)$$

The proof is simple:

$$\begin{aligned}||e||_m &\le c_1^{-1}\,a(e,e)^{1/2} \le c_1^{-1}\,a(u-u_I, u-u_I)^{1/2} \\ &\le \frac{c_2}{c_1}||u-u_I||_m \le c_3\,h^\alpha\,||u||_r \end{aligned} \qquad (7.352)$$

using—in this sequence—the equivalence of the energy norm and the Sobolev norm $||.||_m$, Céa's lemma, once more the equivalence, and finally the estimate (7.349).

To obtain estimates in lower-order norms, $0 \le s \le m$, the Aubin–Nitsche trick [51] is employed, which yields

$$||e||_s = c\,h^\alpha\,||u||_{k+1} \qquad \alpha = \min(k+1-s, 2(k+1-m))\,. \qquad (7.353)$$

For $k = m = 1$ (e.g., linear shape functions in plate problems), the error e of the FE solution becomes

$$||e||_0 = c\,h^2\,||u||_2 \qquad ||e||_1 = c\,h\,||u||_2\,, \qquad (7.354)$$

where the typical pattern of a Taylor series shines through. The constants c are generic quantities independent of u and h.

In the case of a bar—a one dimensional structure—the singularity is due to the load. In plates or slabs singularities typically occur at corner points. If an L-shaped opening is covered with a cloth that is pulled taut by a force H, then under heavy wind pressure p the membrane can be pressed against the vertical edge of the abutment, and the cloth tears apart, because the stress

$$\sigma_n = H\,\nabla w \bullet \boldsymbol{n} = H\,(w_{,x}\,n_x + w_{,y}\,n_y) \qquad \boldsymbol{n} = \text{normal vector} \quad (7.355)$$

becomes infinite. The reason is that the deflection $w(\boldsymbol{x})$ of the membrane

$$-H\,\Delta w = p \quad \text{in } \Omega \quad w = 0 \quad \text{on } \Gamma \tag{7.356}$$

is basically of the form

$$w = k_1\,s_1(r,\varphi) + w_R = k_1\,r^\alpha\,t(\varphi) + w_R \qquad \alpha = 90°/360° = 0.25 < 1\,, \tag{7.357}$$

where k_1 is the stress intensity factor and w_R is the regular part of the solution, which has bounded stresses.

If s_1^h is the FE approximation of the singular function s_1, then the estimate for the L_2-norm of the error is

$$||s_1 - s_1^h||_0 + h^\alpha\,||\nabla(s_1 - s_1^h)||_0 \le c\,h^{2\,\alpha} \tag{7.358}$$

and for the pointwise error

$$||s_1 - s_1^h||_{L_\infty} \le c\,h^\alpha\,. \tag{7.359}$$

Hence the singularity in the solution determines the highest possible convergence rate, and higher-degree elements would not provide a better convergence rate [47].

Interpolation estimates

We want to give a short sketch of the proof[6] of the energy error estimate (1.431), p. 151,

$$||e_i||_E^2 := a(\boldsymbol{e}, \boldsymbol{e})_{\Omega_i} \le c\,\eta_i^2\,. \tag{7.360}$$

Recall that the error $\boldsymbol{e}$ is the displacement field of the structure under the action of the residual forces $\boldsymbol{r}_i = (\boldsymbol{p} - \boldsymbol{p}_h)$ on each element Ω_i, and the jump terms $\boldsymbol{j}_i$ at the edges Γ_i of the elements. The principle of conservation of energy $W_i = W_e$ (Green's first identity) implies that

$$2\,W_i = a(\boldsymbol{e}, \boldsymbol{e}) = \sum_i^n \left\{ \int_{\Omega_i} \boldsymbol{r}_i \bullet \boldsymbol{e}\,d\Omega + \int_{\Gamma_i} \boldsymbol{j}_i \bullet \boldsymbol{e}\,ds \right\} = 2\,W_e\,. \tag{7.361}$$

Because of the *Galerkin orthogonality*, the field $\boldsymbol{I}_h\boldsymbol{e} \in V_h$ that interpolates the field $\boldsymbol{e}$ at the nodes can be subtracted from $\boldsymbol{e}$ without changing the result:

---

[6] For details see e.g. [2]

$$||e||_E^2 := a(e, e) = \sum_i^n \left\{ \int_{\Omega_i} r_i \cdot (e - I_h e) \, d\Omega + \int_{\Gamma_i} j_i \cdot (e - I_h e) \, ds \right\}$$

$$\leq \sum_i^n \left\{ ||r_i||_{L_2, \Omega_i} ||e - I_h e||_{L_2, \Omega_i} + ||j_i||_{L_2, \Gamma_i} ||e - I_h e||_{L_2, \Gamma_i} \right\} . \qquad (7.362)$$

Under suitable assumptions about the FE space V_h, the following *interpolation property* holds:

$$||e - I_h e||_{L_2, \Omega_i} \leq c_1 \, h_i \, ||e||_{E, \Omega_i} \qquad ||e - I_h e||_{L_2, \Gamma_i} \leq c_2 \, h_i^{0.5} \, ||e||_{E, \Omega_i} \qquad (7.363)$$

where h_i is the diameter of the element Ω_i. Hence if (7.362) is divided by $||e||_E$ we have

$$||e||_E \leq c_3 \sum_i^n \left\{ h_i \, ||r||_{L_2, \Omega_i} + h_i^{0.5} \, ||j||_{L_2, \Gamma_i} \right\} . \qquad (7.364)$$

Now a sum $\sum_i^n x_i$ of n terms is always at most $\sqrt{n} \times$ the length of the vector x, or to be precise

$$|\sum_i^n x_i| = |x \cdot 1| \leq ||x|| \, ||1|| = \sqrt{n} \, ||x|| \qquad 1 := [1, 1, \ldots, 1]^T \quad (7.365)$$

so that (the sum consists of $2n$ terms)

$$||e||_E \leq c_3 \sqrt{2n} \left[\sum_i^n \left\{ h_i^2 \, ||r||_{L_2, \Omega_1}^2 + h_i \, ||j||_{L_2, \Gamma_i}^2 \right\} \right]^{1/2} \qquad (7.366)$$

or if both sides are squared [10], [15],

$$||e||_E^2 \leq c_4 \sum_i^n \left[h_i^2 \, ||r_i||_{L_2, \Omega_i}^2 + h_i \, ||j_i||_{L_2, \Gamma_i}^2 \right] . \qquad (7.367)$$

Error estimators

The terms r_i and j_i are *error indicators*. Not to be confused with these error indicators are the so-called *error estimators* ε, which provide upper and lower bounds on the error of an FE solution:

$$c_1 \varepsilon < ||u - u_h||_E < c_2 \varepsilon . \qquad (7.368)$$

and thus can serve as a *stopping* criterion [19]. The significance of such error estimators is that a) they can be calculated unlike the exact solution u, and

b) we trust that there exist constants c_i that bound the error. Nevertheless little is known about the magnitude of these constants.

Hence, the error estimator has the same tendency as the exact error. It is a reliable *substitute* for the unknown error, and it can be calculated. In structural mechanics the standard error indicators η_i are also often used as error estimators; that is the stress discontinuities between elements and the residual forces within elements are considered to be reliable estimators for the error of an FE solution.

One characteristic feature of a good error indicator/estimator is that it is *efficient* in the sense that there exists a constant C independent of the element size so that

$$||e||_E \leq c \sum_i \eta_i \leq C ||e||_E \qquad (7.369)$$

which guarantees that the error indicator mirrors the actual error rather then becoming too pessimistic if h tends to zero.

Recovery based error estimators

Many commercial FE programs offer the option to smooth the raw discontinuous stresses and the error estimation is then simply based on comparing the post-processed stresses σ_{ij}^p with the raw stresses σ_{ij}

$$\eta^2 = \int_\Omega [(\sigma_{xx}^p - \sigma_{xx})^2 + (\sigma_{xy}^p - \sigma_{xy})^2 + (\sigma_{yy}^p - \sigma_{yy})^2]\, d\Omega\,. \qquad (7.370)$$

This seems a very natural approach and it can result in good estimates but pollution may spoil the whole strategy as in Fig. 1.103 on p. 145 where a large but smooth error remained undetected. See also the remarks in Sect. 1.31, p. 152.

Explicit error estimators

These estimators are based on the element residual forces r and the traction discontinuities j and are typically of the form (7.367). They are also called explicit a posteriori estimators, because they are based on readily available data.

Implicit error estimators

Many so-called implicit error estimators are based on Green's first identity which for one element states that

$$G(e, v)_{\Omega_e} = (p - p_h, v)_{\Omega_e} + [r - r_h, v]_{\Gamma_e} - a(e, v)_{\Omega_e} = 0 \qquad (7.371)$$

where the brackets denote the boundary integral. This is motivation to find an approximation to e either by solving a Dirichlet problem on a single element or patch of elements ($u = 0$ on the edge)

$$G(e, v)_{\Omega_e} = (p - p_h, v)_{\Omega_e} - a(e, v)_{\Omega_e} = 0 \qquad v \in V_h(\Omega_e) \quad (7.372)$$

or an "equilibrated" Neumann problem

$$G(e, v)_{\Omega_e} = (p - p_h, v)_{\Omega_e} + [r - r_h, v]_{\Gamma_e} - a(e, v)_{\Omega_e} = 0$$
$$v \in V_h(\Omega_e). \quad (7.373)$$

Equilibrated because the unknown edge forces r must be replaced by approximations $\hat{r}$ which are subject to the equilibrium conditions. The FE solutions of these local problems then serve as error estimators. Because such estimators require the solution of additional problems they are referred to as being of implicit type.

7.12 Important equations and inequalities

For the convenience of the reader we repeat some definitions: in a load case p loads are applied, for example,

$$-EA\,u''(x) = p(x) \qquad 0 < x < l \qquad u(0) = 0, \ N(l) = P \quad (7.374)$$

while in a load case δ displacements are prescribed

$$-EA\,u''(x) = 0 \qquad 0 < x < l \qquad u(0) = 0, \ u(l) = \delta. \quad (7.375)$$

V denotes the trial or test space, that is the set of all possible virtual displacements that are compatible with the support conditions of the structure.

S is the solution space. In a load case p the two spaces coincide, $S = V$, while in a load case δ it is $S = u_\delta \oplus V$ where u_δ is a function with the properties $u_\delta(0) = 0, u_\delta(l) = \delta$, see Sect. 1.7, p. 22. The symbol $\oplus$ means, that any function u in S is the sum of u_δ plus a function $v \in V$.

$V_h \subset V$ is the space spanned by the nodal unit displacements φ_i. That is the generic element of V_h is $v_h = \sum_i v_i\,\varphi_i(x)$.

S_h is the composition of a function u_δ^h plus the space V_h

$$S_h = u_\delta^h \oplus V_h \quad (7.376)$$

where u_δ^h is a displacement field, that satisfies the homogeneous geometric boundary conditions such as $u(0) = 0$ exactly and the inhomogeneous geometric boundary conditions such as $u(l) = \delta$ either exactly ($S_h \subset S$) or approximately ($S_h \not\subset S$).

In a load case p the FE solution lies in V_h and in a load case δ it lies in S_h.

In a load case p the FE solution is an expansion in terms of the nodal unit displacements

$$u_h = \sum_i u_i \, \varphi_i(x) \tag{7.377}$$

and in a load case δ a function u_δ^h is added

$$u_h = u_\delta^h(x) + \sum_i u_i \, \varphi_i(x) = u_\delta^h(x) + u_V^h(x) \tag{7.378}$$

to satisfy the (partially) inhomogeneous geometric boundary conditions

$$u_\delta^h(0) = 0 \qquad u_\delta^h(l) = \delta\,. \tag{7.379}$$

The exact solution u satisfies

$$a(u, v) = p(v) \qquad v \in V\,, \tag{7.380}$$

and the FE solution satisfies

$$a(u_h, v_h) = p(v_h) \qquad v_h \in V_h\,. \tag{7.381}$$

In a load case δ the load p is zero and $u = u_\delta + u_V$, where $u_V \in V$, so that (7.380) is equivalent to

$$a(u_V, v) = -a(u_\delta, v) \qquad v \in V\,, \tag{7.382}$$

and in the same sense in the finite element case, see (7.378),

$$a(u_V^h, v_h) = -a(u_\delta^h, v_h) \qquad v_h \in V_h\,. \tag{7.383}$$

Note that in FE analysis we replace the right-hand side $a(u_\delta^h, \varphi_i)$ by the virtual work done by the negative[7] fixed end forces resulting from u_δ^h, i.e.,

$$a(u_\delta^h, \varphi_i) = p_\delta^h(\varphi_i) = f_i\,. \tag{7.384}$$

An important role in FE analysis plays Green's first identity

$$G(u, \hat{u}) = \int_0^l -EA\,u''\,\hat{u}\,dx + [N\,\hat{u}]_0^l - \int_0^l EA\,u'\,\hat{u}'\,dx = 0 \tag{7.385}$$

which allows to switch at will between external and internal virtual work

$$G(u_h, v_h) = p_h(v_h) - a(u_h, v_h) = 0\,, \tag{7.386}$$

where

---

[7] negative because of actio instead of reactio

$$p_h(v_h) := \int_0^l -EA\,u_h''\,v_h\,dx + [N_h, v_h]_0^l \qquad (7.387)$$

$$a(u_h, v_h) := \int_0^l EA\,u_h'\,v_h'\,dx\,. \qquad (7.388)$$

Note that if u is the exact solution and $\hat{u} \in V$ then

$$G(u, \hat{u}) = \delta\,\Pi(u, \hat{u}) = \left[\frac{d}{d\varepsilon}\,\Pi(u + \varepsilon\,\hat{u})\right]_{\varepsilon=0} \qquad (7.389)$$

is the first variation of the potential energy.

Boundary conditions as $u(0) = 0$ or $u(l) = \delta$ are *essential boundary conditions* and boundary conditions as $N(0) = -P$ or $N(l) = P$ are *natural boundary conditions*. In a problem such as

$$-EA\,u''(x) = p(x) \quad 0 < x < l\,, \quad u(0) = 0 \;\; N(l) = P \qquad (7.390)$$

the essential boundary condition enters the definition of the trial and solution space V

$$V = \{u \in H^1(0, l)\,|\,u(0) = 0\} \qquad (7.391)$$

while the natural boundary condition enters the definition of the virtual external work $p(\hat{u})$ in the associated variational formulation

$$a(u, \hat{u}) = \int_0^l EA\,u'\,\hat{u}'\,dx = \int_0^l p\,\hat{u}\,dx + P \cdot \hat{u}(l) =: p(\hat{u})\;\;\hat{u} \in V\,. \qquad (7.392)$$

It is called a natural boundary condition, because the variational solution satisfies $N(l) = P$.

The boundary integrals in Green's identities are L_2 products (or simply products in 1-D problems) of conjugate boundary terms as $[N\,\hat{u}]_0^l = N(l) \cdot \hat{u}(l) - N(0) \cdot \hat{u}(0)$. A boundary value problem is well posed if of two such conjugate boundary terms one is prescribed and the other is unknown.

Under regular conditions the strain energy product constitutes a norm on V

$$\|u\|_E := \sqrt{a(u, u)} \qquad (7.393)$$

which is equivalent to the Sobolev norm, i.e., there exist constants c_1 and c_2 such that

$$c_1\,\|u\|_m \le \|u\|_E \le c_2\,\|u\|_m\,, \qquad (7.394)$$

and the strain energy product is a continuous bilinear form on $V \times V$

$$|a(u, v)| \le c_3\,\|u\|_m\,\|v\|_m\,, \qquad (7.395)$$

and the virtual work is continuous on V

$$p(v) \leq c \, ||v||_m \, . \tag{7.396}$$

If u is the solution of the load case p then $G(u, u) = 0$ implies

$$a(u, u) = p(u) \tag{7.397}$$

and because of the definition (7.393) and (7.314) we have

$$||u||_E = \sqrt{a(u, u)} = \frac{p(u)}{||u||_E} = \sup_{\substack{v \in V \\ v \neq 0}} \frac{p(v)}{||v||_E} =: ||p||_{-E} \, . \tag{7.398}$$

The strain energy product of the FE solution u_h and a test function $v_h \in V_h$ can be identified with the work done by the forces in a load case p_h acting through v_h

$$a(u_h, v_h) = p_h(v_h) \tag{7.399}$$

where $p_h(v_h)$ is obtained from $a(u_h, v_h)$ by integration by parts, that is

$$p_h(v_h) = G(u_h, v_h) - a(u_h, v_h) \qquad v_h \in V_h \tag{7.400}$$

where $G(u_h, v_h)$ is Green's first identity.
   The FE method

$$u_h \in V_h : \qquad a(u_h, v_h) = p(v_h) \qquad v_h \in V_h \tag{7.401}$$

is therefore equivalent to

$$p_h(v_h) = p(v_h) \quad v_h \in V_h \qquad \textit{Equivalence theorem} \, . \tag{7.402}$$

Let u be the exact solution, u_h the FE solution, and $e = u - u_h$ the error;

- On V

$$a(e, v) = p(v) - a(u_h, v) = p(v) - p_h(v) \qquad v \in V \, . \tag{7.403}$$

- Hence in particular

$$p(u) - p(u_h) = p(e) = a(e, u) = p(u) - p_h(u) \tag{7.404}$$

and as well

$$p(u_h) = p_h(u) \qquad \textit{symmetry} \tag{7.405}$$

which is of course Betti's theorem.

- Let $e = G_0 - G_0^h$ then follows

$$u(\boldsymbol{x}) - u_h(\boldsymbol{x}) = a(G_0 - G_0^h, u) = (\delta_0, u) - (\delta_0^h, u), \qquad (7.406)$$

i.e., the work done by the exact solution u acting through the jump terms and element residuals of the approximate Dirac delta $\delta_0^h = \{j, r\}$ has the same value as the error.

- For test functions $v_h \in V_h \subset V$ the right-hand side in (7.403) is zero because $p(v_h) = p_h(v_h)$, and hence on V_h the error is orthogonal to the test functions

$$a(e, v_h) = 0 \quad v_h \in V_h \qquad \textit{Galerkin orthogonality}. \qquad (7.407)$$

- The residual forces $p - p_h$ are orthogonal to the test functions

$$a(e, v_h) = p(v_h) - p_h(v_h) = 0 \quad v_h \in V_h. \qquad (7.408)$$

- If $p = 0$ as in stability problems, the FE load p_h is orthogonal to each test function

$$p_h(v_h) = 0 \quad v_h \in V_h. \qquad (7.409)$$

- The FE load p_h is orthogonal to the displacement error e

$$a(u_h, e) = p_h(e) = 0. \qquad (7.410)$$

- The unit load cases p_i are orthogonal to the error e

$$a(\varphi_i, e) = p_i(e) = 0. \qquad (7.411)$$

- The FE solution u_h attains the smallest possible value for the strain energy product of the error e on V_h, i.e., the FE solution has the shortest distance in the energy metric from the true solution

$$a(e, e) \le a(u - v_h, u - v_h) \quad v_h \in V_h, \qquad (7.412)$$

because for any $w_h \in V_h$

$$a(e + w_h, e + w_h) = \underbrace{a(e, e)}_{> 0} + 2\underbrace{a(e, w_h)}_{= 0} + \underbrace{a(w_h, w_h)}_{> 0} \qquad (7.413)$$

and choosing $w_h = u_h - v_h$ gives the result. Céa's lemma is based on (7.412), the equivalence (7.394) and the continuity (7.395)

$$c_1 ||e||_m^2 \le a(e, e) = a(e, u - v_h) + a(e, v_h - u_h)$$
$$= a(e, u - v_h) \le c_3 ||e||_m ||u - v_h||_m \qquad (7.414)$$

and therefore

$$||e||_m = ||u - u_h||_m \le \frac{c_3}{c_1} \inf_{v_h \in V_h} ||u - v_h||_m. \qquad (7.415)$$

- In a load case p the internal energy of the FE solution is less than the internal energy of the exact solution (we drop the factor $1/2$ on both sides)

$$a(u_h, u_h) \leq a(u, u) \qquad \text{in a load case } p, \qquad (7.416)$$

because

$$0 < a(u, u) = a(u_h + e, u_h + e)$$
$$= a(u_h, u_h) + 2 \underbrace{a(e, u_h)}_{=0} + \underbrace{a(e, e)}_{>0}. \qquad (7.417)$$

- In a load case δ where $\Pi(u) = 1/2\, a(u, u)$ the internal energy of the FE solution exceeds the internal energy of the exact solution

$$a(u, u) \leq a(u_h, u_h) \qquad \text{in a load case } \delta. \qquad (7.418)$$

The proof was given in Sect. 1.7 on p. 16.

- In a load case p holds

$$a(u, u) = a(u_h, u_h) + a(e, e) \quad \text{'Pythagoras } c^2 = a^2 + b^2 \text{'} \qquad (7.419)$$

or 'the energy of the error is the error in the energy'

$$a(e, e) = a(u, u) - a(u_h, u_h), \qquad (7.420)$$

because

$$a(u, u) = a(u_h - e, u_h - e) = a(u_h, u_h) - 2 \underbrace{a(e, u_h)}_{=0} + a(e, e). \,(7.421)$$

In a load case δ the equation $a(e, u_h) = 0$ is not true, because $u_h = u_\delta^h + u_V^h \notin V_h$, while $a(e, v_h) = 0$ is true if $v_h \in V_h$.

- The potential energy of the FE solution exceeds the potential energy of the exact solution

$$\Pi(u) \leq \Pi(u_h), \qquad (7.422)$$

because

$$\Pi(u_h) = \Pi(u - e) = \frac{1}{2} a(u, u) - a(u, e) + \frac{1}{2} a(e, e) - p(u) + p(e)$$
$$= \Pi(u) \underbrace{-a(u, e) + p(e)}_{G(u,e)=0} + \frac{1}{2} \underbrace{a(e, e)}_{>0}. \qquad (7.423)$$

- The virtual external work done by p acting through u_h is less than the work done acting through u

$$p(u_h) = a(u_h, u_h) < a(u, u) = p(u). \qquad (7.424)$$

- The fact that the residual forces are orthogonal to the φ_i

$$p(\varphi_i) - p_h(\varphi_i) = 0 \qquad \varphi_i \in V_h \tag{7.425}$$

does not suffice to guarantee, that the FE solution u_h interpolates the exact solution at the nodes. This would be true, if the residual forces were orthogonal to the Green's functions $G_0[\boldsymbol{x}_i] = G_0(\boldsymbol{y}, \boldsymbol{x}_i)$ of the nodal displacements $u_i = u(\boldsymbol{x}_i)$

$$p(G_0[\boldsymbol{x}_i]) - p_h(G_0[\boldsymbol{x}_i]) \overset{?}{=} 0 . \tag{7.426}$$

But the nodal Green's functions $G_0[\boldsymbol{x}_i]$ do not lie in V_h

$$G_0(\boldsymbol{y}, \boldsymbol{x}_i) \neq \sum_j u_j(\boldsymbol{x}_i)\, \varphi_j(\boldsymbol{y}) . \tag{7.427}$$

The exception are 1-D problems where—in standard situations—the inclusion $G_0(y, x_i) \in V_h$ is true. The reason is, that (i) the Green's functions $G_0(y, x_i)$ are expansions in terms of homogeneous solutions and (ii) the element shape functions φ_i^e form a complete set of linearly independent homogeneous solutions of the governing equation.

- Let G_0^h the FE approximation of G_0, then holds

$$u_h(\boldsymbol{x}) = \int_\Omega G_0(\boldsymbol{y}, \boldsymbol{x})\, p_h(\boldsymbol{y})\, d\Omega_{\boldsymbol{y}} = \int_\Omega G_0^h(\boldsymbol{y}, \boldsymbol{x})\, p(\boldsymbol{y})\, d\Omega_{\boldsymbol{y}}$$

$$= \int_\Omega u(\boldsymbol{y})\, \delta^h(\boldsymbol{y} - \boldsymbol{x})\, d\Omega_{\boldsymbol{y}} \tag{7.428}$$

where δ^h is the approximate Dirac delta, i.e., that assemblage of external loads that attempts to imitate the action of the true Dirac delta or simply the "right-hand side" of G_0^h.

This means that the FE solution can be written in two ways

$$u_h(\boldsymbol{x}) = \begin{cases} \sum_i \varphi_i(\boldsymbol{x})\, u_i = (\delta_0^h, u) \\ \int_\Omega G_0^h(\boldsymbol{y}, \boldsymbol{x})\, p(\boldsymbol{y})\, d\Omega_{\boldsymbol{y}} = (G_0^h, p) \end{cases} \tag{7.429}$$

so that

$$u_h(\boldsymbol{x}) = \sum_i \varphi_i(\boldsymbol{x})\, u_i = \int_\Omega G_0^h(\boldsymbol{y}, \boldsymbol{x})\, p(\boldsymbol{y})\, d\Omega_{\boldsymbol{y}}$$

$$= \int_\Omega \sum_i \varphi_i(\boldsymbol{y})\, u_j^G(\boldsymbol{x})\, p(\boldsymbol{y})\, d\Omega_{\boldsymbol{y}} = \sum_i u_i^G(\boldsymbol{x})\, f_i$$

$$= \boldsymbol{u}_G^T \boldsymbol{f} = \boldsymbol{u}_G^T \boldsymbol{K}\, \boldsymbol{u} = \boldsymbol{u}_G^T \boldsymbol{K}^T \boldsymbol{u} = \boldsymbol{f}_G^T \boldsymbol{u} \tag{7.430}$$

or in short (see Fig. 1.52 p. 77)

$$u_h(\boldsymbol{x}) = \boldsymbol{u}_G^T \boldsymbol{f} = \boldsymbol{f}_G^T \boldsymbol{u} \,. \qquad (7.431)$$

Note that this holds true for any quantity, $\sigma_h(\boldsymbol{x}) = \ldots$, $V_h(x) = \ldots$ if the nodal vectors $\boldsymbol{u}_G$ and $\boldsymbol{f}_G$ respectively are replaced by the corresponding vectors of the Green's function G_i.

- Betti's theorem and the principle of virtual work, $\delta W_e = \delta W_i$, imply

$$u(\boldsymbol{x}) = \overbrace{\underbrace{p(G_0[\boldsymbol{x}])}_{\delta W_e}}^{Betti} = \underbrace{a(G_0[\boldsymbol{x}], u)}_{\delta W_i} \qquad (7.432)$$

and therefore as well

$$u(\boldsymbol{x}) - u_h(\boldsymbol{x}) = p(G_0[\boldsymbol{x}]) - p_h(G_0[\boldsymbol{x}]) = a(G_0[\boldsymbol{x}], u - u_h) \,. \quad (7.433)$$

- The Galerkin orthogonality implies that

$$a(G_0[\boldsymbol{x}] - G_0^h[\boldsymbol{x}], v_h) = 0 \qquad v_h \in V_h \qquad (7.434)$$

and

$$a(G_0^h[\boldsymbol{x}], u - u_h) = 0 \qquad \text{because } G_0^h[\boldsymbol{x}] \in V_h \,. \qquad (7.435)$$

- Which proves that on V_h the kernel G_0^h is a perfect replacement for the kernel G_0

$$v_h(\boldsymbol{x}) = a(G_0^h[\boldsymbol{x}], v_h) = (\delta_0^h, v_h) \qquad v_h \in V_h \,. \qquad (7.436)$$

Which is also true for higher kernels G_i^h as well.

- All this jointly implies that, see also (1.451) p. 158,

$$\begin{aligned}
u(\boldsymbol{x}) - u_h(\boldsymbol{x}) &= p(G_0[\boldsymbol{x}]) - p_h(G_0[\boldsymbol{x}]) \\
&= p(G_0[\boldsymbol{x}] - G_0^h[\boldsymbol{x}]) - p_h(G_0[\boldsymbol{x}] - G_0^h[\boldsymbol{x}]) \\
&= a(G_0[\boldsymbol{x}] - G_0^h[\boldsymbol{x}], u - u_h) \,,
\end{aligned} \qquad (7.437)$$

and this allows an estimate such as

$$|u(\boldsymbol{x}) - u_h(\boldsymbol{x})| \leq ||G_0[\boldsymbol{x}] - G_0^h[\boldsymbol{x}]||_E \, ||u - u_h||_E \,, \qquad (7.438)$$

where

$$||u||_E := a(u, u)^{1/2} \qquad (7.439)$$

is the energy norm.

- If the Green's function $G_0 = g_0 + u_R$ is split into a fundamental solution g_0 and a regular part u_R and the FE approximation $G_0^h = g_0 + u_R^h$ as well, the previous estimate can be replaced by

$$|u(\boldsymbol{x}) - u_h(\boldsymbol{x})| \leq ||u_R[\boldsymbol{x}] - u_R^h[\boldsymbol{x}]||_E \, ||u - u_h||_E \,. \qquad (7.440)$$

- If the exact solution u lies in V_h, i.e., if $u = u_h$ the error of any approximate Green's function G_i^h is orthogonal to the load case p (the right-hand side of u), see Sect. 1.17, p. 57,

$$p(G_i) - p(G_i^h) = a(G_i - G_i^h, u) = a(G_i - G_i^h, u_h) = 0. \quad (7.441)$$

This means, for example, that even though the Green's function G_1 for the stresses does not lie in V_h, the stresses are exact $\sigma = \sigma^h = p(G_1^h)$.

- The turnstile character of the symmetric strain energy

$$p(\hat{u}) \quad \leftarrow \quad a(u, \hat{u}) \quad \rightarrow \quad \hat{p}(u) \quad (7.442)$$

plays a very important role in FE analysis. Betti's theorem rests on this character

$$W_{1,2} = p(\hat{u}) = a(u, \hat{u}) = \hat{p}(u) = W_{2,1} \quad (7.443)$$

and also Tottenham's equation is an application of Betti's theorem but with a special twist, namely that (7.443) remains true if u and $\hat{u}$ are replaced by the FE solutions u_h and $\hat{u}_h$ of the load cases p and $\hat{p}$ respectively

$$W_{1,2} = p(\hat{u}_h) = a(u, \hat{u}_h) = a(\hat{u}, u_h) = \hat{p}(u_h) = W_{2,1}. \quad (7.444)$$

- Note that the Galerkin orthogonality

$$a(u - u_h, \hat{u}_h) = 0 \quad (7.445)$$
$$a(\hat{u} - \hat{u}_h, u_h) = 0 \quad (7.446)$$

implies

$$a(u, \hat{u}_h) = a(u_h, \hat{u}_h) \quad (7.447)$$
$$a(\hat{u}, u_h) = a(\hat{u}_h, u_h) \quad (7.448)$$

which establishes

$$a(u, \hat{u}_h) = a(\hat{u}, u_h) \quad (7.449)$$

and therewith proves (7.444).

- The FE solution can be written in two ways

$$u_h(x) = \begin{cases} \displaystyle\sum_i \varphi_i(x) \, u_i \\ \displaystyle\int_0^l G_0^h(y, x) \, p(y) \, dy \end{cases} \quad (7.450)$$

or (see Fig. 1.52 p. 77)

$$u_h(x) = \sum_i \varphi_i(x)\, u_i = \int_0^l G_0^h(y, x)\, p(y)\, dy$$

$$= \int_0^l \sum_i \varphi_i(y)\, u_j^G(x)\, p(y)\, dy = \sum_i u_i^G(x)\, f_i$$

$$= \boldsymbol{u}_G^T \boldsymbol{f} = \boldsymbol{u}_G^T \boldsymbol{K}\, \boldsymbol{u} = \boldsymbol{u}_G^T \boldsymbol{K}^T \boldsymbol{u} = \boldsymbol{f}_G^T \boldsymbol{u}\,. \qquad (7.451)$$

This extends naturally to any quantity as for example

$$\sigma_h(x) = \begin{cases} \sum_i \sigma(\varphi_i(x))\, u_i = \boldsymbol{f}_G^T \boldsymbol{u} \\ \int_0^l G_1^h(y, x)\, p(y)\, dy = \boldsymbol{u}_G^T \boldsymbol{f} \end{cases} \qquad (7.452)$$

where $\boldsymbol{u}_G$ and $\boldsymbol{f}_G$ are now the nodal displacements and equivalent nodal forces respectively of the Dirac delta δ_1.

References

1. Adams V, Askenazi A (1998) Building Better Products With Finite Element Analysis. OnWord Press, Santa Fe
2. Ainsworth M, Oden JT (1997) "A posteriori error estimation in finite element analysis", Computer Methods in Appl. Mech. Eng. 142: 1–88
3. Akin JE (1994) Finite Elements for Analysis and Design. Academic Press, London
4. Akin JE (2005) Finite Elements with Error Estimators. Elsevier, Amsterdam
5. Allman DJ (1988) "A quadrilateral finite element including vertex rotations for plane elasticity problem", Int. J. Num. Methods in Eng. 26: 2645–2655
6. Altenbach H, Altenbach J, Kissing W (2004) Mechanics of Composite Structural Elements. Springer-Verlag Berlin Heidelberg New York
7. Andrianov IV, Awrejcewicz J, Manevitch LI (2004) Asympotical Mechanics ot Thin-Walled Structures. Springer-Verlag Berlin Heidelberg New York
8. Apel T (1999) Anisotropic Finite Elements: Local Estimates and Applications Advances in Numerical Mathematics. Teubner Verlag, Stuttgart
9. Arnold DN, Falks RS (1989) "Edge effects in the Reissner–Mindlin plate theory", in: Analytic and Computational Models of Shells (Eds. Noor AK, Belytschko T, Simo JC), A.S.M.E., New York: 71–90
10. Babuška I, Rheinboldt WC (1978) "Error estimates for adaptive finite element computations", SIAM J. Numer. Anal. 4, 15: 736–754
11. Babuška I, Miller AD (1984) "The post-processing approach in the finite element method, I: calculations of displacements, stresses and other higher derivatives of the displacements", Int. J. Num. Methods in Eng. 20: 1085–1109
12. Babuška I, Miller AD (1984) "The post-processing approach in the finite element method, II: the calculation of stress intensity factors", Int. J. Num. Methods in Eng. 20: 1111–1129
13. Babuška I, Miller AD (1984) "The post-processing approach in the finite element method, III: a posteriori error estimation and adaptive mesh selection", Int. J. Num. Methods in Eng. 20: 2311–2324
14. Babuška I (1986) "Uncertainties in Engineering Design. Mathematical Theory and Numerical Experience", in: The Optimal Shape (Eds. Bennett J and Botkin MM) Plenum Press
15. Babuška I, Miller A (1987) "A feedback finite element method with a posteriori error estimation. Part I: the finite element method and some basic properties of the a posteriori error estimator", Computer Methods in Appl. Mech. Eng. 61: 1–40
16. Babuška I (1990) "The problem of modeling the elastomechanics in engineering", Computer Methods in Appl. Mech. Eng. 82: 155–182
17. Babuška I, Pitkäranta J (1990) "The plate paradox for hard and soft simple support", SIAM Journal for Numerical Analysis 21: 551–576

18. Babuška I, Osborn J (1998) Can a Finite Element Method Perform Arbitrarily Badly? TICAM Report 98-19 ICES at the University of Texas, Austin
19. Babuška I, Strouboulis T (2001) The Finite Element Method and its Reliability. Oxford University Press, Oxford
20. Babuška I, Banerjee U, Osborn J (2004) "Generalized Finite Element Mthods: Main ideas, results and perspective", International Journal of Computational Methods 1 (1): 67-103
21. Babuška I, Oden JT (2005) "The reliability of computer predictions: can they be trusted?", Int. J. Numerical Analysis and Modeling 1: 1–18
22. Bangerth W, Rannacher R. (2003) Adaptive Finite Element Methods for Differential Equations. Birkhäuser Verlag Basel Boston Berlin
23. Backstrom G (1998) Deformation and Vibration by Finite Element Analysis. Studentlitteratur, Lund
24. Barthel C (2000) Comparative study of a floor plate. University Kassel
25. Barton G (1989) Elements of Green's Functions and Propagation, Potentials, Diffusion and Waves. Oxford Scientific Publications, Oxford
26. Bathe K-J (1996) Finite Element Procedures. Prentice Hall New Jersey
27. Bathe K-J, Dvorkin EN (1985) "A four-node plate bending element based on Mindlin/Reissner theory and a mixed interpolation", Int. J. Num. Methods in Eng. 21: 367–383
28. Bathe K-J, Chapelle D (2002) The Finite Element Analysis of Shells – Fundamentals. Springer-Verlag Berlin Heidelberg New York
29. Batoz JL, Bathe K-J, Ho LW (1980) "A study of three-node triangular plate bending elements", Int. J. Num. Methods in Eng. 15: 1771–1812
30. Batoz JL (1982) "An explicit formulation for an efficient triangular plate bending element", Int. J. Num. Methods in Eng. 18: 1077–1089
31. Batoz JL, Katili I (1992) "Simple triangular Reissner–Mindlin plate based on incompatible modes and discrete contacts", Int. J. Num. Methods in Eng. 35: 1603-1632
32. Baumann A (2000) Plate analysis with SOFISTIK. TU Munich
33. Bazeley GP, Cheung YK, Irons BM, Zienkiewicz OC (1965) "Triangular elements in plate bending conforming and non-conforming solutions", Proc. (1st) Conf. On Matrix Methods in Struct. Mech. AFFDL TR 66-80: 547–578
34. Becker R, Rannacher R (2001) "An optimal control approach to a posteriori error estimation in finite element methods", Acta Numerica, 1–102
35. Beer G, Watson JO (1992) Introduction to Finite and Boundary Element Methods for Engineers. John Wiley & Sons, Chichester
36. Beer G, Golser H, Jedlitschka G, Zacher P (1999) Coupled Finite Element/Boundary Element Analysis in Rock Mechanics - Industrial Applications Rock Mechanics for Industry (Eds. Amadei, Kranz, Scott & Smeallie) A.A Balkema. Rotterdam
37. Beer G (2000) Programming the Boundary Element Method. John Wiley & Sons, Chichester
38. Belytschko T, Liu WK and Moran B (2000) Nonlinear Finite Elements for Continua and Structures. John Wiley & Sons, New York
39. Belytschko T, Chiapetta T, Bartel RL (1976) "Efficient large-scale non-linear transient analysis by finite elements", Int. J. Num. Methods in Eng. 10: 579–596
40. Belytschko T, Bindemann LP (1987) "Assumed strain stabilization of the 4-node quadrilateral with 1-point quadrature for nonlinear problems", Computer Methods in Appl. Mech. Eng. 88: 311–340

41. Benvenuto E (1981) La Scienza delle Costruzioni e il suo Sviluppo Storico. Sansoni, Firenze, p. 107

42. Bergan PG, Felippa CA (1985) "A triangular membrane element with rotational degrees of freedom", Computer Methods in Appl. Mech. Eng. 50: 25–69

43. Bernadou M (1996) Finite Element Methods for Thin Shell Problems. John Wiley & Sons Chichester, Masson Paris

44. Beskos DE Ed. (1991) Boundary Element Analysis of Plates and Shells. Springer Series in Computational Mechanics, Ed. Atluri SN, Springer-Verlag Berlin Heidelberg New York

45. Betsch P. (1966) Statische und dynamische Berechnungen von Schalen endlicher elastischer Deformationen mit gemischten finiten Elementen. Dissertation, Institut für Baumechanik und Numerische Mechanik, University of Hannover,

46. Bletzinger K-U, Bischoff M, Ramm E (2000) "A unified approach for shear–locking free triangular and rectangular shell finite elements", Comp. & Structures, 73: 321–334.

47. Blum H (1988) "Numerical treatment of corner and crack singularities", in: Finite Element and Boundary Element Techniques from Mathematical and Engineering Point of View. (Eds. Stein E, Wendland W) CISM Lecture Notes 301, Springer-Verlag Wien New York

48. Bonnet M (1995) Boundary Integral Equation Methods for Solids and Fluids. John Wiley & Sons, England

49. Boussinesq J (1885) Application des Potentiels à l'Etude de l'Equilibre et du Mouvement des Solides Elastique. Gauthier-Villars, Paris

50. Braack M, Ern A (2003) "A posteriori control of modeling errors and discretization errors", Multiscale Model. Simul. 1: 221–238

51. Braess D (1997) Finite Elements: Theory, Fast Solvers and Applications in Solid Mechanics. Cambridge University Press, Cambridge

52. Brauer JR (1988) What Every Engineer Should Know About Finite Element Analysis. Marcel Dekker Inc. New York

53. Brenner SC, Scott R (2002) The Mathematical Theory of Finite Element Methods. Springer-Verlag Berlin Heidelberg New York (2nd ed.)

54. Bucalem ML, Bathe KJ (2006) The Mechanics of Solids and Structures. Springer-Verlag Berlin Heidelberg New York

55. Bürge M, Schneider J (1994) Variability in Professional Design. Strct. Engineering Intern. 4: 247–250

56. Carroll WF (1999) A Primer for Finite Elements in Elastic Structures. John Wiley & Sons, New York

57. Case J, Chilver L, Ross CTF (1993) Strength of Materials and Structures With an Introduction to Finite Element Methods. Chapman & Hall, New York (3rd ed.)

58. Chandrupatla TR, Belegundu AD (2001) Introduction to Finite Elements in Engineering. Prentice-Hall, Englewood Cliffs. Pearson Education

59. Chang CH (2005) Mechanics of Elastic Structures with Inclined Members. Springer-Verlag Berlin Heidelberg New York

60. Chen GJ Zhou (1992) Boundary Element Methods. Academic Press, London San Diego

61. Choi KK, Kim NH (2005) Structural Sensitivity Analysis and Optimization 1 - Linear Systems. Springer-Verlag Berlin Heidelberg New York

62. Choi KK, Kim NH (2005) Structural Sensitivity Analysis and Optimization 2 - Nonlinear Systems and Applications. Springer-Verlag Berlin Heidelberg New York

63. Ciarlet PG (1978) The Finite Element Method for Elliptic Problems. North-Holland Amsterdam
64. Ciarlet PG, Lions JL Eds. (1991) Handbook of Numerical Analysis, Volume II: Finite Element Methods. North-Holland, Amsterdam
65. Ciarlet PG, Lions JL Eds. (1996) Handbook of Numerical Analysis, Volume IV: Finite Element Methods (Part 2), Numerical Methods for Solids (Part 2). Elsevier, Amsterdam
66. Ciarlet PG (2005) An Introduction to Differential Geometry with Applications to Elasticity. Springer Verlag
67. Çirak F, Ramm E (2000) "A posteriori error estimation and adaptivity for elastoplasticity using the reciprocal theorem", Int. J. Num. Methods in Eng. 47: 379–393
68. Clough RW, Tocher JL (1965) "Finite element stiffness matrices for analysis of plate bending", Proc. (1st) Conf. On Matrix Methods in Struct. Mech., AFFDL TR 66–80: 515–546
69. Cook RD (1987) "A plane hybrid element with rotational d.o.f. and adjustable stiffness", Int. J. Num. Methods in Eng. 24, 8: 1499–1508
70. Cook RD, Malkus DS, Plesha ME (1989) Concepts and Applications of Finite Element Analysis. John Wiley & Sons, New York (3rd ed.)
71. Cook RD (1995) Finite Element Modeling for Stress Analysis. John Wiley & Sons, New York
72. Crisfield MA (1991) Non-linear Finite Element Analysis of Solids and Structures. 1: Essentials. John Wiley & Sons, Chichester
73. Crisfield MA (1997) Non-linear Finite Element Analysis of Solids and Structures. 2: Advanced Topics. John Wiley & Sons, New York
74. Dautray R, Lions JL (1990) Mathematical Analysis and Numerical Methods for Science and Technology. Springer-Verlag Berlin Heidelberg New York
75. Deif A (1986) Sensitivity Analysis in Linear Systems. Springer-Verlag Berlin Heidelberg New York
76. Desai CS, Kundu T (2001) Introductory Finite Element Method. CRC Press, Boca Raton, FL
77. Dolbow J, Moës N, Belytschko T (2000) "Discontinuous enrichment in finite elements with a partition of unity method", Finite Elements in Analysis and Design, 36: 235–260
78. Dow JO (1999) Finite Element Methods and Error Analysis Procedures: A Unified Approach. Academic Press, San Diego
79. Duddeck F (2002) Fourier BEM. Springer-Verlag Berlin Heidelberg New York
80. Dvorkin EN, Goldschmit MB (2005) Nonlinear Continua. Springer-Verlag Berlin Heidelberg New York
81. Edelsbrunner H (2001) Geometry and Topology for Mesh Generation. Cambridge University Press, Cambridge
82. Elishakoff I, Ren Y (2003) Finite Element Methods for Structures With Large Stochastic Variations. Oxford University Press, Oxford
83. Estep D, Holst M, Larson M (2003) Generalized Green's functions and the effective domain of influence. Preprint 2003-10, Chalmers Finite Element Center, Chalmers University of Technology, Goteborg, Sweden
84. Fagan MJ (1992) Finite Element Analysis: Theory and Practice. John Wiley & Sons, New York
85. Fellin W, Lessmann H, Oberguggenberger M, Vieider R (Eds.) (2005) Analyzing Uncertainty in Civil Engineering. Springer-Verlag Berlin Heidelberg New York

86. Fenner RT (1975) Finite Element Methods for Engineers. The Macmillan Press Ltd., London

87. Fix GJ, Strang G (1969) "Fourier analysis of the finite element method in Ritz-Galerkin theory", Studies in Appl. Math. 48 265–273

88. Frey PJ, George PL (2000) Mesh Generation Application to Finite Elements. Hermes Science Publishing, Oxford, Paris

89. Fujita H. (1955) "Contributions to the theory of upper and lower bounds in boundary value problems", J. Phys. Soc. Japan 10: 1–8

90. Gambhir ML (2004) Stability Analysis and Design of Structures. Springer-Verlag Berlin Heidelberg New York

91. Gao XW, Davies TG (2002) Boundary Element Programming in Mechanics, Cambridge University Press

92. Gaul L, Kögl M, Wagner M (2003) Boundary Element Methods for Engineers and Scientists. Springer-Verlag Berlin Heidelberg New York

93. Germain P (1962) Mècanique des milieux continus. Masson Paris

94. Gerold F (2004) 3D-Modellierung von Gebäuden mit der Methode der Finiten Elemente. Master Thesis, FH Konstanz

95. Gould PL (1985) Finite Element Analysis of Shells of Revolution. Surveys in Structural Engineering and Structural Mechanics 4, Pitman Publishing, New York

96. Grätsch T (2002) L_2-Statik, PhD-thesis University Kassel

97. Grätsch T, Hartmann F (2003) "Finite element recovery techniques for local quantities of linear problems using fundamental solutions", Computational Mechanics, 33:15-21

98. Grätsch T, Hartmann F (2004) "Duality and finite elements", Finite Elements in Analysis and Design, 40: 1005–1020

99. Grätsch T, Bathe KJ, (2006) "Goal-oriented error estimation in the analysis of fluid flows with structural interactions", Computer Methods in Applied Mechanics and Engineering (in press)

100. Grätsch T, Bathe KJ (2005) "Influence functions and goal-oriented error estimation for finite element analysis of shell structures", International Journal for Numerical Methods in Engineering, 63(5):631-788

101. Grätsch T, Bathe KJ (2005) "A posteriori error estimation techniques in practical finite element analysis", Computers & Structures, 83: 235-265

102. Grätsch T, Hartmann F (2006) Pointwise error estimation and adaptivity for the finite element method using fundamental solutions, Computational Mechanics, 37(5): 394-40

103. Grasser E, Thielen G (1991) Hilfsmittel zur Berechnung der Schnittgrößen und Formänderungen von Stahlbetontragwerken, Deutscher Ausschuß für Stahlbeton, Heft 240, 3. Auflage Beuth Berlin

104. Gruttmann F, Sauer R, Wagner W (2000) "Theory and numerics of three-dimensional beams with elastoplastic behaviour", Int. J. Num. Methods in Eng. 48: 1675–1702

105. Gruttmann F, Wagner W (2004) "A stabilized onepoint integrated quadrilateral ReissnerMindlin plate element", Int. J. Num. Methods in Eng. 61: 2273-2295

106. Gruttmann F, Wagner W (2005) "A linear quadrilateral shell element with fast stiffness computation", Comp. Meth. Appl. Mech. Eng. 194: 4279-4300

107. Gruttmann F, Wagner W (2006) "Structural analysis of composite laminates using a mixed hybrid shell element", Comput. Mech. (2006) 37: 479-497

108. Gu Jinshen (1999) Domain Decomposition Methods for Nonconforming Finite Element Discretizations. Nova Science Publishers Inc., Commack, NY

109. Gupta M (1977) "Error in eccentric beam formulation", Int. J. Num. Methods in Eng. 11: 1473–1477
110. Gupta OP (1999) Finite and Boundary Element Methods in Engineering. Balkema, Rotterdam
111. Gupta KK, Meek JL (2000) Finite Element Multidisciplinary Analysis. AIAA, New York
112. Gurtin ME (1972) The Linear Theory of Elasticity. In: Handbook of Physics. Ed. Flügge S, Volume VIa/2 Solid mechancis II. Ed. Truesdell C, Springer-Verlag Berlin Heidelberg New York
113. Haldar A, Guran A, Ayyub BM Eds. (1997) Uncertainty Modeling in Finite Element, Fatigue and Stability of Systems. World Scientific Publ., River Edge, NJ
114. Haldar A and Mahadevan S (2000) Reliability Assessment Using Stochastic Finite Element Analysis. John Wiley & Sons, New York
115. Hartmann F (1985) The Mathematical Foundation of Structural Mechanics. Springer-Verlag Berlin Heidelberg New York
116. Hartmann F (1989) Introduction to Boundary Elements. Springer-Verlag Berlin Heidelberg New York
117. Hartmann F, Jahn P (2001) "Boundary element analysis of raft foundations on piles", Meccanica **36**: 351–366
118. Hjelmstad, KD (2005) Fundamentals of Structural Mechanics. Springer Science + Business Media New York 2nd edition
119. Heyman J (1969) "Hambly's paradox: why design calculations do not reflect real behaviour", Proc. Inst. Civil Eng. 114: 161–166
120. Huebner KHH, Dewhirst DL, Smith DE, Byrom TG (2001) Finite Element Method. John Wiley & Sons, New York
121. Hughes TJR (1987) The Finite Element Method. Prentice-Hall, Englewood Cliffs, New Jersey
122. Hughes TJR (1981) "Finite elements based upon mindlin plate theory with particular reference to the four-node bilinear isoparametric element", Journal of Applied Mechanics, 48: 587–596
123. Hurtado JE (2004) Structural Reliability. Springer-Verlag Berlin Heidelberg New York
124. Irons BM (1976) "The semiloof shell element, finite elements for thin shells and curved members", (Eds. D.G. Ashwell, R.H. Gallagher) John Wiley London, 197–222
125. Irons BM, Razzaque A (1972) "Experience with the patch test for convergence of finite elements", In: The Mathematical Foundations of the Finite Element Method with Applications to Partial Differential Equations. (Aziz AK Ed.) Academic Press, New York, 557–587
126. Jahn P, Hartmann F (1999) "Integral representations for the deflection and the slope of a plate on an elastic foundation", Journal of Elasticity 56: 145–158
127. Jahn P, Hartmann F (2002) Numerical Calculation for Plates on an Elastic Foundation. Preprint University of Kassel
128. Jakobsen B, Rasendahl F, (1994) "The Sleipner platform accident", Structural Engineering International, 3: 190–194
129. Jeyachandrabose C, Kirkhope J (1985) "An alternative formulation for the DKT plate bending element", Int. J. Num. Methods in Eng. 21: 1289–1293
130. Jiang B (1998) The Least-Square Finite Element Method. Springer-Verlag Berlin Heidelberg New York

131. Johnson C (1995) Numerical Solution of Partial Differential Equations by the Finite Element Method. Cambridge University Press

132. Johnson C, Hansbo P (1992) "Adaptive finite element methods in computational mechanics", Computer Methods in Appl. Mech. Eng. 101, North-Holland, Amsterdam

133. Kachanov M, Shafiro B, Tsukrov I (2004) Handbook of Elasticity Solutions. Springer-Verlag Berlin Heidelberg New York

134. Kaliakin VN (2001) Introduction to Approximate Solutions Techniques, Numerical Modeling, and Finite Element Methods. Marcel Dekker Inc., New York

135. Kato T (1953) "On some approximate methods concerning the operaort T^*T", Math. Ann., 126: 253–262

136. Katsikadelis JT (2002) Boundary Elements: Theory and Applications. Elsevier Science

137. Kattan PI, Voyiadjis GZ (2002) Damage Mechanics with Finite Elements practical Applications with Computer Tools. Springer-Verlag Berlin Heidelberg New York

138. Kattan P (2003) MATLAB Guide to Finite Elements: An Interactive Approach. Springer-Verlag Berlin Heidelberg New York

139. Katz C., Stieda (1992) "Praktische FE-Berechnung mit Plattenbalken", Bauinformatik 1: 30–34

140. Katz C, Werner H (1982) "Implementation of nonlinear boundary conditions in finite element analysis", Computers & Structures 15: 299–304

141. Katz C (1995) "Kann die FE-Methode wirklich alles?", FEM 95 - Finite Elemente in der Baupraxis. (Eds. Ramm E, Stein E, Wunderlich W), Ernst & Sohn, Berlin

142. Katz C (1986) "Berechnung von allgemeinen Pfahlwerken", Bauingenieur 61 Heft 12

143. Katz C (1997) "Fliesszonentheorie mit Interaktion aller Stabschnittgrößen bei Stahltragwerken", Stahlbau 66: 205–213

144. Katz C (1996) "Vertrauen ist gut, Kontrolle ist besser", in: Software für Statik und Konstruktion. (Eds. Katz C, Protopsaltis B) Balkema A.A., Rotterdam

145. Kelly DW, Gago JP de SR, Zienkiewicz OC, Babuška I (1983) "A posteriori error analysis and adaptive processes in the finite element method: part I - error analysis", Int. J. Num. Methods in Eng. 19: 1595–1619

146. Kemmler R, Ramm E (2001) "Modellierung mit der Methode der Finiten Elemente", in: Betonkalender 2001 Ernst & Sohn, Berlin, 381–446

147. Knöpke B (1994) "The hypersingular integral equation for bending moments m_{xx}, m_{yy} and m_{xy} of the Kirchhoff plate", Computational Mechanics 14: 1–12

148. Kojic M, Bathe KJ (2005) Inelastic Analysis of Solids and Structures. Springer-Verlag Berlin Heidelberg New York

149. Kotsovos MD, Pavlovic MN (1995) Structural Concrete: Finite Element Analysis for Limit-State Design. Telford, London

150. Krenk S (2001) Mechanics and Analysis of Beams, Columns and Cables. Springer-Verlag Berlin Heidelberg New York (2nd ed.)

151. Kuhn G, Partheymüller P (1997) "Analysis of 3D elastoplastic notch and crack problems using boundary element method", ni Wendland WL Ed., Boundary Element Topics, 99–116. Springer-Verlag Berlin Heidelberg New York

152. Kuhn G, Köhler O (1997) "A field boundary formulation for axisymmetric finite strain elastoplasticity", Proceedings of the IUTAM/IACM-Symposium on

Discretization Methods in Structural Mechanics II, Vienna, Kluwer Academic Publishers

153. Larsson F, Hansbo P, Runesson K (2002) "Strategies for computing goal-oriented a posteriori error measures in nonlinear elasticity", Int. J. Num. Methods in Eng. 55: 879–894

154. Lesaint P (1976) "On the convergence of Wilsons nonconforming element for solving elastic problems", Computer Methods in Appl. Mech. Eng. 7: 1–16

155. Lepi SM (1998) Practical Guide to Finite Elements: A Solid Mechanics Approach. Marcel Dekker Inc. London

156. Li S, Liu WK (2002) "Meshfree and particle methods and their applications", Applied Mechanics Review 55(1): 1–34

157. Liu GR and Quek SS (2003) Finite Element Method: A Practical Course. Butterworth Heinemann, Oxford

158. Lueschen GGG, Bergman LA (1996), "Green's functions for uniform Timoshenko beams", Journal of Sound and Vibration 194(1): 93-102

159. MacNeal RH, Harder RL (1985) "A proposed standard set of problems to test finite element accuracy", Finite Elements in Analysis and Design 1: 3-20

160. MacNeal RH (1994) Finite elements: their design and performance. Dekker, New York

161. Mackie RI (2000) Object Oriented Methods and Finite Element Analysis. Saxe-Coburg Publ., Edinburgh

162. Materna D (2004) Goal-oriented recovery bei nichtlinearen Scheibenproblemen. Diploma Thesis University of Kassel

163. Melenk JM, Babuška I (1996) "The partition of unity finite element method: Basic theory and applications", Computer Methods in Appl. Mech. Eng. 139: 289–314

164. Melosh RJ (1990) Structural Engineering Analysis by Finite Elements. Prentice-Hall, Englewood Cliffs

165. Melzer H, Rannacher R (1980) "Spannungskonzentrationen in Eckpunkten der Kirchhoffschen Platte", Bauingenieur 55: 181–184

166. Mindlin RD (1936) "Force at a point in the interior of a semi-infinite solid", Physics 7: 195–202

167. Mindlin RD (1951) "Influence of rotatory inertia and shear on flexural motions of isotropic elastic plates", ASME Journal of Applied Mechanics, Vol 18: 31–38

168. Moaveni S (1999) Finite Element Analysis: Theory and Application With ANSYS. Prentice-Hall, Upper Saddle River, NJ

169. Moës, N, Dolobow J, Belytschko T (1999) "A finite element method for crack growth without remeshing", Int. J. Num. Methods in Eng. 46: 131–150

170. Mohammed A (2001) Boundary Element Analysis: Theory & Programming. CRC Press

171. Nayfeh AH, Pai PF (2004) Linear and Nonlinear Structural Mechanics. John Wiley & Sons, Chichester

172. Niku-Lari A, Ed. (1986) Structural Analysis Systems-Software, Hardware, Capability, Compatibility, Applications Vol 1. Pergamon Press, Oxford

173. Niku-Lari A, Ed. (1986) Structural Analysis Systems-Software, Hardware, Capability, Compatibility, Applications Vol 2. Pergamon Press, Oxford

174. Nowinski JL (1981) Applications of Functional Analysis in Engineering. Plenum Press New York London

175. Oden JT, Reddy JN (1976) An Introduction to the Mathematical Theory of Finite Elements. John Wiley & Sons New York London

176. Oden JT, Zohdi TI (1997) "Analysis and adaptive modeling of highly heterogeneous elastic structures", Comput. Methods Appl. Mech. Engrg. 148: 367–391
177. Oden JT, Prudhomme S (2001) "Goal-oriented error estimation and adaptivity for the finite element method", Computers and Mathematics with Applications 41: 735–756
178. Oden JT, Prudhomme S (2002) "Estimation of modeling error in computational mechanics", J. Computational Physics 182: 496–515
179. Paraschivoiu M, Peraire J, Patera A (1997) "A posteriori finite element bounds for linear-functional outputs of elliptic partial differential equations", Computer Methods in Applied Mechanics and Engineering, 150: 289–312
180. Pauli W (2000) "Unerwartete Effekte bei nichtlinearen Berechnungen ", Software für Statik und Konstruktion - 3 (Eds. Katz C, Protopsaltis B) A.A.Balkema, Rotterdam Broolfield
181. Peraire J, Patera AT (1998) "Bounds for linear-functional outputs of coercive partial differential equations: Local indicators and adaptive refinement"Advances in Adaptive Computational Methods in Mechanics, (Ed. Ladevèze P, Oden JT), 199-21, Elsevier Amsterdam
182. Perelmuter A, Slivker V (2003) Numerical Structural Analysis. Springer-Verlag Berlin Heidelberg New York
183. Pflanz G (2001) Numerische Untersuchung der elastischen Wellenausbreitung infolge bewegter Lasten mittels der Randelementmethode im Zeitbereich. VDI Fortschritt-Bericht, Reihe 18, Nr. 265
184. Pian THH (1964) "Derivation of element stiffness matrices by assumed stress distribution", AIAA J. 2: 1332–1336
185. Pian THH, Wu CC (2005) Hybrid and Incompatible Finite Element Methods. Chapman & Hall/CRC
186. Pierce NA, Giles MB (2000) "Adjoint recovery of superconvergent functionals from pde approximations", SIAM Review, 42: 247–264
187. Pilkey WD, Wunderlich W (1994) Mechanics of Structures, Variational and Computational Methods. CRC Press Boca Raton, Ann Arbor London Tokyo
188. Piltner R, Taylor RL (1999) "A systematic construction of B-bar functions for linear and non-linear mixedenhanced finite elements for plane elasticity problems", Int. J. Numer. Methods in Eng. 44: 615-639
189. Pitkaeranta J, Matache AM, Schwab C (1999) Fourier mode analysis of layers in shallow shell deformations. Research Report ETH Seminar for Applied Mathematics, Zürich
190. Pomp A (1998) The Boundary-Domain Integral Method for Elliptic Systems - With Application to Shells. Springer-Verlag Berlin Heidelberg New York
191. Portela A, Charafi A (2002) Finite Elements Using Maple A Symbolic Programming Approach. Springer-Verlag Berlin Heidelberg New York
192. Potts D, Zdravkovic L (1999) Finite Element Analysis in Geotechnical Engineering: Volume I - Theory. Telford Publishing, London
193. Potts D, Zdravkovic L (1999) Finite Element Analysis in Geotechnical Engineering: Volume II - Application. Telford Publishing, London
194. Prathap G (1993) The Finite Element Method in Structural Engineering. Solid Mechanics and Its Applications, 24. Kluwer Academic Publ., Dordrecht
195. Qin QH (2000) The Trefftz Finite and Boundary Element Method. WIT Press, Southampton
196. Raamachandran J (2000) Boundary and Finite Elements Theory and Problems CRC Press, Boca Raton, FL

588   References

197. Rajagopalan K (1993) Finite Element Buckling Analysis of Stiffened Cylindrical Shells. Ashgate Publishing Company, Aldershot, Hampshire
198. Ramm, E, Hofmann TJ (1995) Stabtragwerke, in: Der Ingenieurbau, Ed. G. Mehlhorn, Baustatik Baudynamik, Ernst & Sohn, Berlin 1995
199. Rannacher R, Suttmeier FT (1997) "A feed-back approach to error control in finite element methods: application to linear elasticity", Computational Mechanics, 19: 434–446
200. Rannacher R (1998) Error Control in Finite Element Computations. preprint 1998–54, Inst. Angew. Math. University Heidelberg,
201. Rao SS (1999) The Finite Element Method in Engineering. Pergamon Press, Oxford (4th ed.)
202. Reddy BD (1998) Introductory Functional Analysis. With Applications to Boundary Value Problems and Finite Elements. Springer-Verlag Berlin Heidelberg New York
203. Reddy JN (1991) Applied Functional Analysis and Variational Methods in Engineering. Krieger Publishing Company Malabar
204. Roache PJ (1998) Verification and validation in Computational Science and Engineering. Hermosa Publisher, Albuquerque
205. Rössle A, Sändig A-M (2001) Corner singularities and regularity results for the Reissner/Mindlin plate model. Preprint 01/04 of SFB 404 "Mehrfeldprobleme in der Kontinuumsmechanik" at University Stuttgart
206. Rössle A (2000) "Corner singularities and regularity of weak solutions for the two-dimensional lamé equations on domains with angular corners", Journal of Elasticity, 60: 57–75
207. Ross CTF (1990) Finite Element Methods in Engineering Science. Horwood Publishing Ltd, Chichester, UK
208. Ross CTF (1996) Finite Element Programs in Structural Engineering and Continuum Mechanics. Albion Publishing, Chichester
209. Ross CTF (1998) Advanced Applied Finite Element Methods. Horwood Publishing Ltd, Chichester
210. Rubin H (2006) Private communication, TU Vienna
211. Rücker M, Krafczyk M, Rank E (1998) "A parallel p-version FE-approach for structural engineering", Advances in Computational Mechanics with High Performance Computing. Civil-Comp Press, Edinburgh, UK, 73–78
212. Runesson K (2002) "Goal-oriented finite element error control and adaptivity with emphasis on nonlinear material behavior and fracture", 15th Nordic Seminar on Computational Mechanics NSCM 15 (Eds. Lund E, Olhoff N, Stegmann J): 25–32
213. Schwab C (1998) p- and hp-Finite Element Methods. Oxford University Press, Oxford
214. Schwalbe JW (1989) Finite Element Analysis of Plane Frames and Trusses. Krieger Publishing, Melbourne, FL
215. Schwarz HR (1988) Finite Element Methods, Computational Mathematics and Applications Series. Academic Press, London
216. Schenk C, Schuëller G (2005) Uncertainty Assessment of Large Finite Element Systems. Lecture Notes in Applied and Computational Mechanics, 24. Springer-Verlag Berlin Heidelberg New York
217. Simo JC, Rifai MS (1990) "A class of mixed assumed strain methods and the method of incompatible modes", Int. J. Num. Methods in Eng. 29: 1595–1638

218. Simo JC, Armero F (1992) "Geometrically non-linear enhanced strain mixed methods and the method of incompatible modes", Int. J. Num. Methods in Eng. 33: 1413–1449

219. Simo JC, Armero F, Taylor RL (1993) "Improved versions of assumed enhanced strain tri-linear elements for 3D finite deformation problems", Computer Methods in Appl. Mech. Eng. 110: 359–386

220. Sladek V, Sladek J (1998) Singular Integrals in Boundary Element Methods (Advances in Boundary Elements Vol. 3). Computational Mechanics Publication Southampton

221. Snieder R (2004) A guided Tour of Mathematical Methods for the Physical Sciences. Cambridge University Press, Cambridge

222. Spyrakos CC, Raftoyiannis J (1997) Linear and Nonlinear Finite Element Analysis in Engineering Practice. Algor Inc., Pittsburgh, PA

223. Stark RF, Booker JR (1997) "Surface displacements of a non-homogeneous elastic half-space subjected to uniform surface tractions. Part I: loading on arbitrarily shaped areas", Int. J. Num. Analytical Meth. Geomechanics, 21: 361–378

224. Stark RF, Booker JR (1997) "Surface displacements of a non-homogeneous elastic half-space subjected to uniform surface tractions. Part II: loading on rectangular shaped areas", Int. J. Num. Analytical Meth. Geomechanics, 21: 379–395

225. Steele JM (1989) Applied Finite Element Modeling - Practical Problem Solving for Engineers. Marcel Dekker Inc., New York

226. Stein E Ed. (2002) Error-Controlled Adaptive Finite Elements in Solid Mechanics. John Wiley & Sons, New York

227. Stein E, Ohnimus S (1999) "Anisotropic discretization and model-error estimation in solid mechanics by local Neumann problems", Computer Methods in Appl. Mech. Eng. 176: 363–385

228. Stein E, Wendland W Eds. (1988) Finite Element and Boundary Element Techniques from Mathematical and Engineering Point of View. Springer-Verlag Berlin Heidelberg New York

229. Stein E, de Borst R, Hughes TJR Eds. (2004) Encyclopedia of Computational Mechanics, Vol. 1 Fundamentals, Vol. 2 Solids and Structures, Vol. 3 Fluids. Wiley, Chichester

230. Strang G, Fix GJ (1973) An analysis of the finite element method. Prentice Hall, Englewood Cliffs, N.J.

231. Strang G (1991) Calculus. Wellesley-Cambridge Press, Wellesley

232. Strang G (1986) Introduction to Applied Mathematics. Wellesley-Cambridge Press, Wellesley

233. Strang G (2003) Linear Algebra and Its Applications. Saunders (3rd ed.)

234. Stummel F (1980) "The limitations of the patch test", Int. J. Num. Methods in Eng. 15: 177–188

235. Stummel F (1979) "The generalized patch test", SIAM Journal for Numerical Analysis, 16, 3: 449–471

236. Stricklin JA et al (1969) "A rapidly converging triangular plate element", AIAA J. 7: 180–181

237. Sudarshan R, Amaratunga K, Grätsch T (2006) "A combined approach for goal-oriented error estimation and adaptivity using operator-customized finite element wavelets" Int. J. Num. Methods in Eng. 66: 1002–1035

238. Szabó B, Babuška I (1991) Finite Element Analysis. John Wiley & Sons, Inc. New York

239. Tabtabai SMR (1997) Finite Element-Based Elasto-Plastic Optimum Reinforcement Dimensioning of Spatial Concrete Panel. Springer-Verlag Berlin Heidelberg New York

240. Taylor RL, Simo JC (1985) "Bending and membrane elements for analysis of thick and thin shells", Proceedings of the NUMETA '85 Conference: 587–591 (Eds. Middleton J, Pande GN) Swansea, Balkema A.A. Rotterdam

241. Taylor RL, Beresford PJ, Wilson EL (1976) "A non-conforming element for stress analysis", Int. J. Num. Methods in Eng. 10: 1211–1219

242. Teller E, Teller W, Talley W (1991) Conversations on the dark secrets of physics. Plenum Press New York London

243. Tenek LT, Argyris J (1998) Finite Element Analysis for Composite Structures. Kluwer Academic Publ., Dordrecht

244. Topping BHV, Muylle J, Putanowicz R, Cheng B (2000) Finite Element Mesh Generation. Saxe-Coburg Publ., Edinburgh

245. Tottenham H (1970) "Basic Principles", in: Finite Element Techniques in Structural Mechanics. (Eds. Tottenham H, Brebbia C), Southampton University Press, Southampton 1970

246. Trompette P (1992) Structural Mechanics by FEM: Statics and Dynamics. Masson, Paris

247. Vemaganti K (2004) "Modelling error estimation and adaptive modelling of perforated materials", Int. J. Num. Methods in Eng. 59: 1587–1604

248. Verführt R (1996) Review of A Posteriori Error Estimation and Adaptive Mesh-Refinement Techniques. Wiley Teubner

249. Wagner W, Gruttmann, F (2005) "A robust nonlinear mixed hybrid quadrilateral shell element", Int. J. Num. Methods in Eng. 64: 635-666

250. Wahlbin L (1995) Superconvergence in Galerkin Finite Element Methods. Springer-Verlag Berlin Heidelberg New York

251. Whiteman JR Ed. (2000) The Mathematics of Finite Elements and Applications X. Elsevier, Amsterdam

252. Williams ML (1952) Stress singularities resulting from various boundary conditions in angular corners of plates in extension. Jounal of Applied Mechanics, 12: 526–528

253. Wilson E, Taylor RL, Doherty WP, Ghaboussi J (1971) "Incompatible displacement models", Symposium on Numerical Methods, University of Illinois

254. Wolf JP, Song Chongmin (1996) Finite-Element Modeling of Unbounded Media. John Wiley & Sons, Chichester

255. Wriggers P (2001) Nichtlineare Finite-Element-Methoden, Springer-Verlag Berlin Heidelberg

256. Wriggers P (2006) Computational Contact Mechanics. Springer-Verlag Berlin Heidelberg (2nd ed.)

257. Wrobel LC, Aliabadi M (2002) The Boundary Element Method. Vol. 1, Applications in Thermo-Fluids and Acoustics. Vol. 2, Applications in Solids and Structures. John Wiley & Sons Chichester

258. Zienkiewicz OC, Taylor RL, Zhu JZ (2006) Finite Element Method: Volume 1– Its Basis & Fundamentals. Butterworth Heinemann, London

259. Zienkiewicz OC, Taylor RL (2006) Finite Element Method: Volume 2 – For Solid and Structural Mechanics. Butterworth Heinemann, London

260. Zienkiewicz OC, Taylor RL, Nithiarasu P (2006) Finite Element Method: Volume 3 – For Fluid Dynamics. Butterworth Heinemann, London

261. Zienkiewicz OC, Zhu JZ (1987) "A simple error estimator and adaptive procedure for practical engineering analysis", Int. J. Num. Methods in Eng. 24: 337–357
262. Zienkiewicz OC, Zhu JZ (1992) "The superconvergent patch recovery and a posteriori error estimates. Part 1: The recovery technique", Int. J. Num. Methods in Eng. 33: 1331–1364
263. Zienkiewicz OC, Zhu JZ (1992) "The superconvergent patch recovery and a posteriori error estimates, Part 2: Error estimates and adaptivity", Int. J. Num. Methods in Eng. 33: 1365–1382

Index

Printing: Krips bv, Meppel
Binding: Stürtz, Würzburg